Published volumes

Volume 1: *Mémoires ORSTOM* 91: 1-558, 225 figs, 39 pls (1981). ISBN: 2-7099-0578-7.
Volume 2: *Mémoires du Muséum national d'Histoire naturelle* 133 (A): 1-525, 126 figs, 37 pls (1986). ISBN: 2-85653-136-9.
Volume 3: *Mémoires du Muséum national d'Histoire naturelle* 137 (A): 1-254, 82 figs, 9 pls (1987). ISBN: 2-85653-141-5.
Volume 4: *Mémoires du Muséum national d'Histoire naturelle* 143 (A): 1-260, 103 figs, 23 pls (1989). ISBN: 2-85653-150-4.
Volume 5: *Mémoires du Muséum national d'Histoire naturelle* 144 (A): 1-385, 128 figs, 35 pls (1989). ISBN: 2-85653-164-4.
Volume 6: *Mémoires du Muséum national d'Histoire naturelle* 145 (A): 1-388, 190 figs, 4 color plates (1990). ISBN: 2-85653-171-7.
Volume 7: *Mémoires du Muséum national d'Histoire naturelle* 150 (A): 1-264, 587 figs (1991). ISBN: 2-85653-180-6.
Volume 8: *Mémoires du Muséum national d'Histoire naturelle* 151 (A): 1-468, 198 figs (1991). ISBN: 2-85653-186-5.
Volume 9: *Mémoires du Muséum national d'Histoire naturelle* 152 (A): 1-520, 283 figs, 6 color plates (1992). ISBN: 2-85653-191-1.
Volume 10: *Mémoires du Muséum national d'Histoire naturelle* 156: 1-491, 163 figs, 2 color plates (1993). ISBN: 2-85653-206-3.
Volume 11: *Mémoires du Muséum national d'Histoire naturelle* 158: 1-426, 159 figs (1993). ISBN: 2-85653-208-X.
Volume 12: *Mémoires du Muséum national d'Histoire naturelle* 161: 1-569, 269 figs, 11 color plates (1994). ISBN: 2-85653-212-8.
Volume 13: *Mémoires du Muséum national d'Histoire naturelle* 163: 1-517, 132 figs, 4 color plates (1995). ISBN: 2-85653-224-1.
Volume 14: *Mémoires du Muséum national d'Histoire naturelle* 167: 1-647, 987 figs, 3 color plates (1995). ISBN: 2-85653-217-9.
Volume 15: *Mémoires du Muséum national d'Histoire naturelle* 168: 1-539, 205 figs, 6 color plates (1996). ISBN: 2-85653-501-1.
Volume 16: *Mémoires du Muséum national d'Histoire naturelle* 172: 1-667, 432 figs, 2 color plates (1997). ISBN: 2-85653-506-2.
Volume 17: *Mémoires du Muséum national d'Histoire naturelle* 174: 1-213, 93 figs (1997). ISBN: 2-85653-500-3.
Volume 18: *Mémoires du Muséum national d'Histoire naturelle* 176: 1-570, 485 figs, 7 color plates (1997). ISBN: 2-85653-511-9.
Volume 19: *Mémoires du Muséum national d'Histoire naturelle* 178: 1-255, 70 figs, 4 color plates (1998). ISBN: 2-85653-517-8.
Volume 20: *Mémoires du Muséum national d'Histoire naturelle* 180: 1-588, 192 figs, 2 color plates (1999). ISBN: 2-85653-520-8.
Volume 21: *Mémoires du Muséum national d'Histoire naturelle* 184: 1-813, 384 figs, 5 color plates (2000). ISBN: 2-85653-526-7.
Volume 22: *Mémoires du Muséum national d'Histoire naturelle* 185: 1-406, 638 figs, 4 color plates (2001). ISBN: 2-85653-527-5.
Volume 23: *Mémoires du Muséum national d'Histoire naturelle* 191: 1-640, 308 figs, 3 color plates (2004). ISBN: 2-85653-557-7.
Volume 24: *Mémoires du Muséum national d'Histoire naturelle* 193: 1-417, 146 figs, 1 color plate (2006). ISBN: 2-85653-585-2.
Volume 25: *Mémoires du Muséum national d'Histoire naturelle* 196: 1-806, 710 figs, 5 color (2008). ISBN: 978-2-85653-614-8.
Volume 26: *Mémoires du Muséum national d'Histoire naturelle* 200: 1-436, 156 figs, 17 color (2010). ISBN: 978-2-85653-642-1.
Volume 27: *Mémoires du Muséum national d'Histoire naturelle* 204: 1-501, 128 figs, 32 color (2013). ISBN: 978-2-85653-692-6.
Volume 28: *Mémoires du Muséum national d'Histoire naturelle* 207: 1-362, 200 figs, 10 color (2015). ISBN: 978-2-85653-767-1.
Volume 29: *Mémoires du Muséum national d'Histoire naturelle* 208: 1-463, 555 figs, 244 color (2016). ISBN: 978-2-85653-774-9.
Volume 30: *Mémoires du Muséum national d'Histoire naturelle* 212: 1-612, 306 figs, 2 color (2018). ISBN: 978-2-85653-822-7.

Tropical Deep-Sea Benthos

volume 30

Cover photographs
First cover: *Uroptychus australis* (Henderson, 1895), male 6.8 mm, MNHN-IU-2013-2259, Bismarck Sea, 790-808 m.
Spine: *Uroptychus adnatus* n. sp. (carapace and anterior part of abdomen).

ISBN : 978-2-85653-822-7
ISSN : 1243-4442
© Publications Scientifiques du Muséum, Paris, 2018

Tropical Deep-Sea Benthos

volume 30

Mémoires du Muséum national d'Histoire naturelle
Tome 212

Chirostylidae of the Western and Central Pacific:
Uroptychus and a new genus
(Crustacea: Decapoda: Anomura)

Keiji BABA

Kumamoto University
Faculty of Education
2-40-1 Kurokami
Kumamoto 860-8555
Japan
kbaba.kumamoto@gmail.com

Publications Scientifiques du Muséum
Paris

2018

CONTENTS

ABSTRACT

BABA K. 2018 — *Chirostylidae of the Western and Central Pacific:* Uroptychus *and a new genus (Crustacea: Decapoda: Anomura)*, in *Tropical Deep-Sea Benthos 30*. Muséum national d'Histoire naturelle, Paris, 612 p. (Mémoires du Muséum national d'Histoire naturelle; 212). ISBN: 978-2-85653-822-7.

Squat lobsters of the genus *Uroptychus* Henderson, 1888 and a herein proposed new genus *Heteroptychus* (Anomura: Chirostylidae) from the Western and Central Pacific are reported based upon the material now in the collection of the Muséum national d'Histoire naturelle, Paris. The material consists of 3,784 specimens distributed among 152 species in the two genera, including 100 new species of *Uroptychus* and six new species of *Heteroptychus*. *Heteroptychus* n. gen. is established for *Uroptychus scambus* Benedict, 1902 and eight related species including one western Atlantic species. *Uroptychus edwardi* Kensley, 1977 (now transferred to *Heteroptychus*), previously synonymized with *U. scambus*, is resurrected. The number of species of *Uroptychus* from the Indo-West and Central Pacific now stands at 219. One hundred twenty-six species of *Uroptychus* and six of *Heteroptychus* occur in New Caledonia and its vicinity (northward to the Solomon Islands, southward to the Norfolk Ridge, and eastward to Fiji and Tonga). Somewhat lengthy diagnoses are provided for previously known species in order to accurately characterize their specific status, and for some known species and new species, full descriptions are given. A key to species is provided for *Uroptychus* from the Indo-West Pacific and Central Pacific, and for *Heteroptychus* worldwide. For the purpose of elaborating on the specific status of a number of known species that were briefly described, their type materials were examined, and their characters are incorporated in the key to species.

RÉSUMÉ

BABA K. 2018 — *Chirostylidae of the Western and Central Pacific:* Uroptychus *and a new genus (Crustacea: Decapoda: Anomura)*, in *Tropical Deep-Sea Benthos 30*. Muséum national d'Histoire naturelle, Paris, 612 p. (Mémoires du Muséum national d'Histoire naturelle; 212). ISBN: 978-2-85653-822-7.

Chirostylidae du Pacifique Ouest et Central: *Uroptychus* et un nouveau genre (Crustacea: Decapoda: Anomura).
Les galathées du genre *Uroptychus* Henderson, 1888 et du genre *Heteroptychus* n. gen. (Anomura: Chirostylidae) décrites ici sont issues du matériel de la collection du Muséum national d'Histoire naturelle de Paris, provenant du Pacifique Ouest et Central. Ce matériel comprend 3,784 spécimens représentant un total de 152 espèces dont 100 espèces d'*Uroptychus* et 6 espèces d'*Heteroptychus*, nouvelles pour la science. Le genre *Heteroptychus* n. gen. a été établi pour l'espèce *Uroptychus scambus* Benedict, 1902 et 8 espèces relatives dont une espèce de l'Atlantique Ouest. *Uroptychus edwardi* Kensley, 1977 (maintenant placé dans le genre *Heteroptychus*), précédemment synonymisé avec *U. scambus*, est désormais ré-établi. Les espèces d'*Uroptychus* de l'Indo-Ouest Pacifique et Central sont désormais connues de 219 espèces avec un total de 126 espèces d'*Uroptychus* et six *Heteroptychus* présentes en Nouvelle Calédonie et ses environs (du nord, des îles Salomons vers le sud, la ride de Norfolk en passant par l'est des îles Tonga aux îles Fidji). Les diagnoses de certaines espèces déjà établies sont également décrites dans cette étude afin de définir de manière précise leur statut. Les descriptions des espèces nouvelles et déjà connues sont également détaillées. Une clé des espèces du genre *Uroptychus* est fournie pour l'Indo-Ouest Pacifique et Central ainsi que pour les espèces du nouveau genre *Heteroptychus* au niveau mondial. Le matériel type des espèces connues et dont le statut a été révisé, a été examiné et leurs caractères diagnostiques ont été incorporés dans la clé des espèces.

INTRODUCTION

The squat lobster genus *Uroptychus* Henderson, 1888, belonging to the family Chirostylidae, is represented by relatively small species at most 20 mm in postorbital carapace length, with the smallest ovigerous female less than 2 mm. The majority of the species are common in deep-water habitats, usually associated with alcyonacean corals, rarely with antipatharians, pennatulaceans and crinoids, and a few are known from hydrothermal vent sites and a cold seep (Baba & de Saint Laurent 1992; Baba & Williams 1998; Martin & Haney 2005; Baeza 2011; Dong & Li 2015). Shallow-water species are rare but often observed by underwater photographers, in association with alcyonacean corals. Their mimic colorations are displayed online (search *Uroptychus kudayagi*) and in print (Minemizu 2000; Kawamoto & Okuno 2003).

In an earlier paper (Baba 2005), a historical overview of the studies of Indo-Pacific squat lobsters including *Uroptychus* was given, listing all known species. A key to species of the genus from the Indo-West Pacific was also provided, noting that the key was provisional, pending descriptions of forthcoming new species. Baba *et al.* (2008) listed 100 species that were known to occur in the Indo-West Pacific. Since then, additional 21 new species have been described: four from Taiwan (Baba & Lin 2008); six from the Kermadec Islands (Schnabel 2009); one from the Philippines, Taiwan and Japan, and another one from Western Australia (Poore & Andreakis 2011), one from Taiwan (Cabezas *et al.* 2011), five from the continental margin of Western Australia (McCallum & Poore 2013), one from the Macquarie Ridge, New Zealand (Ahyong *et al.* 2015), and two from the northeastern South China Sea (Dong & Li 2015). Prior to the present work, the genus contained 153 species worldwide: 121 from the Indo-West Pacific, five from the eastern Pacific, 22 from the western Atlantic (Baba & Wicksten 2015, 2017a, 2017b) and five from the eastern Atlantic (Baba & Macpherson 2012).

It was mentioned that additional ca.100 species were to be described by K. Baba and K. Schnabel (Baba *et al.* 2008). Part of these new species are now presented here in this paper.

The study material now in the collection of the Muséum national d'Histoire naturelle consists of 3,646 specimens distributed among 150 species, including 106 new species in two genera. *Uroptychus scambus* Benedict, 1902 and eight related species including six new species and one western Atlantic species are placed under the new genus *Heteroptychus*. The Indo-West Pacific and Central Pacific species of *Uroptychus* are now increased from 119 (two of the 121 previously known species are excluded: one relegated to synonym and another one transferred to *Heteroptychus*) to 219. One hundred twenty-six species of *Uroptychus* and six of *Heteroptychus* occur in New Caledonia and its vicinity (including the Solomon Islands, the Chesterfield Islands, Vanuatu, the Norfolk Ridge, Fiji and Tonga), a notable region of high species richness. This collection is one of the largest ever assembled and contributes much to the knowledge of squat lobster species diversity.

The following species or species groups each also possess unique characters worthy of creating new genera, but they stay in *Uroptychus* until more phylogenetic information becomes available including molecular data: *Uroptychus naso* Van Dam, 1933, along with two related species (*U. cyrano* Poore & Andreakis, 2011 and *U. pinocchio* Poore & Andreakis, 2011), characterized by a very narrow dorsal orbital margin and elongate rostrum covering the proximal half of cornea; *U. scandens* Benedict, 1902, *U. articulatus* n. sp., *U. imparilis* n. sp. and *U. parisus* n. sp., all having truncate dactyli of pereopods 2-4; *U. ctenodes* n. sp. featuring a depressed cephalothorax with laterally crested dorsal carapace surface; *U. ciliatus* (Van Dam, 1933), *U. spinirostris* (Ahyong & Poore, 2004), *U. chacei* (Baba, 1986b), *U. numerosus* n. sp., *U. quartanus* n. sp. and *U. senarius* n. sp. share a spinose body and appendages. *Uroptychus diaphorus* shows different spination between the P2 and P4 dactyli (P3 is missing in the sole known specimen but its spination is in all probability the same as that of P4). *Uroptychus inaequalis* n. sp., *U. plautus* n. sp. and *U. pilosus* Baba, 1981 are unusual among the species of *Uroptychus* in having P2-4 with only two terminal spines rather than a row of flexor marginal spines as in the other species of the genus.

Species accounts are provided for each species with novel characters, including length proportions of pereopods 2-4 articles and shape of the pterygostomian flap. Diagnoses are more detailed than those in the previous papers because very subtle differences have proved useful in distinguishing species. Many of the previously reported specimens require reexamination to increase the detail of descriptions. Only a small portion of these have been examined until now and many descriptions were incomplete.

A key to species is provided for *Uroptychus* from the Indo-West and Central Pacific, and for *Heteroptychus* worldwide. Because of the brevity of the description or lack of discriminating characters in some of the previously reported species, their type materials were examined and their key characters are incorporated in the key to species. The following species are not included in the key because their taxonomic status has not been settled: *Uroptychus gracilimanus bidentatus* Doflein & Balss, 1913 from the western Indian Ocean off Somali Republic; *U. nitidus* (A. Milne Edwards, 1880) from the Andaman and Laccadive Seas (Alcock & Anderson 1894; Anderson 1896) and from southeastern South Africa (Barnard 1950; Kensley 1977).

MATERIALS AND METHODS

The materials used in this study, is deposited in the collection of the Muséum national d'Histoire naturelle. This collection was obtained by the following oceanographic cruises led by MNHN to the western and central Pacific, covering the Philippines, Indonesia, Solomon Islands, New Caledonia and vicinity, the Chesterfield Islands, Vanuatu, Norfolk Ridge, Fiji, Tonga and French Polynesia: Aztéque, BATHUS 1-4, BENTHAUS, BERYX 2, BERYX 11, BIOCAL, BIOGEOCAL, BOA 0, BORDAU 1-2, CALSUB, CHALCAL 1-2, CORAIL 2, CORINDON, EBISCO, GEMINI, HALIPRO 1-2, KARUBAR, LAGON, LITHIST, MUSORSTOM 1-10, NORFOLK 1-2, SALOMON 1-2, SANTO, SMIB 1-6, 8, VAUBAN 1978-1979, VOLSMAR (http://expeditions.mnhn.fr/program/tropicaldeep-seabenthos). The materials examined of SANTO and BOA 0 collections do not represent their entire collections, the remaining material awaiting extensive studies.

Comparative materials including types, as well as previously reported specimens that needed confirmation of their identity, were examined during my visits to their repositories or made available on loan from: the Smithsonian Institution, Washington, D.C; the Natural History Museum, London; the Natural History Museum of Denmark, Copenhagen; the Museum für Natuurkunde an der Humboldt-Universität, Berlin; the Queensland Museum, Brisbane; the National Taiwan Ocean University, Keelung; the Kitakyushu Museum of Natural History and Human History, Kitakyushu; the National Museum of Nature and Science, Tsukuba; and the Coastal Branch of the Natural History Museum and Institute, Chiba.

The terminology used in the text generally follows Baba *et al.* (2009, 2011). Carapace regions in *Uroptychus* are not demarcated, but for convenience of description, those used for Galatheidae (Baba *et al.* 2009, 2011) are applied. "Sternum or sternite" used in the description is restricted to the thoracic somites. Pereopod 1 in natural condition is more like mesiodorsal than illustrated in the text. However, what is illustrated in the text is construed as "dorsal." The term "dorsal-ventral" is used for the P2-4 meri, and "extensor-flexor" for the other distal articles. In some specimens of *Heteroptychus*, a suture between the antennal scale and article 2 is visible on the ventral side but not on the dorsal side. This case is regarded as not articulated.

Uroptychus species display a variety of spination on pereopods 2-4, especially distal articles, which may have resulted from adaptation of mobility to commensal hosts. Size and arrangement of the spines on the dactyli, a useful character for discriminating species was often ignored in previous descriptions, and thus a full description of the dactyli for all species is given here. Typical spination is illustrated for the reader to better understand what is described in the text (Figure 1).

The measurements given under the "Material examined" indicate the postorbital carapace length measured from the orbital margin to the posterior margin of the carapace in midline. The breadth of the rostrum is measured between left and right midpoints of each orbital dorsal margin. The measurements taken for the pterygostomian flap and thoracic

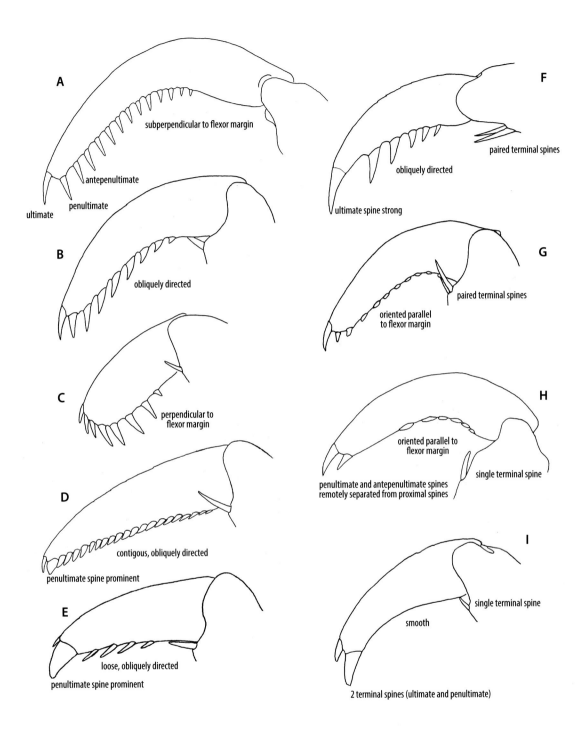

FIGURE 1

Distal part of P2, showing various forms of spination. **A**, *Heteroptychus anouchkae*. **B**, *Uroptychus dejouanneti*. **C**, *U. scandens*. **D**, *U. adiastaltus*. **E**, *U. laurentae*. **F**, *U. shanei*. **G**, *U. disangulatus*. **H**, *U. vandamae*. **I**, *U. inaequalis*.

sternite 4 are as illustrated (Figure 2). Pereopod 1 total length is measured on the ventral midline, P1 article length on the dorsal midline, and pereopods 2-4 articles along the lateral midline; the P2-4 merus breadth is measured between dorsal and ventral margins at its maximum, although this may be called "height" in natural orientation.

The species are categorized in three size groups: small species up to 5.0 mm (or ovigerous females less than 5.0 mm); medium-sized species below 10.9 mm; and large species more than 11.0 mm.

DNA analyses were conducted by Laure Corbari, Sarah Samadi and Marie-Catherine Boisselier of MNHN for a number of samples. The results helped to verify morphology-based identifications, and some suggested rechecking of some provisionally classified specimens was required. Some data clearly separated material into two or three molecularly different species but no morphological differences were found; these cases are noted as such in species accounts.

Coloration of species is described based upon photographs taken by Tin-Yam Chan (Figures 305E, 305I, 306C, 306F, 306H), Pierre Laboute (Figures 305A, 305B, 306D), Jean-Louis Menou (Figures 305C, 305D, 305F, 305H, 306A, 306B, 306E, 306G), Gérard Moutham (Figure 306I), and Neville Coleman (Figure 305G).

Commensal host corals in digital photographs were identified by Yukimitsu Imahara of the Biological Institute of Kuroshio, Wakayama. The identification was provisional, pending direct examination of the material.

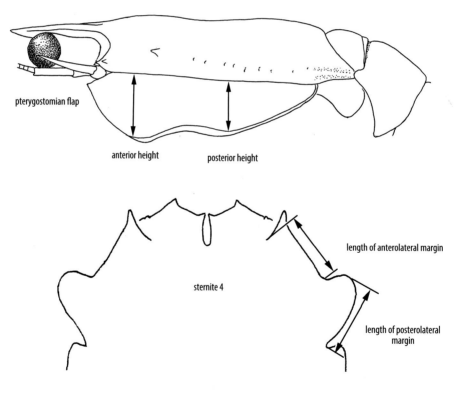

FIGURE 2

Measurements taken for pterygostomian flap and sternite 4.

Abbreviations:

cl	carapace length including rostrum
Mxp1	first maxilliped
Mxp3	third maxilliped
ov.	ovigerous
P1	first pereopod (cheliped)
P2	second pereopod (first walking leg)
P3	third pereopod (second walking leg)
P4	fourth pereopod (third walking leg)
poc	postorbital carapace length

Repositories:

AM	Australian Museum, Sydney, Australia
BMNH	Natural History Museum, London (formerly British Museum of Natural History)
MNHN	Muséum national d'Histoire naturelle, Paris
MZS	Musée Zoologique, Strasbourg
NMV	Museum Victoria, Melbourne
NSMT	National Museum of Nature and Science, Tokyo (formerly National Science Museum, Tokyo)
NTM	Northern Territory Museum, Darwin
NTOU	National Taiwan Ocean University, Keelung
QM	Queensland Museum, Brisbane
SAM	South Australian Museum, Adelaide
SAMC	South African Museum, Cape Town
SNU	Seoul National University, Seoul
TM	Tasmanian Museum and Art Gallery, Hobart, Tasmania
USNM	National Museum of Natural History, Smithsonian Institution, Washington, D.C.
ZLKU	Kitakyushu Museum of Natural History and Human History, Kitakyushu
ZMA	Zoological Museum, Amsterdam (now merged into the Naturalis Biodiversity Center, Leiden)
ZMB	Zoologisches Museum, Zentralinstitut der Humboldt-Universität, Berlin
ZMUC	Zoological Museum, University of Copenhagen, Copenhagen
ZSIC	Zoological Survey of India, Calcutta

SYSTEMATIC ACCOUNT

Superfamily CHIROSTYLOIDEA Ortmann, 1892
Family CHIROSTYLIDAE Ortmann, 1892

Genus *UROPTYCHUS* Henderson, 1888

Diptychus A. Milne Edwards, 1880: 63.
Uroptychus Henderson, 1888: 173 [replacement name for *Diptychus* (junior homonym of *Diptychus* Steindachner, 1866, Pisces)]. — Alcock 1901: 281. — Baba 1988: 17; 2005: 26, 216. — Baba *et al.* 2009: 32. — Macpherson & Baba 2011: 49.

DIAGNOSIS — Carapace dorsally smooth, granulose, with scaly ridges or spines, lateral margin smooth or spinose, anterolateral spine distinct, rarely obsolete. Rostrum narrowly or broadly triangular, flattish, laterally smooth or with small spines. Lateral limit of orbit acuminate, rounded or with small spine. Pterygostomian flap proportionately high from anterior to posterior, rarely very low on posterior half. Anterior margin of sternal plastron distinctly concave, with or without submedian spines and median notch or sinus. Excavated sternum anteriorly ending between bases of Mxp1, with or without median spine or ridge. Antennal scale articulated with or fused to article 2, flagellum of no great length, directed anteriorly, never overreaching tip of P1. Left and right Mxps3 broadly separated from each other, with distal parts accommodated in excavated sternum when folded. P1 spinose or unarmed, ischium with distodorsal spine. P2-4 dactyli with flexor marginal spines of various sizes and arrangements, P4 carpus subequal to, somewhat shorter than, or rarely longer than P3 carpus.

REMARKS — With the description of 100 new species in this paper, the Indo-West and Central Pacific species of the genus are now 219 in number, 125 of which are recorded for the first time from New Caledonia and vicinity, the major region surveyed by the French cruises (see Materials and Methods). Prior to this study, only one species, *Uroptychus amabilis* Baba, 1977, which occurs in shallow water, was known from New Caledonia.

Chirostylus ciliatus Van Dam, 1933 and *Gastroptychus spinirostris* Ahyong & Poore, 2004 have been transferred to *Uroptychus* (Baba 2005). These two species are grouped together with *U. chacei* (Baba, 1986b) and three new species described in this paper, all characterized by the spinose body and appendages, especially maxilliped 3 bearing a strong distomesial spine on the ischium, a feature not observed in other species of *Uroptychus*. These species would better be placed in a different genus, as suggested by molecular data that showed that *U. spinirostris* is placed in a different clade than other species of *Uroptychus* (Bracken-Grissom *et al.* 2013). However, they remain in *Uroptychus*, pending extensive studies.

Uroptychus scambus Benedict, 1902 is now placed in the new genus *Heteroptychus*, together with eight other species including six new species and one western Atlantic species (see below under *Heteroptychus*).

Uroptychus also contains the following aberrant species or species groups, each of which may be shifted to a different genus, but they are retained in the genus until extensive studies have been done, reviewing other related species using molecular data. These include *U. naso* Van Dam, 1933 and two related species; *U. scandens* Benedict, 1902 and three related new species; *U. ctenodes* n. sp.; *U. diaphorus* n. sp.; and *U. inaequalis* n. sp., *U. pilosus* Baba, 1981 and *U. plautus* n. sp. *Uroptychus naso* Van Dam (1933) has recently been reviewed by Poore & Andreakis (2011), with descriptions of two related new species based upon morphological and molecular data. Apparently these species are unusual among the *Uroptychus* species in having very narrow dorsal orbital margin and elongate rostrum, with the mesial half of the eyes concealed beneath the rostrum. *Uroptychus scandens* Benedict, 1902 as defined earlier by having truncate dactyli of P2-4 proved to contain four species as suggested by molecular data (L. Corbari, pers. comm.). *Uroptychus ctenodes* n. sp. is also unique in the genus in having the carapace dorsal surface crested along the lateral margin, with a strong anterolateral spine directly lateral to the rounded lateral limit of the orbit, and the pterygostomian flap very low in the posterior half as in *Heteroptychus* n. gen. *Uroptychus diaphorus* n. sp. shows a different spination between P2 and P4 dactyli (P3 is missing in the sole specimen), the fact suggesting a possible shifting of the species to a different genus as has been applied to *Uroptychodes* Baba, 2004 (see Baba 2004). *Uroptychus inaequalis* n. sp., *U. plautus* n. sp. and *U. pilosus* Baba, 1981 are different from the other species of *Uroptychus* in having only two terminal spines instead of a row of flexor marginal spines.

There still remain problematic species because of their brief accounts: *Uroptychus bacillimanus* Alcock & Anderson, 1899, *U. cavirostris* Alcock & Anderson, 1899, *U. fusimanus* Alcock & Anderson, 1899, *U. indicus* Alcock, 1901, and *U. nigricapillis* Alcock, 1901, all from the Investigator collection and now housed in the Zoological Survey of India, Calcutta. Examination of the types of these species is required to elaborate on their specific status, but for now the material is inaccessible. The relationships between *U. cavirostris* and *U. latirostris* Yokoya, 1933 still have not been settled. The western Indian Ocean material reported under *U. cavirostris* by Tirmizi (1964) is an undescribed species (see under the remarks of *U. latirostris*). *Uroptychus nigricapillis* as appeared in the literature appears to represent a complex of species (see below under this species).

The following species are not included in the key to species from the Indo-West and Central Pacific, because of lack of sufficient morphological details to secure their identity: *Uroptychus nitidus* (A. Milne Edwards, 1880) from off East London (Barnard 1950), off Durban (Kensley 1977), Laccadive Sea [13°47′49″N, 73°7′E (Anonymous 1914)] (Alcock & Anderson 1894), Bay of Bengal [8°44′40″N, 81°20′15″E (Anonymous 1914)] (Anderson 1896). One female specimen of *U. nitidus* collected by the R/V *Meiring Naude* at Station 121 south of Durban in 900-625 m and identified by B. Kensley, now in the collection of the Smithsonian Institution (USNM 1101919), was examined. It looks identical to the material reported by Kensley (1977) from off Durban but is clearly different from *U. nitidus* sensu stricto in having the P2-4 propodi with the terminal spine single, not paired as in the typical form of *U. nitidus* defined by Chace (1942) (Baba & Wicksten 2017a). This is identical to the material reported under *U. gracilimanus* from Madagascar by Baba (1990), as well as the specimens reported under *U. australis* var. *indicus* from Zanzibar by Tirmizi (1964), the identification verified by examination of the material of both (see below under the remarks of *U. psilus* n. sp.). It is apparently an undescribed species and will be described later elsewhere. Actually, true *Uroptychus nitidus* is a western Atlantic element (Baba & Wicksten 2017a), so the Investigator material (Alcock & Anderson 1894; Anderson 1896) will in all probability prove to be a different species.

The taxonomic status of *U. gracilimanus* var. *bidentatus* Doflein & Balss, 1913 from off the east coast of the Somali Republic has not yet been fixed, pending examination of the syntype, ZMB 17483, now in the collection of the Zoologisches Museum, Zentralinstitut der Humboldt-Universität, Berlin.

Uroptychus inclinis Baba, 2005 from the Kai Islands is synonymized with *U. tridentatus* (Henderson, 1885) in this paper.

Key to species from the Indo-West and Central Pacific

1. Dorsal margin of orbit extremely narrow; mesial half of cornea concealed beneath rostrum, not visible in dorsal view **2**
– Dorsal margin of orbit relatively broad; entire cornea visible in dorsal view **4**

2. Sternal plastron 1.1-1.2 × longer than broad. P2 propodus lacking dense tufts of short setae on lateral and mesial faces .... *U. naso* Van Dam, 1933
– Sternal plastron at least 1.4 × as long as broad. P2 propodus with dense tufts of short setae in 2 rows on each of lateral and mesial faces **3**

3. Rostrum with lateral spines on distal half. Posterior branchial margin with 6-7 strong spines widely spaced anteriorly, closely spaced posteriorly .... *U. pinocchio* Poore & Andreakis, 2011
– Rostrum with lateral spines on distal two-thirds. Posterior branchial margin with 9 strong spines evenly spaced by narrow U-shape .... *U. cyrano* Poore & Andreakis, 2011

4. Rostrum very broad, basal breadth at least two-thirds carapace breadth measured at posterior carapace margin.. **5**
– Rostrum basal breadth usually half or less than half, rarely slightly more than half carapace breadth measured at posterior carapace margin **10**

5. Anterior margin of sternite 3 with pair of submedian spines **6**
– Anterior margin of sternite 3 without pair of submedian spines (obsolescent spines may be present) **8**

6. Rostrum distally narrowed, lateral margin somewhat concave. Epigastric region with tubercles behind each eye. Posterior plate of telson laterally and posteriorly convex, not semicircular .... *U. mauritius* Baba, 2005
– Rostrum equilateral broad triangular, lateral margin straight. Epigastric region smooth. Posterior plate of telson semicircular **7**
7. Rostrum half as long as broad, not reaching end of cornea .... *U. simiae* Kensley, 1977
– Rostrum about as long as broad, overreaching cornea .... *U. alcocki* Ahyong & Poore, 2004

8. P1 merus proximally strongly narrowed, distally less so, shaped like a bowling pin. Posterior plate of telson long trianguloid, ending in rounded margin .... *U. yokoyai* Ahyong & Poore, 2004
– P1 merus not narrowed distally. Posterior plate of telson subsemicircular or with emarginate posterior margin [Differences in the following couplet are so slight that examination of the type material of *U. cavirostris* is recommended] **9**

9. Sternite 3 without distinct submedian spines on anterior margin. Telson distinctly emarginate on posterior margin .... *U. cavirostris* Alcock & Anderson, 1899
– Sternite 3 with obsolescent submedian spines on anterior margin. Telson slightly or barely emarginate on posterior margin .... *U. latirostris* Yokoya, 1933

10. P2-4 dactyli truncate **11**
– P2-4 dactyli distally narrowed **14**

11. Antennal scale fused with article 2 .... **12**
– Antennal scale articulated **13**

12. Eyes distally narrowed (cornea narrower than remaining eyestalk) .... *U. scandens* Benedict, 1902
– Eyes subequally broad proximally and distally (cornea as broad as remaining eyestalk) .... *U. parisus* n. sp.

13. Eyes distally narrowed (cornea narrower than remaining eyestalk) .... *U. articulatus* n. sp.
– Eyes subequally broad proximally and distally (cornea slightly broader than remaining eyestalk) ....
.... *U. imparilis* n. sp.

14. Carapace dorsal surface crested laterally, anterolateral spine strong, reaching apex of rostrum. Pterygostomian flap with comb-like row of spines .... *U. ctenodes* n. sp.
– Carapace dorsal surface not crested laterally, anterolateral spine far falling short of apex of rostrum. Pterygostomian flap without comb-like row of spines **15**

15. P2-4 dactyli with 2 terminal (ultimate and antepenultimate) spines only **16**
– P2-4 dactyli with flexor marginal spines (arranged in regular row or separated into distal and proximal groups) **18**

16. P2-4 dactyli with ultimate spine subequal to antepenultimate spine .... *U. pilosus* Baba, 1981
– P2-4 dactyli with ultimate spine more slender than antepenultimate spine **17**

17. Anterolateral corner of carapace rounded, without anterolateral spine. P2-4 dactyli longer than carpi ......
....*U. inaequalis* n. sp.
– Anterolateral corner of carapace angular, produced to small spine. P2-4 dactyli shorter than carpi. ....
.... *U. plautus* n. sp.

18. Spines present on dorsal surface of cardiac and/or branchial region **19**
– No spine on dorsal surface of cardiac and branchial regions **33**

19. P2-4 dactyli with penultimate spine prominent, much greater than (usually >2 ×, rarely >1.5 ×) antepenultimate **20**
- P2-4 dactyli with penultimate spine subequal to or slightly larger than antepenultimate **25**

20. Abdomen with spines on somites 1 and 2 .... *U. abdominalis* n. sp.
– Abdomen unarmed **21**

21. Dorsal spines of carapace strong, especially those on epigastric and anterior cardiac regions ....
.... *U. anoploetron* n. sp.
– Dorsal spines of carapace small **22**

22. Posteriormost of carapace lateral spines largest .... *U. paku* Schnabel, 2009
– Posteriormost of carapace lateral spines distinctly smaller than preceding spine **23**

23. P2-4 propodi with no flexor marginal spine [it is most likely that a terminal pair of spines may have been overlooked] .... *U. sexspinosus* Balss, 1913a
– P2-4 propodi with row of flexor marginal spines, terminal paired **24**

24. Carapace with 3 spines on epigastric region, median one followed behind by 2 spines (1 on posterior gastric, 1 on cardiac region). Antennal article 5 with strong distomesial spine, antennal scale terminating in distal end of article 5 excluding terminal spine. P2-4 dactyli with 7-9 flexor marginal spines .... *U. angustus* n. sp.
– Carapace with numerous small spines on entire dorsal surface. Antennal article 5 unarmed, antennal scale overreaching article 5. P2-4 dactyli with 21 flexor marginal spines ....
.... *U. tracey* Ahyong, Schnabel & Baba, 2015

25. Abdomen unarmed **26**
– Abdomen with spine(s) at least on somite 1 .... **28**

26. P2-4 meri and carpi spinous [characters confirmed by examination of the syntypes, 2 males, 2 ovigerous females, BMNH 1966.2.3.23-26] .... *U. spinimanus* Tirmizi, 1964
– P2-4 meri and carpi without row of spines on dorsal or extensor crest **27**

27. Carapace and P1 very spinous. P2-4 propodi with single terminal spine only on flexor margin [original description; but it may be a pair of spines] .... *U. fusimanus* Alcock & Anderson, 1899
– Carapace and P1 with sparse spines. P2-4 propodi with pair of terminal spines preceded by row of single spines on flexor margin .... *U. setifer* n. sp.

28. Anterior margin of sternite 3 without median notch and submedian spines .... *U. chacei* (Baba, 1986b)
– Anterior margin of sternite 3 with median notch flanked by spine **29**

29. Rostrum proportionately broad distally, with 9 lateral spines .... *U. numerosus* n. sp.
– Rostrum triangularly narrowed distally, with 1-5 lateral spines **30**

30. Rostrum much more than half (0.7-0.8) as long as carapace .... *U. spinirostris* (Ahyong & Poore, 2004)
– Rostrum at most half as long as carapace **31**

31. Sternal plastron broadest on sternite 4, subequally broad between sternites 5 and 7 ... *U. quartanus* n. sp.
– Sternal plastron successively broader posteriorly **32**

32. Small spines on abdomen, somite 3 with a few small spines. P2-4 dactyli with ultimate spine subequal to or slightly larger than penultimate spine .... *U. ciliatus* (Van Dam, 1933)
– Pronounced spines on abdomen, somite 3 with 2 transverse rows of strong spines. P2-4 dactyli with ultimate spine more slender than penultimate spine.. *U. senarius* n. sp.

33. P2 dactylus with flexor marginal spines separated into proximal and distal groups by considerable distance **34**
– P2 dactylus with flexor marginal spines equidistant from one another or somewhat broadly interspaced distally, not remotely separated into proximal and distal groups **40**

34. Pair of strong epigastric spines **35**
– No epigastric spines **37**

35. P2-4 propodi with concave prehensile edge along distal part of flexor margin, distalmost of flexor marginal spines situated near juncture with dactylus, remarkably remote from distal second. Sternite 4 with strong lateral process [characters confirmed by examination of the syntypes, 1 male, 1 ovigerous female, BMNH 1966.2.3.21-22] .... *U. sternospinosus* Tirmizi, 1964
– P2-4 propodi with straight prehensile edge along distal part of flexor margin, flexor marginal spines equidistantly arranged in regular row. Sternite 4 without strong lateral process **36**

36. Antennal scale articulated with article 2 .... *U. jiaolongae* Dong & Li, 2015
– Antennal scale fused with article 2 .... *U. adnatus* n. sp.

37. P2-4 propodi with concave prehensile edge (distal part of flexor margin), distal-most of flexor marginal spines located near juncture with dactylus **38**
– P2-4 propodi with straight prehensile edge (distal part of flexor margin), distal-most of flexor marginal spines remote from juncture with dactylus **39**

38. Epigastric region without elevated ridges. P2 carpus as long as P2 propodus. P1 merus with 1 median spine on distodorsal margin .... *U. thermalis* Baba & de Saint Laurent, 1992

– Epigastric region with pair of elevated ridges behind eyes. P2 carpus 0.8 × as long as P2 propodus. P1 merus unarmed on distodorsal margin .... *U. albus* McCallum & Poore, 2013

39. P2-4 propodi with distalmost flexor marginal spine much more remote from juncture with dactyli than from distal second spine. Sternite 4 having anterolateral margin as long as posterolateral margin ....
.... *U. remotispinatus* Baba & Tirmizi, 1979
– P2-4 propodi with distalmost flexor marginal spine more remote from distal second spine than from juncture with dactyli. Sternite 4 having anterolateral margin much longer than posterolateral margin ....
.... *U. vandamae* Baba, 1988

40. P2 and P4 dactyli different in spination (P2 dactylus with small, loosely arranged spines oriented parallel to flexor margin on P2, P4 dactylus with closely arranged, obliquely directed spines) [P3 not known] ....
.... *U. diaphorus* n. sp.

– P2-4 dactyli with similar spination (flexor marginal spines obliquely or perpendicularly directed, or oriented parallel to margin) **41**

41. P2-4 dactyli with spines oriented parallel to flexor margin (spines may be very small so as to be hardly identified as "oriented parallel") **42**
– P2-4 dactyli with obliquely or perpendicularly directed spines **50**

42. P2-4 propodi with distalmost of flexor marginal spines remote from juncture with dactyli **43**
– P2-4 propodi with distalmost of flexor marginal spines close to juncture with dactyli **44**

43. Anterolateral margin of sternite 4 as long as or slightly shorter than posterolateral margin ....
.... *U. bispinatus* Baba, 1988
– Anterolateral margin of sternite 4 much longer (>1.5 ×) than posterolateral margin .... *U. marcosi* n. sp.

44. Distalmost of flexor marginal spines of P2-4 propodi single, not paired **45**
– Distalmost of flexor marginal spines of P2-4 propodi paired **46**

45. Carapace as long as broad; anterolateral spine strong, overreaching lateral orbital spine. No spine on epigastric region .... *U. brevisquamatus* Baba, 1988
– Carapace longer than broad; anterolateral spine small, not overreaching lateral orbital spine. Pair of spines on epigastric region .... *U. singularis* Baba & Lin, 2008

46. Branchial margins subparallel. P1 ischium with subterminal spine on ventromesial margin ....
.... *U. webberi* Schnabel, 2009
– Branchial margins convex. P1 ischium without distinct subterminal spine on ventromesial margin **47**

47. Lateral limit of orbit unarmed. Antennal scale proportionately broad distally, ending in blunt or rounded tip .... *U. disangulatus* n. sp.
– Lateral limit of orbit with small distinct spine. Antennal scale distally tapering **48**

48. Anterolateral spine of carapace situated directly lateral to lateral orbital spine ....
.... *U. brevirostris* Van Dam, 1933
– Anterolateral spine of carapace situated posterior to position of lateral orbital spine **49**

49. Branchial margin ridged along entire length. Antennal article 2 with very small distolateral spine ....
.... *U. setosipes* Baba, 1981
– Branchial margin ridged along posterior third of length. Antennal article 2 with distinct distolateral spine
.... *U. australis* (Henderson, 1885)

50. P2-4 dactyli with penultimate spine much broader than (usually >2 ×, rarely >1.5 ×) antepenultimate .. **51**
– P2-4 dactyli with penultimate spine subequal to or somewhat broader than antepenultimate **104**

51. Antennal article 5 broadened distally, much broader than antennal scale .... *U. buantennatus* n. sp.
– Antennal article 5 not broadened distally, narrower than antennal scale **52**

52. Anterolateral spine of carapace smaller than or subequal to lateral orbital spine **53**
– Anterolateral spine of carapace distinctly larger than lateral orbital spine **63**

53. Flexor marginal spines of P2-4 dactyli perpendicularly directed (proximal small spines may be oblique) ..
.... **54**
– Flexor marginal spines of P2-4 dactyli obliquely directed **58**

54. Abdominal somite 2 covered with denticle-like small spines. P2-4 meri with row of spines on dorsal crest .... *U. kaitara* Schnabel, 2009
– Abdominal somite 2 smooth on surface. P2-4 meri unarmed or with a few small proximal spines on dorsal crest **55**

55. Posterior branchial region inflated, with a few distinct spines on posterior lateral portion ....
....*U. turgidus* n. sp.
– Posterior branchial region not inflated, with or without denticle-like small spines on posterior lateral portion **56**

56. Anterolateral spine of carapace and lateral orbital spine separated from each other by U-shape in dorsal view .... *U. sarahae* n. sp.
– Anterolateral spine of carapace and lateral orbital spine close to each other, separated by V-shape in dorsal view **57**

57. Carapace dorsal surface with denticle-like small spines on anterior portion (on hepatic and often on epigastric region). Lateral orbital spine of carapace larger than anterolateral spine .... *U. toka* Schnabel, 2009
– Carapace dorsal surface without denticle-like small spines on anterior portion. Lateral orbital spine of carapace subequal to anterolateral spine .... *U. volsmar* n. sp.

58. Carapace and anterior part of abdominal somites covered with denticles .... *U. denticulifer* n. sp.
– Carapace and abdomen not covered with denticles **59**

59. Flexor marginal spines of P2-4 dactyli closely arranged and nearly contiguous to one another ....
....*U. pronus* Baba, 2005
– Flexor marginal spines of P2-4 dactyli loosely arranged **60**

60. Ultimate spine of P2-4 dactyli distinctly broader than antepenultimate .... *U. altus* Baba, 2005
– Ultimate spine of P2-4 dactyli as slender as or more slender than antepenultimate **61**

61. Antennal article 4 unarmed .... *U. laurentae* n. sp.
– Antennal article 4 with distomesial spine **62**

62. Posterior branchial region with dorsolateral projection .... *U. longicheles* Ahyong & Poore, 2004
– Posterior branchial region smooth .... *U. paenultimus* Baba, 2005

63. P2-4 dactyli with flexor marginal spines perpendicularly directed **64**
– P2-4 dactyli with flexor marginal spines obliquely directed **66**

64. Branchial lateral margin of carapace with row of small spines .... *U. yaldwyni* Schnabel, 2009
– Branchial lateral margin of carapace smooth and unarmed **65**

65. Anterolateral spine of carapace very close to lateral orbital spine (nearly contiguous at base in dorsal view). Eyes distally narrowed. Antennal article 2 with distinct distolateral spine .... *U. poorei* n. sp.
– Anterolateral spine of carapace distinctly separated from lateral orbital spine in dorsal view. Eyes slightly swollen distally. Antennal article 2 without distolateral spine [characters of the type material (ZSIC 2340-2350/10) confirmed by K. K. Tiwari, personal comm.] .... *U. bacillimanus* Alcock & Anderson, 1899

66. Carapace lateral margin with anterolateral spine only, no additional spine (tubercle-like small spines may be present) **67**
– Carapace lateral margin with distinct spine(s) in addition to anterolateral spine **72**

67. Carapace distinctly longer than broad. Rostrum not reaching cornea .... *U. minor* n. sp.
– Carapace as long as or shorter than broad. Rostrum overreaching cornea **68**

68. Carapace dorsal surface with tubercle-like very small spines along lateral margin. Pereopods 1-4 spinous, especially P1 carpus covered with small spines; P2-4 meri and carpi with row of dorsal/extensor spines ....
.... *U. obtusus* n. sp.
– Carapace dorsal surface smooth. P1 carpus with distal spines only; P2-4 meri and carpi unarmed **69**

69. P2-4 dactyli with at most 4 loosely arranged spines proximal to prominent penultimate spine ....
.... *U. tomentosus* Baba, 1974

– P2-4 dactyli with more than 9 closely arranged spines proximal to prominent penultimate spine **70**

70. Rostrum as long as broad. P1 merus much shorter than carapace (0.7 ×) .... *U. brevipes* Baba, 1990
– Rostrum longer than broad. P1 merus about as long as or longer than carapace **71**

71. Sternite 4 with anterolateral angle rounded or produced to small spine not reaching anterior end of sternite 3. Carapace pinkish red on anterior part, whitish on remaining carapace, pale on rostrum; abdomen whitish; P1 pale pinkish red, other pereopods much paler .... *U. babai* Ahyong & Poore, 2004
– Sternite 4 with anterolateral angle produced to spine reaching anterior end of sternite 3. Body and appendages reddish .... *U. parilis* Cabezas, Lin & Chan, 2011

72. Carapace lateral margin with 1 spine (may be followed by 1 or 2 very small spines) in addition to anterolateral spine **73**
– Carapace lateral margin with more than 1 spine in addition to anterolateral spine **77**

73. Carapace lateral spine situated at anterior end of branchial margin **74**
– Carapace lateral spine situated at midlength of carapace lateral margin **75**

74. Branchial lateral margins subparallel. Antennal articles 4 and 5 unarmed .... *U. adiastaltus* n. sp.
– Branchial lateral margins convex. Antennal articles 4 and 5 each with distomesial spine ....
.... *U. alius* Baba, 2005

75. Midlateral spine of carapace small. P2-4 dactyli with 5 loosely arranged spines proximal to prominent penultimate spine, ultimate much broader than antepenultimate [characters confirmed by examination of the syntypes, ZMA De 101.693] .... *U. suluensis* Van Dam, 1933
– Midlateral spine of carapace prominent. P2-4 dactyli with more than 10 closely arranged spines, ultimate subequally slender as antepenultimate **76**

76. Antennal flagellum extending far beyond end of P1 merus .... *U. valdiviae* Balss, 1913a
– Antennal flagellum barely reaching end of P1 merus .... *U. raymondi* Baba, 2000

77. Sternite 3 without median notch on anterior margin (very small or ill-defined notch may be present) .. **78**
– Sternite 3 with median notch separating distinct or obsolescent submedian spines on anterior margin ... **81**

78. Carapace 1.7 × broader than long, lateral margin with prominent spine at anterior two-fifths of length, preceded by a few small spines and followed by more than 10 very small spines. Lateral orbital spine well developed but smaller than anterolateral spine. Sternite 3 with shallowly concave anterior margin with tiny median notch .... *U. vulcanus* n. sp.
– Carapace 1.1-1.2 × broader than long, lateral margin with 4 or 5 acute, posteriorly diminishing spines along branchial region. Lateral orbital spine much smaller than anterolateral spine. Sternite 3 with V-shaped anterior margin **79**

79. Carapace dorsal surface with scale-like striae .... *U. strigosus* n. sp.
– Carapace dorsal surface with no scale-like striae on posterior half (small scales may be present on anterior half) **80**

80. Anterolateral margin of sternite 4 about as long as posterolateral margin. Mxp3 merus unarmed ....
.... *U. dentatus* Balss, 1913a
– Anterolateral margin of sternite 4 1.5 × longer than posterolateral margin. Mxp3 merus with distolateral spine and a few small flexor marginal spines .... *U. occultispinatus* Baba, 1988

81. P2 carpus with more than 1 spine on extensor margin **82**
– P2 carpus unarmed or with at most 1 small distal spine on extensor margin **89**

82. Carapace lateral margin with 5 spines (including anterolateral spine) **83**
– Carapace lateral margin with 6 or more spines (including anterolateral spine) **84**

83. Carapace dorsal surface smooth but transverse row of small epigastric spines. P1 merus with field of oblique row of 3 closely arranged spines on mesial proximal surface .... *U. quinarius* n. sp.
– Carapace dorsal surface granulose, without epigastric spines. P1 merus without field of oblique row of 3 closely arranged spines on mesial proximal surface .... *U. vegrandis* n. sp.

84. P2-4 dactyli with 6 loosely arranged flexor marginal spines proximal to pronounced penultimate spine ...
.... *U. japonicus* Ortmann, 1892
– P2-4 dactyli with more than 10 closely arranged (nearly contiguous) flexor marginal spines proximal to pronounced penultimate spine **85**

85. Antennal scale terminating in distal end of antennal article 5. P2 merus with row of spines along ventromesial margin .... *U. nanophyes* McArdle, 1901
– Antennal scale overreaching antennal article 5. P2 merus without row of spines along ventromesial margin **86**

86. Carapace dorsal surface thickly covered with setae; lateral marginal spines small .... *U. echinatus* n. sp.
– Carapace dorsal surface barely or sparsely setose; lateral marginal spines well-developed **87**

87. Row of epigastric spines. Mxp3 ischium with small spine near distal end of flexor margin ....
.... *U. karubar* n. sp.
– Row of epigastric spines absent. Mxp3 ischium without spine near distal end of flexor margin **88**

88. Carapace lateral spines directed anteriorly. Abdominal somite 1 convex from anterior to posterior. Mxp3 ischium not rounded on distal end of flexor margin .... *U. alophus* n. sp.
– Carapace lateral spines directed anterolaterally. Abdominal somite 1 with transverse ridge. Mxp3 ischium rounded on distal end of flexor margin .... *U. longior* Baba, 2005

89. Sternite 4 with posterolateral margin as long as or longer than anterolateral margin **90**
– Sternal 4 with posterolateral margin shorter than anterolateral margin **96**

90. Carapace dorsal surface with tubercles and small spines on anterior and lateral portions ....
.... *U. tuberculatus* n. sp.
– Carapace dorsal surface unarmed or with row of epigastric spines **91**

91. P2-4 dactyli with 4 or 5 loosely arranged spines proximal to prominent penultimate spine ....
.... *U. grandior* n. sp.
– P2-4 dactyli with 12-18 closely arranged spines proximal to prominent penultimate spine **92**

92. Branchial lateral margin with 4 strong spines. P1 merus with oblique row of 3 closely arranged spines on mesioproximal surface **93**
– Branchial lateral margin with 5-7 posteriorly diminishing spines. P1 merus lacking oblique row of 3 closely arranged spines on mesioproximal surface **94**

93. Pterygostomian flap with spine below linea anomurica between second and third lateral spines of carapace. P1 merus with 4 strong ventromesial spines .... *U. floccus* n. sp.
– Pterygostomian flap without spine below linea anomurica between second and third lateral spines of carapace. P1 merus with 2 strong ventromesial spines (occasionally 1 or 2 additional small spines) ....
.... *U. dualis* n. sp.

94. Antennal article 5 with distomesial spine. Antepenultimate spine of P2-4 dactyli long relative to breadth (length-breadth ratio, 3.9-4.8) .... *U. kareenae* n. sp.
– Antennal article 5 unarmed. Antepenultimate spines of P2-4 dactyli short relative to breadth (length-breadth ratio, 2.0-2.9) **95**

95. Gastric region anteriorly sloping down on to rostrum. Antennal scale terminating in distal quarter of article 5. Ultimate spines of P2-4 dactyli as slender as antepenultimate ....*U. levicrustus* Baba, 1988

– Gastric region preceded by depressed rostrum. Antennal scale reaching first or second segment of flagellum. Ultimate spines of P2-4 dactyli more slender than antepenultimate .... *U. depressus* n. sp.

96. P2-4 dactyli with 5-7 slender, loosely arranged spines proximal to pronounced penultimate spine **97**
– P2-4 dactyli with more than 10 slender, closely arranged spines proximal to pronounced penultimate spine
.... **99**

97. Transverse row of spines across epigastric region .... *U. undecimspinosus* Kensley, 1977
– No spine on epigastric region **98**

98. P2-4 dactyli with ultimate flexor marginal spine much broader than antepenultimate ....
....*U. wolffi* Baba, 2005
– P2-4 dactyli with ultimate flexor marginal spine more slender than antepenultimate ....
.... *U. belos* Ahyong & Poore, 2004

99. Eyes strongly narrowed distally .... *U. micrommatus* n. sp.
– Eyes equally broad or somewhat narrowed proximally and distally **100**

100. Field of more than 10 spines across epigastric region .... *U. cardus* Ahyong & Poore, 2004
– No distinct spine on epigastric region **101**

101. Pterygostomian flap anteriorly roundish with small spine. P2 propodus with 15 single spines proximal to pair of terminal spines .... *U. spinulus* n. sp.
– Pterygostomian flap anteriorly angular, produced to spine. P2 propodus with 1-5 single spines proximal to pair of terminal spines **102**

102. Branchial lateral margin with 5 small spines; no spine on hepatic margin. Antennal scale terminating in distal end of article 5 .... *U. dissitus* n. sp.
– Branchial lateral margin with 4 or 5 well-developed, acute spines. Antennal scale overreaching distal end of article 5 .... **103**

103. Branchial lateral margin with 5 well-developed spines. P2 merus with row of small spines on dorsal margin .... *U. worrorra* McCallum & Poore, 2013
– Branchial lateral margin with 4 well-developed spines. P2 merus unarmed on dorsal margin ....
.... *U. crassipes* Van Dam, 1939

104. Eyes 3 × longer than broad .... *U. novaezelandiae* Borradaile, 1916
– Eyes relatively short, at most 2 × longer than broad **105**

105. P2-4 dactyli with ultimate spine more slender than penultimate **106**
– P2-4 dactyli with ultimate spine subequal to or larger than penultimate **153**

106. P2-4 propodi without flexor marginal spines .... *U. patulus* Ahyong & Poore, 2004
– P2-4 propodi with flexor marginal spines **107**

107. P4 propodi with pair of terminal spines only (P2 and P3 may have a few single spines proximal to terminal pair) **108**
– P2-4 propodi with row of flexor marginal spines, terminal single or paired **127**

108. Epigastric spines present. Ultimate spine of P2-4 dactyli slightly more slender than penultimate **109**
– Epigastric spines absent. Ultimate spine of P2-4 dactyli much more slender than penultimate **114**

109. Three well-developed spines on anterior third of carapace lateral margin. Antennal scale overreaching midlength of article 5. Article 4 with distomesial spine **110**
– Two well-developed spines (and 0-2 small spines between) on anterior third of carapace lateral margin. Antennal scale not reaching midlength of article 5. Article 4 unarmed **112**

110. Epigastric region with 2 pairs of spines, median pair small .... *U. mesodme* n. sp.
– Epigastric region with pair of spines **111**

111. Carapace lateral spines directed distinctly anterolaterally. Antennal scale barely reaching end of article 5, article 5 unarmed .... *U. trispinatus* n. sp.
– Carapace lateral spines directed slightly anterolaterally. Antennal scale overreaching article 5, article 5 with distomesial spine .... *U. paraplesius* n. sp.

112. Epigastric region with pair of spines flanked by 2 small spines. Two small but distinct spines between 2 strong anterior spines of carapace lateral margin. Female pleuron of abdominal somite 3 strongly tapering laterally .... *U. clarki* n. sp.
– Epigastric region with pair of spines only. One or two tiny or obsolescent (or obsolete) spines between 2 strong anterior spines of carapace lateral margin. Female pleuron of abdominal somite 3 moderately tapering laterally **113**

113. Antennal article 2 with strong distolateral spine. Posterior branchial margin with 4 or 5 spines ....
.... *U. defayeae* n. sp.
– Antennal article 2 with small distolateral spine. Posterior branchial margin with 1 spine at anterior end .....
.... *U. corbariae* n. sp.

114. Carapace with rounded anterolateral corner, lacking anterolateral spine. P4 merus longer than P3 merus .... *U. cylindropus* n. sp.
– Carapace with anterolateral spine. P4 merus shorter than or subequal to P3 merus **115**

115. Carapace lateral margin without spine other than anterolateral spine **116**
– Carapace lateral margin with spines in addition to anterolateral spine **121**

116. Anterolateral spine of carapace reduced to acuminate angle or very small short spine. Antennal article 2 laterally unarmed .... *U. enriquei* n. sp.
– Anterolateral spine of carapace distinct. Antennal article 2 with distinct distolateral spine **117**

117. Sternite 3 without median notch on anterior margin. Antennal article 4 unarmed .... *U. eratus* n. sp.
– Sternite 3 with median notch on anterior margin. Antennal article 4 with distomesial spine **118**

118. Anterolateral spine of carapace distinctly larger than lateral orbital spine **119**
– Anterolateral spine of carapace small, subequal to lateral orbital spine **120**

119. P2-4 dactyli with 6 perpendicularly directed spine on flexor margin. Mxp3 merus with distinct distolateral spine .... *U. amabilis* Baba, 1977
– P2-4 dactyli with 9 or 10 obliquely directed spines on flexor margin. Mxp3 merus without distolateral spine .... *U. brachycarpus* n. sp.

120. Gastric region with 2 broad prominences. P2-4 dactyli longer than carpi. P4 merus as long as P3 merus .... *U. rutua* Schnabel, 2009
– Gastric region without prominences. P2-4 dactyli as long as carpi. P4 merus 0.8 × as long as P3 merus ......
.... *U. philippei* n. sp.

121. No spine on posterior half of carapace lateral margin **122**
– Spines present on posterior half of carapace lateral margin **123**

122. Carapace lateral margin with 4 spines behind anterolateral spine, anterior 2 small. Antennal scale overreaching antennal article 5 .... *U. zezuensis* Kim, 1972
– Carapace lateral margin with 2 spines behind anterolateral spine, anterior spine occasionally obsolete. Antennal scale barely reaching end of antennal article 5 .... *U. joloensis* Van Dam, 1939

123. Anterolateral spine of carapace smaller than lateral orbital spine .... *U. bertrandi* n. sp.
– Anterolateral spine of carapace larger than lateral orbital spine **124**

124. Antennal article 2 unarmed. Flexor marginal spines of P2-4 dactyli somewhat obliquely directed ....
.... *U. convexus* Baba, 1988
– Antennal article 2 with distinct distolateral spine. Flexor marginal spines of P2-4 dactyli perpendicularly directed **125**

125. Carapace lateral margin with 12-18 small but distinct spines behind anterolateral spine ....
.... *U. spinosior* n. sp.
– Carapace lateral margin with 5 or 6 spines behind anterolateral spine (2 small hepatic marginal, 3 or 4 relatively large branchial marginal) **126**

126. Branchial lateral margin with 4 spines. P1 merus with a few spines on ventral surface. Antennal article 5 with distomesial spine only .... *U. annae* n. sp.
– Branchial lateral margin with 3 spines. P1 merus with row of 5 spines on ventromesial margin and 3 spines on proximal ventral surface. Antennal article 5 with 2 ventromesial spines in addition to distomesial spine ..
.... *U. oxymerus* Ahyong & Baba, 2004

127. Terminal spines of P2-4 propodi single, not paired .... *U. gracilimanus* (Henderson, 1885)
– Terminal spines of P2-4 propodi paired **128**

128. Carapace lateral margin without distinct spine except for anterolateral spine **129**
– Carapace lateral margin spinous **132**

129. Sternite 3 without median notch on anterior margin .... *U. tafeanus* n. sp.
– Sternite 3 with median notch **130**

130. P1 ischium with subterminal spine on ventromesial margin. P2 merus with row of small spines along ventromesial margin .... *U. denticulisquama* n. sp.
– P1 ischium unarmed on ventromesial margin. P2 merus with smooth ventromesial margin **131**

131. Antennal scale extending far beyond eye, reaching second segment of flagellum. Antennal article 5 with distomesial spine .... *U. taylorae* McCallum & Poore, 2013
– Antennal scale not overreaching eye, terminating in distal third of article 5. Antennal article 5 unarmed ...
.... *U. indicus* Alcock, 1901

132. Lateral limit of orbit rounded, lacking lateral orbital spine .... *U. smib* n. sp.
– Lateral orbital spine distinct **133**

133. Anterior margin of sternite 3 without distinct median notch and pair of spines (very small or obsolescent notch may be present) **134**
– Anterior margin of sternite 3 with distinct median notch and/or pair of spines **141**

134. Lateral marginal spines of carapace small (other than anterolateral spine) **135**
– Lateral marginal spines of carapace well-developed (at least one of spines well developed) **136**

135. Rostrum with subapical spine on each side. Antennal articles 4 and 5 each with strong distomesial spine
.... *U. perpendicularis* n. sp.
– Rostrum without subapical spine. Antennal articles 4 and 5 each with obsolescent distomesial spine ....
.... *U. lanatus* n. sp.

136. Branchial margin with 4 spines. Antennal articles 4 and 5 each with strong distomesial spine. P2-4 dactyli with 18-21 spines on flexor margin .... *U. modicus* n. sp.
– Branchial margin with 6 or more spines. Antennal articles 4 and 5 each unarmed or with obsolescent distomesial spine. P2-4 dactyli with 10-14 spines on flexor margin **137**

137. No spine on hepatic lateral margin between anterolateral and anteriormost branchial spine. Antennal scale falling short of apex of rostrum **138**

– Spines present on hepatic lateral margin between anterolateral and anteriormost branchial spines. Antennal scale reaching or overreaching apex of rostrum **139**

138. Dorsal surface of carapace with small spine directly behind rostrum and directly behind anterolateral spine. Sternite 3 with deeply V-shaped anterior margin. P2-4 dactyli shorter than carpi ...... *U. posticus* n. sp.
– No spine behind rostrum and anterolateral spine. Sternite 3 with shallowly V-shaped anterior margin with very small median notch. P2-4 dactyli longer than carpi .... *U. exilis* n. sp.

139. Anterior margin of sternite 3 transverse along median third, not V-shaped. Sternal plastron strongly broadened posteriorly, sternite 7 broadest .... *U. duplex* n. sp.
– Anterior margin of sternite 3 widely V-shaped with very small median notch. Sternal plastron gently broadened posteriorly, sternite 6 broadest **140**

140. Carapace lateral margin with 8 or 9 spines, those on posterior branchial margin narrow at base. Anterolateral spine slightly larger than lateral orbital spine .... *U. macrolepis* n. sp.
– Carapace lateral margin with 12 spines, those on posterior branchial margin broad at base (distally laciniate or bifurcate). Anterolateral spine much larger than lateral orbital spine .... *U. zigzag* n. sp.

141. Anterior margin of sternite 3 with pair of small median spines contiguously placed side by side **142**
– Anterior margin of sternite 3 with distinct median notch separating distinct or obsolescent submedian spines **143**

142. Carapace subtriangular in dorsal view (greatest breadth measured at posterior third, 2.2 × distance between anterolateral spines). Branchial marginal spines short and broad at base (broad triangular). P2-4 meri with spines on dorsal margin .... *U. triangularis* Miyake & Baba, 1967
– Carapace moderately broadened posteriorly (greatest breadth measured at posterior third, 1.8 × distance between anterolateral spines). Branchial marginal spines small. P2-4 meri unarmed on dorsal margin ....
.... *U. palmaris* n. sp.

143. Lateral margin of carapace with small spines (other than anterolateral spine) **144**
– Lateral margin of carapace with well-developed or distinct spine(s) (at most one of spines distinct in low magnification) **145**

144. Anterolateral spine of carapace small, subequal to lateral orbital spine. Rostrum with subapical spine on each side. P1 ischium without subterminal spine on ventromesial margin ....
.... *U. multispinosus* Ahyong & Poore, 2004
– Anterolateral spine of carapace distinctly larger than lateral orbital spine. Rostrum without subapical spine. P1 ischium with subterminal spine on ventromesial margin .... *U. murrayi* Tirmizi, 1964

145. Lateral orbital spine of carapace relatively large, subequal to or slightly smaller than anterolateral spine .
.... **146**
– Lateral orbital spine of carapace much smaller than anterolateral spine **147**

146. Last lateral marginal spine of carapace closer to posterior end than to preceding spine. Antennal scale barely reaching distal end of article 5, article 5 unarmed. P2-4 dactyli with 9 or 10 closely arranged, perpendicularly directed spines proximal to slender ultimate spine .... *U. beryx* n. sp.
– Last lateral marginal spine of carapace much closer to preceding spine than to posterior end. Antennal scale overreaching article 5, article 5 with strong distomesial spine. P2-4 dactyli with 5-7 loosely arranged, obliquely directed spines proximal to slender ultimate spine .... *U. vicinus* n. sp.

147. Antennal scale terminating at most in distal end of article 5 (excluding distal spine); article 5 unarmed or without distinct spine **148**
– Antennal scale overreaching article 5 (excluding distal spine); article 5 with strong distomesial spine **149**

148. Anterolateral margin of sternite 4 with rounded anterior end. P1 ischium smooth on ventromesial margin .... *U. baeomma* n. sp.

– Anterolateral margin of sternite 4 with a few small spines on anterior end. P1 ischium with well-developed subterminal spine on ventromesial margin .... *U. elongatus* n. sp.

149. One of lateral spines of carapace distinct (recognizable under low magnification), other spines small. P1 ischium without subterminal spine on ventromesial margin .... *U. magnipedalis* n. sp.
– Most of lateral spines of carapace well-developed. P1 ischium with subterminal spine on ventromesial margin **150**

150. Hepatic margin (between anterolateral and anteriormost branchial marginal spines) unarmed. Pterygostomian flap with distinct spine between anteriorly produced spine and anterior end of linea anomurica ....
.... *U. crassior* Baba, 1990
– Hepatic margin (between anterolateral and anteriormost branchial marginal spines) spinous. Pterygostomian flap without spine between anteriorly produced spine and anterior end of linea anomurica **151**

151. A few small spines on dorsal surface of hepatic region. P1 merus strongly spiny, with row of 3 obliquely arranged spine on proximal mesial face. P2 merus with row of spines along ventromesial margin ....
.... *U. lumarius* n. sp.
– No spine on dorsal surface of hepatic region. P1 merus weakly spiny, without row of 3 closely arranged spines on proximal mesial face. P2 merus smooth along ventromesial margin **152**

152. P2-4 dactyli strongly narrowed distally, with 6-7 flexor marginal spines including slender terminal. Antennal scale overreaching antennal peduncle by full length of article 5 .... *U. seductus* n. sp.
– P2-4 dactyli proportionately broad distally, with 9 flexor marginal spines including slender terminal. Antennal scale slightly overreaching antennal peduncle (reaching at most second segment of flagellum) ....
.... *U. tridentatus* (Henderson, 1885)

153. P2-4 propodi unarmed on flexor margin **154**
– P2-4 propodi with pair of terminal spines or row of spines on flexor margin **158**

154. Lateral margin of carapace with distinct spines in addition to anterolateral spine **155**
– Lateral margin of carapace with anterolateral spine only **156**

155. P2-4 dactyli with 8 long, obliquely directed spines .... *U. foulisi* Kensley, 1977
– P2-4 dactyli with 19-22 short, perpendicularly directed spines .... *U. crosnieri* Baba, 1990

156. Anterolateral spine of carapace well developed, distinctly overreaching rounded lateral orbital angle ......
.... *U. longvae* Ahyong & Poore, 2004
– Anterolateral spine of carapace small, not overreaching rounded, angular or spiniform lateral orbital angle
.... **157**

157. P2-4 dactyli strongly curving with angle of 90°. P1 fingers directed anterolaterally [Characters verified by examination of type material, 1966.2.3.41-42] .... *U. onychodactylus* Tirmizi, 1964

– P2-4 dactyli moderately curving with angle of 125°. P1 fingers directed straight forward ....
.... *U. setosidigitalis* Baba, 1977

158. P2-4 propodi with pair of terminal spines only on flexor margin .... *U. calcar* Ahyong & Poore, 2004
– P2-4 propodi with row of spines on flexor margin, distalmost single or paired (in some species, P4 propodus with terminal pair only) **159**

159. Anterior margin of sternite 3 without distinct median notch and submedian spines (ill-defined median notch and obsolescent submedian spines may be present) **160**
– Anterior margin of sternite 3 with pair of median spines or well-defined median notch separating distinct or obsolescent spines **179**

160. Carapace lateral margin without distinct spine other than anterolateral spine (fine crenulations or serrations may be present) **161**

175. Branchial marginal spines regularly arranged, subequally spaced. Crista dentata of Mxp3 with distally diminishing denticles .... *U. zeidleri* Ahyong & Poore, 2004
– Branchial marginal spines irregularly arranged, widely spaced between anterior and posterior branchial margins. Crista dentata of Mxp3 with evenly minute denticles **176**

176. Hepatic lateral margin with spinules. Mesial margin of P1 merus with several strong spines. P2-4 propodi each with spines only on distal projection of flexor margin .... *U. insignis* (Henderson, 1885)
– Hepatic lateral margin smooth and unarmed. Mesial margin of P1 merus with strong median spine other than distal spine. P2-4 propodi each with another spine distantly proximal to those on distal projection ....
.... *U. spinulosus* Dong & Li, 2016

177. P2-3 meri with spines on dorsal margin. P1 carpus with row of dorsal spines ....
.... *U. hamatus* Khodkina, in Zarenkov & Khodkina, 1983
– P2-3 meri unarmed on dorsal margin. P1 carpus without row of dorsal spines **178**

178. Rostrum without lateral spines. Sternite 3 having anterior margin medially concave, without median sinus. Antennal articles 4 and 5 unarmed. P1 fingers directed anterolaterally [characters confirmed by examination of male holotype, ZMA De. 101.666] .... *U. xipholepis* Van Dam, 1933
– Rostrum with a few small lateral spines. Sternite 3 having anterior margin V-shaped, with ill-defined median sinus. Antennal articles 4 and 5 each with distomesial spine. P1 fingers directed straight forward ....
.... *U. subsolanus* Ahyong & Poore, 2004

179. Carapace broader than long **180**
– Carapace about as long as or longer than broad **183**

180. Carapace lateral margin with anterolateral spine only **181**
- Carapace lateral margin spinous **182**

181. Anterior margin of sternite 3 shallowly concave, with U-shaped median notch separating small submedian spines. Antennal articles 4 and 5 unarmed. P2 carpus as long as P2 dactylus [characters confirmed by examination of the male holotype, NSMT-Cr. 6177] .... *U. glaber* Baba, 1981
– Anterior margin of sternite 3 broadly V-shaped, with semicircular median notch separating obsolescent submedian spines. Antennal articles 4 and 5 each with small distomesial spine. P2 carpus 1.6 × longer than P2 dactylus [characters confirmed by examination of the ovigerous female holotype, BMNH 1966.2.3.43] ....
.... *U. siraji* Tirmizi, 1964

182. Pterygostomian flap with 2 spines on dorsal margin between anterior terminal spine and anterior end of linea anomurica. Sternite 3 with distinct submedian spines. P2-4 dactyli with 9-11 flexor marginal spines ....
.... *U. paracrassior* Ahyong & Poore, 2004
– Pterygostomian flap smooth on dorsal margin between anterior terminal spine and anterior end of linea anomurica. Sternite 3 without submedian spines. P2-4 dactyli with 6 or 7 flexor marginal spines. ....
.... *U. megistos* n. sp.

183. Row of flexor marginal spines of P2-4 propodi distally ending in single spine **184**
– Row of flexor marginal spines of P2-4 propodi distally ending in pair of spines **188**

184. Carapace dorsal surface and P1 granulose .... *U. soyomaruae* Baba, 1981
– Carapace dorsal surface and P1 smooth (epigastric spines may be present) **185**

185. Pair of spines on epigastric region **186**
– No spine on epigastric region (pair of tuberculate ridges may be present) **187**

186. P2-4 relatively slender: P2 merus 5.2-6.5 × longer than broad. Flexor margins of P2-4 propodi having distalmost spine remote from juncture with dactylus (closer to distal second spine or equidistant between juncture and distal second spine) .... *U. nigricapillis* Alcock, 1901

– P2-4 relatively broad: P2 merus 3.7-4.8 × longer than broad. Flexor margins of P2-4 propodi having distalmost spine very close to juncture with dactylus .... *U. terminalis* n. sp.

187. Carapace longer than broad. Antennal article 2 with distinct distolateral spine .... *U. stenorhynchus* n. sp.
– Carapace as long as broad. Antennal article 2 distolaterally acuminate or with very tiny spine ....
.... *U. dejouanneti* n. sp.

188. Sternite 5 with feebly convex or nearly straight anterolateral margin .... *U. politus* (Henderson, 1885)
– Sternite 5 with distinctly convex anterolateral margin **189**

189. P4 carpus longer than P3 carpus .... *U. longicarpus* n. sp.
– P4 carpus as long as or shorter than P3 carpus **190**

190. Carapace lateral margin with well-developed spine(s) in addition to anterolateral spine **191**
– Carapace lateral margin with anterolateral spine only, unarmed elsewhere (very small spine(s) or denticles may be present) **197**

191. Sternite 4 having anterolateral margin about as long as posterolateral margin ....*U. squamifer* n. sp.
– Sternite 4 having anterolateral margin distinctly longer than posterolateral margin **192**

192. Antennal article 4 with distomesial spine. P1 merus spinous .... *U. boisselierae* n. sp.
– Antennal article 4 unarmed. P1 merus unarmed **193**

193. Epigastric region with several small spines **194**
– Epigastric region unarmed **195**

194. Sternite 4 with anterior breadth (measured between left and right anteriorly produced spines) 0.55 × posterior breadth (greatest breadth between left and right posterolateral lobes); length of anterolateral margin 0.53-0.56 × distance between left and right anteriorly produced spines .... *U. longioculus* Baba, 1990
– Sternite 4 with anterior breadth (measured between left and right anteriorly produced spines) 0.45-0.49 × posterior breadth (greatest breadth between left and right posterolateral lobes); length of anterolateral margin 0.63-0.73 × distance between left and right anteriorly produced spines .... *U. poupini* n. sp.

195. Carapace longer than broad. P1 ischium with small subterminal spine on ventromesial margin ....
.... *U. flindersi* Ahyong & Poore, 2004
– Carapace as long as broad. P1 ischium unarmed on ventromesial margin **196**

196. Cornea slightly inflated. P2-4 meri relatively broad, length-breadth ratio, 3.7-3.9 on P2, 3.4-4.1 on P3, 3.6-3.8 on P4. P2 merus longer than P3 merus, distinctly shorter than (0.8 × length of) carapace ....
.... *U. nebulosus* n. sp.
– Cornea distinctly inflated. P2-4 meri relatively narrow, length-breadth ratio, 5.0-5.6 on P2, 5.0-5.3 on P3, 4.2-4.6 on P4. P2 merus as long as P3 merus, subequal to or very slightly shorter than carapace ....
.... *U. sibogae* Van Dam, 1933

197. Pair of epigastric spines present **198**
– Epigastric spines absent (pair of tuberculate ridges may be present) **201**

198. P2-4 carpi at least slightly shorter than propodi [characters observed in male holotype and in a specimen taken at Lau Back-Arc Basin, 2668 m by Cruise TUIM-06-MV, Dive 140, now in the collection of MNHN] .
.... *U. bicavus* Baba & de Saint Laurent, 1992
– P2-4 carpi less than two-thirds length of propodi **199**

199. Pterygostomian flap with anterior margin sharp angular, ending in distinct spine .... *U. benthaus* n. sp.
– Pterygostomian flap with anterior margin roundish, bearing tiny spine **200**

200. Sternite 4 granulose on surface. P1 palm granulose on ventral surface. P2-4 dactyli with row of plumose setae along extensor margin .... *U. sagamiae* Baba, 2005

– Sternite 4 not granulose on surface. P1 palm smooth on ventral surface. P2-4 dactyli without row of plumose setae along extensor margin .... *U. pollostadelphus* n. sp.

201. P1 ischium with distinct subterminal spine on ventromesial margin **202**

– P1 ischium unarmed or with obsolescent or very small subterminal spine on ventromesial margin **205**

202. Anterolateral spine of carapace not overreaching lateral orbital spine. P1 merus covered with small spines on ventral surface .... *U. psilus* n. sp.

– Anterolateral spine of carapace overreaching lateral orbital spine. P1 merus smooth or granulated on ventral surface **203**

203. Carapace dorsal surface smooth .... *U. granulipes* n. sp.

– Carapace dorsal surface granulose **204**

204. Antennal article 5 with distinct distomesial spine. Dactylus-carpus length ratio 0.6 on P2 and P3, 0.7 on P4 .... *U. brucei* Baba, 1986a

– Antennal article 5 unarmed. Dactylus-carpus length ratio 0.7-0.8 on P2, 0.8-0.9 on P3, 0.9-1.0 on P4 ....
.... *U. maori* Borradaile, 1916

205. Sternite 4 with anterolateral angle strongly produced forward, reaching or overreaching anterior end of sternite 3 .... **206**

– Sternite 4 with anterolateral angle not reaching anterior end of sternite 3 .... **207**

206. Carapace and abdominal somites 1-2 granulated. Submedian spines on anterior margin of sternite 3 contiguous at base .... *U. anatonus* Baba & Lin, 2009

– Carapace and abdominal somites smooth. Submedian spines on anterior margin of sternite 3 separated by distinct notch [characters confirmed by reexamination of the male holotype, USNM 150312] ....
.... *U. acostalis* Baba, 1988

207. Sternite 4 having anterolateral margin about as long as posterolateral margin .... *U. salomon* n. sp.

– Sternite 4 having anterolateral margin distinctly longer than posterolateral margin **208**

208. P1 palm with sharply ridged mesial margin. P2-4 dactyli less than one-third length of propodi ....
.... *U. brachydactylus* Tirmizi, 1994

– P1 palm with roundly ridged mesial margin. P2-4 dactyli distinctly more than one-third length of propodi
.... **209**

209. Carapace 1.2-1.3 × longer than broad .... *U. lacunatus* n. sp.

– Carapace nearly as long as broad (at most 1.1 × longer than broad) **210**

210. Pterygostomian flap anteriorly produced to spine **211**

– Pterygostomian flap anteriorly roundish or bluntly angular, with or without tiny spine at anterior terminus
.... **215**

211. Anterolateral spine of carapace short, not reaching lateral orbital spine .... *U. similis* Baba, 1977

– Anterolateral spine of carapace long, overreaching lateral orbital spine **212**

212. Antennal article 2 acuminate at distolateral angle, lacking distinct spine .... *U. inermis* n. sp.

– Antennal article 2 with well-developed distolateral spine **213**

213. Carapace and pterygostomian flap granular on surface. Anterolateral margin of sternite 4 smooth. P2-4 dactyli subequally long as carpi .... *U. anacaena* Baba & Lin, 2009

– Carapace and pterygostomian flap smooth on surface (pair of granulate ridges on epigastric region may be present in large specimens). Anterolateral margin of sternite 4 irregular. P2-4 dactyli much shorter than carpi **214**

214. Sternite 4 with transverse row of granules on surface. Excavated sternum with angular anterior margin. P2 propodus with 7 or 8 spines along distal half of flexor margin, proximal to pair of terminal spines ....

.... *U. litosus* Ahyong & Poore, 2004

– Sternite 4 with field of granules on surface. Excavated sternum with rounded anterior margin. P2 propodus with 10-12 spines along entire length of flexor margin, proximal to pair of terminal spines ....
.... *U. bardi* McCallum & Poore, 2013

215. P4 merus relatively short, 0.6 × length of P3 merus **216**
– P4 merus relatively long, 0.8-0.9 × length of P3 merus **217**

216. Carapace dorsal surface smooth (epigastric region with pair of granulate ridges in large specimens). No ridge along lateral margin of carapace .... *U. empheres* Ahyong & Poore, 2004
– Carapace dorsal surface sparsely granulated. Lateral margin of carapace ridged along posterior quarter of length [characters confirmed by reexamination of male holotype, USNM 150458] ....
.... *U. comptus* Baba, 1988

217. Antennal article 2 with acuminate distolateral angle, without distinct spine. P4 merus 0.9 × length of P3 merus. Pterygostomian flap without spine at anterior terminus .... *U. septimus* n. sp.
– Antennal article 2 with well-developed distolateral spine. P4 merus 0.8 × length of P3 merus. Pterygostomian flap with small spine at anterior terminus **218**

218. Dorsal surface of carapace with scattered granules on hepatic and branchial regions; epigastric region with pair of granulate ridges. P1 merus with 2 rows of small, low spines on ventral surface ....
.... *U. orientalis* Baba & Lin, 2008
– Dorsal surface of carapace smooth. P1 merus smooth on ventral surface ....
.... *U. jawi* McCallum & Poore, 2013

Uroptychus abdominalis n. sp.

Figures 3, 4, 305A

TYPE MATERIAL — Holotype: **New Caledonia**, Norfolk Ridge. SMIB 3 Stn DW02, 24°54'S, 168°22'E, 537-530 m, 20.V.1987, ♂ 4.9 mm (MNHN-IU-2014-17135). Paratypes: **New Caledonia**, Norfolk Ridge. SMIB 3 Stn DW01, 24°55'S, 168°22'E, 520 m, 20.V.1987, 1 ♂ 4.0 mm, 3 ov. ♀ 3.4-3.8 mm (MNHN-IU-2014-17136). – Stn DW02, station data as for the holotype, 1 ♂ 4.5 mm, 1 ov. ♀ 4.7 mm (MNHN-IU-2014-17137). – Stn DW03, 24°54'S, 168°22'E, 513 m, 20.V.1987, 1 ♂ 4.0 mm (MNHN-IU-2014-17138). – Stn DW05, 24°55'S, 168°22'E, 502-512 m, 21.V.1987, 1 ov. ♀ 3.5 mm (MNHN-IU-2014-17139). CHALCAL 2 Stn CC02, 24°55'S, 168°21'E, 500-610 m, 28.X.1986, 1 ov. ♀ 3.0 mm (MNHN-IU-2014-17140). – Stn DW72, 24°54.5'S, 168°22.3'E, 527 m, 28.X.1986, 2 ♂ 2.6, 4.0 mm, 1 ov. ♀ 4.0 mm (MNHN-IU-2014-17141). – Stn DW73, 24°40'S, 168°38'E, 573 m, 29.X.1986, 2 ov. ♀ 3.6, 4.0 mm (MNHN-IU-2014-17142). NORFOLK 2 Stn DW2056, 24°40.32'S, 168°39.17'E, 573-600 m, 25.X.2003, 1 ♂ 4.7 mm, 1 ♀ 3.2 mm (MNHN-IU-2014-17143). – Stn DW2058, 24°39.76'S, 168°40.43'E, 591-1032 m, 25.X.2003, 1 ov. ♀ 3.9 mm (MNHN-IU-2014-17144). – Stn DW2060, 24°39.84'S, 168°38.50'E, 582-600 m, 25.X.2003, 1 ♂ 4.3 mm (MNHN-IU-2014-17145). – Stn DW2087, 24°56.22'S, 168°21.66'E, 518-586 m, 28.X.2003, 1 ov. ♀ 3.4 mm (MNHN-IU-2014-17146). – Stn CP2088, 24°57.48'S, 168°21.70'E, 627-1089 m, 28.X.2003, 3 ♂ 3.3-4.5 mm, 1 ov. ♀ 4.2 mm (MNHN-IU-2014-17147). – Stn CP2089, 24°44.30'S, 168°08.83'E, 227-230 m, 29.X.2003, 9 ♂ 3.0-5.0 mm, 5 ov. ♀ 3.1-4.3 mm (MNHN-IU-2014-17148).

OTHER MATERIAL EXAMINED — New Caledonia, Norfolk Ridge. LITHIST Stn CP15, 23°40.4'S, 168°15.0'E, 389-404 m, 12.VIII.1999, 2 ov. ♀ 3.7, 3.8 mm (MNHN-IU-2014-17149). BERYX 11 Stn CP08, 24°54'S, 168°21'E, 540-570 m, 15.X.1992, 7 ♂ 3.3-4.5 mm, 4 ov. ♀ 3.1-3.9 mm (MNHN-IU-2014-17150). – Stn DW09, 24°52.10'S, 168°21.95'E, 630-680 m, 15.X.1992, 1 ov. ♀ 3.8 mm (MNHN-IU-2014-17151). – Stn DW10, 24°52.85'S, 168°21.40'E, 560-600 m, 15.X.1992, 1 ♂ 5.1 mm, 1 ov. ♀ 3.8 mm, 1 ♀ 4.1 mm (MNHN-IU-2014-17152). SMIB 4 Stn DW55, 23°21.4'S, 168°04.5'E,

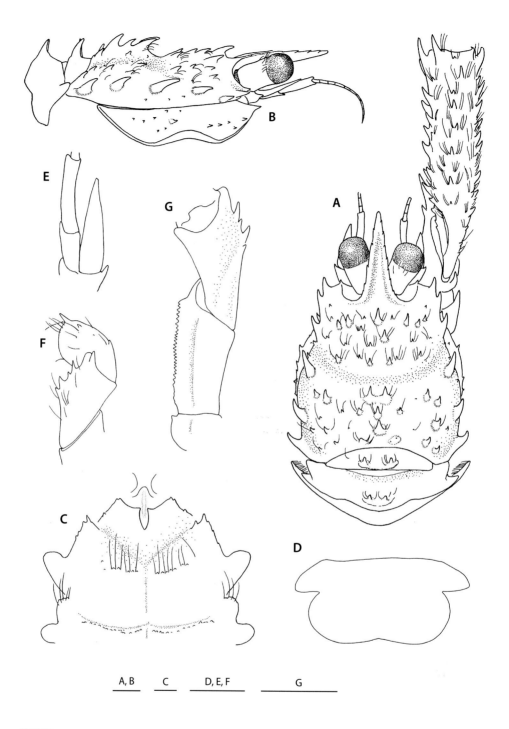

FIGURE 3

Uroptychus abdominalis n. sp., holotype, male 4.9 mm (MNHN-IU-2014-17135). **A**, carapace and anterior part of abdomen, proximal articles of left P1 included, dorsal. **B**, same, P1 omitted, lateral. **C**, anterior part of sternal plastron, excavated sternum included. **D**, telson. **E**, left antenna, ventral. **F**, merus and carpus of left Mxp3, lateral. **G**, same, ventral. Scale bars: 1 mm.

215-260 m, 9.III.1989, 1 ov. ♀ 4.2 mm (MNHN-IU-2014-17153). SMIB 8 Stn DW153, 24°53.55'S, 168°21.33'E, 547-560 m, 27.I.1993, 1 ov. ♀ 3.9 mm (MNHN-IU-2014-17154). NORFOLK 1 Stn DW1691, 24°55'S, 168°21'E, 509-513 m, 23.VI.2001, 2 ♂ 2.8, 4.3 mm, 1 ov. ♀ 3.8 mm (carapace broken) (MNHN-IU-2014-17155). – Stn DW1697, 24°40'S, 168°39'E, 569-616 m, 24.VI.2001, 1 ♀ 3.5 mm (MNHN-IU-2014-17156).

ETYMOLOGY — The Latin *abdominalis* (abdominal) refers to the abdomen of the species that has tergites 1 and 2 each armed with two strong spines placed side by side, a character to separate the species from the close relative, *U. anoploetron* n. sp.

DISTRIBUTION — Norfolk Ridge; 227-1089 m.

SIZE — Males, 2.6-5.1 mm; females, 3.0-4.7 mm; ovigerous females from 3.0 mm.

DESCRIPTION — Small species. *Carapace*: Broader than long (length 0.84-0.95 breadth); greatest breadth 1.5 × distance between anterolateral spines. Dorsal surface somewhat convex, separated into anterior and posterior portions by well depressed concavity along ordinary site of cervical groove, with spines and very sparse setae as figured; gastric region indistinctly separated from hepatic region; epigastric region with 7 spines arranged roughly in transverse row, often with 1-2 additional small spines laterally; median and posterior gastric regions also with small spines; cardiac region delimited in inverse triangle, moderately elevated anteriorly, with 2 strong anterior (placed side by side) and 2 posterior (placed in midline) spines followed by pair of spines located on intestinal region; branchial region also with scattered spines often reduced to several in number and arranged in longitudinal row. Lateral margins convexly divergent posteriorly, armed with 6 spines, often somewhat constricted behind third spine; anterior 3 more or less close to one another; first anterolateral, somewhat overreaching small lateral orbital spine, and moderately distant from that spine; second distinctly more slender than first, closer to first than to third, ventral to level of remainder; third larger than first; fourth to sixth prominent, subequal, equidistant from one another; last (sixth) arising from ridged margin near posterior end. Rostrum long triangular, with interior angle of about 20°, directed slightly dorsally; length 0.6-0.7 × that of remaining carapace, breadth less than half carapace breadth measured at posterior carapace margin; dorsal surface concave; lateral margin with 4-7 denticle-like spines often very small or obsolete. Pterygostomian flap with scattered small spines, anteriorly produced, ending in distinct spine.

Sternum: Excavated sternum with anterior margin strongly convex between Mxps1, surface with relatively sharp ridge in midline. Sternal plastron about as long as broad, lateral extremities gently divergent posteriorly. Sternite 3 moderately depressed; anterolateral angle produced, rarely ending in bifid spine; anterior margin gently concave, with V-shaped, rarely U-shaped median notch flanked by spine distinct, obsolescent or obsolete; lateral margin with proximal small spine. Sternite 4 slightly narrower than sternite 5; with short, nearly straight anterolateral margin anteriorly ending in small spine often accompanying 1 or 2 smaller spines mesial to it, followed by posteriorly diminishing denticles; posterolateral margin about as long as anterolateral margin. Anterolateral margin of sternite 5 strongly convex, 0.6 × as long as posterolateral margin of sternite 4.

Abdomen: Smooth and glabrous. Somites 1 and 2 each with pair of well-developed submedian spines on tergite; somite 2 tergite 2.6-2.7 × broader than long; pleuron having lateral margins well concave, knife-edge-like, sharply produced on anterior and posterior ends. Pleura of somites 3 and 4 laterally tapering. Telson about half as long as broad; posterior plate 1.5 × longer than anterior plate, posterior margin feebly concave or distinctly emarginate.

Eye: 1.7 × longer than broad, overreaching midlength of and barely reaching tip of rostrum. Cornea slightly broader than and much more than half as long as proximally narrowed remaining eyestalk.

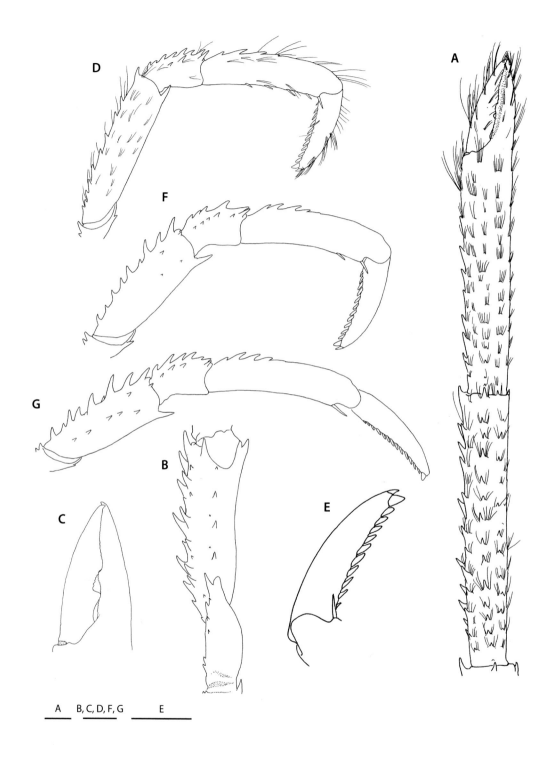

FIGURE 4

Uroptychus abdominalis n. sp., holotype, male 4.9 mm (MNHN-IU-2014-17135). **A**, right P1, proximal articles omitted, dorsal. **B**, left P1, proximal part, ventral. **C**, same, fingers, ventral. **D**, right P2, lateral. **E**, same, distal part, setae omitted, lateral. **F**, right P3, setae omitted, lateral. **G**, right P4, lateral. Scale bars: 1 mm.

Antennule and antenna: Ultimate article of antennular peduncle 2.5-3.2 × longer than high. Antennal peduncle barely or slightly overreaching apex of rostrum. Article 2 with distinct distolateral spine. Antennal scale much broader than article 5, overreaching midlength of, and, falling short of distal end of article 4, rarely with lateral spine at proximal third. Article 4 with strong distomesial spine ventrally. Article 5 unarmed, about twice as long as article 4, breadth at most two-thirds height of ultimate article of antennular peduncle. Flagellum of 9-12 segments far falling short of distal end of P1 merus.

Mxp: Mxp1 with bases broadly separated. Mxp3 basis with 1 denticle (often bidentate) on distal portion of well-convex mesial ridge, bearing 1 or 2 additional small denticles proximal to it in large specimens. Ischium with flexor margin not rounded distally, crista dentata with more than 20 small denticles. Merus 1.7 × longer than ischium, flexor margin well crested, with 2 or 3 spines on distal half of length, often accompanying smaller spine proximal to each; distolateral spine well developed. Carpus with well-developed distolateral spine and often 1 or 2 additional small spines distantly proximal to it.

P1: Slender, spinose, sparsely or moderately setose, 5.6-6.6 × longer than carapace. Ischium with prominent dorsal spine and 4-6 proximally diminishing ventromesial spines, distal one subterminal and well developed. Merus 1.4 × longer than carapace, with 5 rows of spines (1 mesial, 2 dorsal, 1 lateral, 1 ventral) continued onto carpus; mesial spines stronger. Carpus 1.2-1.3 × longer than merus, with lateral row of small spines somewhat dorsal in position, often reduced in number. Palm 4.1-5.4 × longer than broad (with no distinct sexual difference), 0.9 × as long as carpus; mesial margin with spines except on distal two-fifths or more of length; dorsal proximal spines in midline often obsolescent; lateral margin unarmed. Fingers distally slender, slightly crossing, ending in small incurved blunt spine; opposable margins with row of denticles, not spooned, somewhat gaping in large males, proximal process on movable finger usually low and reduced in females and small males, moderate in size in large males; movable finger slightly less than half as long as palm.

P2-4: Relatively broad. Meri successively shorter posteriorly (P3 merus 0.9 × length of P2 merus, P4 merus 0.8-1.0 × length of P3 merus), equally broad on P2-4; length-breadth ratio, 3.5-3.8 on P2, 2.9-3.5 on P3, 2.7-3.2 on P4; P2 merus 0.8 × length of carapace, about as long as P2 propodus; P3 merus 0.8 × length of P3 propodus; P4 merus 0.7 × length of P4 propodus; dorsal margins with 8-10 spines (occasionally distalmost spine with smaller accompanying spine), and 1 distoventral spine somewhat longer than distalmost dorsal spine; lateral surface with another row of several spines present usually on P3, occasionally on P4. Carpi subequal, about 0.3 × as long as propodi on P2-4; extensor margin with 5-7 spines paralleling row of smaller spines on lateral surface. Propodi shorter on P2 than on P3-4; extensor margin with 2-4 spines on proximal portion; flexor margin somewhat concave in lateral view, ending in pair of spines preceded by 3-5 slender spines on P2, 1 on P3, none on P4. Dactyli relatively stout, more than 1.5 × longer than carpi on P2-4, about 0.6 × length of propodi on P2-4; flexor margin slightly curving, with 11-13 obliquely directed movable spines including small ultimate (terminal); penultimate (distal second) prominent, short and broad at base; remaining proximal spines slender, distally blunt, diminishing toward base of article.

Eggs. Ova, 0.9-1.0 mm in diameter; up to 21 eggs carried.

Color. Ovigerous female from Chalcal 2, Stn CC02 (MNHN-IU-2014-17140): Basic color of translucent light purple; carapace and abdomen yellowish; P1 with yellow bands, 2 on each of merus, carpus and propodus, 1 on basi-ischium and dactylus; P2-4 also with yellow bands, 2 on merus and propodus, 1 on carpus, dactylus yellow-tinged.

REMARKS — The species strongly resembles *U. anoploetron* n. sp. in the carapace ornamentation and spination of the pereopods. Their relationships are discussed under the remarks of that species (see below).

Uroptychus adiastaltus n. sp.

Figures 5, 6

TYPE MATERIAL — Holotype: **New Caledonia**, Isle of Pines. BIOCAL Stn DW33, 23°11'S, 167°10'E, 675-680 m, 29.VIII.1985, 1 ov. ♀ 4.3 mm (MNHN-IU-2012-687). Paratypes: **New Caledonia**, Isle of Pines. BIOCAL Stn DW33, 23°11'S, 167°10'E, 675-680 m, 29.VIII.1985, 3 ♂ 3.3-4.9 mm, 7 ov. ♀ 3.7-4.6 mm (MNHN-IU-2014-17157). – Stn DW36, 23°09'S, 167°11'E, 650-680 m, 29.VIII.1985, 1 ♂ 2.8 mm, 2 ov. ♀ 3.5, 3.7 mm (MNHN-IU-2012-685). – Stn DW51, 23°05'S, 167°45'E, 680-700 m, 31.VIII.1985, 13 ♂ 2.3-3.1 mm, 12 ov. ♀ 2.7-3.6 mm, 5 ♀ 2.2-3.9 mm (MNHN-IU-2012-684). MUSORSTOM 4 Stn DW220, 22°58.5'S, 167°38.3'E, 505-550 m, 29.IX.1985, 5 ♂ 3.3-3.7 mm, 2 ov. ♀ 3.5, 3.6 mm, 1 ♀ 3.1 mm (MNHN-IU-2012-686). **New Caledonia**, Norfolk Ridge. NORFOLK 1 Stn DW1666, 23°42'S, 167°43'E, 469-860 m, 20.VI.2001, 1 ov. ♀ 4.9 mm (MNHN-IU-2012-693). NORFOLK 2 Stn DW2027, 23°26.34'S, 167°51.38'E, 465-650 m, 21.X.2003, 2 ♂ 4.9, 5.0 (MNHN-IU-2014-17158). – Stn DW2091, 24°45,36'S, 168°06.24'E, 600-896 m, 29.X.2003, 1 ♂ 3.7 mm (MNHN-IU-2012-690). – Stn DW2101, 23°55.09'S, 167°44.10'S, 700-730 m, 30.X.2003, 1 ♀ 4.0 mm (MNHN-IU-2012-688). – Stn DW2106, 23°53.72'S, 167°41.76'E, 685-757 m, 30.X.2003, 1 ov. ♀ 3.7 mm (MNHN-IU-2012-689).

ETYMOLOGY — From the Greek *adiastaltos* (not clearly defined), alluding to sternite 3 medially bearing a poorly defined sinus, by which the species is readily distinguished from *U. kareenae* n. sp., *U. depressus* n. sp. and *U. levicrustus* Baba, 1988.

DISTRIBUTION — Isle of Pines and Norfolk Ridge, in 465-896 m.

SIZE — Males, 2.3-5.0 mm; females, 2.2-4.9 mm; ovigerous females from 2.7 mm.

DESCRIPTION — Small species. *Carapace*: Slightly broader than long (0.9 × as long as broad); greatest breadth 1.7-1.8 × distance between anterolateral spines. Dorsal surface feebly convex from anterior to posterior, without any distinct border; surface nearly smooth, sparingly with short fine setae. Lateral margins divergent posteriorly to point at anterior end of branchial region, then subparallel or slightly divergent posteriorly; ridged along branchial margin; anterolateral spine well developed, overreaching lateral orbital spine; branchial margin with well-developed spine at anterior end, followed by a few to several granulate ridges visible in lateral view, occasionally 1 or 2 very small spines. Rostrum narrow triangular, with interior angle of 22-24°, length usually slightly less than half, rarely about half that of carapace, breadth about one-third to two-fifths carapace breadth measured at posterior carapace margin; dorsal surface slightly concave. Pterygostomian flap anteriorly moderately angular, ending in distinct spine, smooth on surface.

Sternum: Excavated sternum broad triangular on anterior margin, with weak ridge in midline; sternal plastron slightly broader than long, lateral extremities divergent posteriorly. Sternite 3 well depressed, anterior margin of broad V-shape medially sharply notched, laterally angular. Sternite 4 having anterolateral margin relatively short, three-quarters as long as posterolateral margin, convex, anteriorly ending in blunt corner; anterolateral margin of sternite 5 convex, about half as long as posterolateral margin of sternite 4.

Abdomen: Smooth and nearly glabrous; somite 1 with antero-posteriorly convex transverse ridge. Somite 2 tergite 2.4-2.8 × longer than broad; pleural lateral margins feebly concave and weakly divergent posteriorly. Pleuron of somite 3 posterolaterally blunt angular. Telson half as long as broad or slightly less than so, feebly emarginate or slightly concave on posterior margin, posterior plate 1.2-1.4 × longer than anterior plate.

Eye: Relatively short (1.4 × longer than broad), reaching or slightly overreaching midlength of rostrum, medially somewhat swollen; cornea not inflated, more than half length of remaining eyestalk.

Antennule and antenna: Ultimate article of antennular peduncle 3.4 × longer than high. Antennal article 2 with small lateral spine; antennal scale reaching distal end of antennal article 5; article 4 usually unarmed, occasionally with tubercle-like mesial distoventral spine; article 5 unarmed, 1.4 × length of article 4, breadth about half height of ultimate article of antennule; flagellum of 9-12 segments far falling short of distal end of P1 merus.

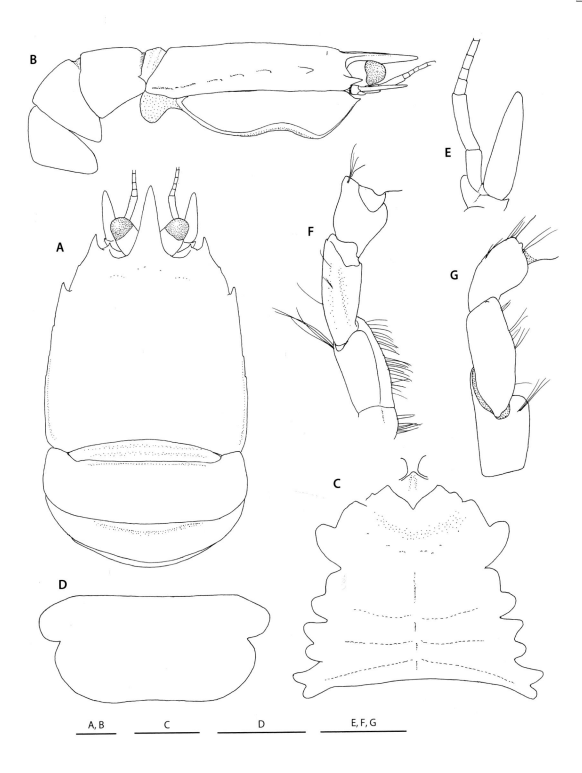

FIGURE 5

Uroptychus adiastaltus n. sp., holotype, ovigerous female 4.3 mm (MNHN-IU-2012-687). **A**, carapace and anterior part of abdomen, dorsal. **B**, same, lateral. **C**, sternal plastron, with excavated sternum, basal parts of Mxp1 included. **D**, telson. **E**, left antenna, ventral. **F**, right Mxp3, ventral. **G**, same, lateral. Scale bars: 1 mm.

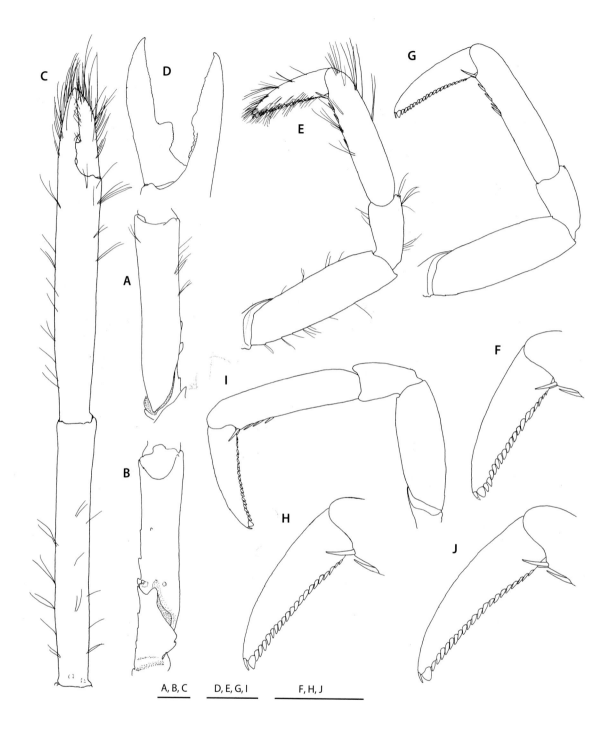

C D G A E I B F H J

A, B, C   D, E, G, I   F, H, J

FIGURE 6

Uroptychus adiastaltus n. sp., holotype, ovigerous female 4.3 mm (MNHN-IU-2012-687). **A**, right P1, proximal part, dorsal. **B**, same, ventral. **C**, same, distal 3 articles, dorsal. **D**, same, fingers, ventral. **E**, left P2, lateral. **F**, same, distal part, setae omitted, lateral. **G**, left P3, setae omitted, lateral. **H**, same, distal part, lateral. **I**, left P4, lateral. **J**, same, distal part, lateral. Scale bars: 1 mm.

Mxp: Mxp1 with bases separated. Mxp3 basis without denticle on mesial ridge. Ischium with distally rounded flexor margin, no distinct denticle on crista dentata. Merus 2.1 × longer than ischium, flattish on mesial face, weakly ridged along flexor margin, distolateral spine obsolescent; flexor margin unarmed or with obsolescent spine somewhat distal to midlength; carpus with tiny distolateral spine.

Pereopods sparsely with soft fine setae.

P1: Slender, 5.5-6.0 × longer than carapace. Ischium dorsally with dorso-ventrally flattened, short spine, ventromesially with strong subterminal spine proximally followed by a few denticles. Merus 1.2-1.3 × longer than carapace, ventromesially with strong or moderate-sized spines: 1 at distal end, 1 at proximal end (these spines not prominent in small specimens), and occasionally additional few small spines between. Carpus 1.4-1.5 × longer than merus. Palm 3.5-4.3 × (males), 5.5-5.7 × (females) longer than broad, subequal to or slightly shorter than carpus. Fingers slightly incurved distally, gaping in large males, not gaping in females and small males, movable finger 0.4-0.5 × length of palm; opposable margin of movable finger with obtuse process of triangular shape having proximal margin oblique; that of fixed finger with prominence or lower process distal to position of opposite process of movable finger.

P2-4: Meri flattish on mesial face, moderately inflated on lateral face, dorsal margin rounded, successively shorter posteriorly (P3 merus 0.9 × length of P2 merus, P4 merus 0.8-0.9 × length of P3 merus), subequally broad on 3 and P4, slightly narrower on P2, length-breadth ratio, 3.5-3.6 on P2, 2.9-3.1 on P3, 2.4-2.6 on P4; P2 merus 0.7-0.8 × length of carapace,1.0-1.2 × longer than P2 propodus; P3 merus about as long as P3 propodus; P4 merus 0.7-0.8 × length of P4 propodus. Carpi subequal, 0.4 × length of propodus on P2, 0.3 × on P3 and P4. Propodi usually subequal on P3 and P4, shorter on P2, rarely successively longer posteriorly; flexor margin straight, ending in pair of long movable spines preceded by 4, 3-4, 2-3 single spines at most on distal half of length on P2, P3, P4, respectively. Dactyli much longer than carpi (1.3-1.5 ×, 1.6-1.9 ×, 1.8-2.0 × longer on P2, P3, P4, respectively), distinctly more than half (0.6 × on P2 and P3, 0.6-0.7 × on P4) as long as propodi, well compressed, relatively broad distally; flexor margin straight, with 16-20 (P2), 18-22 (P3-4) obliquely directed spines nearly contiguous to one another and obscured by setae, ultimate more slender than antepenultimate, penultimate more than 2 × broader than antepenultimate, remaining spines distinctly broader than ultimate spine except for proximal spines; antepenultimate spine 2.0-2.7 × longer than broad.

Eggs. Number of eggs carried, up to 12; size, 0.83 × 0.92 mm - 1.10 × 1.30 mm.

PARASITES — One of the females from BIOCAL Stn DW51 (MNHN-IU-2012-684) bears a rhizocephalan externa on the abdomen.

REMARKS — *Uroptychus adiastaltus* keys out in a couplet together with *U. alius* Baba, 2005 (see above under the key to species), having the anteriormost branchial marginal spine not followed by distinct spines. However, the very characteristic sternite 4, having the anterolateral margin much shorter than the posterolateral margin, links *U. adiastaltus* much more closely to *U. depressus* n. sp., *U. kareenae* n. sp. and *U. levicrustus* Baba, 1988. Their relationships are discussed under the remarks of *U. kareenae*.

Uroptychus adnatus n. sp.

Figures 7, 8

TYPE MATERIAL — Holotype: **Vanuatu**. MUSORSTOM 8 Stn CP1111, 14°51.09'S, 167°14.00'E, 1210-1250 m, 8.X.1994, 1 ♂ 8.9 mm (MNHN-IU-2014-17159).

ETYMOLOGY — From the Latin *adnatus* (joined to, united with), alluding to the antennal scale being fused with the antennal article 2, a character displayed by the new species.

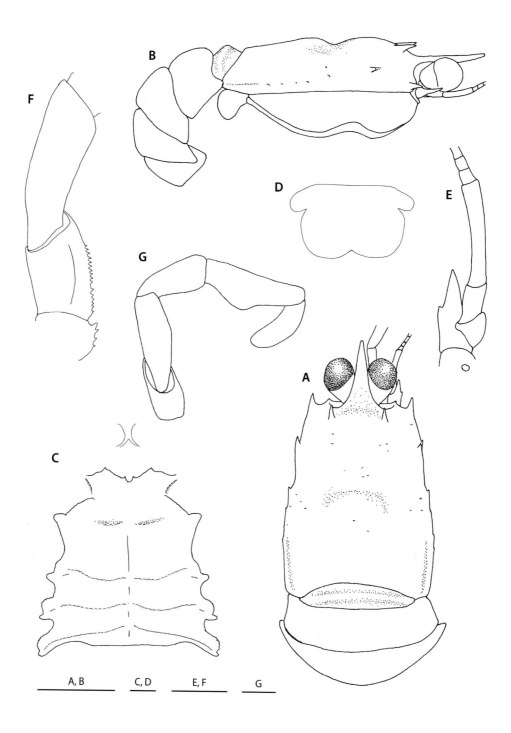

FIGURE 7

Uroptychus adnatus n. sp., holotype, male 8.9 mm (MNHN-IU-2014-17159). **A**, carapace and anterior part of abdomen, dorsal. **B**, same, lateral. **C**, sternal plastron, with excavated sternum and basal parts of Mxp1. **D**, telson. **E**, right antenna, ventral. **F**, right Mxp3, setae omitted, ventral. **G**, same, lateral. Scale bars: A, B, 5 mm; C-G, 1 mm.

DISTRIBUTION — Vanuatu, 1210-1250 m.

DESCRIPTION — *Carapace*: Slightly longer than broad (length 1.1 × breadth); greatest breadth 1.5 × distance between anterolateral spines. Dorsal surface glabrous, with pair of strong epigastric spines and very small tubercles scattered as illustrated; distinct depression between moderately inflated gastric and cardiac regions, boundaries between cardiac and branchial regions indistinct. Lateral margins somewhat divergent posteriorly; anterolateral spine well developed, distinctly overreaching small lateral orbital spine, followed by somewhat smaller spine located at anterior end of branchial region, and some posteriorly diminishing tubercular processes on posterior branchial region; ridged along posterior third of length. Rostrum narrow, with interior angle of 20°, nearly straight, directed slightly dorsally; dorsal surface flattish, smooth and glabrous; length 0.4 × that of remaining carapace, breadth less than half carapace breadth measured at posterior carapace margin. Lateral orbital spine moderately remote from and anterior to level of anterolateral spine. Pterygostomian flap smooth on surface, anterior margin roundish, without distinct spine.

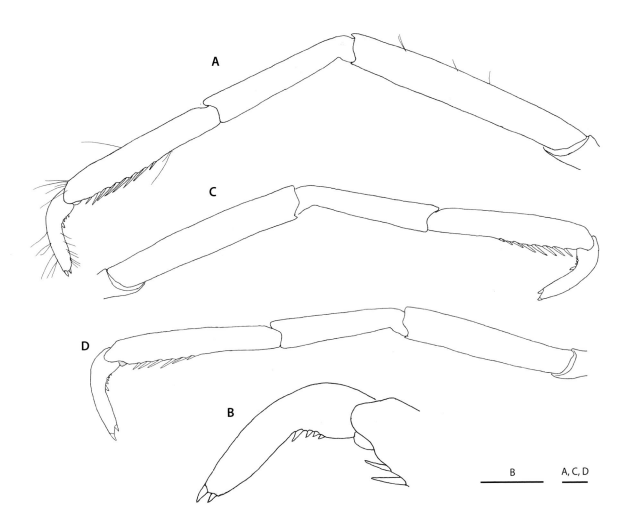

FIGURE 8

Uroptychus adnatus n. sp., male holotype 8.9 mm (MNHN-IU-2014-17159). **A**, left P2, lateral. **B**, same, distal part, setae omitted, lateral. **C**, right P3, lateral. **D**, left P4, lateral. Scale bars: 1 mm.

Sternum: Excavated sternum anteriorly sharp triangular, surface somewhat inflated, with no ridge in midline and no spine in center. Sternal plastron about as long as broad, lateral extremities slightly divergent posteriorly. Sternite 3 moderately depressed, anterior margin of broad V-shape with 2 very small submedian spines separated by broad sinus. Sternite 4 with denticulate transverse ridge preceded by depression (in ventral view), anterolateral margin relatively short, about quarter of greatest breadth, nearly straight, with several small tubercles, posterolateral margin relatively long, nearly as long as anterolateral margin. Sternite 5 anterolateral margin somewhat convex and much shorter than posterolateral margin of sternite 4.

Abdomen: Glabrous. Somite 1 with rounded transverse ridge. Somite 2 tergite 2.7x broader than long; pleuron with posteriorly somewhat divergent lateral margin rounded on anterior and posterior corners. Pleuron of somite 3 posterolaterally blunt. Telson slightly more than half as long as broad; posterior plate 2.3 × length of anterior plate, posterior margin emarginate.

Eye: Short relative to breath (1.4 × longer than broad), distally broadened, proximally narrowed, reaching anterior third of rostrum. Cornea slightly inflated, longer than remaining eyestalk.

Antennule and antenna: Ultimate article of antennule 3.2 × longer than high. Antennal peduncle not reaching rostral tip. Article 2 fused with antennal scale. Antennal scale short, laterally with small spine near base (somewhat distal to ordinary end of article 2), slightly overreaching distal end of article 4 but falling far short of midlength of article 5, 1.7 × broader than article 5. Article 5 2.8 × longer than article 4, breadth slightly less than half height of antennular ultimate article. Flagellum consisting of 19-21 segments.

Mxp: Mxp1 with bases close to each other, but not contiguous. Mxp3 relatively thick and not compressed mesio-laterally. Basis with 4-5 denticles on mesial ridge, distalmost larger than remainder. Ischium distally not rounded on flexor margin, crista dentata with 16 denticles. No spine on merus and carpus. Merus slightly more than 3 × longer than ischium, not crested but rounded along flexor margin.

P1: Missing.

P2-4: Slender, sparsely setose, moderately depressed. Meri successively shorter posteriorly (P3 merus 0.8 × length of P2 merus, P4 merus 0.9 × length of P3 merus), subequally broad on P2-4. Meri with small spine at distal end of flexor margin, and another small spine at distal end of extensor margin on P2 and P3, absent on P4; length-breadth ratio, 7.0 on P2, 6.1 on P3, 5.3 on P4; P2 merus as long as carapace, 1.4 × longer than P2 propodus; P3 merus 1.2 × length of P3 propodus; P4 merus about as long as P4 propodus. Carpi successively shorter posteriorly (P3 carpus 0.9 × length of P2 carpus; P4 carpus 0.9 × length of P3 carpus), relatively long; carpus-propodus length ratio, 0.9 on P2, 0.8-0.9 on P3, 0.7 on P4. Propodi subequal in length on P2 and P3, slightly longer on P4; flexor margin straight, with 8 or 9 spines on P2, 7-8 on P3, 6 on P4, at most on distal two-thirds, distalmost single, equidistant between juncture with dactylus and distal second spine. Dactyli subequal in length on P2-4; dactylus-carpus length ratio, 0.5 on P2 and P3, 0.6 on P4; dactylus-propodus length ratio, 0.4 × on P2-4; strongly curved at proximal third; flexor margin with somewhat inclined short spines in 2 groups remotely separated by sharp crest, distal group of 2 spines (ultimate stronger) and proximal group of 4-5 spines.

REMARKS — This species is very similar in nearly all details to *U. jiaolongae* Dong & Li, 2015 from the South China Sea southwest of Taiwan. They differ only in the antennal scale, which is fused with article 2 in *U. adnatus*, but articulated in *U. jiaolongae*. The same difference also discriminates between *U. scandens* Benedict, 1902 and *U. articulatus* n. sp., its validity being supported by molecular evidence (see below under *U. scandens*). *Uroptychus adnatus* and *U. jiaolongae* were taken at depths of more than 1000 m, but from disjunct localities, and the latter was found in a cold seep community. Although molecular data are not available, I believe *U. adnatus* can be regarded as a distinct species.

Uroptychus adnatus and *U. remotispinatus* Baba & Tirmizi, 1979 share the short antennal scale, sternite 4 with no strong lateral spine, relatively long P2-4 carpi, and the P2 dactylus with remotely separated proximal and distal groups of spines. *Uroptychus adnatus* differs from *U. remotispinatus* in having a pair of epigastric spines, in having the branchial lateral margin with a distinct spine instead of being unarmed on the anterior end, in having the P2-4 propodi with the distalmost of the flexor marginal spines close to instead of considerably remote from the juncture with the dactylus, and the antennal scale fused with article 2.

Uroptychus alcocki Ahyong & Poore, 2004

Figure 9

Uroptychus alcocki Ahyong & Poore, 2004: 15, fig. 2. — Baba 2005: 28, fig. 6. — Poore *et al.* 2011: 328, pl.6, fig. A.
Uroptychus brevirostris — Baba 1973: 117 (Not *U. brevirostris* Van Dam, 1933).
? *Uroptychus cavirostris* — Van Dam 1933: 22, figs 33, 34 (see remarks).

TYPE MATERIAL — Holotype: **Australia**, Southeast of Ballina, New South Wales, female (AM P31412). [not examined]

MATERIAL EXAMINED — **Solomon Islands.** SALOMON 1 Stn DW1742, 11°29.4′S, 159°57.4′E, 366-421 m, 23.IX.2001, 1 ♂ 4.5 mm, 1 ♀ 3.1 mm (MNHN-IU-2014-17160). – Stn DW1854, 9°46.4′S, 160°52.9′S, 229-260 m, 7.X.2001, 1 ♂ 5.3 mm, 1 ♀ 5.2 mm (MNHN-IU-2014-17161). – Stn DW1856, 9°46.4′S, 160°52.3′E, 254-281 m, 7.X.2001, 1 ov. ♀ 5.1 mm (MNHN-IU-2014-17162). **New Caledonia**, Chesterfield Islands. CHALCAL 1 Stn CP08, 19°43.80′S, 158°35.25′E, 348 m, 19.VII.1984, 4 ♂ 5.0-5.5 mm, 2 ov. ♀ 5.0, 5.6 mm (MNHN-IU-2014-17163). MUSORSTOM 5 Stn DW256, 25°18.0′S, 159°52.50′E, 290-300 m, 7.X.1986, 1 ♂ 3.7 mm (MNHN-IU-2014-17164). – Stn DW301, 22°06.90′S, 159°24.60′E, 487-610 m, with coral of Chrysogorgiidae (Calcaxonia), 12.X.1986, 2 ♀ 4.8, 4.9 mm MNHN-IU-2014-17165). – Stn CP307, 22°11.07′S, 159°24.07′E, 350-345 m, with coral Chrysogorgiidae (Calcaxonia), 12.X.1986, 1 ♂ 4.5 mm, 1 ov. ♀ 5.2 mm (MNHN-IU-2014-17166). – Stn CP309, 22°10.20′S, 159°22.80′E, 340 m, 12.X.1986, 1 ov. ♀ 5.0 mm (MNHN-IU-2014-17167). – Stn CP315, 22°25.32′S, 159°27.40′E, 330-335 m, with coral Chrysogorgiidae (Calcaxonia), 13.X.1986, 1 ov. ♀ 6.0 mm (MNHN-IU-2014-17168). – Stn DW337, 19°53.80′S, 158°38.00′E, 412-430 m, with coral Chrysogorgiidae (Calcaxonia), 15.X.1986, 3 ♂ 3.5-5.1 mm, 3 ov. ♀ 5.3-6.1 mm (MNHN-IU-2014-17169). – Stn DC361, 19°52.50′S, 158°38.10′E, 400 m, 19.X.1986, 1 ♂ 3.0 mm (MNHN-IU-2014-17170), 1 ov. ♀ 5.8 mm (MNHN-IU-2014-17171). – Stn DC371, 19°54.85′S, 158°38.17′E, 350 m, *Chrysogorgia* sp. (Calcaxonia, Chrysogorgiidae), 20.X.1986, 1 ♂ 4.1 mm (MNHN-IU-2014-17172). – Stn DC378, 19°53.74′S, 158°38.30′E, 355 m, 20.X.1986, 1 ♀ 4.1 mm (MNHN-IU-2014-17173). **New Caledonia**. BIOCAL Stn DW37, 23°00′S, 167°16′E, 350 m, 30.VIII.1985, 1 ♂ 5.2 mm, 1 ov. ♀ 5.3 mm, 1 ♀ 4.8 mm (MNHN-IU-2014-17175). – Stn DW38, 23°00′S, 167°15′E, 360 m, 30.VIII.1985, 2 ♂ 4.0, 4.5 mm (MNHN-IU-2014-17176). – Stn CP110, 22°12.38′S, 167°06.43′E, 275 m, 9.IX.1985, 1 ov. ♀ 3.8 mm (MNHN-IU-2014-17177). SMIB 1 Stn DW02, 22°51.9′S, 167°13′E, 415 m, 5.II.1986, 1 ♂ 4.9 mm, 5 ov. ♀ 4.9-5.9 mm, 1 ♀ 5.5 mm (MNHN-IU-2014-17178). SMIB 2 Stn DW03, 22°54′S, 167°15′E, 412-428 m, 1.IX.1986, 1 ♂ 4.9 mm, 3 ov. ♀ 4.9-5.4 mm, 1 ♀ 5.4 mm (MNHN-IU-2014-17179). – Stn DW05, 22°56′S, 167°14′E, 398-410 m, 17.IX.1986, 13 ♂ 3.3-5.7 mm, 17 ov. ♀ 4.0-5.3 mm, 4 ♀ 4.2-5.3 mm (MNHN-IU-2014-17180). – Stn DW06, 22°56′S, 167°16′E, 442-460 m, 17.IX.1986, 3 ♂ 5.0-5.4 mm, 1 ov. ♀ 5.6 mm, 1 ♀ 2.9 mm (MNHN-IU-2014-17181). – Stn DW08, 22°53′S, 167°14′E, 435-447 m, 18.IX.1986, 3 ♂ 4.6-4.8 mm, 2 ov. ♀ 4.8, 4.9 mm, 3 ♀ 4.4-5.0 mm (MNHN-IU-2014-17182). – Stn DW14, 22°53′S, 167°13′E, 405-444 m, 18.IX.1986, 1 ♂ 5.3 mm, 1 ov. ♀ 4.7 mm (MNHN-IU-2014-17183). SMIB 3 Stn DW08, 24°45.2′S, 168°08′E, 233 m, 21.V.1987, 1 ♂ 4.3 mm (MNHN-IU-2014-17184). – Stn DW28, 22°47′S, 167°12′E, 394 m, 25.V.1987, 3 ov. ♀ 5.3-5.4 mm, 1 ♀ 5.9 mm (MNHN-IU-2014-17185). – Stn DW30, 22°58′S, 167°22′E, 648 m, 25.V.1987, 1 ♂ 3.8 mm, 2 ov. ♀ 3.7, 4.3 mm (MNHN-IU-2014-17186). MUSORSTOM 4 Stn DW183, 19°01.80′S, 163°25.80′E, 280 m, 18.IX.1985, 1 ov. ♀ 4.2 mm (MNHN-IU-2014-17187). – Stn DW210, 22°44′S, 167°09′E, 340-345 m, 28.IX.1985, 1 ♂ 4.7 mm, 1 ♀ 4.0 mm (MNHN-IU-2014-12784), 1 ♀ 3.4 mm (MNHN-IU-2014-12785). – Stn DW211, 22°46.0′S, 167°09.8′E, 370 m, 28.IX.1985, 1 ov. ♀ 4.0 mm (MNHN-IU-2014-17188). – Stn DW212, 22°47.4′S, 167°10.5′E, 375-380 m, with coral Chrysogorgiidae (Calcaxonia), 28.IX.1985, 6 ♂ 4.2-5.3 mm, 4 ov. ♀ 4.9-5.7 mm (MNHN-IU-2014-17189). – Stn CP213, 22°51.3′S, 167°12.0′E, 405-430 m, with coral Chrysogorgiidae (Calcaxonia), 28.IX.1985, 14 ♂ 4.0-6.0 mm, 21 ov. ♀ 4.6-6.1 mm, 3 ♀ 4.5-5.9 mm (MNHN-IU-2014-17190). – Stn CP214, 22°53.8′S, 167°13.9′E, 425-440 m, 28.IX.1985, 1 ♂ 5.2 mm, 1 ov. ♀ 5.7 mm, 1 ♀ 4.2 mm (MNHN-IU-2014-17191). – Stn DW222, 22°57.6′S, 167°33.0′E, 410-440 m, with coral Chrysogorgiidae (Calcaxonia), 30.IX.1985, 1 ov. ♀ 4.4 mm, 2 ♀ 3.6, 4.6 mm (MNHN-IU-2014-17192), 1 ♂ 5.1 mm MNHN-IU-2014-17193), 1 ♂ 4.3 mm (MNHN-IU-2014-17194), 3 ♂ 4.7-5.0 mm, 2 ov. ♀ 5.0, 5.3 mm, 2 ♀ 4.7, 4.7 mm (MNHN-IU-2014-17195). – Stn DW226, 22°47.2′S, 167°21.6′E, 395 m, 30.IX.1985, 2 ♂ 3.9, 4.0 mm (MNHN-IU-2014-17196). – Stn DW234, 22°15.4′S, 167°08.3′E, 350-365 m, 2.XII.1985, 1 ov. ♀ 3.4 mm (MNHN-IU-2014-17197). BATHUS 1 Stn CP656, 21°13.17′S, 165°53.98′E, 452-460 m, 12.III.93, 1 ♀ 4.9 (MNHN-IU-2014-17198). – Stn CP670, 20°54′S, 165°53′E, 394-397 m, 14.III.1993, 3 ♂ 4.5-4.8 mm (MNHN-IU-2014-17199). – Stn CP701, 20°57′S, 165°35′E, 302-335 m, 18.III.1993, 1 ♂ 4.3 mm, 1 ov. ♀ 3.8 mm, 1 ♀ 5.3 mm (MNHN-IU-2014-17200). – Stn CP710, 21°43.16′S, 166°36.35′E, 320-386 m, 19.III.1993, 1 ♂ 4.0 mm (MNHN-IU-2014-17201). BATHUS 2 Stn DW717, 22°44′S, 167°16′E, 350-393 m, 11.V.1993, 1 ♀ 4.7 mm (MNHN-IU-2014-17202). – Stn DW718, 22°46.70′S, 167°14.45′E, 430-436 m, 11.V.1993, 1 ♀ 4.7 mm (MNHN-IU-2014-17203). – Stn DW729, 22°52.42′S, 167°11.90′E, 400 m, 12.V.1993, 4 ♂ 4.4-5.1 mm, 2 ov. ♀ 4, 8, 4.9 mm, 2 ♀ 4.5, 4.8 mm (MNHN-IU-2014-17204). – Stn CP742, 22°33′S, 166°25′E, 340-470 m, 14.V.1993, 1 ♂ 4.1 mm (MNHN-IU-2014-17205). BATHUS 3 Stn CP805, 23°41′S, 168°01′E, 278-310 m, 27.XI.1993, 1 ♀ 5.2 mm (MNHN-IU-2014-17206). – Stn CP806, 23°42,31′S, 168°00,52′E, 308-312 m, 27.XII.1993, 1 ov. ♀ 4.9 mm (MNHN-IU-2014-17207). – Stn DW829, 23°21′S, 168°02′E, 386-390 m, 29.XI.1993, 1 ♀ 4.4 mm (MNHN-IU-2014-17208). – Stn DW830, 23°20′S, 168°01′E, 361-365 m, 29.XI.1993, 2 ♂ 4.4, 4.9 mm, 1 ov. ♀ 4.1 mm (MNHN-IU-2014-17209).

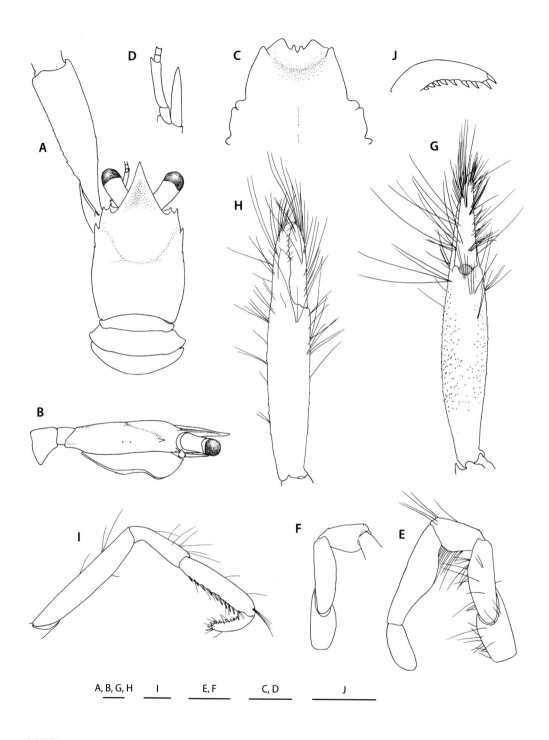

A, B, G, H   I   E, F   C, D   J

FIGURE 9

Uroptychus alcocki Ahyong & Poore, 2003; **A-E**, **G**, **I**, **J**, male 4.5 mm (MNHN-IU-2014-17211); **F**, male 4.8 mm (MNHN-IU-2014-17251); **H**, male 4.3 mm (MNHN-IU-2014-17200). **A**, carapace and anterior part of abdomen, proximal part of left P1 included, dorsal. **B**, same, lateral. **C**, anterior part of sternal plastron. **D**, left antenna, ventral. **E**, left Mxp3, lateral. **F**, right Mxp3, lateral. **G**, left P1, mesial. **H**, same, dorsal. **I**, right P2, lateral. **J**, same, distal part, lateral. Scale bars: 1 mm.

BATHUS 4 Stn CP899, 20°16.68'S, 163°50.26'E, 500-600 m, 3.VIII.1994, 3 ♂ 3.9-5.0 mm, 1 ov. ♀ 4.2 mm (MNHN-IU-2014-17210). – Stn CP900, 20°16.74'S, 163°50.06'E, 580 m, 3.VIII.1994, 1 ♂ 4.5 mm (MNHN-IU-2014-17211). – Stn CP905, 19°02.45'S, 163°15.65'E, 294-296 m, 4.IV.1994, 2 ♂ 4.1, 5.0 mm, 1 ♀ 5.2 mm (MNHN-IU-2014-17212). VAUBAN 1978-1979, no data, 8 ov. ♀ 4.6-6.1 mm, 1 ♀ 5.2 mm (MNHN-IU-2014-17213). – Stn DR15, 22°49'S, 167°12.0'E, 390-395 m, 10.IV.1978, 8 ♂ 4.1-5.4 mm, 3 ov. ♀ 4.3-5.8 mm, 2 ♀ 3.2, 5.0 mm (MNHN-IU-2014-17214). – Stn DR24, 22°48'S, 167°09'E, 355-360 m, on unidentified coral, 13.IV.1978, 1 ♂ 5.4 mm, 1 ♀ 4.5 mm (MNHN-IU-2014-17215). LAGON DW444, 18°15'S, 162°59'E, 300-350 m, 28.II.1985, 1 ♂ 5.9 mm, 1 ov. ♀ 5.7 mm (MNHN-IU-2014-17174). – DW423, 22°46'S, 167°13'E, 405 m, 24.I.1985, 3 ov. ♀ 4.8-5.5 mm, 1 ♀ 2.4 mm (MNHN-IU-2014-17216). **Southern New Caledonia**. SMIB 2 Stn DW05, 22°56'S, 167°14'E, 398-410 m, 17.IX.1986, 1 ♂ 4.9 mm (MNHN-IU-2014-17218). No Stn, 23°00'S, 167°17'E, Dredge, 430 m, 6. IV.1976, coll. A. Michel, 2 ♂ 4.3, 4.4 mm, 1 ♀ 4.6 mm (MNHN-IU-2010-5480). – No Stn, 22°40'S, 167°10'E, 200-350 m, 7-10.X.1986, 1 ♀ 4.5 mm (MNHN-IU-2014-17217). **New Caledonia**, Isle of Pines. 360 m, 13.IV.1978, on hydroids, coll. A. Intes, 3 ov. ♀ 5.3-6.0 mm (MNHN-IU-2014-17219). **New Caledonia**, Norfolk Ridge. CHALCAL 2 Stn CP26, 23°18.15'S, 168°03.58'E, 296 m, on coral Chrysogorgiidae (Calcaxonia), 31.X.1986, 2 ♂ 4.1, 4.5 mm, 2 ov. ♀ 4.5, 6.4 mm, 1 ♀ 4.5 mm (MNHN-IU-2014-17220). – Stn CP 27, 23°15.29'S, 168°04.55'E, 289 m, 3.X.1986, 1 ov. ♀ 4.9 mm MNHN-IU-2014-17221). 3 ♂ 4.6-5.3 mm, 7 ov. ♀ 4.5-5.7 mm, 1 ♀ 4.8 mm (MNHN-IU-2014-17222). – Stn DW81, 23°19.6'S, 168°03.4'E, 311 m, 31.X.1986, 2 ♀ 2.6, 4.6 mm (MNHN-IU-2014-17223). SMIB 5 Stn DW72, 23°42.0'S, 168°00.8'E, 280-400 m, 7.IX.1989, 1 ♀ 5.2 mm (MNHN-IU-2014-17224). – Stn DW87, 22°18.7'S, 168°41.3'E, 335-370 m, 13.IX.1989, 3 ♂ 3.0-3.4 mm, 4 ov. ♀ 4.6-5.8 mm (MNHN-IU-2014-17225). – Stn DW97, 23°01.1'S, 168°18.0'E, 240-300 m, 14.IX.1989, 1 ♂ 3.1 mm (MNHN-IU-2014-17226). – Stn DW101, 23°21.2'S, 168°04.9'E, 270-285 m, 14. IX.1989, 1 ov. ♀ 4.3 mm (MNHN-IU-2014-17227). – DW102, 23°19.6'S, 168°04.7'E, 290-305 m, 14.IX.1989, 2 ♂ 5.1, 5.5 mm, 1 ov. ♀ 3.8 mm (MNHN-IU-2014-17228). – Stn DW103, 23°16'S, 168°04.8'E, 300-315 m, 14.IX.1989, 1 ♂ 5.3 mm, 4 ov. ♀ 4.0-5.1 mm, 1 ♀ 3.2 mm (MNHN-IU-2014-17229). – Stn DW104, 23°15.7'S, 168°04.4'E, 305-335 m, 14.IX.1989, 1 ov. ♀ 4.9 mm (MNHN-IU-2014-17230). – MUSORSTOM 4 no station data, Banc Alis, 250 m, rock, 13.IX.1985, 1 ♂ 3.0 mm, 1 ov. ♀ 3.3 mm (MNHN-IU-2014-17231). SMIB 8 Stn DW181, 23°17.74'S, 168°04.82'E, 311-330 m, 31.I.1993, 1 ov. ♀ 4.7 mm (MNHN-IU-2014-17232). – Stn DW183, 23°18.27'S, 168°04.95'E, 330-367 m, 31.I.1993, 1 ♂ 6.0 mm, 1 ov. ♀ 4.8 mm (MNHN-IU-2014-17233). – Stn DW185, 23°15'S, 168°04.3'E, 311-355 m, 31.I.1993, 1 ♀ 5.1 mm (MNHN-IU-2014-17234). – Stn DW190, 23°18'S, 168°05'E, 305-310 m, 31.I.1993, 2 ov. ♀ 4.7, 5.2 mm (MNHN-IU-2014-17235). – Stn DW197, 22°51.27'S, 168°12.54'E, 414-436 m, 1.II.1993, 3 ov. ♀ 3.3-3.7 mm (MNHN-IU-2014-17236). – Stn DW198, 22°51.6'S, 167°12.4'E, 414-430 m, 1.II.1993, 1 ov. ♀ 5.7 mm (MNHN-IU-2014-17237). NORFOLK 1 Stn DW1651, 23°27.3'S, 167°50.4'E, 276-350 m, ?Acanthogorgiidae (Acanthogorgiidae), 19.VI.2001, 1 ♂ 4.8 mm, 1 ov. ♀ 6.8 mm (MNHN-IU-2014-17238). – Stn DW1653, 23°28'S, 167°51'E, 328-340 m, 19.VI.2001, 1 ♂ 3.3 mm, 1 ov. ♀ 7.2 mm (MNHN-IU-2014-17239). – Stn DW1654, 23°26'S, 167°52'E, 366-560 m, 19.VI.2001, 1 ♂ 5.2 mm, 1 ov. ♀ 5.8 mm (MNHN-IU-2014-17240). – Stn DW1657, 23°26'S, 167°52'E, 305-332 m, 19.VI.2001, 1 ♂ 5.7 mm (MNHN-IU-2014-17241). – Stn CP1671, 23°41'S, 168°00'E, 320-397 m, alcyonacean *Chironephthya* sp. (Alcyoniina: Nidaliidae), 21.VI.2001, 2 ov. ♀ 4.9, 5.3 mm (MNHN-IU-2014-17242). – Stn DW1729, 23°20'S, 168°16'E, 340-619 m, 27.VI.2001, 1 ov. ♀ 6.6 mm (MNHN-IU-2014-17243). – Stn DW1733, 22°56'S, 167°15'E, 427-433 m, 28.VI.2001, 3 ♂ 4.7-4.9 mm, 5 ov. ♀ 4.3-5.3 mm (MNHN-IU-2014-17244). – Stn DW1736, 22°51'S, 167°12'E, 383-407 m, 28.VI.2001, 10 ♂ 3.2-5.9 mm, 7 ov. ♀ 4.3-5.7 mm (MNHN-IU-2014-17245). – Stn DW1737, 22°52'S, 167°12'E, 400 m, with coral Chrysogorgiidae (Calcaxonia), 28.VI.2001, 3 ♂ 3.3-5.3 mm, 6 ov. ♀ 4.2-5.2 mm (MNHN-IU-2014-17246). NORFOLK 2 Stn DW2024, 23°28'S, 167°51'E, 370-371 m, 21.X.2003, 1 ♂ 4.2 mm, 2 ov. ♀ 3.8, 5.1 mm (MNHN-IU-2014-17247). – Stn CP2038, 23°41.86'S, 168°00.28'E, 290-330 m, 23.X.2003, 1 ov. ♀ 4.9 mm (MNHN-IU-2014-17248). – Stn CP2130, 23°15.90'S, 168°13.54'E, 375-427 m, 2.XI.2003, 1 ov. ♀ 5.2 mm (MNHN-IU-2014-17249). – Stn CP2139, 23°00.68'S, 168°22.60'E, 372-393 m, 3.XI.2003, 1 ov. ♀ 5.2 mm (MNHN-IU-2014-17250). – Stn CP2154, 22°49.90'S, 167°12.60'E, 410-423 m, 5.XI.2003, 11 ♂ 4.3-5.3 mm, 18 ov. ♀ 3.7-5.2 mm (MNHN-IU-2014-17251). – Stn CP2160, 22°41.81'S, 167°09.70'E, 313-315 m, 6.XI.2003, 1 ♂ 4.2 mm (MNHN-IU-2014-17252). BERYX 11 Stn CP45, 23°40.27'S, 168°00.95'E, 270-290 m, 20.X.1992, 1 ov. ♀ 5.1 mm (MNHN-IU-2014-17253). **New Caledonia**. Loyalty Ridge. MUSORSTOM 6 Stn DW392, 20°47.32'S, 167°04.60'E, 340 m, with *Acanella* sp. (Calcaxonia: Isididae), 13.II.1989, 1 ov. ♀ 3.3 mm (MNHN-IU-2014-17254). – Stn DW406, 20°40.65'S, 167°06.80'E, 373 m, 15.II.1989, 1 ♀ 6.0 mm (MNHN-IU-2014-17255). – Stn DW407, 20°40.70'S, 167°06.60'E, 360 m, with coral Chrysogorgiidae (Calcaxonia), 15.II.1989, 1 ♂ 4.3 mm, 1 ♀ 3.6 mm (MNHN-IU-2014-17256). – Stn CP409, 20°41.05'S, 167°07.25'E, 385 m, 15.II.1989, 1 ov. ♀ 5.8 mm (MNHN-IU-2014-17257). – Stn CP419, 20°41.65'S, 167°03.70'E, 283 m, 16.II.1989, 1 ov. ♀ 3.6 mm, 1 ♀ 3.6 mm (MNHN-IU-2014-17258). – Stn CP438, 20°23.00'S, 166°20.10'E, 800 m, 18.II.1989, 1 ov. ♀ 3.8 mm (MNHN-IU-2014-17259). – Stn DW472, 21°08.60'S, 167°54.70'E, 300 m, 22.II.1989, 3 ♂ 2.8-5.7 mm, 3 ov. ♀ 3.9-6.2 mm, 1 ♀ 4.0 mm (MNHN-IU-2014-17260). – Stn DW479, 21°09.13'S, 167°54.95'E, 310 m, 22.II.1989, 1 ov. ♀ 5.8 mm (MNHN-IU-2014-17261). **New Caledonia**, Hunter and Matthew Islands. VOLSMAR Stn DW16, 22°25.1'S, 171°40.7'E, 420-500 m, 3.VI.1989, 2 ♂ 4.0, 4.1 mm (MNHN-IU-2014-17274). – Stn DW39, 22°20.5'S, 168°43.5'E, 280-305 m, 8.VI.1989, 1 ♂ 4.8 mm, 2 ov. ♀ 4.2, 4.3 mm (MNHN-IU-2014-17275). **Vanuatu**. MUSORSTOM 8 Stn CP963, 20°20.10'S, 169°49.08'E, 400-440 m, 21.IX.1994, 1 ov. ♀ 4.1 mm (MNHN-IU-2014-17262). – Stn CP982, 19°21.80'S, 169°26.47'E, 408-410 m, 23.IX.1994, 2 ♂ 4.0, 4.7 mm, 2 ov. ♀ 4.9-5.4 mm, 1 ♀ 4.0 mm (MNHN-IU-2014-17263). – Stn CP1024, 17°48.21'S, 168°38.77'E, 335-370 m, 28.IX.1994, 1 ♂ 5.7 mm, 2 ov. ♀ 4.1, 5.5 mm, 1 ♀ 6.4 mm (MNHN-IU-2014-17264). – Stn DW1058, 16°12.03'S, 167°20.80'E, 319 m, with *Acanella sp.* (Calcaxonia: Isididae), 2.X.1994, 1 ♂ (carapace lost) (MNHN-IU-2014-17265). – Stn DW1060, 16°13.82'S, 167°20.80'E, 375- 394 m, 2.X.1994, 1 ♂ 5.5 mm (MNHN-IU-2014-17266). – Stn CP1091, 15°10.24'S, 167°13.01'E, 344-350 m, 6.X.1994, 1 ov. ♀ 4.9 mm (MNHN-IU-2014-17267). – Stn DW1097, 15°05.13'S, 167°10.76'E, 281-288 m, 7.X.1994, 1 ♀ 4.5 mm

(MNHN-IU-2014-17268). – Stn DW1106, 15°05.27'S, 167°11.88'E, 305-314 m, 7.X.1994, 1 ov. ♀ 4.9 mm (MNHN-IU-2014-17269). – Stn CP1137, 15°41.52'S, 167°02.67'E, 360-371 m, 11.X.1994, 1 ov. ♀ 5.3 mm (MNHN-IU-2014-17270). SANTO Stn AT01, 15°32.4'S, 167°16.4'E, 167-367 m, 14. IX.2006, 1 ♂ 4.5 mm (MNHN-IU-2014-17271). – AT08, 15°40.5'S, 167°01.5'E, 366-389 m, 17.IX.2006, 1 ♂ 4.5 mm (MNHN-IU-2014-17272). BOA1 Stn CP2479, 16°45'S, 167°52'E, 350-358 m, 15.IX.2005, 1 ♀ 4.5 mm (MNHN-IU-2014-17273).

DISTRIBUTION — Previously known from New South Wales, Queensland, Tasman Sea, Formosa Strait and Japan, in 64-419 m, and now from the Solomon Islands, Chesterfield Islands, Vanuatu, Loyalty Islands, New Caledonia, Norfolk Ridge and Hunter-Matthew, in 167-780 m.

SIZE — Males, 2.8-6.1 mm; females, 2.4-7.2 mm; ovigerous females from 3.3 mm.

DESCRIPTION — Small to medium-sized species. Usually small but often exceeding 5.0 mm and attaining 7.2 mm. *Carapace*: As long as broad; greatest breadth 1.2-1.3 × distance between anterolateral spines. Dorsal surface smooth, unarmed, somewhat convex in profile, anteriorly gradually lowered toward rostral dorsal concavity. Lateral margins moderately convex on branchial region; anterolateral spine small, directed anteroventrally, barely overreaching smaller lateral orbital spine, followed by larger spine at ordinary end of cervical groove (or anterior end of branchial region), and often followed by 1 or 2 small or obsolescent spines on branchial margin. Rostrum broad triangular, with interior angle of 42-53°, distally sharp; breadth fully two-thirds carapace breadth measured at posterior carapace margin; dorsal surface deeply concave. Lateral orbital spine somewhat anterior to level of anterolateral spine. Pterygostomian flap anteriorly angular, ending in small spine; no spine on surface.

Sternum: Excavated sternum anteriorly produced to spine, having longitudinal ridge in midline (often obsolete). Sternal plastron as long as broad, lateral extremities slightly convexly divergent posteriorly. Sternite 3 strongly depressed, anterior margin deeply excavated, with pair of submedian spines slightly separated from or close to each other, and often incurved, anterolateral angle rounded. Sternite 4 long relative to breadth; anterolateral margins gently divergent, very much longer than posterolateral margin. Anterolateral margin of sternite 5 anteriorly rounded, slightly more than half length of anterolateral margin of, and much longer than posterolateral margin of, sternite 4.

Abdomen: Smooth and glabrous. Somite 1 without transverse ridge. Somite 2 tergite 2.5-2.8 × broader than long, pleural lateral margins barely or feebly concave and strongly divergent posteriorly, ending in blunt tip. Pleuron of somite 3 with rounded lateral end. Telson three-quarters as long as broad; posterior plate semi-elliptical or semicircular, somewhat narrowed posteriorly, not emarginate on posterior margin, length subequal to greatest breadth, twice that of anterior plate.

Eye: Very elongate, slightly falling short of apex of rostrum, broadened distally, mesial margin convex. Cornea inflated, less than half length of remaining ocular peduncle.

Antennule and antenna: Ultimate article of antennule about twice as long as high. Antennal peduncle slender, barely reaching tip of eye. Article 2 with very small distolateral spine. Antennal scale usually slightly falling short of, rarely reaching distal end of article 5, slightly broader than article 5. Article 5 with distinct distomesial spine, length 2.5-3.0 × that of article 4, breadth about half height of ultimate antennular article. Flagellum of 17-24 segments not reaching distal end of P1 merus.

Mxp: Mxp1 with bases nearly contiguous. Mxp3 basis with 3 denticles obscured by stiff setae on well convex mesial ridge. Ischium with 16-23 denticles on crista dentata, flexor margin not rounded distally. Merus twice as long as ischium, well ridged along distal two-thirds of flexor margin but not ridged on proximal third, distolateral spine very small, often obsolete. Carpus with obsolescent distolateral spine and very small spine on proximal part of extensor margin.

P1: With long setae on distal portion, nearly glabrous elsewhere; more or less massive in males, slender in females; length up to 7 × longer than carapace; small tubercles on ventral surface, on merus and carpus in particular, irrespective of sex, but often absent, also occasionally present on palm. Ischium with small dorsal spine, without subterminal spine on ventromesial margin. Merus 1.4-1.5 × longer than carapace. Carpus 1.4-1.5 × longer than merus, dorsally with 1 median spine at anterior end. Palm 3.1-5.3 × (males), 3.9-5.0 × (females) longer than broad, slightly shorter than or subequal to carpus. Fingers distally incurved, crossing, occasionally gaping in large males (with rounded process at midlength of gaping opposable margin of movable finger), movable finger half as long as palm.

P2-4: Slender, moderately compressed; with sparse long setae. Meri successively shorter posteriorly (P3 merus 0.9 × length of P2 merus, P4 merus 0.9 × length of P3 merus), equally broad on P2-4 or very slightly narrower on P2 than on P3; length-breadth ratio, 5.6-6.5 on P2, 5.3-6.2 on P3, 4.5-5.3 on P4; unarmed; P2 merus as long as, occasionally slightly shorter than carapace, 1.6-1.7 × longer than P2 propodus; P3 merus 1.3 × length of P3 propodus; 4 merus 1.0-1.1 × length of P4 propodus. Carpi subequal, 1.3-1.6 × longer than dactyli on P2-4, 0.6 × length of propodus on P2 and P3, 0.5-0.6 × on P4. Propodi successively longer posteriorly; length-breadth ratio, 5.0 on P2-4; flexor margin with pair of distal spines preceded by row of spines along entire length at least on P2 and P3. Dactyli 0.4 × length of propodi on P2-4; curving at proximal third, ending in elongate subtriangular spine preceded by similar but successively diminishing spines, 7 or 8 in number, on flexor margin.

Eggs. Number of eggs carried, 3-30; small eggs (25 in number), 1.03 mm × 1.30 mm - 1.03 × 1.44 mm; large eggs (4 in number) ready to hatch, 1.69 mm × 1.55 mm.

COLOR — The coloration of a specimen from Vanuatu was shown in Poore *et al.* (2011: pl. 6, fig. A). A female 4.5 mm, taken at SANTO Stn AT1, Vanuatu (MNHN-IU-2014-17271): pale pink overall, abdominal pleura and tailfan translucent.

REMARKS — This is one of the most common species in the southwestern Pacific.

The carapace spination, broad rostrum, elongate eyes, and the shapes of antenna, Mxp3, and P2-4 are very similar in *U. alcocki*, *U. cavirostris* Alcock & Anderson, 1899, *U. latirostris* Yokoya, 1933, *U. yokoyai* Ahyong & Poore, 2004 and *U. mauritius* Baba, 2005. However, this species is readily distinguished from the congeners except *U. mauritius* by the sternite 3 that bears a pair of distinct submedian spines rather than being unarmed on the anterior margin. This character suggests that the material reported under *U. cavirostris* by Van Dam (1933) from the Kai Islands is likely referable to *U. alcocki*. *Uroptychus alcocki* differs from *U. mauritius* in having the epigastric region smooth instead of bearing a pair of distinct ridges.

Examination of the material reported under *U. brevirostris* (ZLKU 4882) from the Ryukyu Islands by Baba (1973) discloses that it should be referred to *U. alcocki*. The specimen has a broad rostrum that is much wider than half the breadth of the carapace measured along the posterior margin and a pair of spines on the anterior margin of sternite 3, both characteristic of *U. alcocki*.

The material reported under *U. cavirostris* from the western Indian Ocean (Tirmizi 1964) is an undescribed species, bearing no distinct spine on the anterior branchial margin, the character confirmed by examination of the material now in the Natural History Museum, London (BMNH 1966.2.3.27-40).

Uroptychus alophus n. sp.

Figures 10, 11

TYPE MATERIAL — Holotype: **New Caledonia**, Chesterfield Islands. MUSORSTOM 5 Stn DW272, 24°40.91'S, 159°43.00'E, 500-540 m, 9.X.1986, 1 ♀ 3.7 mm (MNHN-IU-2014-17276). Paratypes: **New Caledonia**, Chesterfield Islands. Station data as for the holotype, 2 ♂ 2.6, 3.4 mm, 1 ov. ♀ 3.0 mm, 2 ♀ 3.5, 3.5 mm (MNHN-IU-2014-17277). **New Caledonia**, Norfolk Ridge. BIOCAL Stn CP52, 23°06'S, 167°47'E, 540-600 m, 31.VIII.1985, 1 ♂ 4.1 mm (MNHN-IU-2014-17278). BATHUS 3 Stn DW817, 23°42'S, 168°16'E, 405-410 m, 28.XI.1993, 1 ♂ 3.3 mm (MNHN-IU-2014-17279). CHALCAL 2 Stn CC01, 24°54.96'S, 168°21.91'E, 500-580 m, 28.X.1986, 1 ov. ♀ 4.7 mm (MNHN-IU-2014-17280). NORFOLK 2 Stn DW2041, 23°40.93'S, 168°01.29'E, 400 m, 23.X.2003, 6 ♂ 2.9-4.8 mm, 2 ov. ♀ 3.2, 3.5 mm (MNHN-IU-2014-17281). – Stn DW2042, 23°40.51'S, 168°00.58'E, 235-245 m, 23.X.2003, 1 ov. ♀ 3.7 mm (MNHN-IU-2014-17282). – Stn CP2083, 24°53.23'S, 168°21.86'E, 530-540 m, 28.X.2003, 3 ♂ 4.0-4.4 mm (MNHN-IU-2014-17283).

FIGURE 10

Uroptychus alophus n. sp., holotype, female 3.7 mm (MNHN-IU-2014-17276). **A**, carapace and anterior part of abdomen, dorsal. **B**, same, lateral. **C**, sternal plastron, excavated sternum included. **D**, telson. **E**, right antenna, ventral. **F**, left Mxp3, setae omitted, ventral. **G**, same, lateral. Scale bars: 1 mm.

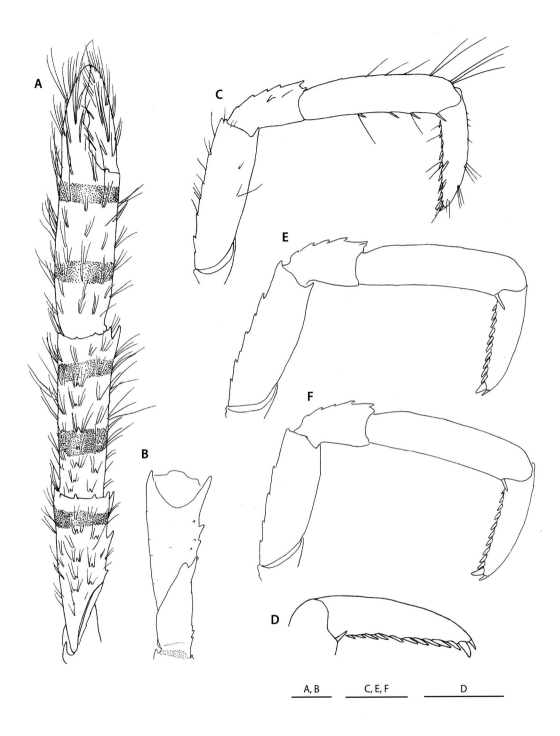

FIGURE 11

Uroptychus alophus n. sp., holotype, female 3.7 mm (MNHN-IU-2014-17276). **A**, left P1, dorsal. **B**, right P1, proximal part, ventral. **C**, right P2, lateral. **D**, same, distal part, setae omitted, lateral. **E**, right P3, lateral. **F**, right P4, lateral. Scale bars: 1 mm.

ETYMOLOGY — From the Greek *lophos* (ridge) with suffix *a-*, meaning without ridge, referring to the absence of sharp transverse ridge on the first abdominal tergite by which the species is readily distinguished from the related species, *U. longior* Baba, 2005.

DISTRIBUTION — Chesterfield Islands and Norfolk Ridge; 235-600 m.

SIZE — Males, 2.6-4.8 mm; females 3.0-4.7 mm; ovigerous female from 3.0 mm.

DESCRIPTION — Small species. *Carapace*: Somewhat broader than long (0.9 × as long as broad); greatest breadth measured between bases of last lateral marginal spines, 1.6 × distance between anterolateral spines. Dorsal surface polished, somewhat convex from anterior to posterior, without distinct groove; with sparse short setae. Lateral margins weakly convex, usually with 8 (rarely 9) spines; first anterolateral, reaching tip of smaller lateral orbital spine; second and third small; remainder acute, posteriorly diminishing, fourth situated at anterior end of branchial region, mesially accompanying 1 or 2 small spines, last followed by ridged margin. Rostrum broad triangular, with interior angle of 25-28°, horizontal; length about half that of remaining carapace (two-thirds in paratype MNHN-IU-2014-17282), breadth half carapace breadth measured at posterior carapace margin; dorsal surface somewhat concave; lateral margin with a few distinct or indistinct small spines distally (7 spines in one of male paratypes MNHN-IU-2014-17281). Lateral orbital spine relatively close to but distinctly anterior to level of anterolateral spine. Pterygostomian flap with anterior margin produced to acute spine followed by a few spines along anterior part of linea anomurica.

Sternum: Excavated sternum blunt triangular on anterior margin, bearing distinct (often sharp) ridge in midline of surface. Sternal plastron slightly broader than long, lateral extremities gently divergent posteriorly. Sternite 3 moderately depressed; anterior margin weakly concave, with broad U-shaped median sinus separating 2 small spines; anterolateral angle rounded or sharply angular. Sternite 4 with anterolateral margin slightly longer than posterolateral margin, moderately or slightly convex, or straight, anteriorly ending in pronounced process occasionally with accompanying small spine directly lateral and/or mesial to it. Anterolateral margin of sternite 5 moderately convex, about as long as posterolateral margin of sternite 4.

Abdomen: Smooth. Somite 1 moderately convex from anterior to posterior. Somite 2 tergite 2.3-2.7 × broader than long; pleuron rounded on anterolateral and posterolateral ends, somewhat concave on lateral margin. Pleuron of somite 3 laterally blunt. Telson 0.4-0.5 × as long as broad; posterior plate distinctly or slightly emarginate on posterior margin, length 1.2-1.4 × that of anterior plate.

Eye: about 2 × longer than broad, overreaching midlength of rostrum, lateral and mesial margins subparallel. Cornea not dilated, length about half or more than half that of remaining eyestalk.

Antennule and antenna: Antennular ultimate article 2.8-3.3 × longer than high. Antennal peduncle distinctly overreaching cornea, not reaching apex of rostrum. Article 2 with small, distinct spine. Antennal scale 1.5-1.7 × broader than article 5, reaching or slightly overreaching tip of distal spine of article 5, reaching or slightly falling short of apex of rostrum. Distal 2 articles each with strong distomesial ventral spine; article 5 1.7-2.0 × longer than article 4, breadth 0.7-0.8 × height of antennular ultimate article. Flagellum of 8-11 segments barely reaching distal end of P1 merus.

Mxp: Mxp1 with bases separated. Mxp3 basis with 1 denticle near distal end of mesial ridge. Ischium with 14-23 denticles on crista dentata; flexor margin sharply crested, rounded distally. Merus 2.1 × longer than ischium, flattish and smooth on mesial face; flexor margin well ridged with a few to several small, occasionally obtuse spines on distal half; distolateral spine distinct, occasionally accompanying additional small spine. Carpus with distolateral spine and a few small spines on extensor margin.

P1: 4 × longer than carapace; relatively massive and setose. Ischium dorsally with sharp, procurved distal spine occasionally followed by small spine, ventromesially with well-developed, subterminal spine occasionally distantly followed by a few small proximal spines. Merus mesially with a few spines of moderate size, laterally unarmed, dorsally with 2 or 3 rows of spines usually small, often obsolete, ventrally with 2 rows of spines and well-developed distomesial and distolateral spines; length subequal to that of carapace. Carpus subequal to or slightly longer than merus, dorsally with 2-3 rows of spines continued from merus, ventrally with well-developed distomesial and distolateral spines. Palm 2.1-2.5 × longer than broad, about as long as carpus, somewhat depressed, unarmed. Fingers curved slightly ventrally, relatively broad, distally ending in slightly incurved spine, crossing when closed; denticulate opposable margins sinuous (proximal obtuse process of movable finger fitting to between 2 eminences on immovable finger when closed), not spooned; movable finger 0.6-0.7 × length of palm.

P2-4: Compressed mesio-laterally, moderately setose, broad relative to length, meri in particular. Meri successively shorter posteriorly (P3 merus 0.9-1.0 × length of P2 merus, P4 merus 0.8-0.9 × length of P3 merus), subequal in breadth on P2-4; length-breadth ratio, 2.8-3.2 on P2, 2.6-2.9 on P3, 2.3-2.5 on P4; P2 merus 0.7 × length of carapace, 0.7-0.9 × length of P2 propodus; P3 merus 0.8 × length of P3 propodus; P4 merus 0.7-0.8 × length of P4 propodus; extensor margin with row of spines continued onto carpus, distinct on P2 and P3, obsolescent on P4; ventrolateral margin with 1 or 2 distal spines on P2-4. Carpi subequal, carpus-propodus length ratio, 0.4 on P2 and P3, 0.3-0.4 on P4; extensor margin with row of small spines, distalmost somewhat larger, with smaller or subequal-sized accompanying spine lateral to it (another row of a few spines lateral to extensor row in male paratype of MNHN-IU-2014-17281). Propodi slightly longer on P3 than on P2 and P4 or subequal on P3 and P4 and somewhat shorter on P2; extensor margin with a few proximal spines distinct on P2 and P3 but occasionally obsolete; flexor margin ending in pair of spines preceded by 1-3 spines on P2, 0 or 1 on P3-4. Dactyli shorter on P2 than on P3 and P4, subequal on P3 and P4; distinctly longer than carpi (dactylus-carpus length ratio, 1.5 on P2, 1.6 on P3 and P4) and more than half length of propodi (dactylus-propodus length ratio, 0.6-0.7 on P2 and P3, 0.6 on P4), proportionately broad; flexor margin nearly straight, with 9-13 (usually 9) spines obliquely directed and close (nearly contiguous) to each other, ultimate slender, smaller than antepenultimate, penultimate prominent, about twice as broad as antepenultimate.

Eggs. Up to 25 eggs carried; size, 0.80 mm × 0.94 mm - 1.15 mm × 1.19 mm.

Color. P1 with orangish bands in preservative, 1 on merus, 2 on each of carpus and palm (Figure 11A).

REMARKS — The arrangements of spines on the carapace and P2-4, the shape of sternal plastron, and the long antennal scale are very similar to those of *U. longior* Baba, 2005. However, the new species has a low instead of sharp transverse ridge on the abdominal somite 1; P1 is much broader and shorter, with the merus as long as instead of distinctly longer than the carapace; the P1 carpus bears a pronounced instead of small distomesial spine.

The rostrum having a few very small lateral spines, the carapace lateral margin spinose, and the P2-4 dactyli well compressed mesio-laterally, with obliquely directed spines on the flexor margin suggest that *U. alophus* n. sp. is close to *U. nanophyes* McArdle, 1901. The new species is distinguished from that species by more pronounced spines on the carapace lateral margin; the anterolateral spine of sternite 4 is very short instead of reaching the anterior end of sternite 3; and no row of ventromesial spines is present on the P2 merus.

Uroptychus angustus n. sp.

Figures 12, 13

TYPE MATERIAL — Holotype: **Tonga**. BORDAU 2 Stn DW1602, 20°49'S, 174°57'W, 263-320 m, 15.VI.2000, ov. ♀ 1.9 mm (MNHN-IU-2014-17284).

ETYMOLOGY — From the Latin *angustus* (narrow), referring to the ultimate article of antennular peduncle that is as broad as the antennal peduncle, characteristic of the new species. This article is usually much broader than the antennal peduncle in other species.

DISTRIBUTION — Tonga; 263-320 m.

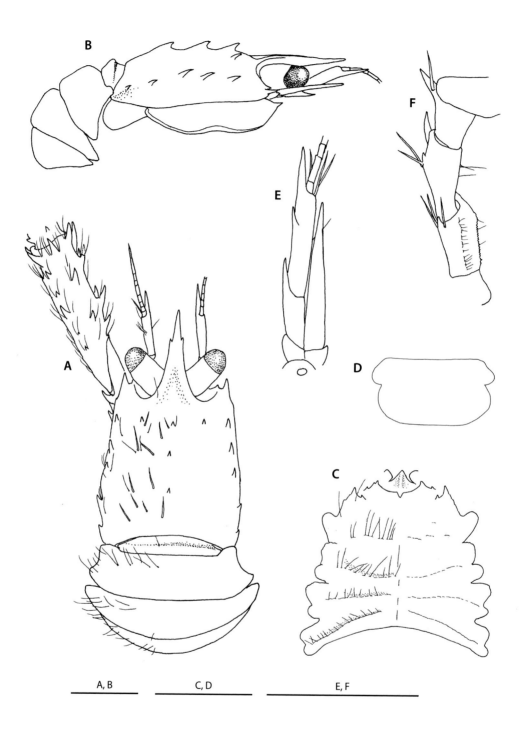

A, B   C, D   E, F

FIGURE 12

Uroptychus angustus n. sp., holotype, ovigerous female 1.9 mm (MNHN-IU-2014-17284). **A**, carapace and anterior abdomen (setae omitted on right half), proximal part of left P1 included, dorsal. **B**, same, lateral. **C**, sternal plastron, with excavated sternum, basal parts of Mxp1 included. **D**, telson. **E**, left antenna, ventral. **F**, right Mxp3, distal articles omitted, ventral. Scale bars: 1 mm.

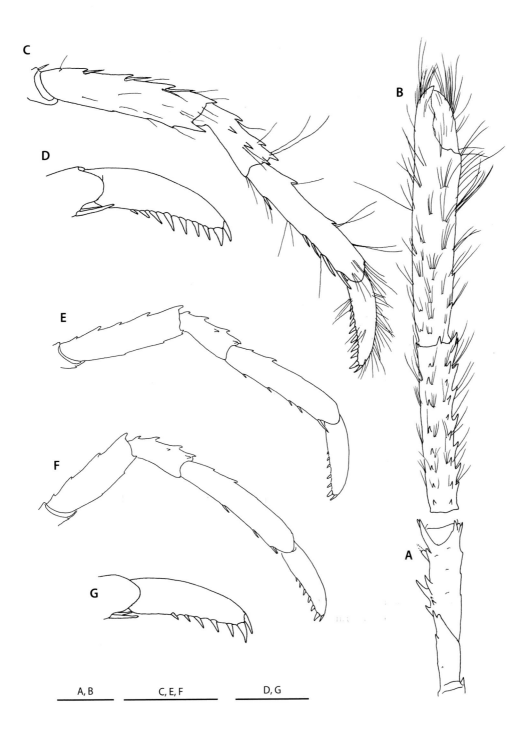

FIGURE 13

Uroptychus angustus n. sp., holotype, ovigerous female 1.9 mm (MNHN-IU-2014-17284). **A**, left P1, proximal part, setae omitted, ventral. **B**, same, proximal part omitted, dorsal. **C**, right P2, lateral. **D**, same, distal part, setae omitted, lateral. **E**, right P3, setae omitted, lateral. **F**, right P4, lateral. **G**, same, distal part, lateral. Scale bars: A, B, C, E, F, 1 mm; D, G, 0.5 mm.

DESCRIPTION — Small species. *Carapace*: slightly broader than long; greatest breadth 1.4 × distance between anterolateral spines. Dorsal surface moderately convex from side to side and anterior to posterior, with weak depression between gastric and cardiac regions, bearing 3 spines in midline (1 on median gastric region, 1 on posterior gastric region, 1 on cardiac region), anteriormost of these flanked by 1 spine on each side; with sparse coarse setae. Lateral margins slightly convex and slightly divergent posteriorly, bearing 6 relatively large spines, first anterolateral, overreaching small lateral orbital spine, remaining spines nearly equidistant on branchial margin, second to fourth somewhat dorsal to level of first, sixth (last) somewhat posterior to midpoint between fifth and posterior end of lateral margin; small tubercle at midlength between first and second spines. Rostrum sharp triangular, with interior angle of 20°, directed slightly dorsally; length slightly more than half that of carapace, breadth half carapace breadth measured at posterior carapace margin; dorsal surface concave, lateral margin with small spine at anterior quarter. Lateral orbital spine about at same level as and close to anterolateral spine. Pterygostomian flap smooth on surface, anteriorly produced to sharp spine.

Sternum: Excavated sternum with sharp ridge in midline of surface, anterior margin triangular, reaching distal end of basal article of Mxp1. Sternal plastron slightly broader than long, lateral extremities gently divergent posteriorly. Sternite 3 moderately depressed, anterior margin moderately concave, with shallow V-shaped median notch, without submedian spines; anterolateral corner angular. Sternite 4 with interrupted transverse ridge bearing coarse setae, anterolateral margin nearly straight, anteriorly ending in strong spine mesially accompanying small tubercle-like spine, length subequal to that of posterolateral margin. Sternite 5 having anterolateral margin strongly convex, shorter than posterolateral margin of sternite 4.

Abdomen: With sparse coarse setae. Somite 1 with transverse ridge. Somite 2 tergite 2.2 × broader than long, with setae transversely arranged on anterior portion; pleural lateral margins concave and strongly divergent posteriorly, posterior end bluntly angular. Pleura of somites 3 and 4 laterally tapering. Telson half as long as broad; posterior plate 1.5 × longer and slightly narrower than anterior plate, posterior margin nearly transverse, not emarginate.

Eye: Slightly overreaching midlength of rostrum, distally narrowed, length 1.9 × breadth; lateral margin slightly convex proximally. Cornea not dilated, half as long as remaining eyestalk.

Antennule and antenna: Ultimate article of antennule 4.0 × longer than high. Antennal peduncle nearly reaching apex of rostrum. Article 2 with small lateral spine. Antennal scale as broad as article 5, slightly falling short of distal end of article 5. Article 4 with strong distomesial spine. Article 5 about twice as long as article 4, with strong, elongate distomesial spine distinctly overreaching rostral tip and another strong ventromesial spine at midlength, breadth slightly more than height of antennular ultimate article. Flagellum consisting of 7 segments, falling short of distal end of P1 merus.

Mxp: Mxp1 with bases broadly separated. Mxp3 basis lacking denticles on mesial ridge. Ischium having flexor margin with rounded distal end mesial to small spine flanked by a few long setae; crista dentata with about 20 very small denticles. Merus 1.7 × longer than ischium, well compressed; mesial face flattish or slightly concave; flexor margin cristiform, angular at distal third, bearing distinct spine and long setae there (absent on right side); with prominent distolateral spine. Carpus also with prominent distolateral spine.

P1: Right P1 much shorter and narrower than left, probably regenerated. Left P1 5.3 × longer than carapace, slender, subcylindrical except for palm and fingers, moderately setose on all articles. Ischium dorsally with sharp strong spine, ventromesially with somewhat smaller subterminal spine proximally followed by 2 much smaller spines (obsolescent on right P1). Merus 1.2 × longer than carapace; 4 rows of spines (2 dorsal, 1 mesial, 1 mesioventral) continued on to carpus. Carpus 1.2 × longer than merus. Palm slightly broadened distally, 4.3 × longer than broad, 1.2 × longer than carpus; mesial margin with 4 spines on proximal two-thirds (unarmed on right P1). Fingers directed somewhat laterally, distally somewhat incurved, crossing when closed; opposable margins sinuous; movable finger one-third length of palm, opposable margin with process of moderate size proximal to midlength, fitting to opposing concave margin of fixed finger when closed.

P2-4: Setose like P1; P3 and P4 much shorter on right than on left, presumably regenerated (each article of P3-4 on right side about 0.8 × that on left side except for propodus). Meri moderately compressed, successively diminishing posteriorly (P3 merus 0.9 × length of P2 merus, P4 merus 0.8 × length of P3 merus), equally broad on P2-4; length-breadth ratio, 5.4

on P2, 4.6 on P3, 4.1 on P4; dorsal margin not crested but rounded, with 5 spines on P2-4; ventrolateral margin ending in acute spine on P2-4; ventromesial margin with 3 remotely separated small spines on P2, unarmed on P3-4; P2 merus slightly shorter than carapace, slightly longer than P2 propodus; P3 merus 0.9 × length of P3 propodus; P4 merus 0.8 × length of P4 propodus. Carpi successively slightly shorter posteriorly (P3 carpus 0.97 length of P2 carpus, P4 carpus 0.94 length of P3 carpus); carpus-propodus length ratio, 0.46 on P2, 0.48 on P3, 0.43 on P4; extensor margin with row of 4 spines paralleling row of 3 spines on lateral surface. Propodi subequal on P2-3, longer on P4 (posteriorly shorter on right side); extensor margin with 2 or 3 spines; flexor margin ending in pair of terminal spines preceded by 4 or 5 relatively broadly separated movable spines on P2, 4 on P3, 2-4 on P4. Dactyli well compressed, subequal in length on P2-4; dactylus-carpus length ratio, 1.1 on P2, 1.2 on P3, 1.3 on P4; dactylus-propodus length ratio, 0.5 on P2-4; flexor margin slightly concave in lateral view, with row of 7-9 loosely arranged, sharp spines diminishing toward base of article, penultimate about twice as broad as antepenultimate as well as ultimate, closer to ultimate than to antepenultimate.

Eggs. Two eggs carried, measuring 0.72 × 0.69 mm.

REMARKS — *Uroptychus angustus* keys out in a couplet together with *U. tracey* Ahyong, Schnabel and Baba, 2015 (see above under the key to species), but they are morphologically largely different. In *U. angustus*, the carapace dorsal surface bears fewer spines, the antennal article 5 bears a strong distomesial spine instead of being unarmed, and the P2-4 dactyli bear 7-9 loosely arranged instead of 21 closely arranged flexor marginal spines. The spination of the carapace dorsal surface is similar to that of *U. setifer* n. sp. Their relationships are discussed under the accounts of that species (see below). The spinose carapace and pereopods 1-4, unarmed abdomen, and the long posterolateral margin of sternite 4 link the species to *U. anoploetron* n. sp., but the carapace in *U. angustus* has much smaller, fewer spines and the flexor marginal spines of the P2-4 dactyli are loosely arranged, not contiguous to one another as in *U. anoploetron*.

Uroptychus angustus and *U. buantennatus* n. sp. (see below) possess the antennal article 5 that is broader than the height of the ultimate article of antennule, an unusual character among the members of the genus. However, this article is as broad as the antennal scale in *U. angustus*, whereas it is much broader in *U. buantennatus*.

Uroptychus annae n. sp.

Figures 14, 15

Uroptychus tridentatus — Baba 2005, 61 (part), fig. 21 (not *U. tridentatus* (Henderson, 1885)).

TYPE MATERIAL — Holotype: **Vanuatu**. MUSORSTOM 8 Stn DW1100, 15°04.72'S, 167°09.99'E, 258-265 m, 10.X.1994, 1♀ 4.2 mm (MNHN-IU-2014-17285). Paratypes. **Vanuatu**. MUSORSTOM 8 Stn CP1099, 15°05.39'S, 167°10.51'E, 275-284 m, 7.X.1994, 1 ♂ 4.0 mm (MNHN-IU-2014-17286). – Stn CP1018, 17°52.88'S, 168°25.08'E, 300-301 m, 27.IX.1994, 1 ♂ 4.1 mm (MNHN-IU-2014-17287). – Stn CP1017, 17°52.80'S, 168°26.20'E, 294-295 m, 27.IX.1994, 1 ♂ 3.8 mm (MNHN-IU-2014-17288). **New Caledonia**. Loyalty Islands. CALSUB PL03, 20°36'S, 167°13'E, 2465-2885 m, 22.II.1989, 1 ♀ 1.8 mm (MNHN-IU-2014-17289). MUSORSTOM 6 Stn DW482, 21°21.50'S, 167°46.80'E, 375 m, 23.II.1989, 1 ♀ 3.7 mm (MNHN-IU-2014-17290). New Caledonia. EBISCO Stn DW2530, 22°48.8'S, 159°23.0'E, 338-343 m, 9.X.2005, 1 ov. ♀ 3.5 mm (MNHN-IU-2014-17291). LAGON. Stn DW393, 22°46'S, 167°04'E, 265 m, 22.I.1985, 1 ov. ♀ 3.2 mm (MNHN-IU-2014-17299). **New Caledonia**. Norfolk Ridge. AZTÈQUE Stn CH01, 23°16.7'S, 168°04.7'E, 290-460 m, 12.II.1990, 1 ♂ 4.1 mm (MNHN-IU-2014-17292 [= Ga 4612]) (incorporated in Baba 2005). CHALCAL 2 Stn CP27, 23°15.29'S, 168°04.55'E, 289 m, 3.X.1986, 2 ov. ♀ 3.7, 3.9 mm (MNHN-IU-2014-17293). SMIB 5 Stn DW101, 23°21.2'S, 168°04.9'E, 270 m, 14.IX.1989, 1 ov. ♀ 3.5 mm (MNHN-IU-2014-17294). SMIB 8 DW182, 23°19.28'S, 168°04.82'E, 314-340 m, 31.I.1993, 1 ♂ 3.1 mm (MNHN-IU-2014-17295). BATHUS 3 Stn CP804, 23°41.40'N, 168°00.42'E, 244-278 m, 27.XI.1993, 1 ov. ♀ 4.3 mm (MNHN-IU-2014-17296). – Stn CP806, 23°42,31'S, 168°00,52'E, 27.XII.1993, 308-312 m, 1 ♂ 3.7 mm, 1 ov. ♀ 3.8 mm, 1 ♀ 4.6 mm (MNHN-IU-2014-17297). Norfolk 1 Stn DWCP1671, 23°41'S, 168°00'E, 320-397 m,

21.VI.2001, 1 ov. ♀ 3.8 mm (MNHN-IU-2014-17298[= Ga 4613]) (incorporated in Baba 2005). **New Caledonia**, Hunter and Matthew Islands. VOLSMAR Stn DW39, 22°20.5'S, 168°43.5'E, 305 m, 8.VI.1989, 1 ♂ 2.3 mm, 2 ov. ♀ 3.0, 3.2 mm, 2 ♀ 3.1, 4.0 mm (MNHN-IU-2014-17300).

ETYMOLOGY — Named for Anna W. McCallum for her contributions to the knowledge of western Australian Chirostylidae.

DISTRIBUTION — Vanuatu, Loyalty Islands, New Caledonia, Norfolk Ridge, Grand Récif du Sud, and Hunter-Matthew Islands; 248-460 m (the depth record of CALSUB Dive 03 from the Loyalty Islands, 60-600 m, is not cited here).

SIZE — Male, 2.3-4.1 mm; females, 1.8-4.6 mm; ovigerous females from 3.0 mm.

DESCRIPTION — Small species. Body and appendages sparingly with long fine setae. *Carapace*: Slightly shorter than (0.90-0.95) broad; greatest breadth 1.5-1.7 × distance between anterolateral spines. Dorsal surface unarmed, slightly convex from anterior to posterior. Lateral margins convex or convexly somewhat divergent posteriorly, usually with 7 spines: first anterolateral, overreaching lateral orbital spine; second and third much smaller, situated ventral to level of other spines; fourth to seventh usually acute, placed on branchial region, seventh situated at point one-third from posterior end, occasionally followed by obsolescent spine. Rostrum triangular, with interior angle of 20-30°, bearing subterminal small spine on each side; dorsal surface concave; length slightly less than half that of remaining carapace, breadth about half carapace breadth measured at posterior carapace margin. Lateral orbital spine small, located directly mesial to anterolateral spine. Pterygostomian flap anteriorly sharp angular, produced to sharp spine, surface with small spines on anterior half, anterior half distinctly higher than posterior half.

Sternum: Excavated sternum with distinct ridge in midline arising from center, leading anteriorly to bluntly produced anterior margin between bases of Mxp1. Sternal plastron slightly longer than broad, lateral extremities subparallel. Sternite 3 depressed well, anterior margin shallowly concave, with relatively narrow, deep median notch without flanking spine; posterolateral margin much more than half length of anterolateral margin. Anterolateral margins of sternite 5 subparallel or slightly convergent posteriorly, about as long as posterolateral margin of sternite 4.

Abdomen: Somite 2 tergite 2.4-2.7 × broader than long; pleuron posterolaterally blunt. Pleuron of somite 3 posterolaterally bunt. Telson half as long as broad; posterior plate 1.1-1.3 × longer than anterior plate, posterior margin feebly convex or feebly concave.

Eyes: Elongate, 1.6 × longer than broad, distally narrowed, overreaching midlength of rostrum; cornea more than half length of remaining eyestalk.

Antennule and antenna: Ultimate article of antennular peduncle 2.5-3.0 × longer than high. Antennal peduncle overreaching cornea. Article 2 with acute lateral spine. Antennal scale fully reaching or overreaching opposite peduncle. Distal 2 articles each with strong distomesial spine; article 5 1.5 × longer than article 4, breadth 0.7-0.9 × height of ultimate article of antennule. Flagellum of 9-12 segments not reaching distal end of P1 merus.

Mxps: Mxp1 with bases broadly separated. Mxp3 basis without denticles on mesial ridge. Ischium with small spine lateral to rounded distal end of flexor margin, crista dentata with about 12 denticles. Merus 1.8-2.2 × longer than ischium, flattish, with distolateral spine and 2 (rarely 1) small spines on distal third of flexor margin. Carpus with distolateral spine and 1-3 spines on extensor surface.

P1: Relatively massive in males, 4.9-5.5 × (males), 5.0-5.4 × (females) longer than carapace. Ischium with strong dorsal spine and distinct subterminal spine on ventromesial margin. Merus with a few small ventral spines proximally, short ventromesial and ventrolateral spines distally, length 1.1-1.3 × that of carapace. Carpus 1.0-1.2 × length of merus, dorsal surface with 3 proximal tubercles. Palm 2.7-3.3 × (males), 2.7-4.0 × (females) longer than broad, length 1.1-1.3 × longer than carpus. Fingers distally narrowed and slightly incurved, slightly gaping in males; movable finger 0.4-0.5 × length of palm, opposable margin with subtriangular blunt proximal process fitting to opposite groove proximal to low eminence of fixed finger when closed; opposable margin of fixed finger sinuous.

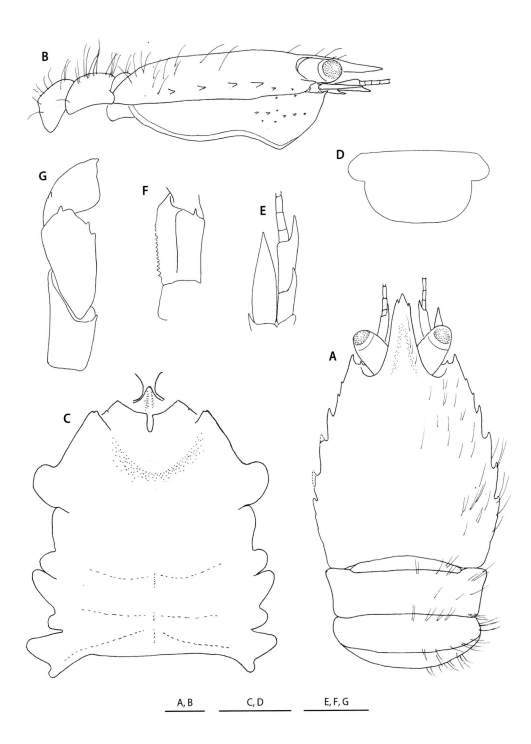

A, B   C, D   E, F, G

FIGURE 14

Uroptychus annae n. sp., holotype, ♀ 4.2 mm (MNHN-IU-2014-17285). **A**, carapace and anterior part of abdomen, dorsal. **B**, same, lateral. **C**, sternal plastron, with excavated sternum and basal parts of Mxp1. **D**, telson. **E**, right antenna, ventral. **F**, left Mxp3, ventral. G, same, lateral. Scale bars: 1 mm.

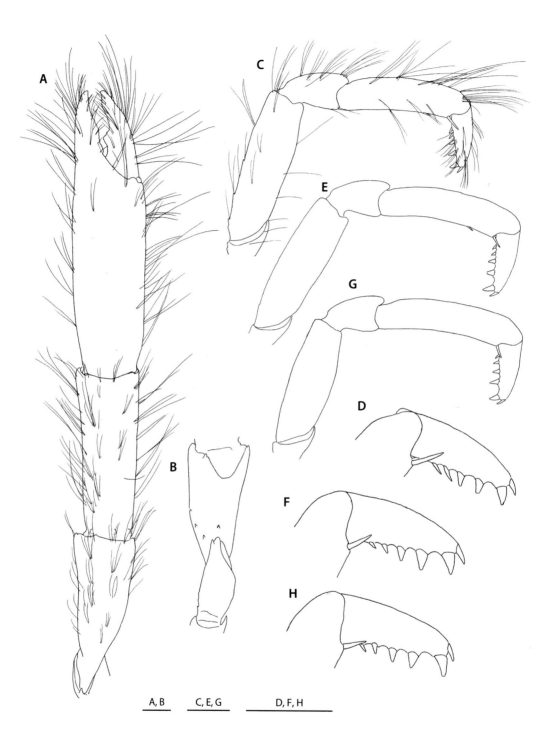

A, B   C, E, G   D, F, H

FIGURE 15

Uroptychus annae n. sp., holotype, ♀ 4.2 mm (MNHN-IU-2014-17285). **A**, left P1, dorsal. **B**, same, proximal part, setae omitted, ventral. **C**, right P2, lateral. **D**, same, distal part, lateral. **E**, right P3, lateral. **F**, same, distal part, lateral. **G**, right P4, lateral. **H**, same, distal part, lateral. Scale bars: 1 mm.

P2-4: Meri compressed mesio-laterally and relatively broad, mesial face flattish, successively shorter posteriorly (P3 merus 0.9 × length of P2 merus, P4 merus 0.8-0.9 × length of P3 merus), subequally broad on P2-4; length-breadth ratio, 3.1-3.6 on P2, 3.0-3.6 on P3, 2.6-3.5 on P4; dorsal margin with a few small proximal spines distinct on P2, occasionally obsolescent on P3, obsolete on P4, ventrolateral margin distally lobe-like, not produced to spine; P2 merus 0.8-0.9 × length of carapace, slightly longer (1.1 x) than P2 propodus; P3 merus as long as or very slightly shorter than P3 propodus; P4 merus 0.8-0.9 × length of P4 propodus. Carpi relatively short, subequal (slightly longer on P2 than on P3 and P4 or successively slighter shorter posteriorly), unarmed, length less than half that of propodus. Propodi successively shorter posteriorly, successively longer posteriorly or subequal on P3 and P4 and slightly shorter on P2; flexor margin slightly concave, with pair of terminal spines preceded by 1-3 single spines (often distal pair only on P4). Dactyli subequal in length on P2-4 or slightly shorter on P2 than on P3 and P4, slightly longer than carpi (dactylus-carpus length ratio, 1.1 on P2, 1.2 on P3, 1.1-1.3 on P4); flexor margin nearly straight, bearing slender terminal spine preceded by 3 triangular spines proximally diminishing and perpendicular to margin, and additional 1 or 2 (rarely 3) slender, somewhat inclined proximal spines; penultimate slightly larger than antepenultimate.

Eggs. Up to 18 eggs carried; size, 0.83 mm × 0.88 mm - 0.81 mm × 0.92 mm.

REMARKS — *Uroptychus tridentatus* (Henderson, 1885) has been discussed in my earlier paper (Baba, 2005), in which some of the present material was included. The occurrence of another specimen from the Philippines (MNHN-IU-2014-16995; see *Uroptychus tridentatus*) that is more like the non-intact type material of *U. tridentatus*, suggests that the New Caledonian material discussed earlier and listed above from the collection of the Paris Museum would better be placed in a different species. This is now named *U. annae* n. sp.

Uroptychus annae is distinguished from *U. tridentatus* (Henderson, 1885) by the following differences: the carapace lateral margin has the posteriormost spine more remote from the posterior end (situated at point one-third instead of quarter from the end); the P2-4 propodi have fewer flexor spines other than the distal pair (1-3 instead of 5 or 6); the P2-4 dactyli have fewer flexor spines (6 instead of 8 other than the slender terminal spine); and the P2-4 meri have the ventrolateral margin distally ending in a lobe-like instead of acuminate process. *Uroptychus annae* resembles *U. oxymerus* Ahyong & Baba, 2004 from northwestern Australia, from which it is distinguished by the following details: the carapace lateral margin bears 7 instead of 6 spines (4 instead of 3 on the branchial margin), the second and third much smaller, and the last located at the posterior third instead of slightly posterior to the midlength; the antennal article 5 bears a distomesial spine only instead of additional small spines on the mesial margin; and the P1 merus is spineless on the ventral mesial margin, instead of beraring a row of five spines. In addition, the small spine mesial to the fourth lateral spine of the carapace is distinct in *U. oxymerus*, absent in *U. annae*.

Uroptychus anoploetron n. sp.

Figures 16, 17, 305B

TYPE MATERIAL — Holotype: **New Caledonia**, Norfolk Ridge. CHALCAL 2 Stn DW73, 24°39.9'S, 168°38.1'E, 573 m, 29.X.1986, 1 ♂ 4.2 mm (MNHN-IU-2014-17301). Paratypes: **New Caledonia**, Norfolk Ridge. CHALCAL 2 Stn DW73, station data as for the holotype, 2 ♂ 2.3, 3.6 mm, 1 ov. ♀ 2.7 mm (MNHN-IU-2014-17302). – Stn DW74, 24°40.36'S, 168°38.38'E, 650 m, 29.X.1986, 1 ♂ 2.9 mm, 1 ov. ♀ 3.1 mm (MNHN-IU-2014-17303). – Stn DW75, 24°39.31'S, 168°39.67'E, 600 m, 29.X.1986, 4 ♂ 2.5-3.8 mm, 2 ov. ♀ 3.3, 3.7 mm, 2 ♀ 3.7, 4.2 mm (MNHN-IU-2014-17304). BATHUS 2 Stn CP735, 23°01'S, 166°56'E, 530-570 m, 13.V.1993, 1 ov. ♀ 3.0 mm (MNHN-IU-2014-17305). NORFOLK 1 Stn DW1694, 24°40'S, 168°39'E, 575-589 m, 24.VI.2001, 1 ov. ♀ 3.6 mm (MNHN-IU-2014-17306). – Stn DW1699, 24°40'S, 168°40'E, 581-600 m, 24.VI.2001, 5 ov. ♀ 2.9-3.5 mm (MNHN-IU-2014-17307). – Stn DW1700, 24°40'S, 168°40'E, 605-752 m, 24.VI.2001, 1 ♂ 3.2 mm (MNHN-IU-2014-17308). NORFOLK 2 Stn DW2056, 24°40.32'S, 168°39.17'E, 573-600 m, 25.X.2003, 1 ov. ♀ 2.8 mm (MNHN-IU-2014-17309). – Stn DW2057, 24°40.10'S, 168°39.34'E, 555-565 m,

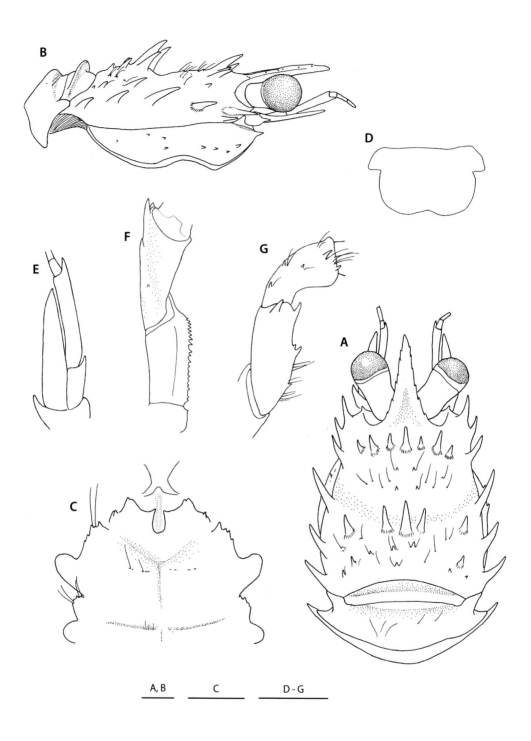

FIGURE 16

Uroptychus anoploetron n. sp., **A-C**, **E-G**, holotype, male 4.2 mm (MNHN-IU-2014-17301); **D**, male paratype 3.1 mm (MNHN-IU-2014-17311). **A**, carapace and anterior part of abdomen, dorsal. **B**, same, lateral. **C**, anterior part of sternal plastron, with excavated sternum, basal parts of Mxp1 included. **D**, telson. **E**, right antenna, ventral. **F**, right Mxp3, ventral. **G**, same, merus and carpus, lateral. Scale bars: 1 mm.

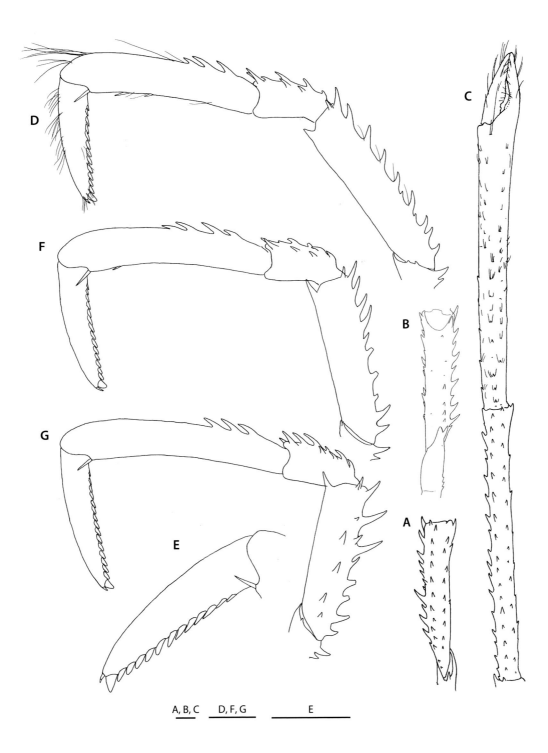

FIGURE 17

Uroptychus anoploetron n. sp., holotype, male 4.2 mm (MNHN-IU-2014-17301). **A**, right P1, proximal part, dorsal. **B**, same, ventral. **C**, same, proximal part omitted, dorsal. **D**, left P2, lateral. **E**, same, distal part, setae omitted, lateral. **F**, left P3, lateral. **G**, left P4, setae omitted, lateral. Scale bars: 1 mm.

25.X.2003, 2 ♂ 3.9, 4.4 mm, 1 ov. ♀ 4.0 mm (MNHN-IU-2014-17310). – Stn DW2059, 24°39.96'S, 168°39.84'E, 564 m, 25.X.2003, 1 ♂ 3.1 mm (MNHN-IU-2014-17311). – Stn DW2060, 24°39.84'S, 168°38.50'E, 582-600 m, 25.X.2003, 1 ♂ 3.4 mm (MNHN-IU-2014-17312). – Stn CP2062, 24°40.05'S, 168°39.70'E, 560-572 m, 25.X.2003, 1 ov. ♀ 3.4 mm (MNHN-IU-2014-17313).

ETYMOLOGY — The specific name is a noun in apposition from the Greek *anoplos* (unarmed) and *etron* (abdomen), for the unarmed abdominal somites by which the species is distinguished from the congener *U. abdominalis* n. sp. described above.

DISTRIBUTION — Norfolk Ridge; 530-650 m.

SIZE — Males, 2.3-4.4 mm; females, 2.7-4.2 mm; ovigerous females from 2.7 mm.

DESCRIPTION — Small species. *Carapace*: Slightly broader than long (0.9 × as long as broad); greatest breadth 1.4 × distance between anterolateral spines. Dorsal surface nearly horizontal in profile, with gastric and cardiac regions bordered by transverse depression, cardiac region moderately inflated; epigastric region with transverse row of 5-7 spines: median strong spine flanked by 3 spines, lateral-most occasionally very small or absent; mesogastric region unarmed, occasionally with a few to several smaller spines and very sparse setae; cardiac, intestinal and branchial regions also with spines: 3 transversely arranged anterior cardiac spines usually prominent, followed by smaller spines on intestinal region; branchial region with 3 spines arranged in longitudinal row paralleling lateral margin, posteriorly decreasing in size (anterior second and posteriormost often obsolescent) and often with additional few small spines mesial to row. Lateral margins usually divergent posterolaterally, occasionally somewhat convex on branchial region, armed with 6 acute spines: first (anterolateral) spine extending far beyond lateral orbital spine; first to third subequal, smaller than remainder, closer to each other than fourth to sixth; second distinctly ventral to level of remainder; fourth to sixth prominent, remotely equidistant, situated on branchial region, last (sixth) located near posterior end. Rostrum straight, directed slightly dorsally, sharply triangular, with interior angle of 20°; length distinctly more than half that of remaining carapace, breadth half carapace breadth measured at posterior carapace margin; dorsal surface moderately concave; lateral margin with 3-6 small spines (distinct in specimens having a few spines, obsolescent in those having more spines). Lateral orbital spine small, moderately remote from anterolateral spine and situated anterior to level of that spine. Pterygostomian flap anteriorly ending in sharp spine, surface with small spines arranged roughly in mid-longitudinal row.

Sternum: Excavated sternum anteriorly ending in broad triangular or convex margin between bases of Mxp1, with well-developed ridge in midline. Sternal plastron slightly broader than long, somewhat broadened posteriorly. Sternite 3 moderately depressed, anterior margin moderately concave, with deep U-shaped, rarely narrowly V-shaped median notch flanked by small incurved spine; anterolateral angle produced. Sternite 4 broad relative to length, about as broad as sternite 5; anterolateral margin about as long as posterolateral margin, anteriorly ending in spine of moderate size, occasionally reduced to small size and followed by a few small spines. Sternite 5 having anterolateral margin strongly convex and much shorter than posterolateral margin of sternite 4.

Abdomen: Smooth, barely setose. Somite 1 with cristiform transverse ridge. Somite 2 tergite 2.3-2.5 × broader than long; tergite with feeble anterior transverse ridge occasionally bearing a few very small spines and a few setae, pleuron anterolaterally strongly produced and sharply angular, posterolaterally bluntly angular, lateral margins knife-edge-like, strongly concave and strongly divergent posteriorly. Pleuron of somite 3 tapering to blunt tip. Telson about half as long as broad or slightly more than so; posterior plate twice as long as anterior plate, posterior margin distinctly emarginate.

Eye: Relatively large, 1.8 × longer than broad, overreaching midlength of, but falling short of apex of rostrum, distally broadened, lateral and mesial margins concave. Cornea dilated, about half as long as remaining eyestalk.

Antennule and antenna: Ultimate article of antennular peduncle 3.5 × longer than high. Antennal peduncle slender, reaching rostral tip. Article 2 with strong distolateral spine. Antennal scale 1.5 × broader than article 5, overreaching midlength of, but not reaching distal end of peduncle, rarely with lateral spine near midlength. Distal 2 articles each with well-developed distomesial spine; article 5 slightly more than twice as long as article 4, breadth two-thirds height of ultimate article of antennule. Flagellum consisting of 9-12 segments, not reaching distal end of P1 merus.

Mxp: Mxp1 with bases broadly separated. Mxp3 basis with 1 denticle often obsolete on distal part of mesial ridge. Ischium with 15-20 small denticles on crista dentata, flexor margin not rounded distally. Merus 1.7 × longer than ischium, concavely smooth on mesial face; lateral surface with distal spine; flexor margin sharply ridged, bearing 2-6 spines of irregular sizes on distal two-thirds of length, distalmost usually strong. Carpus with distinct distolateral spine, and occasionally with additional scattered small spines on lateral surface.

P1: 6-9 × longer than carapace, slender, subcylindrical, with sparse setae. Ischium with prominent dorsal spine, ventromesially with well-developed subterminal spine and a few tubercles on proximal portion. Merus and carpus very spinose. Merus 1.5-1.9 × longer than carapace, with 5 rows of spines (2 dorsal, 1 mesial, 1 lateral, 1 ventromesial) continued on to carpus, mesial spines larger than remainder. Carpus 1.3-1.7 × longer than merus. Palm 7-8 × longer than broad, as long as or slightly shorter than carpus, with a few to several mesial marginal spines (usually on proximal half) paralleling very small tubercle-like dorsal spines occasionally absent; ventral surface with row of 3 or 4 tubercles on proximal ventromesial portion. Fingers somewhat gaping in large specimens, not gaping in females and small males, slightly crossing distally; opposable margins slightly sinuous; when gaping, opposable margin of movable finger convex on proximal half bearing small notch somewhat proximal to its midlength; movable finger one-third length of palm or less than so.

P2-4: Relatively broad. Meri successively shorter posteriorly (P3 merus 0.9 × length of P2 merus, P4 merus 0.9 × length of P3 merus), equally broad on P2-4; length-breadth ratio, 4.0 on P2, 3.0-4.0 on P3 and P4; P2 merus as long as or slightly shorter than carapace as well as P2 propodus; P3 merus 0.8 × length of P3 propodus; P4 merus 0.7 × length of P4 propodus; dorsal margin with 8-11 spines on P2, 7-9 on P3, 5-8 on P4; ventrolaterally with terminal spine; lateral surface with another few small spines paralleling dorsal row on P4. Carpi subequal, with row of 4 marginal spines on extensor margin and another 2 or 3 smaller ones on lateral surface; length about half that of dactylus and 0.3 × that of propodus on P2-4. Propodi longer on P3 and P4 than on P2; extensor margin with 2-4 spines proximally; flexor margin somewhat concave in lateral view, ending in pair of spines preceded by 1 or 2 spines on P2 and P3, none on P4. Dactyli successively longer posteriorly, subequal on P2-4 or slightly longer on P3 than on P2 and P4, length 0.6 × that of propodus on P2-P3, 0.6-0.7 × on P4, distally relatively broad; flexor margin nearly straight, with 13-15 spines on P2, additional 1 or 2 spines on P3 and P4; all of these spines elongate, obliquely directed and nearly contiguous to one another, ultimate spine more slender than antepenultimate, penultimate broadest, 2 × broader than antepenultimate, remaining spines proximally diminishing.

Eggs. Up to 15 eggs carried; size, 1.0 mm × 1.2 mm.

Color. Ovigerous female paratype (MNHN-IU-2014-17302) from CHALCAL 2, Stn DW73: Carapace and abdomen pale grayish yellow, spines reddish at base; pereopods pale purple in basic color and yellow-tinged, with 2 reddish bands on merus, carpus, propodus.

REMARKS — The carapace ornamentation, spination of P1-P4, shape of the sternal plastron and the color pattern displayed by the species are very similar to those of *U. abdominalis* n. sp. *Uroptychus anoploetron* is differentiated from that species by the following: the P1-4 are more slender and the carapace spines are more pronounced; the abdominal somite 1 and 2 are unarmed, instead of bearing a pair of strong submedian spines; the tergite of abdominal somite 1 is sharply ridged transversely rather than convex from anterior to posterior; the antennal article 5 bears a distinct distomesial spine instead of being unarmed. These two species have the similar color pattern but the colorations are different: bands on the pereopods are reddish in *U. anoploetron*, yellowish in *U. abdominalis*.

Uroptychus articulatus n. sp.

Figures 18, 19

TYPE MATERIAL — Holotype: **Vanuatu**. MUSORSTOM 8 Stn DW1070, 15°37'S, 167°16'E, 184-190 m, 4.X.1994, 1 ov. ♀ 2.8 mm (MNHN-IU-2013-8569). Paratypes: **New Caledonia**. Loyalty Ridge. MUSORSTOM 6 Stn DW417, 20°14.80'S, 167°03.65'E, 283 m, with ?*Chrysogorgia* sp. (Calcaxonia, Chrysogorgiidae), 16.IX.1989, 1 ♂ 2.3 mm (MNHN-IU-2014-17314). – Stn CP454, 21°00.60'S, 167°26.50'E, 260 m, 20.II.1989, 3 ♂ 1.8-2.2 mm, 2 ov. ♀ 2.0, 2.2 mm (MNHN-IU-2013-8566). **New Caledonia**. BATHUS 1 Stn CP669, 20°57.3'S, 165°35.3'E, 255-280 m, 14.III.1993, 1 ov. ♀ 2.3 mm (MNHN-IU-2014-17315). **Vanuatu**. MUSORSTOM 8 Stn CP963, 20°20.10'S, 169°49.08'E, 400-440 m, 21.IX.1994, 1 ov. ♀ 3.2 mm (MNHN-IU-2014-17316). – Stn DW967, 20°19.45'S, 169°52.87'E, 295-334 m, 21.IX.1994, 1 ov. ♀ 2.9 mm (MNHN-IU-2014-17317). – Stn CP1017, 17°52.80'S, 168°26.20'E, 294-295 m, 27.IX.1994, 4 ♂ 2.3-3.1 mm, 2 ov. ♀ 3.0, 3.3 mm (MNHN-IU-2013-8567). – Stn CP1018, 17°52.88'S, 168°25.08'E, 300-301 m, 27.IX.1994, 2 ov. ♀ 2.8, 3.2 mm, 1 ♀ 3.4 (with rhizocephalan parasite) (MNHN-IU-2014-17318). – Stn CP1023, 17°47.60'S, 168°48.83'E, 321 m, 28.IX.1994, 2 ♂ 2.5, 3.1 mm (MNHN-IU-2013-12280 & MNHN-IU-2013-12281). – Stn CP1024, 17°48.21'S, 168°38.77'E, 335-370 m, 28.IX.1994, 1 ov. ♀ 3.0 mm, 1 ♀ 3.3 mm (MNHN-IU-2013-12282 & MNHN-IU-2013-12283).

ETYMOLOGY — From the Latin *articulatus* (articulated), in reference to the articulated antennal scale, by which the species is differentiated from *U. scandens*.

DISTRIBUTION — New Caledonia, Vanuatu and Loyalty Ridge, in 184-334 m.

SIZE — Males, 1.8-3.1 mm; females, 2.0-3.4 mm; ovigerous females from 2.0 mm.

DESCRIPTION — Small species. *Carapace*: Broader than long (0.8-0.9 × as long as broad); greatest breadth 1.4-1.6 × distance between anterolateral spines. Dorsal surface somewhat convex from anterior to posterior, setose on lateral portion and on epigastric region, with very small spines on epigastric, hepatic and anterior lateral branchial regions, those on hepatic region somewhat larger. Lateral margins feebly or barely concave between hepatic and branchial regions, convex along branchial region, with row of small spines; anterolateral spine larger than others, directed somewhat anterolaterally. Rostrum narrow triangular, directed straight forward, with interior angle of 25-30°; lateral margins slightly concave, with several obsolescent denticles; dorsal surface somewhat concave; length subequal to or slightly larger than (1.1 ×) breadth, 0.4 × that of remaining carapace, breadth about half carapace breadth measured at posterior carapace margin; lateral orbital spine very small, somewhat anterior to level of anterolateral spine. Pterygostomian flap anteriorly angular, produced to strong sharp spine followed by a few spines along dorsal margin anterior to linea anomurica and another few spines along ventral margin; surface with small spines on anterior half.

Sternum: Excavated sternum with slightly convex anterior margin; surface smooth, without ridge and central spine, with setae along anterior margin. Sternal plastron 0.8-0.9 × as long as broad; lateral extremities slightly convex, sternite 6 broadest. Sternite 3 slightly depressed in ventral view, anterior margin broadly and deeply excavated in subsemicircular shape, with pair of small median spines contiguous at base or separated by small notch, occasionally with a single median spine instead, laterally rounded or angular, occasionally with a few denticles. Sternite 4 having anterolateral margin feebly convex, anteriorly ending in anteriorly directed spine occasionally followed by a few small spines, posterolateral margin 0.8 × as long as anterolateral margin. Anterolateral margin of sternite 5 as long as posterior margin of sternite 4.

Abdomen. Somite 1 gently convex from anterior to posterior. Somite 2 tergite 2.4-2.7 × broader than long, pleuron slightly concavely divergent posteriorly, ending in rounded margin. Pleura of somites 3-5 laterally rounded. Telson half as long as broad; posterior plate with posterior margin slightly emarginate, concave or feebly convex, length 1.9-2.5 × that of anterior plate.

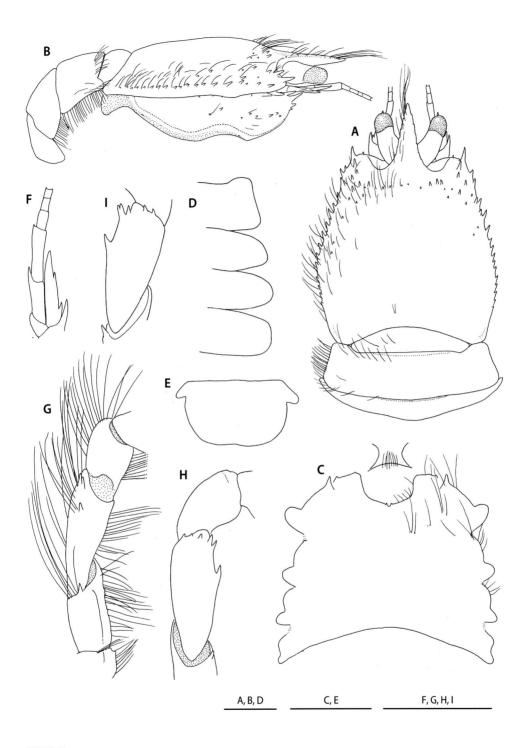

FIGURE 18

Uroptychus articulatus n. sp., holotype, ovigerous female 2.8 mm (MNHN-IU-2013-8569). **A**, carapace and anterior part of abdomen, dorsal. **B**, same, lateral. **C**, sternal plastron, with excavated sternum and basal parts of Mxp1. **D**, right pleura of abdominal somites 2-5, dorsolateral. **E**, telson. **F**, left antenna, ventral. **G**, right Mxp3, ventral. **H**, same, setae omitted. **I**, same, lateral. Scale bars: 1 mm.

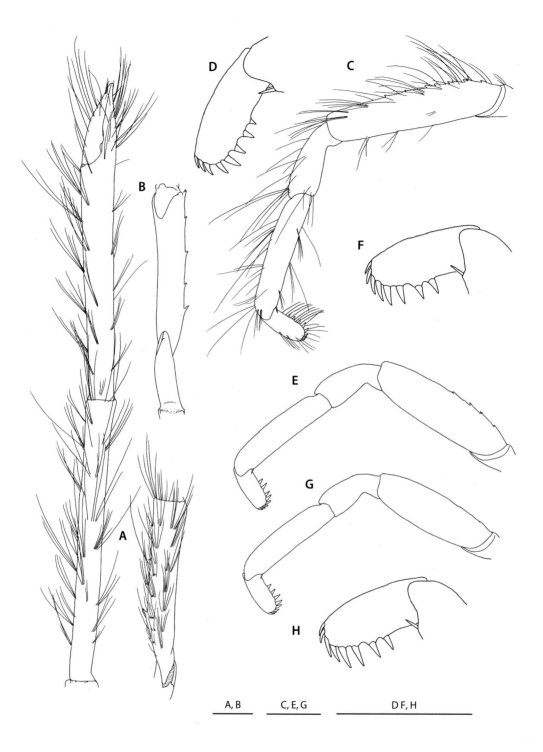

FIGURE 19

Uroptychus articulatus n. sp., holotype, ovigerous female 2.8 mm (MNHN-IU-2013-8569). **A**, right P1, dorsal. **B**, same, proximal part, ventral. **C**, left P2, lateral. **D**, same, distal part, setae omitted, lateral. **E**, left P3, setae omitted, lateral. **F**, same, distal part. **G**, left P4, lateral. **H**, same, distal part, lateral. Scale bars: 1 mm.

Eyes: Elongate, 1.8-2.1 × longer than broad, falling short of apex of rostrum, distinctly broader proximally than distally, greatest breadth 1.2-1.3 × larger than that of cornea, lateral margin convex, mesial margin concave. Cornea slightly inflated, 0.4-0.5 × length of remaining eyestalk.

Antennule and antenna: Ultimate article of antennular peduncle 3.3-3.6 × longer than high. Antennal peduncle reaching apex of rostrum. Article 2 with strong distolateral spine reaching midlength of article 4. Antennal scale 1.2 × broader than article 5, terminating in or overreaching midlength of, and not reaching point two-thirds length of article 5, laterally with 0-2 spines. Articles 4 with distinct distomesial spine. Article 5 with very small, obsolescent distomesial spine, length 1.3-1.8 × that of article 4, breadth 0.5 × height of ultimate article of antennule. Flagellum of 8-11 segments reaching midlength to distal quarter of P1 merus, apical seta somewhat longer than flagellum, reaching or overreaching distal end of P1 merus.

Mxp: Mxp1 with bases broadly separated. Mxp3 with long setae. Basis with 1 or 2 distal denticles on mesial ridge. Ischium with small spine lateral to distal end of flexor margin, crista dentata with a few to several obsolescent denticles. Merus more than twice (2.3 ×) length of ischium, ridged along flexor margin; with 2-4 spines on distal third of flexor margin and 1 or 2 small distolateral spines. Carpus unarmed.

P1: Slender, subcylindrical, 6.6-7.4 × (males), 7.0-7.6 × (females) longer than carapace, bearing long setae. Ischium dorsally with basally broad, short, depressed spine accompanying much smaller spine laterally, unarmed ventromesially. Merus 1.5-1.6 × longer than carapace, bearing 2 rows of spines (mesial row of somewhat larger spines subparalleling another row of denticle-like, bifurcate spines directly dorsolateral to it). Carpus 1.3-1.5 × longer than merus, unarmed. Palm 4.1-5.0 × (males), 6.3-6.9 × (females) as long as broad, 0.8-0.9 × as long as carpus, very slightly broadened distally. Fingers relatively narrow distally, gaping in proximal third strongly in large males, slightly in females and small males; opposable margins fitting to each other when closed in distal two-thirds; movable finger 0.4 × (rarely 0.5 x) length of palm, with obtuse process (more pronounced in males than in females) at midpoint of gaping portion.

P2-4: With long setae like P1. Meri successively shorter posteriorly (P3 merus 0.8 × length of P2 merus, P4 merus 0.8-0.9 × length of P3 merus), slightly broader on P3 than on P2 and P4; length-breadth ratio, 4.5-5.0 on P2, 3.5-3.8 on P3, 3.1-3.4 on P4; dorsal margin with 4-8 small spines on proximal two-thirds on P2, 3-7 spines on P3, unarmed on P4; P2 merus 1.1-1.2 × length of carapace, 1.3-1.4 × length of P2 propodus; P3 merus 1.2 × length of P3 propodus; P4 merus 1.1 × length of P4 propodus. Carpi subequal on P3 and P4 and longer on P2 or successively shorter posteriorly; carpus-propodus length ratio, 0.5 on P2-P4. Propodi successively shorter posteriorly; flexor margin straight in lateral view, with pair of short, slender terminal spines only. Dactyli subequal in length on P2-4, shorter than carpus, dactylus-carpus length ratio, 0.6 on P2, 0.7 on P3 and P4; dactylus-propodus length ratio, 0.3 on P2-P4; truncate, bearing 6-8 slender spines obscured by setae, 3 of these located on terminal (distal) margin, remainder perpendicular to flexor margin, terminal spines somewhat smaller than flexor marginal spines.

Eggs. Number of eggs carried, 3-26; size, 0.58 × 0.63 mm - 0.63 × 0.81 mm.

REMARKS — The species is grouped together with *Uroptychus scandens* Benedict, 1902, *U. imparilis* n. sp. and *U. parisus* n. sp., by having truncate P2-4 dactyli. Characters distinguishing these four species are very slight and are outlined under the account of *U. scandens* (see below).

The male from MUSORSTOM 6 Stn DW417 (MNHN-IU-2014-17314) was collected together with chrysogorgiid corals, but their commensal relationship cannot be assumed at present.

Uroptychus australis (Henderson, 1885)

Figures 20-23

Diptychus australis Henderson, 1885: 420 (part; specimens from Port Jackson and off Banda Island [Not specimens from north of the Kermadec Islands]; see below for designation of lectotype).

Uroptychus australis — Henderson 1888: 179 (part), pl. 21: figs 4, 4a-4c (specimens from Port Jackson and off Banda). — Ahyong & Poore 2004: 18, fig. 3. — Baba 2005: 224 (designation of lectotype: ♂, BMNH 88:33, Challenger Stn 164). — Poore *et al.* 2008: 17 (fig.).

Not *Uroptychus australis* var. *indicus* — Van Dam 1933: 18, figs 25-28 (see under *U. vandamae*).

Not *Uroptychus australis* var. *indicus* — Tirmizi 1964: 394 (undescribed species).

TYPE MATERIAL — Lectotype: **Australia**, Port Jackson. CHALLENGER Stn 164, 39°13'S, 151°38'E, 410 fms (746 m), 1 ♂ 4.4 mm (BMNH 88:33). [Examined]. Paralectotypes: same data as for the lectotype, 1 ov. ♀ 7.2 mm, 1 ♀ 7.5 mm (BMNH 88:33). [Examined]. Indonesia, off Banda. CHALLENGER Stn 194, 4°31'00"S, 129°57'20"E, 360 fms (655 m), 1 ♂ 4.8 mm, 1 ♀ 5.0 mm (BMNH 88:33). [Examined].

OTHER MATERIAL EXAMINED — **Solomon Islands**. SALOMON 1 Stn DW1827, 9°59.1'S, 161°05.8'E, 804-936 m, 4.X.2001, 1 ov. ♀ 7.0 mm (MNHN-IU-2014-17319). SALOMON 2 Stn CP2197, 8°24.2'S, 159°22.5'E, 897-1057 m, 24.X.2004, 1 ♂ 6.8 mm, 2 ov. ♀ 6.2, 6.5 mm (MNHN-IU-2014-17320). – Stn CP2230, 6°27.8'S, 156°24.3'E, 837-945 m, 29.X.2004, 18 ♂ 5.1-7.7 mm, 24 ov. ♀ 5.9-7.5 mm, 8 ♀ 4.1-6.5 mm (MNHN-IU-2014-17321). – Stn CP2253, 7°26.5'S, 156°15.0'E, 1200-1218 m, 2.XI.2004, 1 ♀ 5.4 mm (MNHN-IU-2014-17322). – Stn CP2289, 8°19.6'S, 160°01.9'E, 660-854 m, 7.XI.2004, 2 ♂ 5.8, 6.4 mm, 2 ov. ♀ 5.3, 5.6 mm, 2 ♀ 4.3, 4.8 mm (MNHN-IU-2014-17323). – Stn CP2297, 9°08.8'S, 158°16.0'E, 728-777 m, 8.XI.2004, 4 ♂ 4.3-7.5 mm, 3 ov. ♀ 6.0-6.9 mm, 2 ♀ 5.1, 6.7 mm (MNHN-IU-2014-17324). **Wallis and Futuna Islands**. MUSORSTOM 7 Stn CP550, 12°15'S, 177°28'W, 800-810 m, *Chrysogorgia* sp. (Calcaxonia, Chrysogorgiidae), 18.V.1992, 1 ♂ 6.8 mm (MNHN-IU-2014-17325), 1 ♂ 5.7 mm (MNHN-IU-2014-17326). – Stn CP551, 12°15'S, 177°28'W, 791-795 m, 18.V.1992, 1 ♀ 6.5 mm (MNHN-IU-2014-17327), 1 ♀ 7.5 mm (MNHN-IU-2014-17328). – Stn CC553, 12°17'S, 177°28'W, 780-794 m, with corals of Chrysogorgiidae (Calcaxonia), 18.V.1992, 1 ♂ 7.5 mm (MNHN-IU-2014-17329), 1 ov. ♀ 7.5 mm (MNHN-IU-2014-17330). **Vanuatu**. MUSORSTOM 8 Stn CP992, 18°52.34'S, 168°55.16'E, 748-775m, 24.IX.1994, 2 ♂ 5.4, 6.0 mm (MNHN-IU-2014-17331). – Stn CP993, 18°48.78'S, 168°54.04'E 780-783 m 24.IX.1994, 3 ♂ 5.4-6.4 mm, 1 ov. ♀ 7.0 mm, 2 ♀ 6.5, 7.2 mm (MNHN-IU-2014-17332). – Stn CC996, 8°52.41'S, 168°55.73'E, 764-786 m, 24.IX.1994, 2 ♂ 5.8, 6.5 mm, 1 ov. ♀ 6.7 mm (MNHN-IU-2014-17333). – Stn CP1036, 18°01.00'S, 168°48.20'E, 920-950 m, 29.IX.1994, 1 ♂ 6.2 mm, 1 ov. ♀ 7.1 mm (MNHN-IU-2014-17334). – Stn CP1074, 15°48.42'S, 167°24.27'E, 775-798 m, 4.X.1994, 1 ov. ♀ 6.9 mm (MNHN-IU-2014-17335). – Stn CP1080, 15°57.30'S, 167°27.73'E, 799-850 m, 5.X.1994, 7 ♂ 6.3-7.8 mm, 7 ov. ♀ 6.3-7.1 mm, 2 ♀ 7.1, 7.2 mm (MNHN-IU-2014-17336). – Stn DW1128, 16°02.14'S, 166°38.39'E, 778-811 m, 10.X.1994, 1 ov. ♀ 6.0 mm (MNHN-IU-2014-17337). SANTO Stn AT62, 15°41.5'S, 167°58.0'E, 830-918 m, 4.X.2006, 1 ov. ♀ 7.7 mm (MNHN-IU-2014-17338). **Tonga**. BORDAU 2 Stn CP1565, 20°58'S, 175°16'W, 869-880 m, 9.VI.2000, 2 ♂ 6.7, 7.7 mm, 2 ov. ♀ 6.3, 6.9 mm, 1 ♀ 6.9 mm (MNHN-IU-2014-17339). – Stn DW1588, 18°40'S, 173°52'W, 630-710 m, 13.VI.2000, 1 ♀ 5.8 mm (MNHN-IU-2014-17340). – Stn CP1600, 20°48'S, 174°52'W, 902-907 m, Chrysogorgiidae gen. sp. (Calcaxonia), 15.VI.2000, 4 ♂ 3.7-6.0 mm, 1 ov. ♀ 5.7 mm, 1 ♀ 4.2 mm (MNHN-IU-2014-17341). – Stn CP1613, 23°03'S, 175°47'W, 331-352 m, 17.VI.2000, 1 ♀ 7.1 mm (MNHN-IU-2014-17342). – Stn CP1625, 23°28'S, 176°22'W, 824 m, 19.VI.2000, 2 ov. ♀ 6.2, 6.9 mm (MNHN-IU-2014-17343). **New Caledonia**. BATHUS 4 Stn CP951, 20°31.44'S, 164°54.97'E, 960 m, 10.VIII.1994, 1 ov. ♀ 6.7 mm (MNHN-IU-2014-17344). BATHUS 2 Stn CP742, 22°33'S, 166°25'E, 340-470 m, 14.V.1993, 1 ♂ 5.1 mm (MNHN-IU-2014-17345). HALIPRO 1 Stn CH876, 23°10'S, 166°49'E, 870-1000 m, 31.III.1994, 1 ♀ 7.8 mm (MNHN-IU-2014-17346). **New Caledonia**, Loyalty Ridge. MUSORSTOM 6 Stn CP438, 20°23.00'S, 166°20.10'E, 800 m, 18.II.1989, 2 ♂ 7.3, 7.7 mm, 4 ov. ♀ 5.7-6.0 mm (MNHN-IU-2014-17347). – Stn CP427, 20°23.35'S, 166°20.00'E, 800 m, 17.VII.1989, 1 ov. ♀ (carapace broken) (MNHN-IU-2014-17348). BIOGEOCAL Stn CP290, 20°36.91'S, 167°03.34'E, 760-920m, 27.IV.1987, 1 ♂ 7.0 mm, 1 ov. ♀ 6.9 mm, 2 ♀ 4.6, 6.2 mm (MNHN-IU-2014-17349). **New Caledonia**, Loyalty Basin. BIOGEOCAL Stn CP232, 21°33.81'S, 166°27.07'E, 760-790 m, 12.IV.1987, 1 ♂ 8.1 mm, 1 ♀ 7.2 mm (MNHN-IU-2014-17350). **New Caledonia**, Hunter and Matthew Islands. VOLSMAR Stn DW04, 22°24.7'S, 171°49.0'E, 825-850 m, 1.VI.1989, 2 ♂ 4.6, 6.9 mm, 1 ov. ♀ 6.0 mm (MNHN-IU-2014-17352). – Stn CP26, 22°22.8'S, 171°21.4'E, 915-980 m, 4.VI.1989, 1 ♂ 6.5 mm, 2 ♀ 4.5, 6.0 mm (MNHN-IU-2014-17353). **New Caledonia**, Norfolk Ridge. BATHUS 3 Stn CP844, 23°06'S, 166°46'E, 908 m, 1.XII.1993, 1 ♂ 7.9 mm, 4 ov. ♀ 5.8-6.3 mm, 2 ♀ 5.3, 5.8 mm (MNHN-IU-2014-17354). BIOCAL Stn CP31, 23°08'S, 166°51'E, 850 m, 29.VIII.1985, 1 ov. ♀ 7.3 mm (MNHN-IU-2014-17355), 1 ov. ♀ 7.1 mm (MNHN-IU-2014-17356). NORFOLK 2 Stn DW2054, 23°39.62'S, 168°15.17'E, 736-800 m, 24.X.2003, 1 ♂ 6.8 mm, 1 ♀ 8.4 mm (MNHN-IU-2014-17357). – Stn DW2066, 25°16.90'S, 168°55.11'E, 834-870 m, 26.X.2003, 3 ♂ 4.7-6.3 mm (MNHN-IU-2014-17358). – Stn DW2080, 25°20.40'S, 168°18.74'E, 764-816 m, 27.X.2003, 2 ♂ 2.7 mm, carapace broken, 2 ov. ♀ 5.3, 5.7 mm (MNHN-IU-2014-17359). **New Caledonia**, Lord Howe Rise. No cruise name, Stn Dr 2, 379-391 m, X.1999, 1 ♀ 6.9 mm (MNHN-IU-2014-17360). **Kiribati**. 750 m, IV.1987, coll. Crutz, 1 ov. ♀ 7.1 mm (MNHN-IU-2014-17351).

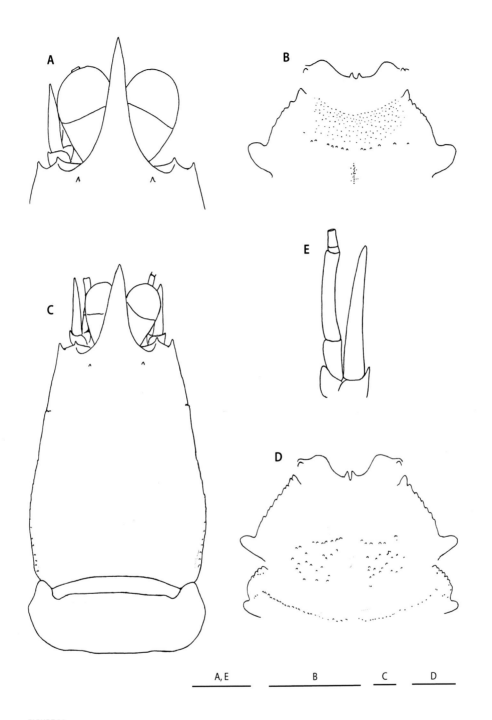

FIGURE 20

Uroptychus australis (Henderson, 1885), types from *Challenger* Stn 164. **A**, **B**, lectotype, male 4.4 mm (BMNH 88:33); **C-E**, paralectotype, female 7.5 mm (BMNH 88:33). **A**, anterior part of carapace, dorsal. **B**, anterior part of sternal plastron. **C**, carapace and anterior part of abdomen, dorsal. **D**, anterior part of sternal plastron. **E**, left antenna, ventral. Scale bars: 1 mm.

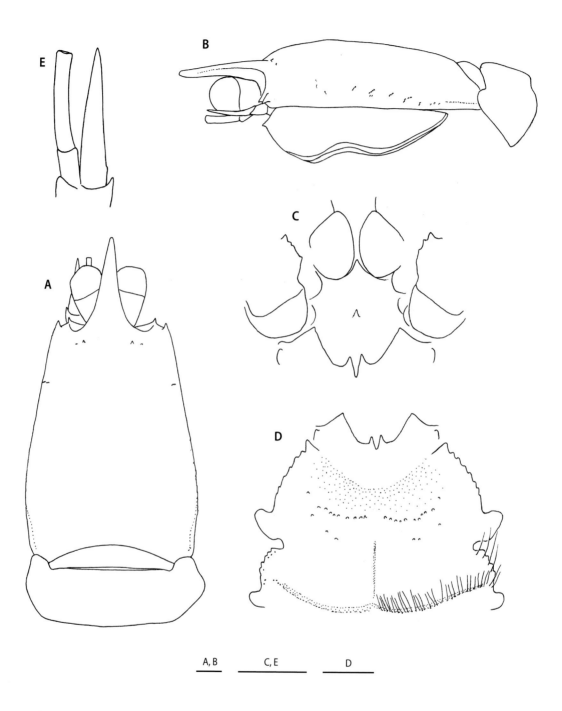

FIGURE 21

Uroptychus australis (Henderson, 1885), paralectotype, ovigerous female 7.2 mm, from *Challenger* Stn 164 (BMNH 88:33). **A**, carapace and anterior part of abdomen, dorsal. **B**, same, lateral. **C**, excavated sternum, basal parts of Mxps 1-3 included. **D**, anterior part of sternal plastron. **E**, left antennal peduncle, ventral. Scale bars: 1 mm.

DISTRIBUTION — Western Australia, New South Wales, Victoria and Tasmania, Makassar Strait, off Banda, in 458-1150 m; and now Solomon Islands, Wallis and Futuna Islands, Vanuatu, Tonga, Loyalty Islands, New Caledonia, Kiribati, Hunter-Matthew, Norfolk Ridge, Lord Howe Rise, in 331-1218 m.

SIZE — Males, 3.7-8.1 mm; females, 4.1-7.5 mm; ovigerous females from 5.3 mm.

DESCRIPTION — Medium-sized species. *Carapace*: Slightly longer than broad (length 1.1 × breadth); greatest breadth 1.6 × distance between anterolateral spines. Dorsal surface smooth and glabrous, without depression between gastric and cardiac regions; gastric region somewhat inflated in profile, distinctly elevated from level of rostrum, bearing pair of small epigastric spines or tubercles often obsolete. Lateral margins somewhat convexly divergent posteriorly, with ridge along posterior third or posterior half of length; anterolateral spine small, somewhat larger than, separated well from, and barely reaching tip of, lateral orbital spine. Rostrum sharp, narrow triangular, straight horizontal, with interior angle of 18-20°; length less than half that of remaining carapace, breadth half carapace breadth measured at posterior carapace margin; dorsal surface flattish or slightly concave. Pterygostomian flap anteriorly ending in small spine, smooth on surface.

Sternum: Excavated sternum with small central process often absent on somewhat elevated ridge in midline, anteriorly sharply produced. Sternal plastron gradually broadened posteriorly, length about four-fifths breadth. Sternite 3 moderately or strongly depressed from level of sternite 4, anterior margin deeply excavated, with pair of submedian spines, laterally rounded or angular. Sternite 4 with anterolateral margin slightly or moderately convex, moderately produced anteriorly, followed by posteriorly diminishing tubercles, posterolateral margin short, at most half as long as anterolateral margin. Anterolateral margin of sternite 5 well convex and tuberculose, about as long as posterolateral margin of sternite 4.

Abdomen: Smooth and glabrous. Somite 1 convex from anterior to posterior. Somite 2 tergite 2.1-2.4 × broader than long; pleuron anterolaterally rounded, posterolaterally bluntly angular, lateral margins strongly divergent posteriorly. Pleuron of somite 3 blunt angular posterolaterally. Telson slightly more than half (0.6) as long as broad; posterior plate 1.5-2.2 × longer than anterior plate, posterior margin distinctly emarginate.

Eye: Relatively large, 1.5 × longer than broad, proximally narrowed, overreaching midlength of, and barely reaching apex of rostrum. Cornea dilated, length more than half that of remaining eyestalk.

Antennule and antenna: Antennular ultimate article 2.0-2.5 × longer than high. Antennal peduncle reaching or slightly overreaching eyes. Article 2 with distinct distolateral spine. Antennal scale relatively slender, reaching or slightly falling short of distal end of article 5. Distal 2 articles unarmed; article 5 2.2-2.7 × longer than article 4, breadth half or less than half height of antennular ultimate article; flagellum of 13-15 (8 in male 2.7 mm) segments not reaching distal end of P1 merus.

Mxp: Mxp1 with bases close to each other or nearly contiguous. Mxp3 scarcely setose laterally. Basis with 4-7 denticles on somewhat convex, well cristate mesial ridge. Ischium with 14-20 denticles on crista dentata, flexor margin distally not rounded. Merus 2 × longer than ischium, unarmed, flexor margin not sharply crested but with rounded ridge, mesial face not flattish. No spine on carpus.

P1: With sparse long setae, very setose on fingers; scattered small tubercles on ventral surface of merus, often on carpus and palm; length 5-6 × that of carapace in both sexes. Ischium with very short distodorsal spine, ventromesial margin with or without tubercles on proximal portion or along entire length. Merus 1.1-1.2 × longer than carapace. Carpus 1.3-1.5 × longer than merus, unarmed. Palm having lateral and mesial margins subparallel, about 3 × longer than broad, slightly shorter than carpus. Fingers slightly crossing in males, not crossing in females; largely gaping in males, not gaping in females and small males; movable finger in males with strong, bifid process on gaping portion, that in female with prominent or low proximal process fitting into distinct or indistinct concavity on opposing fixed finger when closed; movable finger at most half length of palm.

P2-4: Slender and compressed mesio-laterally, with long setae on distal portions of merus, carpus and entire propodus and dactylus. Meri with smooth surface, dorsal crest with very small, often obsolescent denticles on proximal half on P2, on proximal third on P3, distally bearing fine setae; length-breadth ratio, 5.0 on P2, 4.0-5.0 on P3, 3.0-4.0 on P4;

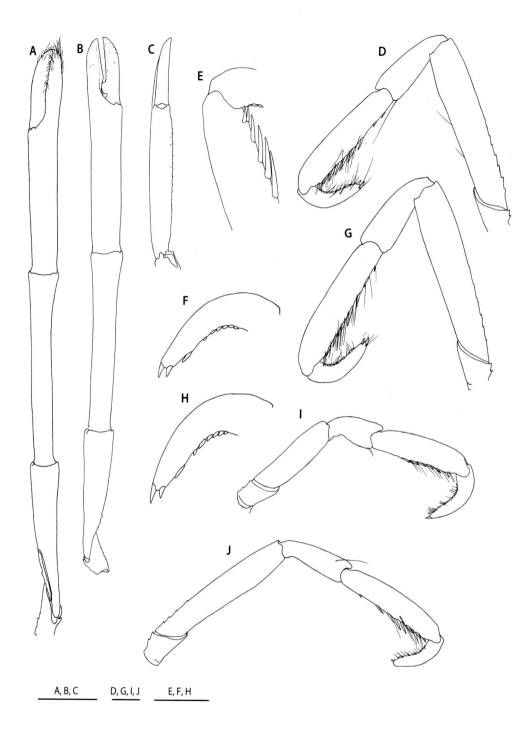

A, B, C   D, G, I, J   E, F, H

FIGURE 22

Uroptychus australis (Henderson, 1885). Detached pereopods, *Challenger* Stn 164 (BMNH 88:33). **A**, right P1, dorsal. **B**, left P1, dorsal. **C**, same, distal articles, mesial. **D**, left P2, lateral. **E**, same, distal part, mesial. **F**, same, dactylus, setae omitted, lateral. **G**, left P3, lateral. **H**, same, dactylus, lateral. **I**, right P4, lateral. **J**, right P2 or P3. Scale bars: A-C, 5 mm; D-J, 1 mm.

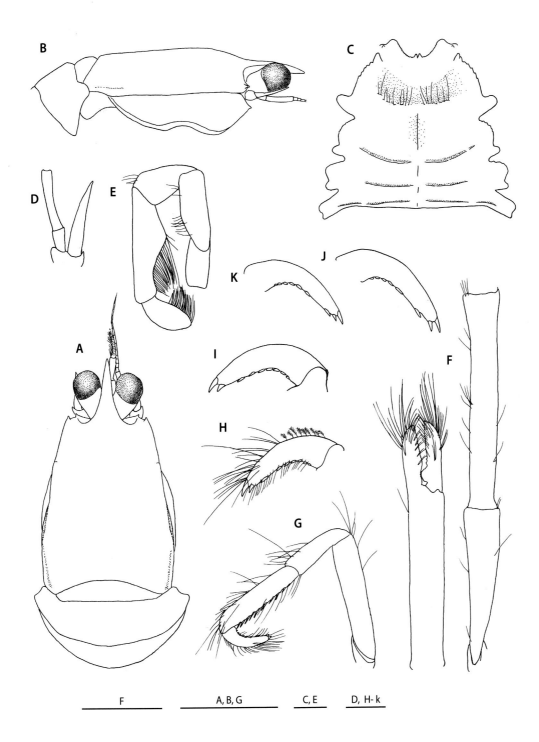

F A, B, G C, E D, H- k

FIGURE 23

Uroptychus australis (Henderson, 1885). **A-I**, ovigerous female 7.3 mm (MNHN-IU-2014-17355). **J**, **K**, ovigerous female 7.1 mm (MNHN-IU-2014-17356). **A**, carapace and anterior part of abdomen, dorsal. **B**, same, lateral. **C**, sternal plastron. **D**, left antenna, ventral. **E**, left Mxp3, lateral. **F**, left P1, dorsal. **G**, left P2, lateral. **H**, same, distal part, lateral. **I**, left P3, distal part, setae omitted, lateral. **J**, right P3, distal part, setae omitted, lateral. **K**, right P4, distal part, setae omitted, lateral. Scale bars: A, B, G, F, 5 mm; C-E, H-K, 1 mm.

P2 merus subequal to P3 merus in length, subequal to or slightly shorter than carapace, 1.2-1.3 × longer than P2 propodus; P4 merus much narrower than (ca. 0.6 x) P3 merus, length 0.5-0.6 × that of P3 merus, 0.8-0.9 × that of P4 propodus. P2-3 carpi subequal; P4 carpus short, 0.6-0.7 × length of P3 carpus; 0.5-0.6 × length of propodus on P2, 0.4-0.5 × on P3 and P4. Propodi longest on P3, shortest on P4, bearing long setae; flexor margin nearly straight, with pair of distal spines somewhat distant from juncture with dactylus, preceded by movable spines (6-9 in number on P2, 5-8 on P3, 4-6 on P4) along entire length to distal three-quarters of length. Dactyli rather slender, strongly or moderately curved proximally, dactylus-carpus length ratio, 0.7-0.8 on P2, 0.8-0.9 on P3, 1.0-1.1 on P4; dactylus-propodus length ratio, 0.4 on P2-3, 0.5 on P4; extensor margin often with short plumose setae on proximal two-thirds; flexor margin with 8-9 spines, distal 2 nearly terminal in position, first (ultimate) largest, other proximal spines contiguously oriented parallel to flexor margin, third (antepenultimate) rather remote from second (penultimate) and equidistant between second and fourth, occasionally closer to second, remaining spines close to one another and diminishing toward base of article.

Eggs. Up to 30 eggs carried, 1.17 mm × 1.31 mm - 1.37 mm × 1.53 mm.

Color. A Western Australian specimen was illustrated by Poore *et al.* (2008).

PARASITES — Rhizocephalan externae on one female from BORDAU 2 Stn CP1565 (MNHN-IU-2014-17339), one female from Stn BORDAU 2 CP1588 (MNHN-IU-2014-17340) and the female from HALIPRO 1 Stn CH876 (MNHN-IU-2014-17346). One male from SALOMON 2 Stn CP2230 (MNHN-IU-2014-17321) is infected by a bopyrid isopod, as also are one male and two females from SALOMON 2 Stn CP2297 (MNHN-IU-2014-17324), the male from MUSORSTOM 7 Stn CP550 (MNHN-IU-2014-17326) and the female from CP551 (MNHN-IU-2014-17327).

REMARKS — The type material of *Uroptychus australis* (Henderson, 1885) now in the collection of the Natural History Museum, London, was collected from four different localities: *Challenger* Stn 164 off Port Jackson, Stn 194 off Banda, Stns 170 and 171 in the Kermadec Islands. Examination of the material discloses that it includes three different species. All the material (1 male, 1 ovigerous female, and 1 non-ovigerous female) from Stn 164 and part (1 male and 1 non-ovigerous female) from Station 194 are identical and referred to *U. australis*. Hence the lectotype was assigned to the male from Station 164 (Baba 2005: 224). The materials from Stns 170 and 171 are identified as *U. terminalis* n. sp. Part of the material from Stn 194 (1 ovigerous female) is referred to *U. empheres* Ahyong & Poore, 2004 (see below).

The general features of the carapace and the P2-4 dactyli bearing spines oriented parallel to the flexor margin suggest that *U. australis* is close to *U. brevirostris* Van Dam, 1933 (see above) and *U. setosipes* Baba, 1981, from Japan. Characters distinguishing *U. australis* from *U. brevirostris* are discussed under the remarks of *U. brevirostris*. *Uroptychus australis* has a short P4 merus, which is 0.5-0.6 times as long as the P3 merus rather than being slightly shorter as in many of the other species. *Uroptychus setosipes* has a short P4 merus (confirmed by examinination of the holotype), sharing with *U. australis* a characteristic arrangement of the anterolateral spine of the carapace situated clearly posterior to the lateral orbital spine. However, the branchial margin in *U. setosipes* is sharply ridged along the entire length and the antennal article 2 bears a very small instead of distinct distolateral spine, the characters to mention the obvious differences from *U. australis*.

The ovigerous female paratype of *U. vandamae* from Makassar Strait was noted to be different from both the holotype and male paratype of that species in the spination of the P2-4 dactyli (Baba, 1988: 52). Re-examination of the types disclosed that this ovigerous paratype is referable to *U. australis* (see below under *U. vandamae*).

The material reported under *U. australis* var. *indicus* from Zanzibar area by Tirmizi (1964) is identical to the material of *U. gracilimanus* from Madagascar (Baba, 1990), but this species is different from *U. gracilimanus* (and *U. dejouanneti* n. sp.) in having the P2-4 carpi long relative to dactyli and the propodi with the different location of the flexor terminal spine, representing an undescribed species (Baba, 2005: 36). Additional material referable to this species has been collected from the Mozambique Channel by Mainbaza Station CP3138. This species will be described elsewhere.

Uroptychus babai Ahyong & Poore, 2004
Figures 24-27, 305C

Uroptychus babai Ahyong & Poore, 2004: 22, fig. 4. — Poore *et al.* 2011: 328, pl. 6: fig. D.
Uroptychus granulatus — Baba 1990: 923, fig. 9 (Not *U. granulatus* Benedict, 1902).
Not *Uroptychus babai*: Baba *et al.* 2009: 38, figs 30-31 (= *U. parilis* Cabezas, Lin & Chan, 2012).

TYPE MATERIAL — Holotype: **Australia**, E of Broken Bay, 33°31-34'S, 152°02-04'E, 905-914 m, male (AM P26782). [not examined].

MATERIAL EXAMINED — **New Caledonia**, Chesterfield Islands. MUSORSTOM 5 Stn CP324, 21°15.01'S, 157°51.33'E, 970 m, 14.X.1986, 1 ♀ 9.2 mm (MNHN-IU-2014-16300), 1 ov. ♀ 9.4 mm (MNHN-IU-2014-16301). **Solomon Islands**. SALOMON 2 Stn CP2181, 8°49.9'S, 159°39.8'E, 645-840 m, 22.X.2004, 1 ♂ 9.2 mm (MNHN-IU-2014-16302). – Stn CP2189, 8°19.6'S, 160°01.9'E, 660-854 m, 23.X. 2004, 5 ♂ 6.8-11.8 mm, 1 ov. ♀ 13.0 mm, 4 ♀ 7.5-12.5 mm (MNHN-IU-2014-16303). – Stn CP2193, 8°23.9'S, 159°26.6'E, 362-432 m, 24.XI.2004, 1 ♂ 7.8 mm, 1 ♀ 7.7 mm (MNHN-IU-2014-16304). – Stn CP2241, 6°55.3'S, 156°21.2'E, 815-1000 m, 30.X.2004, 1 ♀ 9.0 mm (MNHN-IU-2014-16305). – Stn CP2269, 7°45.1'S, 156°56.3'E, 768-890 m, 4.XI.2004, 1 ♂ 5.2 mm (MNHN-IU-2014-16306). – Stn CP2272, 8°56.2'S, 157°44.1'E, 380-537 m, 5.XI.2004, 1 ♀ 11.3 mm (MNHN-IU-2014-16307). – Stn CP2289, 08°36'S, 157°28'E, 627-623 m, 07.XI.2004, 4 ♂ 3.1-5.2 mm (MNHN-IU-2014-16308). – No station number, XI.2004, 1 ♂ 7.8 mm, 1 ♀ 3.5 mm (MNHN-IU-2014-16309).

DISTRIBUTION — Previously known from Madagascar, New South Wales, in 880-1152 m; and now from Chesterfield Islands and Solomon Islands, in 362-1000 m.

SIZE — Males, 3.1-11.8 mm; females, 3.5-13.0 mm, ovigerous females from 9.4 m.

DESCRIPTION — Large species. Body and appendages sparsely or thickly covered with fine setae (usually thick in large specimens). *Carapace*: Slightly broader than long (length 0.8-0.9 × breadth); greatest breadth 1.8 × distance between anterolateral spines. Dorsal surface granulose in large specimens, less so in small specimens, moderately convex from anterior to posterior, with very weak (in small specimens) or distinct (in large specimens) depression suggesting cervical groove, anteriorly smoothly continued on to rostrum. Lateral margins convex, with short, oblique granulate ridges: one at anterior end of branchial region well elevated, rarely representing tiny spine, another ridge behind posterior branchial region usually visible in dorsal view, and others discernible under high magnification; ridged along posterior half; anterolateral spine well developed, distinctly overreaching lateral orbital spine. Rostrum broad sharp triangular, with interior angle of 30-35°, somewhat upcurved distally; length 0.4-0.6 × that of remaining carapace (greater in small specimens than in large specimens), breadth less than half carapace breadth measured at posterior carapace margin; dorsal surface slightly concave. Lateral orbital spine small, occasionally reduced to acuminate angle, moderately remote from and slightly anterior to level of anterolateral spine. Pterygostomian flap with granulate short ridges or fine granules supporting setae on surface, anterior margin angular, produced to small sharp spine.

Sternum: Excavated sternum with convex anterior margin between Mxp1, with low ridge in midline. Sternal plastron slightly broader than long, lateral extremities gently divergent posteriorly. Sternite 3 strongly depressed from level of sternite 4; anterior margin deeply or moderately emarginate in broad V-shape, without submedian spines; lateral margin with small spine near lateral end. Sternite 4 with anterolateral margin somewhat or moderately convex anteriorly, anterior end rounded, angular or produced to anteriorly directed spine falling short of anterior end of sternite 3; posterolateral margin 0.6-0.8 × length of anterolateral margin.

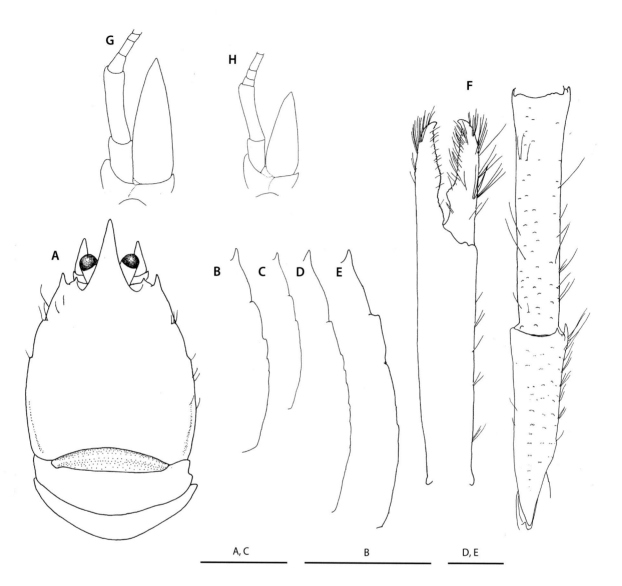

FIGURE 24

Uroptychus babai Ahyong & Poore, 2004. **A**, **B**, **F**, female 9.2 mm from Chesterfield Islands (MNHN-IU-2014-16300); **C**, male, 6.8 mm from Solomon Islands (MNHN-IU-2014-16303); **D**, male 10.7 mm (MNHN-IU-2014-16303); **E**, male 11.8 mm (MNHN-IU-2014-16303); **G**, male 9.9 mm from Madagascar (MNHN-IU-2014-12824); **H**, female 7.8 mm (MNHN-IU-2014-12824). **A**, carapace and anterior part of abdomen, dorsal. **B-E**, Lateral margin of carapace, dorsomesial. **F**, left P1, dorsal. **G**, **H**, left antenna, ventral. Scale bars: A-F, 5 mm; G, H, 1 mm.

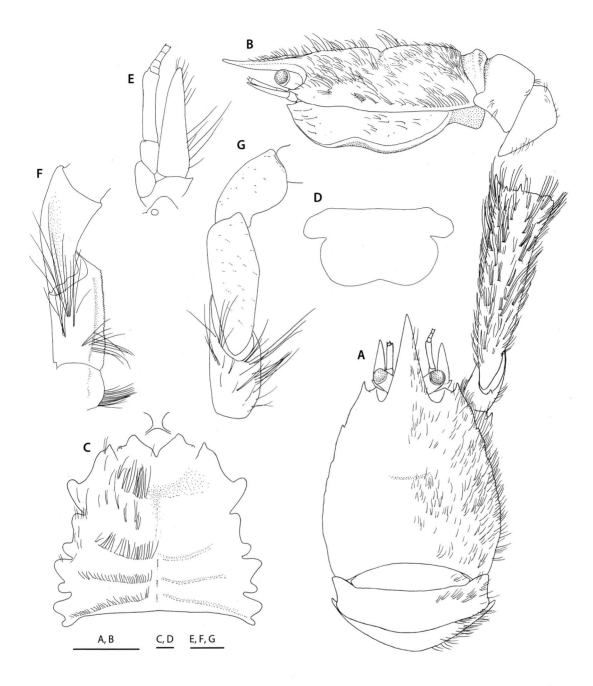

FIGURE 25

Uroptychus babai Ahyong & Poore, 2004, male 11.8 mm (MNHN-IU-2014-16303). **A**, carapace and anterior part of abdomen, proximal part of right P1 included, dorsal. **B**, same, lateral. **C**, sternal plastron, with excavated sternum and basal parts of Mxp1. **D**, telson. **E**, left antenna, ventral. **F**, right Mxp3, ventral. **G**, same, lateral. Scale bars: A, B, 5 mm; C-G, 1 mm.

Abdomen: Somite 1 with well-elevated, rounded transverse ridge. Somite 2 tergite 2.6-2.7 × broader than long, pleuron with concavely divergent lateral margin posteriorly ending in rounded corner. Pleuron of somite 3 tapering. Telson about half as long as broad; posterior plate emarginate on posterior margin, length 1.2-1.6 × that of anterior plate.

Eye: Relatively small, 1.5-1.8 × longer than broad, terminating in or slightly overreaching midlength of rostrum; lateral and mesial margins subparallel. Cornea not dilated, length much more than half that of remaining eyestalk.

Antennule and antenna: Ultimate article of antennular peduncle short relative to height in small specimens, long in large specimens (breadth-height ratio, 2.5-3.1 in specimens 3.2-7.6 mm, 4.5-4.9 in specimens 12.0-12.8 mm). Antennal peduncle relatively slender. Article 2 with small lateral spine. Antennal scale 1.4-2.5 × broader than article 5, varying from slightly falling short of to terminating in distal end of article 5, rarely reaching proximal third segment of antennal flagellum. Distal 2 articles unarmed (in large specimens, each with very tiny tubercle-like ventral distomesial spine); article 5 1.6-2.1 × length of article 4; breadth 0.4-0.5 × height of ultimate antennular article. Flagellum consisting of 22-23 segments slightly falling short of or overreaching distal end of P1 merus (13 or 14 segments overreaching distal end of P1 merus in males 3.1-4.3 mm).

Mxp: Mxp1 with bases broadly separated. Mxp3 basis without distinct denticles on mesial ridge. Ischium with 26-47 tiny denticles on crista dentata, flexor margin not rounded distally. Merus 2.4 × longer than ischium, unarmed, flexor ridge not cristate, moderately rounded. Carpus unarmed.

P1: 3.2-6.6 × longer than carapace (usually shorter in small specimens than in large specimens), relatively slender; with simple fine setae more numerous in large specimens. Ischium with basally broad dorsal spine, ventromesially unarmed. Merus granulate dorsally and ventrally, with 2 distoventral spines and 1 or 2 small, often obsolescent distodorsal spines; length 0.9-1.3 × that of carapace (shorter in small specimens<4 mm). Carpus also granulate like merus, 1.2-1.6 × longer than merus (shorter in small specimens, longer in larger specimens), with distomesial and distolateral spines on ventral surface and 1 distodorsal spine (often obsolete). Palm more slender in females than in males, length-breadth ratio, 3.3-6.9 in males and 4.0-9.2 × in females, lateral and mesial margins subparallel, dorsal granulation more weak than on carpus; length subequal to that of carpus. Fingers slightly curving ventrally, distally incurved, crossing when closed (in small specimens, slightly incurved, not gaping, opposable margins nearly straight or with low eminence on movable finger); movable finger with obtuse proximal process fitting to narrow longitudinal groove on opposite fixed finger when closed, length 0.3-0.5 × that of palm (short in large specimens).

P2-4: Relatively slender, subcylindrical on meri, somewhat compressed mesio-laterally on propodi. Meri unarmed, successively shorter posteriorly (P3 merus 0.8-0.9 × length of P2 merus; P4 merus 0.8-0.9 × length of P3 merus), subequally broad on P2-4; length-breadth ratio, 4-5 on P2, 4 on P3, 3-4 on P4; P2 merus 0.7-1.0 × length of carapace (0.7-0.8 × in specimens<5.2 mm), 1.0-1.3 × length of P2 propodus (shorter in small specimens); P3 merus as long as P3 propodus; P4 merus 0.7-0.9 × length of P4 propodus; dorsal margin with obsolescent denticles on P2. Carpi subequal or slightly longer on P2 (P3 carpus 0.9 × length of P2 carpus), shorter than dactyli, 0.4 × length of propodi on P2, 0.3-0.4 × on 3 and P4. Propodi curving, shorter on P2 than on P3 and P4, subequal on P3 and P4 or slightly longer on P4 than on P3; flexor margin with pair of distal spines only. Dactyli relatively stout, slightly curved; length 0.4-0.6 × that of propodus on P2-4; flexor margin ending in slender spine preceded by much broader penultimate and 12-13 close (nearly contiguous), obliquely directed, proximally diminishing spines on P2, 12-14 on P3, 13-16 on P4, all spines obscured by dense setae; penultimate broadest, antepenultimate broader than ultimate, half as broad as penultimate.

Eggs. About 40 eggs carried by largest female (12.8 mm); size, 1.11 mm × 1.33 mm - 1.28 mm × 1.55 mm.

Color. The coloration of the female (MNHN-IU-2014-16300) from MUSORSTOM 5 Stn CP324 is exactly the same as that of the figure of the holotype from eastern Australia provided by Poore *et al.* (2011): pinkish red on anterior part of carapace, paler on rostrum, whitish on remaining carapace and abdomen; P1 pale pinkish red, other pereopods much paler.

REMARKS — Specimens >10 mm are usually very setose on the body and appendages. Small specimens <5.2 mm are very sparsely setose, as also are some of large specimens (male 9.2 mm, MNHN-IU-2014-16302; females 9.2 mm, MNHN-IU-2014-16300; ovigerous female 9.4 mm, MNHN-IU-2014-16301). The specimens as illustrated in Figures 25-26 look

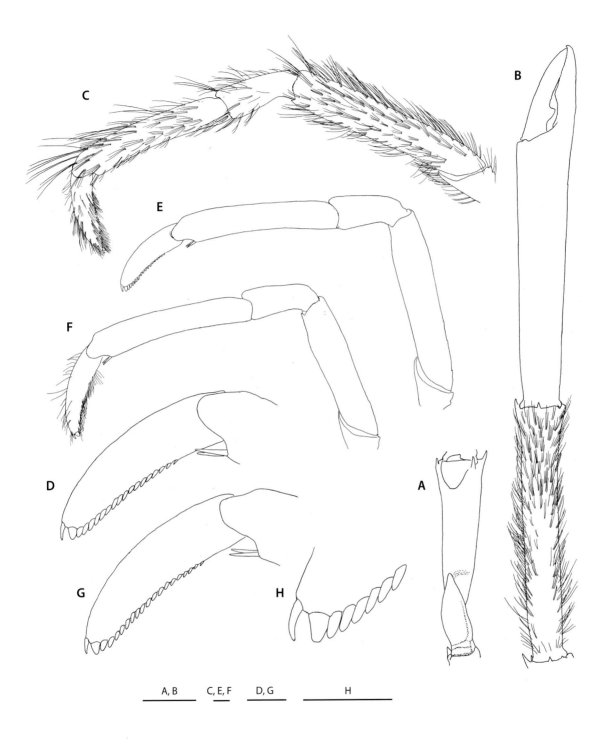

FIGURE 26

Uroptychus babai Ahyong & Poore, 2004, male 11.8 mm (MNHN-IU-2014-16303). **A**, right P1, proximal part, ventral. **B**, same, carpus, palm and fingers, setae omitted on palm and fingers, dorsal. **C**, left P2, lateral. **D**, same, distal part, lateral. **E**, left P3, setae omitted, lateral. **F**, left P4, setae omitted on merus, carpus and propodus, lateral. **G**, same, distal part, lateral. **H**, same, distal part enlarged. Scale bars: 1 mm.

very much like *U. parilis* Cabezas, Lin & Chan, 2012. The holotype of *U. parilis* from Taiwan was first referred to *U. babai* by Baba *et al.* (2009) but subsequently described as new. Cabezas *et al.* (2012) noted that the major differences between the two species are found on the carapace lateral margin (almost smooth in *U. parilis*, crenulated in *U. babai*), thoracic sternite 4 (anterolaterally produced to a distinct spine in *U. parilis*, lacking spine in *U. babai*), and the antennal scale (falling short of to slightly overreaching the distal end of antennal article 5 in *U. parilis*, distinctly overreaching article 5 in *U. babai*).

The present specimens seem to cover all these differences. Two distinct ridges visible in dorsal view on the lateral margin of the carapace are usually small in large specimens so as to be seen as smooth but discernible in all the specimens examined (Figure 24A-E), as well as in the holotype of *U. parilis* illustrated in Baba *et al.* (2009: fig. 31a). Sternite 4 is also variable, even in several specimens from one sampling station (Figure 27F, G, H); the anterolateral spine as diagnosed for *U. parilis* is also present in some of the specimens examined as well as in one of the specimens of *U. babai* from Madagascar (MNHN-Ga 2234), although not so strong to reach the anterior end of sternite 3 as in the holotype of *U. parilis* (see Figure 27). The antennal scale varies from falling short of to overreaching the distal end of the antennal article 5 (Figure 24G, H).

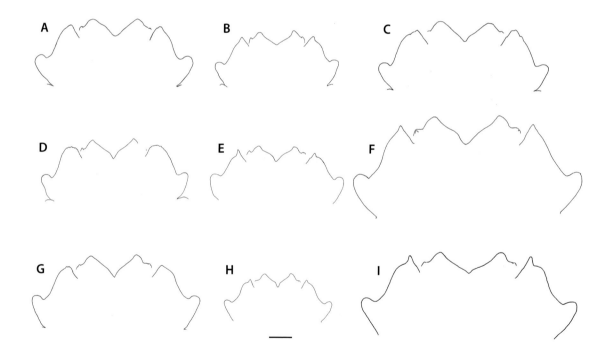

FIGURE 27

Uroptychus babai Ahyong & Poore, 2004, sternites 3-4. **A**, male 9.9 mm (MNHN-IU-2014-12824). **B**, female 7.8 mm (MNHN-IU-2014-12824). **C**, ovigerous female 9.4 mm (MNHN-IU-2014-16301). **D**, female 9.2 mm (MNHN-IU-2014-16300). **E**, female 7.7 mm (MNHN-IU-2014-16304). **F**, male 11.3 mm (MNHN-IU-2014-16303). **G**, female 9.3 mm (MNHN-IU-2014-16303). **H**, male 6.8 mm (MNHN-IU-2014-16303). **I**, female 11.3 mm (MNHN-IU-2014-16307). Scale bar: 1 mm.

P1 appears longer relative to the carapace in *U. parilis*, measuring 6 times longer than the carapace (Cabezas *et al.* 2012), whereas about 4 times as long in *U. babai* (Ahyong & Poore, 2004, fig. 4a). However, the present specimens (poc 3.1-13.0 mm) show a wide, age related variation in P1-carapace length ratios: 3.3-4.8 in small specimens (3.2-5.2 mm), 5.0-6.8 in large specimens (6.8-13.0 mm). An exceptional case is two large females (9.2, 9.4 mm) from the Chesterfield Islands (MNHN-IU-2014-16300 & MNHN-IU-2014-16301), which have ratios 4.5 and 4.7. P1 in *U. parilis* looks much more slender than in *U. babai*, as illustrated by Cabezas *et al.* (2012: fig. 1A) and Ahyong & Poore (2004: fig. 4A). This difference appears to be sex-related, as usual in other species of *Uroptychus*. In the present material, the P1 palm is 3.3-6.9 (males), 4.0-9.2 (females) times longer than broad, the length-breadth ratio being greater in large specimens. However, the females from the Chesterfield Island the ratios are 4.0 and 4.3, whereas three females from SALOMON 2 Station CP2189 (MNHN-IU-2014-16303) of nearly the same size as the Chesterfield specimens show the ratios 8.7-9.2 and the largest female 5.6. The ratio is thus widely variable.

In conclusion, the morphological differences noted by Cabezas *et al.* (2012) do not seem to be applicable especially to small specimens. The only definite difference resides on coloration: orange red overall including appendages in *U. parilis* versus reddish on the anterior part of the carapace, whitish on remaining carapace and abdomen in *U. babai*. Molecular data would clarify their systematic status.

Uroptychus baeomma n. sp.
Figures 28, 29

TYPE MATERIAL — Holotype: **New Caledonia**, Loyalty Islands. BIOCAL Stn DW83, 20°35'S, 166°54'E, 460 m, 6.IX.1985, ov. ♀ 9.8 mm (MNHN-IU-2014-16310). Paratypes: Station data as for the holotype, 1 ♂ 7.5 mm, 1 ♀ 5.0 mm (MNHN-IU-2014-12786). **New Caledonia**, MUSORSTOM 4 Stn CP216, 22°59.5'S, 167°22.0'E, 490-515 m, 29.IX.1985, 1 ♂ 7.6 mm, 3 ov. ♀ 9.7-10.0 mm, 1 ♀ 10.1 mm (MNHN-IU-2014-16311). – Stn CP238, 22°13.0'S, 167°14.0'E, 500-510 m, 2.X.1985, 1 ♂ 7.5 mm, 1 ov. ♀ 7.6 mm (MNHN-IU-2014-16312). **New Caledonia**, Hunter and Matthew Islands. VOLSMAR Stn DW16, 22°25'S, 171°41'E, 420-500 m, 03.VI.1989, 1 ♂ 4.7 mm, 1 ♀ 5.8 mm (MNHN-IU-2014-16313). **Vanuatu**. MUSORSTOM 8 Stn CP974, 19°21.51'S, 169°28.26'E, 492-520 m, 22.IX.1994, 2 ov. ♀ 7.5, 7.8 mm (MNHN-IU-2014-16314). – Stn CP983, 19°21.61'S, 169°27.76'E, 480-475 m, 23.IX.1994, 1 ov. ♀ 7.3 mm (MNHN-IU-2014-16315).

ETYMOLOGY — The specific name is a noun in apposition from the Greek *baios* (small) and *omma* (eye), referring to relatively small corneae of the species.

DISTRIBUTION — Loyalty Islands, New Caledonia, Hunter and Matthew Islands, and Vanuatu; 460-520 m.

SIZE — Males, 4.7-7.6; females, 5.0-10.0 mm; ovigerous females from 7.3 mm.

DESCRIPTION — Medium-sized species. *Carapace*: Slightly broader than long (0.9 × as long as broad); greatest breadth 1.9-2.0 × distance between anterolateral spines. Dorsal surface smooth with scattered setae, slightly convex from anterior to posterior, with or without laterally diminishing cervical groove. Lateral margins well convex, with 5 well-developed spines on anterior two-thirds of length: first anterolateral, overreaching small lateral orbital spine, located at level of that spine; remaining 4 strong, situated on branchial region, directed anteriorly, somewhat diminishing posteriorly, third spine rarely preceded by small spine, last spine followed by distinct ridge leading to posterior end; 1-3 very small tubercle-like spines often obsolescent between first (anterolateral) and second spines. Rostrum broad triangular, with interior angle of 37°, slightly deflected ventrally, length about half or less than half that of remaining carapace, breadth less than half carapace breadth measured at posterior carapace margin; dorsal surface concave, ventral surface horizontal;

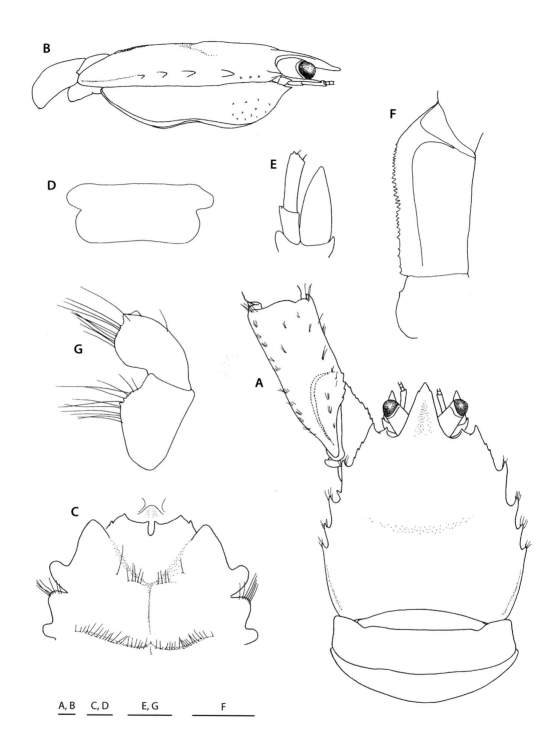

FIGURE 28

Uroptychus baeomma n. sp., holotype, ovigerous female 9.8 mm (MNHN-IU-2014-16310). **A**, carapace and anterior part of abdomen, proximal part of left P1 included, dorsal. **B**, same, lateral. **C**, anterior part of sternal plastron, with excavated sternum and basal parts of Mxp 1. **D**, telson. **E**, left antenna, ventral. **F**, left Mxp3, ventral. **G**, same, merus and carpus, lateral. Scale bars: 1 mm.

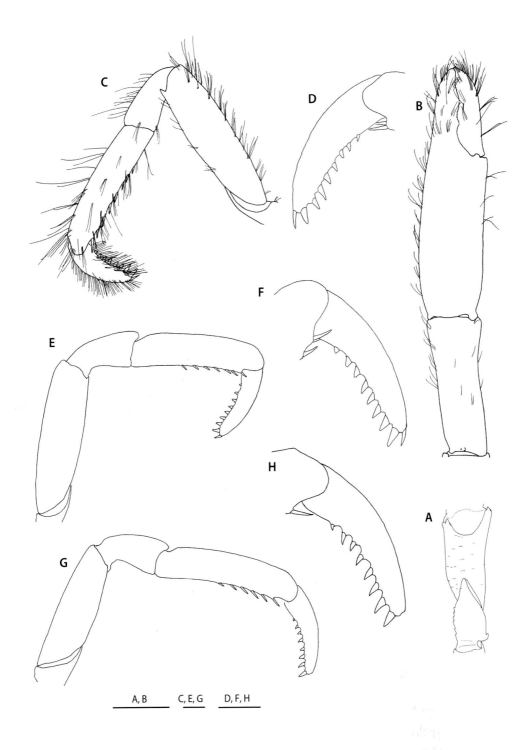

A, B   C, E, G   D, F, H

FIGURE 29

Uroptychus baeomma n. sp., holotype, ovigerous female 9.8 mm (MNHN-IU-2014-16310). **A**, left P1, proximal part, ventral. **B**, same, distal part, dorsal. **C**, left P2, lateral. **D**, same, distal part, setae omitted. **E**, right P3, lateral. **F**, same, distal part, lateral. **G**, right P4, lateral. **H**, same, distal part, lateral. Scale bars: A, B, 5 mm; C-H, 1 mm.

lateral margin with subapical spine usually very small, often obsolescent. Pterygostomian flap with scattered small spines or tubercles on anterior portion, anterior margin sharply angular, terminating in distinct spine.

Sternites: Excavated sternum with strongly convex anterior margin between Mxp1, surface with weak broad longitudinal ridge in midline. Sternal plastron slightly broader than long; lateral extremities moderately divergent posterolaterally. Sternite 3 strongly depressed; anterior margin shallowly concave, with narrow, U-shaped median sinus flanked by distinct or indistinct spine, anterolaterally angular; lateral margin with small spine or process at midlength. Sternite 4 with anterolateral margin nearly entire, somewhat convex, anterolateral angle minutely denticulate, rounded or somewhat angular; posterolateral margin relatively short, about two-thirds of anterolateral margin. Anterolateral margin of sternite 5 strongly convex anteriorly, about as long as posterolateral margin of sternite 4.

Abdomen: Smooth, glabrous. Somite 1 feebly convex from anterior to posterior on dorsal surface. Somite 2 tergite 2.3-2.5 × broader than long; pleuron with lateral margin slightly concave, anterolateral and posterolateral margins rounded. Pleuron of somite 3 laterally blunt. Telson 0.4 × as long as broad; posterior plate 1.2-1.4 × longer than anterior plate, posterior margin feebly convex, nearly transversal or feebly concave.

Eye: Relatively small, moderately elongate, about 2 × longer than broad, overreaching midlength of rostrum, slightly narrowed distally or with lateral and mesial margins subparallel. Cornea not inflated, about half as long as remaining eyestalk.

Antennule and antenna: Ultimate article of antennular peduncle 2.6-3.5 × longer than high. Antennal peduncle relatively slender and short, overreaching cornea, not reaching apex of rostrum. Article 2 with sharp distolateral spine. Antennal scale nearly reaching or slightly falling short of distal end of article 5, breadth distinctly more than twice that of article 5. Articles 4 and 5 each with small distomesial spine, that of article 4 often larger. Article 5 1.5-2.0 × longer than article 4, breadth 0.7-0.8 × height of ultimate antennular article. Flagellum of 20-22 segments barely reaching distal end of P1 merus.

Mxp: Mxp1 with bases broadly separated. Mxp3 basis with 1 distal denticle on mesial ridge. Ischium having flexor margin sharply ridged, distally rounded; crista dentata with 23-30 denticles proximally obsolescent, medially distinct, distally smaller. Merus short relative to breadth, 1.6 × longer than ischium, flattish on mesial face, sharply ridged along flexor margin, occasionally with small distolateral spine; extensor and flexor margins subparallel in distal half, flexor margin with very small tubercles or denticles distal to midlength. Carpus unarmed.

P1: 3.8-4.4 × longer than carapace, massive, moderately or sparsely setose. Ischium dorsally with basally broad, flattened, short blunt spine, ventromesially with row of short, tubercle-like processes or small spines, without distinct subterminal spine. Merus nearly as long as or slightly shorter than carapace, dorsally with distomesial spine distinct in small specimens, obsolescent or absent in large specimens, mesially with small (often distinct) spines, tubercles or granules on proximal portion, ventrally with distomesial and distolateral spines, distomesial one large and acute in small specimens. Carpus distally broadened, dorsally with 3 small blunt processes in transverse row on proximal portion, ventrally with distomesial and distolateral spines, both obtuse and small; length 1.0-1.2 × that of merus. Palm moderately depressed, 1.6-2.6 × longer than broad, 1.2-1.4 × longer than carpus; mesial margin with rounded ridge well developed in large specimens, distally diminishing in small specimens. Fingers distally curved ventrally, short relative to breadth, distally incurved, crossing when closed; opposable margin of fixed finger sinuous, that of movable finger with proximal process fitting to opposite longitudinal narrow groove on fixed finger when closed; movable finger 0.5-0.6 × as long as palm.

P2-4: Mesio-laterally compressed, relatively broad, bearing sparse short fine setae. Meri successively shorter posteriorly (P3 merus 0.9 × length of P2 merus, P4 merus 0.8 × length of P3 merus), slightly narrower on P4 than on P2 and P3; length-breadth ratio, 3.0-3.5 on P2, 3.1-3.3 on P3, 3.2 on P4; dorsally with 3 or 4 small spines or denticles often obsolescent on P2, obsolete on P3 and P4, ventrally with small distolateral spine; P2 merus much shorter than (three-quarters) carapace, 1.1 × longer than P2 propodus; P3 merus as long as P3 propodus; P4 merus 0.8 × length of P4 propodus. Carpi subequal to or slightly shorter than length of dactyli (carpus-dactylus length ratio, 0.9 on P2 and on P3, 0.8 on P4 in holotype), dactylus-propodus length ratio, 0.4-0.5 on P2, 0.4 on P3, 0.3-0.4 on P4; unarmed. Propodi shorter on P2 than on P3-4; flexor margin nearly straight, ending in pair of spines preceded by 4-6, 3-6, 1-5 slender spines on P2, P3 and P4, respectively. Dactyli shorter on P2 than on P3 and P4, very setose along flexor and extensor margins; length 0.4-0.5 × that of propodus on P2-4; flexor margin slightly curving, ending in slender terminal spine preceded by 8-10 loosely arranged, slightly inclined, sharp triangular spines gradually diminishing toward base of article.

Eggs. Number of eggs carried, 3-40; size, 1.16 mm × 1.00 mm - 1.25 mm × 1.48 mm.

REMARKS — *Uroptychus baeomma* is very close to *U. elongatus* n. sp. and *U. modicus* n. sp. Their relationships are discussed under the remarks of these species (see below).

The species also resembles *U. crassipes* Van Dam, 1939 in the carapace spination and the shape of the sternal plastron. However, it is differentiated from that species by the absence instead of presence of a pronounced subterminal spine on the ventromesial margin of P1 ischium; the antennal articles 4 and 5 each have a very tiny instead of strong terminal spine; the antennal scale terminates short of instead of distinctly overreaching the tip of article 5; and the P2-4 dactyli bear a slender terminal spine preceded by relatively sharp, slightly inclined, loosely arranged spines, instead of a broad penultimate spine preceded by slender, distinctly obliquely directed, closely arranged, spines.

Uroptychus bardi McCallum & Poore, 2013

Figure 30

Uroptychus bardi McCallum & Poore, 2013: 158, fig.4.

TYPE MATERIAL — Holotype: **Western Australia**, off Cape Leveque (14°33.43'S, 121°20.38'E - 14°32.76'S, 121°19.65'E), 924-1101 m, male (NMV J63754). [not examined].

MATERIAL EXAMINED — **Wallis and Futuna Islands**. MUSORSTOM 7 Stn DW539, 12°27'S, 177°27'E, 700 m, 17.V.1992, 1 ♂ 10.0 mm (MNHN-IU-2014-16316). – Stn DW615, 14°27'S, 177°26'W, 700-750 m, with Chrysogorgiidae gen. sp. (Calcaxonia), 27.V.1992, 1 ♂ 12.5 mm (MNHN-IU-2014-16317); 1 ov. ♀ 11.8 mm (MNHN-IU-2014-16318). **Vanuatu**. MUSORSTOM 8 Stn CP1074, 15°48.42'S, 167°24.27'E, 775-798 m, 4.X.1994, 1 ov. ♀ 9.1 mm (MNHN-IU-2014-16319).

DISTRIBUTION — Western Australia off Cape Leveque, 924-1101 m, and now Wallis and Futuna Islands (SW Pacific) and Vanuatu, 700-798 m.

SIZE — Males, 10.0-12.5 mm; ovigerous females, 9.1-11.8 mm.

DESCRIPTION — Large species. *Carapace*: Slightly broader than long (0.9 × as long as broad); greatest breadth 1.9 × distance between anterolateral spines. Dorsal surface glabrous, moderately convex from anterior to posterior, without distinct border between gastric and cardiac regions; epigastric region feebly or barely granulose. Lateral margins convexly divergent posteriorly (greatest breadth of carapace measured at posterior third), with row of short, oblique, granulate ridges along branchial region; anterolateral spine well developed, extending far beyond much smaller lateral orbital spine, situated slightly posterior to level of, but moderately remote from that spine. Rostrum sharp triangular, with interior angle of 25°; length about one-third postorbital carapace length, breadth less than half carapace breadth measured at posterior carapace margin; dorsal surface flattish, ventral surface straight horizontal. Pterygostomian flap anteriorly angular, ending in small but distinct spine; surface smooth.

Sternum: Excavated sternum with convex anterior margin bearing 1 or 2 small median spines, surface with or without small spine in center. Sternal plastron 0.8-0.9 × as log as broad, posteriorly broadened. Sternite 3 strongly depressed, anterior margin deeply excavated in V-shape, with 2 submedian spines separated by narrow notch, anterolateral corner sharp angular or rounded. Sternite 4 long relative to breadth; surface with tubercles on posterior portion; anterolateral margin strongly produced forward, slightly falling short of anterior end of sternite 3; posterolateral margin half length

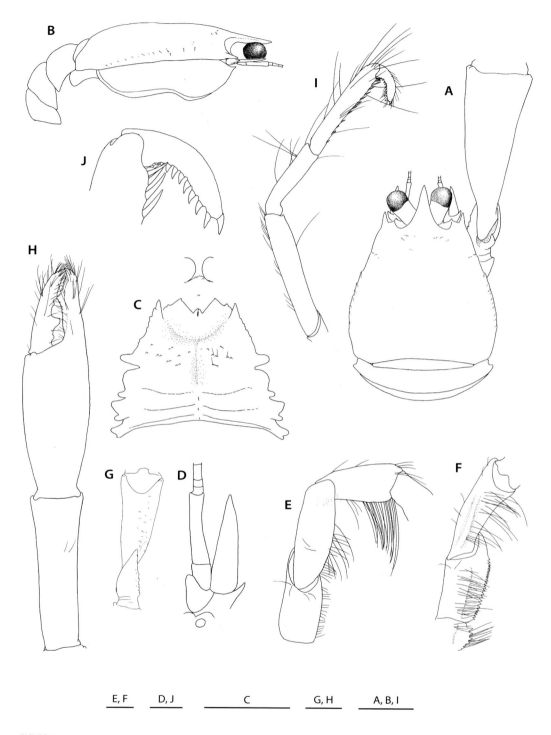

FIGURE 30

Uroptychus bardi McCallum & Poore, 2013, male 12.5 mm (MNHN-IU-2014-16317). **A**, carapace and anterior part of abdomen, proximal part of right P1 included, dorsal. **B**, same, lateral. **C**, sternal plastron, with excavated sternum and basal parts of Mxp1. **D**, left antenna, ventral. **E**, right Mxp3, lateral. **F**, same, ventral. **G**, right P1, proximal part, ventral. **H**, same, dorsal. **I**, right P2, lateral. **J**, same, distal part, setae omitted, lateral. Scale bars: A, B, G, H, I, 5 mm; C, D, E, F, J, 1 mm.

of anterolateral margin. Anterolateral margin of sternite 5 strongly convex, much longer than posterolateral margin of sternite 4.

Abdomen: Smooth and glabrous. Somite 1 without transverse ridge. Somite 2 tergite 2.5-2.6 × broader than long; pleural lateral margins concavely divergent posteriorly, anterior end rounded, posterior end angular. Telson half as long as broad; posterior plate moderately concave on posterior margin, length 1.5-1.6 × that of anterior plate.

Eye: Long relative to breadth (length 1.8 × breadth), reaching distal quarter to tip of rostrum; mesial margin somewhat concave. Cornea slightly dilated, broader than remaining eyestalk, length more than half that of remaining eyestalk.

Antennule and antenna: Ultimate article of antennule 2.3-2.6 × longer than high. Antennal peduncle slender, reaching rostral tip. Article 2 with strong distolateral spine. Antennal scale much broader than peduncle, nearly or barely reaching distal end of article 5. Distal 2 articles unarmed; article 5 2.7 × longer than article 4, breadth about half height of ultimate antennular article. Flagellum of 22-30 segments falling short of distal end of P1 merus.

Mxp: Mxp1 with bases close to each other but not contiguous. Mxp3 barely setose on lateral surface. Basis having mesial ridge with 5-7 denticles diminishing toward proximal end of article. Ischium with 24-26 denticles on crista dentata, flexor margin distally not rounded. Merus and carpus unarmed; merus 2.2 × longer than ischium, relatively slender, flexor margin moderately ridged.

P1: 5.1-5.8 × longer than carapace, relatively massive, glabrous except for fingers. Ischium dorsally with well-developed curved spine, ventromesially with tubercle-like spines, without subterminal spine. Merus 1.2-1.4 × longer than carapace, ventral surface with tubercles in longitudinal row and blunt low distal spine near each of mesial and lateral margins. Carpus 1.2-1.3 × longer than merus. Palm 2.1-2.6 × longer than broad, slightly shorter than carpus. Fingers distally incurved, crossing when closed, gaping in males in proximal two-thirds, fitting when closed in distal half, not gaping in females; length more than half (0.53-0.67) that of palm; opposable margin of movable finger with strong blunt median process on gaping portion in males, disto-proximally broad process on proximal third in females (and on right P1 in smaller male).

P2-4: Relatively slender, compressed mesio-laterally, with sparse long setae. Meri successively shorter posteriorly (P3 merus 0.9 × length of P2 merus; P4 merus 0.8 × length of P3 merus), subequal in breadth on P2-4, dorsal margin smooth; length-breadth ratio, 6.0 on P2, 5.0-6.0 on P3, 5.0 on P4; P2 merus as long as carapace, 1.2 × longer than P2 propodus; P3 merus subequal to P3 propodus; P4 merus 0.90-0.96 × length of P4 propodus. Carpi relatively long, 0.6 × length of propodus on P2 and P3, 0.5 × on P4; P3 carpus subequal to or slightly shorter than P2 carpus, P4 carpus 0.8-0.9 × length of P3 carpus. Propodi successively slightly shorter posteriorly or subequal on P2 and P3 and shortest on P4; flexor margin slightly curving, ending in pair of spines preceded by 10-12 spines on entire length on P2, 8 or 9 on P3, 6 on P4. Dactylus relatively short; dactylus-propodus length ratio, 0.30-0.36 on P2-4; dactyls-carpus length ratio, 0.5 on P2, 0.6 on P3, 0.7 on P4; strongly curving at proximal third; flexor margin with 11-14 loosely arranged, obliquely directed, sharp, proximally diminishing spines.

Eggs. Number of eggs carried, 6-23; size, 1.48 mm × 1.56 mm - 1.60 mm × 1.77 mm.

REMARKS — The species strongly resembles *U. litosus* Ahyong & Poore, 2004. In addition to the differences given by McCallum & Poore (2013), *U. bardi* is distinguished from that species by the following: the P2 propodus bears 10-12 instead of 7-8 spines proximal to a pair of terminal spines, along the entire length instead of the distal half of the flexor margin; the excavated sternum anteriorly ends in a rounded margin, not angular as in *U. litosus*; and the P4 merus is as broad as instead of distinctly narrower than the P3 merus.

Uroptychus belos Ahyong & Poore, 2004

Figure 31

Uroptychus belos Ahyong & Poore, 2004: 25, fig. 5.

TYPE MATERIAL — Holotype: **Australia**, Britannia Seamount, SE of Brisbane, Tasman Sea, 419 m holotype, female (AM P65830). [not examined].

MATERIAL EXAMINED — **New Caledonia**, Chesterfield Islands. MUSORSTOM 5 Stn DW337, 19°53.80′S, 158°38.00′E, 412-430 m, gorgonian corals, 15.X.1986, 1 ov. ♀ 4.0 mm (MNHN-IU-2014-16320). **New Caledonia**. MUSORSTOM 4 Stn DW222, 22°57.6′S, 167°33.0′E, 410-440 m, 30.IX.1985, 1 ♀ 4.2 mm (MNHN-IU-2014-16321). BATHUS 2 Stn DW719, 22°48′S, 167°15′E, 444-445 m, 11 .V.1993, 1 ov. ♀ 3.7 mm (MNHN-IU-2014-16322). **New Caledonia**, Norfolk Ridge. LITHIST Stn DW13, 23°45.0′S, 168°16.7′E, 400 m, 12.VIII.1999, 1 ov. ♀ 3.3 mm (MNHN-IU-2014-16323). NORFOLK 1 Stn DW1654, 23°28′S, 167°52′E, 366-560 m, 19.VI.2001, 1 ♂ 3.4 mm (MNHN-IU-2014-16324). – Stn DW1733, 22°56′S, 167°15′E, 427-433 m, 28.VI.2001, 1 ♀ 4.3 mm (MNHN-IU-2014-16325). NORFOLK 2 Stn DW2156, 22°54′S, 167°15′E, 468-500 m, 05.XI.2003, 1 ♂ 3.6 mm (MNHN-IU-2014-16326).

DISTRIBUTION — Tasman Sea, and now Chesterfield Islands, New Caledonia and Norfolk Ridge; 366-560 m.

SIZE — Males, 3.4-3.6 mm; females, 3.3-4.2 mm; ovigerous females from 3.3 mm.

DESCRIPTION — Small species. *Carapace*: Distinctly broader than long (0.8 × as long as broad); greatest breadth 1.6 × distance between anterolateral spines. Dorsal surface with sparse short setae, weakly convex from anterior to posterior, without groove and depression. Lateral margins convexly divergent posteriorly, bearing 5 spines: first anterolateral, overreaching small lateral orbital spine, situated directly lateral to and somewhat distant from that spine, subequal in size to second but smaller than third, second somewhat posterior to midpoint between first and third, second to fifth equidistant, fourth smaller than second, fifth (last) very small or rudimentary; no ridge along posterior lateral margin. Rostrum elongate triangular, with interior angle of 28-30°, ending in sharp point, dorsally concave moderately, length varying from slightly less than half to more than half that of carapace, breadth half carapace breadth measured at posterior carapace margin. Lateral orbital angle ending in small spine. Pterygostomian flap smooth on surface, anteriorly angular, ending in distinct spine.

Sternum: Excavated sternum with distinct longitudinal ridge in midline, anterior margin nearly transverse or slightly convex. Sternal plastron as long as or slightly longer than broad, lateral extremities subparallel between sternites 4-6, sternite 7 broadest. Sternite 3 weakly depressed, anterior margin shallowly excavated with deep or shallow U- or V-shaped median notch but no submedian spines, lateral margin smooth, spineless. Sternite 4 with nearly straight or slightly concave anterolateral margin about 1.5 × longer than posterolateral margin. Anterolateral margin of sternite 5 anteriorly convex, about as long as posterolateral margin of sternite 4

Abdomen: With sparse short setae. Somite 1 smooth on surface. Somite 2 tergite 2.5 × broader than long; pleuron posterolaterally blunt, lateral margin gently concave and moderately divergent posteriorly. Pleuron of somite 3 somewhat angular laterally. Telson about half as long as broad; posterior plate nearly as long as anterior plate, posterior margin somewhat or distinctly emarginate.

Eye: Slender, subcylindrical, about twice as long as broad, slightly overreaching midlength of rostrum. Cornea not dilated, half as long as remaining eyestalk.

Antennule and antenna: Ultimate antennular article 2.7-2.9 × longer than high. Antennal peduncle overreaching cornea, not reaching rostral tip. Article 2 with small distolateral spine. Antennal scale much broader than article 5, varying from slightly falling short of to slightly overreaching distal end of article 5. Article 4 with distoventral spine small, blunt or often obsolete. Article 5 unarmed or with very small distal spine discernible under high magnification; length 1.4-1.6 × that of article 4, breadth slightly more than half height of ultimate antennular article. Flagellum consisting of 12 or 13 segments, reaching distal end of P1 merus.

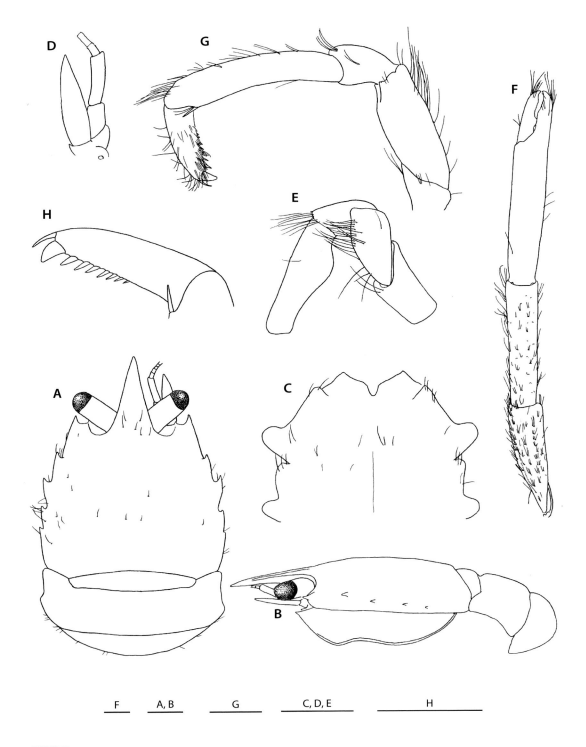

F   A, B   G   C, D, E   H

FIGURE 31

Uroptychus belos Ahyong & Poore, 2004, ovigerous female 4.0 mm (MNHN-IU-2014-16320). **A**, carapace and anterior part of abdomen, dorsal. **B**, same, lateral. **C**, anterior part of sternal plastron. **D**, right antenna, ventral. **E**, left Mxp3, lateral. **F**, right P1, dorsal. **G**, left P3, lateral. **H**, same, distal part, setae omitted, lateral. Scale bars: 1 mm.

Mxp: Mxp1 with bases broadly separated. Mxp3 with long setae on lateral faces of ischium and merus. Basis with proximally rounded mesial ridge bearing no denticle. Ischium with tuft of long setae lateral to rounded distal end of flexor margin; denticles on crista dentata very small. Merus 1.7 × longer than ischium, flattish on mesial face; distolateral spine very small, often obsolescent; flexor margin with long setae and 2 very small, often obsolete spines. Carpus unarmed.

P1: 4.3-4.4 × longer than carapace in females (missing in males); subcylindrical but palm more or less depressed; with fine setae particularly along mesial margins of merus and carpus, and on fingers, less setose on palm. Ischium with small distodorsal spine, without subterminal spine on ventromesial margin. Merus slightly longer than carapace. Merus and carpus each with pair of very small distoventral spines (distomesial and distolateral). Carpus slightly longer than merus. Palm about 4 × longer than broad, 1.1-1.2 × longer than carpus. Fingers with longer setae distally, distally somewhat deflected ventrally, not gaping, distally incurved, crossing when closed; movable finger 0.4 × as long as palm, opposable margin with dorsoventrally depressed process at proximal third fitting into narrow longitudinal groove on opposite fixed finger when closed.

P2-4: Moderately compressed, sparsely setose, more setose on dactyli. Meri unarmed, successively decreasing in size posteriorly (P3 merus 0.9 × length of P2 merus, P4 merus 0.9 × length of P3 merus), equally broad on P2-4 or slightly narrower on P4 than on P3; length-breadth ratio, 3.3-3.5 on P2, 2.9-3.5 on P3, 2.5-3.0 on P4; P2 merus 0.6-0.7 × length of carapace, 0.8-0.9 × (rarely subequal to) length of P2 propodus; P3 merus 0.8-0.9 × length of P3 propodus; P4 merus 0.7-0.8 × length of P4 propodus. Carpi subequal on P2-4 or slightly shorter on P4, length about one-third that of propodi on P2-4. Propodi successively longer posteriorly or slightly shorter on P2 than on P3 and P4; flexor margin somewhat concave in lateral view, with pair of terminal spines only. Dactyli stout, about half as long as propodi and much longer than carpi (dactylus-carpus length ratio, 1.5 on P2, 1.6 on P3, 1.7 on P4); flexor margin nearly straight, with 9-10 more or less loosely arranged spines, ultimate somewhat more slender and longer than antepenultimate, penultimate pronouncedly broad, remaining spines much smaller than penultimate, moderately inclined, successively diminishing toward base of article.

Eggs. Number of eggs carried, 4; size, 1.00 mm × 1.08 mm.

REMARKS — The material generally agrees with the description of *U. belos* Ahyong and Poore, 2004 from the Tasman Sea. The distolateral spine and the flexor marginal spines of the Mxp3 merus, originally described as distinct, are usually barely discernible, and very tiny in one of the material examined (MNHN-IU-2014-16324); also usually very small or obsolete in the present material is the distolateral spine of the antennal article 5, and rarely absent is the distolateral spine of article 4. These spines are small but distinct in the type.

The carapace ornamentation, elongate eyes, elongate sternal plastron, and spination of P2-4 dactyli are very similar to those of *U. grandior* n. sp. Their relationships are discussed under that species (see below).

Uroptychus benthaus n. sp.
Figures 32, 33

TYPE MATERIAL — Holotype: **French Polynesia**, Austral Islands. BENTHAUS Stn DW1873, 29°00'S, 140°15'W, 456-813 m, with ?Chrysogorgiidae gen. sp. (Calcaxonia), 4.XI.2002, ♂ 4.6 mm (MNHN-IU-2014-16327). Paratypes: **French Polynesia**, Austral Islands. BENTHAUS, collected with holotype, 1 ♂ 6.4 mm, 1 ov. ♀ 6.2 mm (MNHN-IU-2014-16328). – Stn DW1894, 27°40.1'S, 144°21.5'W, 100 m, 8.XI.2002, 1 ♀ 8.3 mm (MNHN-IU-2014-16329). – Stn DW1956, 23°18.4'S, 149°27'W, 600-990 m, 18.XI.2002, 1 ♂ 4.9 mm (MNHN-IU-2014-16330).

ETYMOLOGY — Named for the cruise Benthaus, by which the type material was collected; used as a noun in apposition.

DISTRIBUTION — French Polynesia; 100-813 m.

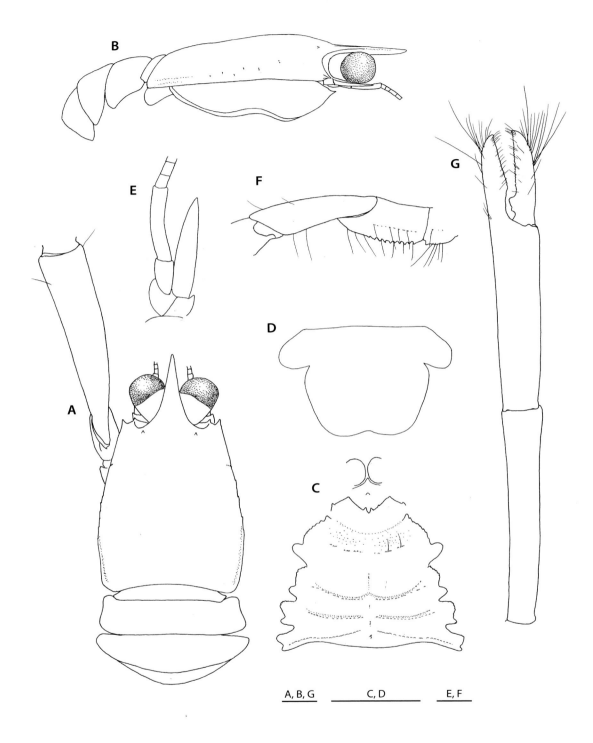

FIGURE 32

Uroptychus benthaus n. sp., holotype, male 4.6 mm (MNHN-IU-2014-16327). **A**, carapace and anterior part of abdomen, proximal part of left P1 included, dorsal. **B**, same, lateral. **C**, sternal plastron, with excavated sternum and basal parts of Mxp1. **D**, telson. **E**, left antenna, ventral. **F**, left Mxp3, ventral. **G**, left P1, proximal part omitted, dorsal. Scale bars: 1 mm.

SIZE — Males, 4.6-6.4 mm; females, 6.2-8.3 mm; ovigerous females from 6.2 mm.

DESCRIPTION — Medium-sized species. *Carapace*: 1.1 × longer than broad; greatest breadth 1.5 × distance between anterolateral spines. Dorsal surface smooth, glabrous, moderately convex from anterior to posterior, without distinct depression or groove. Epigastric region with pair of very small spines. Lateral margins convex or convexly divergent posteriorly; anterolateral spine small, situated somewhat posterior to level of but terminating at tip of smaller lateral orbital spine; branchial lateral margin with short protuberant ridge at anterior end, followed by obsolescent granules, weakly ridged along posterior third. Rostrum narrow triangular, with interior angle of 20°, straight horizontal, length half that remaining carapace, breadth subequal to carapace breadth measured at posterior carapace margin; dorsal surface flattish. Pterygostomian flap smooth on surface, anteriorly angular, ending in small sharp spine.

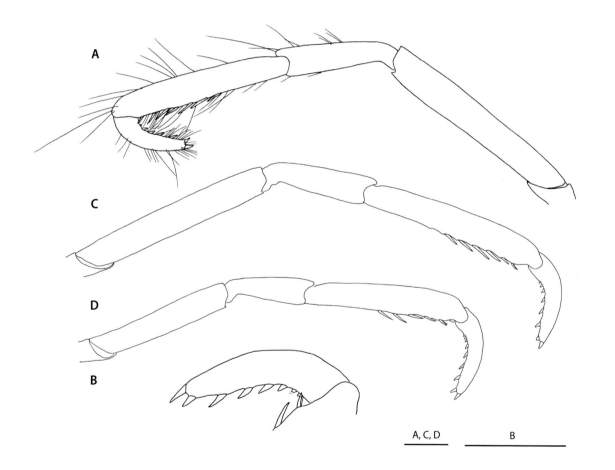

FIGURE 33

Uroptychus benthaus n. sp., holotype, male 4.6 mm (MNHN-IU-2014-16327). **A**, left P2, lateral. **B**, same, distal part, setae omitted, lateral. **C**, right P3, setae omitted, lateral. **D**, right P4, setae omitted, lateral. Scale bars: 1 mm.

Sternum: Excavated sternum anteriorly produced, ending in spine; surface with small spine in center. Sternal plastron slightly shorter than broad, posteriorly broadened. Sternite 3 well depressed; wide V-shaped anterior margin bearing pair of submedian spines separated by shallow V-shaped notch, laterally angular. Sternite 4 with denticulate, setiferous interrupted transverse ridge; anterolateral margin convex, anteriorly ending in short spine followed by row of denticles, length slightly more than 1.5 × that of posterolateral margin. Anterolateral margin of sternite 5 strongly convex anteriorly and denticulate, about as long as posterolateral margin of sternite 4.

Abdomen: Smooth and glabrous. Somite 1 moderately convex from anterior to posterior. Somite 2 tergite 2.3-2.8 × broader than long; pleuron posterolaterally blunt, lateral margin gently concave and moderately divergent posteriorly. Pleuron of somite 3 bluntly angular laterally. Telson 0.7 × as long as broad; posterior plate sharply emarginate on posterior margin, length twice that of anterior plate.

Eye: Broad relative to length, 1.3-1.4 × longer than broad, proximally slightly narrower, overreaching midlength of rostrum. Cornea slightly broader than and nearly as long as remaining eyestalk.

Antennule and antenna: Ultimate article of antennule 2.1-2.7 × longer than high. Antennal peduncle slightly overreaching cornea. Article 2 with small distolateral spine. Antennal scale slightly falling short of distal end of article 5, breadth 1.2 × that of article 4. Distal 2 articles unarmed. Article 5 2.0-2.2 × length of article 4, breadth less than half height of ultimate antennular article. Flagellum of 12 segments not reaching distal end of P1 merus.

Mxp: Mxp1 with bases close to each other. Mxp3 barely setose on lateral surface. Basis with 1-3 denticles on mesial ridge. Ischium with 10-14 denticles on crista dentata; flexor margin not rounded distally. Merus 2.3 × longer than ischium, moderately thick mesio-laterally, flexor margin unarmed, not sharply ridged. No spine on carpus.

P1: Relatively slender, 4.8-5.5 × longer than carapace; surface smooth and glabrous except for setose fingers. Ischium with short, triangular dorsal spine, ventromesially unarmed. Merus 1.1-1.3 × longer than carapace. Carpus 1.3-1.4 × longer than merus. Palm 4.0-4.1 × (males), 3.7-4.6 × (females) longer than broad, 0.8-0.9 × length of carpus. Fingers gaping in males, not gaping in females, slightly incurved at tip; in females, opposable margin of movable finger with very low proximal process fitting to opposing concavity when closed; in males, opposable margins nearly straight in distal half, gaping in proximal half, bearing process as figured; length of movable finger 0.51-0.54 × that of palm.

P2-4: Moderately compressed, relatively slender, with sparse long setae but meri glabrous. P2 merus slightly shorter than carapace, 1.1 × length of P2 propodus; P3 merus subequal to P2 merus in length and breadth or slightly shorter, subequal to length of P3 propodus; P4 merus 0.7 × length and 0.9 × breadth of P3 merus, 0.8-0.9 × length of P4 propodus; length-breadth ratio, 6.6-6.8 on P2, 6.1-7.0 on P3, 4.8-5.1 on P4; ventrolateral margin with very small terminal spine. Carpi subequal in length on P2 and P3, shorter on P4 (P4 carpus 0.8 × length of P3 carpus); length 0.5-0.6 × that of propodus on P2, 0.5 × on P3 and P4. Propodus longest on P3, shortest on P4; flexor margin ending in pair of spines preceded by 7-9 slender spines on distal half on P2, 6 or 7 on P3, 5 or 6 on P4. Dactyli proximally curved, 0.7 × length of carpi on P2 and P3, 0.95 × on P4, 0.4 × length of propodi on P2-4; flexor margin with 8 or 9 loosely arranged, obliquely directed, proximally diminishing spines, penultimate closer to ultimate than to antepenultimate.

Eggs. Number of eggs carried, 9; size, 1.30 mm × 1.77 mm - 1.33 mm × 1.58 mm.

REMARKS — The species resembles *U. sagamiae* Baba, 2005 and *U. pollostadelphus* n. sp. in having a pair of epigastric spines, in having the antennal articles 4 and 5 unarmed, and in having the P2-4 propodi with a row of flexor marginal spines distally ending in paired spines. *Uroptychus benthaus* is readily distinguished from *U. sagamiae* by the pterygostomian flap that is sharply angular anteriorly and produced to a distinct spine, instead of being roundish with a tiny spine; and P2-4 are more slender, *e.g.*, the P2 merus being at least 6.6 instead of 3.7 times longer than broad. The relationships with *U. pollostadelphus* are discussed under the remarks of that species (see below).

Uroptychus bertrandi n. sp.

Figures 34, 35

TYPE MATERIAL — Holotype: **New Caledonia**. VAUBAN 1978-1979 Stn DR15, 22°49'S, 167°12'E, 390-395 m, 10.IV.1978, ov. ♀ 2.9 mm (MNHN-IU-2013-8521). Paratypes: Collected together with holotype, 1 ov. ♀ 3.2 mm, 1 ♀ (with exuviae) 3.2 mm (MNHN-IU-2014-1633). **New Caledonia**, Loyalty Ridge. MUSORSTOM 6 Stn CP467, 21°05.13'S, 167°32.11'E, 575 m, 21.II.1989, 1 ♂ 2.3 mm (MNHN-IU-2013-8530). **New Caledonia**, Norfolk Ridge. NORFOLK 1 Stn DW1651, 23°27.3'S, 167°50.4'E, 276-350 m, 19.VI.2001, 2 ov. ♀ 2.9, 3.1 mm (MNHN-IU-2013-8522). – Stn DW1653, 23°28'S, 167°51'E, 328-340 m, 19.VI.2001, 1 ♂ 3.0 mm (MNHN-IU-2013-8523). – Stn DW1654, 23°28'S, 167°52'E, 366-560 m, 19.VI.2001, 1 ov. ♀ 3.2 mm (MNHN-IU-2013-8524).

ETYMOLOGY — Named for Bertrand Richer de Forges for his enormous efforts in the field work.

DISTRIBUTION — New Caledonia and Norfolk Ridge; 276-560 m.

SIZE — Male, 2.3-3.0 mm; females, 2.9-3.2 mm; ovigerous females from 2.9 mm.

DESCRIPTION — Small species. *Carapace*: Broader than long (0.8 × as long as broad); greatest breadth 1.7-1.8 × distance between anterolateral spines. Dorsal surface well convex from side to side and from anterior to posterior, without distinct groove or depression, covered with short fine setae, sparingly bearing denticles on hepatic, anterior branchial and lateral epigastric regions. Lateral margins moderately convex, somewhat divergent posteriorly, with row of denticles or small spines; anterolateral spine small, located slightly posterior to level of lateral orbital spine, distinctly separated from that spine in dorsal view. Rostrum triangular, with interior angle of 35°, nearly horizontal; length 0.9 × breadth, about 0.4 × that of remaining carapace, breadth less than half carapace breadth measured at posterior carapace margin; dorsal surface distinctly depressed along midline; lateral margin with obsolescent spine near tip. Lateral orbital spine larger than anterolateral spine. Pterygostomian flap covered with denticles or tubercle-like small spines on surface, anteriorly angular, ending in small spine.

Sternum: Excavated sternum with anterior margin convex between bases of Mxp1, surface with weak ridge in midline on anterior half. Sternal plastron 0.9 × as long as broad, lateral extremities slightly divergent between sternites 4 and 7. Sternite 3 shallowly depressed, with anterior margin gently emarginate with U-shaped median sinus flanked by small spine; anterolateral end rounded or angular. Sternite 4 with relatively short anterolateral margin anteriorly rounded with or without a few denticles; posterolateral margin slightly shorter than anterolateral margin. Anterolateral margin of sternite 5 anteriorly somewhat convex, about as long as posterolateral margin of sternite 4.

Abdomen: Setose like carapace. Somite 1 convex from anterior to posterior. Somite 2 tergite 2.3-2.7 × broader than long; pleuron posterolaterally rounded, lateral margin weakly concave and slightly divergent posteriorly. Pleura of somites 3 and 4 laterally rounded. Telson slightly less than half as long as broad; posterior plate 1.0-1.4 × longer than anterior plate, feebly concave on posterior margin.

Eye: Broad relative to length (1.4-1.6 × longer than broad), reaching at most distal fifth of rostrum, strongly inflated proximally, especially along mesial margin. Cornea not inflated, half as long as remaining eyestalk.

Antennule and antenna: Ultimate article of antennule 2.8-3.3 × longer than high. Antennal peduncle relatively slender, reaching rostral tip. Article 2 with distinct distolateral spine. Antennal scale lanceolate, much broader (1.5-2.0 x) than article 5, reaching or slightly falling short of midlength of article 5. Article 4 with small distomesial spine. Article 5 unarmed, 1.1-1.3 × longer than article 4, breadth about half height of antennular ultimate article. Flagellum of 12-14 segments slightly falling short of or nearly reaching distal end of P1 merus.

Mxp: Mxp1 with bases broadly separated. Mxp3 moderately setose on lateral surface. Basis without denticles on convex mesial ridge. Ischium having flexor margin not rounded distally; crista dentata with 25 denticles distally diminishing.

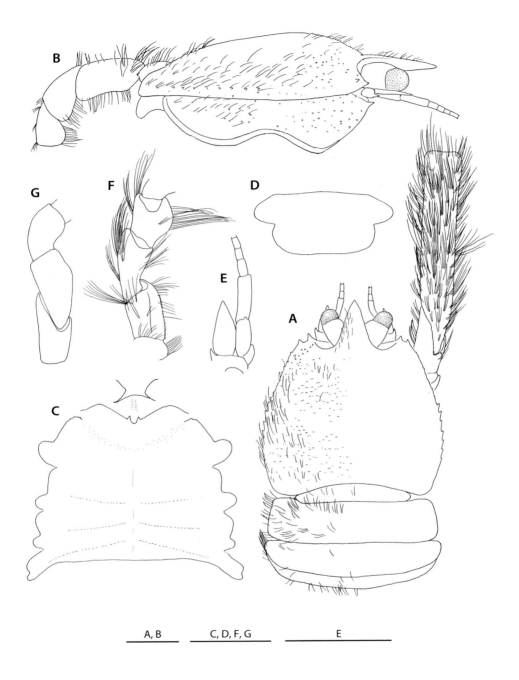

FIGURE 34

Uroptychus bertrandi n. sp., holotype, ovigerous female 2.9 mm (MNHN-IU-2013-8521). **A**, carapace and anterior part of abdomen, proximal part of right P1 included, dorsal. **B**, same, lateral. **C**, sternal plastron, with excavated sternum and basal parts of Mxp1. **D**, telson. **E**, right antenna, ventral. **F**, right Mxp3, ventral. **G**, same, setae omitted, lateral. Scale bars: 1 mm.

FIGURE 35

Uroptychus bertrandi n. sp., holotype, ovigerous female 2.9 mm (MNHN-IU-2013-8521). **A**, left P1, merus and proximal articles omitted, dorsal. **B**, same, proximal part, setae omitted, ventral. **C**, right P2, lateral. **D**, same, distal part, setae omitted, lateral. **E**, right P3, setae omitted, lateral. **F**, same, distal part, setae omitted, lateral. **G**, right P4, setae omitted, lateral. **H**, same, distal part, setae omitted, lateral. Scale bars: 1 mm.

Merus 2.3 × longer than ischium, usually unarmed, occasionally with small or obsolescent spine at distal third of flexor margin, length 2.1-2.3 × that of ischium, flexor margin not sharply ridged. Carpus unarmed.

P1: Slender, subcylindrical, with fine setae thickly, unarmed except for ischium bearing short flattish, laciniate dorsal spine; length 7.4-7.6 × (males) or 6.7-7.5 × (females) that of carapace. Merus 1.4-1.6 × longer than carapace. Carpus 1.4-1.6 × longer than merus. Palm slightly depressed (0.8 × as high as broad), 0.9 × length of carpus, 4.9-5.0 × (males) or 6.0-7.4 × (females) longer than broad. Fingers slightly gaping, ending in slightly incurved spine, not spooned; movable finger 0.4 × (rarely 0.3 x) length of palm, opposable margin with 2 small subtriangular processes, proximal larger, situated proximal to position of opposing low prominence at midlength of fixed finger.

P2-4: Moderately compressed, with soft fine setae. Meri successively shorter posteriorly (P3 merus 0.9-1.0 × length of P2 merus, P4 merus 0.9 × length of P3 merus), equally broad on P2 and P3, very slightly narrower on P4 than on P3; length-breadth ratio, 4.4 on P2, 4.1-4.3 on P3, 3.8-4.0 on 4; P2 merus 0.9-1.0 × length of carapace, 1.1 × length of P2 propodus; P3 merus subequal to length of P3 propodus; P4 merus 0.9 × length of P4 propodus; extensor margin with a few eminences or denticles often obsolete on P2, obsolete on P3 and P4. Carpi subequal, unarmed, length less than half that of propodi (carpus-propodus length ratio, 0.33-0.35 on P2, 0.31-0.33 on P3, 0.29-0.32 on P4). Propodi shorter on P2 than on subequal P3 and P4; flexor margin with pair of movable slender terminal spines only. Dactyli relatively broad distally in lateral view, longer than carpus (dactylus-carpus length ratio, 1.1-1.3 on P2 and P3, 1.2-1.4 on P4), less than half that of propodi (dactylus-propodus length ratio, 0.37-0.43 on P2, 0.35-0.42 on P3, 0.36-0.41 on P4); flexor margin nearly straight, with 6-9 (usually 7) spines often obscured by setae, ultimate short, slender, and very close to strongest penultimate spine preceded by 4-7 (usually 5) loosely arranged spines successively diminishing proximally and nearly perpendicular to flexor margin but proximal-most slightly inclined, antepenultimate spine three-quarters as broad as penultimate.

Eggs. Number of eggs carried, 7-15; size, 0.75 × 0.67 mm - 0.77 × 0.92 mm.

REMARKS — The combination of the following characters link the species to *U. rutua* Schnabel, 2009, *U. sarahae* n. sp. and *U. toka* Schnabel, 2009: the carapace with denticle-like small spines around the hepatic region, the anterolateral spine small, subequal to or smaller than the lateral orbital spine, the P2-4 dactyli with loosely arranged, perpendicularly directed flexor marginal spines, and the antenna with a short antennal scale ending at most in the midlength of article 5. *Uroptychus bertrandi* is readily distinguished from all of these species by the unique eyes that are noticeably inflated proximally. When viewed dorsally, the anterolateral spine of the carapace and lateral orbital spines are distinctly separated in *U. bertrandi*, *U. rutua* and *U. sarahae*, contiguous at the base in *U. toka*. The broad prominences on the gastric region as distinct in *U. rutua* are absent in *U. bertrandi*, *U. sarahae* and *U. toka*.

Uroptychus beryx n. sp.

Figures 36, 37

TYPE MATERIAL — Holotype: **New Caledonia**, Norfolk Ridge. BERYX 11 Stn CH49, 23°45'S, 168°17'E, 400-460 m, 21.X.1992, ov. ♀ 2.5 mm (MNHN-IU-2013-8514).

ETYMOLOGY — Named for the BERYX 11 cruise by which the new species was collected; used as a noun in apposition.

DISTRIBUTION — Norfolk Ridge; 400-460 m.

DESCRIPTION — Small species. *Carapace*: Slightly longer than broad (length 1.06 × breadth); greatest breadth 1.7 × distance between anterolateral spines. Dorsal surface feebly convex from anterior to posterior, without distinct groove, with sparse short setae, unarmed except or a few spinules on hepatic region. Lateral margins slightly divergent posteriorly, with row of about 10 small lateral spines along entire length; anterolateral spine slightly falling short of tip of lateral orbi-

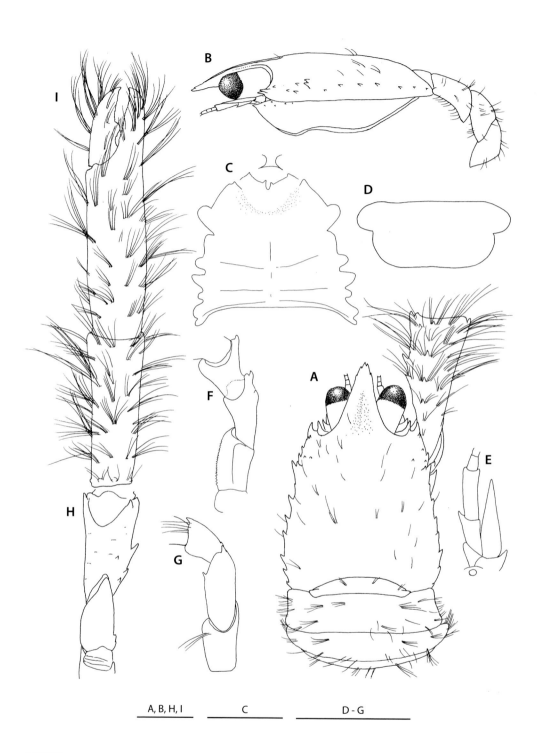

A, B, H, I — C — D - G

FIGURE 36

Uroptychus beryx n. sp., holotype, ovigerous female 2.5 mm (MNHN-IU-2013-8514). **A**, carapace and anterior part of abdomen, proximal part of right P1 included, dorsal. **B**, same, lateral. **C**, sternal plastron, with excavated sternum and basal parts of Mxp1. **D**, telson. **E**, left antenna, ventral. **F**, left Mxp3, setae omitted, ventral. **G**, same, lateral. **H**, right P1, proximal part, setae omitted, ventral. **I**, same, distal part, dorsal. Scale bars: 1 mm.

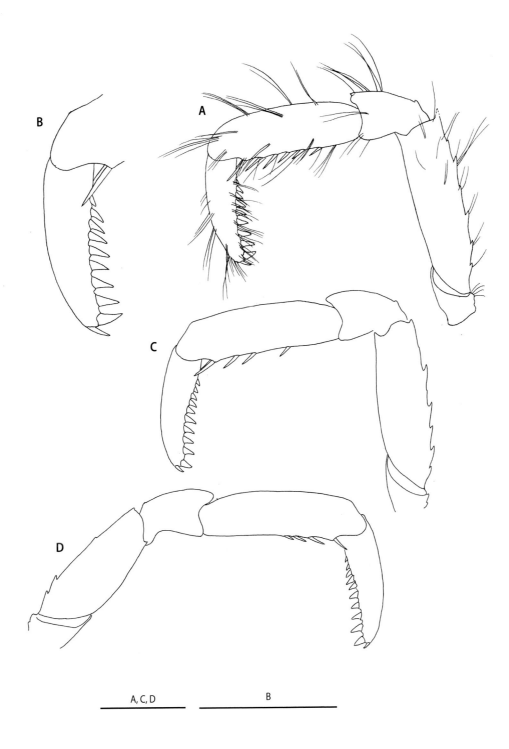

FIGURE 37

Uroptychus beryx n. sp., holotype, ovigerous female 2.5 mm (MNHN-IU-2013-8514). **A**, left P2, lateral. **B**, same, distal part, setae omitted, lateral. **C**, left P3, setae omitted, lateral. **D**, right P4, lateral. Scale bars: 1 mm.

tal spine; spine at anterior end of branchial region well developed, preceded by 1 or 2 small spines, with accompanying smaller spine dorsomesial to it, followed by 7 spines (first of these much smaller and ventral to level of remainder, last closer to posterior end than to preceding). Rostrum slightly deflected ventrally, deeply excavated dorsally, with 2 subapical spines (proximal obsolescent) on lateral margin; length about half that of carapace, breadth somewhat more than half carapace breadth measured at posterior carapace margin. Lateral orbital spine well developed, slightly smaller than anterolateral spine, both situated at same level and close to each other, but not contiguous in dorsal view.

Sternum: Excavated sternum with anterior margin transverse between bases of Mxp1, surface feebly ridged in midline. Sternal plastron slightly broader than long, somewhat broader posteriorly. Sternite 3 shallowly depressed; anterior margin well emarginate, with U-shaped median notch flanked by small but distinct spine, laterally sharp angular. Sternite 4 with slightly convex anterolateral margin anteriorly angular or with small spine, posterolateral margin slightly shorter than anterolateral margin. Anterolateral margin of sternite 5 anteriorly convex, about as long as posterolateral margin of sternite 4.

Abdomen: Smooth, with sparse tufts of setae. Somite 1 feebly convex from anterior to posterior. Somite 2 tergite 2.3 × broader than long; pleural lateral margins concavely divergent posteriorly, posterolateral end somewhat angular. Pleura of somites 3 and 4 laterally angular. Telson slightly less than half as long as broad; posterior plate somewhat emarginate on posterior margin, length two-thirds length of anterior plate.

Eye: Relatively broad (1.6 × longer than broad), with subparallel lateral and mesial margins, overreaching midlength of rostrum, falling short of rostral tip. Cornea not dilated, half as long as remaining eyestalk.

Antennule and antenna: Ultimate article of antennular peduncle 2.8 × longer than high. Antennal peduncle ending in distal margin of cornea. Article 2 with strong distolateral spine. Antennal scale tapering, barely reaching distal end of article 5, breadth 1.3 × that of article 4. Article 4 with distinct distomesial spine. Article 5 unarmed, 1.5 × longer than article 5, breadth three-quarters height of ultimate antennular article. Flagellum consisting of 12 (left) or 9 (right) segments, reaching distal end of P1 merus.

Mxp: Mxp1 with bases broadly separated. Mxp3 basis without denticles on mesial ridge. Ischium flexor margin rounded distally, crista dentata with 30 small denticles diminishing toward distal end. Merus 2 × longer than ischium, moderately ridged along flexor margin bearing small spine at distal third; distolateral spine distinct. Carpus also with distolateral spine.

P1-4 with long soft setae. *P1*: 4.0 × longer than carapace. Ischium dorsally with basally broad, depressed, short spine, ventrally with 2 denticles proximally, without subterminal spine. Merus about as long as carapace, bearing spines (1 dorsal row in midline, 1 dorsomesial row, a few mesial spines, another few small spines on proximo-ventral surface close to mesial margin, and ventral distomesial and distolateral spines). Carpus slightly longer than merus, dorsally with 4 tubercle-like processes directly distal to juncture with merus. Palm with subparallel lateral and mesial margins, 3.0 × longer than broad, 1.2 × longer than carpus. Fingers ending in slightly incurved blunt spine; movable finger half as long as palm, opposable margin with median process; opposing margin of fixed finger nearly straight.

P2-4: Relatively broad and thick, sparse setose. Meri subequal in length on P2 and P3, P4 merus 0.9 × as long as and 0.9 × as broad as P3 merus; length-breadth ratio, 2.8-3.2 on P2, 2.8-2.9 on P3, 2.8 on P4; P2 merus 0.7 × as long as carapace and subequal to length of P2 propodus; P3 merus 0.8 × length of P3 propodus; P4 merus 0.7 × length of P4 propodus; dorsal crest with 7 (left) or 5 (right) spines on P2, 6 on P3, 2 or 3 on P4, one of these situated at distal end; ventrolateral margin with small but distinct terminal spine. Carpi slightly shorter posteriorly (P3 carpus 0.96 × length of P2 carpus, P4 carpus 0.95 × length of P3 carpus), much shorter than dactyli; carpus-propodus length ratio, 0.37 on P2, 0.36 on P3, 0.34 on P4; extensor margin with 1 proximal and 1 distal spine. Propodi subequal; flexor margin ending in pair of spines preceded by 4 or 5 spines on P2, 3 or 4 on P3, 3 on P4. Dactyli relatively stout, ending in slender spine preceded by 9 or 10 sharp strong spines nearly perpendicular to slightly curving flexor margin and gradually smaller proximally (proximal spines somewhat oblique); dactylus-carpus length ratio, 1.4 on P2, 1.6 on P3, 1.7 on P4, dactylus-propodus length ratio, 0.6 on P2-4.

Eggs. Number of eggs carried, 7; size, 0.90 mm × 1.02 mm.

REMARKS — The new species resembles *U. multispinosus* Ahyong & Poore, 2004 and *U. vicinus* n. sp., in having the anterolateral spine slightly larger than or subequal to the lateral orbital spine and in having the anterior margin of sternite 3 with median notch. *Uroptychus beryx* differs from *U. multispinosus* in the following features: the rostrum is as long as instead of longer than broad; the antennal scale is slightly broader instead of more than twice broader than article 5, falling short of instead of extending far beyond the apex of article; the antennal article 5 is unarmed instead of bearing a distinct distal spine; the P2-4 meri bear spines instead of being unarmed on the dorsal crest; and the P2-4 dactyli bear more numerous and more closely arranged flexor spines (9 or 10 instead of 6 or 7 spines that are more loosely arranged). The relationships with *U. vicinus* are discussed under that species (see below).

Uroptychus bispinatus Baba, 1988

Figure 38

Uroptychus bispinatus Baba, 1988: 25, fig. 9. — Baba *et al.* 2009: 40, figs 32-33. — Poore *et al.* 2011: 328, pl. 6, fig. E.

TYPE MATERIAL — Holotype: **Indonesia**, Molucca Sea between Halmahera and northern Sulawesi, ALBATROSS Stn 5614, 2013 m, female (USNM 150311). [not examined].

MATERIAL EXAMINED — Fiji Islands. BORDAU 1 Stn CP1458, 17°22'S, 179°28'W, 1216-1226 m, 5.III.1999, 3 ♂ 5.3-5.6 mm, 2 ov. ♀ 5.4, 5.8 mm (MNHN-IU-2014-16332).

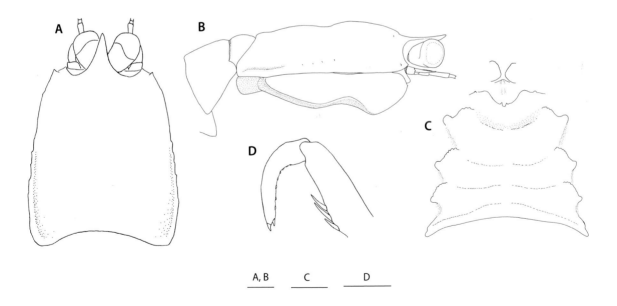

A, B   C   D

FIGURE 38

Uroptychus bispinatus Baba, 1988, **A**, **D**, male 5.5 mm; **B**, **C**, ovigerous female 5.4 mm (MNHN-IU-2014-16332). **A**, carapace, dorsal. **B**, carapace and anterior part of abdomen, lateral. **C**, sternal plastron, with excavated sternum and basal parts of Mxp1. **D**, distal part of left P2, lateral. Scale bars: 1 mm.

DISTRIBUTION — Molucca Sea between Halmahera and northern Sulawesi and Taiwan, and now Fiji Islands; 1173-2013 m.

DIAGNOSIS — Carapace as long as broad, greatest breadth 1.6-1.9 × distance between anterolateral spines; unarmed and smooth dorsally (epigastric spines vestigial); lateral margins ridged, with feeble crenulations (in dorsal view) along branchial region; anterolateral spine very small, distinctly posterior to level of small lateral orbital spine or acuminate lateral limit of orbit. Rostrum short triangular, with interior angle of 26-30°, dorsally flattish, length at most one-third that of remaining carapace, breadth less than half carapace breadth measured at posterior carapace margin. Pterygostomian flap anteriorly roundish with very small spine. Excavated sternum with convex anterior margin followed by longitudinal ridge in midline. Sternal plastron slightly broader than long; lateral extremities divergent posteriorly; sternite 3 depressed well, anterior margin deeply excavated in semicircular shape, with submedian spines flanking small median sinus; sternite 4 having anterolateral margins strongly divergent posteriorly, about as long as or slightly shorter than posterolateral margin. Anterolateral margin of sternite 5 convexly strongly divergent posteriorly, as long as anterolateral margin of sternite 4. Abdominal somite 1 without transverse ridge; somite 2 tergite 2.4-2.5 × broader than long; pleural lateral margins barely or slightly concave and posteriorly divergent, ending in bluntly angular tip; pleuron of somite 3 posterolaterally bluntly angular. Telson slightly more than half as long as broad; posterior plate 1.2-1.8 × longer than anterior plate, posterior margin slightly convex or slightly concave, not distinctly emarginate. Eyes relatively broad, 1.4 × longer than broad, distally broadened, proximally narrowed, barely or fully reaching, or slightly overreaching rostral tip. Distal article of antennular peduncle about twice as long as high. Antennal peduncle overreaching apex of rostrum; article 2 without distinct spine; antennal scale varying from slightly overreaching article 4 to terminating in midlength of article 5; distal 2 articles unarmed; breadth of article 5 one-third height of ultimate article of antennule; flagellum of 13-14 segments slightly falling short of distal end of P1 merus. Mxp1 with bases close to each other, not contiguous. Mxp3 basis with 1 distal denticle on mesial ridge; ischium with flexor margin not rounded distally, crista dentata with 3-7 loosely arranged denticles; merus not flattened, rather thick and unarmed, length 2.5-2.7 × that of ischium. P1 slender; ischium dorsally with antero-posteriorly compressed, basally broad, short spine; no spine elsewhere; merus 1.0-1.1 × longer than carapace; carpus 1.2-1.4 × longer than merus; palm 3.6-3.7 × (males), 4.4-5.6 × (females) longer than broad, 0.8 × length of carpus; fingers relatively broad distally, spooned on prehensile face, not crossing, length 0.5-0.6 × length of palm. P2-4 slender; meri with flattish lateral and mesial surfaces, unarmed on dorsal margin, successively shorter posteriorly (P3 merus 0.9 × length of P2 merus, P4 merus 0.9 × length of P3 merus), subequally broad on P2-4; P2 merus 0.8-0.9 × length of carapace, 1.3-1.4 × length of P2 propodus; P3 merus 1.1-1.2 × length of P3 propodus; P4 merus 0.9-1.1 × length of P4 propodus; carpi successively slightly shorter posteriorly; carpus-propodus length ratio, 0.7-0.8 on P2, 0.7 on P3, 0.6-0.7 on P4; propodi subequal or successively slightly longer posteriorly; flexor margin inflated at midlength, with 2 or 3 (usually 2) movable spines close to each other and located directly distal to midlength and remote from distal end of article; dactyli much shorter than carpi (dactylus-carpus length ratio, 0.6 on P2, 0.7 on P3 and P4), about half length of propodi (dactylus-propodus length ratio, 0.4 on P2, 0.4-0.5 on P3, 0.5 on P4); relatively slender, flexor margin strongly curving at proximal third, with 2 distal spines of moderate size (ultimate larger) preceded by 8 very small spines oriented parallel to flexor margin, all obscured by setae.

Eggs. Number of eggs carried, 8-10; size, 1.60 mm × 1.70 mm - 1.65 × 1.80 mm.

Color. A specimen from Taiwan was illustrated in Baba *et al.* (2009) and Poore *et al.* (2011).

Parasites. One of the males examined bears a rhizocephalan externa.

REMARKS — A slight difference between the type and the present material is noted: the excavated sternum bears a central spine on the surface in the type instead of a longitudinal ridge in the present material as well as in the material from Taiwan (Baba *et al.* 2009).

The P2-4 dactyli bear thick setae along the flexor margin by which the small spines are obscured, as shown in Baba *et al.* (2009). The presence of these spines was also confirmed in the type material.

The small anterolateral spine of the carapace and the P2-4 dactylar spines oriented parallel to the flexor margin link the species to *U. australis* (Henderson, 1885) and *U. setosipes* Baba, 1981 from Japan. *Uroptychus bispinatus* differs from both in having the pterygostomian flap anteriorly rounded instead of produced, in having the sternite 4 anterolateral margin as long as or slightly shorter than instead of twice as long as the posterolateral margin, in having the P4 merus 0.9 × instead of 0.6 × as long as the P3 merus, and in having the P2-4 propodi with two or three flexor marginal spines remotely distant from the juncture with the dactylus instead of a pair of terminal spines preceded by a row of spines.

The species also resembles *U. remotispinatus* Baba & Tirmizi, 1979 in having a short antennal scale, in having the P2-4 dactyli with the ultimate spine distinctly larger than the penultimate, and in having the P2-4 propodi with the distalmost of flexor spines considerably remote from the juncture with the dactyli. Their relationships were discussed by Baba *et al.* (2009).

The coloration was illustrated by Baba *et al.* (2009) based upon the material from Taiwan.

Uroptychus boisselierae n. sp.
Figures 39, 40

TYPE MATERIAL — Holotype: **Vanuatu**. MUSORSTOM 8 Stn CP1088, 15°09.23'S, 167°15.13'E, 425-455 m, 6.X.1994, ♂ 7.5 mm (MNHN-IU-2014-16333). Paratypes: **Vanuatu**. MUSORSTOM 8 Stn CP982, 19°21.80'S, 169°26.47'E, 408-410 m, 23.IX.1994, 1 ♂ 4.0 mm, 1 ♀ 6.6 mm (MNHN-IU-2014-16334).

ETYMOLOGY — Named for Marie-Catharine Boisselier of MNHN for her help with molecular analyses.

DISTRIBUTION — Vanuatu; 408-455 m.

DESCRIPTION — Medium sized-species. *Carapace*: As long as broad; greatest breadth 1.6 × distance between anterolateral spines. Dorsal surface polished and glabrous, somewhat convex from anterior to posterior, with shallow depression between gastric and cardiac regions; epigastric region with pair of denticulate ridges; a few denticle-like spines mesial to anterior end of branchial lateral margin. Lateral margins slightly convexly divergent posteriorly, ridged along posterior fifth of length; 2 strong spines: first anterolateral, moderately distant from, distinctly posterior to level of, and overreaching small lateral orbital spine; second situated at anterior end of branchial margin, preceded by 1 or 2 tiny spines, followed by a few to several small spines or tubercle-like processes. Rostrum sharp triangular, with interior angle of 30-35°; length slightly less than half that of carapace, breadth more than half carapace breadth measured at posterior carapace margin; dorsal surface concave, ventral surface horizontal, lateral margins straight. Pterygostomian with anterior margin more or less roundish ending in small spine; surface smooth.

Sternum: Excavated sternum with small spine in center, anterior margin broad triangular. Sternal plastron slightly shorter than broad, with posteriorly divergent lateral extremities. Sternite 3 strongly depressed; anterior margin deeply excavated, with narrow U-shaped median notch separating 2 well-developed submedian spines. Sternite 4 having anterolateral margin convex, anteriorly ending in short blunt spine or process followed by a few posteriorly diminishing small spines; posterolateral margin half as long as anterolateral margin. Anterolateral margins of sternite 5 convexly somewhat divergent posteriorly, 1.3 × longer than posterolateral margin of sternite 4.

Abdomen: Smooth and barely setose. Somite 1 transversely ridged. Somite 2 tergite 2.7-2.9 × broader than long; pleuron posterolaterally angular, lateral margin concavely moderately divergent posteriorly. Pleuron of somite 3 with angular posterolateral end. Telson about half as long as broad; posterior plate 1.5 × longer than anterior plate, laterally expanded and lobed, posterior margin deeply emarginate.

Eye: Elongate, 2 × as long as broad, overreaching midlength of rostrum; mesial margin somewhat convex proximally. Cornea not dilated, length slightly more than half that of remaining eyestalk.

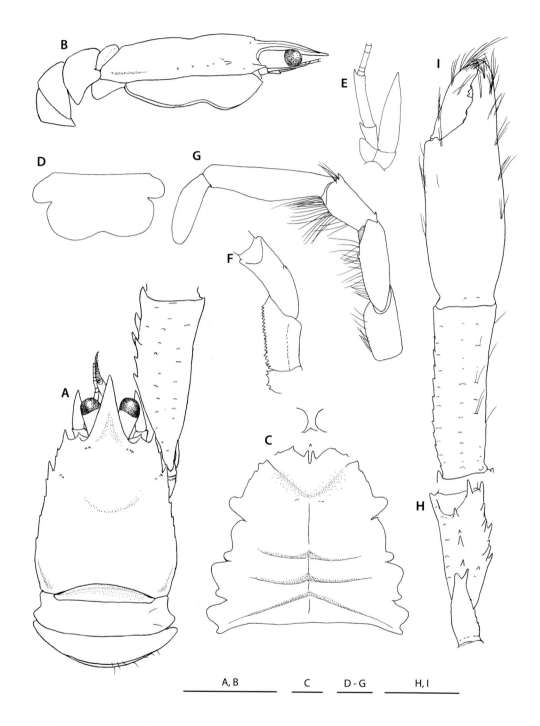

FIGURE 39

Uroptychus boisselierae n. sp., holotype, male 7.5 mm (MNHN-IU-2014-16333). **A**, carapace and anterior part of abdomen, proximal part of P1 included, dorsal. **B**, same, lateral. **C**, sternal plastron, with excavated sternum and basal parts of Mxp1. **D**, telson. **E**, left antenna, ventral. **F**, left Mxp3, setae omitted, ventral. **G**, same, lateral. **H**, right P1, proximal part, setae omitted, ventral. **I**, same, dorsal. Scale bars: A, B, H, I, 5 mm; C-G, 1 mm.

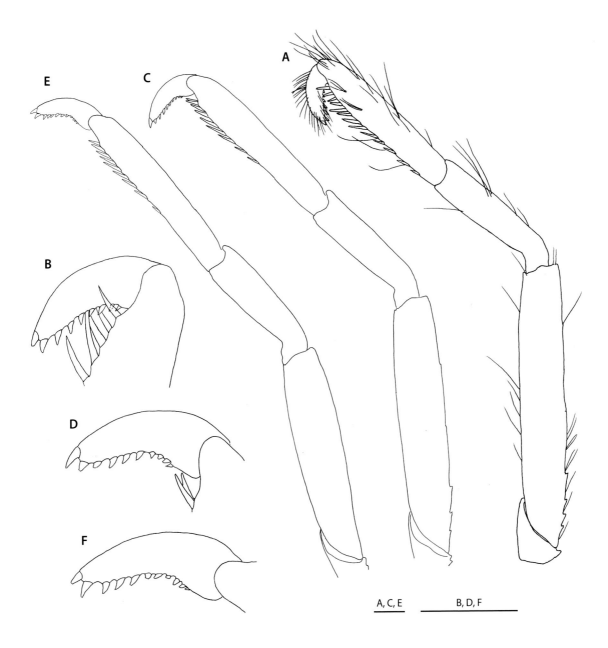

FIGURE 40

Uroptychus boisselierae n. sp., holotype, male 7.5 mm (MNHN-IU-2014-16333). **A**, left P2, lateral. **B**, same, distal part, setae omitted, lateral. **C**, left P3, setae omitted, lateral. **D**, same, distal part, setae omitted, lateral. **E**, left P4, setae omitted, lateral. **F**, same, distal part, setae omitted, lateral. Scale bars: 1 mm.

Antennule and antenna: Ultimate article of antennule 2.2-2.7 × longer than high. Antennal peduncle slightly overreaching cornea. Article 2 with very small distolateral spine. Antennal scale twice as broad as article 5, reaching or slightly falling short of distal end of article 5. Articles 4 and 5 each with distomesial spine. Article 5 2.3 × longer than article 4, breadth half height of antennular ultimate article. Flagellum consisting of 18-24 segments, reaching distal end of P1 merus.

Mxp: Mxp1 with bases close to each other. Mxp3 barely setose on lateral face. Basis with 3 or 4 denticles on mesial ridge. Ischium having flexor margin not rounded distally, crista dentata with 18-25 denticles. Merus 2 × longer than ischium; distolateral spine small; flexor margin sharply ridged, bearing small spine distal to midlength. Carpus with small distolateral spine and another small spine on extensor proximal margin.

P1: Somewhat massive in larger male, 3.8 × (smaller male), 5.2 × (larger male), 4.7 × (female) longer than carapace; barely setose on proximal articles, sparingly setose on carpus and palm, more setose on fingers. Ischium dorsally with strong spine, ventromesially with strong subterminal spine. Merus 1.1-1.2 × longer than carapace, polished, with spines of good size: 1 dorsal distomesial, 1 ventral distolateral, 1 ventral distomesial, 3 ventral in longitudinal row, 3-5 ventromesial, and 2-3 mesial (spination weak in holotype). Carpus ventrally flattish, dorsally with denticulate short ridges, ventrolaterally with tubercle-like spines, mesially with row of small spines; length 1.3 × (0.95 × in male paratype) that of merus. Palm moderately depressed, 2.3-2.6 × longer than broad, subequal to length of carpus, ridged on mesial and lateral margins, with row of 8 or 9 small spines (continued on proximal part of fixed finger) along ventrolateral margin (not visible in dorsal view) in male holotype and female paratype, obsolescent in male paratype. Fingers distally incurved, crossing when closed; movable finger 0.5-0.6 × as long as palm, opposable margin with pronounced process proximal to opposite eminence on fixed finger.

P2-4: Slender, somewhat compressed mesio-laterally, sparsely with simple long setae. Meri successively shorter posteriorly (P3 merus 0.9 × length of P2 merus (P2 and P3 meri subequal in male paratype), P4 merus 0.8-0.9 × length of P3 merus), equally broad on P2-4; length-breadth ratio, 5.9-6.2 on P2, 5.7-6.0 on P3, 4.4-4.8 on P4; dorsal margin not cristate but rounded, with several small spines on proximal portion on P2 and P3 (obsolescent in male paratype); P2 merus subequal to (holotype and female paratype) or slightly shorter than (male paratype) carapace, 1.5-1.6 × longer than P2 propodus; P3 merus 1.3 × length of P3 propodus; P4 merus subequal to or slightly longer than P4 propodus. Carpi unarmed; P3 carpus slightly longer than P2 carpus (subequal in smaller male), P4 carpus subequal to P3 carpus; carpus-propodus length ratio, 0.7-0.8 on P2, 0.6-0.7 on P3, 0.6 on P4; carpus-dactylus length ratio, 2.2 on P2, 2.4 on P3, 2.6 on P4. Propodi successively slightly shorter posteriorly; flexor margin slightly convex on distal half, with pair of terminal spines preceded by 7-12 spines along nearly entire length on P2, 7-8 spines on P3, 6-8 spines on P4. Dactyli subequal in length on P2 and P3 and longer on P4 or shortest on P2 and subequal on P3 and P4; length 0.4 × that of carpus and 0.3 × that of propodus on P2-4; extensor margin with plumose setae; flexor margin strongly curving at proximal third, with row of 9-10 sharp, somewhat obliquely directed spines successively diminishing toward base of article and obscured by setae; ultimate and penultimate spines subequal and nearly as broad as antepenultimate.

REMARKS — The spination of the carapace lateral margin and P2-4, propodi and dactyli in particular, and the shape of sternite 3 in the new species resemble those of *U. longicarpus* n. sp. (see below). *Uroptychus longicarpus* is unique among the species of the genus in having the P4 carpus distinctly longer than the P3 carpus. *Uroptychus boisselierae* differs from *U. longicarpus* in having the epigastric region of the carapace with a pair of denticulate ridges instead of being smooth; the anterolateral margin of sternite 4 is twice instead of three to four times the length of posterolateral margin; the antennal article 4 bears a distinct distomesial spine instead of being unarmed; the corneas are not inflated as in *U. longicarpus*; the P1 merus is spinose instead of granulose on the ventral surface, and the ischium bears a distinct subterminal spine instead of being unarmed on the ventromesial margin; and the P2-4 dactyli bears the ultimate spine subequal to instead of larger than the penultimate spine.

Uroptychus brachycarpus n. sp.

Figures 41, 42

TYPE MATERIAL — Holotype: **New Caledonia**, Norfolk Ridge. SMIB 4 Stn DW61, 23°01'S, 167°22'E, 520-550 m, 10.III.1989, with Chrysogorgiidae gen. sp. (Calcaxonia), sex indet., 7.5 mm (MNHN-IU-2014-16335). Paratypes: **New Caledonia**, Norfolk Ridge. MUSORSTOM 4 Stn CP216, 22°59.5'S, 167°22.0'E, 490-515 m, with Chrysogorgiidae gen. sp. (Calcaxonia), 29.IX.1985, 2 ♂ 6.3, 7.2 mm, 1 ov. ♀ 8.7 mm (MNHN-IU-2014-16338). SMIB 8 Stn DW194, 22°59.6'S, 168°22.5'E, 491 m, 1.II.1993, 1 ♀ 8.3 mm (MNHN-IU-2014-16339). BIOCAL Stn CP52, 23°06'S, 167°47'E, 540-600 m, 31.VIII.1985, 1 ♂ 6.3 mm, 1 ov. ♀ 6.9 mm, 1 ♀ 4.4 mm (MNHN-IU-2014-16340). **Indonesia**, Kai Islands. KARUBAR Stn CP05, 5°49'S, 132°18'E, 296-299 m, 22.X.1991, with ?*Acanella* sp. (Calcaxonia: Isididae), 1 ♂ 6.2 mm, 1 ov. ♀ 7.5 mm (MNHN-IU-2014-16336). **Vanuatu.** MUSORSTOM 8 Stn CP1088, 15°09.23'S, 167°15.13'E, 425-455 m, 6.X.1994, 3 ov. ♀ 7.7-8.8 mm, 1 ♀ 6.5 mm (MNHN-IU-2014-16337).

ETYMOLOGY — From the Greek *brachys* (short) and *carpos* (carpus, wrist), referring to short carpi of P2-4, by which the species is distinguished from the related species *U. brucei*.

DISTRIBUTION — South of New Caledonia, Isle of Pines, Vanuatu, and Kai Islands (Indonesia); 296-600 m.

SIZE — Males, 6.2-7.2 mm; females, 4.4-8.8 mm; ovigerous females from 6.9 mm.

DESCRIPTION — Medium-sized species. *Carapace*: Slightly broader than long (0.9 × as long as broad); greatest breadth 1.8 × distance between anterolateral spines. Dorsal surface smooth and glabrous, slightly convex from anterior to posterior, with weak or indistinct depression between gastric and cardiac regions; covered with small circular spots (visible but not hollowed out), and with granulations along lateral margins, more distinctly on anterior part in large specimens. Lateral margins convexly divergent posteriorly, bearing granules or denticles on anterior third, ridged along branchial region; anterolateral spine relatively large, overreaching small lateral orbital spine. Rostrum elongate triangular, with interior angle of 30°, dorsally excavated and somewhat deflected ventrally; lateral margins feebly convex anteriorly; length 0.3-0.5 that of remaining carapace, breadth less than half carapace breadth measured at posterior carapace margin. Lateral orbital spine situated at same level as, and more or less remote from, anterolateral spine. Pterygostomian flap anteriorly sharp angular, produced to small spine; surface granulose.

Sternum: Excavated sternum with convex anterior margin between bases of Mxp1, surface with rounded ridge in midline. Sternal plastron 0.9 × as long as broad, lateral extremities divergent posteriorly. Sternite 3 depressed well, anterolaterally sharp angular, anterior margin shallowly concave, with V-shaped median notch flanked by small spine. Sternite 4 having anterolateral margins nearly straight or anteriorly convex, anterior end rounded, often with granules or denticles; posterolateral margin short, about half as long as anterolateral margin. Anterolateral margin of sternite 5 anteriorly strongly convex, slightly more than 1.5 × longer than posterolateral margin of sternite 4.

Abdomen: Smooth and almost glabrous. Somite 1 convex from anterior to posterior, without transverse ridge. Somite 2 tergite 2.4-2.8 × broader than long; lateral margins somewhat concavely divergent and posteriorly rounded. Pleuron of somite 3 posterolaterally blunt. Telson half as long as broad; posterior plate nearly transverse or slightly emarginate on posterior margin, length subequal to or slightly more than that of anterior plate.

Eye: Relatively elongate, slightly less than 2 × longer than broad, slightly broadened proximally, overreaching midlength of rostrum. Cornea not inflated, more than half that of remaining eyestalk.

Antennule and antenna: Ultimate article of antennular peduncle 3.0-3.3 × longer than high. Antennal peduncle overreaching cornea, not reaching apex of rostrum. Article 2 with short, sharp distolateral spine. Antennal scale more than twice as broad as article 5, slightly falling short of (rarely reaching) distal end of article 5. Article 4 with small distomesial

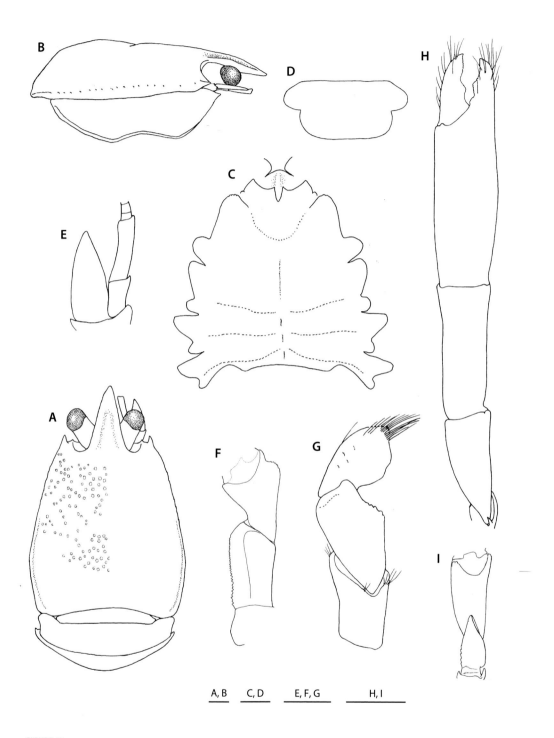

A, B   C, D   E, F, G   H, I

FIGURE 41

Uroptychus brachycarpus n. sp., holotype, sex indet. 7.5 mm (MNHN-IU-2014-16335). **A**, carapace and abdomen, small spots on right side omitted, dorsal. **B**, same, abdomen excluded, lateral. **C**, sternal plastron, with excavated sternum and basal parts of Mxp1. **D**, telson. **E**, right antenna, ventral. **F**, left Mxp3, ventral. **G**, right Mxp3, lateral. **H**, left P1, dorsal. **I**, same, proximal part, ventral. Scale bars: A-G, 1 mm; H, I, 5 mm.

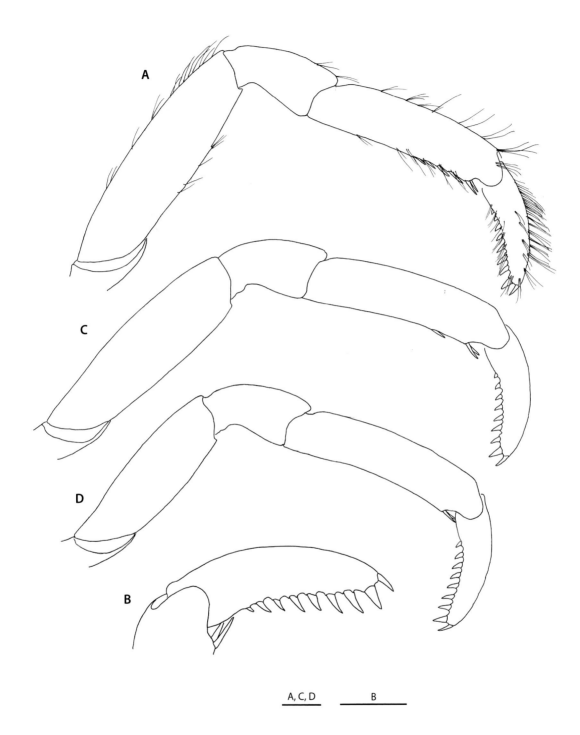

FIGURE 42

Uroptychus brachycarpus n. sp., holotype, sex indet. 7.5 mm (MNHN-IU-2014-16335). **A**, right P2, lateral. **B**, same, distal part, setae omitted, lateral. **C**, right P3, setae omitted, lateral. **D**, right P4, setae omitted, lateral. Scale bars: 1 mm.

spine. Article 5 unarmed, 1.6-1.7 × longer than article 4, breadth 0.7 × height of ultimate article of antennule. Flagellum of 15-22 segments barely reaching distal end of P1 merus.

Mxp: Mxp1 with bases broadly separated. Mxp3 basis smooth on mesial ridge. Ischium with distally diminishing small denticles on crista dentata; flexor margin sharply ridged, distally rounded. Merus 2.2 × longer than ischium, compressed well, mesially flattish, flexor margin angular at point one-third from distal end, with 3-5 tubercle-like spines distal to angular portion. Carpus unarmed.

P1: Massive, unarmed, nearly glabrous except for fingers; length 4.4-4.9 × that of carapace. Ischium dorsally with basally broad, antero-posteriorly flattish, distally sharp spine, ventromesially with row of low processes on proximal portion, without distinct subterminal spine. Merus 1.0-1.1 × longer than carapace, weakly granulose on proximal portions of ventral and mesial surfaces. Carpus 1.0-1.4 × longer than merus. Palm 2.2-2.4 × (males), 2.8-3.5 × (females) longer than broad, 1.0-1.3 × longer than carpus. Fingers depressed, sparingly with relatively short setae, not crossing distally; movable finger with subtriangular process on opposable median margin fitting to opposing concavity of fixed finger when closed; length about half or slightly less than half that of palm.

P2-4: Broad relative to length, well compressed mesio-laterally, setose barely on meri and carpi, sparsely on propodi, thickly on dactyli, setae relatively short. Meri and carpi unarmed. Meri successively shorter posteriorly (P3 merus 0.9 × length of P2 merus, P4 merus 0.8 × length of P3 merus), equally broad on P2-4; length-breadth ratio, 3.4-3.8 on P2, 3.2-4.0 on P3, 3.0-3.5 on P4; P2 merus 0.8 × length of carapace, 1.2 × (rarely 1.1 ×) longer than P2 propodus; P3 merus 1.1 × length of P3 propodus; P4 merus 0.8 × length of P4 propodus. Carpi successively slightly shorter posteriorly (P3 carpus 0.92-0.96 × length of P2 carpus, P4 carpus 0.92-0.96 × length of P3 carpus), less than half length of propodi (0.41-0.48 on P2, 0.37-0.41 on P3, 0.33-0.40 on P4). Propodi shorter on P2 than on P3 and P4, subequal on P3 and P4; flexor margin ending in pair of movable spines preceded by 3-5 spines on P2, 1-3 on P3, 0 or 1 on P4. Dactyli about half length of propodi, slightly (at most 1.1 x) longer than or subequal to carpi on P2-4; setae on extensor margin plumose; flexor margin slightly curving, with 9 or 10 well-developed, loosely arranged, proximally diminishing spines, all subtriangular, slightly obliquely directed, ultimate much more slender than penultimate, antepenultimate broader than ultimate and narrower than penultimate.

Eggs. About 50 eggs carried; size, 1.09 mm × 1.21 mm - 1.16 mm × 1.26 mm.

Parasites. The holotype bears an externa of rhizocephalan parasite on the abdomen.

REMARKS — This species keys out in a couplet together with *U. amabilis* Baba, 1977 from New Caledonia, sharing the carapace lateral margin with a well-developed anterolateral spine only, sternite 3 with a median notch on the anterior margin, the ultimate spine of the P2-4 more slender than the penultimate, and the P4 merus with a pair of terminal spines only on the flexor margin. However, they differ from each other in the spination of the P2-4 dactyli. *Uroptychus brachycarpus* has on the flexor margin, excepting the slender ultimate spine, a row of 8 or 9 long triangular, obliquely directed, proximally diminishing spines, whereas these spines in *U. amabilis* are less numerous (5 in number), short triangular and perpendicularly directed, the penultimate and antepenultimate being relatively broad. In addition, the Mxp3 merus in *U. brachycarpus* bears tubercle-like small spines only on the distal half of the flexor margin, whereas in *U. amabilis* this article bears a few distinct flexor marginal spines and a well-developed distolateral spine.

Uroptychus brachycarpus also resembles *U. denticulisquama* n. sp. in the carapace shape, in having a broad antennal scale, and in the spination of the P2-4 dactyli. Differences between these species are discussed under the remarks of *U. denticulisquama* (see below).

Uroptychus brachydactylus Tirmizi, 1964

Figures 43-45

Uroptychus brachydactylus Tirmizi, 1964: 399, fig. 19.

TYPE MATERIAL — Holotype: **South Arabian coast**, JOHN MURRAY Stn 42, 1415 m, male (BMNH 1966.2.3.20). [examined].

MATERIAL EXAMINED — **New Caledonia**, Isle of Pines. BIOGEOCAL Stn CP232, 21°33.81'S, 166°27.07'E, 760-790 m, 12.IV.1987, 1 ♂ 14.7 mm (MNHN-IU-2014-16341).

DISTRIBUTION — South Arabian coast and now Isle of Pines; 760-1415 m.

DESCRIPTION — Large species. *Carapace*: 1.1 × longer than broad; greatest breadth 2.0 × distance between anterolateral spines. Dorsal surface glabrous, feebly granulose, weakly convex from anterior to posterior, bearing very weak depression between gastric and cardiac regions. Lateral margins convexly divergent posteriorly, with row of short granulate ridges, and ridged along posterior fourth of length; anterolateral spine small, reaching distal end of antennal article 2, relatively remote from, slightly posterior to level of, and slightly overreaching small lateral orbital spine. Rostrum elongate sharp triangular, with interior angle of 23°; length slightly more than one-third that of carapace, breadth less than half carapace breadth measured at posterior carapace margin; dorsal surface flattish. Pterygostomian flap granulose, anteriorly roundish, with small spine.

Sternum: Excavated sternum anteriorly ending in small distinct spine, surface with small central spine. Sternal plastron as long as broad, lateral extremities posteriorly divergent. Sternite 3 with deep depression expanded posteriorly on to sternite 4; anterior margin deeply excavated, with pair of submedian spines separated by narrow deep notch. Sternite 4 granulose on surface; anterolateral margin longer than distance between anterolateral angles of sternite 3, 1.6 × length of posterolateral margin, anteriorly ending in distinct spine. Anterolateral margin of sternite 5 anteriorly convex, 1.4 × longer than posterolateral margin of sternite 4.

Abdomen: Smooth and glabrous. Somite 1 well convex from anterior to posterior. Somite 2 tergite 2.9 × broader than long; pleuron posterolaterally bluntly angular, lateral margin strongly concave. Pleuron of somite 3 posterolaterally angular. Telson about half as long as broad; posterior plate somewhat emarginate on posterior margin, length 1.4 × that of anterior plate.

Eye: Elongate, overreaching midlength of rostrum, laterally and mesially concave proximal to cornea; length about twice breadth. Cornea more than half length of remaining eyestalk.

Antennule and antenna: Ultimate article of antennular peduncle 3 × longer than high. Antennal peduncle slightly overreaching cornea. Article 2 with small distolateral spine. Antennal scale slightly more than twice as broad as opposite peduncle, terminating in distal quarter of article 5, overreaching cornea. Distal 2 articles slender, unarmed; article 5 2.6 × longer than article 4, breadth half height of ultimate article of antennule. Flagellum of 18 segments barely reaching distal end of P1 merus.

Mxp: Mxp1 with bases close to each other but not contiguous. Mxp3 basis with 4 denticles on somewhat convex mesial ridge, distalmost largest. Ischium with flexor margin not rounded distally, crista dentata with 26 or 27 denticles distally diminishing. Merus 2.4 × longer than ischium, relatively narrow, moderately thick mesio-laterally, unarmed, flexor margin ridged but not sharply crested. No spine on carpus.

P1: Relatively massive, setose distally. Ischium with low, broad, dorso-ventrally depressed, plate-like process dorsally, without spines on ventromesial margin. Merus covered with short granulate ridges, length slightly more than that of carapace. Carpus somewhat granulose, 1.4 × longer than merus, subcylindrical but somewhat ridged along dorsomesial and dorsolateral margins, distally broadened. Palm 2.4 × longer than broad, slightly shorter than carpus, smooth and

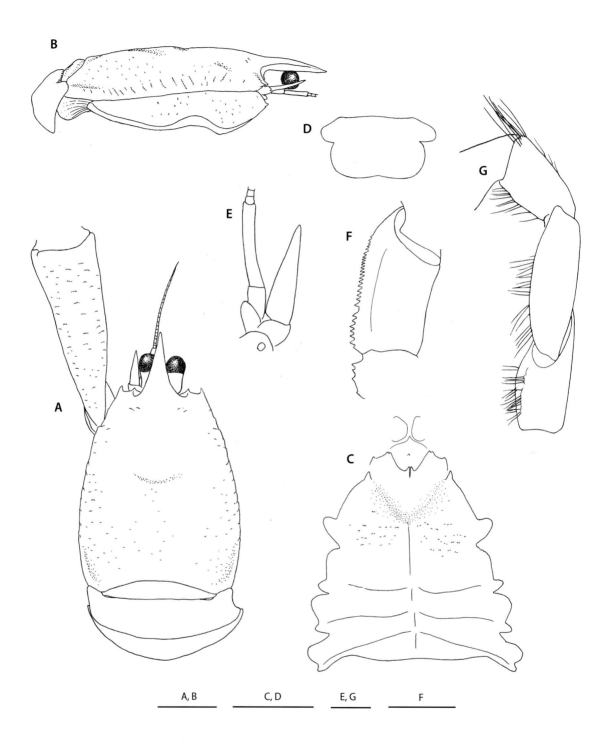

FIGURE 43

Uroptychus brachydactylus Tirmizi, 1964, male 14.7 mm (MNHN-IU-2014-16341). **A**, carapace and anterior part of abdomen, proximal part of left P1 included, dorsal. **B**, same, lateral. **C**, sternal plastron, with excavated sternum and proximal parts of Mxp1. **D**, telson. **E**, left antenna, ventral. **F**, left Mxp3, ventral. **G**, same, lateral. Scale bars: A-D, 5 mm; E-G, 1 mm.

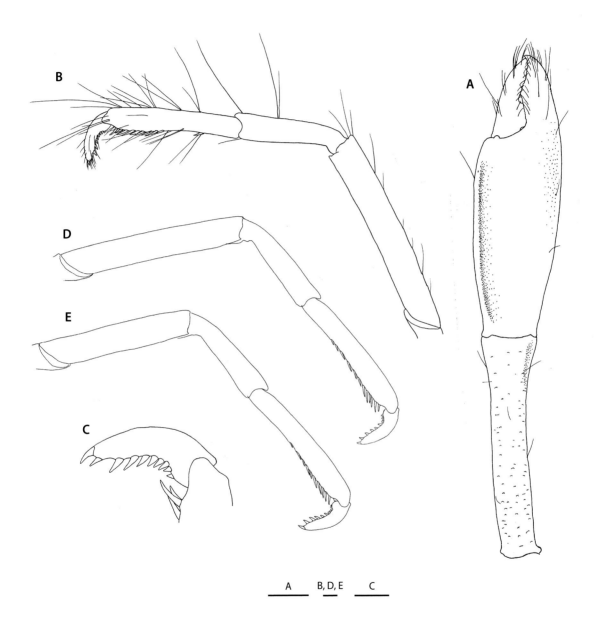

FIGURE 44

Uroptychus brachydactylus Tirmizi, 1964, male 14.7 mm (MNHN-IU-2014-16341). **A**, right P1, proximal articles omitted, dorsal. **B**, left P2, lateral. **C**, same, distal part, setae omitted, lateral. **D**, right P3, setae omitted, lateral. **E**, right P4, lateral. Scale bars: A, 5 mm; B-E, 1 mm.

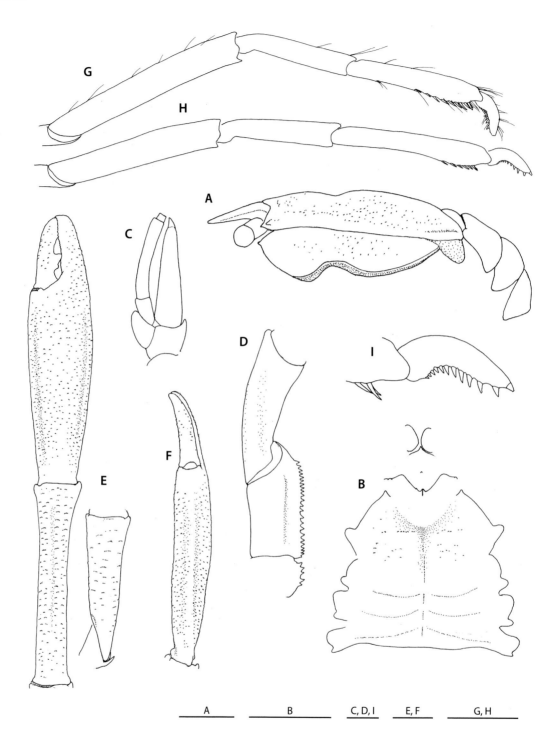

FIGURE 45

Uroptychus brachydactylus Tirmizi, 1964, holotype, male 15.2 mm (BMNH 1966.2.3.20). **A**, carapace and anterior part of abdomen, lateral. **B**, sternal plastron, with excavated sternum and basal parts of Mxp1. **C**, left antenna, ventral. **D**, right Mxp3, ventral. **E**, right P1, dorsal. **F**, same, mesial. **G**, right P2, lateral. **H**, right P4, setae omitted, lateral. **I**, same, distal part. Scale bars, A, B, E-H, 5 mm; C, D, I, 1 mm.

glabrous, depressed well, distally broadened, greatest breadth 2.2 × that of carpus measured at midlength; dorsal and ventral surfaces convex, ridged along lateral and mesial margins, mesial margin crested. Fingers distally crossing, ending in short sharp spine; opposable margins fitting to each other in distal half (in distal third in holotype), gaping in proximal half; movable finger with broad low process on gaping portion, length about half that of palm.

P2-4: Slender, moderately compressed, with sparse long setae. Meri successively shorter posteriorly (P3 merus 0.9 × length of P2 merus; P4 merus 0.8 × length of P3 merus), equally broad on P2-4; length-breadth ratio, 6.9 on P2, 6.7 on P3, 5.3 on P4; P2 merus slightly shorter than carapace, 1.5 × longer than P2 propodus, dorsal margin unarmed, ventrolateral margin with small distal spine; P3 merus 1.2 × length of P3 propodus; P4 merus as long as P4 propodus. Carpi successively shorter posteriorly (P3 carpus 0.9 × length of P2 carpus, P4 carpus 0.9 × length of P3 carpus); length 0.7 × that of propodus on P2, 0.6 × on P3 and P4. Propodi subequal on P2-4 or subequal on P2-3 and longer on P4; flexor margin ending in pair of spines preceded by 9 or 10 spines on P2, 11 on P3, 7 or 9 on P4. Dactyli subequal on P2-4, 0.3 × length of propodi, 0.4 × length of carpi on P2 and P3, 0.5 × on P4; distally slender, strongly curving at proximal third; flexor margin with 10 or 11 (12 on P4 in holotype) somewhat inclined, sharp, proximally diminishing spines.

REMARKS — The penultimate of the flexor marginal spines of dactylus is smaller than the antepenultimate on the left P2, but those on the other appendages including P3 and P4 are larger, presumably regenerated after being broken.

The material generally agrees with the description of the male holotype of *U. brachydactylus* Tirmizi, 1964 previously known from the western Indian Ocean. However, the following slight differences are found by examination of the holotype (BMNH 1966.2.3.20; Figure 45). In the present material, the antennal scale falls short of instead of reaching the tip of article 5, and article 2 bears a tiny rather than distinct distolateral spine; the P1 palm and fingers are smooth instead of noticeably granulose; the P4 propodus bears flexor marginal spines along the distal three-fifths of the length instead of the distal quarter; and the posterolateral margin of sternite 4 is relatively short, 0.8-0.9 instead of 1.3 times as long as the anterolateral margin of sternite 5. The last mentioned difference may indicate this specimen is a separate species, as is shown by *U. longioculus* Baba, 1990 from Madagascar and *U. poupini* n. sp. from Fiji, Tonga and New Caledonia, the separation of which are based only on the length-breadth ratio of sternite 4 and supported by DNA analyses (see below under *U. poupini*). The present material is provisionally placed in *U. brachydactylus*, awaiting more material to allow genetic analyses.

Uroptychus brachydactylus resembles *U. brucei* Baba, 1986a. Their relationships are discussed under that species (see below).

Uroptychus brevirostris Van Dam, 1933

Figure 46

Uroptychus brevirostris Van Dam, 1933: 20, figs 29-32. — Van Dam 1940: 96. — Baba 2005 (synonymies, key).
Not *Uroptychus brevirostris*: Baba 1973: 117 (= *U. alcocki* Ahyong & Poore, 2004).

TYPE MATERIAL — Holotype: **Indonesia**, Sulu Archipelago, 5°43.5'N, 119°40'E, 522 m female (ZMA De. 101.694). [examined].

MATERIAL EXAMINED — **Philippines**. MUSORSTOM 3 Stn CP133, 11°58'N, 121°52'E, 334-390 m, with Chrysogorgiidae gen. sp. (Calcaxonia), 5.VI.1985, 1 ov. ♀ 5.4 mm (MNHN-IU-2014-16342), 1 ♂ 4.0 mm, 1 ov. ♀ 6.0 mm, 1 sp. (sex indet.) 2.4 mm (MNHN-IU-2014-16343).

DISTRIBUTION — Sulu Archipelago and Java Sea, in 41-520 m, and now northeastern Panay, Philippines, in 334-390 m.

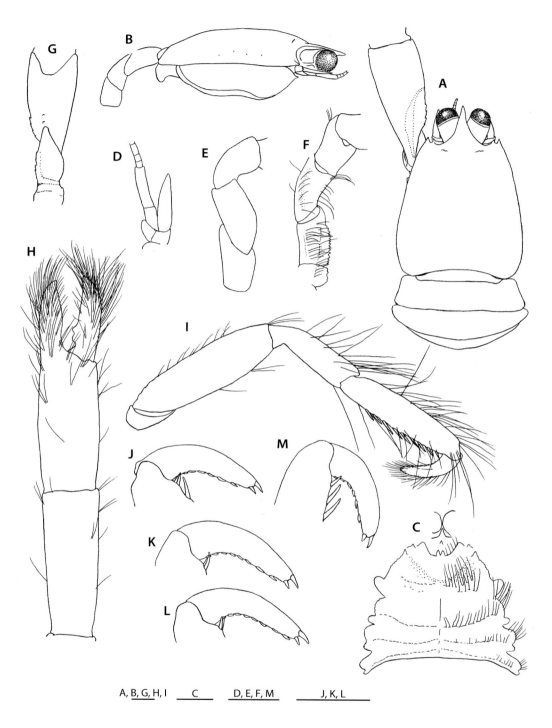

A, B, G, H, I   C   D, E, F, M   J, K, L

FIGURE 46

Uroptychus brevirostris Van Dam, 1933, **A-L**, ovigerous female 5.4 mm (MNHN-IU-2014-16342); **M**, holotype, female (ZMA De. 101.694). **A**, carapace and anterior part of abdomen, proximal part of left P1 included, dorsal. **B**, same, lateral. **C**, sternal plastron, with excavated sternum and basal parts of Mxp1. **D**, left antenna, ventral. **E**, right Mxp3, setae omitted, lateral. **F**, same, ventral. **G**, left P1, proximal part, ventral. **H**, same, dorsal. **I**, right P2, lateral. **J**, same, distal part, setae omitted, lateral. **K**, right P3, distal part, lateral. **L**, right P4, distal part, lateral. **M**, right P2, distal part, lateral. Scale bars: 1 mm.

DESCRIPTION — Medium-sized species. *Carapace*: As long as broad; greatest breadth 1.8 × distance between anterolateral spines. Dorsal surface well convex from side to side and anterior to posterior, without depression between gastric and cardiac regions, smooth and glabrous. Epigastric region with pair of denticulate short ridges. Lateral margins weakly divergent posteriorly, convex posteriorly, with well-developed anterolateral spine overreaching small lateral orbital spine. Rostrum subtriangular, with interior angle of 22-30°, dorsally flattish, ventrally straight horizontal; length 0.3 × that of remaining carapace, breadth about half carapace breadth measured at posterior carapace margin; lateral margins slightly concave. Lateral orbital spine much smaller than, located slightly anterior to level of, and more or less close to, anterolateral spine. Pterygostomian flap anteriorly somewhat roundish, ending in small spine; surface smooth.

Sternum: Excavated sternum strongly produced anteriorly, ending in sharp spine; surface with spine in center. Sternal plastron slightly broader than long, gradually broadened posteriorly; relatively long setae on transverse ridges and lateral margins of sternites 5-7. Sternite 3 deeply depressed, anterior margin deeply emarginate, with pair of submedian spines contiguous at base, without median notch. Sternite 4 with more or less convex anterolateral margin anteriorly produced to short spine followed by granules, posterolateral margin short, about half length of anterolateral margin. Anterolateral margins of sternite 5 convexly divergent posteriorly, about as long as posterolateral margin of sternite 4.

Abdomen: Nearly glabrous and smooth. Somite 1 with antero-posteriorly convex. Somite 2 tergite 2.5 × broader than long; pleural lateral margin slightly concavely divergent posteriorly, ending in rounded terminus. Pleuron of somite 3 posterolaterally blunt. Telson slightly more than half as long as broad; posterior plate 1.7 × longer than anterior plate, posterior margin distinctly emarginate.

Eye: Relatively broad (length 1.6 × breadth), slightly broadened proximally, slightly falling short of apex of rostrum. Cornea not dilated, more than half length of remaining eyestalk.

Antennule and antenna: Antennular ultimate article 1.7-2.0 × longer than high. Antennal peduncle terminating in distal corneal margin. Article 2 with small distolateral spine. Antennal scale slightly broader than article 5, slightly falling short of or reaching distal end of peduncle. Articles 4 and 5 unarmed; article 5 1.5-1.9 × longer than article 4, breadth half height of ultimate article of antennule. Flagellum consisting of 12-14 segments (9 in smallest specimen), not reaching distal end of P1 merus.

Mxp: Mxp1 with bases nearly contiguous. Mxp3 smooth, glabrous on lateral surface. Basis with 4 denticles on mesial ridge. Ischium not rounded at distal end of flexor margin; crista dentata with 14-15 denticles. Merus and carpus unarmed; merus 1.9 × longer than ischium.

P1: Massive, smooth, very setose on fingers, glabrous on ischium and merus; length 4.2-4.5 × that of carapace. Ischium with basally broad, short dorsal spine, ventromesial margin with row of denticle-like spines on proximal half, lacking subterminal spine. Merus slightly (1.1 ×) longer than carapace, mesially bearing several tubercle-like spines on proximal portion, ventrally a few proximal tubercles. Carpus 1.1-1.2 × longer than merus. Palm 2.2 × longer than broad, subequal to or slightly shorter than (0.9) carpus. Fingers very setose, distally not sharply pointed, not distinctly incurved; opposable margin of movable finger with depressed, broad proximal process bearing 2 obtuse teeth; movable finger 0.6 × length of palm.

P2-4: Smooth, relatively broad, with long setae especially numerous on mesial faces of propodi and dactyli; P4 much shorter and narrower than P2 and P3. Meri strongly compressed mesio-laterally and distally narrowed, very slightly shorter on P3 than on P2, subequally broad on P2 and P3, smallest on P4 (P4 merus 0.7 × length and 0.7 × breadth of P3 merus); length-breadth ratio, 3.5-3.6 on P2, 3.4-3.5 on P3, 3.5-3.6 on P4 (in smallest specimen: 4.9, 4.7, 4.4 respectively); P2 merus 0.9 × length of carapace, 1.1 × length of P2 propodus; dorsal margin with very small denticles or granules on proximal third on P2, obsolete on P3 and P4; P3 merus as long as P3 propodus; P4 merus 0.9 × length of P4 propodus. Carpi subequal in length on P2 and P3, much shorter on P4 (P4 carpus 0.7-0.8 × length of P3 carpus), 0.6 × length of propodus on P2-3, 0.5-0.6 × on P4; unarmed. Propodi subequal in length on P2 and P3, shortest on P4; flexor margin straight, with pair of terminal spines preceded by row of 7 or 8 spines on P2-3 (6 on smallest specimen), 6 or 5 on P4. Dactyli proportionately broad proximally and distally, length 0.4 × that of propodi on P2 and P3, 0.4-0.5 × on P4, and 0.6-0.7 × length of carpi on P2, 0.7 × on P3, 0.8 × on P4; strongly curving at proximal quarter; flexor margin with 9 or 10 spines, obscured by setae in large specimens, ultimate longest, penultimate about as broad as or slightly narrower than ultimate, remaining spines proximally diminishing and oriented parallel to flexor margin.

Eggs. Number of eggs carried, 11 in larger female; size, 0.58 × 0.59 mm - 0.60 × 0.66 mm (yolky); 14 eggs in smaller female; size, 1.06 × 0.97 mm.

REMARKS — The specimens agree quite well with the original description of *U. brevirostris* Van Dam, 1933, except that the rostral margin is somewhat concave. Examination of the type material shows that the P2-4 dactyli bear flexor spines oriented parallel to the dactylar margin, not as illustrated by Van Dam (1933: fig. 32) (see Figure 46M). The presence or absence of the epigastric ridges in the type has not been confirmed.

Uroptychus brevirostris strongly resembles *U. disangulatus* n. sp. in the shape of the carapace, pterygostomian flap, sternal plastron and P2-4 dactyli. Their relationships are discussed under the remarks of that species (see below).

Uroptychus brevirostris somewhat resembles *U. australis* (Henderson, 1885) in the spination of P2-4 dactyli and in having short P4s, from which it is readily distinguished by the following differences: the carapace is as long as instead of distinctly longer than broad, and the rostrum is about as long as instead of much longer than broad; the anterolateral spine of the carapace is closer to instead of moderately remote from the lateral orbital spine, when viewed from dorsal side; P1 is more massive, with the palm short relative to breadth (length 2.3 times breadth in *U. brevirostris*, more than 3 times in *U. australis*); and the P4 merus is 0.7-0.8 instead of at most 0.6 times the length of P3 merus.

Uroptychus brevisquamatus Baba, 1988

Figures 47, 48

Uroptychus brevisquamatus Baba, 1988: 28, fig. 10.

TYPE MATERIAL — Holotype: **Indonesia**, off southern Obi, ALBATROSS Stn 5635, 732 m, female (USNM 150319). [examined].

MATERIAL EXAMINED — **Solomon Islands**. SALOMON 2 Stn CP2230, 6°27.8'S, 156°24.3'E, 837-945 m, 29.X.2004, 1 ♀ 7.3 mm (MNHN-IU-2014-16344). **Vanuatu**. MUSORSTOM 8 Stn DW1128, 16°02.14'S, 166°38.39'E, 778-811 m, 10.X.1994, 1 ov. ♀ 7.2 mm (MNHN-IU-2014-16345). **Wallis and Futuna Islands**. MUSORSTOM 7 Stn DW548, 12°23'S, 177°24'W 700-740 m, 17.V.1992, 1 ♀ 7.7 mm (MNHN-IU-2014-16347). – Stn DW635, 13°49S, 179°56'E, 700-715 m, 30.V.1992, 1 ♀ 6.3 mm (carapace broken) (MNHN-IU-2014-16348). **New Caledonia**, Loyalty Ridge. BATHUS 3 Stn DW776, 24°44'S, 170°08'E, 770-830 m, 24.XI.1993, 2 ♂ 4.3, 7.3 mm, 1 ov. ♀ 7.0 mm (MNHN-IU-2014-16349). – Stn DW778, 24°43'S, 170°07'E, 750-760 m, 24.XI.1993, 7 ♂ 3.8-7.8 mm, 6 ov. ♀ 6.3-7.8 mm, 3 ♀ 3.6-8.3 mm (MNHN-IU-2014-16350). **New Caledonia**, Norfolk Ridge. NORFOLK 2 Stn DW2047, 23°43.04'S, 168°01.92'E, 759-807 m, 23.X. 2003, 1 ♂ 7.9 mm, 1 ov. ♀ 7.0 mm (MNHN-IU-2014-16351). – Stn DW2107, 23°53'S, 167°41'E, 742-820 m, 30.X.2003, 1 ♀ 7.2 mm (MNHN-IU-2014-16346). – Stn DW2113, 23°45.17'S, 168°17.99' E, 888-966 m, 31.X.2003, 1 ♂ 6.7 mm (MNHN-IU-2014-16352).

DISTRIBUTION — Off southern Obi, and now Solomon Islands, Wallis and Futuna Islands, Vanuatu, Norfolk Ridge and Loyalty Ridge; 700-966 m.

SIZE — Males, 3.8-7.9 mm; females, 3.6-8.3 mm; ovigerous females from 6.3 mm.

DESCRIPTION — Medium-sized species. *Carapace*: Nearly as long as broad; greatest breadth 1.9 × distance between anterolateral spines. Dorsal surface moderately convex from anterior to posterior, with weak depression on ordinary site of cervical groove; gastric region more elevated than level of rostrum; anterior part of carapace including rostrum with sparse, very short, fine setae discernible under high magnification. Lateral margins convexly divergent, with finely

A, B   C, G, H   D, E, F   I, J

FIGURE 47

Uroptychus brevisquamatus Baba, 1988, female 7.7 mm (MNHN-IU-2014-16347). **A**, carapace and anterior part of abdomen, proximal part of right P1 included, dorsal. **B**, same, abdomen omitted, lateral. **C**, anterior part of cephalothorax, showing eye and orbit, dorsal. **D**, sternal plastron, with excavated sternum and basal parts of Mxp1. **E**, telson. **F**, right antenna, ventral. **G**, right Mxp3, ventral. **H**, left Mxp3, lateral. **I**, right P1, proximal part, ventral. **J**, same, proximal part omitted, dorsal. Scale bars: A-H, 1 mm; I, J, 5 mm.

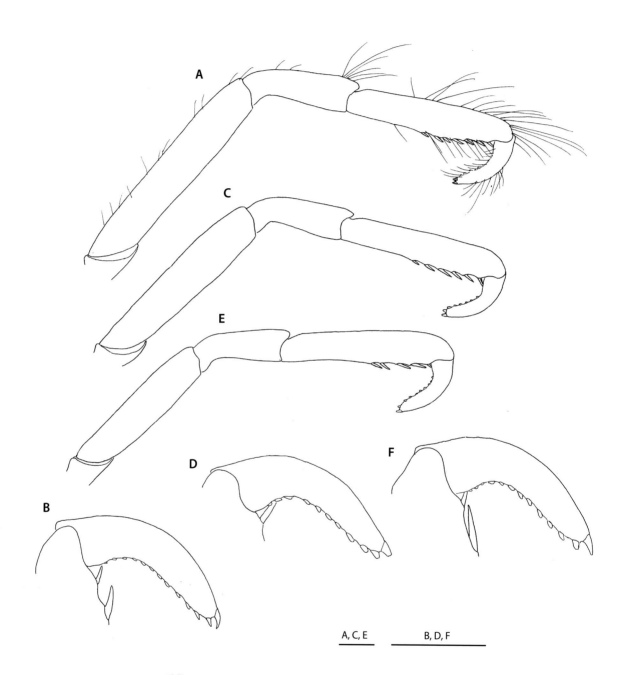

FIGURE 48

Uroptychus brevisquamatus Baba, 1988, female 7.7 mm (MNHN-IU-2014-16347). **A**, right P2, lateral. **B**, same, distal part, setae omitted, lateral. **C**, right P3, lateral. **D**, same, distal part, lateral. **E**, right P4, lateral. **F**, same, distal part, lateral. Scale bars: 1 mm.

granulate short ridges at anterior end of branchial region; ridged along posterior half; anterolateral spine relatively stout, reaching distal end of antennal article 2, situated at level of, but distinctly overreaching small lateral orbital spine Rostrum slightly overreaching cornea, bluntly triangular, with interior angle of 26-30° (rarely 35°), somewhat deflected ventrally; length less than one-third postorbital carapace length, breadth less than half carapace breadth measured at posterior carapace margin; dorsal surface flattish or slightly concave. Pterygostomian flap smooth on surface, anteriorly roundish with very small spine.

Sternum: Excavated sternum anteriorly sharp triangular, produced between bases of Mxp1, surface somewhat ridged in midline, with small spine in center. Sternal plastron slightly broader than long; lateral extremities gently divergent posteriorly. Sternite 3 depressed well, anterior margin deeply emarginate, bearing 2 submedian spines separated by narrow notch. Sternite 4 denticulate on anterolateral margin, posterolateral margin relatively long, more than half (0.6-0.7) as long as anterolateral margin. Anterolateral margin of sternite 5 granulated, gently convexly divergent posteriorly, somewhat longer than posterolateral margin of sternite 4.

Abdomen: Nearly glabrous and smooth. Somite 1 strongly convex from anterior to posterior. Somite 2 tergite 2.3-2.8 × broader than long; pleuron having lateral margin strongly concave, divergent posteriorly, posterolaterally bluntly angular. Pleuron of somite 3 posterolaterally blunt angular. Telson half as long as broad; posterior plate 1.4-1.6 × longer than anterior plate, somewhat emarginate on posterior margin.

Eye: Broad relative to length (length 1.6 × breadth), lateral margin convex, mesial margin concave. Cornea slightly dilated, more than half as long as remaining eyestalk.

Antennule and antenna: Ultimate article of antennular peduncle 2.0-2.5 × longer than high. Antennal peduncle slightly overreaching cornea. Article 2 with short stout spine. Antennal scale twice (rarely 1.5 x) as broad as article 5, terminating in or somewhat overreaching midlength of article 5. Distal 2 articles unarmed; article 5 1.7-2.0 × longer than article 4, breadth half or less than half height of ultimate antennular article. Flagellum consisting of 14-18 segments, not reaching distal end of P1 merus.

Mxp: Mxp1 with bases very close to each other or nearly contiguous. Mxp3 basis with 3 or 4 denticles on mesial ridge. Ischium with 21-23 denticles on crista dentata, flexor margin not rounded distally. Merus 3 × longer than ischium, relatively thick mesio-laterally, flattish on mesial face, unarmed. Carpus unarmed.

P1: More or less massive, length about 4 × that of carapace; smooth (occasionally somewhat granulose on ventral surface of merus), fingers setose. Ischium dorsally with strong curved spine, ventromesially with short subterminal spine. Merus as long as or slightly shorter than carapace (0.8 × in holotype). Carpus 1.2-1.3 × longer than merus. Palm 2.4-3.0 × longer than broad, subequal to or slightly shorter than carpus. Fingers ending in slightly incurved tip, gaping in large males, not gaping in females and small males; movable finger half length of palm or slightly more; opposable margin with low bilobed process at proximal third.

P2-4: Moderately compressed mesio-laterally, with sparse long setae, particularly on carpi and propodi. Meri successively shorter posteriorly (P3 merus 0.9 × length of P2 merus; P4 merus 0.8 × length of P3 merus), equally broad on P2-4 (occasionally slightly narrower on P4 than on P3); dorsal margin unarmed; length-breadth ratio, 4.3-4.8 on P2, 3.9-4.4 on P3, 3.4-3.9 on P4; P2 merus 0.8 × length of carapace, 1.2-1.3 × length of P2 propodus; P3 merus subequal to length of P3 propodus; P4 merus 0.9-1.1 × length of P4 propodus. Carpi successively slightly shorter posteriorly (P3 carpus 0.9 × length of P2 carpus, P4 carpus 0.9 × length of P3 carpus); carpus-propodus length ratio, 0.5-0.6 on P2 and P3, 0.5 on P4, much longer than dactyli (carpus-dactylus length ratio, 1.6 on P2, 1.5 on P3, 1.4 on P4. Propodi subequal on P2-3 (longer on P4) or P3-4 (longer on P2), more than twice as long as dactyli; flexor margin with 6-8 slender spines on distal half, distalmost single, slightly proximal to juncture with dactylus and mesial to midline in flexor view. Dactyli proportionately broad distally, strongly curving at proximal quarter, length 0.3-0.4 × that of propodus; dactylus-carpus length ratio, 0.6 on P2, 0.7 on P3 and P4; flexor margin 10-12 loosely arranged, relatively short spines, but ultimate spine longer and broader than penultimate, remaining spines successively smaller proximally, obliquely directed, not directed parallel to margin, and obscured by dense setae.

Eggs. Up to 25 eggs carried; size, 1.50 mm × 1.60 mm - 1.30 mm × 1.60 mm.

PARASITES — The female from NORFOLK 2 Stn DW2107 (MNHN-IU-2014-16346) bears a rhizocephalan externa.

REMARKS — In the type, the P2 merus is shorter relative to the carapace than in the present material (0.7 versus 0.8), as is also the P4 merus relative to the P4 propodus (0.8 versus 0.9-1.1); and the lateral limit of the orbit is rounded instead of bearing a small spine. These differences are regarded as individual variation.

This material is assigned to *U. brevisquamatus* based on the following characters: the anterolateral spine of the carapace is relatively stout; the pterygostomian flap bears a roundish anterior margin with a tiny spine; the eyes are relatively short bearing a concave mesial margin; the antennal scale at most slightly overreaches the midlength of antennal article 5; the distalmost of the flexor marginal spines of the P2-4 propodi is single, not paired, all also confirmed by examination of the holotype.

The species resembles *U. disangulatus* n. sp. and *U. webberi* Schnabel, 2009 from the Kermadec Ridge in the shapes of the carapace and sternal plastron, and in having a short and broad antennal scale and small inclined spines on the flexor margin of the P2-4 dactyli. The relationships with *U disangulatus* are discussed under the remarks of that species (see below). *Uroptychus brevisquamatus* shares the roundish anterior margin of the pterygostomian flap with *U. webberi*, from which it clearly differs in having the P2-4 propodi with a single instead of paired terminal spines on the flexor margin; and the carapace lateral margins are convexly divergent rather than subparallel.

Uroptychus brucei Baba, 1986
Figure 49

Uroptychus brucei Baba, 1986a: 1, figs 1, 2. — Ahyong & Poore 2004: 14 (key). — Baba 2005: 219 (key), 225 (synonymy).

TYPE MATERIAL — Holotype: **Australia**, off Western Australia, 17°59.4'S, 118°18.4'E, 406-416 m, male (NTM Cr. 000604). [not examined].

MATERIAL EXAMINED — **Indonesia**, Tanimbar Islands. KARUBAR Stn CC40, 7°46'S, 132°31'E, 443-468 m, 28.X.1991, 1 ♂ 13.7 mm, 1 ♀ 15.8 mm (MNHN-IU-2014-16353).

DISTRIBUTION — Northwest Australia, and now Tanimbar Islands, Indonesia; 406-468 m.

DESCRIPTION — Large species. *Carapace*: About as long as broad; greatest breadth 1.6 × distance between anterolateral spines. Dorsal surface somewhat granulose and unarmed, slightly convex from anterior to posterior, with moderate depression suggesting cervical groove. Lateral margins with small tubercles or granules; somewhat elevated ridge at end of anterior branchial margin; ridged along posterior third; anterolateral spine well developed, situated posterior to level of but somewhat overreaching much smaller lateral orbital spine, separated from that spine by twice basal breadth. Rostrum sharp triangular, with interior angle of 30°, length less than half that of carapace, breadth less than half carapace breadth measured at posterior carapace margin; dorsal surface concave. Pterygostomian flap finely granulose, anteriorly roundish, with small spine.

Sternum: Excavated sternum anteriorly triangular or ending in small spine, surface with obsolescent ridge in midline. Sternal plastron slightly shorter than broad, lateral extremities gently divergent posteriorly. Sternite 3 strongly depressed, anterior margin excavated in shallow V-shape, with 2 submedian spines flanking deep narrow sinus. Sternite 4 having anterolateral margin smoothly convex, bluntly produced on anterior end, 1.4 × longer than posterolateral margin. Anterolateral margin of sternite 5 strongly convex, about as long as posterolateral margin of sternite 4.

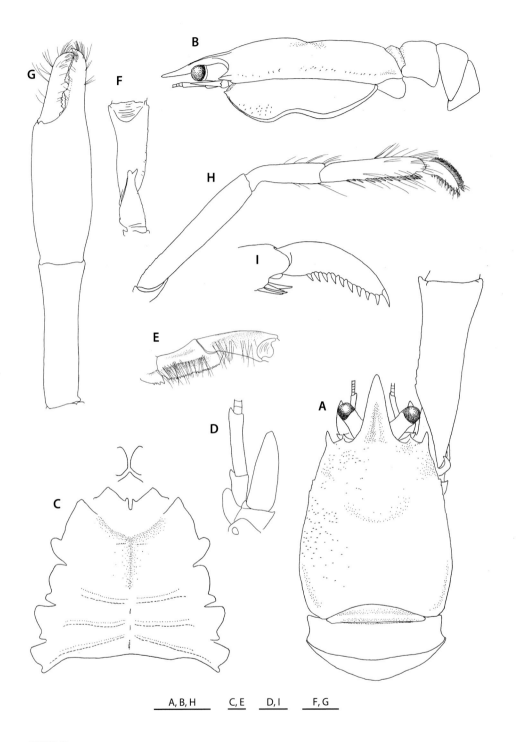

A, B, H   C, E   D, I   F, G

FIGURE 49

Uroptychus brucei Baba, 1986a, male 13.7 mm (MNHN-IU-2014-16353). **A**, carapace and anterior part of abdomen, proximal part of right P1 included, dorsal. **B**, same, lateral. **C**, sternal plastron, with excavated sternum and basal parts of Mxp1. **D**, left antenna, ventral. **E**, right Mxp3, ventral. **F**, right P1, proximal part, ventral. **G**, same, dorsal. **H**, right P2, lateral. **I**, same, distal part, setae omitted. Scale bars: A, B, F, G, H, 5 mm; C-E, I, 1 mm.

Abdomen: Somite 1 well convex from anterior to posterior, hence transversely ridged. Somite 2 tergite 2.1-2.5 × broader than long; pleuron posterolaterally blunt, lateral margin concavely divergent posteriorly. Pleuron of somite 3 posterolaterally blunt angular. Telson 0.4-0.5 × as long as broad, posterior plate 1.5 × longer than anterior plate, somewhat emarginate on posterior margin.

Eye: Elongate and slender, 2.2-2.6 × longer than broad, slightly overreaching midlength of rostrum, lateral and mesial margins somewhat concave medially. Cornea one-third as long as remaining eyestalk.

Antennule and antenna: Ultimate article of antennule 2.9-3.3 × longer than high. Antennal peduncle distinctly overreaching cornea. Article 2 with small lateral spine. Antennal scale relatively broad, overreaching midlength of, but falling short of distal end of article 5. Article 4 with small distomesial spine. Article 5 2.1-2.5 × longer than article 4, distomesially with distinct spine; breadth 0.6 × height of ultimate antennular article. Flagellum of 43-47 segments overreaching P1 merus.

Mxp: Mxp1 with bases close to each other but not contiguous. Mxp3 basis with 4 or 5 proximally diminishing denticles. Ischium half as long as merus, laterally flattish, mesially excavated longitudinally, flexor margin straight, distally not rounded, crista dentata with 27-30 denticles. Merus 2 × longer than ischium, well ridged along flexor margin, mesial face flattish, with long setae. Carpus unarmed.

P1: Relatively massive, barely setose except for fingers, length 5.1 × (male), 4.8 × (female) that of carapace. Ischium dorsally with strong depressed spine, ventrally with well-developed subterminal spine on mesial margin. Merus tuberculose ventrally and mesially (female), with more pronounced tubercles and small spine about at midlength in line with subterminal spine of ischium (male), length 1.1 × that of carapace. Carpus 1.2-1.3 × longer than merus, somewhat tuberculose ventromesially. Palm 2.3 × longer than broad, subequally long as carpus, ridged moderately along lateral margin, sharply along mesial margin. Fingers distally incurved, crossing when closed, somewhat gaping on proximal three-quarters; movable finger about half as long as palm, opposable margin with 2 processes (1 at middle, 1 at proximal portion) interspersed by a few small processes.

P2-4: Relatively slender, unarmed on merus and carpus. Meri equally broad on P2-4; P2 merus subequal to or very slightly longer than P3 merus as well as carapace, 1.4 × longer than P2 propodus; P3 merus 1.2 × length of P3 propodus; P4 merus 0.8-0.9 × length of P3 merus, subequal to length of P4 propodus; length-breadth ratio, 6.1-6.8 on P2, 5.6-6.5 on P3, 4.6-5.4 on P4. Carpi subequal in length on P2 and P3 or slightly shorter (0.9) on P3 than on P2, P4 carpus 0.9 × length of P3 carpus; carpus-propodus length ratio, 0.6-0.7 on P2, 0.6 on P3, 0.5 on P4; carpus-dactylus length ratio, 1.8 on P2, 1.7 on P3, 1.5 on P4. Propodi subequal in length on P2-4, or slightly shorter on P2 than on P3 and P4; flexor margin ending in pair of spines preceded by row of 11-14 spines on nearly entire to distal two-thirds length on P2, 8-11, 8-9 spines on distal half on P3 and P4 respectively. Dactyli about one-third as long as propodi on P2-4; dactylus-carpus length ratio, 0.6 on P2 and P3, 0.7 on P4; flexor margin well curved, setose, with 9-12 sharp, somewhat inclined, loosely arranged spines subequally broad but somewhat smaller proximally; extensor margin fringed with plumose setae at least on median three-fifths of length.

REMARKS — The ornamentation of the carapace and short dactyli of P2-4 displayed by the species are very much like those of *U. brachydactylus* Tirmizi, 1964 (see above). However, *Uroptychus brucei* is readily distinguished from that species by the P1 ischium that bears a well-developed subterminal spine instead of being unarmed on the ventromesial margin; the P2-4 dactyli are not so strongly curved as in *U. brachydactylus*, bearing a fringe of plumose setae along the extensor margin; and the distal two articles of the antennal peduncle each bear a distinct distomesial spine, instead of being unarmed.

Uroptychus buantennatus n. sp.

Figures 50, 51

TYPE MATERIAL — Holotype: **New Caledonia**, Hunter and Matthew Islands. VOLSMAR Stn DW20, 22°21'S, 171°24'E, 460-500 m, 03.VI.1989, with corals of Antipatharia (Hexacorallia), ov. ♀ 3.0 mm (MNHN-IU-2014-16354). Paratypes: same station as for the holotype, same host, 1 ♂ 2.5 mm (MNHN-IU-2014-16355). **New Caledonia**, Norfolk Ridge. NORFOLK 1 Stn DW1704, 23°45'S, 168°16'E, 400-420 m, 25.VI.2001, 1 ♂ 2.5 mm (MNHN-IU-2014-16356).

ETYMOLOGY — From the Latin *bu* (prefix meaning large, huge) plus *antenna* plus *tus* (suffix denoting possession), referring to the relatively large size of the antennal peduncle displayed by the species.

DISTRIBUTION — Hunter-Matthew Islands and Norfolk Ridge; 400-500 m.

DESCRIPTION — Small species. *Carapace*: 1.3 × broader than long; greatest breadth 1.8 × distance between anterolateral spines. Dorsal surface smooth and unarmed, slightly convex from anterior to posterior, without distinct depression in profile. Lateral margins armed with 5 spines, divergent posterolaterally toward last spine and convergent behind it; first anterolateral, medium-sized, more or less close to and barely overreaching lateral orbital spine; second much smaller than first (absent on left side on male paratype), distinctly anterior to midpoint between first and third, third largest, situated at midlength; fourth somewhat smaller than third, about at point one-third from posterior end; last very small or obsolete. Rostrum nearly horizontal, narrow triangular, with interior angle of 15-19°, overreaching distal end of P1 merus, length subequal to postorbital carapace length, breadth less than half carapace breadth measured at posterior carapace margin; dorsal surface concave; lateral margin entire. Lateral orbital spine subequal to or slightly smaller than anterolateral spine. Pterygostomian flap smooth on surface, anteriorly angular, ending in distinct spine.

Sternum: Excavated sternum subtriangular on anterior margin, surface with ridge in midline. Sternal plastron 1.4 × broader than long, lateral extremities posteriorly divergent. Sternite 3 moderately depressed, anterior margin of broad V-shape (medially rounded in male paratype, MNHN-IU-2014-16356), anterolaterally sharply angular. Sternite 4 with anterolateral margin convex, bearing a few small blunt spines anteriorly, posterolateral margin two-thirds length of anterolateral margin. Anterolateral margin of sternite 5 anteriorly convex, about as long as posterolateral margin of sternite 4.

Abdomen: Smooth and glabrous. Somite 1 smooth on surface, feebly convex from anterior to posterior. Somite 2 tergite 2.5-2.8 × broader than long; pleuron anterolaterally blunt angular, lateral margins strongly concave and posteriorly produced to angular point. Pleuron of somite 3 moderately angular on lateral end. Telson about half as long as broad; posterior plate feebly concave on posterior margin, length 1.5 × that of anterior plate.

Eye: 1.7-1.8 × longer than broad, not reaching midlength of rostrum; lateral and mesial margins subparallel. Cornea not dilated, length about half that of remaining eyestalk.

Antennule and antenna: Ultimate article of antennular peduncle 3.3-3.5 × longer than high. Antennal peduncle broad relative to length, article 5 in particular, overreaching midlength of rostrum. Article 2 with sharp distolateral spine. Antennal scale slender, slightly narrower than article 4, reaching distal end of article 5. Article 4 with small distomesial spine. Article 5 broadened distally, with very strong distomesial spine (mesially with small accompanying spine in holotype); greatest breadth 1.5 × that of antennal scale, 1.5 × height of antennular ultimate article; length 1.3-1.4 × that of article 4. Flagellum consisting of 7-8 segments, slightly overreaching distal ends of both rostrum and P1 merus.

Mxp: Mxp1 with bases more or less close to each other but distinctly separated. Mxp3 barely setose on ischium and merus. Basis with a few obsolescent denticles on mesial ridge. Ischium relatively thick, with about 20 denticles on crista dentata, flexor margin not rounded distally. Merus 1.8 × longer than ischium, flexor margin not cristate but rounded, with very small spine at distal quarter of length; distolateral spine small. Carpus with small distolateral spine.

P1: 4.5 × longer than carapace, massive, subcylindrical except for palm and fingers, sparsely bearing short, fine setae along lateral and mesial margins. Ischium dorsally with short blunt spine, ventromesially with no subterminal spine

FIGURE 50

Uroptychus buantennatus n. sp., **A-C**, **E-G**, holotype, ovigerous female 3.0 mm (MNHN-IU-2014-16354); **D**, paratype, male 2.5 mm (MNHN-IU-2014-16356); **E**, paratype, male 2.5 mm (MNHN-IU-2014-16355). **A**, carapace and anterior part of abdomen, proximal part of right P1 included, dorsal. **B**, same, lateral. **C**, anterior part of sternal plastron. **D**, sternal plastron, with excavated sternum and basal parts of Mxp1; **E**, telson. **F**, left antenna, ventral. **G**, left Mxp3, ventral. **H**, same, lateral. Scale bars: 1 mm.

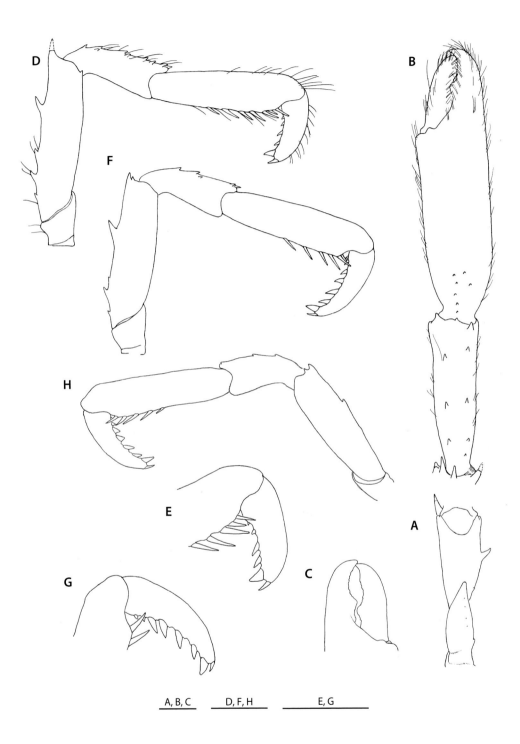

FIGURE 51

Uroptychus buantennatus n. sp., holotype, ovigerous female 3.0 mm (MNHN-IU-2014-16354). **A**, right P1, proximal part, ventral. **B**, same, proximal part omitted, dorsal. **C**, same, distal part, ventral. **D**, right P2, lateral. **E**, same, distal part, setae omitted. **F**, right P3, setae omitted, lateral. **G**, same, distal part, setae omitted, lateral. **H**, left P4, lateral. Scale bars: 1 mm.

(additional small spine at point proximal quarter in holotype); row of very small ventral spines along mesial margin. Merus as long as carapace, armed with 4 terminal spines (2 dorsal and 1 dorsolateral much pronounced, ventromesial smaller and blunt, no ventrolateral) plus strong mesial spine somewhat distal to midlength (2 additional small spines proximal to 2 dorsal terminal on left side in holotype); length subequal to that of carapace. Carpus 1.3 × longer than merus, with 2 rows of dorsal spines. Palm 2.5 × longer than broad, 1.3 × longer than carpus, 2.1-2.5 × longer than movable finger; dorsal surface with several small spines on proximal portion. Fingers broad relative to length, strongly incurved distally, crossing when closed, not gaping; fixed finger having opposable margin with dorsoventrally depressed, low triangular process distal to opposite proximal process of movable finger; movable finger slightly falling short of distal end of fixed finger, length half that of palm.

P2-4: Sparsely setose, meri successively shorter posteriorly (P3 merus 0.91-0.95 × length of P2 merus, P4 merus 0.80-0.87 × length of P3 merus), equally broad on P2 and P3, slightly narrower (0.9) on P4 than on P3; length-breadth ratio, 3.5-3.9 on P2, 3.3-3.6 on P3, 3.0-3.3 on P4; dorsal margin with 4-5 proximally diminishing spines on P2 and P3, 0 or 2 on P4; no spine on ventrolateral margin; P2 merus 0.8 × as long as carapace, about as long as or slightly longer than P2 propodus; P3 merus 0.8-0.9 × length of P3 propodus; P4 merus 0.9 × length of P4 propodus. P3 carpus subequal to or slightly longer than P4 carpus, 0.8-0.9 × length of P2 carpus, carpus-propodus length ratio, 0.54-56 on P2, 0.43-0.45 on P3, 0.42 on P4; extensor margin with row of small spines, 6 on P2, 3 on P3 and P4 (spines obsolescent on P3 and P4 in paratype of MNHN-IU-2014-16355). Propodi subequal; flexor margin somewhat convex around distal quarter, ending in pair of movable spines preceded by 5 (P2) or 3-4 (P3, P4) elongate slender spines at least on distal half. Dactyli somewhat curved, longer on P3 and P4 than on P2; slightly longer than carpi on P2, as long as or slightly longer on P3, subequal to or slightly shorter on P4; flexor margin with 7 or 8 loosely arranged spines, ultimate slender, remaining spines subtriangular, moderately inclined, gradually diminishing toward base of article, penultimate much broader, twice as broad as antepenultimate on P2, less than so on P3 and P4.

Eggs. Eggs carried 4 in number; size, 1.0 mm × 1.1 mm.

REMARKS — The combination of the following characters easily distinguishes the new species from the other members of the genus: the long rostrum is subequal to the length of carapace, and the relatively stout antennal peduncle has article 5 distally noticeably broadened with a strong terminal spine, and the massive P1. The antennal article 5 in the new species is broader than the height of antennular ultimate article as in *U. angustus* n. sp. (see above) but *U. buantennatus* is readily distinguished from that species by having no spine on the carapace dorsal surface, by having the sternite 3 anterior margin V-shaped without median notch, and by having P1-4 broader relative to length, with P1 ischium unarmed on the ventromesial margin.

The species is somewhat similar to *U. triangularis* Miyake & Baba, 1967 in the spination of the carapace and P2-4 meri. In addition to the above-mentioned differences, *U. buantennatus* is distinguished from that species by the P2-4 carpi bearing distinct spines (rarely obsolescent on P3-4) instead of being unarmed on the extensor margin; and the anterior margin of sternite 3 is V-shaped instead of semicircular.

Uroptychus ciliatus (Van Dam, 1933)

Figure 52

Chirostylus ciliatus Van Dam, 1933: 12, figs 17-19.
Uroptychus ciliatus — Baba 2005: 33, fig. 9. — Baba *et al.* 2009: 42, figs 34-35. — Poore *et al.* 2011: 326, pl. 6, fig. F. — McCallum & Poore 2013: 160, figs 5, 12B.

TYPE MATERIAL — Holotype: **Indonesia**, Kur Island, Kei Islands, 204 m, female, (ZMA De. 101.696). [not examined].

MATERIAL EXAMINED — **Indonesia**, Kai Islands. KARUBAR Stn CP16, 5°17'S, 122°50'E, 315-349 m, 24.X.1991, 1 ♂ 7.0 mm (MNHN-IU-2014-16357). – Stn CP25, 5°30'S, 132°52'E, 336-346 m, 26.X.1991, 1 ov. ♀ 7.5 mm (MNHN-IU-2014-16358).

DISTRIBUTION — Western Australia, Kai Islands and Taiwan; in 204-439 m.

DIAGNOSIS — Body and appendages spinose. Carapace broader than long; greatest breadth 1.7 × distance between anterolateral spines; gastric, cardiac and branchial regions well inflated; anterolateral spine overreaching small lateral orbital spine. Rostrum narrow and elongate, with a few to several spines dorsally and marginally, breadth less than one-third carapace breadth measured at posterior carapace margin. Excavated sternum with subtriangular anterior margin, surface with sharp cristate ridge in midline on anterior half. Sternal plastron 0.7 × as long as broad, markedly broadened posteriorly; sternite 3 with shallow V-shaped anterior margin bearing pair of submedian spines; sternite 4 with 2 pairs of spines on anterolateral margin and transverse row of small spines on posterior surface, posterolateral margin slightly shorter than anterolateral margin. Anterolateral margin of sternite 5 with or without small distal spine,

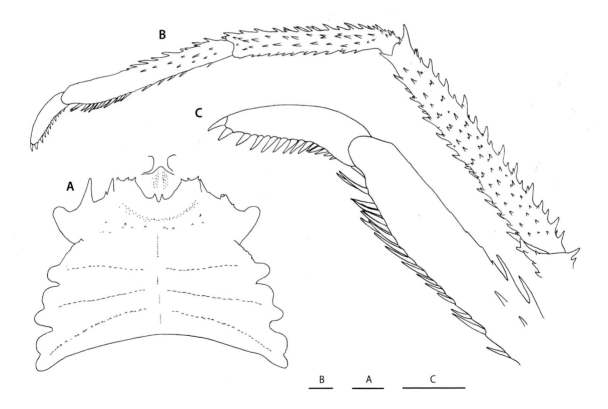

FIGURE 52

Uroptychus ciliatus (Van Dam, 1933), male 7.0 mm (MNHN-IU-2014-16357). **A**, sternal plastron, with excavated sternum and basal parts of Mxp1. **B**, left P2, setae omitted, lateral. **C**, same, distal part, setae omitted, lateral. Scale bars: 1 mm.

distinctly shorter (0.7) than posterolateral margin of sternite 4. Abdominal somite 1 with median spine flanked by small spine; somite 2 tergite 2.5-2.9 × broader than long; pleuron anterolaterally produced, ending in blunt tip, posterolaterally strongly produced and tapering to sharp point; pleuron of somite 3 also tapering; somite 2 with anterior and posterior transverse rows each of 4 spines; somite 3 with 4 anterior and 2 posterior spines, somites 4-5 with 2 anterior and 4 posterior spines; somite 6 more spinose, posterior margin with spines. Protopod of uropod smooth on mesial margin; endopod 1.8-2.0 × longer than broad. Telson 0.6 × as long as broad, posterior plate 1.6 × longer than anterior plate, posterior margin subsemicircular (male) or slightly concave (female). Antennal article 2 with strong distolateral spine; antennal scale slightly falling short of distal end of article 5, with or without a few lateral spines; article 3 with small distolateral spine; article 5 somewhat longer than article 4, with small distolateral (dorsal) and strong distomesial (ventral) spine; article 4 also with strong distomesial spine; flagellum of 16-21 segments falling short of distal end of P1 merus. Mxp1 with bases broadly separated. Mxp3 basis with obsolescent denticles on mesial ridge; ischium with 15-17 small denticles on crista dentata; pronounced distolateral spine on each of ischium, merus and carpus; merus 1.7-1.8 × longer than ischium. Pereopods spinose. P1 merus 1.5-1.6 × carapace length; palm as long as or slightly shorter than carpus, about 5 × longer than broad; movable finger 0.4 × length of palm. P2-3 meri subequal in length and breadth or P3 merus very slightly shorter than P2 merus, P4 merus four-fifths to three-quarters as long as and slightly narrower than P3 merus; P2 merus 1.4 × longer than carapace, 1.4-1.5 × length of P2 propodus; P4 merus 1.1-1.2 × length of P4 propodus; P2-3 carpi subequal, P4 carpus 0.8 × length of P3 carpus; carpus-propodus length ratio, 0.9 on P2, 0.8-0.9 on P3, 0.6-0.7 on P4; carpus-dactylus length ratio, 3.0 on P2-4; flexor margin with pair of terminal spines preceded by row of movable slender spines in close, zigzag arrangement in distal half on P2-3, in distal third on P4, and additional similar spines more remote from one another on remaining proximal portion; dactyli subequal, 0.3 × length of carpi, flexor margin with proximally diminishing spines, ultimate subequal to or slightly larger than penultimate.

Eggs. Number of eggs carried, 20 (excluding exuviae on right pleopods); size, 1.10 mm × 1.10 mm - 1.10 mm × 1.17 mm.

Color. A female from Taiwan was illustrated by Baba *et al.* (2009) and Poore *et al.* (2011), and an ovigerous female from Western Australia by McCallum & Poore (2013).

REMARKS — The species resembles *U. spinirostris* (Ahyong & Poore, 2004) in the spinose body and appendages. In addition to the differences between the two species noted by Baba (2005), *U. ciliatus* is distinguished from that species by the following: the bases of Mxp1 are broadly separated rather than close to each other; and the ultimate of the flexor marginal spines of P2-4 dactyli is subequal to or slightly larger instead of smaller than the penultimate.

Uroptychus ciliatus is also close to *U. quartanus* n. sp. described in this paper. Their relationships are discussed under the account of that new species (see below).

Uroptychus clarki n. sp.
Figures 53, 54

TYPE MATERIAL — Holotype: **New Caledonia**, Norfolk Ridge. NORFOLK 1 Stn DW1697, 24°39'S, 168°38'E, 569-616 m, 24.VI.2001, 1 ov. ♀ 5.5 mm (MNHN-IU-2013-8575).

ETYMOLOGY — Named after Paul F. Clark for his friendship and help in arranging loans of material.

DESCRIPTION — Medium-sized species. *Carapace:* Broader than long (0.7 × as long as broad); greatest breadth 2.0 × distance between anterolateral spines. Dorsal surface smooth, somewhat convex from anterior to posterior, with feeble depression between gastric and cardiac regions; ridged along posterior fifth of lateral margin; epigastric region with pair of short stout spines flanked by 2 small spines placed side by side. Lateral margins concave or weakly constricted at distal third; anterior part (hepatic and anterior branchial margins) gently divergent posteriorly, with 4 spines: first

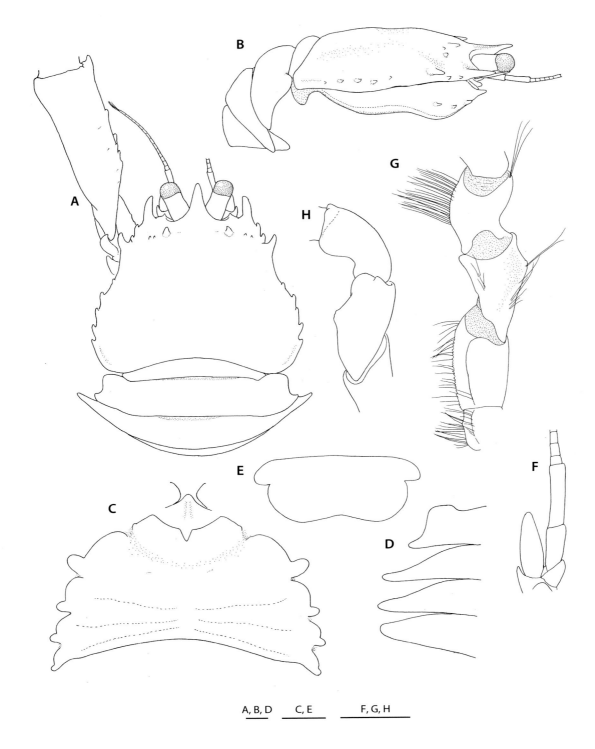

FIGURE 53

Uroptychus clarki n. sp., holotype, ovigerous female 5.5 mm (MNHN-IU-2013-8575). **A**, carapace and anterior part of abdomen, proximal part of left P1 included, dorsal. **B**, same, lateral. **C**, sternal plastron, with excavated sternum and basal parts of Mxp1. **D**, left pleura of abdominal somites 2-5, dorsolateral. **E**, telson. **F**, right antenna, ventral. **G**, left Mxp3, ventral. **H**, same, setae omitted, lateral. Scale bars: 1 mm.

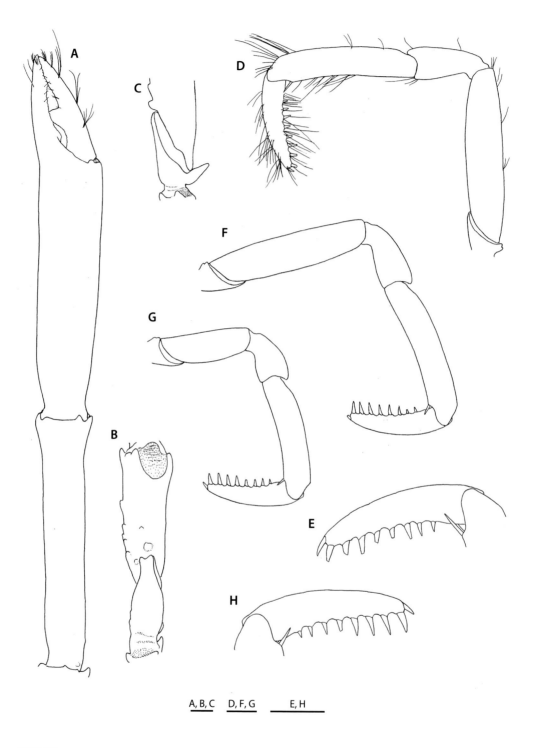

FIGURE 54

Uroptychus clarki n. sp., holotype, ovigerous female 5.5 mm (MNHN-IU-2013-8575). **A**, left P1, dorsal. **B**, same, proximal part, ventral. **C**, same, lateral. **D**, left P2, lateral. **E**, same, distal part, setae omitted, lateral. **F**, right P3, lateral. **G**, right P4, lateral. **H**, same, distal part, lateral. Scale bars: 1 mm.

anterolateral, strong, directed straight forward, reaching cornea, second and third small, situated at level of first in lateral view, directed anteriorly, fourth strong but smaller than first, directed anterolaterally; posterior part (posterior branchial margin) convex, with 5 short spines, first spine remotely separated from fourth spine on anterior part, last spine followed by ridge. Rostrum narrow triangular, with interior angle of 29°, directed slightly dorsally on dorsal surface, strongly so on ventral surface; length 0.8 × breadth, 0.3 × that of remaining carapace; dorsal surface somewhat concave; breadth much less than half carapace breadth measured at posterior carapace margin. Lateral orbital spine small, situated somewhat anterior to level of anterolateral spine. Pterygostomian flap anteriorly angular, produced to strong spine, surface with 3 spines arranged roughly in longitudinal row on anterior part; height of posterior half about 0.4 × that of anterior half.

Sternum: Excavated sternum sharply ridged in midline on surface, anterior margin produced triangularly between bases of Mxp1. Sternal plastron 1.6 × broader than long, lateral extremities convexly divergent behind sternite 4, sternite 7 slightly narrower than sternite 6. Sternite 3 shallowly depressed; anterior margin moderately concave, with V-shaped median notch, with no submedian spines. Sternite 4 short relative to breadth; anterolateral margin convex, anteriorly rounded, length 2.0 × that of posterolateral margin. Sternite 5 with convex anterolateral margin 1.3 × longer than posterolateral margin of sternite 4.

Abdomen: Somites short relative to breadth. Somite 1 well convex from anterior to posterior. Somite 2 tergite 3.4 × broader than long, pleuron well produced posterolaterally. Pleuron of somite 3 tapering to blunt terminus. Telson 0.5 × as long as broad; posterior plate 1.5 × longer than anterior plate, posterior margin distinctly emarginate.

Eye: Elongate, 2.0 × longer than broad, reaching apex of rostrum, feebly narrowed medially. Cornea slightly more than half length of remaining eyestalk.

Antennule and antenna: Ultimate article of antennular peduncle 3.0 × longer than high. Antennal article 2 with strong distolateral spine. Antennal scale proportionately broad, 1.6 × broader than article 5, ending in point proximal third of length of article 5. Article 4 unarmed. Article 5 also unarmed, length 1.4 × that of article 4, breadth 0.5 × height of ultimate article of antennule. Flagellum of 15 segments not reaching distal end of P1 merus.

Mxp: Mxp1 with bases broadly separated. Mxp3 basis with 1 obsolescent denticle on mesial ridge. Ischium having flexor margin distally rounded, crista dentata with about 20 tiny denticles. Merus twice as long as ischium, mesial face flattish; flexor margin roundly ridged on proximal half, sharply ridged on distal half, with blunt eminence distal to midlength; laterally with short blunt distal spine. Carpus unarmed.

P1: 6.7 × longer than carapace, smooth and barely setose except for fingers. Ischium with strong dorsal spine; ventromesial margin with subterminal spine very short and blunt, not reaching distal end of ischium, proximally with obsolescent protuberances. Merus with 2 rows of short blunt spines along mesial margin (2 dorsomesial spines, 5 ventromesial spines), and a few low blunt processes or spines on ventral surface directly distal to ischium; ventral distomesial spine short and blunt; length 1.2 × that of carapace. Carpus 1.5 × length of merus, distally with dorsomesial spine. Palm unarmed, slightly broadened distally, 3.8 × longer than broad. Fingers relatively slender distally, sparingly setose, moderately gaping in proximal three-quarters, not crossing distally; fixed finger directed slightly laterally; movable finger 0.5 × as long as palm, opposable margin with prominent median process (distal margin of process perpendicular to opposable margin) on gaping margin; no longitudinal groove on ventromesial face of fixed finger.

P2-4: Relatively slender, well compressed mesio-laterally, with sparse simple setae, numerous on dactyli. Ischia with 2 short dorsal spines. Meri successively shorter posteriorly (P3 merus 0.9 × length of P2 merus, P4 merus 0.6 × length of P3 merus); subequally broad on P2 and P3, slightly narrower on P4; dorsal margin unarmed; length-breadth ratio, 4.6 on P2, 4.3 on P3, 3.0 on P4; P2 merus 0.8 × as long as carapace, 1.1 × longer than P2 propodus; P3 merus subequal to length of P3 propodus, P4 merus 0.7 × length of P4 propodus. Carpi successively shorter posteriorly, carpus-propodus length ratio, 0.5 on P2, 0.4 on P3, 0.3 and P4. Propodi subequal in length on P2 and P3, shortest on P4; flexor margin slightly concave in lateral view, with pair of slender terminal spines only. Dactyli subequal in length on P2-4; dactylus-carpus length ratio, 1.3 on P2, 1.6 on P3, 2.0 on P4; dactylus-propodus length ratio, 0.6 on P2 and P3, 0.7 on P4; flexor margin slightly curving, with 9 sharp spines proximally diminishing, loosely arranged and nearly perpendicular to margin, ultimate spine shorter and more slender than penultimate, penultimate somewhat broader than antepenultimate.

Eggs. Number of eggs carried, 38; size, 0.97 × 1.12 mm - 1.08 × 1.10 mm.

REMARKS — The carapace shape including the presence of a pair of epigastric spines, the pterygostomian flap with spines on the anterior surface and the spination of P2-4, especially the arrangement of spines on dactyli, are similar in *U. clarki* n. sp., *U. corbariae* n. sp., *U. defayeae* n. sp., *U. mesodme* n. sp., *U. paraplesius* n. sp. and *U. trispinatus* n. sp. These species also share a well depressed cephalothorax especially in the posterior half: the pterygostomian flap is posteriorly lowered, the height of posterior half 0.4-0.5 times that of anterior half. The first three species differ from the latter three species in having the carapace lateral margin with irregular spines (lacking three strong spines) on the anterior portion and in having the antennal article 4 unarmed and the antennal scale barely reaching the midlength of article 5. Characters to distinguish *U. clarki* from *U. corbariae* and *U. defayeae* are outlined in the accounts of the latter two species (see below).

Uroptychus convexus Baba, 1988

Figure 55

Uroptychus convexus Baba, 1988: 32, fig. 12. — Poore *et al.* 2011: 328, pl. 6, fig. G.

TYPE MATERIAL — Holotype: **Indonesia**, between Cebu and Bohol, 265 m, female (USNM 150320). [not examined].

MATERIAL EXAMINED — Indonesia, Kai Islands. KARUBAR Stn DW13, 5°26'S, 132°38'E, 417-425 m, 24.X.1991, 2 ♂ 2.5, 2.9 mm, 1 ♀ 2.7 mm (MNHN-IU-2014-16359). – Stn DW18, 5°18'S, 133°01'E, 205-212 m, 24.X.1991, 3 ♂ 2.3-2.4 mm, 4 ♀ 2.2-2.7 mm (MNHN-IU-2014-16360).

DISTRIBUTION — Philippines between Cebu and Bohol, and now Kai Islands; in 205-425 m.

SIZE — Males, 2.3-2.8 mm; females, 2.2-2.7 mm.

DIAGNOSIS — Small species. Cephalothorax relatively high. Carapace broader than long (0.8 × as long as broad), greatest breadth 1.5 × distance between anterolateral spines; dorsal surface unarmed, strongly convex from side to side and anterior to posterior, without distinct border between gastric and cardiac regions. Lateral margins weakly convexly divergent posteriorly, with 7 spines: first anterolateral, overreaching lateral orbital spine; second situated at anterior end of anterior branchial region, somewhat posterior to midpoint between first and third; third to seventh situated on posterior branchial region, posteriorly diminishing, third largest. Rostrum short, less than half length of remaining carapace, broad, equilateral triangular, with interior angle of 45°, dorsally concave; breadth less than half carapace breadth measured at posterior carapace margin. Lateral orbital spine small, very slightly anterior to level of anterolateral spine. Pterygostomian flap anteriorly angular, ending in sharp spine, unarmed on surface. Excavated sternum with convex anterior margin, surface with weak ridge in midline; sternal plastron as long as broad, lateral extremities slightly divergent posteriorly; sternite 3 moderately depressed, anterior margin gently concave, with deep, broad median notch separating obsolescent submedian spines, anterolaterally sharp angular; sternite 4 having anterolateral margin convex, anteriorly bearing a few small spines, length 1.3-1.7 × that of posterolateral margin. Anterolateral margin of sternite 5 anteriorly strongly convex, about as long as or slightly longer than posterolateral margin of sternite 4. Abdominal somite 2 tergite 2.5-2.6 × broader than long; pleuron posterolaterally rounded, lateral margins slightly convergent posteriorly; pleuron of somite 3 laterally blunt. Telson half as long as broad; posterior plate 1.0-1.4 × longer than anterior plate, emarginate on posterior margin. Eyes 2 × longer than broad, proximally broad, distally narrowed, reaching or overreaching rostral tip; cornea about half as long as remaining eyestalk or less than so. Ultimate article of antennular peduncle 2.7-2.9 × longer than high. Antennal article 2 without lateral spine; antennal scale overreaching midlength of but never reaching distal end of article 5; article 4 with ventral distomesial spine; article 5 unarmed, 1.3-1.6 × longer than article 4, breadth about half height of ultimate article of antennule; flagellum of 9-10 segments nearly or barely reaching distal end of P1 merus. Mxp1 with bases broadly separated. Mxp3 basis with a few obsolescent denticles on mesial ridge; ischium with distally rounded flexor margin, crista dentata with about 25 denticles; merus 2.0 × longer than ischium, flexor margin sharply ridged along distal two-thirds, bearing a few (usually 2) denticle-like small spines distal to midlength. P1 massive, sparsely setose; ischium with sharp curved dorsal spine, ventromesially with distinct subterminal spine; merus subequal to or slightly shorter than carapace, with a few well-developed mesial and ventral spines; carpus as long as merus; palm 1.8-2.1 × longer than broad, 1.1-1.3 × longer than carpus; fingers distally incurved, crossing when closed, opposable margins not spooned. P2-4 meri and carpi relatively thick mesio-laterally, dorsal margin with several small denticle-like spines distinct on P2, obsolescent on P3, obsolete on P4; P2 merus 0.7-0.9 × length of carapace, subequal in length and breadth to P3 merus, as long as P2 propodus; P4 merus 0.9-1.0 × length of, 0.7-0.8 × breadth of P3 merus, 0.8-0.9 × length of P4 propodus; carpi subequal, each less than half length of propodus; propodi subequal in length on P3 and P4, shorter on P2; flexor margin with pair of distal spines only; dactyli slightly shorter than or subequal to carpus on P2-4, slightly less than half as long as propodi, ending in slender spine preceded by 4-5 somewhat obliquely directed, loosely arranged spines on straight flexor margin, penultimate and antepenultimate strong, remaining proximal spines slender.

Color. A specimen in color from the Philippines was illustrated by Poore *et al.* (2011).

REMARKS — The specimens agree well with the type of *U. convexus.* Ovigerous females have not been collected.

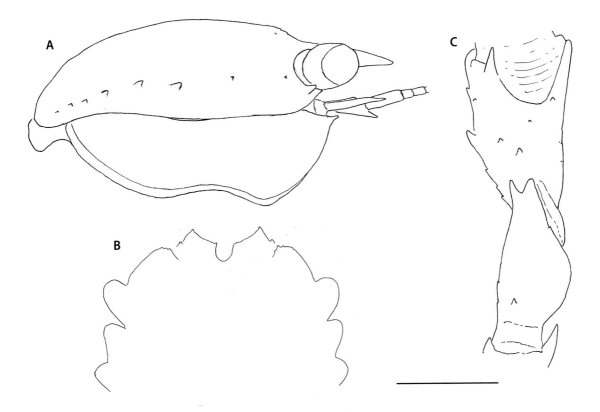

FIGURE 55

Uroptychus convexus Baba, 1988, male 2.9 mm (MNHN-IU-2014-16359). **A**, carapace, lateral. **B**, anterior part of sternal plastron. **C**. left P1, proximal part, ventral. Scale bar: 1 mm.

Uroptychus corbariae n. sp.
Figures 56, 57

TYPE MATERIAL — Holotype: **Indonesia**, Kai Islands. KARUBAR Stn CP19, 5°15'S, 133°01'E, 605-576 m, on gorgonian coral, 25.X.1991, ♂ 5.3 mm (MNHN-IU-2014-16361). Paratype: Collected with holotype, 1 ov. ♀ 5.4 mm (MNHN-IU-2014-16362).

DISTRIBUTION — Kai Islands, Indonesia, 605-576 m.

ETYMOLOGY — Dedicated to Laure Corbari of MNHN for her help and support.

DESCRIPTION — Medium-sized species. *Carapace*: Broader than long (0.8-0.9 × as long as broad); greatest breadth 1.4-1.9 × distance between anterolateral spines. Dorsal surface smooth, somewhat convex from anterior to posterior, feebly depressed between gastric and cardiac regions, bearing pair of low blunt epigastric spines; ridged along posterior third of lateral margin. Lateral margins gently constricted at anterior third; anterior part (anterior to constrictions) gently divergent posteriorly, with 2 well-developed spines, first anterolateral spine directed straight forward or slightly anteromesially, overreaching base of antennal scale, second short, situated at anterior end of anterior branchial region, nearly smooth without distinct spine between first and second; posterior part (posterior to constrictions or posterior branchial margin) gently convex, bearing short blunt spine at anterior end of posterior branchial margin or at midlength of carapace lateral margin, followed by obsolescent crenulations. Rostrum narrow triangular, directed somewhat dorsally or straight horizontal, with interior angle of 21°; slightly longer (male) or slightly shorter (female) than broad; dorsal surface concave; length 0.4 × (male) or 0.3 × (female) that of remaining carapace; breadth much less than half carapace breadth measured at posterior carapace margin. Lateral orbital spine small, slightly anterior to level of anterolateral spine. Pterygostomian flap anteriorly angular, produced to strong spine, with 3-5 small spines on anterior surface; height of posterior half slightly more than half that of anterior half.

Sternum: Excavated sternum with sharp ridge in midline, anterior margin produced triangularly between bases of Mxp1. Sternal plastron 1.4 or 1.5 × broader than long, lateral extremities convexly divergent behind sternite 4, sternite 7 slightly narrower than sternite 6. Sternite 3 shallowly depressed; anterior margin moderately concave, with deep, narrow U-shaped median notch. Sternite 4 broad relative to length; anterolateral margin anteriorly convex, length 1.4-1.6 × that of posterolateral margin. Sternite 5 with convex anterolateral margin as long as posterolateral margin of sternite 4.

Abdomen: Somites short relative to breadth. Somite 1 with antero-posteriorly convex transverse ridge. Somite 2 tergite 2.9 × longer than broad, pleuron posterolaterally produced more bluntly on male than on female. Pleura of somites 3-4 laterally more rounded in male than in female. Telson 0.4 × as long as broad; posterior plate 1.3-1.6 × longer than anterior plate, posterior margin slightly emarginate.

Eye: Elongate, 1.8 × longer than broad, reaching distal quarter of rostrum, slightly narrowed medially. Cornea slightly dilated, much more than half length of remaining eyestalk.

Antennule and antenna: Ultimate article of antennular peduncle 3.1-3.3 × longer than high. Antennal peduncle reaching or overreaching apex of rostrum. Article 2 with short distolateral spine. Antennal scale proportionately broad, distally rounded or blunt, 1.5 × broader than article 5, slightly overreaching article 4 or reaching proximal third of length of article 5. Article 4 unarmed. Article 5 also unarmed, length 1.4 × that of article 4, breadth 0.5 × height of ultimate article of antennule. Flagellum of 14 segments not reaching distal end of P1 merus.

Mxp: Mxp1 with bases broadly separated. Mxp3 basis without denticle on mesial ridge. Ischium with flexor margin distally rounded, crista dentata with about 20 minute (obsolescent) denticles. Merus twice as long as ischium, mesial face flattish; flexor margin roundly ridged on proximal half, sharply ridged on distal half, bearing 1 or 2 very short, blunt spines on distal third; lateral face with short blunt distal spine. Carpus unarmed.

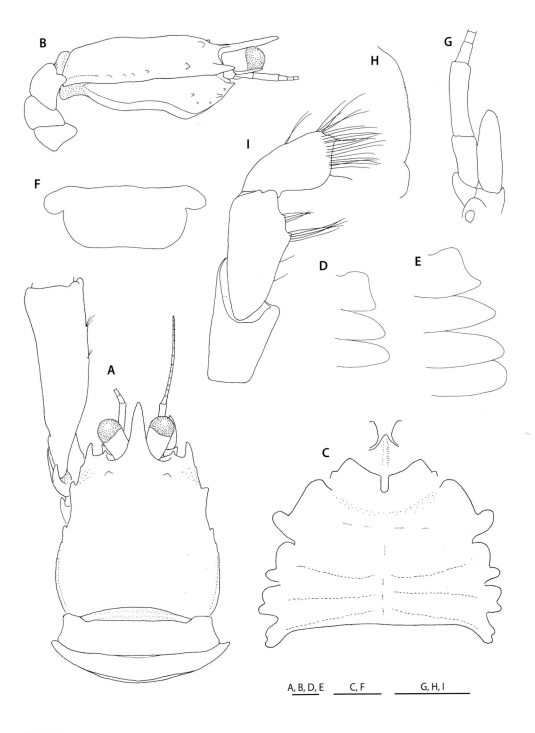

FIGURE 56

Uroptychus corbariae n. sp., **A-D**, **F-I**, holotype, male 5.3 mm (MNHN-IU-2014-16361); **E**, paratype, ovigerous female 5.4 mm (MNHN-IU-2014-16362). **A**, carapace and anterior part of abdomen, proximal part of left P1 included, dorsal. **B**, same, lateral. **C**, sternal plastron, with excavated sternum and basal parts of Mxp1. **D**, right pleura of abdominal somites 2-4, dorsolateral. **E**, right pleura of abdominal somites 2-5, dorsolateral. **F**, telson. **G**, left antenna, ventral. **H**, right crista dentata, ventral. **I**, right Mxp3, lateral. Scale bars: 1 mm.

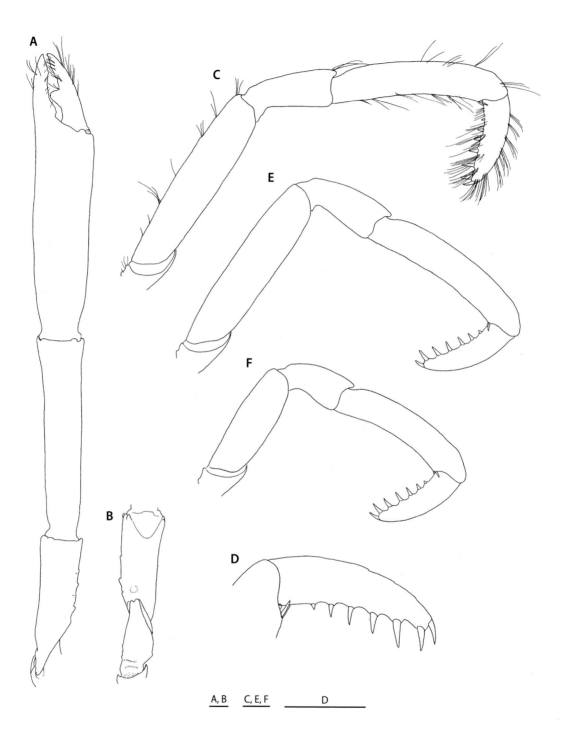

FIGURE 57

Uroptychus corbariae n. sp., holotype, male 5.3 mm (MNHN-IU-2014-16361). **A**, left P1, dorsal. **B**, same, proximal part, ventral. **C**, right P2, lateral. **D**, same, distal part, setae omitted, lateral. **E**, right P3, setae omitted, lateral. **F**, right P4, setae omitted, lateral. Scale bars: 1 mm.

P1: 5.3 × (female), 6.4 (male) longer than carapace, smooth and barely setose except for fingers. Ischium with strong dorsal spine with or without small accompanying spine proximally; ventromesial margin with subterminal spine very short and blunt, not reaching distal end of ischium, proximally with obsolescent protuberances. Merus with 2-4 obsolescent, low, blunt spines along mesial margin, blunt distomesial spine on dorsal surface, and blunt distomesial spine on ventral surface; length 1.3 or 1.4 × that of carapace. Carpus 1.4 × length of merus, with no spine laterally and mesially. Palm unarmed, depressed, 0.6 × as high as broad, lateral and mesial margins subparallel, somewhat narrowed proximally; 4.4-3.5 × as long as broad. Fingers relatively slender distally, sparingly setose, moderately gaping in proximal two-thirds in male, not gaping in female, distally not clearly crossing; fixed finger directed laterally in male, not distinctly so in female; movable finger 0.4 × as long as palm, opposable margin with prominent process (distal margin of process perpendicular to opposable margin in male) fitting into narrow longitudinal groove on opposite ventromesial face of fixed finger when closed.

P2-4: Relatively slender, well compressed, sparsely with simple setae, more setose on dactyli. Ischia with 2 short dorsal spines. Meri subequal in length on P2 and P3 or very slightly shorter on P3 than on P2, shortest on P4 (P4 merus 0.6 × length of P3 merus); subequally broad on P2 and P3, very slightly narrower on P4; dorsal margin unarmed; length-breadth ratio, 4.3 (male) or 4.8 (female) on P2, 4.3 on P3, 2.9 (male) or 3.1 (female) on P4; P2 merus 0.9 × (male) or 0.8 × (female) length of carapace, subequal to or very slightly longer than P2 propodus; P3 merus subequal to length of P3 propodus. P4 merus 0.7 × (female) or 0.8 × (male) length of P4 propodus. Carpi subequal in length on P2 and P3 (female) or slightly longer on P2 than on P3 (male), shortest on P4 (P4 carpus 0.7 × (female) or 0.8 × (male) length of P3 carpus), 0.5 × length of propodus on P2, 0.4 × (male) or 0.5 × (female) on P3, slightly less than 0.4 × on P4. Propodi subequal on P2 and P3, shortest on P4; flexor margin slightly concave in lateral view, with pair of slender terminal spines only. Dactyli subequal in length on P2-4; dactylus-carpus length ratio, 1.0 on P2, 1.0-1.1 on P3, 1.4-1.5 on P4; dactylus-propodus length ratio, 0.5 on P2 and P3, 0.5-0.6 on P4; flexor margin slightly curving, with 7 sharp spines loosely arranged, proximally diminishing and nearly perpendicular to margin, ultimate spine slightly more slender than penultimate.

Eggs. Number of eggs carried, 4; size, 1.08 × 1.46 mm - 1.25 × 1.42 mm.

REMARKS — *Uroptychus corbariae* resembles *U. clarki* n. sp. and *U. defayeae* n. sp. (see their similarities under the account of *U. clarki*).

Uroptychus corbariae is distinguished from *U. clarki* by lacking spines between the first and second strong spines on the carapace lateral margin; the posterior branchial margin is crenulated (at most bearing a small spine) instead of bearing 4 or 5 spines; the epigastric region bears a pair of spines not flanked by two small spines; the antennal article 2 has a small distolateral spine, not pronounced as in *U. clarki*. The relationships with *U. defayeae* are discussed under the remarks of that species (see below).

The combination of the carapace lateral marginal spines restricted to the anterior half and the ultimate of the P2-4 dactylar spines more slender than the penultimate links this species to *U. joloensis* Van Dam, 1939 and *U. zezuensis* Kim, 1972. However, *Uroptychus corbariae* is more distant from both than from the above-mentioned five congeners in having a pair of blunt, distinct epigastric spines and in having the penultimate and antepenultimate of the P2-4 dactylar spines relatively slender, not so strong as in *U. joloensis* and *U. zezuensis*.

Uroptychus crassipes Van Dam, 1939

Figure 58

Uroptychus crassipes Van Dam, 1939: 392, fig. 1. — Baba 1988: 35. — Baba 2005: 35, 225. — Baba *et al.* 2008: 31.

TYPE MATERIAL — Holotype: **Indonesia**, Kei Islands, 5°29′S, 132°27′E, 290 m, male (ZMUC-CRU 6124). [not examined].

MATERIAL EXAMINED — **Indonesia**, Kai Islands. KARUBAR Stn CP09, 5°23′S, 132°29′E, 368-389 m, 23.X.1991, 5 ♂ 4.0-8.5 mm, 4 ov. ♀ 6.1-9.0 mm, 2 ♀ 5.0, 6.0 mm (MNHN-IU-2014-16363). – Stn CP16, 5°17′S, 132°50′E, 315-349 m, 24.X.1991, 1 ♂ 7.2 mm, 1 ov. ♀ 6.8 mm (MNHN-IU-2014-16364). Tanimbar Islands. KARUBAR Stn CP69, 8°42′S, 131°53′E, 356-368 m, 2.XI.1991, 2 ♂ 5.7, 6.1 mm, 1 ov. ♀ 6.6 mm, 1 ♀ 7.0 mm (MNHN-IU-2014-16365). – Stn CP83, 09°23′S, 131°00′E, 285-297 m, 4.XI.1991, 1 ov. ♀ 4.7 mm (MNHN-IU-2014-16366). **Philippines**. MUSORSTOM 2 Stn CP36, 13°31′N, 121°24′E, 569-595 m, 24.XI.1980, 3 ov. ♀ 4.9-5.4 mm (MNHN-IU-2014-16367). – Stn CP46, 13°27′N, 122°18′E, 445-520 m, 26.XI.1980, 2 ♀ 3.8, 5.3 mm (MNHN-IU-2014-16368).

DISTRIBUTION — Kai and Tanimbar Islands (Indonesia), and off eastern Mindoro (Philippines); 285-595 m.

SIZE — Males, 4.0-8.5 mm, females, 3.8-9.0 mm; ovigerous females from 4.7 mm.

DIAGNOSIS — Medium-sized species. Body and pereopods covered with fine setae, especially long on pereopods. Carapace 0.8 × as long as broad, greatest breadth 1.7 × distance between anterolateral spines; lateral margins convex, bearing 7 spines: first anterolateral, relatively small, slightly overreaching lateral orbital spine; second and third much smaller than first; third occasionally obsolete; fourth to seven acute, situated on branchial region. Rostrum sharp triangular, with interior angle of 20-25°, breadth less than half carapace breadth measured at posterior carapace margin; dorsal surface concave, lateral margin with small subapical spine on each side. Pterygostomian flap anteriorly angular, produced to sharp spine. Excavated sternum having anterior margin convex, surface with low longitudinal ridge in midline; sternal plastron slightly shorter than broad, sternites successively broader posteriorly; sternite 3 strongly depressed, anterior margin shallowly concave, with pair of submedian spines separated by narrow or relatively broad notch; sternite 4 with nearly straight or slightly concave anterolateral margin about 1.5 × longer than posterolateral margin. Anterolateral margin of sternite 5 anteriorly produced bluntly or roundly, length distinctly more than that of posterolateral margin of sternite 4. Abdominal somite 1 with antero-posteriorly convex setiferous transverse ridge. Somite 2 tergite 2.6 × broader than long; pleuron posterolaterally blunt angular, lateral margin slightly concave and somewhat divergent posteriorly. Pleuron of somite 3 bluntly angular on posterolateral margin. Telson about half as long as broad; posterior plate as long as anterior plate, feebly concave or moderately emarginate on posterior margin. Eyes elongate (2.0-2.6 × longer than broad), slightly overreaching midlength of rostrum, mesial and lateral margins subparallel; cornea not dilated, length more than half that of remaining eyestalk. Ultimate article of antennular peduncle 3.7-4.0 × longer than high. Antennal article 2 with small distolateral spine; antennal scale overreaching distal end of article 5, reaching subapical spine of rostrum; articles 4 and 5 each with strong distomesial spine, article 5 1.5-1.7 × longer than article 4, breadth 0.7-0.8 × height of ultimate article of antennule, flagellum of 20-24 segments distinctly overreaching distal end of P1 merus. Mxp1 with bases broadly separated. Mxp3 basis having convex mesial ridge without denticle; ischium with distally rounded flexor margin, crista dentata with more than 30 denticles; merus and carpus each with distinct distolateral spine; merus 1.9 × longer than ischium, flexor margin with a few small spines on distal third; carpus with distolateral spine. P1 massive, with short transverse ridges supporting setae; ischium with 1 strong dorsal and 1 ventromesial subterminal spine; merus as long as or slightly longer than carapace, with terminal (including 2 dorsal terminal) and mesial marginal spines; carpus with terminal (including 2 dorsal terminal) spines, length subequal to that of merus and 0.8-0.9 × that of palm; palm 2.1-2.6 × longer than broad; fingers 0.5-0.6 × length of palm, distally crossing, opposable margins sinuous, proximal process of

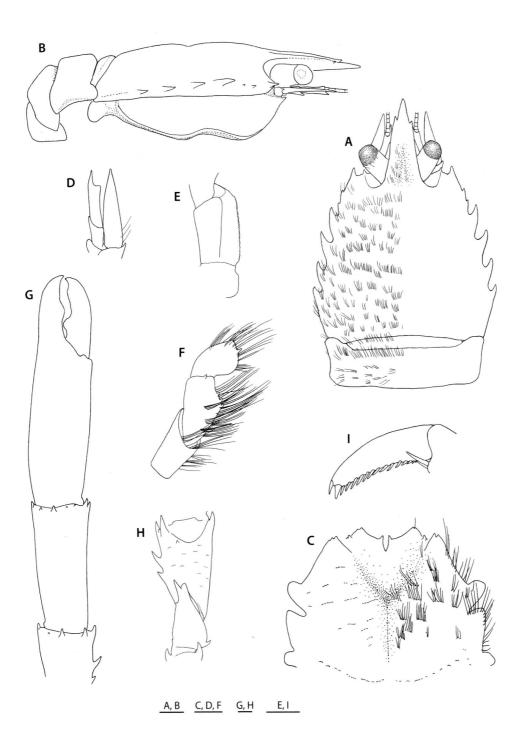

FIGURE 58

Uroptychus crassipes Van Dam, 1939; **A**, **B**, male 6.1 mm (MNHN-IU-2014-16365); **C-F**, **I**, male 8.5 mm (MNHN-IU-2014-16363); **G**, **H**, male 7.2 mm (MNHN-IU-2014-16363). **A**, carapace and anterior part of abdomen, dorsal. **B**, same, setae omitted, lateral. **C**, anterior part of sternal plastron. **D**, left antenna, ventral. **E**, right Mxp3, setae omitted, ventral. **F**, same, lateral. **G**, left P1, setae (and supporting ridges) omitted, dorsal. **H**, same, proximal part, setae omitted, ventral. **I**, left P2, distal part, setae omitted, lateral. Scale bars: 1 mm.

movable finger fitting to between 2 low eminences of fixed finger. P2-4 with setiferous short ridges like those on P1, meri somewhat compressed mesio-laterally, ventrolaterally bearing distolateral spine; successively slightly shorter posteriorly or often subequal on P2-3; P2 merus 0.7-0.8 × length of carapace, slightly longer than P2 propodus; P4 merus 0.8-0.9 × as long as and 0.9 × as broad as P3 merus, 0.8 × length of P4 propodus; carpi successively slightly shorter posteriorly, half as long as propodus on P2, nearly half or slightly less than half as long on P3, much less than so on P4; extensor margin with terminal spine accompanied with another small spine lateral to it; propodal flexor margin with pair of terminal spines preceded by 1-5 spines on P2, 0-4 on P3 and P4; dactyli longer than carpi (dactylus-carpus length ratio, 1.1-1.3 on P2, 1.2-1.5 on P3, 1.4-1.5 on P4), flexor margin with obliquely directed spines close to one another and obscured by dense setae, penultimate prominent, 2 × as broad as antepenultimate, remainder slender, ultimate subequal to antepenultimate.

Eggs. Up to about 60 eggs carried; size, 0.86 mm × 0.98 mm - 0.85 mm × 1.00 m.

REMARKS — The specimens from the Kai Islands bear setiferous short ridges on the carapace dorsal surface so as to resemble *U. micrommatus* n. sp. that was taken together at KARUBAR Station CP09. However, the following characteristics are distinctive of *U. crassipes*: the rostrum is strongly excavated along the dorsal midline, bearing a distinct subterminal spine on each side; the anterior margin of sternite 3 has a pair of distinct submedian spines; the P1 ischium bears a stronger subterminal spine on the ventromesial margin, and the merus and carpus bear terminal spines on the dorsal side; and the eyes are subcylindrical, elongate and uniform in breadth. *Uroptychus crassipes* also resembles *U. occultispinatus* Baba, 1988, and *U. worrorra* McCallum & Poore 2013. *Uroptychus crassipes* differs from *U. occultispinatus* in having small but distinct spines on the distodorsal margins of the P1 merus and carpus, in having sharper spines on the branchial lateral margin, and in having a well-developed subterminal spine on the ventromesial margin of the P1 ischium. *Uroptychus worrorra* is characterized by having five branchial lateral spines (four in *U. crassipes*), smooth lateral rostral margins (bearing a subapical spine in *U. crassipes*), P1 slender rather than massive, with the carpus longer than the palm (shorter in *U. crassipes*), and the P2 merus bearing a row of small dorsal spines (unarmed in *U. crassipes*).

Uroptychus ctenodes n. sp.
Figures 59, 60

TYPE MATERIAL — Holotype: **Indonesia**, Tanimbar Islands. KARUBAR Stn CP45, 7°54'S, 132°47'E, 302-305 m, 29.X.1991, ov. ♀ 4.6 mm (MNHN-IU-2014-16369).

ETYMOLOGY — From the Greek *ktenodes* (comb-like), alluding to a comb-like arrangement of spines on the lateral margin of the carapace and pterygostomian flap, characteristic of the species.

DISTRIBUTION — Tanimbar Islands, Indonesia; 302-305 m.

DESCRIPTION — Small species. *Carapace*: Strongly broadened posteriorly, 1.7 × broader than long (0.6 × as long as broad); greatest breadth 2.1 × distance between anterolateral spines. Dorsal surface convex on gastric and cardiac regions, covered with very fine short setae discernible under high magnification; posterior half laterally crested. Lateral margins nearly straight and strongly divergent posteriorly in anterior half; anterolateral spine strong, extending forward to level of rostral tip; posterior half convexly convergent posteriorly, with row of 7 or 8 strong blunt spines, all anterolaterally directed. Rostrum broad triangular, as long as broad, with interior angle of 73°, horizontal, flattish on dorsal surface, not reaching distal end of eye; length 0.2 × that of remaining carapace, breadth less than half carapace breadth measured at posterior carapace margin. Lateral limit of orbit rounded, not angular, directly leading to anterolateral spine. Pterygostomian flap low in posterior half (posterior half about 0.3 × as high as anterior half), anteriorly somewhat angular, produced to small

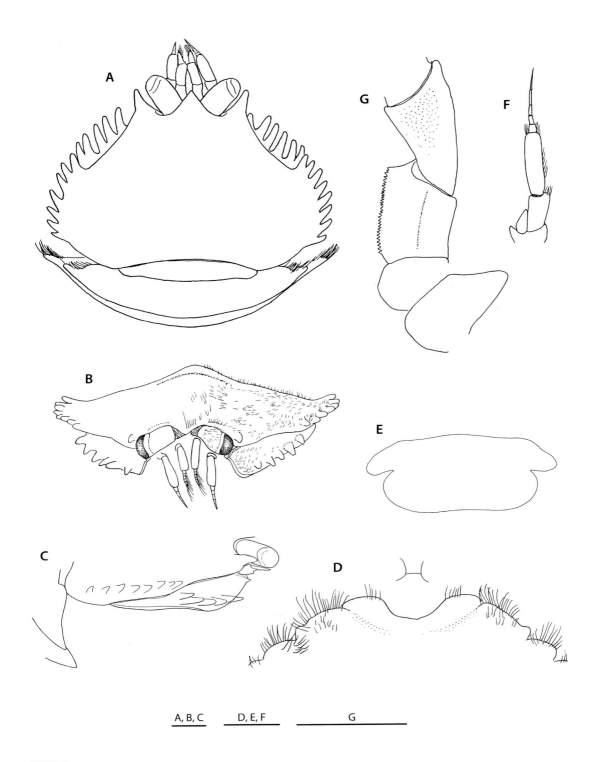

A, B, C   D, E, F   G

FIGURE 59

Uroptychus ctenodes n. sp., holotype, ovigerous female 4.6 mm (MNHN-IU-2014-16369). **A**, carapace and anterior part of abdomen, fine setae omitted, dorsal. **B**, same, fine setae on right side omitted, anterior. **C**, same, dorsal part omitted, lateral. **D**, anterior part of sternal plastron, with excavated sternum and basal parts of Mxp1. **E**, telson. **F**, right antenna, ventral. **G**, left Mxp3, ventral. Scale bars: 1 mm.

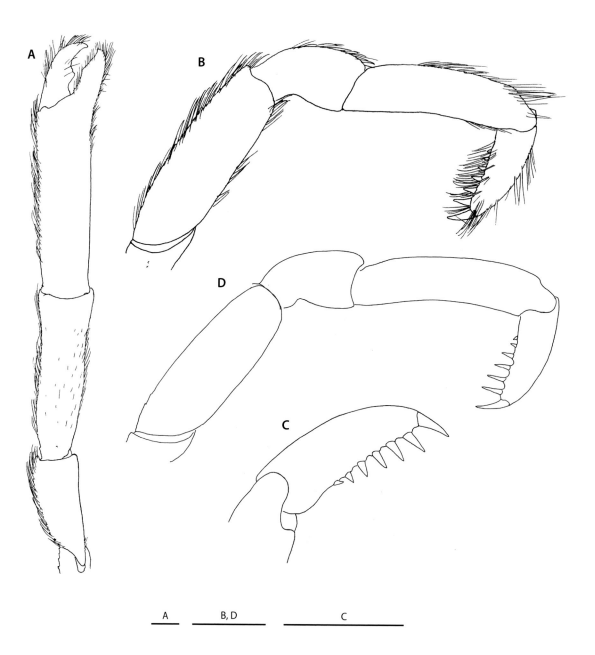

FIGURE 60

Uroptychus ctenodes n. sp., holotype, ovigerous female 4.6 mm (MNHN-IU-2014-16369). **A**, right P1, dorsal. **B**, right P2, lateral. **C**, same, distal part, setae omitted, lateral. **D**, right P3, setae omitted, lateral. Scale bars: 1 mm.

but distinct spine; surface with row of 5 or 6 strong blunt spines arranged along lower margin: in dorsal view, this row in line with carapace lateral spines representing comb-like arrangement.

Sternum: Excavated sternum with anterior margin nearly transverse between bases of Mxp1, surface strongly excavated, without median ridge and central spine. Sternal plastron strongly depressed posteriorly in ventral view, setose laterally; lateral margins of somites 4-6 strongly divergent. Sternite 3 shallowly depressed; anterior margin deeply excavated, representing semicircular shape, without median notch and submedian spines, laterally rounded. Sternite 4 with convex anterolateral margin, posterolateral lobe not pronounced. Anterolateral margins of sternite 5 convex and strongly divergent posteriorly, length much more than that of posterolateral margin of sternite 4.

Abdomen: Smooth, covered with short fine setae. Somite 1 smoothly convex from anterior to posterior, without elevated transverse ridge. Somite 2 tergite 3.1 × broader than long; pleuron anterolaterally rounded, posterolaterally tapering, lateral margin concave and strongly divergent posterolaterally. Pleuron of somites 3 also tapering. Telson 0.4 × as long as broad, posterior plate feebly concave on posterior margin, length 1.2 × that of anterior plate.

Eye: Elongate, 2 × longer than broad, with lateral and mesial margins subparallel. Cornea not dilated, length about half that of remaining eyestalk.

Antennule and antenna: Ultimate article of antennular peduncle 2.8 × longer than high. Antennal peduncle extending far beyond eye. Article 2 with stout distolateral spine. Antennal scale slightly more than half as broad as article 5, slightly overreaching midlength of article 4. Articles 4 and 5 unarmed, with soft fine setae on distal portion; article 5 1.8 × longer than article 4, breadth 0.9 × height of antennular ultimate article. Flagellum consisting of 5 segments, very short, nearly as long as article 5.

Mxp: Mxp1 with bases broadly separated. Mxp3 with relatively long soft setae particularly on lateral surface of merus and distal part of lateral surface of ischium. Coxa ventrolaterally produced to blunt, depressed process. Basis without denticles on mesial ridge. Ischium with about 30 small denticles on crista dentata; flexor margin not rounded distally. Merus 2.2 × longer than ischium, flexor margin not cristate but roundly ridged. Carpus unarmed.

P1: 4 × longer than carapace; tubercle-like spines on proximal mesial portion of merus and along ventromesial margin of ischium. Ischium with small blunt dorsal spine, unarmed ventrally and ventromesially. All articles with short fine setae thick along mesial margin. Merus 0.8 × length of carapace. Carpus 1.5 × longer than merus. Palm 3.4 × longer than broad, 1.2 × longer than carpus. Fingers directed anterolaterally, relatively broad and moderately depressed; movable finger distally incurved, length 0.4 × that of palm, opposable margin with triangular (dorsal view) median process; opposable margin of fixed finger with eminence distal to position of opposite process on movable finger.

P2-4: P2-P3 relatively thick mesio-laterally, with soft fine setae. P4 missing. Meri slightly shorter on P3 than on P2, subequal in breadth on P2 and P3, length-breadth ration, 3.0 on P2, 2.8 on P3; dorsal margin proximally with 2 small tubercle-like spines only discernible under high magnification; P2 merus 0.6 × length of carapace, slightly longer than P2 propodus; P3 merus as long as P3 propodus. Carpi subequal in length on P2 and P3, slightly less than half length of propodi (carpus-propodus length ratio, 0.45 on P2, 0.43 on P3), nearly as long as dactyli on P2 and P3. Propodi somewhat longer on P3 than on P2; flexor margin unarmed. Dactyli slightly longer on P3 than on P2, with straight flexor margin ending in strong spine preceded by 6 loosely arranged, sharp, proximally diminishing spines nearly perpendicular to margin (proximal-most spine obliquely directed), proximal spines obscured by setae.

Eggs. Eggs carried 22 in number; size, 1.16 mm in diameter.

REMARKS — The species is unique in *Uroptychus* in the following particulars: the anterolateral spine is extremely strong, situated directly lateral to the unarmed lateral limit of the orbit; the dorsal surface of the carapace is crested laterally in the posterior half; the antenna bears a very small antennal scale much narrower than article 5 and a very short, five-segmented flagellum as long as article 5, and the pterygostomian flap is very low in the posterior half and bears a row of spines that is, when viewed dorsally, continued on to the row of lateral carapace spines. This species could be placed in a different genus but it stays in *Uroptychus* awaiting a discovery of additional material that allows molecular analyses.

Uroptychus cylindropus n. sp.

Figures 61, 62

TYPE MATERIAL — Holotype: **New Caledonia**, Norfolk Ridge. SMIB 8 Stn DW156, 24°46'S, 168°08'E, 275-300 m, 28.I.1993, ov. ♀ 5.3 mm (MNHN-IU-2014-16370). Paratypes: **New Caledonia**, Loyalty Islands. MUSORSTOM 6 Stn DW483, 21°19.80'S, 167°47.80'E, 600 m, 23.II.1989, 1 ♂ 4.2 mm (MNHN-IU-2014-16371). **New Caledonia**, Norfolk Ridge. NORFOLK 1 Stn DW1689, 24°54'S, 168°23'E, 600-620 m, 23.VI.2001, 1 ♀ 5.5 mm (MNHN-IU-2014-16372). NORFOLK 2 Stn DW2064, 25°16.59'S, 168°55.64'E, 609-691 m, 26.X.2003, 1 ov. ♀ 4.8 mm (MNHN-IU-2014-16373). – Stn CP2089, 24°44'S, 168°09'E, 227-230 m, 29.X.2003, 4 ♂ 4.9-5.7 mm, 1 ov. ♀ 4.9 mm, 1 ♀ 5.4 mm (MNHN-IU-2014-16374).

ETYMOLOGY — The specific name is a noun in apposition from the Greek *kylindros* (a cylinder) and *pous* (a foot), referring to subcylindrical pereopods of the species.

DISTRIBUTION — Loyalty Islands and Norfolk Ridge; 227-691 m.

SIZE — Males, 4.2-5.7 mm, females 4.8-5.5 mm; ovigerous females from 4.8 mm.

DESCRIPTION — Medium-sized species. *Carapace*: As long as broad; greatest breadth 1.8 × distance between ordinary places of anterolateral spines. Dorsal surface with very fine setae discernible only under high magnification, convex from side to side, with very weak depression lateral to cardiac region, dorsal midline very slightly convex in profile continuing onto rostral dorsum. Lateral margins anteriorly rounded, with no distinct anterolateral angle and no anterolateral spine, somewhat divergent posteriorly to point one-third from anterior end, then convexly divergent further posteriorly. Rostrum elongate sharp triangular, with interior angle of 15°, nearly horizontal but slightly arched in profile, 2.5 × longer than broad, 0.7-0.8 × length of remaining carapace, breadth less than half carapace breadth measured at posterior carapace margin; dorsal surface flattish proximally, somewhat convex from side to side, with very fine short setae. Lateral limit of orbit with small spine. Pterygostomian flap anteriorly angular, produced to small, often blunt spine; surface smooth.

Sternum: Excavated sternum blunt triangular or convex on anterior margin, and smooth on surface. Sternal plastron slightly shorter than broad; sternite 5 as broad as or slightly narrower than sternite 4, narrower than sternit 6. Sternite 3 moderately depressed; anterior margin deeply excavated, representing broad V-shape, with no median notch and no median spines, anterolateral angle rounded. Sternite 4 with anterolateral margin well convex, without spine; posterolateral margin as long as anterolateral margin. Anterolateral margin of sternite 5 stongly convex, two-thirds length of posterolateral margin of sternite 4.

Abdomen: Smooth, nearly glabrous, long relative to breadth. Somite 1 antero-posteriorly convex, without distinct transverse ridge. Somite 2 tergite 1.9-2.2 × broader than long; lateral margin concavely divergent posteriorly, ending in rounded terminus. Pleuron of somite 3 bluntly produced. Telson slightly more than half as long as broad; posterior plate 1.3-1.4 × longer than anterior plate, posterior margin slightly concave or transverse medially.

Eye: Relatively short (length 1.6-1.7 × breath), falling short of proximal third of rostrum, mesial margin slightly concave, lateral margin slightly convex. Cornea not dilated, length about half that of remaining eyestalk.

Antennule and antenna: Ultimate antennular article 3.5-4.0 × longer than high. Antennal peduncle terminating in midlength of rostrum. Article 2 without lateral spine. Antennal scale slightly broader than article 5, falling short of distal end of article 4. Distal 2 articles unarmed; article 5 subequal to or slightly longer than article 4, breadth less than half height of antennular ultimate article. Flagellum consisting of 7-9 segments, not reaching rostral tip and distal end of P1 merus.

Mxp: Mxp1 with bases broadly separated. Mxp3 basis without denticles on mesial ridge. Ischium without distinct denticles on crista dentata, flexor margin not rounded distally. Merus 2.1 × longer than ischium, relatively thick mesiolaterally, unarmed. Carpus also unarmed.

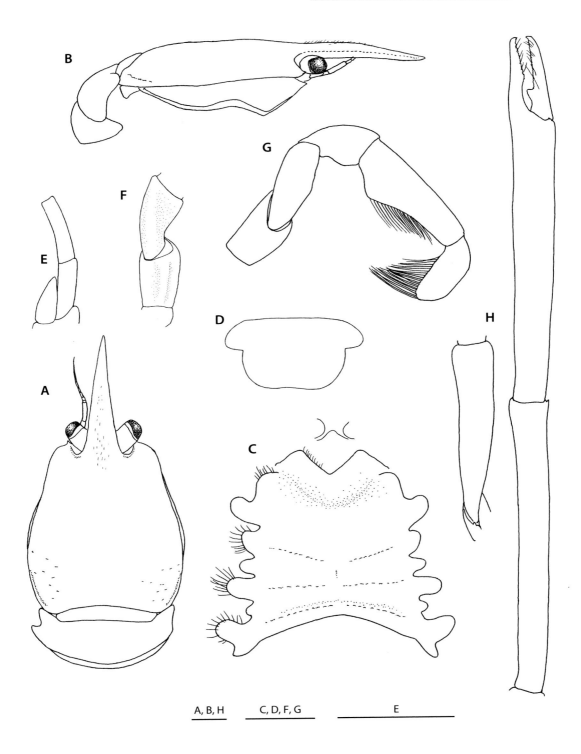

A, B, H   C, D, F, G   E

FIGURE 61

Uroptychus cylindropus n. sp., holotype, ovigerous female 5.3 mm (MNHN-IU-2014-16370). **A**, carapace and anterior part of abdomen, dorsal.
B, same, lateral. **C**, sternal plastron, with excavated sternum and basal parts of Mxp1. **D**, telson. **E**, right antenna, ventral. **F**, right Mxp3, ventral.
G, same, lateral. **H**, left P1, dorsal. Scale bars: 1 mm.

P1: Slender (more so in females than in males), subcylindrical on merus and carpus, 5.0-5.6 × longer than carapace, covered with very short fine setae. Ischium with very small, often obsolescent dorsal spine, unarmed elsewhere. Merus 1.1-1.2 × longer than carapace. Carpus 1.3-1.4 × (males), 1.6-1.8 × (females) longer than merus. Palm 3.8-4.0 × (males), 7.0-9.0 × (females) longer than broad, slightly longer than or as long as carpus, moderately depressed. Fingers gaping (strongly gaping in males), setose on gaping portions and ventral surface, ending in small incurved spine, and distally slightly crossing, not spooned; movable finger with subtriangular process at one-third from proximal end (or at midlength of opposable margin in largely gaping fingers), length 0.3 × (females) or 0.4 × (males) that of palm; in females, fixed finger with opposable margin denticulate on distal third of length, fitting to opposite denticulate margin of movable finger when closed.

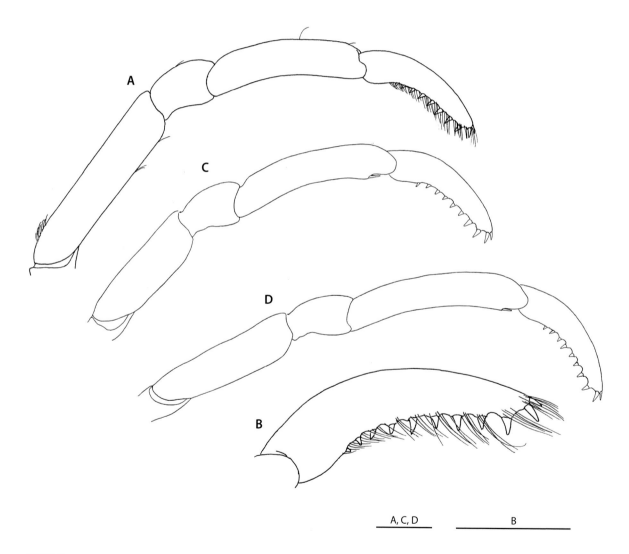

A, C, D   B

FIGURE 62

Uroptychus cylindropus n. sp., holotype, ovigerous female 5.3 mm (MNHN-IU-2014-16370). **A**, right P2, lateral. **B**, same, distal part, lateral. **C**, right P3, lateral. **D**, right P4, lateral. Scale bars: 1 mm.

P2-4: Relatively short, subcylindrical, meri and carpi with sparse short fine setae. Meri unarmed, longest on P2, shortest on P3, equally broad on P2-4; length-breadth ratio, 4.5-4.7 on P2, 3.2-3.7 on P3, 3.9-4.2 on P4; P2 merus 0.6-0.7 × length of carapace, 1.1-1.2 × length of P2 propodus; P3 merus 0.7 × as long as P2 merus, 0.7 × length of P3 propodus; P4 merus 1.1-1.3 × length of P3 merus, 0.8-0.9 × length of P4 propodus. Carpi subequal on P2-4, unarmed; length about one-third that of propodi. Propodi longer on P4 than on P2 and P3; length 4.5-5.4 × breadth on P2, 4.6-5.8 × on P3, 5.4-5.9 × on P4; somewhat curving along flexor margin, sparsely setose; flexor margin with single terminal spine occasionally obsolete. Dactyli moderately curved, length about two-thirds that of propodus (in holotype, dactylus-propodus length ratio, 0.72-0.73 on P2, 0.61-0.66 on P3, 0.60-0.63 on P4), and 2 × longer than carpi on P2-4; flexor margin very setose, with 9-10 loosely arranged, sharp spines nearly perpendicular to margin, proximally diminishing, ultimate much more slender and shorter than penultimate.

Eggs. Number of eggs carried, 8-11; size, 1.04 mm × 1.08 mm - 1.32 mm × 1.56 mm.

REMARKS — Pereopod 1 in the new species shows sex-related differences: the palm length-breadth ratio is greater in females than in males (7.0-8.8 versus 3.8-3.9); the carpus-merus length ratio is greater in females than in males (1.7-1.8 versus 1.3-1.4), as also is the dactylus-palm length ratio greater in males than in females (0.4 versus 0.3).

The species is differentiated from all other known species of the genus by the absence of the anterolateral spine of the carapace, and by the very long rostrum, subcylindrical pereopods, and the P3 merus that is much shorter than the P2 merus as well as P4 merus.

Uroptychus defayeae n. sp.

Figures 63, 64

TYPE MATERIAL — Holotype: **Vanuatu**. MUSORSTOM 8 Stn CP975, 19°23.60'S, 169°28.93'E, 566-536 m, 22.IX.1994, 1 ♂ 5.0 mm (MNHN-IU-2014-16375). Paratypes: **New Caledonia**, Chesterfield Islands. MUSORSTOM 5 Stn CP387, 20°53.41'S, 160°52.12'E, 650-660 m, 22.X.1986, 1 ♂ 3.4 mm (MNHN-IU-2014-16376). **New Caledonia**, Norfolk Ridge. CHALCAL 2 Stn DW73, 24°39.9'S, 168°38.1'E, 573 m, 29.X.1986, 1 ov. ♀ 4.5 mm (MNHN-IU-2014-16377). NORFOLK 2 Stn DW2077, 25°20.63'S, 168°18.53'E, 666-1000 m, 27.X.2003, 1 ♂ 3.5 mm (MNHN-IU-2014-16378).

ETYMOLOGY — Name for Danielle Defaye of MNHN for her help during my stays at the Paris Museum.

DISTRIBUTION — Chesterfield Islands, Vanuatu, and Norfolk Ridge; 536-1000 m.

DESCRIPTION — Small to medium-sized species. *Carapace*: Broader than long (0.8 × as long as broad); greatest breadth 1.8-1.9 × distance between anterolateral spines. Dorsal surface smooth, somewhat convex from anterior to posterior, with very shallow depression between gastric and cardiac regions; somewhat ridged along posterior third of lateral margin; epigastric region with pair of small spines (left spine absent in paratype MNHN-IU-2014-16376); small spine mesial to second stout lateral marginal spine (absent in paratypes MNHN-IU-2014-16376 and MNHN-IU-2014-16378). Lateral margins gently constricted at anterior third; anterior part (anterior to constrictions) gently divergent posteriorly, with 2 well-developed spines: first anterolateral, directed forward, or inclined slightly mesially or slightly laterally, slightly overreaching antennal scale; second shorter than first, placed at anterior end of anterior branchial region; 2 tiny spines present or obsolescent between first and second; posterior part (posterior to constriction or posterior branchial margin) gently convex, with 4 or 5 short, posteriorly diminishing spines, anteriormost of these remotely separated from second spine on anterior part. Rostrum broad or narrow triangular, directed slightly dorsally, with interior angle of 22° (paratypes) or 30° (holotype); length 0.8 × (holotype) or subequal to (paratypes) breadth, and 0.4 × that of remaining carapace; dorsal surface concave; breadth less than half that of posterior carapace margin. Lateral orbital spine small, situated at

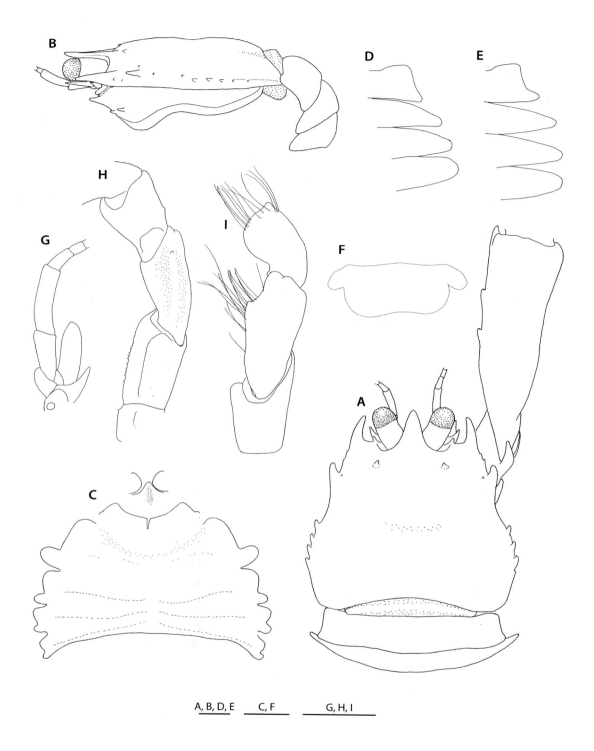

A, B, D, E   C, F   G, H, I

FIGURE 63

Uroptychus defayeae n. sp., **A-D**, **F-I**, holotype, male 5.0 mm (MNHN-IU-2014-16375); **E**, ovigerous female paratype 4.5 mm (MNHN-IU-2014-16377). **A**, carapace and anterior part of abdomen, proximal part of right P1 included, dorsal. **B**, same, lateral. **C**, sternal plastron, with excavated sternum and basal parts of Mxp1. **D**, right pleura of abdominal somites 2-5, dorsolateral. **E**, same. **F**, telson. **G**, left antenna, ventral. **H**, left Mxp3, setae omitted, ventral. **I**, same, lateral. Scale bars: 1 mm.

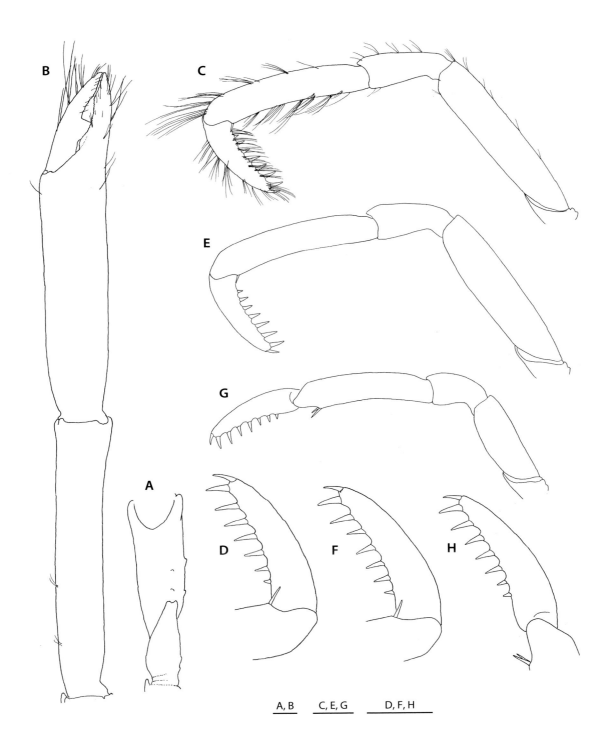

FIGURE 64

Uroptychus defayeae n. sp., holotype, male 5.0 mm (MNHN-IU-2014-16375). **A**, right P1, proximal part, ventral. **B**, same, distal three articles, dorsal. **C**, left P2, lateral. **D**, same, distal part, setae omitted, lateral. **E**, left P3, setae omitted, lateral. **F**, same, distal part, setae omitted, lateral. **G**, left P4, setae omitted, lateral. **H**, same, distal part, setae omitted, lateral. Scale bars: 1 mm.

same level as anterolateral spine. Pterygostomian flap anteriorly angular, produced to strong spine, surface with 2-5 spines on anterior part; height of posterior half 0.5 × that of anterior half.

Sternum: Excavated sternum with sharp ridge in midline leading onto anterior margin produced triangularly between bases of Mxp1. Sternal plastron 1.4 × broader than long, lateral extremities convexly divergent behind sternite 4, sternite 7 slightly narrower than sternite 6. Sternite 3 shallowly depressed; anterior margin moderately concave, with deep narrow or broad V-shaped median notch without flanking spine. Sternite 4 short relative to breadth; anterolateral margin anteriorly convex, length 1.5 × that of posterolateral margin. Sternite 5 with convex anterolateral margin subequal to or slightly longer than posterolateral margin of sternite 4.

Abdomen: Somites short relative to breadth. Somite 1 well convex from anterior to posterior. Somite 2 tergite 3.1-3.3 × broader than long; pleuron posterolaterally produced more strongly in female than in males. Pleura of somite 3-5 laterally rounded, with no great difference between sexes. Telson 0.4-0.5 × as long as broad; posterior plate 1.0-1.3 × longer than anterior plate, posterior margin slightly or moderately emarginate.

Eye: Elongate, 1.8-1.9 × longer than broad, reaching or slightly overreaching apex of rostrum, slightly broadened proximally, mesial margin slightly concave medially. Cornea very slightly dilated, 0.7-0.8 × length of remaining eyestalk.

Antennule and antenna: Ultimate article of antennular peduncle 2.6-2.9 × longer than high. Antennal peduncle extending far beyond apex of rostrum. Article 2 with strong distolateral spine. Antennal scale proportionately broad, distally blunt, 1.3 × broader than article 5, slightly overreaching article 4, not reaching midlength of article 5. Article 4 unarmed or with very tiny tubercle-like process distomesially. Article 5 unarmed, length 1.3-1.4 × that of article 4, breadth 0.5 × height of ultimate article of antennule. Flagellum of 12-15 segments not reaching distal end of P1 merus.

Mxp: Mxp1 with bases broadly separated. Mxp3 basis with obsolescent distal denticle on mesial ridge. Ischium having flexor margin distally rounded, crista dentata with 10-20 small or obsolescent denticles. Merus twice as long as ischium, mesial face flattish; flexor margin roundly ridged on proximal half, sharply ridged on distal half, bearing 1 or 2 very short blunt spine at distal third; distolateral spine obsolescent. Carpus unarmed.

P1: 6.9-7.2 × (males), 6.2 × (female) longer than carapace, smooth and barely setose except for fingers. Ischium with strong dorsal spine with or without small accompanying spine proximally, ventromesial margin with subterminal spine very short and blunt, not overreaching distal end of ischium, and a few obsolescent proximal protuberances. Merus with 2 rows each of a few short blunt spines (1 dorsomesial, 1 ventromesial), ventral distomesial spine blunt and short; length 1.4-1.5 × (males), 1.3 × (female) that of carapace. Carpus 1.5-1.6 × length of merus, distally with dorsomesial spine, no spine elsewhere. Palm unarmed, depressed, two-thirds as high as broad, slightly broader distally or with subparallel lateral and mesial margins, somewhat narrowed proximally; 3.5-4.5 × (males), 4.0 × (female) as long as broad. Fingers relatively slender distally, sparingly setose, moderately gaping in largest male and female, not gaping in smaller males, not clearly crossing when closed; fixed finger directed slightly laterally, opposable margin sinuous; movable finger 0.4 × as long as palm, opposable margin with prominent process (distal margin of process perpendicular to opposable margin); no longitudinal groove on ventromesial face of fixed finger.

P2-4: Relatively slender, well compressed mesio-laterally, with sparse simple setae, more setose on dactyli. Ischia with 2 short dorsal spines placed distally and proximally. Meri subequal in length on P2 and P3, shortest on P4 (P4 merus 0.5-0.6 × length of P3 merus); subequally broad on P2 and P3 or slightly broader on P3 than on P2, slightly narrower on P4; dorsal margin unarmed; length-breadth ratio, 4.4-4.7 on P2, 4.0-4.4 on P3, 2.7-3.1 on P4; P2 merus 0.9-1.0 × length of carapace, subequal to or very slightly longer than P2 propodus; P3 merus subequal to length of P3 propodus. P4 merus 0.7 × length of P4 propodus. Carpi subequal in length on P2 and P3 or slightly longer on P2 than on P3, shortest on P4 (P4 carpus 0.7 × length of P3 carpus), carpus-propodus length ratio, 0.4 on P2 and P3, 0.3-0.4 on P4. Propodi longest on P3, shortest on P4; flexor margin slightly concave in lateral view, with pair of slender terminal spines only. Dactyli subequal in length on P2-4; dactylus-carpus length ratio, 1.1-1.3 on P2, 1.2-1.4 on P3, 1.7-1.8 on P4; dactylus-propodus length ratio, 0.5 on P2 and P3, 0.6 on P4; flexor margin slightly curving, with 7-9 sharp spines loosely arranged, proximally diminishing and nearly perpendicular to margin, ultimate spine slightly but consistently more slender than penultimate.

Eggs. Number of eggs carried, 25; size, 1.08 × 1.00 mm - 1.13 × 1.17 mm.

Color in preservative. Reddish bands on P1 as in *U. trispinatus*, but no bands on fingers.

REMARKS — *Uroptychus defayeae* resembles *U. clarki* n. sp. and *U. corbariae* n. sp. (see their similarities under *U. clarki*). Morphological differences between *U. clarki* and *U. defayeae* are small. In *U. defayeae* the carapace dorsal surface has a pair of epigastric spines only, instead of bearing additional spines flanking the epigastric pair as in *U. clarki*; the strong anterolateral spine of the carapace is followed by two obsolescent instead of small but distinct spines; the crista dentata bears denticles on the median part, not along the entire length as in *U. clarki*; the distolateral spine of the antennal article 2 is much stronger. *Uroptychus defayeae* differs from *U. corbariae* in having the posterior branchial margin with 4 or 5 spines instead of 1 spine (followed by a few short oblique ridges), in having the antennal article 2 with a strong instead of very small distolateral spine, and in having more slender pleura of abdominal somites 3-4.

Uroptychus dejouanneti n. sp.
Figures 65, 66

Uroptychus gracilimanus Baba 1969: 45, figs 3, 4; 1988: 35; 2005: 36. — Baba *et al.* 2009: 44, figs 36-37 (not *U. gracilimanus* (Henderson, 1885)).

TYPE MATERIAL — Holotype: **Indonesia**, Tanimbar Islands. KARUBAR Stn CP38, 7°40'S, 132°27'E, 620-666 m, with *Acanella* sp. (Isididae) and *Thouarella* sp. (Primnoidae), 28.X.1991, ov. ♀ 6.9 mm (MNHN-IU-2013-8538). Paratypes: **Indonesia**, Kai Islands. KARUBAR Stn CP19, 5°15'S, 133°01'E, 605-576 m, on gorgonacean, 25.X.1991, 2 ov. ♀ 6.9 mm, 7.5 mm (MNHN-IU-2013-8535). – Stn CP20, 5°15'S, 132°59'E, 769-809 m, Chrysogorgiidae gen. sp. (Calcaxonia), 25.X.1991, 3 ♂ 6.8, 6.8, 7.6 mm, 6 ov. ♀ 7.2-7.8 mm (MNHN-IU-2013-8536). – Stn CC21, 5°14'S, 133°00'E, 688-694 m, 25.X.1991, 2 ♂ 7.6, 7.8 mm, 4 ov. ♀ 6.0-8.0 mm, 2 ♀ 4.8, 6.8 mm (MNHN-IU-2013-8537). **Indonesia**, Tanimbar Islands. KARUBAR Stn CP38, 7°40'S, 132°27'E, 620-666 m, *Acanella* sp. (Isididae) + *Thouarella* sp. (Primnoidae), 28.X.1991, 5 ♂ 5.7-7.8 mm, 5 ov. ♀ 6.7-7.6 mm, 3 ♀ 5.6-7.8 mm, 1 ♂ 7.6 mm (MNHN-IU-2014-10140, MNHN-IU-2014-10141, MNHN-IU-2014-10142, MNHN-IU-2014-16379). – Stn CP39, 7°47'S, 132°26'E, 477-466 m, 28.X.1991, 1 ♂ 8.6 mm (MNHN-IU-2013-8539). – Stn CC57, 8°19'S, 131°53'E, 603-620 m, 31.X.1991, 3 ♂ 5.0-6.2 mm, 1 ov. ♀ 6.2 mm, 1 ♀ 7.2 mm (MNHN-IU-2013-8540). – Stn CC58, 8°19'S, 132°02'E, 457-461 m, 31.X.1991, 1 ♂ 7.1 mm (MNHN-IU-2013-8541).

ETYMOLOGY — Named after Jean-Françoise Dejouannet, artist at the Paris Museum, for his friendship.

DISTRIBUTION — Indonesia (Kai Islands, Tanimbar Islands, Molucca Sea off west coast of Halmahera), Philippines (east of Zamboanga), Taiwan and East China Sea; 441-1060 m.

SIZE — Males 5.0-8.6 mm; females, 4.8-8.0 mm; ovigerous females from 6.0 mm.

DESCRIPTION — Medium-sized species. *Carapace*: As long as broad; greatest breadth 1.6-1.8 × distance between anterolateral spines. Dorsal surface smooth, glabrous and unarmed, anteriorly somewhat inflated in profile, with or without faint depression between gastric and cardiac regions. Lateral margins somewhat convex and divergent posteriorly, with row of obsolescent short ridges along anterior half of branchial margin; ridged along posterior half; anterolateral spine small, varying from barely to fully reaching tiny lateral orbital spine where present, situated distinctly posterior to level of that spine. Rostrum 0.8-1.0 × as long as broad, triangular with interior angle of 24-30°, straight horizontal or curving slightly ventrally; length 0.3-0.4 × that of remaining carapace, breadth slightly less than half carapace breadth measured at posterior carapace margin; lateral margins slightly concave; dorsal surface slightly concave at base. Lateral orbital spine small, often obsolescent or absent. Pterygostomian flap anteriorly roundish, bearing very small spine.

FIGURE 65

Uroptychus dejouanneti n. sp., holotype, ovigerous female 6.9 mm (MNHN-IU-2013-8538). **A**, carapace and anterior part of abdomen, dorsal. **B**, same, lateral. **C**, anterior part of cephalothorax, showing anterolateral spine, anterior part of pterygostomian flap and antenna, lateral. **D**, sternal plastron, with excavated sternum and basal parts of Mxp1. **E**, telson. **F**, right antenna, ventral. **G**, right Mxp3, setae omitted, ventral. **H**, same, lateral. Scale bars: 1 mm.

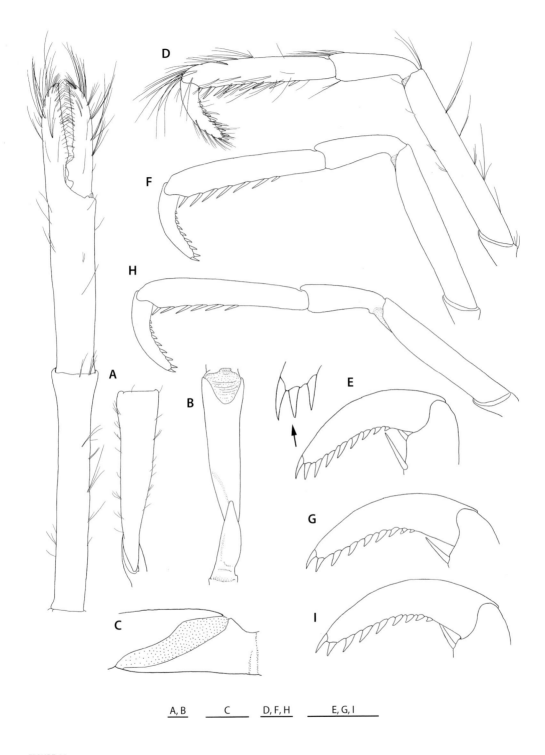

A, B   C   D, F, H   E, G, I

FIGURE 66

Uroptychus dejouanneti n. sp., holotype, ovigerous female 6.9 mm (MNHN-IU-2013-8538). **A**, left P1, dorsal. **B**, same, proximal part, ventral. **C**, same, lateral. **D**, left P2, lateral. **E**, same, distal part, setae omitted, lateral. **F**, left P3, setae omitted, lateral. **G**, same, distal part, setae omitted, lateral. **H**, left P4, setae omitted, lateral. **I**, same, distal part, setae omitted, lateral. Scale bars: 1 mm.

Sternum: Excavated sternum anteriorly broad triangular, ending in small spine, surface with spine in center. Sternal plastron 0.9-1.0 × as long as broad, lateral extremities between sternites 4 and 7 divergent posteriorly. Sternite 3 moderately depressed, with well-excavated anterior margin bearing 2 distinct submedian spines flanking narrow and shallow or U-shaped sinus, laterally angular. Sternite 4 having convex anterolateral margin, with short spines on anterior half (anterior 2 often of good size and others obsolescent), about 1.6-1.8 × length of posterolateral margin. Anterolateral margins of sternite 5 somewhat convexly divergent posteriorly, slightly longer than posterolateral margin of sternite 4.

Abdomen: Tergites smooth and glabrous. Somite 1 without transverse ridge. Somite 2 tergite 2.3-2.5 × broader than long; pleuron posterolaterally ending in rounded terminus, lateral margin somewhat concavely divergent posteriorly. Pleura of somites 3 and 4 laterally rounded. Telson very slightly more than half (0.53-0.58 x) as long as broad; posterior plate moderately emarginate or feebly concave on posterior margin, length 1.6-1.8 × that of anterior plate.

Eye: Short relative to breadth (1.7-1.8 × longer than broad), slightly concave on mesial margin, slightly convex on lateral margin, slightly falling short of rostral tip. Cornea not dilated, about half length of remaining eyestalk.

Antennule and antenna: Ultimate article of antennular peduncle 2.3-2.7 × longer than high. Antennal peduncle overreaching cornea, reaching rostral tip. Article 2 distolaterally acuminate or with very tiny spine. Antennal scale 1.5-1.7 × broader than article 5, usually slightly falling short of midlength of, rarely reaching point three-quarters of article 5. Distal 2 articles unarmed; article 5 1.6-1.8 × longer than article 4, breadth slightly smaller than height of ultimate antennular article. Flagellum of 18-20 segments fully or barely reaching distal end of P1 merus.

Mxp: Mxp1 with bases close to each other. Mxp3 spineless on merus and carpus. Basis with 3 denticles on mesial ridge. Ischium with 20-28 denticles on crista dentata, flexor margin not rounded distally. Merus 2.3 × longer than ischium, relatively thick mesio-laterally, flexor margin sharply ridged.

P1: Slender (male palm in large specimens massive, broader than distance between anterolateral spines of carapace), unarmed, sparsely setose on fingers, glabrous elsewhere; length 4.7-5.5 × (males), 4.9-5.3 × (females) that of carapace. Ischium with basally broad, low and blunt dorsal spine often obsolescent, ventrally unarmed. Merus 1.1-1.3 × longer than carapace. Carpus 1.3-1.4 × longer than merus. Palm 2.2-3.2 × (males; slender in small specimens), 3.6-3.8 × (females) longer than broad, 0.7-0.9 × as long as carpus. Fingers proportionately broad, gaping in proximal half in males, not gaping in females, ending in incurved tips, feebly crossing; movable finger 0.6-0.7 × length of palm; opposable margin of movable finger with broad proximal process low in females, high and 2-toothed in males.

P2-4: Slender, with sparse long setae. Meri successively shorter posteriorly (P3 merus 0.9 × length of P2 merus, P4 merus 0.9 × length of P3 merus), breadths subequal on P2-4; length-breadth ratio 5.7-6.4 on P2, 4.9-5.6 on P3, 4.5-5.5 on P4; dorsal and ventrolateral margins unarmed; P2 merus 0.8-0.9 × length of carapace, 1.2 × length of P2 propodus; P3 merus as long as P3 propodus; P4 merus 0.9 × length of P4 propodus. Carpi subequal or shorter on P2 than on subequal P3 and P4; carpus-propodus length ratio, 0.4-0.5 on P2 and P3, 0.4 on P4. Propodi successively slightly longer posteriorly; flexor margin nearly straight, with 6-8 basally articulated, long spines on two-thirds to four-fifths of length on P2, 6 or 7 spines on P3, 5 or 6 spines more distal on P4, terminal spine single, located very close to distal end and mesial to flexor midline. Dactyli curving at middle, length 0.4 × length of propodi on P2-4; dactylus-carpus length ratio, 0.8-0.9 on P2, 0.8-1.0 on P3, 1.0 on P4; flexor margin with 8-10 sharp, obliquely directed spines, distal 4 or 5 subequal, remaining spines successively diminishing toward base of article.

Eggs. Up to 40 eggs carried; size, 0.92 × 1.25 mm - 1.01 × 1.29 mm.

Color. A specimen from Taiwan reported under *U. gracilimanus* was illustrated by Baba *et al.* (2009).

PARASITES — Three males of the specimens from KARUBAR Stn CC21, CC57 and CC58 and one female from Stn CP38 each bear an externa of rhizocephalan parasite.

REMARKS — This species is so similar to *U. gracilimanus* (Henderson, 1885) that careful examination should be paid to the spination of the P2-4 dactyli: the ultimate of the flexor spines in the new species is subequal to, instead of distinctly narrower than, the penultimate spine as in *U. gracilimanus* (see below). The material reported under *U. gracilimanus*

(Henderson, 1885) by Baba (1969; 1988) and Baba *et al.* (2009) are now referred to *U. dejouanneti. Uroptychus dejouanneti* resembles *U. stenorhynchus* n. sp. Their relationships are discussed under that species (see below).

Molecular data provided by L. Corbari (personal comm.) suggest that the specimens MNHN-IU-2014-10141 and IU-2014-10142 are different from each other at a specific level (the genetic divergence between the two specimens is 7.2 % (COI)), but no morphological differences were found.

Uroptychus denticulifer n. sp.
Figures 67, 68

TYPE MATERIAL — Holotype: **Solomon Islands**. SALOMON 1 Stn DW1854, 9°46.4'S, 160°52.9'E, 229-260 m, 7.X.2001, ov. ♀ 2.5 mm (MNHN-IU-2014-16380). Paratype: **Vanuatu**. MUSORSTOM 8 Stn DW1060, 16°14'S, 167°21'E, 394-375 m, 2.X.1994, 1 ♂ 2.3 mm (MNHN-IU-2014-16381).

ETYMOLOGY — From the Latin *denticulus* (dim. of *dens*, tooth) and *fer* (to bear), referring to the denticles covering the body.

DISTRIBUTION — Solomon Islands and Vanuatu; 229-397 m.

DESCRIPTION — Small species. *Carapace*: 0.8-0.9 × as long as broad; greatest breadth 1.7 × distance between anterolateral spines. Dorsal surface covered with denticles, inflated on lateral gastric, cardiac and lateral branchial regions, with deep groove separating gastric and cardiac regions; anterolateral spine small, not reaching tip of lateral orbital spine, followed by convex lateral margin leading to ordinary end of cervical groove, then followed by inflation on anterior branchial region and convexly divergent posterior branchial margin. Rostrum subtriangular, with interior angle of 22°, distally roundish, not tapering; dorsal surface concave; length about half that of remaining carapace. Lateral orbital spine larger than or subequal to anterolateral spine, situated directly mesial to (at same level as) and more or less close to that spine. Pterygostomian flap covered with denticles, anteriorly roundish, with small spine.

Sternum: Excavated sternum with convex anterior margin, moderately ridged in midline. Sternal plastron slightly longer than broad, sternite 4 as broad as sternite 5, slightly broader than sternite 6, sternite 7 broader than sternite 4. Sternite 3 shallowly depressed, weakly concave on anterior margin bearing narrow, deep median sinus without flanking spine. Sternite 4 with anterolateral margin rounded at anterior end bearing a few obsolescent denticles; posterolateral margin longer than anterolateral margin. Anterolateral margin of sternite 5 strongly convex and posteriorly divergent, about half as long as posterolateral margin of sternite 4.

Abdomen: Somites long relative to breadth, bearing denticles on somites 1 and 2 (and on anterior part of somite 3 in holotype), smooth elsewhere. Somite 1 convex from anterior to posterior. Somite 2 tergite 1.9-2.0 × broader than long; pleural margin feebly concave and slightly divergent posteriorly, posterolateral end rounded. Pleuron of somite 3 laterally rounded. Telson 0.4-0.5 × as long as broad; posterior plate slightly shorter than anterior plate, posterior margin slightly concave.

Eye: Broad relative to length (length 1.4-1.5 × breadth), slightly overreaching midlength of rostrum; lateral and mesial margins slightly or distinctly convex, dorsal surface with or without sparse denticles. Cornea not inflated, length less than half that of remaining eyestalk.

Antennule and antenna: Ultimate article of antennular peduncle 2.0 × longer than high. Antennal peduncle reaching apex of rostrum. Article 2 fused with antennal scale, without lateral spine. Antennal scale slightly overreaching distal end of article 4, with small blunt spines along lateral margin. Article 4 ventrally with a few denticles on terminal margin at juncture with article 5. Article 5 1.6 × length of article 4, unarmed; breadth 0.6-0.7 × height of antennular ultimate article. Flagellum of 6 segments about as long as antennal peduncle, not reaching distal end of P1 merus.

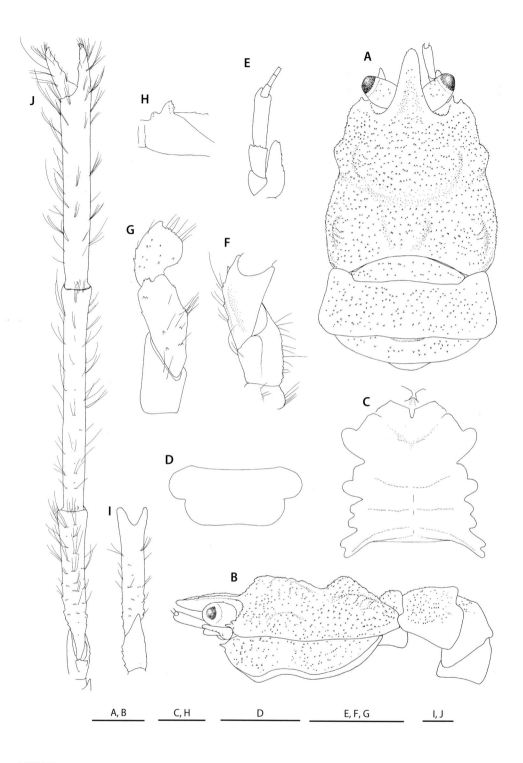

FIGURE 67

Uroptychus denticulifer n. sp., holotype, ovigerous female 2.5 mm (MNHN-IU-2014-16380). **A**, carapace and anterior part of abdomen, dorsal. **B**, same, lateral. **C**, sternal plastron, with excavated sternum and basal parts of Mxp1. **D**, telson. **E**, left antenna, ventral. **F**, right Mxp3, ventral. **G**, same, lateral. **H**, right P1, ischium, lateral. **I**, same, proximal part, ventral. **J**, same, dorsal. Scale bars: 1 mm.

Mxp: Mxp1 with bases moderately separated. Mxp3 sparsely setose on lateral surface. Basis with obsolescent denticles on mesial ridge. Ischium with distally rounded flexor margin; crista dentata with obsolescent denticles. Merus 2.8 × longer than ischium, not cristate but roundly ridged along flexor margin, with short distolateral process distally bearing a few denticles; with scattered small spines or denticles on lateral surface and a few denticles on distal half of flexor margin. Carpus also with scattered denticles dorsally and laterally.

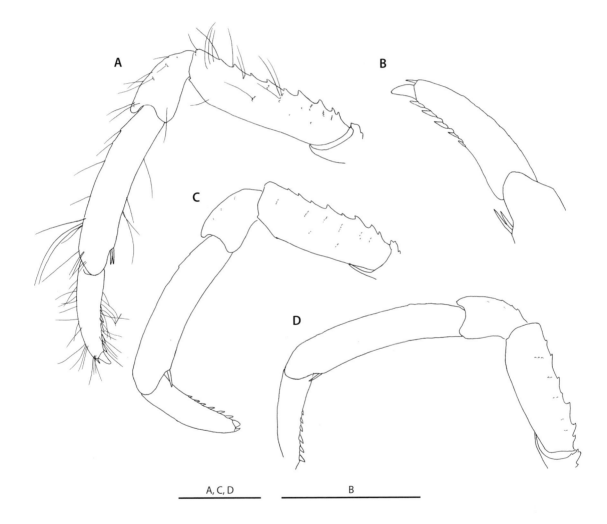

FIGURE 68

Uroptychus denticulifer n. sp., holotype, ovigerous female 2.5 mm (MNHN-IU-2014-16380). **A**, left P2, lateral. **B**, same, distal part, setae omitted, lateral. **C**, left P3, setae omitted, lateral. **D**, left P4, setae omitted, lateral. Scale bars: 1 mm.

P1: 6.9 × longer than carapace, slender, sparsely with soft setae. Ischium with truncate, broad, depressed dorsal process, ventromesially feebly denticulate, lacking subterminal spine. Merus 1.4-1.5 × longer than carapace, mesially with small, blunt spines on proximal half, dorsally and ventrally with denticulate ridges supporting setae. Carpus 1.3-1.5 × longer than merus, with obsolescent denticulate ridges, unarmed. Palm 6.7 × longer than broad, 0.9 × length of carpus. Fingers somewhat incurved distally, not spooned along opposable margins; opposable margin with 2 low processes on movable finger, low median eminence on fixed finger; movable finger 0.3 × length of palm.

P2-4: Relatively thick mesio-laterally, sparsely setose, setae relatively long. Meri successively slightly broader posteriorly or subequally broad, successively shorter posteriorly (P3 merus 0.8 × length of P2 merus, P4 merus 0.9 × length of P3 merus); length 3.8-4.0 × breadth on P2, 3.1-3.3 × on P3, 2.8-3.0 × on P4; P2 merus 0.7-0.8 × length of carapace, as long as P2 propodus; P3 merus 0.8 × length of P3 propodus; P4 merus 0.70-0.75 × length of P4 propodus; dorsal margin with row of short spines, lacking spine at distal end; lateral surface with short denticulate ridges supporting setae. Carpi subequal, about one-third length of propodi (0.30-0.33 on P2, 0.28-0.33 on P3, 0.30-0.31 on P4); lateral surface with a few denticulate ridges supporting setae paralleling extensor margin. Propodi subequal in length on P2 and P3 and longer on P4 or shortest on P2 and subequal on P3 and P4; flexor margin slightly concave in lateral view, with pair of terminal spines only. Dactyli about 1.5 × length of carpi on P2-4 (1.4 × on P2, 1.6 × on P3, 1.5 × on P4), about half as long as propodi on P2-4; flexor margin slightly curving, with prominently broad, long penultimate spine preceded proximally by 5 or 6 loosely arranged, obliquely directed, slender spines; ultimate spine much more slender than antepenultimate.

Eggs. Five eggs carried; size, 0.76 m × 0.88 mm - 0.77 mm × 0.92 mm.

REMARKS — The new species strongly resembles *U. kaitara* Schnabel, 2009 from the Kermadec Ridge in having the body covered with denticles (on the carapace, pterygostomian flap and at least anterior two somites of the abdomen), in having the anterolateral spine of the carapace small and subequal to or smaller than the lateral orbital spine, and in the shape of sternite, especially the posterolateral margin about as long as or longer than the anterolateral spine. It is differentiated from that species by the following: the antennal scale is fused instead of articulated with the antennal article 2; the antennal article 5 is 1.6 times longer than instead of being subequal to article 4; and the P2-4 dactyli bear flexor marginal spines obliquely directed instead of perpendicular to the margin.

Uropytchus denticulisquama n. sp.

Figures 69, 70

TYPE MATERIAL — Holotype: **New Caledonia**. MUSORSTOM 4 Stn CP216, 22°59.5'S, 167°22.0'E, 490-515 m, gorgonacean coral, 29.IX.1985, ♂ 6.3 mm (MNHN-IU-2014-16382). Paratypes: **New Caledonia**. MUSORSTOM 4 Stn CP216, collected with holotype, 1 ov. ♀ 6.3 mm (MNHN-IU-2014-16383). BATHUS 2 Stn CP735, 23°01'S, 166°56'E, 530-570 m, 13.V.1993, 1 ♂ 4.1 mm (MNHN-IU-2014-16384), 1 ov. ♀ 4.8 mm (MNHN-IU-2014-16385). **New Caledonia**, Isle of Pines. SMIB 2 Stn DC26, 22°59'S, 167°23'E, 500-535 m, 21.IX.1986, 1 ov. ♀ 7.0 mm. (MNHN-IU-2014-16386).

ETYMOLOGY — From the Latin *denticulus* (dim. of *dens*, tooth) and *squama* (scale), a noun in apposition, referring to the antennal scale with denticles on the lateral margin.

DISTRIBUTION — New Caledonia and Isle of Pines; 490-570 m.

SIZE — Males, 4.1-6.3 mm; ovigerous females, 4.8-7.0 mm.

DESCRIPTION — Medium-sized species. *Carapace*: About as long as broad; greatest breadth 1.5-1.7 × distance between anterolateral spines. Dorsal surface slightly convex from anterior to posterior, with or without feeble depression between

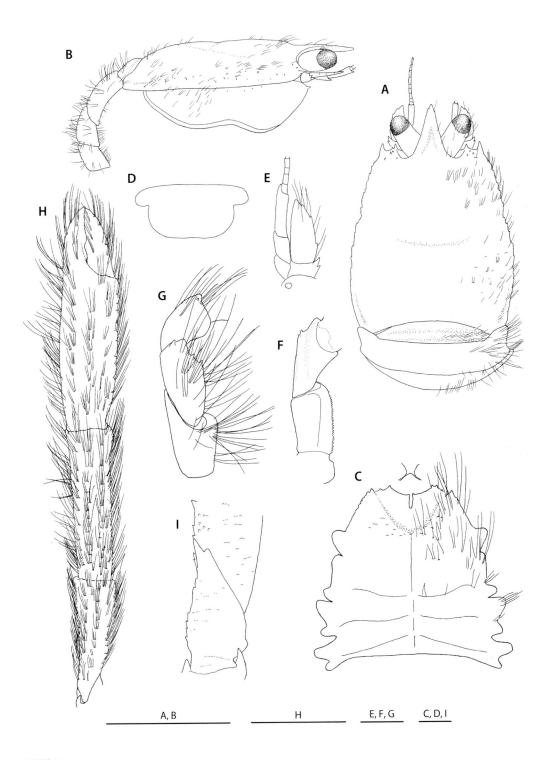

FIGURE 69

Uroptychus denticulisquama n. sp., holotype, male 6.3 mm (MNHN-IU-2014-16382). **A**, carapace and anterior part of abdomen, dorsal. **B**, same, lateral. **C**, sternal plastron, with excavated sternum and basal parts of Mxp1. **D**, telson. **E**, left antenna, ventral. **F**, right Mxp3, setae omitted, ventral. **G**, same, lateral. **H**, left P1, dorsal. **I**, same, proximal part, setae omitted, ventral. Scale bars: A, B, H, 5 mm; C-G, I, 1 mm.

gastric and cardiac regions and between gastric and branchial regions, smooth but feebly or sparsely tuberculose along lateral margins (denticles on hepatic region usually present), sparsely setose. Lateral margins somewhat convexly divergent posteriorly; anterolateral spine short, terminating in or slightly overreaching tip of lateral orbital spine, followed by denticles or tubercles (rarely 2 small spines, one at anterior end of branchial region and another at anterior end of posterior branchial region). Rostrum narrow triangular, with interior angle of 25-30°, nearly straight horizontal or directed slightly dorsally; length 0.3-0.4 × postorbital carapace length, breadth less than half carapace breadth measured at posterior carapace margin; dorsal surface slightly concave; lateral margin with a few obsolescent denticles near tip.

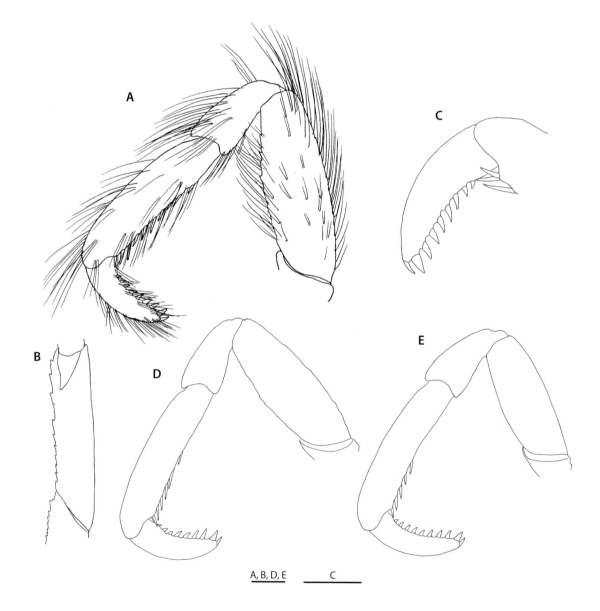

FIGURE 70

Uroptychus denticulisquama n. sp., holotype, male 6.3 mm (MNHN-IU-2014-16382). **A**, left P2, lateral. **B**, same, merus, setae omitted, ventral. **C**, same, distal part, setae omitted, lateral. **D**, left P3, setae omitted, lateral. **E**, left P4, setae omitted, lateral. Scale bars: 1 mm.

Lateral orbital spine small, slightly anterior to level of and moderately remote from anterolateral spine. Pterygostomian flap anteriorly angular, produced to small spine, surface somewhat granulose, with sparse short setae.

Sternum: Excavated sternum subtriangularly produced anteriorly, with ridge in midline. Sternal plastron as long as or very slightly shorter than broad; lateral extremities gently divergent posteriorly. Sternite 3 distinctly depressed, anterior margin shallowly concave, with pair of tiny or obsolescent submedian spines separated by narrow or somewhat broad U-shaped sinus. Sternite 4 having anterolateral margin nearly straight, anteriorly ending in 1 or 2 small spines, length about twice that of posterolateral margin. Anterolateral margins of sternite 5 subparallel, anteriorly rounded, length 1.4 × that of posterolateral margin of sternite 4.

Abdomen: Smooth, sparsely setose. Somite 1 convex from anterior to posterior, forming rounded transverse ridge. Somite 2 tergite 3.6 × broader than long; pleural lateral margin concave, weakly divergent posteriorly, ending in angular tip. Pleuron of somite 3 posterolaterally angular. Telson half as long as broad; posterior plate distinctly or feebly emarginate on posterior margin, length 1.1-1.5 × that of anterior plate.

Eye: Elongate, 2.0 × longer than broad, barely reaching rostral tip; mesial margin somewhat concave proximally. Cornea not dilated, about half as long as remaining eyestalk.

Antennule and antenna: Ultimate article of antennular peduncle 2.7-3.2 × longer than high. Antennal peduncle reaching or slightly overreaching cornea. Article 2 with 1 or 2 lateral spines. Antennal scale relatively broad, fully twice as broad as article 5, reaching or slightly falling short of distal end of article 5; lateral margin with a few small spines and relatively long setae. Article 4 with small distomesial spine. Article 5 with small or obsolescent distomesial spine; length 1.5-1.7 × that of article 4, breadth 0.6 × height of ultimate article of antennule. Flagellum composed of 16-19 segments, fully or barely reaching distal end of P1 merus.

Mxp: Mxp1 with bases broadly separated. Mxp3 setose on lateral surface, setae long. Basis with a few obsolescent denticles on mesial ridge. Ischium with about 30 small denticles on crista dentata, flexor margin distally rounded. Merus 1.9 × longer than ischium, flattish on mesial face, bearing distolateral spine very small or obsolete; flexor margin well ridged with a few small spines on distal third, rounded on proximal two-thirds. Carpus with tiny or obsolescent distolateral spine.

P1: 4.6-5.5 × longer than carapace, covered with distally softened setae, those on carpus and merus arising from short scale-like ridges. Ischium dorsally armed with relatively short, basally broad, depressed spine, ventromesially with strong subterminal spine proximally followed by very small denticle-like spines obscured by setae. Merus 0.9-1.3 × length of carapace, mesially tuberculose, with well-developed mesial distoventral spine and several small spines on proximal mesial surface. Carpus 1.2-1.3 × longer than merus, mesially tuberculose, ventrally with short blunt distomesial and distolateral spines. Palm 2.5-2.6 × (male), 2.8-3.1 × (females) longer than broad, subequal to length of carpus. Fingers broad relative to length, distally somwewhat incurved, weakly crossing when closed; movable finger half as long as palm, opposable margin with bluntly triangular proximal process fitting to opposite longitudinal concavity ventromesial to between 2 prominences of fixed finger when closed.

P2-4: Thickly covered with long setae, on ventral and ventromesial surfaces in particular. Meri moderately compressed mesio-laterally, successively diminishing posteriorly (P3 merus 0.8-0.9 × length of P2 merus, P4 merus 0.8 × length of P3 merus), successively very slightly narrower posteriorly; length-breadth ratio, 2.9-4.0 on P2, 2.7-3.0 on P3, 2.6-3.3 on P4; several eminences on proximal half of dorsal crest and denticle-like small spines along ventromesial margin, both only on P2; P2 merus 0.8-0.9 × length of carapace, 1.1-1.2 × length of P2 propodus; P3 merus 0.9 length of P3 propodus; P4 merus 0.8 × length of P4 propodus. P2 carpus slightly longer than or subequal to P3 carpus, P4 carpus subequal to or slightly shorter than P3 carpus, each 0.4 × length of respective propodus. Propodi subequal on P3 and P4 and slightly shorter on P2, subequal on P2-4, or subequal on P2-3 and shorter on P4 than on P3; flexor margin nearly straight on P2, slightly curving on P3 and P4, with pair of terminal spines preceded by 5-9 slender movable spines at most on distal two-thirds on P2 and P3, 4-6 on P4. Dactyli moderately curving, as long as carpi on P2, 1.0-1.1 × longer on P3, 1.2-1.3 × longer on P4, and 0.4-0.5 × as long as propodi on P2-4; flexor margin with 9-10 sharp, somewhat inclined, loosely arranged spines, ultimate spine shorter and much more slender than penultimate, penultimate broader than antepenultimate.

Eggs. Number of eggs carried, 14-25 but normal number probably more; size, 1.16 mm × 1.32 mm - 1.32 mm × 1.49 mm.

REMARKS — The ovigerous female paratype (MNHN-IU-2014-16386) from SMIB 2 Stn DC26 has a very short rostrum barely reaching midlength of eyes so that the proximal limit of eye is visible in a dorsal aspect but the other characters agree quite well with those of the others. This rostrum is presumably broken.

The species resembles *U. brachycarpus* n. sp. in the shapes of the carapace, Mxp3, antenna and P2-4 dactyli. However, *U. brachycarpus* has no lateral spines on the antennal scale; the thoracic appendages (Mxp3 and P1-4) are less setose; the P2 merus lacks a row of ventromesial marginal spines; and the P4 propodus bears a pair of terminal spines only; all distinctive differences from *U. denticulisquama*.

Uroptychus denticulisquama somewhat resembles *U. indicus* Alcock, 1901, one of the poorly known species (see Figure 217). However, the carapace in *U. indicus* is distinctly longer than broad and smooth on the lateral margin, the antennal articles 4 and 5 are unarmed, the antennal scale lacks lateral spines, and P2-4 look more slender.

Uroptychus depressus n. sp.
Figures 71, 72

TYPE MATERIAL — Holotype: **New Caledonia**, Norfolk Ridge. CHALCAL 2 Stn CP22, 24°40'S, 168°39'E, 650-750 m, 29.X.1986, ♀ 4.5 mm (MNHN-IU-2012-691). Paratypes: **Wallis and Futuna Islands**. MUSORSTOM 7 Stn DW578, 13°08'S, 176°16'W, 640-730 m, 22.V.1992, 1 ♂ 3.6 mm (MNHN-IU-2014-16387). **Tonga**. BORDAU 2 Stn DW1553, 20°42'S, 174°54'W, 650-676 m, 6.VI.2000, 1 ♂ 3.0 mm (MNHN-IU-2014-16388).

ETYMOLOGY — From the Latin *depressus* (= depressed), alluding to the depressed rostral surface clearly bordering the gastric and rostral regions.

DISTRIBUTION — Wallis and Futuna Islands, Tonga and Norfolk Ridge; 640-750 m.

DESCRIPTION — Small species. *Carapace*: Slightly broader than long (0.9 × as long as broad), greatest breadth 1.7 × distance between anterolateral spines. Dorsal surface glabrous, nearly horizontal or weakly convex from anterior to posterior, with feeble depression between gastric and cardiac regions; epigastric region with transverse row of small spines (median spine flanked by 2 denticles placed side by side), preceded by depressed rostrum. Lateral margin posteriorly divergent along hepatic region, slightly convexly divergent along branchial region; anterolateral spine well developed, slightly overreaching lateral orbital spine; branchial margin with 5-7 posteriorly diminishing spines, anteriormost as large as or slightly larger than anterolateral spine, last spine followed by distinct ridge; small spine between and ventral to level of anterolateral and anteriormost branchial marginal spines. Small spine mesial to anteriormost branchial spine. Rostrum narrow subtriangular, with interior angle of 18°, nearly horizontal, dorsal surface excavated; lateral margin unarmed or with 2 tiny denticle-like spines on distal portion; length half that of remaining carapace. Lateral orbital spine small, situated slightly anterior to level of anterolateral spine. Pterygostomian flap smooth or with a few small spines on anterior surface, anteriorly more or less angular, ending in distinct spine.

Sternum: Excavated sternum anteriorly triangular, surface with weak ridge in midline. Sternal plastron 0.9-1.0 × as long as broad, lateral extremities somewhat divergent posteriorly. Sternite 3 moderately depressed, anterior margin concave, laterally sharp angular, with narrow U-shaped median sinus flanked by small or obsolescent spine; lateral margin with small blunt spine. Sternite 4 with anterolateral margin straightly divergent posteriorly, anteriorly angular or produced; posterolateral margin longer than anterolateral margin.

Abdomen: Smooth and glabrous. Somite 1 convex from anterior to posterior, forming rounded transverse ridge. Somite 2 tergite 2.4 × broader than long; pleural lateral margin feebly convex, slightly divergent posteriorly, anterior and posterior ends both rounded. Pleural lateral margin of somite 3 blunt on posterior end. Telson slightly less than half as

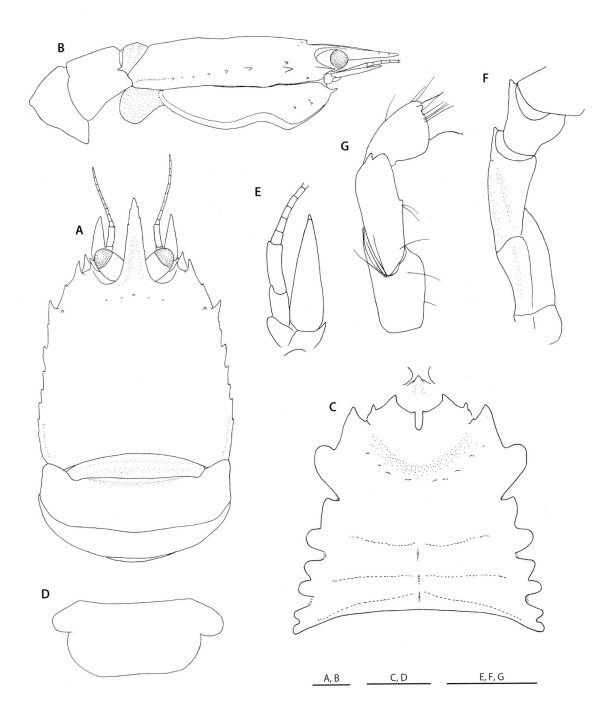

FIGURE 71

Uroptychus depressus n. sp., holotype, female 4.5 mm (MNHN-IU-2012-691). **A**, Carapace and anterior part of abdomen, dorsal. **B**, same, lateral. **C**, sternal plastron, with excavated sternum and basal parts of Mxp1. **D**, telson. **E**, left antenna, ventral. **F**, right Mxp3, setae omitted, ventral. **G**, same, lateral. Scale bars: 1 mm.

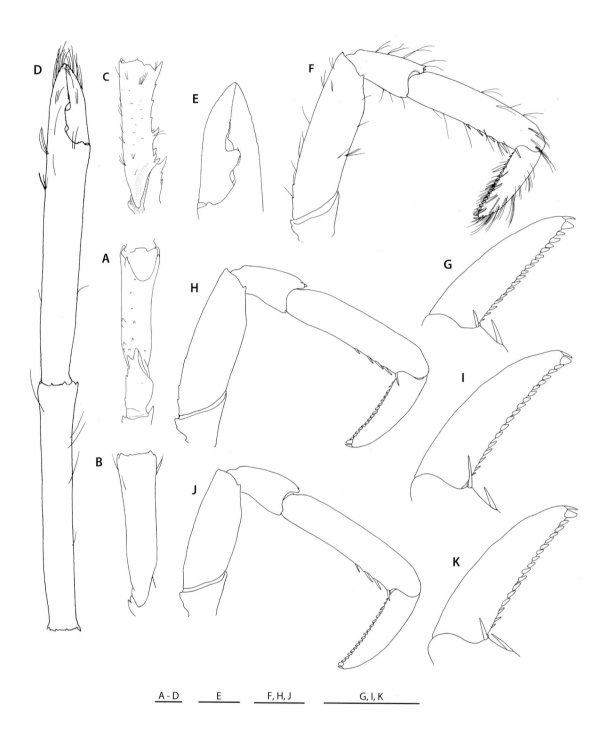

FIGURE 72

Uroptychus depressus n. sp., holotype, female 4.5 mm (MNHN-IU-2012-691). **A**, left P1, proximal part, ventral. **B**, same, dorsal. **C**, same, mesial. **D**, same, proximal part omitted, dorsal. **E**, same, fingers, ventral. **F**, right P2, lateral. **G**, same, distal part, setae omitted, lateral. **H**, right P3, setae omitted, lateral. **I**, same, distal part, setae omitted, lateral. **J**, right P4, setae omitted, lateral. **K**, same, distal part, setae omitted, lateral. Scale bars: 1 mm.

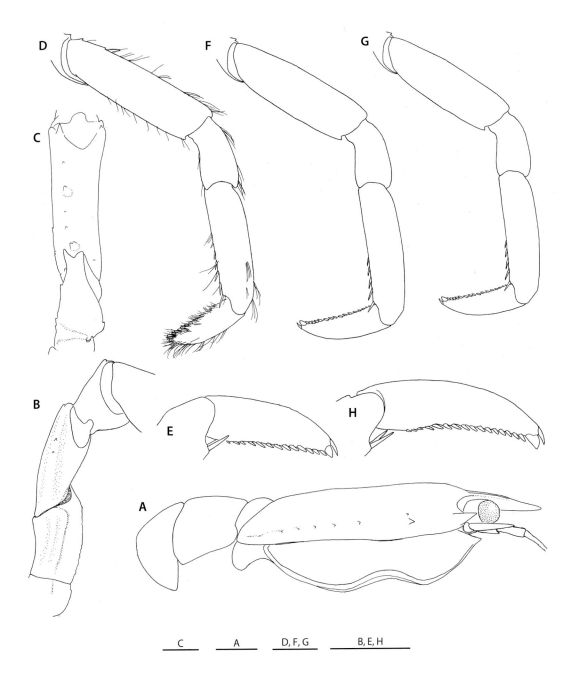

FIGURE 73

Uroptychus levicrustus Baba, 1988, holotype, ovigerous female 4.7 mm (USNM 150309). **A**, carapace and anterior part of abdomen, lateral. **B**, right Mxp3, ventral. **C**, P1, proximal articles, ventral. **D**, right P2, lateral. **E**, same, distal part, setae omitted, lateral. **F**, right P3, setae omitted, lateral. **G**, right P4, setae omitted, lateral. **H**, same, distal part, lateral. Scale bars: 1 mm.

long as broad; posterior plate feebly concave on posterior margin, length 1.3-1.4 × that of anterior plate, greatest breadth not more than distance between left and right constrictions of telson.

Eye: Short relative to breadth (length 1.4 × breath), barely reaching or slightly overreaching midlength of rostrum, narrowed proximally and distally, mesial and lateral margins convex, cornea more than half length of remaining eyestalk.

Antennule and antenna: Ultimate article of antennule 3.0-3.1 × longer than high. Antennal peduncle overreaching eye, falling short of apex of rostrum. Article 2 of antennal peduncle with distinct lateral spine. Antennal scale overreaching article 5, reaching proximal first or second segment of flagellum, breadth more than twice that of article 5. Article 4 with short distomesial spine. Article 5 unarmed, length 1.3-1.4 × that of article 4, breadth half height of antennular ultimate article. Flagellum consisting of 9-11 segments, falling short of end of P1 merus, length less than that of rostrum.

Mxp: Mxp1 with bases broadly separated. Mxp3 basis without denticles on mesial ridge. Ischium with tuft of long setae lateral to rounded distal end of flexor margin, no distinct denticles on crista dentata. Merus 2.1 × longer than ischium, flexor margin somewhat roundly ridged, bearing small blunt spine distal to midlength; distolateral spine on each of merus and carpus.

P1: Slender, smooth and polished, sparsely setose except for fingers; length 5.5-6.0 × that of carapace. Ischium with distinct dorsal spine, ventromesially with strong subterminal spine followed by 2 distal and 2 much smaller proximal spines. Merus with 3 tiny dorsal spines on distal margin, ventrally with small spines not numerous but moderate in number, ventrolaterally with small distal spine, ventromesially with 3 well-developed spines, distal spine terminal and strongest, median spine at point one-third from proximal end; length 1.2 × that of carapace. Carpus as long as or slightly shorter than palm, with 2 small ventral (ventomesial and ventrolateral) and 2 dorsal (on distal margin) spines. Palm 3.7 × (males) or 5.9 × (female) longer than broad, 3 × longer than fingers. Fingers gaping in proximal half, feebly crossing at tip, not spooned; opposable margin of movable finger with prominent blunt proximal process of subtriangular shape fitting to opposing concavity of fixed finger when closed.

P2-4: Relatively thick mesio-laterally, sparsely setose on meri and carpi, moderately so on distal part of propodi and entire dactyli. Meri successively shorter posteriorly (P3 merus 0.9 × length of P2 merus, P4 merus 0.8 × length of P3 merus), subequally broad on P2-4; length-breadth ratio, 3.2-3.8 on P2, 2.8-3.1 on P3, 2.4-2.5 on P4; dorsal margin not crested, rounded; ventrolateral margin ending in small spine. P2 merus 0.7 × length of carapace, subequal to length of P2 propodus; P3 merus 0.9 × length of P3 propodus; P4 merus 0.7-0.8 × length of P4 propodus. Carpi subequal, length less than half that of propodus (0.4 on P2 and P3, 0.3 on P4). Propodi subequal on P2 and P3, longer on P4; flexor margin straight, ending in pair of slender movable spines preceded by row of 4 or 5 (P2 and P3), 3 or 4 (P4) single movable spines at most on distal half. Dactyli successively longer posteriorly, length 1.4 × that of carpus on P2, 1.7 × on P3, 1.9 × on P4, 0.5 × length of propodus on P2, 0.6 × on P3 and P4; flexor margin slightly curving, with row of 18 or 19 obliquely directed spines obscured by setae, ultimate slender, penultimate prominent, slightly more than 2 × broader than antepenultimate, remaining proximal spines including antepenultimate close to each other, subequal to ultimate in breadth; antepenultimate spine 2.7-2.9 × longer than broad.

REMARKS — This new species is very similar to *U. levicrustus* Baba, 1988 from the Moluccas and *U. kareenae* n. sp. (see below) in having a row of branchial marginal spines, in having the anterior margin of sternite 3 with a median sinus separated by obsolescent submedian spines, and in the spination of P2-4. *Uroptychus depressus* is distinguished from *U. levicrustus* by having the epigastric region anteriorly preceded by the depressed rostrum, instead of being gently sloping down on to the rostrum (Figure 73); the eyes are medially swollen and narrowed distally and proximally, instead of having subparallel lateral and mesial margins; the antennal scale overreaches the first segment of the antennal flagellum, instead of barely reaching the distal end of antennal article 5; and the ultimate spines of the P2-4 dactyli are more slender than instead of as slender as the antepenultimae. The relationships with *U. kareenae* are discussed under the remarks of that species (see below).

Uroptychus diaphorus n. sp.

Figures 74, 75

TYPE MATERIAL — Holotype: **Tonga**. BORDAU 2 Stn DW1595, 19°03'S, 174°19'W, 523-806 m, 14.VI.2000, ov. ♀ 2.8 mm (MNHN-IU-2014-16389).

ETYMOLOGY — From the Greek *diaphoros* (different), alluding to an unusually slender P2 especially distal two articles.

DISTRIBUTION — Tonga; 523-806 m.

DESCRIPTION — Small species. *Carapace*: 1.2 × as broad as long (0.8 × as long as broad); greatest breadth measured around posterior end, 1.5 × distance between anterolateral spines. Dorsal surface slightly convex from anterior to posterior, with very feeble depression between gastric and cardiac regions; smooth, with very sparse setae. Lateral margins slightly convex and divergent posteriorly, bearing 6 spines: first anterolateral, small, not reaching tip of lateral orbital spine; second larger than first, subequal to third, located at anterior end of anterior branchial margin, equidistant between first and third; third to sixth on posterior branchial region; last one small (obsolescent on right side), followed by ridge. Rostrum sharp triangular with interior angle of 26°, straight horizontal in profile; length 0.6 × postorbital carapace length, breadth less than half carapace breadth measured at posterior carapace margin; dorsal surface moderately concave, lateral margin entire. Lateral orbital spine larger than and relatively close to anterolateral spine. Pterygostomian flap anteriorly somewhat roundish, produced to small spine, surface smooth.

Sternum: Excavated sternum strongly depressed, surface with weak ridge in midline, anterior margin bluntly subtriangular between bases of Mxp1. Sternal plastron slightly broader than long, lateral extremities somewhat divergent posteriorly. Sternite 3 moderately depressed, anterior margin moderately concave, with narrow V-shaped median notch flanked by very small spine, anterolaterally angular. Sternite 4 having anterolateral margin nearly straight and entire, anteriorly ending in triangular process, posterolateral margin half as long as anterolateral margin. Anterolateral margins of sternite 5 subparallel, anteriorly rounded, length 1.3 × that of posterolateral margin of sternite 4.

Abdomen: Smooth, very sparsely setose. Somite 1 moderately convex from anterior to posterior, without ridge. Somite 2 tergite 2.2 × broader than long; pleural lateral margin weakly concave and weakly divergent posteriorly, posterolaterally rounded. Pleuron of somite 3 posterolaterally bluntly angular. Telson slightly less than half as long as broad; posterior plate with rounded posterior margin, length 1.5 × that of anterior plate.

Eye: Terminating in anterior third length of rostrum, lateral and mesial margin somewhat convex. Cornea not dilated, slightly shorter than remaining eyestalk.

Antennule and antenna: Antennular ultimate article 2.8 × longer than high. Antennal peduncle reaching distal end of cornea. Article 2 without spine on lateral margin. Antennal scale 1.4 × broader than article 4, reaching first segment of flagellum. Article 4 with distinct distomesial spine. Article 5 as long as article 4, bearing small distomesial spine; breadth slightly more than half height of antennular ultimate article. Flagellum consisting of 14 segments, falling short of distal end of P1 merus.

Mxp: Mxp1 with bases broadly separated. Mxp3 having basis smooth on mesial ridge. Ischium with distally rounded flexor margin accompanying a few setae lateral to it; crista dentata without denticles. Merus 2 × longer than ischium, well compressed, mesially flattish, distolateral spine distinct; flexor margin with a few small spines distal to angular point one-third from distal end. Carpus unarmed.

P1: Left P1 missing. Right P1 5.5 × longer than carapace, slender, smooth and polished, sparsely setose but fingers more setose. Ischium with strong, sharp dorsal spine, ventrally unarmed but 2 tubercular processes on proximal mesial margin. Merus 1.3 × longer than carapace, mesially with strong spine at midlength and 2 tiny spinules, ventrally bearing distomesial and distolateral spines, both small but distomesial larger. Carpus 1.2 × longer than merus, ventrally bearing distomesial and distolateral spines, dorsally bearing 2 tubercular processes directly distal to juncture with merus. Palm 3.7 × longer than broad, slightly longer than carpus, somewhat depressed. Fingers somewhat incurved distally, crossing

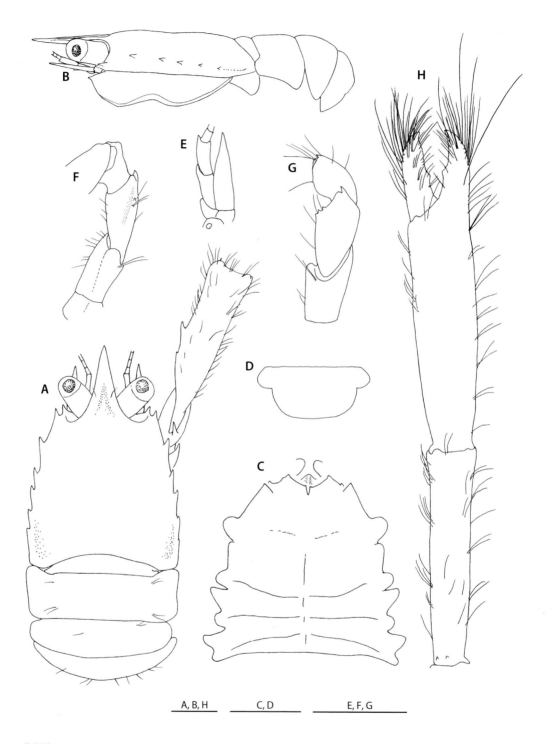

A, B, H C, D E, F, G

FIGURE 74

Uroptychus diaphorus n. sp., holotype, ovigerous female 2.8 mm (MNHN-IU-2014-16389). **A**, carapace and anterior part of abdomen, proximal part of right P1 included, dorsal. **B**, same, lateral. **C**, sternal plastron, with excavated sternum and basal parts of Mxp1. **D**, telson. **E**, left antenna, ventral. **F**, left Mxp3, ventral. **G**, same, lateral. **H**, right P1, dorsal. Scale bars: 1 mm.

when closed; movable finger one-third length of palm; opposable margin of movable finger with low, subtriangular (in dorsal view) process proximal to similarly low process placed at midlength of fixed finger.

P2-4: P3 missing on both sides. P2 and P4 rather flattish, setose like P1. P2 much more slender than P4, especially distal 2 articles; merus 5.9 × longer than broad, slightly longer than carapace, unarmed; carpus 0.4 × length of propodus, slightly shorter than dactylus; propodus distally narrowed, 6.2 × longer than broad at proximal portion of greatest breadth; flexor margin ending in pair of spines preceded by 1 spine; length about twice that of dactylus; dactylus moderately curved, flexor margin with 8 spines, ultimate slender, penultimate pronouncedly broad, remaining spines small (discernible under high magnification), oriented parallel to margin and loosely arranged. P4 merus 3.5 × longer than broad, 0.7 × as long as and 1.2 × broader than P2 merus, 0.9 × length of P4 propodus; carpus as long as that of P2, slightly less than half length of dactylus; propodus and dactylus longer and broader than on P2; propodus 2.8 × length of carpus, flexor margin ending in pair of spines preceded by 1 spine; dactylus 2 × length of carpus, 0.7 × length of propodus, well compressed, flexor margin somewhat curved, with strongly inclined spines distinctly larger than those on P2, penultimate 2 × broader than antepenultimate, proximally preceded by 11 blunt spines nearly contiguous to one another, ultimate more slender than antepenultimate.

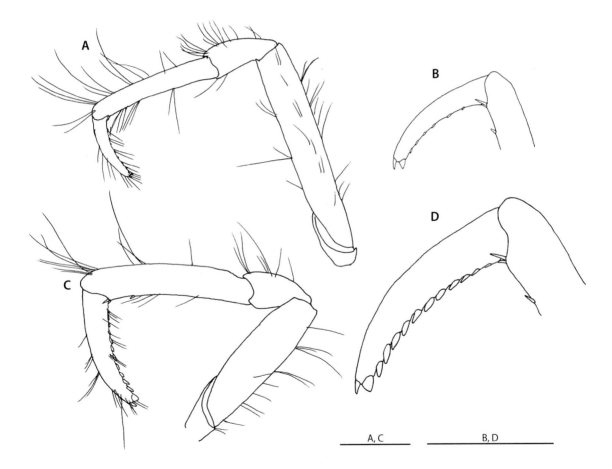

FIGURE 75

Uroptychus diaphorus n. sp., holotype, ovigerous female 2.8 mm (MNHN-IU-2014-16389). **A**, left P2, lateral. **B**, same, distal part, setae omitted, lateral. **C**, left P4, lateral. **D**, same, distal part, setae omitted, lateral. Scale bars: 1 mm.

Eggs. One egg carried on left half of abdomen, nearly ready to hatch, 1 exuvia on right side; size, 1.28 mm × 1.40 mm.

REMARKS — The new species is unique among the species of *Uroptychus* in having the propodus and dactylus of P2 much more slender and shorter than those of P4, with the dactylar spination different between P2 and P4 (P3 is missing in this species): the flexor marginal spines proximal to the pronounced penultimate spine are small, oriented parallel to the flexor margin and loosely arranged on P2, whereas these spines are larger, strongly inclined and nearly contiguous to one another on P4. The slender P2, which is different from P3 and P4 in lacking flexor marginal spines on the dactylus, is one of the diagnostic characters of *Uroptychodes* Baba, 2004. This species may likely be shifted to a different genus but is provisionally placed under *Uroptychus* awaiting molecular analysis.

The new species has the lateral carapace margin bearing a row of spines, the anterolateral spine of the carapace is smaller than the lateral orbital spine, the penultimate of the P2-4 dactylar spines is twice as broad as the antepenultimate, and the ultimate of the same is more slender than the penultimate, all characters shared by *U. pronus* Baba, 2005. *Uroptychus diaphorus* differs from that species in having the antennal scale overreaching the antennal article 5 instead of barely reaching the tip of that article, in having both the P2 and P4 meri unarmed instead of bearing a row of spines along the dorsal margin, and more distinctly in having a different spination between P2 and P4 (see above).

Uroptychus disangulatus n. sp.
Figures 76, 77

TYPE MATERIAL — Holotype: **French Polynesia**, Austral Islands. BENTHAUS Stn DW1898, 27°34.3'S, 144°26.6'W, 580-820 m, 8.XI.2002, ov. ♀ 8.0 mm (MNHN-IU-2014-16390). Paratypes: **French Polynesia**, Austral Islands. Station data as for the holotype, 1 ♂ 4.7 mm (MNHN-IU-2014-16391). – Stn DW2001, 22°26.6'S, 151°20.1'W, 200-550 m, with Chryso-gorgiidae gen. sp. (Calcaxonia), 23.XI.2002, 1 ♂ 8.1 mm (MNHN-IU-2014-16392). **New Caledonia**, Hunter and Matthew Islands. VOLSMAR Stn DW05, 22°26'S, 171°46'E, 620-700 m, 01.VI.1989, 1 ♀ 7.8 mm (MNHN-IU-2014-16393). **New Caledonia**, Norfolk Ridge. NORFOLK 2 Stn DW2027, 23°26.34'S, 167°51.38'E, 465-650 m, 21.X.2003, 2 ♂ 7.1, 8.6 mm, 1 ov. ♀ 9.9 mm (MNHN-IU-2014-16394).

ETYMOLOGY — From the Latin *dis* (without) plus *angulatus* (having angles), alluding to the absence of lateral orbit spine, a character to distinguish the species from *U. brevisquamatus*.

DISTRIBUTION — Hunter-Matthew, Norfolk Ridge and French Polynesia; in 200-820 m.

DESCRIPTION — Medium-sized species. *Carapace*: As long as broad; greatest breadth 1.7 × distance between anterolateral spines. Dorsal surface smooth or somewhat granulose, moderately convex from anterior to posterior, with depression between gastric and cardiac regions; epigastric region with pair of denticulate ridges. Lateral margins convex posteriorly, with somewhat elevated, short granulate ridge at anterior end of branchial region, and ridged along posterior third or posterior half; anterolateral spine relatively large, overreaching lateral limit of orbit without spine. Rostrum anteriorly ending in blunt or rounded margin, with interior angle of 22-27°, somewhat overreaching eyes; length at most one-third postorbital carapace length, breadth less than half carapace breadth measured at posterior carapace margin; dorsal surface flattish. Lateral limit of orbit rounded, not angular, lacking spine. Pterygostomian flap anteriorly roundish, ending in small spine; surface somewhat granulose.

Sternum: Excavated sternum with low ridge in midline or weak tubercle-like process in center, anterior margin sharp or blunt triangular. Sternal plastron moderately broadened posteriorly, slightly shorter than broad. Sternite 3 depressed well, anterior margin deeply excavated, with pair of small submedian spines separated by narrow or small V-shaped median notch; lateral margin with distinct spine at midlength. Sternite 4 with anterolateral margin convex and anteriorly ending

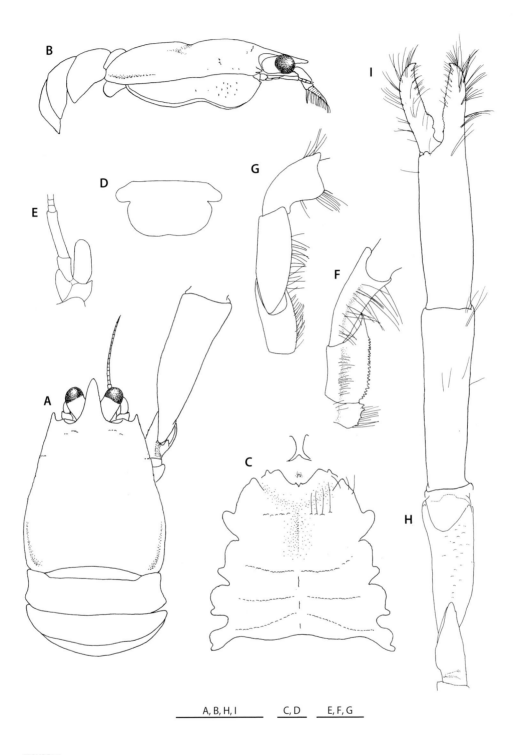

A, B, H, I   C, D   E, F, G

FIGURE 76

Uroptychus disangulatus n. sp., holotype, ovigerous female 8.0 mm (MNHN-IU-2014-16390). **A**, carapace and anterior part of abdomen, proximal part of P1 included, dorsal. **B**, same, lateral. **C**, sternal plastron, with excavated sternum and basal parts of Mxp1. **D**, telson. **E**, left antenna, ventral. **F**, right Mxp3, ventral. **G**, same, lateral. **H**, right P1, proximal part, ventral. **I**, same, proximal part omitted, dorsal. Scale bars: A, B, H, I, 5 mm; C-G, 1 mm.

in roundish terminus; posterolateral margin relatively long, more than half length of anterolateral margin. Anterolateral margin of sternite 5 convex well, about as long as posterolateral margin of sternite 4.

Abdomen: Smooth, barely setose. Somite 1 strongly convex from anterior to posterior. Somite 2 tergite 2.3-2.5 × broader than long; lateral margin strongly divergent, posterolateral corner bluntly angular. Pleura of sternites 3 and 4 posterolaterally angular. Telson half as long as broad; posterior plate emarginate on posterior margin, length 1.3-1.6 × that of anterior plate.

Eye: Long relative to breadth, about twice as long as broad, reaching distal quarter of rostrum; mesial margin concave proximal to cornea. Cornea slightly more than half length of remaining eyestalk.

Antennule and antenna: Ultimate article of antennular peduncle 2.1-2.5 × longer than high. Antennal peduncle slightly overreaching cornea. Article 2 produced to short, stout spine on distal lateral margin. Antennal scale relatively broad (1.7-2.0 × as broad as article 4), roundly truncate or ending in blunt tip, varying from overreaching to barely reaching midlength of antennal article 5, not reaching its distal end. Distal 2 articles unarmed; article 5 2.0 × longer than article 4, breadth slightly less than half height of ultimate antennular article. Flagellum consisting of 13-17 segments, reaching or barely reaching distal end of P1 merus.

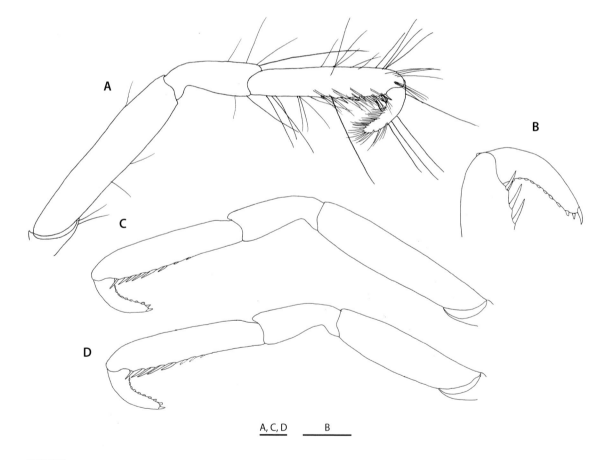

FIGURE 77

Uroptychus disangulatus n. sp., holotype, ovigerous female 8.0 mm (MNHN-IU-2014-16390). **A**, right P2, lateral. **B**, same, distal part, setae omitted, lateral. **C**, left P3, setae omitted, lateral. **D**, left P4, lateral. Scale bars: 1 mm.

Mxp: Mxp1 with bases close to each other, slightly separated or nearly contiguous. Mxp3 unarmed on merus and carpus. Basis with 3-4 denticles on mesial ridge. Ischium short relative to merus, flexor margin not rounded distally, crista dentata with 22-26 denticles. Merus 2.4 × longer than ischium, thick mesio-laterally, with rounded ridge along flexor margin. No spine on carpus.

P1: Relatively massive, 4.0-4.4 × longer than carapace, barely setose except for fingers, bearing tubercles on ventral surface of ischium, merus and carpus (smooth in smallest male). Ischium dorsally with basally broad, depressed, relatively short distal spine, ventrally with rudimentary subterminal spine on mesial margin. Merus subequal to or slightly longer than carapace, ventrally weakly granulose bearing distomesial and distolateral spines occasionally obsolescent. Carpus 1.0-1.2 × longer than merus. Palm 2.2-2.6 × longer than broad, as long as carpus. Fingers distally incurved, crossing when closed, slightly gaping in males, not gaping in females; opposable margin of movable finger with low proximal process (bidentate in males); that of fixed finger sinuous in proximal third in males, nearly straight in females; movable finger 0.6 × as long as palm.

P2-4: Unarmed on meri and carpi, distal 2 articles with sparse long setae. Meri moderately compressed mesio-laterally, distally narrowed, successively shorter posteriorly (P3 merus 0.9 × length of P2 merus, P4 merus 0.8-0.9 × length of P3 merus), subequal in breadth on P2-4; length-breadth ratio, 4.8-5.3 on P2, 4.2-4.7 on P3, 3.4-3.8 on P4; P2 merus 0.9-1.0 × length of carapace, 1.3-1.4 × length of P2 propodus; P3 merus 1.2 × length of P3 propodus; P4 merus 0.8-0.9 × length of P4 propodus. Carpi subequal or successively shorter posteriorly (P3 carpus 0.9-1.0 × length of P2 carpus, P4 carpus 0.9 × length of P3 carpus), carpus-propodus length ratio, 0.5-0.6 on P2, 0.5 on P3, 0.4-0.5 on P4. Propodi subequal on P2-4 or somewhat longer on P4 than on P2 and P3; flexor margin nearly straight, terminating in pair of spines preceded by 6-9 spines on P2, 6-8 on P3, 4-6 on P4. Dactyli slender relative to propodi, subequal, about one-third (max. 0.4) length of propodi; dactylus-carpus length ratio, 0.6-0.7 on P2 and P3, 0.7-0.8 on P4; flexor margin with 11 moderately loosely arranged spines, ultimate longest, penultimate and antepenultimate subequal and shorter than ultimate, remaining spines small, blunt and oriented parallel to flexor margin, all obscured by setae.

Eggs. Number of eggs carried, 4-13; size, 1.48 mm × 1.65 mm - 1.55 mm × 1.68 mm; 4 ova carried by the specimen from Norfolk 2, Stn DW2027 nearly ready to hatch.

REMARKS — The ocular peduncles in the specimens from the Norfolk Ridge (MNHN-IU-2014-16394) are much more slender than in the other specimens, and the carapace is finely granulose in the female (9.9 mm) and the smaller male (7.1 mm), faintly so in the larger male (8.6 mm).

This species has a carapace bearing an anterolateral spine only on the lateral margin and the P2-4 dactyli bearing small spines oriented parallel to the flexor margin, characters also possessed by *U. australis* (Henderson, 1885) and *U. brevirostris* Van Dam, 1933. *Uroptychus disangulatus* is distinguished from these species by the lateral limit of the orbit, which is rounded and unarmed instead of produced to a distinct spine; the antennal scale, which is broader and distally blunt or rounded instead of distally tapering; and the flexor marginal spines of the P2-4 dactyli, which are much shorter. In addition, *U. disangulatus* differs from *U. australis* in the P4 merus-P3 merus length ratio which is 0.8-0.9 instead of 0.6, and in having the anterolateral spine located very slightly posterior to the lateral orbital angle instead of being more posterior. *Uroptychus brevirostris* has broader P2-4 meri, *i.e.*, the length-breadth ratio on P2 merus being 3.5-3.6 instead of more than 4.8 as in *U. australis* and *U. disangulatus*.

The new species also strongly resembles *U. brevisquamatus* Baba, 1988 (see above) and *U. webberi* Schnabel, 2009 from the Kermadec Ridge, in the shape of the carapace, sternal plastron, and in the spination of the P2-4 dactyli. In *U. brevisquamatus*, the lateral orbital spine is distinct; the P1 ischium bears a distinct subterminal spine rather than being unarmed on the ventromesial margin; and the terminal of the flexor marginal spines of P2-4 propodi is single, not paired as in the present new species. *Uroptychus disangulatus* differs from *U. webberi* in the following particulars: the lateral limit of the orbit is rounded instead of ending in a spine; the carapace lateral margins are convex instead of subparallel along the branchial region; the Mxp3 ischium bears 22-26 instead of 12 denticles on the crista dentata; and the P1 ischium bears a rudimentary instead of small but distinct subterminal spine on the ventromesial margin.

Uroptychus dissitus n. sp.

Figures 78, 79

TYPE MATERIAL — Holotype: **Vanuatu**. MUSORSTOM 8 Stn DW1060, 16°14'S, 167°21'E, 394-375 m, 2.X.1994, ♂ 2.4 mm (MNHN-IU-2014-16395).

ETYMOLOGY — From the Latin *dissitus* (distant), alluding to the anterior second of the carapace lateral marginal spines that is remotely separated from the first anterolateral spine.

DISTRIBUTION — Vanuatu; 397-375 m.

DESCRIPTION — *Carapace*: Slightly broader than long (0.9 × as long as broad); greatest breadth 1.4 × distance between anterolateral spines. Dorsal surface smooth, bearing scattered short setae, nearly horizontal from anterior to posterior, with feebly depression between gastric and cardiac regions. Lateral margins slightly convex and slightly divergent posteriorly, bearing 6 spines: first spine anterolateral, reaching tip of smaller lateral orbital spine, and close to that spine; second to sixth situated on branchial margin, posteriorly diminishing, second nearly as large as first, placed at anterior end of branchial margin and remotely equidistant between first and third; last followed by 1 or 2 tubercles. Rostrum sharp triangular, with interior angle of 22°; dorsal surface moderately concave; length about half length of carapace, breadth half carapace breadth measured at posterior carapace margin. Pterygostomian flap nearly smooth but scattered very tiny tubercles supporting short setae on surface, anterior margin angular, produced to small spine.

Sternum: Excavated sternum anteriorly triangular, bearing ridge in midline. Sternal plastron about as long as broad; lateral extremities subparallel between sternites 4-7. Sternite 3 shallowly depressed, anterior margin weakly concave, with narrow U-shaped median sinus without flanking spine, anterolaterally angular. Sternite 4 having anterolateral margin nearly straight, anteriorly bidentate; posterolateral margin slightly shorter than anterolateral margin. Anterolateral margin of sternite 5 convex anteriorly, length 0.8 × that of posterolateral margin of sternite 4.

Abdomen: With very sparse short setae. Somite 1 moderately convex from anterior to posterior. Somite 2 tergite 2.8 × broader than long; pleuron anterolaterally blunt angular, posterolaterally rounded, lateral margin somewhat concave, slightly divergent posteriorly. Pleuron of somite 3 bluntly angular on posterolateral terminus. Telson slightly less than half as long as broad; posterior plate with convex posterior margin, length 1.2 × that of anterior plate.

Eye: Slightly falling short of rostral tip, 1.6 × longer than broad, lateral and mesial margins slightly convex. Cornea not dilated, more than half as long as remaining eyestalk.

Antennule and antenna: Ultimate antennular article 2.9 × longer than high. Antennal peduncle overreaching eye, nearly reaching rostral tip. Article 2 with acute short distolateral spine. Antennal scale slightly overreaching article 5; breadth about 1.7 × that of article 5. Articles 4 and 5 each with distomesial spine. Article 5 1.3 × longer than article 4, breadth slightly less than half height of antennular ultimate article. Flagellum consisting of 10-11 segments, barely reaching distal end of P1 merus.

Mxp: Mxp1 with bases close to each other. Mxp3 basis smooth, without denticles on mesial ridge. Ischium having flexor margin distally rounded; crista dentata with obsolescent denticles. Merus 2.0 × longer than ischium, broad relative to length, flattish, and concave on mesial face; flexor margin sharply cristate and convex on distal third, bearing a few obsolete denticles. Small but distinct distolateral spine on each of merus and carpus.

P1: 5.5 × longer than carapace, smooth, moderately setose but glabrous on ventral surface, setae arising from short transverse or scale-like granulate ridges. Ischium with long, sharp dorsal spine, ventromesially with strong subterminal spine. Merus 1.2 × length of carapace, dorsally with a few distal spines near juncture with carpus, and row of small spines along dorsomesial margin, mesially with 3-4 strong, proximally diminishing spines on proximal half of length; ventrally with sparse granules supporting setae, and small distolateral and strong distomesial spines. Carpus also with 2 distoventral (mesial and lateral) spines, 3 small distodorsal spines, a few tubercles distal to dorsal juncture with merus; length slightly

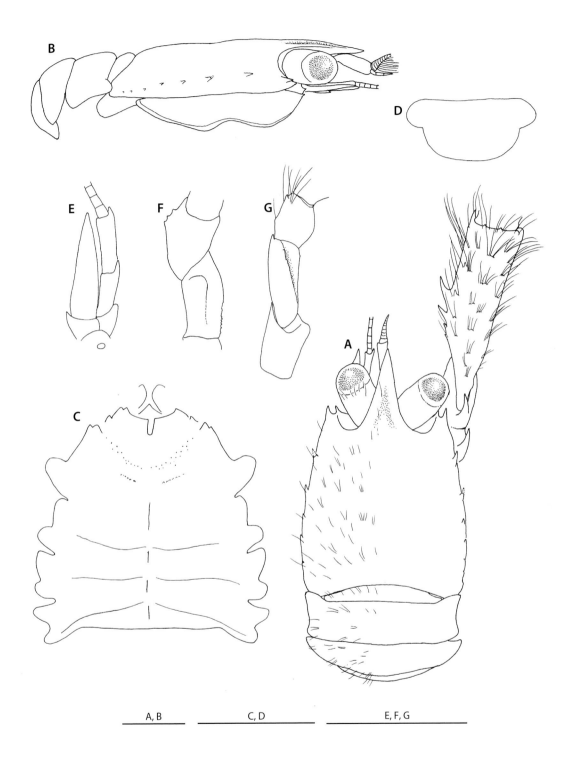

FIGURE 78

Uroptychus dissitus n. sp., holotype, male 2.4 mm (MNHN-IU-2014-16395). **A**, carapace and anterior part of abdomen, setae omitted on right half, proximal part of right P1 included, dorsal. **B**, same, setae omitted, lateral. **C**, sternal plastron, with excavated sternum and basal parts of Mxp1. **D**, telson. **E**, right antenna, ventral. **F**, right Mxp3, ventral. **G**, same, lateral. Scale bars: 1 mm.

more than that of merus. Palm unarmed, somewhat massive but moderately depressed, somewhat broader distally, 2.7-2.8 × longer than broad, 1.2 × longer than carpus. Fingers gaping, distally incurved, crossing when closed; movable finger half as long as palm, opposable margin angular at distal quarter, with process of moderate size proximal to midlength, opposing margin of fixed finger somewhat sinuous.

P2-4: Setose like P1, moderately compressed mesio-laterally. Meri dorsally unarmed, ventrally with distolateral spine, successively shorter posteriorly (P3 merus 0.95 × length of P2 merus, P4 merus 0.8 × length of P3 merus), subequally broad on P2-4; length-breadth ratio, 4.6 on P2, 3.7 on P3, 3.0 on P4; P2 merus slightly shorter than carapace, 1.2 × longer than P2 propodus; P3 merus as long as P3 propodus; P4 merus 0.8 × length of P4 propodus. Carpi subequal; each carpus

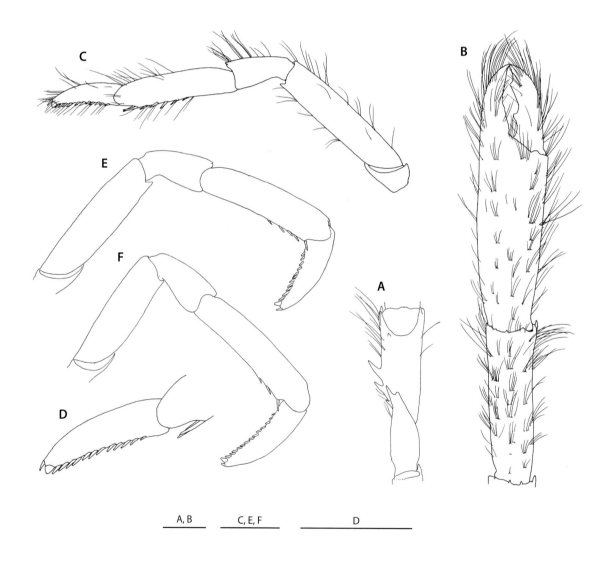

FIGURE 79

Uroptychus dissitus n. sp., holotype, male 2.4 mm (MNHN-IU-2014-16395). **A**, left P1, proximal part, ventral. **B**, same, distal part, dorsal. **C**, left P2, lateral. **D**, same, distal part, setae omitted, lateral. **E**, right P3, setae omitted, lateral. **F**, right P4, lateral. Scale bars: 1 mm.

less than half as long as propodus (carpus-propodus length ratio, 0.45 on P2, 0.40 on P3, 0.37 on P4), distinctly shorter than dactylus. Propodi shortest on P2, subequal on P3 and P4; flexor margin ending in pair of spines located at juncture with dactylus, preceded by 4 or 5 spines on P2 and P3, 2 or 3 on P4. Dactyli subequal on P3 and P4 and shorter on P2; dactylus-propodus length ratio, 0.6 on P2, 0.5 on P3 and P4; 1.4 × longer than carpi on P2-4 (1.2 × longer on left P2); flexor margin nearly straight, with 13-15 obliquely directed, closely arranged spines obscured by setae, penultimate spine more than twice breadth of other spines and distally blunt, remainder slender, distal spines nearly contiguous to one another.

REMARKS — The carapace shape and the spination of the P2-4 dactyli of the species are very similar to those of *U. spinulus* n. sp. Their relationships are discussed under that species (see below).

Uroptychus dualis n. sp.

Figures 80, 81

TYPE MATERIAL — Holotype: **New Caledonia**. MUSORSTOM 4 Stn DW156, 18°54'S, 163°19'E, 530 m, 15.IX.1985, ♂ 3.4 mm (MNHN-IU-2014-16396). Paratypes: **New Caledonia**. MUSORSTOM 4 Stn DW156, 18°54.00'S, 163°18.80E, 530 m, 15.IX.1985, 2 ♂ 3.0-3.4 mm, 1 ov. ♀ 4.2 mm (MNHN-IU-2014-16397). – Stn DW162, 18°35'S, 163°10'E, 535 m, 16.IX.1985, 1 ♂ 3.7 mm, 1 ov. ♀ 4.0 mm (MNHN-IU-2014-16398). – Stn CP194, 18°52'S, 163°21'E, 550 m, 19.IX.1985, 1 ov. ♀ 4.7 mm (MNHN-IU-2014-16399), 24 ♂ 3.5-5.4 mm, 22 ov. ♀ 3.4-5.0 mm, 3 ♀ 3.5-3.7 mm (MNHN-IU-2014-16400). – Stn DW197, 18°51'S, 163°21'E, 560 m, 20.IX.1985, 29 ♂ 2.6-5.1 mm, 19 ov. ♀ 2.7-5.1 mm, 5 ♀ 2.5-4.4 mm (MNHN-IU-2014-16401). – Stn CP199, 18°51'S, 163°14'E, 600 m, 20.IX.1985, 1 ♂ 3.2 mm (MNHN-IU-2014-16402).

OTHER MATERIAL EXAMINED — **New Caledonia**, Chesterfield Islands. MUSORSTOM 5 Stn DW355, 19°36.43'S, 158°43.41'E, 580 m, 18. X.1986, 1 ♂ 2.7 mm (MNHN-IU-2014-16404). **New Caledonia**. BIOCAL Stn DW38, 23°00'S, 167°15'E, 360 m, 30.VIII.1985, 1 ♂ 1.7 mm (MNHN-IU-2010-5435). – Stn DW77, 22°15'S, 167°15'E, 440 m, 5. IX.1985, 1 ov. ♀ 3.3 mm (MNHN-IU-2014-16407). **New Caledonia**, Isle of Pines. MU-SORSTOM 4 Stn CP216, 22°59.5'S, 167°22.0'E, 490-515 m, 29.IX.1985, 1 ov. ♀ 3.4 mm (MNHN-IU-2014-16408). – Stn DW221, 22°58.6'S, 167°36.8'E, 535-560 m, 29.IX.1985, 9 ♂ 2.5-4.2 mm, 6 ov. ♀ 2.5-3.8 mm (MNHN-IU-2014-16409). – Stn DW222, 22°57.6'S, 167°33.0'E, 410-440 m, 30.IX.1985, 1 ♂ 2.8 mm (MNHN-IU-2014-16410). BERYX 11 Stn DW27, 23°37.25'S, 167°41.20'E, 460-470 m, 18.X.1992, 9 ♂ 3.6-5.0 mm, 7 ov. ♀ 3.9-5.0 mm, 1 ♀ 3.9 mm (MNHN-IU-2014-16411). BIOCAL Stn CP52, 23°06'S, 167°47'E, 540-600 m, 31.VIII.1985, 1 ♀ 3.2 mm (MNHN-IU-2014-16412). SMIB 2 Stn DC26, 22°59'S, 167°23'E, 500-535 m, 21. IX.1986, 1 ♀ 3.2 mm (MNHN-IU-2014-16413). SMIB 3 Stn DW12, 23°42'S, 167°41'E, 470-470 m, 22 .V.1987, 2 ♂ 4.5, 5.0 mm, 1 ov. ♀ 4.6 mm, 1 ♀ 4.3 mm (MNHN-IU-2014-16414). – Stn DW24, 22°58.7'S, 167°21.1'E, 535 m, 24.V.1987, 1 ♂ 4.1 mm, 1 ov. ♀ 3.6 mm (MNHN-IU-2014-16415). **New Caledonia**, Norfolk Ridge. CHALCAL 2 Stn DW76, 23°40.5'S, 167°45.2'E, 470 m, 3.X.1986, 2 ♂ 4.2, 4.4 mm (MNHN-IU-2014-16416). – Stn DW77, 23°38.35'S, 167°42.68'E, 435 m, 30.X.1986, 1 ♀ 2.0 mm (MNHN-IU-2014-16417). BERYX 11 Stn DW38, 23°37.53'S, 167°39.42'E, 550-690 m, 19.X.1992, 2 ♂ 4.1, 4.8 mm, 2 ov. ♀ 4.9, 5.0 mm (MNHN-IU-2014-16418). LITHIST Stn CP02, 23°37.1'S, 167°41.1'E, 442 m, 10.VIII.1999, 1 ov. ♀ 4.2 mm (MNHN-IU-2014-16419). – Stn CP03, 23°37.0'S, 167°41.5'E, 447 m, 10.VIII.1999, 1 ov. ♀ 4.5 mm, 1 ♀ 4.5 mm (MNHN-IU-2014-16420). NORFOLK 1 Stn CP1655, 23°26'S, 167°51'E, 680 m, 19.VI.2001, 1 ♂ 3.5 mm (MNHN-IU-2014-16421). – Stn DW1659, 23°37'S, 167°41'E, 449-467 m, 20.VI.2001, 1 ♂ 5.0 mm, 4 ov. ♀ 3.9-4.9 mm (MNHN-IU-2014-16422). – Stn DW1662, 23°38'S, 167°42'E, 462-491 m, 20.VI.2001, 2 ov. ♀ 4.1, 4.4 mm (MNHN-IU-2014-16423). – Stn DW1666, 23°42'S, 167°44'E, 469-860 m, 20.VI.2001, 7 ♂ 3.8-5.2 mm, 6 ov. ♀ 4.0-4.7 mm (MNHN-IU-2014-16424). – Stn DW1667, 23°40'S, 168°01'E, 237-250 m, 21.VI.2001, 1 ♂ 5.1 mm (MNHN-IU-2014-16425). – Stn CP1671, 23°41'S, 168°00'E, 320-397 m, on *Chironephthya* sp. (Nidaliidae: Alcyonacea), 21.VI.2001, 1 ov. ♀ 3.7 mm (MNHN-IU-2014-16426). – Stn DW1707, 23°43'S, 168°16'E, 381-493 m, 25.VI.2001, 1 ♂ 4.0 mm, 2 ov. ♀ 4.8, 4.9 mm (MNHN-IU-2014-16427). – Stn DW1722, 23°18'S, 168°01'E, 540m, 26.VI.2001, 3 ov. ♀ 4.2-4.7 mm, 3 ♀ 2.6-4.0 mm (MNHN-IU-2014-16428). – Stn DW1732, 23°21'S, 168°16'E, 347-1063 m, 27.VI.2001, 1 ♂ 3.7 mm (MNHN-IU-2014-16429). NORFOLK 2 Stn DW2024, 23°27.92'S, 167°50.90'E, 370-371 m, 21.X.2003, 2 ♂ 3.0, 3.2 mm, 1 ov. ♀ 4.0 mm, 2 ♀ 4.5, 4.3 mm (MNHN-IU-2014-16430). – Stn DW2031, 23°38.83'S, 167°44.01'E, 440-440 m, 22.X.2003, 4 ov. ♀ 4.3-5.4 mm (MNHN-IU-2014-16431). – Stn DW2034, 23°40.64'S, 167°41.36'E, 485-505 m, 22.X.2003, 3 ♂ 3.4-5.4 mm, 2 ov. ♀ 4.2, 4.5 mm (MNHN-IU-2014-16432). – Stn DW2035, 23°39.82'S, 167°40.43'E, 515-540 m, 22.X.2003, 1 ov. ♀ 5.0 mm, 1 ♀ 5.3 mm (MNHN-IU-2014-16433). – Stn DW2049, 23°42.88'S, 168°15.43'E, 470-621 m, 24.X.2003, 1 ♂ 4.1 mm, 1 ov. ♀ 3.8 mm, 2 ♀ 3.2, 3.6 mm (MNHN-IU-2014-16434). – Stn DW2056, 24°40'S, 168°39'E, 573-600 m, 25.X.2003, 2 ov. ♀ 4.7, 4.8 mm (MNHN-IU-2014-16435). – Stn DW2132, 23°17.30'S, 168°13.56'E, 405-455 m, 2.XI.2003, 1 ♂ 4.1 mm, 1 ♀ 2.7 mm (MNHN-IU-2014-16436). – Stn DW2136, 23°00.73'S, 168°22.68'E,

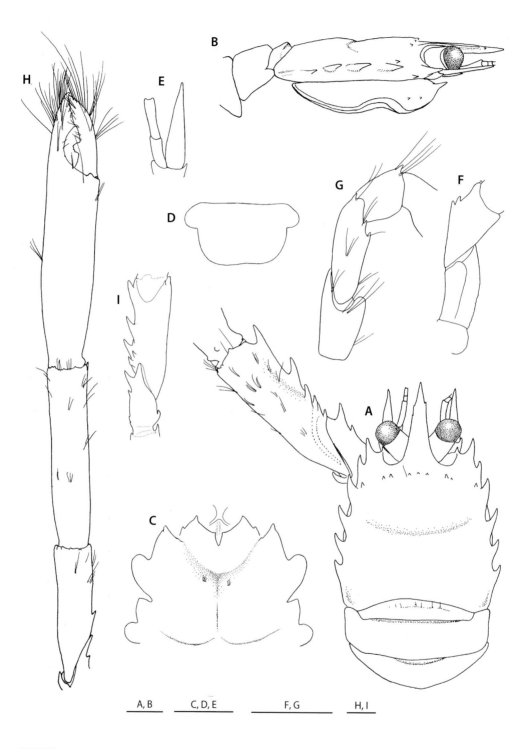

A, B   C, D, E   F, G   H, I

FIGURE 80

Uroptychus dualis n. sp., holotype, male 3.4 mm (MNHN-IU-2014-16396). **A**, carapace and anterior part of abdomen, proximal articles of left P1 included, dorsal. **B**, same, lateral. **C**, anterior part of sternal plastron, with excavated sternum and basal parts of Mxp1. **D**, telson. **E**, left antenna, ventral. **F**, right Mxp3, ventral. **G**, same, lateral. **H**, left P1, dorsal. **I**, same, proximal part, setae omitted, ventral. Scale bars: 1 mm.

402-410 m, 3.XI.2003, 1 ♂ 3.0 mm, 1 ov. ♀ 3.6 mm (MNHN-IU-2014-16437). **Wallis and Futuna Islands**. MUSORSTOM 7 DW525, 13°11′S, 176°15′W, 500-600 m, 13.V.1992, 1 ov. ♀ 4.0 mm (MNHN-IU-2014-16403). **Vanuatu**. MUSORSTOM 8 Stn DW988, 19°16′S, 169°24′E, 372-466 m, 23.IX.1994, 1 ov. ♀ 3.3 mm, 1 ♀ 3.8 mm (MNHN-IU-2014-16405). – Stn DW1003, 18°49′S, 168°59′E, 327-200 m, 25.IX.1994, 1 ♂ 3.7 mm (MNHN-IU-2014-16406).

ETYMOLOGY — From the Latin *dualis* (of two) in reference to two strong ventromesial spines on the P1 merus by which the species can be separated from *U. floccus*, *U. quinarius*, and *U. lumarius*, the latter three species bearing three or four spines.

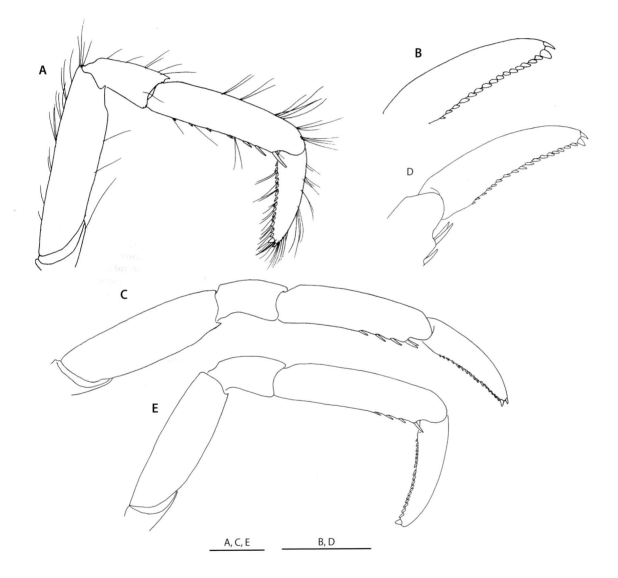

A, C, E   B, D

FIGURE 81

Uroptychus dualis n. sp., holotype, male 3.4 mm (MNHN-IU-2014-16396). **A**, right P2, lateral. **B**, same, distal part, setae omitted, lateral. **C**, right P3, setae omitted, lateral. **D**, same, distal part, lateral. **E**, right P4, lateral. Scale bars: 1 mm.

DISTRIBUTION — Wallis and Futuna Islands, Chesterfield Islands, Vanuatu, New Caledonia, Isle of Pines and Norfolk Ridge; 200-860 m.

SIZE — Males, 2.5-5.4 mm; females, 2.0-5.4 mm; ovigerous females from 2.5 mm.

DESCRIPTION — Small species. *Carapace*: Slightly broader than long (0.8 × as long as broad); greatest breadth 1.4 × distance between anterolateral spines. Dorsal surface slightly convex from anterior to posterior, with depression (occasionally obsolescent) suggesting cervical groove; epigastric region with transverse row of 7 or 8 small spines preceded by depressed rostrum, laterally leading to small spine dorsal to third lateral marginal spine. Lateral margins slightly convex, with 6 sharp spines: first anterolateral, well developed, extending far beyond small lateral orbital spine; second ventral to level of others, much smaller; third to sixth strong, last (sixth) followed by ridge. Rostrum sharp triangular, with interior angle of ca. 20° (18-23°), horizontal; length 0.4-0.6 × postorbital carapace length (relatively long in small specimens), breadth less than half carapace breadth measured at posterior carapace margin; lateral margin with a few very small spines distally but often obsolescent or absent; dorsal surface moderately concave. Lateral orbital spine situated directly mesial to but remote from anterolateral spine. Pterygostomian flap with anterior margin produced to strong spine; surface with a few spines small or obsolete, often well developed, situated on anterior median portion, without spine below linea anomurica between second and third lateral spines.

Sternum: Excavated sternum blunt triangular between bases of Mxp1, surface with longitudinal ridge in midline. Sternal plastron somewhat broader than long, with gently divergent lateral extremities. Sternite 3 depressed well, anterior margin strongly concave, with median notch of narrow V-shape flanked by obsolescent spine, anterolateral end sharply angular, lateral margin with small spine. Sternite 4 with anterolateral margin relatively short, slightly convex, entire, with anterior end angular or produced to small spine; posterolateral margin distinctly longer than anterolateral margin. Anterolateral margin of sternite 5 anteriorly angular or rounded, about half as long as posterolateral margin of sternite 4.

Abdomen: Smooth and glabrous. Somite 1 with distinct transverse ridge. Segment 2 tergite 2.6-2.8 × broader than long; pleural lateral margin weakly concave and weakly divergent posteriorly, posterolaterally rounded. Pleuron of somite 3 laterally blunt. Telson barely or fully half as long as broad; posterior plate 1.3-1.6 × longer than anterior plate, posterior margin feebly to distinctly concave.

Eye: Somewhat elongate (length 1.8 × breadth), slightly overreaching midlength of rostrum, somewhat broadened proximally. Cornea not dilated, more than half as long as remaining eyestalk.

Antennule and antenna: Ultimate antennular article relatively slender, 3.2-3.6 × longer than high. Antennal peduncle overreaching cornea and falling short of rostral tip. Article 2 with sharp distolateral spine. Antennal scale distinctly overreaching article 5, at least reaching second segment of flagellum, breadth about twice that of article 5. Article 4 with small, often obsolescent distomesial spine. Article 5 unarmed, 1.2-1.5 × longer than article 4, breadth 0.5-0.6 × height of ultimate antennular article. Flagellum of 11-12 segments falling short of distal end of P1 merus.

Mxp: Mxp1 with bases close to each other but not contiguous. Mxp3 basis without denticles on mesial ridge. Ischium with distally rounded flexor margin; crista dentata with denticles very small, obsolescent or nearly absent. Merus and carpus each with distolateral spine; merus 2.2 × longer than ischium, mesial face flattish (slightly concave), lateral face convex, flexor margin sharply ridged, with a few spines around distal third of length.

P1: More than 5 × longer than carapace (5.4-6.1 × in males, 5.7-6.0 × in females). Ischium with strong, curved, compressed distodorsal spine, and ventromesially with strong subterminal spine proximally followed by a few diminishing spines. Merus 1.3-1.5 × longer than carapace, with 2 strong mesioventral spines and very often additional 1 or 2 small or obsolete spines proximal to these, and field of 3 small spines at mesial proximal portion. Carpus 1.4-1.7 × longer than merus; distoventral spines on each side. Palm 3.5-4.2 × (males), 4.3-6.6 × (females) longer than broad, as long as or slightly longer than carpus. Fingers gaping in large specimens, more strongly so in males than in females, fitting on distal third, distally slightly incurved, crossing when closed, not spooned; movable finger with median process on gaping opposable margin, length 0.3-0.4 × that of palm.

P2-4: Moderately depressed mesio-laterally, sparsely or moderately setose; meri successively shorter posteriorly (P3 merus 0.8-0.9 × length of P2 merus, P4 merus 0.8-0.9 × length of P3 merus), ventrolateral margin with small terminal spine, dorsal margin unarmed; length-breadth ratio, 4.0-5.3 on P2, 3.6-4.5 on P3, 3.0-3.3 on P4; P2 merus nearly as long as or slightly shorter than carapace, 1.2-1.5 × longer than P2 propodus, slightly more slender than P3-4 meri; P3 merus subequal to length of P3 propodus; P4 merus 0.8 × length of P4 propodus, as broad as P3 merus. Carpi subequal on P2-4, or slightly shorter on P2 than on P3-4; each distinctly less than half length of propodus (carpus-propodus length ratio, 0.34-0.36 on P2, 0.30-0.35 on P3, 0.29-0.33 on P4); extensor margin with small terminal spine distinct on P2, occasionally obsolete on P3, absent on P3. Propodi successively longer posteriorly, flexor margin nearly straight in lateral view, ending in pair of spines (placed at distal end) preceded by 6 spines on P2, 3-4 on P3, 2-3 on P4, at most on distal half of length. Dactyli successively longer posteriorly, more slender and more setose on P2 than on P3 and P4; much longer than carpi (dactylus-carpus length ratio, 1.5 on P2, 1.7 on P3, 1.8 on P4), and 0.5-0.6 × as long as propodi on P2-4; flexor margin straight, with 17-20 closely arranged, obliquely directed, short spines, penultimate about twice as broad as antepenultimate, ultimate more slender and longer than antepenultimate, all obscured by thick setae on P2.

Eggs. Number of eggs carried up to 16; size, 1.08 mm × 1.17 mm - 1.40 mm × 1.20 mm.

PARASITES — One of the females from NORFOLK 2, Stn DW2024 (MNHN-IU-2014-16430) bears a rhizocephalan externa.

REMARKS — This species is common around New Caledonia and the Norfolk Ridge.

The new species is grouped together with *U. floccus* n. sp., *U. lumarius* n. sp., and *U. quinarius* n. sp. (see below) by having a field of three closely arranged spines on the mesioproximal face of P1 merus. The spination of the carapace and P2-4 dactyli and the shape of the antenna displayed by *U. dualis* are very much like those of *U. floccus* n. sp. The relationships with these congeners are discussed under the remarks of the respective species (see below).

Uroptychus duplex n. sp.

Figures 82, 83

TYPE MATERIAL — Holotype: **New Caledonia**, Norfolk Ridge. CHALCAL 2 Stn DW73, 24°39.9'S, 168°38.1'E, 573 m, 2.X.1986, ov. ♀ 4.7 mm (MNHN-IU-2011-5923). Paratypes: **New Caledonia**, Norfolk Ridge. Station data as for the holotype, 1 ♂ 4.4 mm (MNHN-IU-2011-5924). NORFOLK 2 Stn DW2091, 24°45'S, 168°06'E, 600-896 m, 29.X.2003, 1 ov. ♀ 5.1 mm (MNHN-IU-2011-5922). **New Caledonia**, Isle of Pines. BIOCAL Stn CP52, 23°06'S, 167°47'E, 540-600 m, 31.VIII.1985, 1 ♂ 3.9 mm (MNHN-IU-2011-5921).

ETYMOLOGY — The specific name is a noun in apposition from the Latin *duplex* (double) for the flexor marginal spines of the P2-4 propodi which are arranged in double rows, not in a single row as in most of the other species.

DISTRIBUTION — Isle of Pines and Norfolk Ridge; in 573 m. 320-600 m.

DESCRIPTION — Small species. *Carapace*: Much broader than long (0.6-0.7 × as long as broad); greatest breadth measured between last lateral spines situated at posterior quarter of lateral margin, 1.9-2.2 × distance between anterolateral spines. Dorsal surface well convex from anterior to posterior; anterior surface with sparse short setae, hepatic region with several scattered small spines. Lateral margins strongly convexly divergent posteriorly, with row of spines of irregular sizes as figured, first anterolateral, strong, overreaching smaller lateral orbital spine. Rostrum narrow triangular, with interior angle of 22-28°, slightly upcurved, length less than half that of remaining carapace, breadth about one-third carapace breadth measured at posterior carapace margin; dorsal surface concave. Lateral orbital spine distantly mesial to and

FIGURE 82

Uroptychus duplex n. sp., holotype, ovigerous female 4.7 mm (MNHN-IU-2011-5923). **A**, carapace and anterior part of abdomen, proximal articles of right P1 included, dorsal. **B**, same, lateral. **C**, sternal plastron, with excavated sternum and basal parts of Mxp1. **D**, telson. **E**, left antenna, ventral. **F**, left Mxp3, setae omitted, ventral. **G**, same, lateral. **H**, right P1, proximal part, setae omitted, ventral. **I**, same, dorsal. Scale bars: 1 mm.

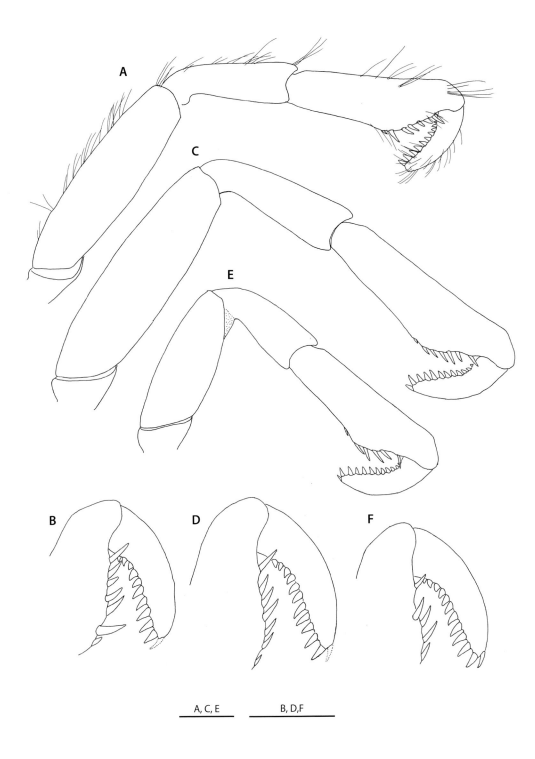

FIGURE 83

Uroptychus duplex n. sp., holotype, ovigerous female 4.7 mm (MNHN-IU-2011-5923). **A**, right P2, lateral. **B**, same, distal part, setae omitted, lateral. **C**, right P3, lateral. **D**, same, distal part, lateral. **E**, right P4, lateral. **F**, same, distal part, lateral. Scale bars: 1 mm.

very slightly anterior to anterolateral spine. Pterygostomian flap anteriorly angular, produced to distinct spine, surface with denticle-like very small spines on anterior half; height of posterior half 0.4 × that of anterior half.

Sternum: Excavated sternum subtriangular between bases of Mxp1, surface with low ridge in midline. Sternal plastron slightly less than twice as broad as long, depressed (in ventral view) on median parts of sternites 6 and 7; lateral extremities of sternites 4-7 convexly divergent posteriorly. Sternite 3 very short relative to breadth, not depressed; anterior margin broadly and shallowly excavated, transverse or very feebly concave on median third, without median notch. Sternite 4 with short lateral margin anteriorly rounded, posterolateral expansion very short, directed laterally, not anterolaterally. Anterolateral margin of sternite 5 strongly convex, much longer than posterolateral margin of sternite 4.

Abdomen: Smooth and glabrous, broad relative to length. Somite 1 without transverse ridge. Somite 2 tergite 2.6-2.8 × broader than long; pleuron anterolaterally rounded, posterolaterally angular, lateral margin feebly concavely divergent posteriorly. Pleura of somites 3 and 4 tapering to sharp point. Telson [0.2]-0.3 × as long as broad in female, 0.4 × in male; posterior plate 1.1-1.2 × that of anterior plate in males, 1.4-1.6 × in females, posterior margin broadly concave or feebly convex.

Eye: 1.5-1.6 × longer than broad, slightly broadened proximally, slightly overreaching midlength of rostrum. Cornea not inflated, more than half as long as remaining eyestalk.

Antennule and antenna: Ultimate article of antennular peduncle 3 × longer than high. Antennal peduncle distinctly overreaching cornea, falling short of apex of rostrum. Article 2 with sharp lateral spine. Antennal scale 2.0 × broader than article 5, slightly overreaching apex of rostrum, overreaching antennal peduncle by more than half length of article 5; lateral margin setiferous, bearing 1 or 2 small spines. Distal 2 articles unarmed; article 5 1.5 × longer than article 4, breadth 0.6 × height of antennular ultimate article. Flagellum consisting of 14-15 segments, falling short of distal end of P1 merus.

Mxp: Mxp1 with bases close to each other but not contiguous. Mxp3 laterally with fine plumose setae on ischium, merus and carpus. Basis without denticles on mesial ridge. Ischium with more than 30 very small denticles on crista dentata, flexor margin rounded distally. Merus 1.6-1.9 × longer than ischium, moderately compressed, distolaterally with distinct spine; flexor crest with 2 small spines distal to midlength. Carpus with small distolateral spine.

P1: 4.8-5.6 (females), 5.8-5.9 (males) × longer than carapace; with soft setae sparingly on fingers and along mesial margin. Ischium with basally broad, depressed, triangular dorsal process, ventromesially with row of 4 or 5 tubercle-like small spines on proximal two-thirds, without subterminal spine. Merus mesially with obsolescent denticles on proximal portion, length slightly more than that of carapace. Carpus 1.4-1.5 (males), 1.3-1.5 (females) × longer than merus, unarmed. Palm 3.7-3.8 (males), 3.8-4.2 (females) × longer than broad, 1.0-1.2 (females), 1.0-1.3 (males) × longer than carpus, more or less depressed. Fingers distally incurved and weakly crossing; movable finger 0.4 (females), 0.3 (males) × as long as palm, opposable margin with low subtriangular proximal process fitting into longitudinal groove proximal to median eminence on opposing fixed finger when closed.

P2-4: Sparsely setose, relatively broad, and unarmed except for distal 2 articles. Meri broad relative to length, largest on P3, smallest on P4, P2 merus 0.9 × P3 merus in length and breadth, P4 merus 0.5-0.6 × P3 merus in length, 0.8 × in breadth; P2 merus as long as or slightly shorter than carapace, 1.2 × length of P2 propodus; P3 merus 1.0-1.1 × length of P3 propodus; P4 merus 0.8 × length of P4 propodus; length-breadth ratio, 2.9-3.3 on P2, 2.8-3.4 on P3, 2.5 on P4. Carpi longest on P3, shortest on P4 (P3 carpus 1.1 × longer than P2 carpus, P4 carpus 0.7 × length of P3 carpus), carpus-propodus length ratio, 0.6 on P2-4. Propodi longer on P3 than on P2 and P4; extensor margin nearly straight; flexor margin inflated around point one-third from distal end, subprehensile with dactylus on distal third of length, bearing double rows of slender mesial spines, terminal spines paired. Dactyli distinctly shorter than carpi (dactylus-carpus length ratio, 0.7 on P2 and P3, 0.9 on P4); flexor margin moderately curved at proximal quarter, with row of 12 or 13 slightly inclined, subtriangular spines, ultimate more slender than penultimate, remaining spines gradually diminishing toward base of article.

Eggs. Number of eggs carried, up to 16; size, 1.58 mm × 1.38 mm.

REMARKS — *Uroptychus duplex* is very similar to *U. macrolepis* n. sp. and *U. zigzag* n. sp. in the spination of the carapace lateral margin, in having the pterygostomian flap very low on the posterior half, in the shape of the antenna, in having the P3 merus distinctly longer than the P2 merus, and in the spination of the P2-4 dactyli. The relationships are discussed under the remarks of these congeners (see below).

Uroptychus echinatus n. sp.

Figures 84, 85

TYPE MATERIAL — Holotype: **New Caledonia**, Hunter and Matthew Islands. VOLSMAR Stn DW05, 22°26'S, 171°46'E, 620-700 m, 01.VI.1989, ov. ♀ 4.9 mm (MNHN-IU-2014-16438). Paratypes: Collected with the holotype, 1 ♂ 3.2 mm, 1 ♀ 3.5 mm (MNHN-IU-2014-16439).

ETYMOLOGY — From the Latin *echinatus* (spiny), alluding to the spiny propodi of the walking legs.

DISTRIBUTION — Hunter and Matthew Islands; 700 m.

DESCRIPTION — Small species. *Carapace*: Slightly shorter than broad; greatest breadth 1.6 × distance between antero-lateral spines. Dorsal surface convex from anterior to posterior, thickly or moderately setose, with a few scattered small spines on hepatic and lateral epigastric regions. Lateral margins convex, ridged along posterior fourth of length, bearing about 6 small spines: first anterolateral, slightly overreaching much smaller lateral orbital spine, well separated from and slightly posterior to level of that spine, followed by a few very tiny spines along hepatic margin and 5 distinct spines along branchial margin. Rostrum with interior angle of ca. 25°, nearly straight horizontal; length about half that of remaining carapace, breadth about half carapace breadth measured at posterior carapace margin; lateral margin somewhat concave, with 1 or 2 small spines near tip; dorsal surface concave and setose. Pterygostomian flap moderately setose, with tiny or obsolescent spines on anterior surface; anterior margin angular, produced to small spine.

Sternum: Excavated sternum broadly convex on anterior margin, with ridge in midline on anterior half surface. Sternal plastron as long as broad, gradually broadened posteriorly. Sternite 3 strongly depressed; anterior margin moderately excavated with small or relatively broad median notch without (holotype) or with (paratypes) small flanking spine, lateral terminus sharp angular. Sternite 4 with relatively short, nearly straight anterolateral margin ending anteriorly in strong or medium-sized spine with accompanying smaller spine lateral or mesial to it; surface with short striae supporting long setae; posterolateral margin slightly shorter than anterolateral margin. Anterolateral margins of sternite 5 subparallel, anteriorly rounded and tuberculose, about as long as posterolateral margin of sternite 4

Abdomen: Less setose than carapace. Somite 1 gently convex from anterior to posterior, with thick setae in transverse line. Somite 2 tergite 2.6-2.7 × broader than long; pleural lateral margin moderately concavely divergent, posterolaterally blunt. Pleuron of somite 3 laterally blunt. Telson half as long as broad; posterior plate concave on posterior margin, length 1.2-1.4 × that of anterior plate.

Eye: 1.6-1.7 × longer than broad, slightly overreaching midlength of rostrum, dorsally bearing setae proximal to cornea, somewhat broadened proximally. Cornea not dilated, more than half as long as remaining eyestalk.

Antennule and antenna: Antennular ultimate article 3.0-3.5 × longer than high. Antennal peduncle extending far beyond cornea, reaching subapical spine of rostrum. Article 2 with acute lateral spine. Antennal scale more than 1.5 × broader than article 5, overreaching distal end of article 5 by two proximal segments of flagellum. Distal 2 articles each with acute distomesial spine. Article 5 1.5 × longer than article 4, breadth three-fifths to three-quarters height of antennular ultimate article. Flagellum consisting of 8-10 segments, reaching distal end of P1 merus.

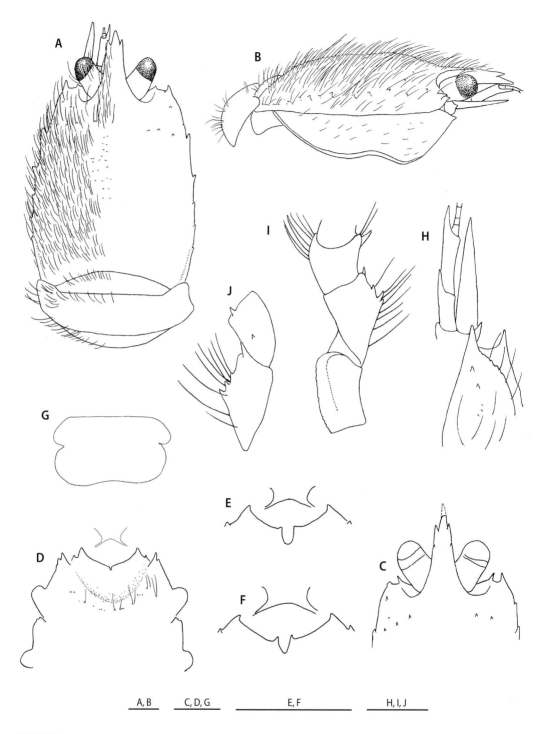

FIGURE 84

Uroptychus echinatus n. sp., **A**, **B**, **D**, **G-J**, holotype, ovigerous female 4.9 mm (MNHN-IU-2014-16438); **C**, **E**, male paratype 3.2 mm (MNHN-IU-2014-16439), **F**, female paratype 3.5 mm (MNHN-IU-2014-16439). **A**, carapace and anterior part of abdomen, setae on right half omitted, dorsal. **B**, same, lateral. **C**, anterior part of carapace, setae omitted, dorsal. **D**, anterior part of sternal plastron, with excavated sternum and basal parts of Mxp1. **E**, same. **F**, same. **G**, telson. **H**, anterior part of cephalothorax, showing anterior part of pterygostomian flap and antenna, ventrolateral. **I**, left Mxp3, ventral. **J**, same, merus and carpus, lateral. Scale bars: 1 mm.

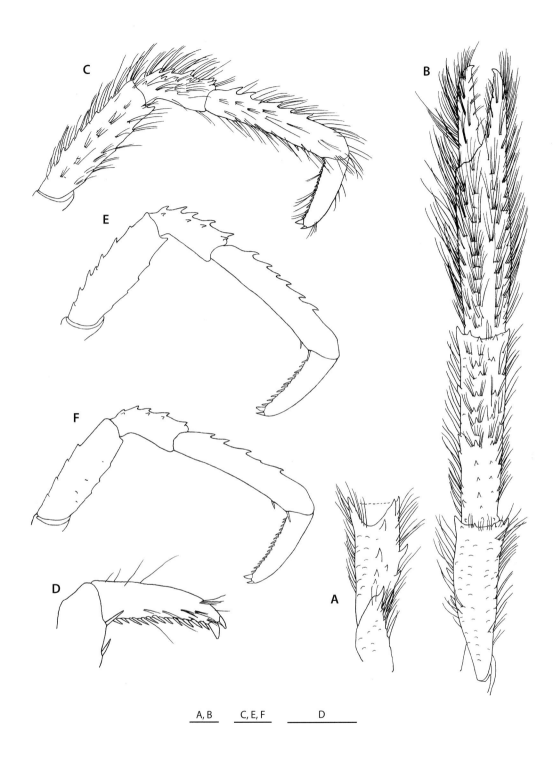

FIGURE 85

Uroptychus echinatus n. sp., holotype, ovigerous female 4.9 mm (MNHN-IU-2014-16438). **A**, right P1, proximal part, ventral. **B**, same, dorsal. **C**, right P2, lateral. **D**, same, distal part. **E**, right P3, setae omitted, lateral. **F**, right P4, setae omitted, lateral. Scale bars: 1 mm.

Mxp: Mxp1 with bases broadly separated. Mxp3 with relatively long, sparse setae other than mesial brushes. Basis with 1 or 2 obsolescent denticles on mesial ridge. Ischium with very small or obsolescent denticles on crista dentata, flexor margin distally rounded. Merus relatively short, 1.5 × longer than ischium, flattened, with well-developed distolateral spine and 2 or 3 spines distal to midlength of flexor margin. Carpus with 1 distinct distolateral and 1 small midlateral spine.

P1: 4.0-4.3 × longer than carapace, with thick simple long setae arising from scale-like granulate ridges. Ischium dorsally with strong spine, ventromesially with strong sharp subterminal spine proximally followed by successively diminishing small spines. Merus subequal to carapace in length, bearing rows of both ventral and mesial spines; terminal margin (at junction with carpus) with a few small dorsal spines and well-developed mesial and lateral spines on ventral and dorsal sides. Carpus 1.2 × longer than merus, dorsally with 3 rows of spines. Palm unarmed, 2.9-3.2 × longer than broad, 0.9-1.0 × length of carpus. Fingers not gaping, distally incurved, crossing when closed; opposable margins finely denticulate, that of movable finger with low prominence (on proximal third) fitting to opposing longitudinal narrow groove on fixed finger when closed; movable finger slightly more than half length of palm.

P2-4: Relatively short, with granulate scale-like ridges supporting coarse setae. Meri equally broad on P2-4, moderately compressed mesio-laterally, very setose like P1; P3 merus subequal to or slightly shorter than P2 merus; P4 merus 0.9 length of P3 merus; dorsal margins with row of small spines; ventrolateral margins distally ending in distinct spine proximally followed by granulate eminences; length-breadth ratio, 3.8-3.9 on P2, 3.6-3.8 on P3, 3.1-3.2 on P4; P2 merus 0.8 × length of carapace, 0.9-1.0 × length of P2 propodus; P3 merus 0.8 × length of P3 propodus; P4 merus 0.7-0.8 × length of P4 propodus. P3 carpus 0.92-0.96 × length of P2 carpus; P4 carpus 0.91-0.95 × length of P3 carpus; extensor margin with row of spines paralleling another row of smaller spines on dorsolateral surface; carpus-propodus length ratio about 0.4 on P2-4. Propodi slightly longer on P2 and P4 than on P3 or subequal on P3 and P4 and shorter on P2; extensor margin with row of 5-6 acute immovable spines, unarmed on distal third; flexor margin ending in pair of movable spines preceded by 1 or 2 similar spines. Dactyli subequal, proportionately broad; dactylus-carpus length ratio, 1.2 on P2, 1.3 on P3, 1.4 on P4, dactylus-propodus length ratio, 0.5-0.6 on P2 and P3, 0.5 on P4; flexor margin straight, with 13 spines on P2, 14 or 15 on P3 and P4, all not contiguous to one another and obliquely directed, ultimate slender, penultimate more than 2 × broader than antepenultimate, remaining spines slender, antepenultimate somewhat narrower than ultimate.

Eggs. Eggs carried 3 in number; size, 1.10 mm × 1.30 mm.

REMARKS — The new species resembles *U. multispinosus* Ahyong & Poore, 2004 (see below) in having small spines along the carapace lateral margin, the trifid apex of the rostrum, the antenna bearing a strong distomesial spine on each of articles 4 and 5 and the antennal scale overreaching article 5. However, *U. echinatus* is readily distinguished from that species by the following differences: the P1 ischium bears a strong subterminal spine instead of being unarmed on the ventromesial margin; the flexor margins of P2-4 dactyli bear strongly inclined, slender spines proximal to the pronounced penultimate spine, instead of proximally diminishing, somewhat inclined, subtriangular spines; and the extensor margins of carpi and propodi bear a row of spines instead of being smooth and unarmed. In addition, the carapace is thickly setose, instead of sparsely so.

The combination of the following characters links the species to *U. alophus* n. sp., *U. karubar* n. sp., *U. longior* Baba, 2005, *U. nanophyes* McArdle, 1901 and *U. tracey* Ahyong, Schnabel & Baba, 2015: the P2-4 dactyli having the flexor margin with more than 10 closely arranged spines proximal to the pronounced penultimate spine, the P2-4 carpi with a few to several extensor marginal spines, and the carapace lateral margin with at least six spines. *Uroptychus nanophyes* is readily distinguished from all the others by having a row of ventrolateral spines on the P2-4 meri and by the antennal scale not overreaching the antennal article 5. *Uroptychus echinatus* differs from the rest of the species in having a very setose carapace with small lateral spines.

Uroptychus elongatus n. sp.
Figures 86, 87, 305D

TYPE MATERIAL — Holotype: **Vanuatu**. MUSORSTOM 8 Stn CP1027, 17°53.05'S, 168°39.35'E, 550-571 m, 28.IX.1994, ov. ♀ 4.3 mm (MNHN-IU-2014-16440). Paratypes: **Vanuatu**. MUSORSTOM 8 Stn CP1026, 17°50.35'S, 168°39.33'E, 437-504 m, 28.IX.1994, 2 ♂ 4.6, 5.0 mm, 1 ov. ♀ 4.8 mm, 1 ♀ 5.4 mm (MNHN-IU-2014-16441).

ETYMOLOGY — From the Latin *elongatus* (prolonged), referring to elongate eyes, one of the characteristic features of the new species.

DISTRIBUTION — Vanuatu; in 437-571 m.

SIZE — Males, 4.6-5.0 mm; females, 4.3-5.4 mm; ovigerous females from 4.3 mm.

DESCRIPTION — Small species. *Carapace*: 0.9 × as long as broad; greatest breadth 1.7 × distance between anterolateral spines. Dorsal surface well convex from anterior to posterior, without distinct depression or groove, smooth, sparsely or barely setose along lateral margin. Lateral margins somewhat convex and slightly or moderately divergent posteriorly, bearing 6 or 7 spines: first anterolateral, relatively stout, overreaching much smaller lateral orbital spine, subequal to third and fourth; second very small, located on hepatic margin; remaining spines on branchial region posteriorly diminishing; sixth followed by eminence or additional small spine and ridge leading to posterior end of margin. Rostrum sharp triangular, with interior angle of 32-35°, nearly horizontal, dorsally excavated; length about one-third that of remaining carapace, breadth less than half carapace breadth measured at posterior carapace margin. Lateral orbital spine situated slightly anterior to level of anterolateral spine, separated from anterolateral spine by basal breadth of that spine. Pterygostomian flap high relative to length, anteriorly angular, produced to small sharp spine; tubercle-like small processes occasionally obsolescent on anterior half of surface.

Sternum: Excavated sternum strongly produced anteriorly between bases of Mxp1, ending in blunt tip, surface with ridge in midline. Sternal plastron slightly broader than long, lateral extremities somewhat divergent posteriorly. Sternite 3 well depressed; anterior margin weakly concave, with subovate or narrow deep median notch flanked by incurved spine, laterally sharp angular. Sternite 4 with anterolateral margin nearly straight, anteriorly ending in small spine usually accompanying 2 smaller spines lateral and mesial to it; posterolateral margin 0.7 × length of anterolateral margin.

Abdomen: With sparse or moderately dense setae. Somite 1 gently convex from anterior to posterior, without transverse ridge. Somite 2 tergite 2.3 × broader than long; pleural lateral margin concavely divergent posteriorly, posterolaterally rounded. Telson half as long as broad; posterior plate 1.3 × longer than anterior plate, posterior margin feebly emarginate or weakly concave.

Eye: Elongate, 2.0-2.2 × longer than broad, reaching or slightly overreaching rostral tip, slightly broadened proximally. Cornea not dilated, half as long as remaining eyestalk.

Antennule and antenna: Ultimate article of antennular article slightly more than 3 × longer than high. Antennal peduncle reaching rostral tip. Article 2 with small lateral spine. Antennal scale reaching distal end of article 5, breadth 1.7 × that of article 5. Article 4 with distomesial spine. Article 5 unarmed, length 1.5 × that of article 4, breadth slightly more than half height of ultimate antennular article. Flagellum consisting of 15-17 segments, slightly falling short of distal end of P1 merus.

Mxp: Mxp1 with bases separated. Mxp3 with tufts of setae on lateral surfaces of ischium and merus. Basis with obsolescent distal denticle or unarmed on mesial ridge. Ischium thick, flexor margin distally rounded, crista dentata with row of very small denticles. Merus 1.8 × longer than ischium, concavely smooth on mesial face, bearing distinct distolateral spine and 1 or 2 obsolescent spines at point one-third from distal end of sharply ridged flexor margin. Carpus unarmed.

Pereopods sparingly with long, distally softened setae.

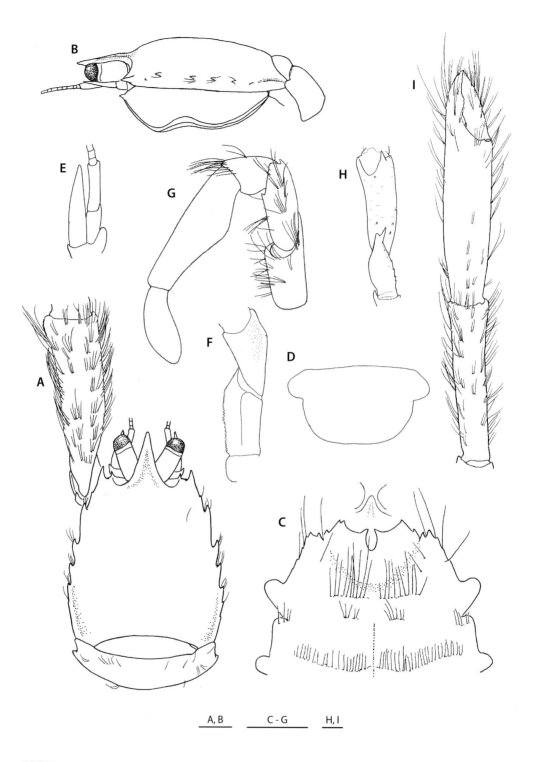

FIGURE 86

Uroptychus elongatus n. sp., holotype, ovigerous female 4.3 mm (MNHN-IU-2014-16440). **A**, carapace and anterior part of abdomen, proximal part of left P1 included, dorsal. **B**, same, lateral. **C**, anterior part of sternal plastron, with excavated sternum and basal parts of Mxp1. **D**, telson. **E**, right antenna, ventral. **F**, left Mxp3, setae omitted, ventral. **G**, same, setae omitted from distal 2 articles, lateral. **H**, left P1, proximal part, setae omitted, ventral. **I**, same, dorsal. Scale bars: 1 mm.

P1: Relatively slender and subcylindrical, palm somewhat massive especially in males; length 4.7-6.0 × postorbital carapace length. Ischium dorsally bearing strong distal spine, ventromesially well-developed subterminal spine proximally followed by row of tubercle-like spines. Merus 1.1-1.2 × longer than carapace, ventrally bearing well-developed distomesial spine and a few small spines on mesioproximal portion. Carpus 1.2-1.4 × longer than merus, with small ventral distomesial and distolateral spines. Palm 2.9 × (male; missing in another male) or 3.4-4.2 × (females) longer than broad, as long as carpus. Fingers slightly incurved distally, slightly gaping in males, not gaping in females; opposable margin of fixed finger sinuous, that of movable finger with proximal blunt process fitting to longitudinal groove on opposite margin of fixed finger when closed; movable finger 0.4-0.5 × length of palm.

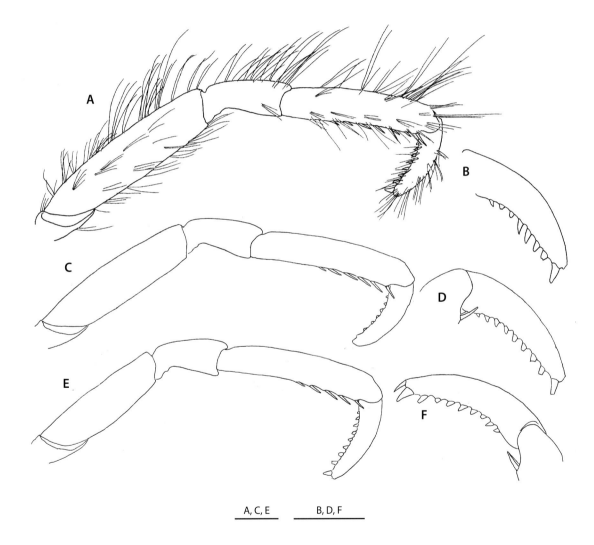

FIGURE 87

Uroptychus elongatus n. sp., holotype, ovigerous female 4.3 mm (MNHN-IU-2014-16440). **A**, right P2, lateral. **B**, same, distal part, setae omitted, lateral. **C**, right P3, lateral. **D**, same, distal part, lateral. **E**, right P4, lateral. **F**, left P4, distal part, lateral. Scale bars: 1 mm.

P2-4: Moderately compressed mesio-laterally. Meri and carpi unarmed. Meri successively shorter posteriorly (P3 merus 0.8-0.9 × length of P2 merus, P4 merus 0.8-0.9 × length of P3 merus), equally broad on P2-4; length-breadth ratio, 3.8-4.2 on P2, 3.6-3.8 on P3, 3.0-3.2 on P4; P2 merus 0.9 × length of carapace, 1.2 × length of P2 propodus; P3 merus 0.9 × length of P3 propodus; P4 merus 0.7-0.8 × length of P4 propodus. Carpi successively slightly shorter posteriorly (P3 carpus 0.9 × length of P2 carpus, P4 carpus 0.9 × length of P3 carpus), slightly shorter than dactyli; length 0.5 × that of propodus on P2, 0.4 × on P3 and P4. Propodi subequal on P2-4 or successively slightly shorter posteriorly; flexor margin straight, ending in pair of spines preceded by row of 6-8 spines on distal two-thirds on P2 and P3, 4 spines on distal half on P4. Dactyli somewhat curving, ending in slender spine preceded by 9-11 sharp, somewhat oblique, loosely arranged spines, ultimate more slender and shorter than antepenultimate, penultimate broader than antepenultimate; dactylus-propodus length ratio, 0.4 on P2, 0.4-0.5 on P3 and P4.

Eggs. Number of eggs carried, 4-19; size, 1.13 mm × 1.03 mm to 1.30 mm × 1.25 mm.

Color. Holotype, MNHN-IU-2014-16440: Base color pale brown, carapace, P1 and P4 darker, abdomen translucent.

REMARKS — The elongate eyestalks and the spination of the carapace and P2-4 are similar in *U. elongatus*, *U. baeomma* n. sp. (see above) and *U. exilis* n. sp. (see below). *Uroptychus elongatus* is distinguished from *U. baeomma* by the P1 ischium that has a well-developed instead of obsolete subterminal spine on the ventromesial margin; and the P3 merus is shorter than instead of subequal to the P2 merus. The relationships with *U. exilis* is discussed under the remarks of that species (see below).

The elongate eyestalks and the spination of the P2-4 dactyli as displayed by this species are also possessed by *U. denticulisquama* n. sp. (see above). However, *U. elongatus* is easily distinguished from that species by having a row of distinct spines instead denticles on the carapace lateral margin; the antennal scale is laterally unarmed instead of bearing small spines; and the P2 merus is unarmed along the ventromesial margin instead of bearing a row of small spines.

Uroptychus empheres Ahyong & Poore, 2004

Figures 88, 89

Diptychus australis Henderson, 1885: 420 (part).
Uroptychus australis — Henderson 1888: 179 (part).
Uroptychus empheres — Ahyong & Poore, 2004: 34, fig. 8.

TYPE MATERIAL — Holotype: **Australia**, "Andys" Seamount, Tasmania, 800 m, male (NMV J52864). [not examined].

MATERIAL EXAMINED — **Indonesia**, Kai Islands. KARUBAR Stn CP19, 5°15'S, 133°01'E, 605-576 m, on gorgonacean, 25.X.1991, 1 ♂ 6.4 mm, 1 ♀ 5.1 mm (MNHN-IU-2014-16442). **Indonesia**, off Banda. CHALLENGER Stn 194, 4°31'0"S, 129°57'20"E, 360 fms [659 m], 1 ov. ♀ 7.4 mm (BMNH 1888:33) [syntype of *U. australis* (Henderson, 1885). **Solomon Islands**. SALOMON 1 Stn CP1751, 09°10'S, 159°53'E, 749-799 m, 25.IX.2001, 2 ♂ 4.9, 7.3 mm, 1 ov. ♀ 5.8 mm (MNHN-IU-2014-16443). – Stn CP1839, 10°16'S, 161°40'E, 575-624 m, 05.X.2001, 1 ov. ♀ 6.7 mm (MNHN-IU-2014-16444). SALOMON 2 Stn CP2213,7°38.7'S, 157°42.9'E, 495-650 m, 26.X.2004, 1 ♂ 6.5 mm, 1 ♀ 5.5 mm (MNHN-IU-2014-16445). – Stn CP2214, 7°41.6'S, 157°43.8'E, 550-682 m, 26.X.2004, 1 ov. ♀ 6.4 mm (MNHN-IU-2014-16446). – Stn DW2238, 06°53'S, 156°21'E, 470-443 m, 30.X.2004, 1 ♂ 7.3 mm, 1 ov. ♀ 6.2 mm (MNHN-IU-2014-16447). – Stn CP2245, 7°43.1'S, 156°26.0'E, 582-609 m, 1.XI.2004, 1 ov. ♀ 6.9 mm (MNHN-IU-2014-16448). – Stn CP2264, 7°52.4'S, 156°51.0'E, 515-520 m, 3.XI.2004, 1 ♂ 8.7 mm (MNHN-IU-2014-16449). – Stn CP2267, 7°48.0'S, 156°52.0'E, 590-600 m, 4.XI.2004, 1 ov. ♀ 6.9 mm (MNHN-IU-2014-12787). – Stn CP2269, 7°45.1'S, 156°56.3'E, 768-890 m, 4.XI.2004, 1 ♂ 6.2 mm (MNHN-IU-2014-16450). – Stn CP2297, 9°08.8'S, 158°16.0'E, 728-777 m, 8.XI.2004, 2 ♂ 5.0, 7.7 mm (MNHN-IU-2014-16451). **Tonga**. BORDAU 2 Stn CP1529, 21°13'S, 174°58'W, 688-710 m, 3.VI.2000, 2 ♂ 6.8, 8.1 mm, 2 ov. ♀ 6.0, 8.2 mm (MNHN-IU-2014-12788). – Stn DW1589, 18°39'S, 173°54'W, 281 m, 13.VI.2000, 1 ♂ 9.0 mm (MNHN-IU-2014-12789). – Stn CP1625, 23°28'S, 176°22'W, 824 m, 19.VI.2000, 1 ♂ 4.1 mm, 1 ov. ♀ 7.5 mm, 3 ♀ 4.1-5.6 mm (MNHN-IU-2014-12790). **New Caledonia**, Loyalty Ridge. BATHUS 3 Stn DW778, 24°43'S, 170°07'E, 750-760 m, 24.XI.1993, 1 ♂ 5.8 mm (MNHN-IU-2014-16452), 1 ♀ 6.8 mm (MNHN-IU-2014-16453). – Stn DW792, 23°46'S, 169°49'E, 730-735 m, 26.XI.1993, 1 ov. ♀ 7.3 mm (MNHN-IU-2014-16454). **New Caledonia**, Norfolk Ridge. NORFOLK 2 Stn DW2102, 23°56'S, 167°44'E, 700-715 m, 30.X.2003, 1 ov. ♀ 8.4 mm (MNHN-IU-2014-16455).

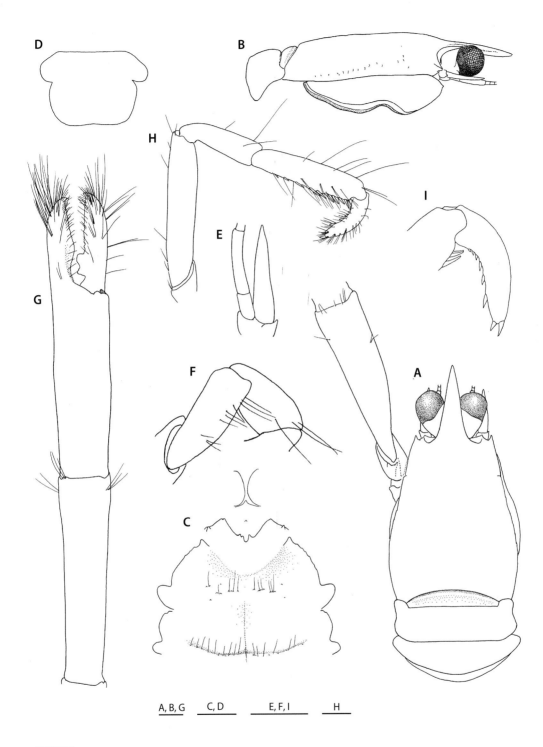

FIGURE 88

Uroptychus empheres Ahyong & Poore, 2004, male 5.8 mm (MNHN-IU-2014-16452). **A**, carapace and anterior part of abdomen, proximal part of left P1 included, dorsal. **B**, same, lateral. **C**, anterior part of sternal plastron, with excavated sternum and basal parts of Mxp1. **D**, telson. **E**, left antenna, ventral. **F**, right Mxp3, merus and carpus, lateral. **G**, left P1, proximal articles omitted. **H**, right P2, lateral. **I**, same, distal part, setae omitted, lateral. Scale bars: 1 mm.

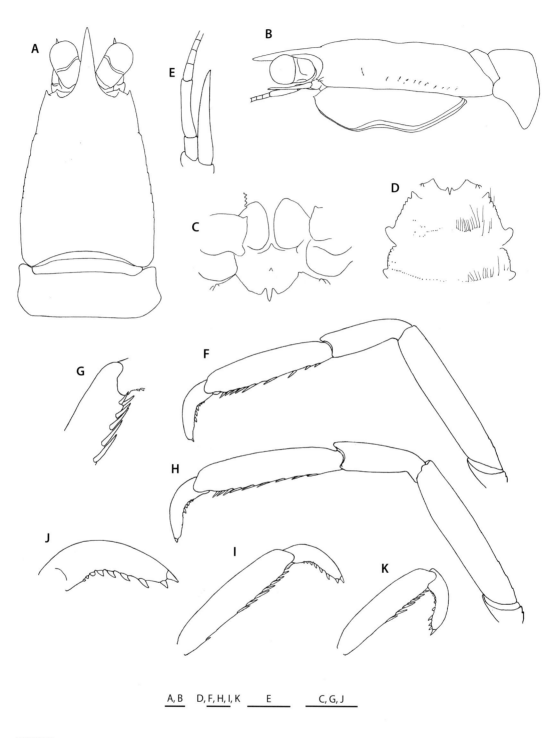

A, B   D, F, H, I, K   E   C, G, J

FIGURE 89

Uroptychus empheres Ahyong & Poore, 2004, ovigerous female 7.4 mm (BMHN 1888:33), Challenger Stn 194. **A**, carapace and anterior part of abdomen, dorsal. **B**, same, lateral. **C**, excavated sternum with basal parts of Mxps 1-3. **D**, anterior part of sternal plastron. **E**, left antenna, ventral. **F**, left P2, lateral. **G**, same, propodus, mesial. **H**, left P3, lateral. **I**, right P3, propodus and dactylus, lateral. **J**, same, dactylus, lateral. **K**, right P4, propodus and dactylus, lateral. Scale bars: 1 mm.

DISTRIBUTION — Tasmania, and now off Banda, Kai Islands, Loyalty Islands, Norfolk Ridge, Tonga and Solomon Islands; 281-890 m.

SIZE — Males, 4.1-9.0 mm; females, 4.1-8.4 mm; ovigerous females from 5.8 mm.

DESCRIPTION — Medium-sized species. *Carapace*: 1.1 × longer than broad; greatest breadth 1.6-1.7 × distance between anterolateral spines. Dorsal surface feebly convex, with or without feeble depression between gastric and cardiac region; pair of epigastric scales composed of granules occasionally absent. Lateral margins posteriorly convex, with elevated small ridge at anterior end of branchial region, followed by row of oblique short ridges so as to be seen as granulose in dorsal view; anterolateral spines small, reaching or slightly falling short of tip of lateral orbital spine. Rostrum relatively narrow and elongate, with interior angle of 24-28°, straight horizontal or feebly deflected ventrally; about half or slightly less than half as long as remaining carapace, breadth slightly more than half carapace breadth measured at posterior carapace margin; dorsal surface flattish. Lateral limit of orbit acuminate or ending in very small spine laterally followed by oblique margin leading to anterolateral spine. Pterygostomian flap smooth on surface, anteriorly roundish with very small spine.

Sternum: Excavated sternum anteriorly ending in small acute process between bases of Mxp1, surface with small spine in center. Sternal plastron slightly broader than or as broad as long, with lateral extremities divergent posteriorly. Sternite 3 well depressed, anterior margin deeply emarginate, bearing 2 submedian spines flanking narrow median notch. Sternite 4 having anterolateral margin somewhat convex and denticulate, anteriorly ending in short, blunt spine; posterolateral margin less than half length of anterolateral margin. Anterolateral margin of sternite 5 convex, slightly longer than posterolateral margin of sternite 4.

Abdomen: Tergites smooth and glabrous. Somite 1 with rounded transverse ridge. Somite 2 tergite 2.3-2.7 × broader than long; pleural lateral margin concavely divergent posteriorly, rounded on posterolateral terminus. Pleuron of somite 3 posterolaterally blunt. Telson three-fifths as long as broad; posterior plate about 1.5 × longer than anterior plate, posterior margin slightly concave or distinctly emarginate.

Eye: Broad relative to length (1.4-1.5 × longer than broad), slightly broadened distally, overreaching midlength of rostrum. Cornea slightly dilated, about as long as remaining eyestalk.

Antennule and antenna: Ultimate article of antennular peduncle 2.2-2.5 × longer than high. Antennal peduncle slightly overreaching cornea. Article 2 with very small distolateral spine. Antennal scale 1.3-1.8 × as broad as article 5, reaching or slightly overreaching distal end of article 5. Distal 2 articles unarmed; article 5 2.2-2.8 × length of article 4, breadth barely or nearly half height of ultimate antennular article. Flagellum of 15-17 segments not reaching distal end of P1 merus.

Mxp: Mxp1 with bases very close to each other. Mxp3 barely setose on lateral surface. Basis with 1-3 denticles on mesial ridge. Ischium with 13-18 denticles on crista dentata, flexor margin not rounded distally. Merus 2.0-2.2 × longer than ischium, thick mesiolaterally, unarmed; flexor margin roundly ridged, not cristate. Carpus unarmed.

P1: Slender, unarmed except for very small distodorsal spine of ischium, smooth on dorsal surface, granulose on ventral surface (not on fingers) in large specimens; length 5.0-5.4 × that of carapace. Merus 1.1-1.3 × longer than carapace, ventromesial and ventral surface with row of short spines often tubercle-like or obsolescent. Carpus 1.2-1.4 × longer than merus. Palm 3-5 × longer than broad (slightly more slender in females), 0.8-0.9 × length of carpus, ventral surface granulose. Fingers proportionately broad, terminating in small incurved spine and distally crossing when closed, ; movable finger with broad two-toothed proximal process, length about half that of palm; fixed finger having opposable margin with low prominence distal to position of opposite process on movable finger.

P2-4: With long setae sparsely or sparingly. Meri subequal in length on P2 and P3 or slightly longer on P3 than on P2, shortest on P4 (P4 merus 0.6 × length of P3 merus), moderately compressed mesio-laterally, unarmed; length-breadth ratio, 5.0-5.2 on P2, 4.8-5.3 on P3, 3.6-4.5 on P4; P2 merus 0.8-1.0 × length of carapace, 1.2-1.3 × length of P2 propodus; P3 merus 1.1-1.2 × length of P3 propodus; P4 merus much narrower than P2-3 meri, length 0.8-0.9 × that of P4 propodus. Carpi subequal in length on P2 and P3 or slightly longer on P3 than on P2 and shortest on P4; carpus-propodus length ratio, 0.5-0.6 on P2, 0.5 on P3 and P4. Propodus longest on P3, subequal on P2 and P4 or shorter on P4 than on P2; flexor margin straight, with pair of terminal spines slightly proximal to juncture with dactylus, preceded by row of 6-9 spines on P2, 5-7 spines on P3, 4-5 spines on P4, distalmost of these unpaired spines equidistant between pair of spines and distal second spine. Dactyli about one-third length of propodus on P2 and P3, 0.4-0.5 × on P4; 0.6-0.7 × length of carpi on P2-4; strongly curved at proximal third; flexor margin with 8-9 obliquely directed, loosely arranged, sharp spines, ultimate spine distinctly longer and broader than penultimate, penultimate as broad as antepenultimate and close to ultimate, antepenultimate more remote from penultimate than from distal quarter.

Eggs. Number of eggs carried 3-25; size, 1.30 mm × 1.50 mm - 1.67 mm × 1.92 mm.

REMARKS — As mentioned under *U. australis* (see above), the ovigerous female, one of the syntypes of *U. australis* from Challenger Stn 194, is referred to *U. empheres*.

Uroptychus empheres is so closely related to *U. australis* (Henderson, 1885) and *U. terminalis* n. sp. described below, in the shape of the carapace and in having the much smaller P4 that Henderson (1885, 1888) could not discriminate among these three species, which are actually mixed in the type material of *U. australis* (see above under *U. australis*). *Uroptychus empheres* is readily distinguished from *U. australis* by the flexor marginal spines of the P2-4 dactyli that are obliquely directed instead of orientated parallel to the margin. The relationships between *U. empheres* and *U. terminalis* are discussed under the remarks of the latter species. The species is also very close to *U. comptus* in nearly all features. Their relationships were discussed by Ahyong & Poore (2004).

Uroptychus enriquei n. sp.
Figures 90, 91

TYPE MATERIAL — Holotype: **New Caledonia**, Norfolk Ridge. BERYX 11 Stn CP53, 23°48'S, 168°17'E, 540-950 m, 21.X.1992, ov. ♀ 10.8 mm (MNHN-IU-2011-5953). Paratypes: **Philippines**. MUSORSTOM 3 Stn CP105, 13°52'N, 120°30'E, 398-417 m, 1.VI.1985, 1 ov. ♀ 7.9 mm (MNHN-IU-2011-5957). **Solomon Islands**. SALOMON 2 Stn CP2227, 6°37.2'S, 156°12.7'E, 508-522 m, 28.X.2004, 1 ♀ 7.0 mm (MNHN-IU-2011-5956). **New Caledonia**, Norfolk Ridge. SMIB 3, Stn DW21, 22°59.2'S, 167°19.0'E, 525 m, 24.V.1987, 1 ♂ 9.5 mm (MNHN-IU-2011-5954). MUSORSTOM 4 Stn CP216, 22°59.5'S, 167°22.0'E, 490-515 m, with Chrysogorgiidae gen. sp. (Calcaxonia), 29.IX.1985, 1 ♂ 8.7 mm (MNHN-IU-2011-5955).

ETYMOLOGY — The species is named for Enrique Macpherson who has contributed greatly to the knowledge of squat lobsters.

DISTRIBUTION — Philippines, Solomon Islands and Norfolk Ridge; 398-950 m.

SIZE — Males, 8.7, 9.5 mm; females, 7.0-10.8 mm; ovigerous females from 7.9 mm.

DESCRIPTION — Medium-sized species. *Carapace:* Somewhat broader than long (0.9 × as long as broad); greatest breadth 2.0 × distance between anterolateral angles. Dorsal surface moderately convex from anterior to posterior, with slight depression between gastric and cardiac regions; sparingly with long plumose setae. Lateral margins convexly divergent posteriorly, bearing row of granulate, oblique, short ridges so as to be seen as granulose in dorsal view; anterolateral corner acuminate or very minutely spiniform, close to and distinctly posterior to level of acuminate lateral limit of orbit. Rostrum triangular, with interior angle of 30°, directed somewhat dorsally; length slightly less than half that of remaining carapace, breadth about one-third carapace breadth measured at posterior carapace margin; dorsal surface flattish. Pterygostomian flap anteriorly angular, ending in acuminate point, not produced to distinct spine; surface spineless.

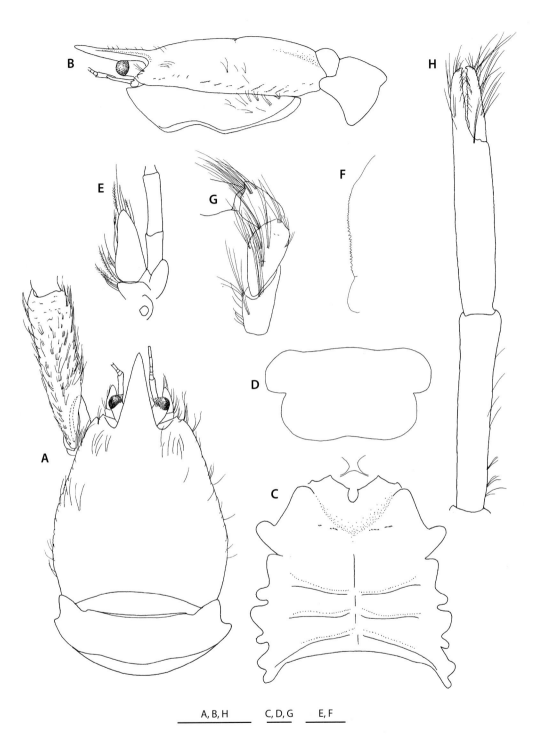

A, B, H   C, D, G   E, F

FIGURE 90

Uroptychus enriquei n. sp., holotype, ovigerous female 10.8 mm (MNHN-IU-2011-5953). **A**, carapace and anterior part of abdomen, proximal part of left P1 included, dorsal. **B**, same, lateral. **C**, sternal plastron, with excavated sternum and basal parts of Mxp1. **D**, telson. **E**, right antenna, ventral. **F**, left Mxp3, mesial ridges of basis and ischium, setae omitted, lateral. **G**, same, lateral. Scale bars: A, B, H, 5 mm; C-G, 1 mm.

Sternum: Excavated sternum with convex anterior margin between bases of Mxp1, surface with rounded ridge in midline. Sternal plastron as long as broad. Sternite 3 shallowly depressed, anterior margin moderately concave, with semi-ovate median notch flanked by obsolescent spine. Sternite 4 as broad as sternite 5; anterolateral margin anteriorly rounded, length 1.5 × that of posterolateral margin. Sternite 6 slightly broader than sternite 5, as broad as sternite 7. Sternite 5 having anterolateral margin anteriorly rounded, about as long as posterolateral margin of sternite 4.

Abdomen: Tergites smooth, with very sparse fine soft setae. Dorsal surface of somite 1 well convex from anterior to posterior. Somite 2 tergite 2.3-2.6 × broader than long; pleural lateral margins concavely divergent posteriorly, rounded on posterior terminus. Pleuron of somite 3 with blunt lateral margin. Somite 4 with rounded lateral margin. Telson half as long as broad; posterior plate distinctly emarginate on posterior margin, length 1.0-1.2 × that of anterior plate.

Eye: Small, 1.8 × longer than broad, proximally broader, bearing plumose setae proximal to cornea (very thick on paratype from Philippines), terminating in midlength of rostrum. Cornea not dilated, length about half that of remaining eyestalk.

Antennule and antenna: Ultimate article of antennular peduncle 3.3-4.0 × longer than high. Antennal peduncle overreaching eyes by full length of article 5. Article 2 laterally blunt, bearing long setae. Antennal scale about 2 × as broad as article 5, ending in midlength of article 5, bearing long plumose setae marginally. Article 4 unarmed or with tiny distomesial spine. Article 5 unarmed, 1.5 × longer than article 4, breadth more than half (0.7) height of ultimate antennular article. Flagellum of 16-19 segments reaching or slightly falling short of distal end of P1 merus.

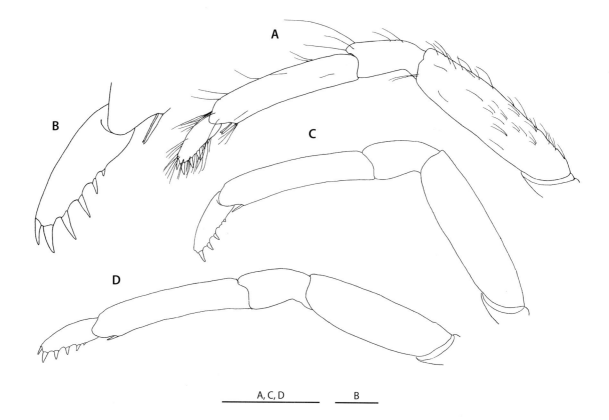

FIGURE 91

Uroptychus enriquei n. sp., holotype, ovigerous female 10.8 mm (MNHN-IU-2011-5953). **A**, left P2, lateral. **B**, same, distal part, setae omitted, lateral. **C**, left P3, lateral. **D**, left P4, lateral. Scale bars: A, C, D, 5 mm; B, 1 mm.

Mxp: Mxp1 with bases broadly separated. Mxp3 with long soft setae on lateral surface. Basis without denticles on mesial ridge. Ischium with more than 30 very small, distally diminishing denticles (barely discernible under high magnification on left side in holotype) on crista dentata; flexor margin not rounded distally. Merus 2.3 × longer than ischium, flattish on mesial face, flexor margin somewhat roundly ridged, without spine. Carpus unarmed.

P1: 5-6 × longer than carapace, unarmed except for ischium bearing bluntly truncate, antero-posteriorly compressed process on distodorsal margin, sparsely setose but palm almost glabrous, fingers more setose. Merus 1.1-1.3 × longer than carapace. Carpus 1.2-1.4 × longer than merus. Palm 3-4 × longer than broad, slightly longer than or subequal to carpus in males, slightly shorter in females; fingers slightly incurved distally, crossing when closed, opposable margins without process in females, with distinct median process on movable finger in males; movable finger 0.4-0.5 × length of palm.

P2-4: Compressed mesio-laterally, with sparse long setae, unarmed on meri and carpi. Meri successively shorter posteriorly (P3 merus 0.9 × length of P2 merus, P4 merus 0.8-0.9 × length of P3 merus), equally broad on P2-4; length-breadth ratio, 4.1-4.3 on P2, 3.0-3.7 on P3, 3.2-3.8 on P4; P2 merus subequal to or slightly shorter than carapace, 1.1-1.2 × longer than P2 propodus; P3 merus subequal to length of P3 propodus; P4 merus 0.9 × length of P4 propodus. Carpi subequal in length on P2 and P3 or successively slightly shorter posteriorly (P3 carpus 0.93-1.00 × length of P2 carpus, P4 carpus 0.93-0.95 × length of P3 carpus); carpus-propodus length ratio, 0.36-0.41 on P2, 0.37-0.38 on P3, 0.35-0.36 on P4. Propodi successively slightly longer posteriorly; flexor margin nearly straight, with pair of distal spines only. Dactyli subequal to carpi (dactylus-carpus length ratio, 0.8-1.0 on P2, 0.9-1.0 on P3, 1.0 on P4), about one-third length of propodi (0.33-0.35 on P2 and P3, 0.34-0.35 on P4); flexor margin straight, ending in slender spine preceded by 5-6 acute, elongate spines diminishing toward base of article (all obscured by setae on P2 and less so on P3 and P4 in Philippine material), ultimate and penultimate spines close to each other, others rather distant from one another, not perpendicular to margin but somewhat obliquely directed, ultimate much shorter than penultimate, penultimate slightly broader than antepenultimate.

Eggs. Number of eggs carried, about 40; size, 1.00 mm × 1.16 mm - 1.44 mm × 1.27 mm.

REMARKS — The specimen from the Philippines is more setose on the body and appendages than the other specimens.

Uroptychus enriquei resembles *U. onychodactylus* Tirmizi, 1964 from the Maldives and *U. setosidigitalis* Baba, 1977 from off Midway Island, in the carapace shape, especially bearing a reduced anterolateral spine. However, these congeners can be distinguished by: sternite 3 which has a deeply excavated anterior margin lacking median notch; the P2-4 dactyli which are strongly curved and much longer relative to the propodi, with closely arranged, more numerous, shorter flexor marginal spines; and the P2-4 propodi without flexor marginal spines.

Uroptychus enriquei also resembles *U. tomentosus* Baba, 1974 from the east coast of New Zealand, in the carapace shape, in having sternite 4 with the anteriorly rounded anterolateral margins, in having a short antennal scale, and in the spination of the P2-4 dactyli. In *U. tomentosus*, however, the anterolateral spine of the carapace is distinct and the penultimate of the flexor marginal spines of P2-4 dactyli is pronounced, measuring about twice as broad as the antepenultimate, both distinctive differences from *U. enriquei*.

Uroptychus eratus n. sp.

Figures 92, 93

TYPE MATERIAL — Holotype: **Philippines**. MUSORSTOM 1 Stn CP03, 14°01'N, 120°15'E, 183-185 m, 19.III.1976, ov. ♀ 1.8 mm (MNHN-IU-2013-8515).

ETYMOLOGY — Front the Greek *eratos* (lovely), alluding to the small lovely species.

DISTRIBUTION — Off southeastern Luzon, Philippines; 183-185 m.

DESCRIPTION — Small species. *Carapace*: Broader than long (length 0.75 × breadth); greatest breadth 1.9 × distance between anterolateral spines. Dorsal surface smooth, somewhat convex from anterior to posterior, smoothly continued on to rostrum. Lateral margins smoothly divergent posteriorly, convex on posterior fourth; anterolateral spine of moderate size (missing on left side, presumably broken), overreaching much smaller lateral orbital spine. Rostrum sharp triangular, with interior angle of 32°; length 0.4 × that postorbital carapace length, breadth less than half carapace breadth measured at posterior carapace margin; dorsal surface slightly concave; ventral surface horizontal. Lateral orbital spine small, situated directly mesial to but separated from anterolateral spine by basal breadth of latter spine. Pterygostomian flap anteriorly angular, ending in small spine, surface smooth.

Sternum: Excavated sternum with convex anterior margin, surface with ridge in midline. Sternal plastron 1.2 × broader than long, lateral extremities successively broader posteriorly. Sternite 3 shallowly depressed, anterior margin gently concave without median notch and submedian spines. Sternite 4 with anterolateral margin slightly convex, anteriorly produced to small spine; posterolateral margin short, one-third length of anterolateral margin. Anterolateral margin of sternite 5 strongly convex, 1.7 × longer than posterolateral margin of sternite 4.

Abdomen: Tergites smooth and glabrous. Somite 1 somewhat convex from anterior to posterior, without transverse ridge. Somite 2 tergite 1.9 × broader than long; pleural lateral margins weakly concave and moderately divergent posteriorly, with rounded posterolateral terminus. Pleuron of sternite 3 with blunt lateral end. Telson 0.3 × as long as broad; posterior plate feebly concave; length 0.6 × that of anterior plate.

Eye: 1.5 × longer than broad, distally narrowed, slightly falling short of apex of rostrum. Cornea not dilated, length one-third that of remaining eyestalk.

Antennule and antenna: Ultimate article of antennular peduncle 1.9 × longer than high. Antennal peduncle reaching distal margin of cornea. Article 2 with well-developed distolateral spine. Antennal scale slightly broader than article 5, terminating in distal end of article 5, slightly overreaching eye. No spine on articles 4-5; article 5 1.3 × length of article 4, breadth 0.6 × height of antennular ultimate article. Flagellum consisting of 6 segments, slightly overreaching P1 merus.

Mxp: Mxp1 with bases broadly separated. Mxp3 basis with a few obsolescent denticles on mesial ridge. Ischium with very small numerous denticles on crista dentata; flexor margin not rounded distally. Merus 2.1 × longer than ischium, flexor margin not cristate but somewhat roundly ridged with 2 close tiny spines distal to point two-thirds of length; distolateral spine small but distinct. Carpus with small distolateral spine and eminence at proximal part of extensor margin.

P1: Right P1 missing. Left P1 with sparse short setae, length 3.4 × that of carapace. Ischium with short, subtriangular dorsal spine, ventromesially bearing a few obsolescent denticles on proximal half, subterminal spine absent. Merus three-quarters length of carapace; distoventral (mesial and lateral) spines distinct, and a few scattered small spines on ventral surface. Carpus 1.1 × longer than merus; ventrally bearing obsolescent distomesial and distolateral spines, unarmed elsewhere. Palm relatively high dorsoventrally, 2.3 × longer than broad, 1.6 × longer than carpus. Fingers broad relative to length, depressed, distally incurved, crossing when closed, and curving ventrally; movable finger half as long as palm, opposable margin with bluntly subtriangular (in dorsal view) process proximal to midlength; opposing margin of fixed finger sinuous.

P2-4: Right P2 and left P3 missing. Right P4 shorter than left P4, especially dactylus much shorter (0.7 x), presumably regenerated. Meri unarmed, sparsely setose, successively shorter posteriorly (P3 merus 0.9 × length of P2 merus; P4 (left) merus 0.9 × length of P3 merus), successively slightly narrower posteriorly; length-breadth ratio, 3.4 on P2, 3.1 on P3, 3.0 on P4; P2 merus three-quarters length of carapace, about as long as P2 propodus; P3 merus 0.8 × length of P3 propodus; P4 merus 0.7 × length of P4 propodus. Carpi successively shorter posteriorly, 0.4 × length of propodus on P2-4. Propodi subequal; flexor margin straight, with pair of terminal spines preceded by 2 movable spines on P2, none on P3 and P4. Dactyli successively longer posteriorly, length half that of propodus on P2-4; dactylus-carpus length ratio, 1.0 on P2, 1.2 on P3 and P4; flexor margin feebly curved, with loosely arranged, subperpendicularly directed (slightly inclined), sharp spines, ultimate somewhat narrower than penultimate on P2, slightly so on P3 and P4, preceded by proximally diminishing spines (8 on left P2, broken on P3, 10 on left P4, 6 on right P4), penultimate spine largest.

Eggs. Two eggs carried; size, 0.84 × 0.94 mm - 0.90 × 0.96 mm.

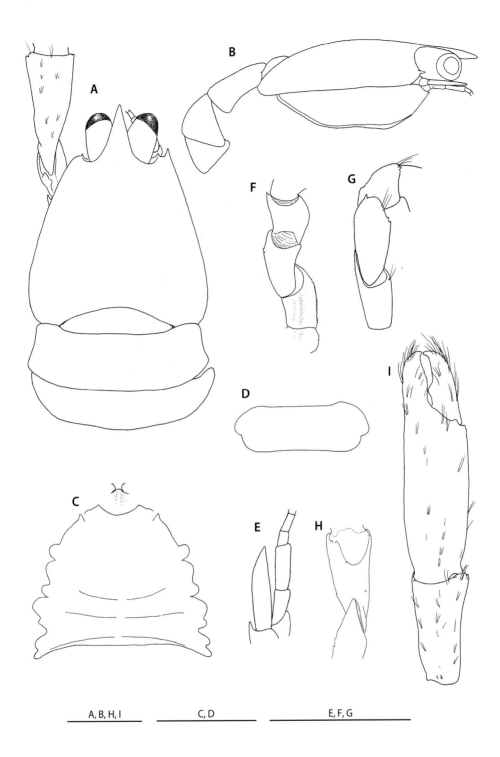

FIGURE 92

Uroptychus eratus n. sp., holotype, ovigerous female 1.8 mm (MNHN-IU-2013-8515). **A**, carapace and anterior part of abdomen, proximal part of left P1 included, dorsal. **B**, same, lateral. **C**, sternal plastron, with excavated sternum and basal parts of Mxp1. **D**, telson. **E**, right antenna, ventral. **F**, right Mxp3, setae omitted, ventral. **G**, same, lateral. **H**, left P1, proximal part, setae omitted, ventral. **I**, same, proximal part omitted, dorsal. Scale bars: 1 mm.

REMARKS — The carapace that is distinctly broader than long and posteriorly broadened, with well-developed antero-lateral spines links the species to *U. patulus* Ahyong & Poore, 2004 from southeastern Australia. However, *U. eratus* is clearly different from that species in having: the sternite 3 anterior margin gently instead of strongly excavated; the eyes reaching nearly the apex of the rostrum instead of terminating in the midlength; the antennal article 2 bearing a disto-lateral spine instead of being unarmed; and the P2-4 dactyli much shorter, at least half as long as the propodi instead of much more than so, with the flexor margin slightly curving instead of strongly so, and more noticeably bearing 9-11 instead of 20 closely arranged spines. In addition, *U. patulus* is a medium-sized species, the largest ovigerous female recorded being 12.6 mm (postorbital carapace length, 8.3 mm).

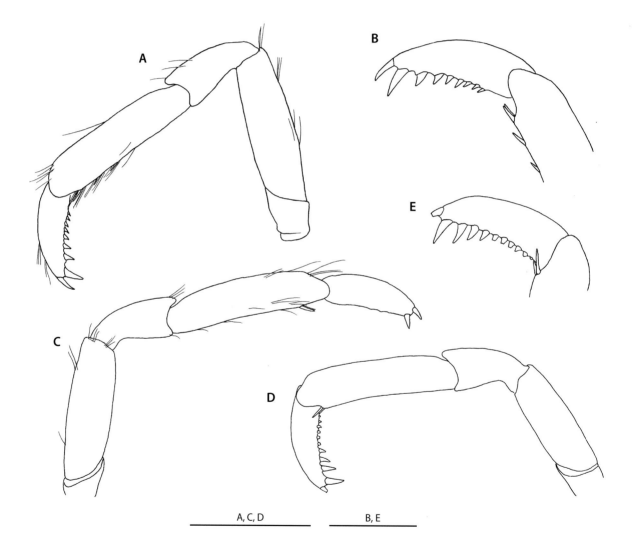

FIGURE 93

Uroptychus eratus n. sp., holotype, ovigerous female 1.8 mm (MNHN-IU-2013-8515). **A**, left P2, lateral. **B**, same, distal part, setae omitted, lateral. **C**, right P3, setae omitted, lateral. **D**, left P4, setae omitted, lateral. **E**, same, distal part, setae omitted, lateral. Scale bars: A, C, D, 1 mm; B, E, 0.5 mm.

Uroptychus exilis n. sp.

Figures 94, 95

TYPE MATERIAL — Holotype: **New Caledonia**, Isle of Pines. MUSORSTOM 4 Stn DW222, 22°57.6'S, 167°33.0'E, 410-440 m, 30.IX.1985, ♀ 3.1 mm (MNHN-IU-2010-1793).

ETYMOLOGY — From the Latin *exilis* (small, slender) referring to slender eyes of the species.

DISTRIBUTION — Isle of Pines; 410-440 m.

DESCRIPTION — Small species. *Carapace*: Broader than long (0.8 × as long as broad); greatest breadth 1.7 × distance between anterolateral spines. Dorsal surface feebly convex on gastric region, without distinct groove; with sparse short setae. Lateral margins divergent posteriorly to point one-fifth from posterior end, then convergent, bearing 7 spines: first anterolateral, slightly falling short of tip of lateral orbital spine; second larger than first, situated at anterior end of anterior branchial region, equidistant between first and fourth, with accompanying small spine dorsomesial to it on left side only; third very small; fourth largest, situated at anterior end of posterior branchial margin, followed by 4 posteriorly diminishing spines, last rudimentary. Rostrum elongate triangular, with interior angle of 34°, 0.4 × as long as remaining carapace, breadth slightly less than half carapace breadth measured at posterior carapace margin; dorsal surface concave, ventral surface straight horizontal. Lateral orbital spine subequal to anterolateral spine, situated directly mesial to and moderately remote from that spine. Pterygostomian flap unarmed on surface, anteriorly angular, produced to small spine.

Sternum: Excavated sternum with convex anterior margin, smooth on surface. Sternal plastron 1.2 × as broad as long, successively broader posteriorly. Sternite 3 shallowly depressed, anterior margin excavated in broad V-shape, with small median notch. Sternite 4 having anterolateral margin relatively short, nearly straight, anteriorly ending in 2 small spines placed side by side, posterolateral margin half as long as anterolateral margin. Anterolateral margin of sternite 5 moderately convexly divergent posteriorly, 1.3 × longer than posterolateral margin of sternite 4

Abdomen: With sparse fine setae. Somite 1 smooth on surface, without ridge. Somite 2 tergite 2.4 × broader than long; pleuron posterolaterally bluntly angular, lateral margins moderately concave and divergent posteriorly. Pleuron of somite 3 tapering, with blunt lateral terminus. Telson 0.4 × as long as broad; posterior plate distinctly emarginate on posterior margin, length four-fifths that of anterior plate.

Eye: Elongate, more than 2 × longer than broad, distally narrowed, reaching distal fifth of rostrum. Cornea not dilated, less than half as long as remaining eyestalk.

Antennule and antenna: Ultimate article of antennule 2.7 × longer than high. Antennal peduncle reaching distal end of cornea. Article 2 with small but distinct distolateral spine. Antennal scale reaching end of first segment of flagellum, breadth 1.8 × that of article 5. Article 4 unarmed. Article 5 1.5 × longer than article 4, bluntly produced distomesially without distinct spine, breadth 0.7 × height of antennular ultimate article. Flagellum of 11 segments overreaching distal end of P1 merus by 2 distal segments of flagellum.

Mxp: Mxp1 with bases broadly separated. Mxp3 barely setose on lateral surface. Basis without denticles on mesial ridge. Ischium with very small denticles on crista dentata, flexor margin distally rounded. Merus 1.5 × longer than ischium, sharply ridged along distal half of flexor margin bearing small spine at distal third. Carpus unarmed.

P1: Unequal, right one slightly larger, left one possibly regenerated. Right P1 3.5 × longer than carapace; very setose, setae plumose, moderate in length. Ischium with distinct dorsal spine and small subterminal spine on ventromesial margin. Merus 0.9 × length of carapace, ventrally bearing small distomesial and distolateral spines, dorsally with distomesial spine, terminal margin in juncture with carpus with a few dorsal denticles. Carpus 0.9 × length of merus, ventrally bearing distomesial and distolateral spines. Palm 2.4 × longer than broad, unarmed, length 1.2 × that of carpus. Fingers gently curving ventrally, distally incurved, crossing when closed; opposable margins denticulate, that of movable finger with blunt

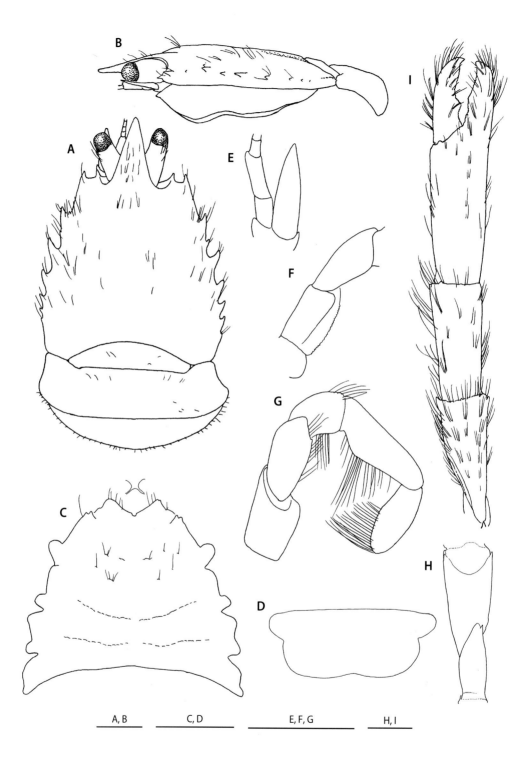

FIGURE 94

Uroptychus exilis n. sp., holotype, female 3.1 mm (MNHN-IU-2010-1793). **A**, carapace and anterior part of abdomen, dorsal. **B**, same, lateral. **C**, sternal plastron, with excavated sternum and basal parts of Mxp1. **D**, telson. **E**, left antenna, ventral. **F**, right Mxp3, setae omitted, ventral. **G**, same, lateral. **H**, right P1, proximal part, setae omitted, ventral. **I**, same, dorsal. Scale bars: 1 mm.

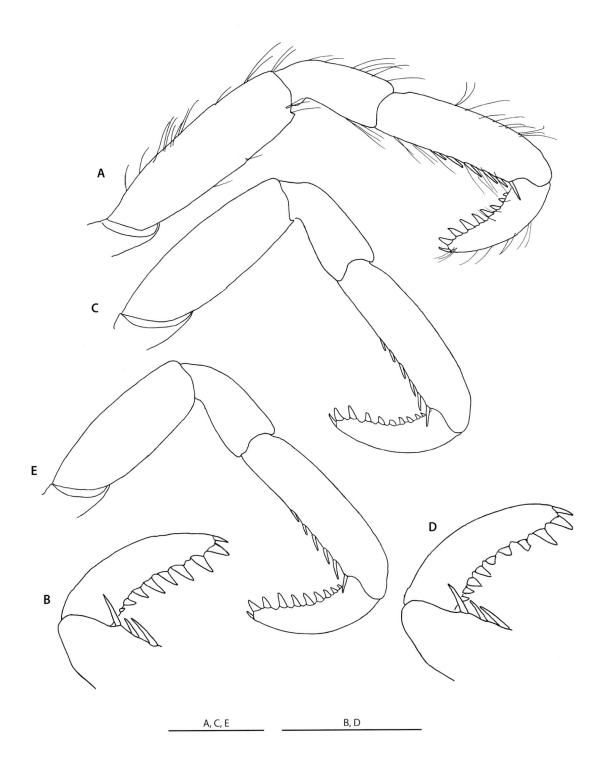

FIGURE 95

Uroptychus exilis n. sp., holotype, female 3.1 mm (MNHN-IU-2010-1793). **A**, right P2, lateral. **B**, same, distal part, setae omitted, lateral. **C**, right P3, setae omitted, lateral. **D**, same, distal part, setae omitted, lateral. **E**, right P4, setae omitted, lateral. Scale bars: 1 mm.

subtriangular process at proximal third, that of fixed finger sinuous, bearing low prominence slightly distal to position of opposite process on movable finger; movable finger 0.6 × length of palm.

P2-4: Moderately compressed mesio-laterally and setose. Meri successively shorter posteriorly (P3 merus 0.9 × length of P2 merus; P4 merus 0.8 × length of P3 merus), equally broad on P2-4; length-breadth ratio, 3.3 on P2, 2.9 on P3, 2.5 × on P4; P2 two-thirds as long as carapace, 1.1 × length of P2 propodus; P3 merus 0.9 × length of P3 propodus; P4 merus 0.8 × length of P4 propodus; dorsal margin unarmed, ventral margin with small distal spine on P2 and P3, unarmed on P4. Carpi subequal on P3 and P4, P3 carpus 0.9 × length of P2 carpus; length 0.5 × that of propodus on P2, 0.45 × on P3-4. Propodi successively slightly longer posteriorly; flexor margin nearly straight, ending in pair of movable spines preceded by row of 5 spines. Dactyli successively slightly longer posteriorly, moderately curving, 0.5 × length of propodi on P2 and P3, 0.6 × on P4, 1.1 × length of carpi on P2-4; flexor margin somewhat curving, with 10 loosely arranged, slightly obliquely directed spines, ultimate much more slender and shorter than penultimate, remaining spines sharp subtriangular, gradually diminishing toward base of article.

REMARKS — *Uroptychus exilis* most closely resembles *U. posticus* in the spination of the carapace lateral margin, the shape of the antenna and the spination of P2-4. Distinguishing characters of these species are outlined under the remarks of the latter species (see below). The new species also resembles *U. baeomma* n. sp. and *U. elongatus* n. sp. (see above), in the elongate eyestalks and the spination of the P2-4 dactyli. *Uroptychus exilis* is distinguished from these congeners by the following: the anterolateral spine of the carapace is subequal to instead of much larger than the lateral orbital spine; sternite 3 has the anterior margin excavated in a broad V-shape bearing a small median notch, instead of being shallowly excavated with 2 submedian spines flanking a subovate or narrow notch. In addition, *U. exilis* differs from *U. baeomma* in having the antennal scale distinctly overreaching, instead of at most ending in, the distal end of article 5. *Uroptychus exilis* also differs from *U. elongatus* in having the P1 ischium with a small instead of well-developed subterminal spine on the ventromesial margin.

Uroptychus floccus n. sp.

Figures 96, 97

TYPE MATERIAL — Holotype: **New Caledonia**. BATHUS 1 Stn CP710, 21°43'S, 166°36'E, 320-386 m, 19.III.1993, ov. ♀ 3.1 mm (MNHN-IU-2014-16456). Paratypes: **New Caledonia**. BATHUS 1, station data as for the holotype, 1 ov. ♀ 3.9 mm (MNHN-IU-2014-16457). – Stn CP711, 21°43'S, 166°36'E, 315-327 m, 19.III.1993, 1 ♂ 4.2 mm (MNHN-IU-2014-16458). **New Caledonia**, Norfolk Ridge. NORFOLK 1 Stn DW1651, 23°27.3'S, 167°50.4'E, 276-350 m, 19.VI.2001, 1 ♂ 4.1 mm, 1 ov. ♀ 4.3 mm (MNHN-IU-2014-16459). – Stn DW1652, 23°26.1'S, 167°50.3'E, 290-378 m, 23.VI.2001, 1 ♂ 4.8 mm (MNHN-IU-2014-16460). – Stn DW1653, 23°28'S, 167°51'E, 328-340 m, 19.VI.2001, 1 ♂ 4.1 mm, 6 ov. ♀ 3.3-5.2 mm, 2 ♀ 3.7, 4.0 mm (MNHN-IU-2014-16461). – Stn DW1654, 23°28'S, 167°52'E, 366-560 m, 19.VI.2001, 3 ♂ 2.4-3.4 mm, 1 ov. ♀ 3.3 mm, 5 ♀ 2.7-3.9 mm (MNHN-IU-2014-16462). – Stn CP1655, 23°26'S, 167°51'E, 680 m, 19.VI.2001, 1 ♀ 5.0 mm (MNHN-IU-2014-16463). – Stn DW1657, 23°28'S, 167°52'E, 305-332 m, 19.VI.2001, 1 ov. ♀ 4.0 mm, 1 ♀ 4.0 mm (MNHN-IU-2014-16464), 2 ov. ♀ 3.9, 4.3 mm (MNHN-IU-2014-16465). – Stn DW1659, 23°37'S, 167°41'E, 449-467 m, 20.VI.2001, 3 ♂ 3.4-4.8 mm, 1 ov. ♀ 4.7 mm, 2 ♀ 3.9, 4.3 mm (MNHN-IU-2014-16466). – Stn DW1667, 23°40'S, 168°01'E, 237-250 m, 21.VI.2001, 1 ov. ♀ 3.2 mm (MNHN-IU-2014-16467). – Stn CP1669, 23°41'S, 168°01'E, 302-325 m, 21.VI.2001, 3 ♂ 3.2-3.7 mm, 3 ov. ♀ 3.3-4.1 mm (MNHN-IU-2014-16468). – Stn CP1671, 23°42'S, 168°01'E, 397-320 m, 21.VI.2001, on *Chironephthya* sp. (Nidaliidae: Alcyonacea), 2 ♂ 3.2, 3.4 mm (MNHN-IU-2014-16469). – Stn CP1672, 23°41'S, 168°01'E, 324-267 m, 21.VI.2001, 1 ov. ♀ 2.9 mm (MNHN-IU-2014-16470). – Stn DW1674, 23°40'S, 168°00'E, 245-253 m, 21.VI.2001, 1 ov. ♀ 3.7 mm (MNHN-IU-2014-16471). – Stn DW1675, 24°45'S, 168°09'E, 231-233 m, 22.VI.2001, 1 ♂ 3.7 mm (MNHN-IU-2014-16472). – Stn DW1704, 23°45'S, 168°16'E, 400-420 m, 25.VI.2001, 1 ov. ♀ 3.9 mm (MNHN-IU-2014-16473), 3 ♂ 2.8-3.6 mm, 7 ov. ♀ 3.6-3.8 mm, 5 ♀ 2.7-3.8 mm (MNHN-

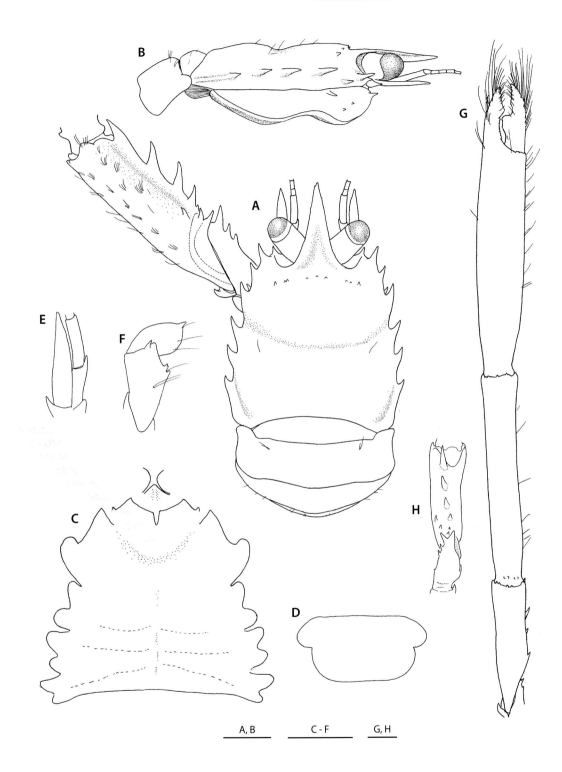

FIGURE 96

Uroptychus floccus n. sp., holotype, ovigerous female 3.1 mm (MNHN-IU-2014-16456). **A**, carapace and anterior part of abdomen, proximal part of left P1 included, dorsal. **B**, same, lateral. **C**, sternal plastron, with excavated sternum and basal parts of Mxp1. **D**, telson. **E**, right antenna, ventral. **F**, right Mxp3 (merus and carpus), lateral. **G**, left P1, dorsal. **H**, same, proximal part, ventral. Scale bars: 1 mm.

IU-2014-16474). – Stn CP1706, 23°44'S, 168°17'E, 383-394 m, 25.VI.2001, 2 ov. ♀ 4.0, 4.5 mm (MNHN-IU-2014-16475). – Stn DW1707, 23°43'S, 168°16'E, 381-493 m, 25.VI.2001, 2 ♂ 4.4 mm (carapace broken), 1 ov. ♀ 3.4 mm (MNHN-IU-2014-16476). – Stn DW1712, 23°23'S, 168°02'E, 180-250 m, 26.VI.2001, 2 ♂ 3.7, 4.3 mm, 3 ov. ♀ 3.3-3.8 mm, 2 ♀ 2.8, 3.8 mm (MNHN-IU-2014-16477). – Stn CP1713, 23°22'S, 168°02'E, 204-216 m, 26.VI.2001, 3 ♂ 3.1-4.8 mm, 1 ♀ 3.7 mm (MNHN-IU-2014-16478). – Stn CP1714, 23°22'S, 168°03'E, 257-269 m, 26.VI.2001, 2 ♂ 4.4, 4.5 mm, 2 ov. ♀ 3.6, 3.8 mm (MNHN-IU-2014-16479). – Stn DW1723, 23°18'S, 168°15'E, 266-267 m, 27.VI.2001, 2 ♂ 3.3, 3.4 mm, 1 ov. ♀ 3.1 mm (MNHN-IU-2014-16480). – Stn DW1729, 23°20'S, 168°16'E, 340-619 m, 27.VI.2001, 4 ♂ 3.3-4.2 mm, 5 ov. ♀ 3.5-4.6 mm, 3 ♀ 3.5-4.6 mm (MNHN-IU-2014-16481). – Stn DW1729, 23°20'S, 168°16'E, 340-619 m, 27.VI.2001, 2 ♂ 3.2, 3.7 mm, 3 ov. ♀ 4.0-4.5 mm (MNHN-IU-2014-16482). – Stn DW1732, 23°21'S, 168°16'E, 347-1063 m, 27.VI.2001, 1 ♂ 4.7 mm (MNHN-IU-2014-16483).

OTHER MATERIAL EXAMINED — **New Caledonia**. MUSORSTOM 4 Stn CP155, 18°54'S, 163°19'E, 500-570 m, 15.IX.1985, 1 ov. ♀ 4.7 mm (MNHN-IU-2014-16484). – Stn DW163, 18°33.8'S, 163°11.5'E, 350 m, 16.IX.1985, 1 ov. ♀ 3.6 mm (MNHN-IU-2014-16485). BATHUS 4 Stn DW924, 18°54.85'S, 163°24.34'E, 344-360 m, 7.VIII.1994, 1 ♂ 4.0 mm, 2 ov. ♀ 3.3, 3.9 mm (MNHN-IU-2014-16486). – Stn DW929, 18°51.55'S, 163°23.27'E, 502-516 m, 7.VIII.1994, 5 ♂ 3.5-5.1 mm, 2 ov. ♀ 4.3, 4.6 mm, 1 ♀ 5.0 mm (MNHN-IU-2014-16487). – Stn DW931, 18°55.38'S, 163°24.36'E, 360-377 m, 7. VIII.1994, 2 ov. ♀ 3.8, 4.7 mm (MNHN-IU-2014-16488). LAGON DW1148, 19°06.5'S, 163°30.1'E, 215-220 m, 28.X.1989, 1 ov. ♀ 2.8 mm (MNHN-IU-2014-16489). HALIPRO 1 Stn CP851, 21°43'S, 166°37'E, 314-364 m, 19.III.1994, 1 ov. ♀ 3.8 mm (MNHN-IU-2014-16490). BATHUS 2 Stn DW717, 22°44'S, 167°16'E, 350-393 m, 11.V.1993, 1 ♂ 3.4 mm, 2 ov. ♀ 3.1, 3.2 mm, 1 ♀ 3.2 mm (MNHN-IU-2014-16491). SMIB 2 Stn DW06, 22°55'S, 167°16'E, 442-460 m, 17.IX.1986, 1 ov. ♀ 4.2 mm (MNHN-IU-2014-16492). BIOCAL Stn DW37, 23°00'S, 167°16'E, 350 m, 30.VIII.1985, 2 ov. ♀ 3.3, 3.8 mm (MNHN-IU-2014-16493). – Stn DW38, 23°00'S, 167°15'E, 360 m, 30.VIII.1985, 2 ov. ♀ 4.0, 4.5 mm (MNHN-IU-2014-16494). MUSORSTOM 4 Stn DW222, 22°57.6'S, 167°33.0'E, 410-440 m, 30.IX.1985, 3 ov. ♀ 3.2-3.9 mm (MNHN-IU-2014-16495). – Stn DW227, 22°46'S, 167°20'E, 320 m, 30.IX.1985, 1 ♂ 3.3 mm (MNHN-IU-2014-16496). LAGON Stn DW380, 22°29,6'S, 167°11,8'E, 60 m, 22.I.1985, 1 ov. ♀ 3.5 mm (MNHN-IU-2014-16497). – Stn DW389, 22°44'S, 167°04'E, 274-274 m, 22.I.1985, 1 ov. ♀ 4.1 mm (MNHN-IU-2014-16498). **New Caledonia**, Loyalty Islands. CALSUB PL16, 20°37,8'S, 167°02,7'E, 825-370 m, 7.III.1989, 1 ov. ♀ 2.8 mm (MNHN-IU-2014-16499). MUSORSTOM 6 Stn CP401, 20°42.15'S, 167°00.35'E, 270 m, 14.II.1989, 1 ♂ 3.7 mm (MNHN-IU-2014-16500). – Stn DW472, 21°08.60'S, 167°54.70'E, 300 m, 22 II. 1989, 3 ov. ♀ 3.3-3.7 mm (MNHN-IU-2014-16501). **New Caledonia**, Hunter and Matthew Islands. VOLSMAR Stn DW39, 22°20'S, 168°44'E, 280-305 m, 08.VI.1989, 1 ♂ 3.9 mm, 1 ♀ 3.8 mm (MNHN-IU-2014-16502). **New Caledonia**, Isle of Pines. CALSUB PL21, 22°45'S, 167°09'E, 344-330 m, 12.III.1989, 1 ♀ (carapace broken) (MNHN-IU-2014-16503). – PL19, 22°46'S, 167°20'E, 416-404 m, 10.III.1989, on *Stylaster* sp. (Hydrozoa: Stylasteridae), 1 ♂ 2.8 mm, 1 ♀ 3.0 mm (MNHN-IU-2014-16504). **New Caledonia**, Norfolk Ridge. CHALCAL 2 Stn CH08, 23°13'S, 168°03'E, 300 m, 31.X.1986, unidentified coral host, 1 ov. ♀ 3.6 mm (MNHN-IU-2014-16505). – Stn CP26, 23°18.15'S, 168°03.58'E, 296 m, unidentified coral host, 31.X.1986, 2 ♂ 2.8, 4.2 mm (MNHN-IU-2014-16506). – Stn CP27, 23°15.29'S, 168°04.55'E, 289 m, 31.X.1986, 1 ov. ♀ 3.3 mm (MNHN-IU-2014-16507). – Stn CP27, 23°15.29'S, 168°04.55'E, 289 m, 31.X.1986, 2 ♂ 3.5, 4.2 mm, 1 ov. ♀ 4.0 mm (MNHN-IU-2014-16508). – Stn DW76, 23°40.5'S, 167°45.2'E, 470 m, 30.X.1986, 1 ♂ 2.8 mm (MNHN-IU-2014-16509). – Stn DW78, 23°41'S, 168°00'E, 233-360 m, 30.X.1986, 1 ♂ 3.6 mm, 3 ov. ♀ 3.2-4.0 mm (MNHN-IU-2014-16510). – Stn DW79, 23°40.5'S, 168°00.1'E, 243 m, 30.X.1986, 2 ov. ♀ 2.4, 3.6 mm (MNHN-IU-2014-16511). – Stn DW81, 23°19.6'S, 168°03.4'E, 311 m, 31.X.1986, 1 ♂ 4.0 mm, 4 ov. ♀ 2.8-3.9 mm, 2 ♀ 2.6, 3.4 mm (MNHN-IU-2014-16512). – Stn DW82, 23°13.68'S, 168°04.27'E, 304 m, 31.X.1986, 2 ov. ♀ 2.9, 3.6 mm (MNHN-IU-2014-16513). LITHIST Stn CP03, 23°37.0'S, 167°41.5'E, 447 m, 10.VIII.1999, 1 ov. ♀ 4.0 mm (MNHN-IU-2014-16514). – Stn CP15, 23°40.4'S, 168°15.0'E, 389-404 m, 12.VIII.1999, 1 ♂ 2.8 mm, 1 ov. ♀ 3.7 mm, 1 ♀ 4.1 mm (MNHN-IU-2014-16515). – Stn DW13, 23°45.0'S, 168°16.7'E, 400 m, 12.VIII.1999, 2 ov. ♀ 3.4, 3.6 mm (MNHN-IU-2014-16516), 6 ♂ 2.9-4.0 mm, 11 ov. ♀ 3.0-4.0 mm (MNHN-IU-2014-16517), 2 ♂ 3.0, 3.4 mm, 1 ov. ♀ 3.8 mm (MNHN-IU-2014-16518). – Stn CP16, 23°43.2'S, 168°16.2'E, 379-391 m, 12.VIII.1999, 3 ov. ♀ 3.2-3.4 mm (MNHN-IU-2014-16519). SMIB 4 Stn DW51, 23°40'S, 168°01'E, 245-260 m, 09.III.1989, 1 ♀ 4.1 mm (MNHN-IU-2014-16520). – Stn DW53, 23°39'S, 168°00'E, 250-270 m, 09.III.1989, 1 ♂ 3.7 mm, 2 ov. ♀ 3.3, 4.0 mm (MNHN-IU-2014-16521). – Stn DW55, 23°21'S, 168°05'E, 215-260 m, 09.III.1989, 2 ♂ 3.3, 3.6 mm (MNHN-IU-2014-16522). SMIB 5 Stn DW70, 23°41'S, 168°01'E, 260-270 m, 07. IX.1989, 1 ♂ 3.8 mm (MNHN-IU-2014-16523). – Stn DW72, 23°42'S, 168°01'E, 280-400 m, 07.IX.1989, 1 ov. ♀ 3.5 mm (MNHN-IU-2014-16524). – Stn DW76, 23°41'S, 168°01'E, 240-280 m, 07.IX.1989, 2 ♂ 3.6, 4.1 mm, 2 ov. ♀ 3.5, 4.0 mm (MNHN-IU-2014-16525). – Stn DW85, 22°20'S, 168°42'E, 240-260 m, 13.IX.1989, 2 ov. ♀ 3.5, 3.8 mm (MNHN-IU-2014-16526). – Stn DW87, 22°18'S, 168°41'E, 335-370 m, 13.IX.1989, 1 ♂ 2.1 mm (MNHN-IU-2014-16527). – Stn DW92, 22°20'S, 168°42'E, 255-280 m, 13.IX.1989, 1 ♂ 3.5 mm, 2 ov. ♀ 3.5, 4.5 mm (MNHN-IU-2014-16528). – Stn DW97, 23°02'S, 168°18'E, 240-300 m, 14.IX.1989, 1 ♂ 3.5 mm (MNHN-IU-2014-16529). – Stn DW98, 23°02'S, 168°16'E, 320-335 m, 14.IX.1989, 2 ♂ 2.6, 2.8 mm, 4 ov. ♀ 2.3-4.4 mm, 4 ♀ 3.1-3.7 mm (MNHN-IU-2014-16530). – Stn DW101, 23°21'S, 168°05'E, 285-270 m, 14.IX.1989, 6 ♂ 2.8-4.3 mm, 4 ov. ♀ 3.3-4.0 mm, 2 ♀ 2.5, 3.6 mm (MNHN-IU-2014-16531). – Stn DW102, 23°19'S, 168°05'E, 290-305 m, 14.IX.1989, 5 ov. ♀ 3.3-4.2 mm (MNHN-IU-2014-16532). – Stn DW103, 23°16'S, 168°04'E, 300-315 m, 14.IX.1989, 2 ♂ 3.8, 4.3 mm, 5 ov. ♀ 3.5-4.0 mm, 4 ♀ 3.0-4.4 mm (MNHN-IU-2014-16533).

SMIB 8 Stn DW175, 23°41'S, 168°01'E, 235-240 m, 29.I.1993, 1 ♂ 4.0 mm (MNHN-IU-2014-16534). – Stn DW179, 23°45.87'S, 168°16.95'E, 400-405 m, 30.I.1993, 1 ov. ♀ 3.3 mm (MNHN-IU-2014-16535). – Stn DW181, 23°17.74'S, 168°04.82'E, 311-330 m, 31 I.1993, 2 ov. ♀ 3.8, 4.0 mm, 1 ♀ 3.2 mm (MNHN-IU-2014-16536). – Stn DW182, 23°19'S, 168°05'E, 330-314 m, 31.I.1993, 1 ov. ♀ 3.5 mm (MNHN-IU-2014-16537). – Stn DW190, 23°18'S, 168°05'E, 305-310 m, 31.I.1993, 1 ov. ♀ 4.2 mm (MNHN-IU-2014-16538). HALIPRO 2 BT94, 23°33'S, 167°42'E, 448-880 m, 24.XI.1996, 1 ♀ 3.7 mm (MNHN-IU-2014-16539). BATHUS 3 Stn CH801, 23°39.40'S, 168°00.50'E, 270-300 m, 27.XI.1993, 2 ov. ♀ 3.9, 4.2 mm, 1 ♀ 4.2 mm (MNHN-IU-2014-16540). – Stn CH802, 23°41'S, 168°00'E, 237-550 m, 27.XI.1993, 1 ♂ 4.2 mm, 3 ov. ♀ 3.9-4.0 mm, 1 ♀ 3.7 mm (MNHN-IU-2014-16541). – Stn CP804, 23°41'S, 168°00'E, 244-278 m, 27.XI.1993, 1 ov. ♀ 3.9 mm (MNHN-IU-2014-16542). – Stn CP805, 23°41'S, 168°01'E, 278-310 m, 27.XI.1993, 7 ♂ 2.8-3.8 mm, 2 ov. ♀ 2.9, 4.1 mm, 1 ♀ 3.1 mm (MNHN-IU-2014-16543), 1 ♂ 3.7 mm (MNHN-IU-2014-16544). – Stn DW807, 23°40'S, 167°59'E, 420-438 m, 27.XI.1993, 2 ov. ♀ 3.4, 4.1 mm (MNHN-IU-2014-16545). – Stn CP811, 23°41'S, 168°15'E, 383-408 m, 28.XI.1993, 1 ♂ 4.3 mm, 3 ov. ♀ 3.5-3.8 mm, 1 ♀ 3.5 mm (MNHN-IU-2014-16546). – Stn CP813, 23°45'S, 168°17'E, 410-415 m, 28.XI.1993, 1 ♂ 2.8 mm (MNHN-IU-2014-16547), 1 ov. ♀ 3.8 mm (MNHN-IU-2014-16548). – Stn DW817, 23°42'S, 168°16'E, 405-410 m, 28.XI.1993, 6 ov. ♀ 3.2-3.8 mm, 1 ♀ 2.2 mm (MNHN-IU-2014-16549). – Stn DW818, 23°44'S, 168°16'E, 394-401 m, 28.XI.1993, 2 ♂ 3.0, 3.1 mm, 4 ov. ♀ 2.8-3.5 mm, 1 ♀ 2.4 mm (MNHN-IU-2014-16550). – Stn DW819, 23°45'S, 168°16'E, 478-486 m, 28.XI.1993, 1 ♀ 4.3 mm (MNHN-IU-2014-16551). – Stn DW830, 23°20'S, 168°01'E, 361-365 m, 29.XI.1993, 2 ov. ♀ 2.7, 4.0 mm (MNHN-IU-2014-16552), 16 ♂ 2.2-4.6 mm, 12 ov. ♀ 3.9-4.3 mm, 5 ♀ 2.8-4.4 mm (MNHN-IU-2014-16553), 2 ov. ♀ 2.7, 4.0 mm (MNHN-IU-2014-16554). BERYX 11 Stn DW40, 23°41.48'S, 168°00.65'E, 240-300 m, 20.X.1992, 5 ♂ 2.9-3.6 mm, 1 ♀ 3.1 mm (MNHN-IU-2014-16555). – Stn CP45, 23°40.27'S, 168°00.95'E, 270-290 m, 20.X.1992, 1 ov. ♀ 3.3 mm (MNHN-IU-2014-16556). – Stn CP46, 23°42.00'S, 168°01.25'E, 300-350 m, 20.X.1992, 2 ♂ 4.7, 5.2 mm (MNHN-IU-2014-16557). – Stn CP51, 23°45'S, 168°17'E, 390-400 m, 21.X.1992, 1 ov. ♀ 4.2 mm (MNHN-IU-2014-16558). NORFOLK 2 Stn DW2023, 23°27.04'S, 167°50.72'E, 282-297 m, 21.X.2003, 1 ♂ 4.4 mm, 1 ov. ♀ 3.5 mm (MNHN-IU-2014-16559). – Stn DW2024, 23°27.92'S, 167°50.90'E, 370-371 m, 21.X.2003, 5 ♂ 2.3-4.4 mm, 3 ov. ♀ 3.4-3.5 mm, 3 ♀ 3.4-4.0 mm (MNHN-IU-2014-16560). – Stn CP2029, 23°38.88'S, 167°44.05'E, 438-445 m, 22.X.2003, 1 ov. ♀ 4.0 mm (MNHN-IU-2014-16561). – Stn DW2032, 23°39.14'S, 167°43.39'E, 420-450 m, 22.X.2003, 2 ♂ 3.6, 3.7 mm, 3 ov. ♀ 3.9-4.2 mm (MNHN-IU-2014-16562). – Stn DW2033, 23°39.13'S, 167°43.15'E, 430-450 m, 22.X.2003, 2 ♂ 4.4, 4.8 mm, 4 ov. ♀ 4.1-4.7 mm, 1 ♀ 3.0 mm (MNHN-IU-2014-16563). – Stn CP2138, 23°01'S, 168°23'E, 396-405 m, 03.XI.2003, 1 ♀ 3.6 mm (MNHN-IU-2014-16564). – Stn DW2034, 23°40.64'S, 167°41.36'E, 485-505 m, 22.X.2003, 1 ov. ♀ 4.1 mm (MNHN-IU-2014-16565). – Stn DW2049, 470-621 m, 23°42.88'S, 168°15.43'E, 24.X.2003, 1 ov. ♀ 4.0 mm (MNHN-IU-2014-16566). – Stn CP2050, 23°42.17'S, 168°15.72'E, 377-377 m, 24.X.2003, 1 ♂ 2.8 mm (MNHN-IU-2014-16567). – Stn DW2052, 23°42.29'S, 168°15.27'E, 473-525 m, 24.X.2003, 2 ♂ 3.3, 3.6 mm, 2 ov. ♀ 3.2, 3.3 mm (MNHN-IU-2014-16568). – Stn DW2108, 23°46.52'S, 168°17.12'E, 403-440 m, 31.X.2003, 2 ♂ 2.3-2.7 mm, 2 ov. ♀ 2.9, 3.2 mm (MNHN-IU-2014-16569). – Stn DW2109, 23°47.46'S, 168°17.04'E, 422-495 m, 31.X.2003, 3 ♂ 3.0-3.7 mm, 4 ov. ♀ 3.0-4.2 mm, 2 ♀ 2.9, 3.3 mm (MNHN-IU-2014-16570). – Stn DW2110, 23°48.34'S, 168°16.81'E, 493-850 m, 31.X.2003, 2 ♂ 2.5, 3.9 mm, 3 ov. ♀ 3.5-4.2 mm, 1 ♀ 4.0 mm (MNHN-IU-2014-16571). – Stn DW2124, 23°18'S, 168°15'E, 260-270 m, 2.XI.2003, 1 ov. ♀ 3.3 mm (MNHN-IU-2014-16572). – Stn DW2125, 23°16.64'S, 168°14.00'E, 275-348 m, 2.XI.2003, 1 ov. ♀ 3.2 mm (MNHN-IU-2014-16573). – Stn DW2126, 23°15.60'S, 168°13.50'E, 398-550 m, 2.XI.2003, 2 ♂ 3.6-3.8 mm, 5 ov. ♀ 3.7-3.9 mm (MNHN-IU-2014-16574). – Stn DW2135, 23°01.61'S, 168°21.35'E, 295-330 m, 3.XI.2003, 1 ♂ 3.1 mm, 1 ov. ♀ 3.2 mm (MNHN-IU-2014-16575). – Stn CP2139, 23°01'S, 168°23'E, 372-393 m, 03.XI.2003, 5 ♂ 3.4-4.3 mm, 3 ov. ♀ 3.7-4.3 mm, 1 ♀ 3.4 mm (MNHN-IU-2014-16576). – Stn DW2142, 23°00.51'S, 168°16.90'E, 550-550 m, 3.XI.2003, 1 ♀ 3.4 mm (MNHN-IU-2014-16577).

ETYMOLOGY — From the Latin *floccus* (tuft of hairs), suggesting the setose dactyli of P2-4 characteristic of the new species.

DISTRIBUTION — New Caledonia, Loyalty Ridge, Isle of Pines, Hunter and Matthew islands and Norfolk Ridge; 180-880 m.

SIZE — Males, 2.1-5.1 mm; females, 2.3-5.2 mm; ovigerous females from 2.3 mm.

DESCRIPTION — Small species. *Carapace*: Slightly broader than long (0.9 × as long as broad); greatest breadth 1.4-1.5 × distance between anterolateral spines. Dorsal surface smooth, somewhat convex from anterior to posterior, with transverse depression suggesting cervical groove; epigastric region with transverse row of 9 small spines preceded by depressed rostrum. Lateral margins somewhat convex, with 6 acute spines: first anterolateral, directed anterolaterally, terminating in or slightly overreaching tip of much smaller lateral orbital spine; second smaller than first, located on hepatic region, ventral to level of other spines; third to sixth on branchial region, all strong, much larger than second and equidistantly arranged; last spine followed by ridge. Rostrum sharp triangular, with interior angle of 20-30°, straight horizontal; length 0.5-0.6 × postorbital carapace length, breadth half carapace breadth measured at posterior carapace margin; dorsal surface concave; lateral margin often with a few small denticles distally. Lateral orbital spine situated anterior to level of anterolateral spine. Pterygostomian flap anteriorly roundish, with sharply produced anterior spine followed by 2 small spines usually distinct, rarely obsolescent on anterior median surface, and another spine usually present (very rarely absent) directly below linea anomurica between second and third marginal spines of carapace.

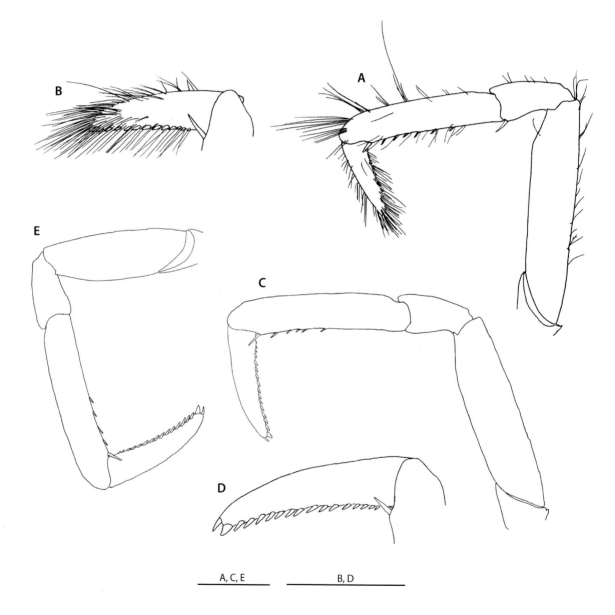

A, C, E   B, D

FIGURE 97

Uroptychus floccus n. sp., holotype, ovigerous female 3.1 mm (MNHN-IU-2014-16456). **A**, left P2, lateral. **B**, same, distal part, lateral. **C**, left P3, setae omitted, lateral. **D**, same, distal part, lateral. **E**, left P4, lateral. Scale bars: 1 mm.

Sternum: Excavated sternum bluntly produced anteriorly between bases of Mxp1 but often broad triangular, surface with ridge in midline on anterior half. Sternal plastron slightly broader than long, lateral extremities gently divergent posteriorly. Sternite 3 depressed well; anterior margin moderately convex, with deep U- or V-shaped median notch without flanking spine, laterally sharp angular. Sternite 4 with anterolateral margin convex, anteriorly sharp angular, posterolateral margin slightly longer than anterolateral margin. Anterolateral margin of somite 5 anteriorly rounded, slightly more than half as long as posterolateral margin of sternite 4.

Abdomen: Barely setose, somites relatively long. Somite 1 dorsally strongly convex from anterior to posterior. Somite 2 tergite 2.6-2.7 × broader than long; pleuron posterolaterally rounded, lateral margin concave and gently divergent posteriorly. Pleuron of somite 3 posterolaterally rounded. Telson nearly half as long as broad or slightly less than so; posterior plate 1.1-1.3 × longer than anterior plate, posterior margin nearly transverse or slightly convex.

Eye: Somewhat elongate (length 1.8 × breadth), slightly or distinctly broadened proximally, slightly overreaching midlength of rostrum. Cornea not dilated, more than half length of remaining eyestalk.

Antennule and antenna: Ultimate antennular article 3.3-3.8 × longer than high. Antennal peduncle overreaching cornea, slightly falling short of rostral tip. Article 2 with strong distolateral spine. Antennal scale distally sharp, terminating in or slightly overreaching distal end of article 5, occasionally bearing small spine on proximal lateral margin, breadth slightly more than 1.5 × that of article 5. Article 4 with distinct distomesial spine. Article 5 1.2 × longer than article 4, distomesially bearing small, often obsolescent spine, breadth two-thirds to half height of ultimate antennular article. Flagellum of 12-15 (usually 12 or 13) segments barely reaching distal end of P1 merus.

Mxp: Mxp1 with bases close to each other but not contiguous. Mxp3 basis usually unarmed, occasionally with 2 or 3 obsolescent denticles on mesial ridge. Ischium with distally rounded flexor margin, crista dentata with obsolescent denticles. Merus and carpus each with distolateral spine; merus 1.8-1.9 × longer than ischium, mesial face flattish, flexor margin with 2 small spines distal to point two-thirds of length.

P1: Slender, more than 6 × longer than carapace in large specimens; sparingly with short soft setae on merus and carpus, sparsely on palm, and more numerous longer setae on fingers. Ischium dorsally bearing basally broad, compressed, strong spine, ventromesially bearing strong subterminal spine proximally followed by a few to a number of proximally diminishing spines. Merus with row of 4 basally broad, equidistant ventromesial spines and field of 3 smaller mesioproximal spines; length 1.3-1.6 × that of carapace. Carpus 1.5-1.8 × longer than merus. Palm 3.6-4.5 × (males), 4.6-7.2 × (females) longer than broad, nearly as long as or slightly longer than carpus. Fingers gaping in large specimens of both sexes, tapering and slightly incurved distally, crossing when closed; movable finger 0.3-0.4 × length of palm, opposable margin with broad, low process on proximal half in both sexes.

P2-4: Slender, well compressed mesio-laterally. Meri successively shorter posteriorly (P3 merus 0.9 × length of P2 merus, P4 merus 0.8-0.9 × length of P3 merus); length-breadth ratio, 5.3-6.0 on P2, 4.0-4.9 on P3, 3.0-3.8 on P4; dorsal margins unarmed; P2 merus slightly shorter than, occasionally as long as or slightly longer than carapace, slightly narrower than P3-4 meri, 1.3-1.7 × longer than P2 propodus; P3 merus subequal to length of P3 propodus; P4 merus as broad as P3 merus, 0.8 × length of P4 propodus. P2-4 carpi subequal or P2 carpus slightly shorter; carpus-propodus length ratio, 0.4 on P2, 0.3-0.4 on P3 and P4. Propodi successively longer posteriorly, more slender on P2 than on P3 and P4; flexor margin straight, ending in pair of terminal spines preceded by 5-6 spines on P2, 3-5 spines on P3-4. Dactyli much more slender and shorter on P2 than on P3 and P4, length 0.6-0.7 × that of propodus on P2 and P3, 0.5-0.7 × on P4, and 1.3 × that of carpus on P2, 1.7-1.9 × on P3, 1.7-1.8 × on P4; flexor margin nearly straight, with 14-17 spines on P2, 17-19 spines on P3 and P4, penultimate much broader (about 2 × broader than antepenultimate), remaining spines slender, strongly inclined, close to one another, obscured by dense setae particularly on P2.

Eggs. Number of eggs carried, 5-22; size, 0.81 mm × 0.86 mm - 1.09 × 1.22 mm.

PARASITES — A rhizocephalan externa on the antennule was observed in the following specimens: 1 ♂ (MNHN-IU-2014-16576), 1 ov. ♀ and 1 ♀ (MNHN-IU-2014-16464), 1 ov. ♀ (MNHN-IU-2014-16562), 1 ov. ♀ (MNHN-IU-2014-16563), 1 ♂ (MNHN-IU-2014-16571).

REMARKS — This is one of the most common species around New Caledonia. The new species strongly resembles *U. dualis* n. sp. (see above) in the carapace spination, in having a field of three spines closely and obliquely arranged on the mesioproximal face of the P1 merus, and in having the very setose P2 dactylus. *Uroptychus floccus* can be differentiated from that species by the pterygostomian flap that bears a spine directly below the linea anomurica between the second and third carapace marginal spines, instead of lacking that spine; the anterolateral spine of the carapace overreaches the lateral orbital spine only slightly instead of greatly; the P2 dactylus is more slender and more setose than in *U. dualis*; and the P1 merus bears four well-developed instead of two strong ventromesial spines (accompanied by one or two small proximal spines).

The spine below the linea anomurica, on the pterygostomian flap, is usually distinct in *U. floccus*, which helps separate the species from *U. dualis* n. sp. However, the spine is absent in the specimens registered under MNHN-IU-2014-16565, MNHN-IU-2014-16575, and two of the five specimens of the lot MNHN-IU-2014-16562.

Uroptychus gracilimanus (Henderson, 1885)

Figures 98-100

Diptychus gracilimanus Henderson, 1885: 420.
Uroptychus gracilimanus — Henderson 1888: 181, pl. 21, figus 5, 5a, 5b. — Tirmizi 1964: 392, figs 6-9. — Baba 1990: fig. 8a.
Not *U. gracilimanus* Baba 1969: 45, figs 3, 4; 1988: 35. — Baba *et al.* 2009: 44, figs 36-37; 2005: 36 (= *U. dejouanneti* n. sp.). — Baba 1990: 941, figs 8b (= undescribed species). — Doflein & Balss 1913: 134 (part) (1 ov. ♀ from Valdivia St. 250 = *U. remotispinatus* Baba & Tirmizi, 1979; 1 ♂ from Valdivia St. 245 = undescribed species). — Ahyong & Poore 2004: 40, fig. 10 (= *U. nigricapillis* Alcock, 1901). — Poore *et al.* 2008: 17 (fig.) (= *U. taylorae* McCallum & Poore, 2013).

Identification questioned:
Uroptychus gracilimanus — Parisi 1917: 3.
Uroptychus gracilimanus var. *bidentatus* — Doflein & Balss, 1913: 20.

TYPE MATERIAL — Holotype: **Australia**, Port Jackson, CHALLENGER Stn 164, 410 fms (750 m), ov. female (BMNH 1888:33). [examined].

MATERIAL EXAMINED — **Australia**, Port Jackson. CHALLENGER Stn 164, 410 fms (750 m), 1 ov. ♀ 7.9 mm, holotype (BMNH 1888: 33). **Solomon Islands**. SALOMON 1 Stn CP1792, 9°15.4'S, 160°08.9'E, 477-505 m, 30.IX.2001, 1 ♂ 5.4 mm, 3 ov. ♀ 6.2-7.6 mm, 1 ♀ 7.0 mm (MNHN-IU-2013-8542). – Stn CP1793, 9°13.4'S, 160°07.8'E, 505-510 m, 30.IX.2001, 2 ♂ 7.1, 7.4 mm, 1 ov. ♀ 6.5 mm (MNHN-IU-2013-8543). – Stn CP1794, 9°16.1'S, 160°07.7'E, 494-504 m, 30.IX.2001, 1 ov. ♀ 7.2 mm (MNHN-IU-2013-8544). – Stn CP1858, 9°37.0'S, 160°41.7'E, 435-461 m, 7.X.2001, 1 ♂ 6.0 mm (MNHN-IU-2011-5914), 1 ♂ 6.1 mm (MNHN-IU-2011-5915). – No data, 1 ♂ 6.7 mm (MNHN-IU-2013-8545). SALOMON 2 Stn CP2176, 9°09.4'S, 158°59.2'E, 600-875 m, 21.X.2004, 1 ♂ 4.9 mm, 1 ♀ 5.7 mm (MNHN-IU-2013-8546). – Stn CP2213,7°38.7'S, 157°42.9'E, 495-650 m, 26.X.2004, 1 ♂ 7.1 mm, 1 ov. ♀ 7.2 mm (MNHN-IU-2013-8547). – Stn DW2238, 06°53'S, 156°21'E, 470-443 m, 30.X.2004, 1 ov. ♀ 6.2 mm (MNHN-IU-2013-8548). – Stn CP2245, 7°43.1'S, 156°26.0'E, 582-609 m, 1.XI.2004, 1 ov. ♀ 6.1 mm (MNHN-IU-2013-8549). – Stn CP2246, 7°42.6'S, 156°24.6'E, 664-682 m, 1.XI.2004, 1 ov. ♀ 6.7 mm, 1 ♀ 4.9 mm (MNHN-IU-2013-8550). – Stn CP2247, 7°44.9'S, 156°24.7'E, 686-690 m, 1.XI.2004, 2 ♂ 5.7, 6.8 mm, 3 ov. ♀ 5.4-6.2 mm (MNHN-IU-2013-8551). – Stn CP2253, 7°26.5'S, 156°15.0'E, 1200-1218 m, 2.XI.2004, 1 ov. ♀ 7.5 mm (MNHN-IU-2013-8552). – Stn CP2267, 7°48.0'S, 156°52.0'E, 590-600 m, 4.XI.2004, 1 ♂ 7.1 mm (MNHN-IU-2013-8553). – Stn CP2269, 7°45.1'S, 156°56.3'E, 768-890 m, 4.XI.2004, 1 ♂ 6.3 mm, 1 ov. ♀ 5.5 mm (MNHN-IU-2013-8554). – Stn CP2289, 08°36'S, 157°28'E, 627-623 m, 07.XI.2004, 1 ♂ 6.4 mm, 1 ♀ 5.4 mm (MNHN-IU-2013-8555). **Vanuatu**. MUSORSTOM 8 Stn CP993, 18°48.78'S, 168°54.04'E, 780-783 m, 24.IX.1994, 1 ov. ♀ 6.5 mm, 2 ♀ 6.6, 8.0 mm (MNHN-IU-2013-8556). – Stn DW1064, 16°16'S, 167°21'E, 459-459 m, 2.X.1994, 1 ♂ 6.0 mm (MNHN-IU-2011-5913). **New Caledonia**, Loyalty Ridge. BATHUS 3 Stn DW786, 23°54.46'S, 169°49.15'E, 699-715 m, 25.XI.1993, 1 ♂ 4.9 mm (MNHN-IU-2011-5911).). **New Caledonia**, Isle of Pines. BATHUS 2 CP738, 23°02'S, 166°56'E, 558-647 m, 13.V.1993, 1 ♀ 5.1 mm (MNHN-IU-2011-5912). **New Caledonia**, Norfolk Ridge. BATHUS 3

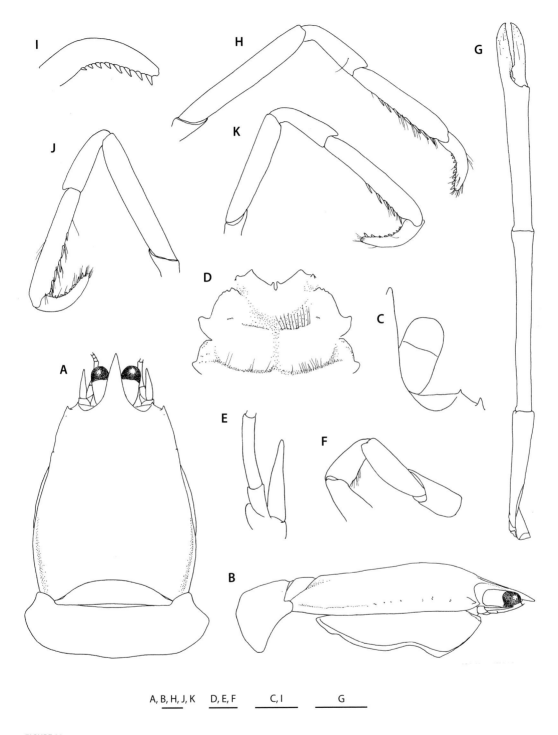

FIGURE 98

Uroptychus gracilimanus (Henderson, 1885), holotype, ovigerous female 7.9 mm (BMNH 1888:33). **A**, carapace and anterior part of abdomen, dorsal. **B**, same, lateral. **C**, anterior part of carapace, showing lateral limit of orbit, anterolateral spine and eye, right, dorsal. **D**, anterior part of sternal plastron. **E**, left antenna, ventral. **F**, left Mxp3, distal articles omitted, lateral. **G**, left P1, dorsal. **H**, right P2, lateral. **I**, same, distal part, setae omitted, lateral. **J**, left P3, lateral. **K**, right P4, lateral. Scale bars: A-F, H-K, 1 mm; G, 5 mm.

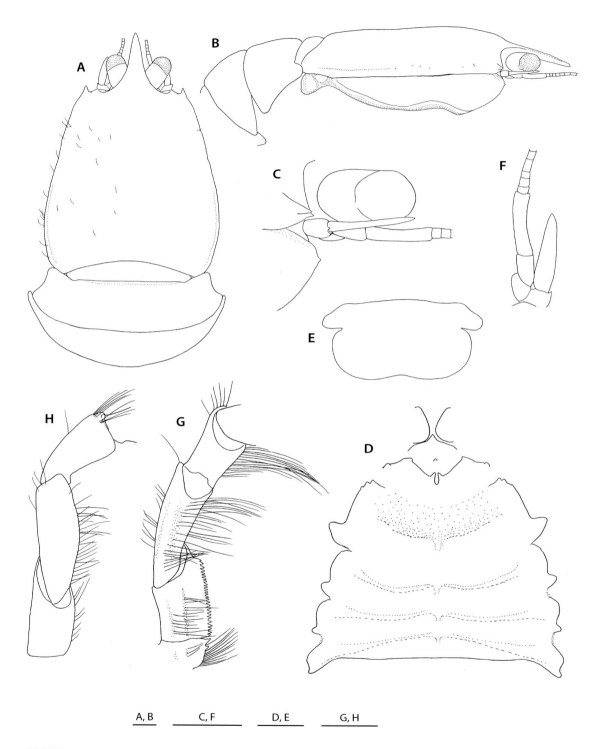

FIGURE 99

Uroptychus gracilimanus (Henderson, 1885), ovigerous female 7.4 mm (MNHN-IU-2013-8560). **A**, carapace and anterior part of abdomen, dorsal. **B**, same, lateral. **C**, anterior part of cephalothorax, left, showing lateral orbital spine, anterior part of pterygostomian flap, eye, and antenna, lateral. **D**, sternal plastron, with excavated sternum and basal parts of Mxp1. **E**, telson. **F**, left antenna, ventral. **G**, right Mxp3, ventral. **H**, same, lateral. Scale bars: 1 mm.

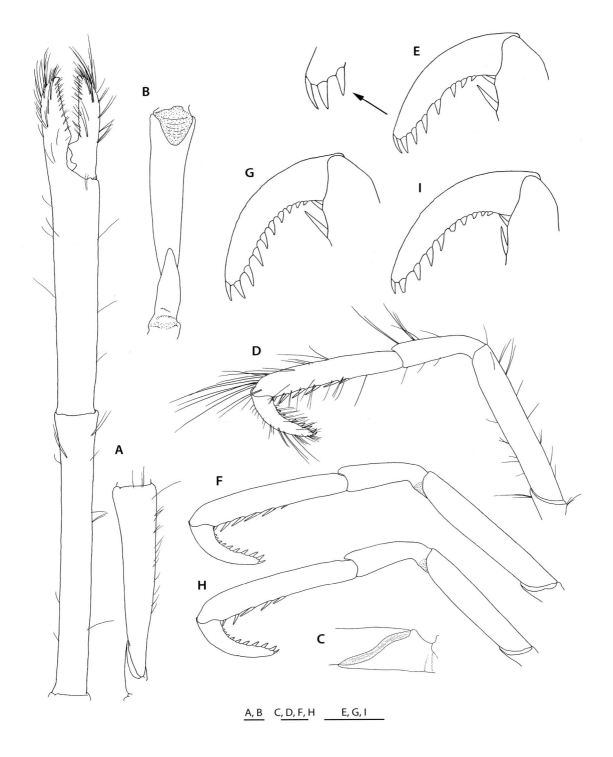

FIGURE 100

Uroptychus gracilimanus (Henderson, 1885), ovigerous female 7.4 mm (MNHN-IU-2013-8560). **A**, left P1, dorsal. **B**, same, proximal part, ventral. **C**, same, lateral. **D**, left P2, lateral. **E**, same, distal part, setae omitted, lateral. **F**, left P3, setae omitted, lateral. **G**, same, distal part, lateral. **H**, left P4, setae omitted, lateral. **I**, same, distal part, lateral. Scale bars: 1 mm.

Stn CP831, 23°04'S, 166°56'E, 650-658 m, 30.XI.1993, 1 ov. ♀ 6.7 mm (MNHN-IU-2013-8557). – Stn CP832, 23°03'S, 166°54'E, 650-659 m, 30.XI.1993, 2 ov. ♀ 7.0, 7.5 mm (MNHN-IU-2013-8558). – Stn CC848, 23°02'S, 166°53'E, 680-700 m, 1.XII.1993, 1 ♂ 6.9 mm (MNHN-IU-2013-8559). HALIPRO 1 Stn CH872, 23°02'S, 166°52'E, 620-700 m, 30.III.1994, 1 ov. ♀ 7.4 mm (MNHN-IU-2013-8560). – Stn CH873, 23°02'S, 166°54'E, 640-680 m, 30.III.1994, 1 ♀ 6.0 mm (MNHN-IU-2013-8561).

DISTRIBUTION — Port Jackson and Zanzibar, 421-750 m; and now Solomon Islands, Vanuatu, Isle of Pines, and Norfolk Ridge; in 435-1218 m.

SIZE — Males, 4.9-7.4 mm; females, 4.9-7.9 mm; ovigerous females from 5.4 mm.

DESCRIPTION — Medium-sized species. *Carapace*: As long as broad; greatest breadth 1.6-1.7 × distance between anterolateral spines. Dorsal surface smooth, glabrous and unarmed, anteriorly inflated in profile, with or without faint depression between gastric and cardiac regions. Lateral margins convexly divergent posteriorly, with row of obsolescent denticulate short ridges along anterior half; ridged along posterior half; anterolateral spine small, varying from barely to fully reaching tiny lateral orbital spine where present, situated distinctly posterior to position of that spine. Rostrum triangular with interior angle of 25-30°, horizontal or directed slightly ventrally; length 0.3-0.4 × postorbital carapace length, breadth much less than half carapace breadth measured at posterior carapace margin; lateral margins often concave; dorsal surface slightly concave at base. Lateral orbital spine small, often obsolescent or absent. Pterygostomian flap anteriorly roundish, bearing very small spine, surface smooth.

Sternum: Excavated sternum anteriorly broad triangular, ending in sharp spine, surface with spine in center. Sternal plastron about as long as broad, lateral extremities between sternites 4 and 7 straight divergent posteriorly. Sternite 3 moderately depressed, with well excavated anterior margin bearing 2 small submedian spines flanking narrow, shallow sinus. Sternite 4 having anterolateral margin convex with a few small spines on anterior half (anterior 2 often pronounced and others obsolescent), about 1.5 × length of posterolateral margin. Anterolateral margins of sternite 5 moderately convexly divergent posteriorly or with lateral margins nearly subparallel, slightly longer than posterolateral margin of sternite 4.

Abdomen: Tergites smooth and glabrous. Somite 1 dorsally convex from anterior to posterior, without transverse ridge. Somite 2 tergite 2.0-2.5 × broader than long; pleuron posterolaterally rounded, lateral margin somewhat concavely divergent posteriorly. Pleuron of somite 3 laterally blunt. Telson about half as long as broad; posterior plate distinctly or feebly emarginate, feebly concave or feebly convex on posterior margin, length 1.5-2.0 × that of anterior plate.

Eye: Short relative to breadth (about 1.6 × longer than broad), slightly concave on mesial margin, slightly convex on lateral margin, ending in between midlength and point four-fifths of rostrum. Cornea not dilated, about half length of remaining eyestalk.

Antennule and antenna: Ultimate article of antennular peduncle 2.1-2.5 × longer than high. Antennal peduncle overreaching cornea, barely reaching rostral tip. Article 2 with tiny lateral spine. Antennal scale 1.3-1.4 × broader than article 5, terminating in or overreaching midlength of article 5, never reaching its distal end. Distal 2 articles unarmed; article 5 2.0-2.5 × longer than article 4, breadth one-third to half height of ultimate antennular article. Flagellum of 16-20 segments nearly or barely reaching distal end of P1 merus.

Mxp: Mxp1 with bases close to each other. Mxp3 spineless on merus and carpus. Basis with 3 or 4 denticles on mesial ridge. Ischium with 20-30 denticles on crista dentata, flexor margin not rounded distally. Merus 2.3 × longer than ischium, relatively thick mesio-laterally, flexor margin sharply ridged. Carpus unarmed.

P1: Slender (male palm in large specimens massive, broader than distance between anterolateral spines of carapace), sparsely setose on fingers, glabrous elsewhere; length 4.9-5.9 × (males), 4.1-7.5 × (females) that of carapace. Ischium with basally broad, low and blunt dorsal spine often obsolescent; ventrally unarmed. Merus 1.2-1.3 × longer than carapace. Carpus 1.3-1.5 × longer than merus. Palm with length-breadth ratios of 2.1-4.0 (males; slender in small specimens) and 3.7-5.0 (females), length 0.7-0.9 × that of carpus. Fingers proportionately broad, ending in incurved tips, feebly crossing;

movable finger 0.5-0.7 × (males), 0.4-0.6 × (females) length of palm; opposable margin of movable finger with broad proximal process low in females, high and 2-toothed in males.

P2-4: Slender. Meri successively shorter posteriorly (P3 merus 0.9 × length of P2 merus, P4 merus 0.9 × length of P3 merus), breadths subequal on P2-4 or slightly narrower on P2 than on P3; length-breadth ratio, 5.6-6.9 on P2, 5.0-6.5 on P3, 4.0-5.4 on P4; dorsal and ventral margins unarmed; P2 merus 0.8-0.9 × length of carapace, 1.1-1.3 × length of P2 propodus; P3 merus subequal to length of P3 propodus; P4 merus 0.8-0.9 × length of P4 propodus. Carpi subequal or successively shorter posteriorly; carpus-propodus length ratio, 0.5-0.6 on P2, 0.4-0.5 on P3, 0.4 on P4. Propodi subequal or successively slightly longer posteriorly; flexor margin nearly straight, with 6-9 long spines on distal two-thirds to four-fifths on P2, 5-8 spines more distal on P3-4, terminal spine single, located very close to distal end and mesial to flexor midline. Dactyli curving at middle, length 0.4-0.5 × that of propodus on P2-4; dactylus-carpus length ratio, 0.8-0.9 (rarely 1.0) on P2, 0.9-1.0 on P3 and P4; flexor margin with 8-10 sharp spines diminishing toward base of article, ultimate spine more slender than penultimate spine.

Eggs. Up to about 40 eggs carried; size, 1.10 mm × 1.20 mm - 1.33 mm × 1.41 mm. Holotype with 30 eggs, 1.7 mm in diameter.

REMARKS — The holotype has all the P2-4 detached from the body, with the dactyli broken at the tip. Material in the Australian Museum collected from NE of Wollongong and E of Broken Bay, the vicinity of Port Jackson, was examined at my request by Shane Ahyong. His illustrations show that the ultimate of the flexor spines of the dactyli is more slender than the penultimate, a consistent character in all the above-listed material that separates the species from *U. dejouanneti* n. sp. *Uroptychus dejouanneti* is described based upon material from Indonesia (see above). It has the ultimate and penultimate spines subequal, and is identical with the previously reported material under *G. gracilimanus* from the East China Sea (Baba 1969), Taiwan (Baba *et al.* 2009), Indonesia in the Molucca Sea (Baba 1988), and the Philippines off Zamboanga (Baba, 2005).

Examination of the John Murray material from Zanzibar (Tirmizi 1964) shows that it has been correctly identified, perfectly fitting the species definition herein proposed. The specimens from Madagascar reported by Baba (1990) represent an undescribed species and have been removed from the synonymy and part of the Valdivia collection has been transferred to *U. remotispinatus* (see Baba 2005: 36).

Uroptychus gracilimanus resembles *U. brevisquamatus* Baba, 1988 in the carapace shape, from which it is readily differentiated by the following: the P2-4 propodi bear a single instead of paired terminal spines on the flexor margin, the flexor margins of the P2-4 dactyli bear elongate triangular instead of very short spines; and the antennal article 2 bears a tiny instead of strong lateral spine.

All the material of *U. gracilimanus* reported by Ahyong & Poore (2004) from southeastern Australia and Tasmania agrees well with *U. nigricapillis* Alcock, 1901.

According to molecular data provided by L. Corbari (personal comm.), the two specimens from Salomon 1 Stn CP1858 (MNHN-IU-2011-5914 and 5915; divergence of 5.7 %) are different from each other at a specific level, and one of these is also genetically different from the other specimens from the Solomon Islands, Vanuatu, Loyalty Ridge and Isle of Pines. However, I find no clear morphological differences between them.

Uroptychus grandior n. sp.

Figures 101, 102

TYPE MATERIAL — Holotype: **New Caledonia**. LAGON Stn CP1153, 18°58.l4'S, 163°23.0'E, 330 m, 29.IX.1989, ♂ 2.4 mm (MNHN-IU-2011-5929). Paratypes: **New Caledonia**, Loyalty Islands. CALSUB PL11, 20°52,5'S, 167°03'E, 681-140 m, 28.II.1989, 1 ♀ 1.6 mm (MNHN-IU-2011-5928). **New Caledonia**, Norfolk Ridge. SMIB 4 Stn DW54, 23°41'S, 168°00'E, 230-235 m, 09.III.1989, 1 ov. ♀ 2.3 mm (MNHN-IU-2011-5925). NORFOLK 1 Stn DW1657, 23°28'S, 167°52'E, 305-

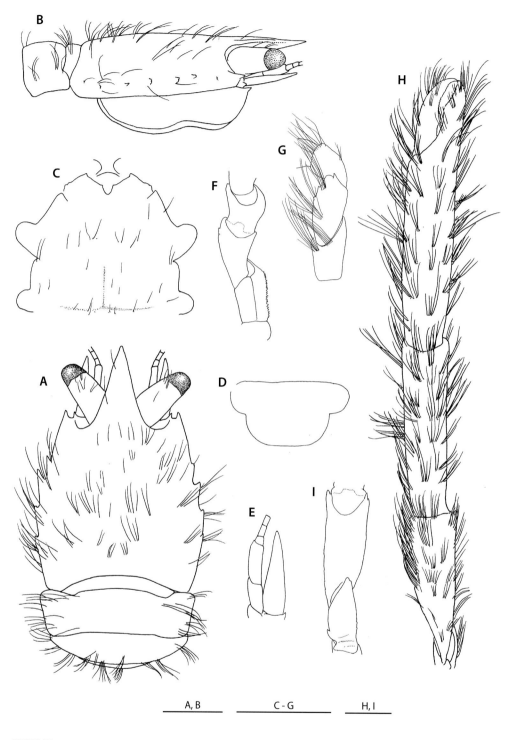

FIGURE 101

Uroptychus grandior n. sp., holotype, male 2.4 mm (MNHN-IU-2011-5929). **A**, carapace and anterior part of abdomen, dorsal. **B**, same, lateral. **C**, anterior part of sternal plastron, with excavated sternum and basal parts of Mxp1. **D**, telson. **E**, left antenna, ventral. **F**, right Mxp3, setae omitted, ventral. **G**, left Mxp3, lateral. **H**, right P1, dorsal. **I**, same, proximal part, setae omitted, ventral. Scale bars: 1 mm.

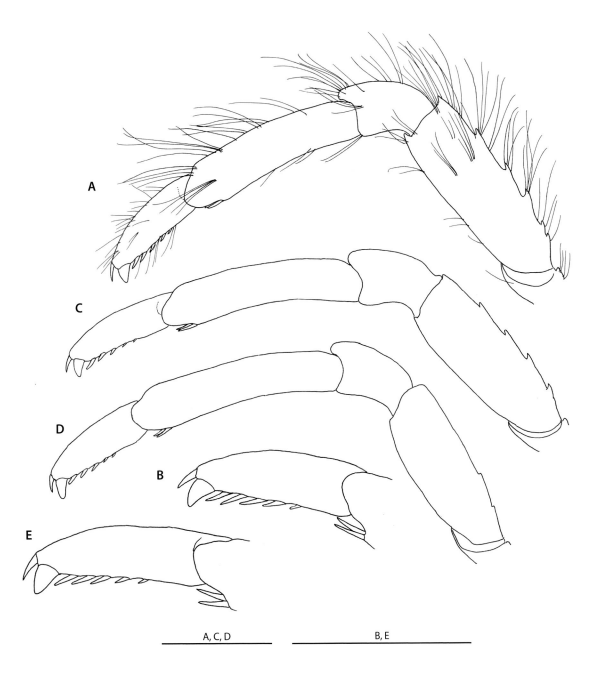

FIGURE 102

Uroptychus grandior n. sp., holotype, male 2.4 mm (MNHN-IU-2011-5929). **A**, left P2, lateral. **B**, same, distal part, setae omitted, lateral. **C**, left P3, setae omitted, lateral. **D**, left P4, setae omitted, lateral. **E**, same, distal part, setae omitted. Scale bars: 1 mm.

332 m, 19.VI.2001, 1 ♂ 2.2 mm (MNHN-IU-2011-5927). – Stn CP1682, 24°43'S, 168°10'E, 331-379 m, 22.VI.2001, 1 ♂ 2.4 mm (MNHN-IU-2011-5926).

ETYMOLOGY — From the Latin *grandior* (larger), alluding to the unusually large distal second spine on the flexor margin of the P2-4 dactyli, one of the discriminating characters of the species.

DISTRIBUTION — New Caledonia, Loyalty Islands and Norfolk Ridge; in 220-379 m.

SIZE — Males, 2.2-2.4 mm; females, 1.6-2.3 mm; ovigerous female, 2.3 mm.

DESCRIPTION — Small species. Body and appendages with distally softened plumose setae.

Carapace: Slightly broader than long (0.8 × as long as broad); greatest breadth 1.5 × distance between anterolateral spines. Dorsal surface feebly convex or nearly straight horizontal from anterior to posterior, anteriorly smoothly continued on to rostrum; moderately setose, setae arising from short striae. Lateral margins convex, with 5 spines: first anterolateral, close to and directly lateral to lateral orbital spine, not overreaching that spine; second as large as first, remote from first and closer to third, occasionally preceded by eminence or 1 or 2 small spines; third to fifth situated on branchial margin; third larger than second and fourth; fifth much smaller than preceding two, followed by obsolescent spine or eminence and even additional small spine. Rostrum broad triangular, with interior angle of 32°, distally sharp, nearly straight horizontal, dorsal surface flattish; length slightly more than half that of carapace, breadth somewhat more than half carapace breadth measured at posterior carapace margin. Lateral orbital spine much smaller than anterolateral spine. Pterygostomian flap anteriorly angular, produced to small spine, surface smooth.

Sternum: Excavated sternum with convex anterior margin, bearing weak ridge in midline on surface (barely discernible in smallest specimen). Sternal plastron 1.2 × longer than broad, with subparallel lateral extremities, sternite 6 somewhat narrower than remainder. Sternite 3 shallowly depressed; broad V-shaped anterior margin with median sinus of broad U-shape, with no distinct flanking spine, laterally angular. Sternite 4 with anterolateral margin nearly straight, angular at anterior terminus; posterolateral margin relatively long, subequal to anterolateral margin in length. Anterolateral margin of sternite 5 somewhat convex, length 0.8 × that of posterolateral margin of sternite 4.

Abdomen: Smooth. Somite 1 without transverse ridge. Somite 2 tergite 2.1-2.3 × broader than long; pleuron posterolaterally blunt, lateral margin feebly concave and subparallel. Pleuron of somite 3 with rounded lateral margin. Telson about half as long as broad; posterior plate as long as anterior plate, posterior margin feebly convex.

Eye: Elongate, about twice as long as broad, distally narrowed, overreaching midlength of but barely reaching apex of rostrum. Cornea not dilated, less than half as long as remaining eyestalk.

Antennule and antenna: Ultimate article of antennular peduncle 2.5-3.3 × longer than high. Antennal peduncle reaching distal end of eyes. Article 2 with small distolateral spine. Antennal scale slightly less than twice as broad as article 5, distally sharp, extending far beyond article 5, falling short of rostral tip. Article 4 with well-developed distomesial spine. Article 5 with very small or distinct distomesial spine; length 1.5 × that of article 4, breadth 0.7 × height of antennular ultimate article. Flagellum consisting of 10-12 segments, slightly falling short of distal end of P1 merus.

Mxp: Mxp1 with bases broadly separated. Mxp3 with tufts of plumose setae on distal part of lateral margin of ischium, lateral face of merus and distal part of lateral face of carpus. Basis without distinct denticles on mesial ridge. Ischium with flexor margin distally rounded, crista dentata with small denticles diminishing distally. Merus 2.2 × longer than ischium, with distolateral spine, mesial face flattish, flexor margin sharply ridged with a few very small spines often obsolescent on distal third. Carpus unarmed.

P1: Relatively slender, subcylindrical, with tufts of relatively long setae, length 5.0-5.5 × that of carapace. Ischium dorsally bearing strong distal spine, ventrally unarmed. Merus 1.2-1.3 × longer than carapace, ventrally bearing distomesial and distolateral spines. Carpus 1.1-1.2 × longer than merus, ventrally with 2 distal spines. Palm 3.5-4.3 × longer than broad, 1.0-1.2 × longer than carpus. Fingers slightly curving laterally, distally incurved, crossing when closed, not gaping;

movable finger 0.4 × length of palm, opposable margin with low, denticulate median process fitting to opposite narrow longitudinal groove on fixed finger when closed.

P2-4: Moderately compressed mesio-laterally. Meri relatively broad, successively slightly shorter posteriorly (P3 merus 0.9 × length of P2 merus, P4 merus 0.9-1.0 × length of P3 merus), subequally broad on P2-4; length-breadth ratio, 3.1-3.3 on P2, 3.0-3.3 on P3, 2.8 on P4; dorsal crest with 4-5 short, acute spines along entire length on P2 and P3, a few obsolescent spines or eminences on proximal part on P4; ventrolateral margin with terminal spine distinct on P2, much smaller on P3, obsolescent on P4; P2 merus 0.7 length of carapace, about as long as P2 propodus; P3 merus 0.9 × length of P3 propodus; P4 merus 0.8 × length of P4 propodus. Carpi subequal, unarmed; carpus-propodus length ratio, 0.3-0.4 on P2 and P3, 0.3 on P4. Propodi subequal on P2 and P3 and longer on P4 or subequal on P3 and P4 and shorter on P2; flexor margin straight, with pair of distal spines only. Dactyli proportionately broad, about half as long as propodi; dactylus-carpus length ratio, 1.6 on P2, 1.5 on P3, 1.6 on P4; flexor margin nearly straight, bearing 6-7 spines, ultimate slender and curved, penultimate prominent, distinctly longer than remainder, about 3 × broader than antepenultimate, remaining spines slender, strongly inclined but not contiguous to one another, successively shorter toward base of article.

Eggs. Number of eggs carried, 5; size, 1.1 mm × 1.1 mm.

REMARKS — *Uroptychus grandior* strongly resembles *U. belos* Ahyong & Poore, 2004 in the spination of the carapace and P2-4 dactyli and in the elongate eyes. The two species each key out in a remotely different couplet (see under the key to species), due to the difference of length ratio of the anterolateral margin to the posterolateral margin of sternite 4. *Uroptychus grandior* can be distinguished from *U. belos* by the following differences: the eyes are distally narrowed instead of subequally broad proximally and distally; the dorsal crests of the P2-4 meri each bear a row of spines instead of being unarmed; the flexor marginal spines of the P2-4 dactyli are strongly oblique instead of gently inclined; and the antennal scale distinctly overreaches instead of terminating at most in the distal end of article 5.

Uroptychus granulipes n. sp.
Figures 103, 104

TYPE MATERIAL — Holotype: **New Caledonia**, Norfolk Ridge. NORFOLK 2 Stn DW2060, 24°39.84'S, 168°38.50'E, 582-600 m, 25.X.2003, ♀ 7.0 mm (MNHN-IU-2014-16578).

ETYMOLOGY — From the Latin *granulus* (a small grain) plus *pes* (foot), alluding the granulose P1 merus and carpus.

DISTRIBUTION — Norfolk Ridge; 582-600 m.

DESCRIPTION — *Carapace*: As long as broad; greatest breadth 1.6 × distance between anterolateral spines. Dorsal surface smooth and glabrous, slightly convex from anterior to posterior, with very feeble depression between gastric and cardiac regions. Lateral margins somewhat convex posteriorly, finely denticulate on anterior branchial margin, feebly ridged along posterior portion behind pterygostomian flap; anterolateral spine prominent, directed forward, situated directly lateral to (not posterior to level of) smaller lateral orbital spine, extending far beyond that spine. Rostrum directed somewhat ventrally, elongate triangular, with interior angle of 27°, slightly less than half length of rostrum, breadth half carapace breadth measured at posterior carapace margin; dorsal surface moderately concave. Pterygostomian flap anteriorly roundish, produced to sharp spine, surface smooth.

Sternum: Excavated sternum anteriorly produced to spine, slightly ridged in midline on surface. Sternal plastron slightly broader than long, successively broader posteriorly. Sternite 3 depressed well; anterior margin deeply excavated with deep, narrow median sinus flanked by sharp, incurved spine, lateral end sharply produced straight forward. Sternite 4 having

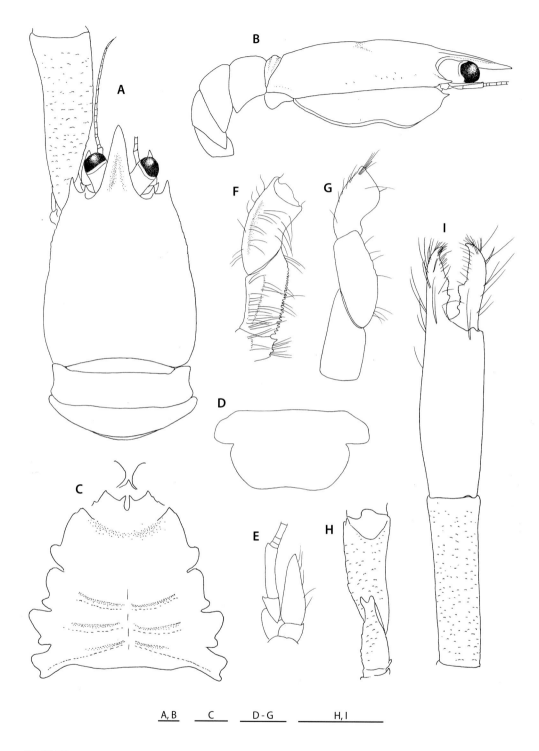

FIGURE 103

Uroptychus granulipes n. sp., holotype, female 7.0 mm (MNHN-IU-2014-16578). **A**, carapace and anterior part of abdomen, proximal part of left P1 included, dorsal. **B**, same, lateral. **C**, sternal plastron, with excavated sternum and basal parts of Mxp1. **D**, telson. **E**, left antenna, ventral. **F**, right Mxp3, ventral. **G**, same, lateral. **H**, left P1, proximal part, ventral. **I**, same, dorsal. Scale bars: A-G, 1 mm; H, I, 5 mm.

anterolateral margin convex, anteriorly angular, slightly more than 1.5 × length of posterolateral margin. Anterolateral margin of sternite 5 anteriorly convex, 1.2 × longer than posterolateral margin of sternite 4.

Abdomen: Tergites smooth, barely setose. Somite 1 strongly convex from anterior to posterior. Somite 2 tergite 2.8 × broader than long; pleuron anterolaterally and posterolaterally blunt angular, lateral margin weakly concave and slightly divergent posteriorly. Pleura of somites 3 and 4 laterally angular. Telson half as long as broad; posterior plate concave on posterior margin, length 1.4 × that of anterior plate.

Eye: Slightly less than twice as long as broad, somewhat overreaching midlength of rostrum, mesial and lateral margins slightly concave. Cornea slightly inflated, more than half as long as remaining eyestalk.

Antennule and antenna: Ultimate article of antennular peduncle 2.3 × longer than high. Antennal peduncle relatively slender, slightly overreaching cornea. Article 2 with small distolateral spine. Antennal scale about twice as broad as article 5, slightly falling short of distal end of article 5, lateral margin with sparse setae. Article 4 with distinct distomesial spine.

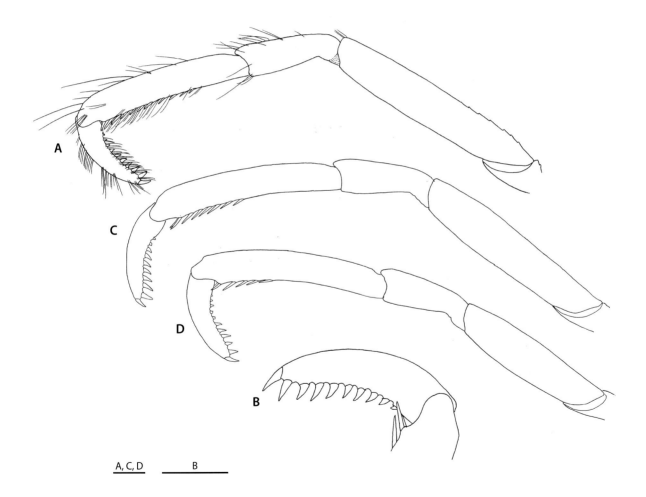

A, C, D   B

FIGURE 104

Uroptychus granulipes n. sp., holotype, female 7.0 mm (MNHN-IU-2014-16578). **A**, left P2, lateral. **B**, same, distal part, setae omitted, lateral. **C**, left P3, setae omitted, lateral. **D**, left P4, setae omitted, lateral. Scale bars: 1 mm.

Article 5 unarmed, length slightly more than twice as long as article 4, breadth 0.7 × height of antennular ultimate article. Flagellum consisting of 22-24 segments, reaching distal end of P1 merus.

Mxp: Mxp1 with bases separated. Mxp3 barely setose on lateral surface. Basis with 3 denticles on mesial ridge. Ischium with 25 distally diminishing denticles on crista dentata, flexor margin not rounded distally. Merus unarmed, 1.8 × as long as ischium; flexor margin not cristate but somewhat roundly ridged, with a few eminences. Carpus unarmed.

P1: Relatively massive, bearing short, fine setae on fingers, barely setose elsewhere, granulose on ischium, merus and carpus. Ischium dorsally with basally broad, long spine, ventromesially with well-developed subterminal spine. Merus 1.1 × longer than carapace. Carpus 1.2 × longer than merus. Palm 2.8 × longer than broad, as long as carpus. Fingers slightly curving ventrally, distally strongly incurved, crossing when closed; movable finger half as long as palm, opposable margin with 2 low processes proximal to midlength; opposable margin of fixed finger with 2 prominences, proximal one fitting to between 2 processes on movable finger.

P2-4: Slender, sparsely with fine setae on distal articles. Meri moderately compressed mesio-laterally, successively shorter posteriorly (P3 merus 0.9 × length of P2 merus, P4 merus 0.8 × length of P3 merus), equally broad on P2-4; length-breadth ratio, 5.4-5.9 on P2, 4.9-5.4 on P3, 3.8-4.2 on P4; P2 merus as long as carapace, 1.3 × longer than P2 propodus; P3 merus about as long as P3 propodus; P4 merus 0.8 × length of P4 propodus; dorsal margin with obsolescent denticles on proximal half. Carpi subequal on P2 and P3, P4 carpus 0.9 × length of P3 carpus; length 0.5 × that of propodus on P2 and P3, 0.4 × on P4. Propodi subequal in length on P3 and P4 and slightly shorter on P2; flexor margin ending in pair of spines preceded by 9 or 10 spines on P2, 6 on P3, 4 on P4. Dactyli shorter than carpi (dactylus-carpus length ratio, 0.8 on P2, 0.9 on P3 and P4), 0.4 × as long as propodi on P2-4; flexor margin curving, setose, with 11-12 slightly inclined, loosely arranged triangular spines successively diminishing toward proximal end of article; extensor margin with fringe of setae.

REMARKS — The new species is very close to *U. inermis* n. sp. in nearly all aspects. Their relationships are discussed under the remarks of that species (see below).

The shape of the carapace which bears strong anterolateral spines, and the spination of P2-4 displayed by this species resemble those of *U. brucei* Baba, 1986a and *U. maori* Borradaile, 1916. *Uroptychus granulipes* is differentiated from *U. brucei* by the carapace dorsal surface that is smooth instead of granulose; the anterolateral spine is situated directly lateral to instead of posterior to the position of lateral orbital spine; the carapace lateral margin is almost smooth instead of distinctly ridged along the posterior portion; the P1 merus and carpus are granulose instead of smooth; the antennal article 5 is unarmed instead of bearing a distinct spine; and the dactylus-carpus length ratio is 0.8 on P2, 0.9 on P3 and P4, instead of 0.6 on P2 and P3, 0.7 on P4. *Uroptychus granulipes* is different from *U. maori* (see below) in having the carapace dorsal surface smooth instead of granulose, bearing a ridge along the posterior part of lateral margin, and in having the pterygostomian flap smooth instead of granulated on the surface.

Uroptychus imparilis n. sp.

Figures 105, 106

Uroptychus scandens Baba 1981: 132 (not *Uroptychus scandens* Benedict, 1902).

TYPE MATERIAL — Holotype: **Philippines**. MUSORSTOM 1 Stn CP40, 13°57.4'N, 120°27.8'E, 287-265 m, 24.III.1976, ov. ♀ 3.8 mm (MNHN-IU-2013-8565). Paratypes: **Philippines**. MUSORSTOM 1, collected with holotype 1 ♂ 3.2 mm, 1 ov. ♀ 3.6 mm (MNHN-IU-2014-10143 & MNHN-IU-2014-10144), 1 ov. ♀ 4.0 mm (MNHN-IU-2014-16579). MUSORS-TOM 2 Stn CP75, 13°51'N, 120°30'E, 300-330 m, 1.XII.1980, 1 ov. ♀ 3.2 mm (MNHN-IU-2014-16580). **Indonesia**, Kai Islands. KARUBAR Stn CP06, 5°49'S, 132°21'E, 298-287 m, 22.X.1991, 1 ♂ 3.5 mm (MNHN-IU-2014-16582). **Vanuatu**. MUSORSTOM 8 Stn CP1025, 17°49.01'S, 168°39.37'E, 385-410 m, 28.IX.1994, 1 ov. ♀ 3.0 mm (MNHN-IU-2014-16583).

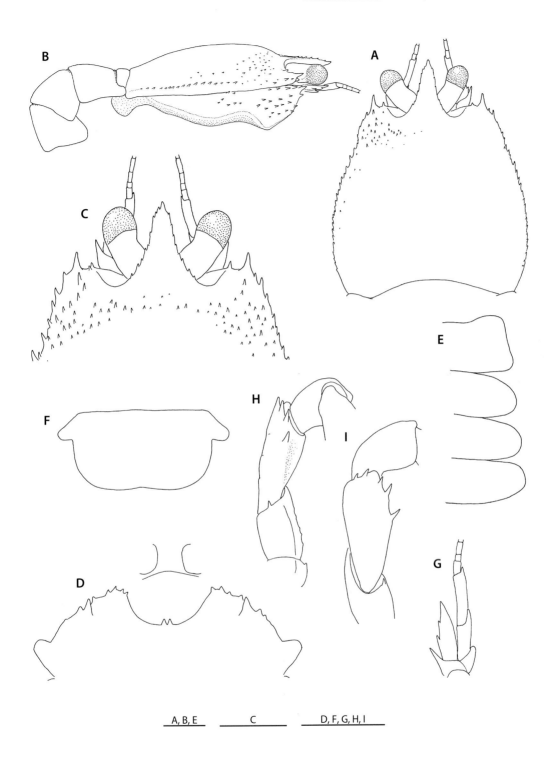

A, B, E   C   D, F, G, H, I

FIGURE 105

Uroptychus imparilis n. sp., holotype, ovigerous female 3.8 mm (MNHN-IU-2013-8565). **A**, carapace, dorsal. **B**, same, anterior part of abdomen included, lateral. **C**, anterior part of carapace, dorsal. **D**, anterior part of sternal plastron, with excavated sternum and basal parts of Mxp1. **E**, right pleura of abdominal somites 2-5, dorsolateral. **F**, telson. **G**, right antenna, ventral. **H**, right Mxp3, ventral, setae omitted. **I**, same, lateral. Scale bars: 1 mm.

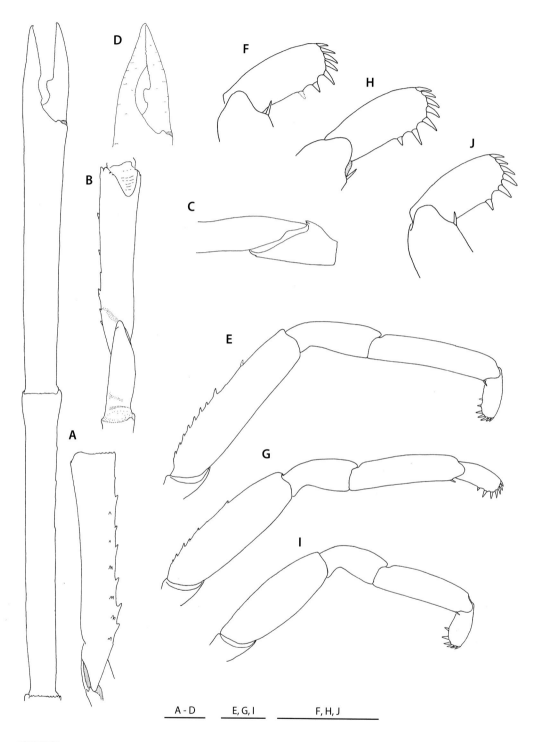

FIGURE 106

Uroptychus imparilis n. sp.; **A-C**, **E-I**, holotype, ovigerous female 3.8 mm (MNHN-IU-2013-8565); **D**, male paratype 3.2 mm (MNHN-IU-2014-10143). **A**, left P1, setae omitted, dorsal. **B**, same, proximal part, ventral. **C**, same, lateral. **D**, left fingers, setae omitted, dorsal. **E**, right P2, setae omitted, lateral. **F**, same, distal part, lateral. **G**, right P3, lateral. **H**, same, distal part, lateral. **I**, right P4, lateral. **J**, same, distal part, lateral. Scale bars: 1 mm.

OTHER MATERIAL EXAMINED — **Japan**, Tosa Bay. On pennatulacean, 24.XII.1959, K. Sakai coll., 1 ov. ♀ (carapace broken), 1 ♀ 3.6 mm (ZLKU 7466); 150 m, 31.I.1959, K. Kurohara coll., 1 ♂ 3.6 mm (ZLKU 5871); IV.1960, K. Sakai coll., 1 ov. ♀ 3.2 mm (ZLKU 14393). Southern Kyushu, SOYO-MA-RU Stn 72d, 31°13.6′N, 129°58.5′E, 310 m, 12.II.1959, 2 ♂ 2.7, 3.1 mm, 1 ♀ 3.2 mm (NSMT). Hachijo-jima, SOYO-MARU Stn B3, 33°06.5′N, 140°04.8′E, 490-495 m, 12.XII.1963, 1 ♂ 3.7 mm (NSMT).

ETYMOLOGY — From the Latin *imparilis* (different), suggesting that the species is different from the closest congener *U. parisus* n. sp. (see below).

DISTRIBUTION — Philippines, Indonesia, Vanuatu and Japan; 150-495 m.

SIZE — Males, 2.7-3.7 mm; females, 3.0-3.8 mm; ovigerous females from 3.0 mm.

DESCRIPTION — Small species. *Carapace*: Broader than long (0.8 × as long as broad); greatest breadth 1.8 × distance between anterolateral spines. Dorsal surface setose laterally, with very small spines on epigastric, hepatic and anterior lateral branchial regions, those on hepatic region somewhat larger. Lateral margins convexly divergent posteriorly, occasionally slightly constricted between hepatic and branchial regions, bearing row of small spines; anterolateral spine much larger than others, directed straight forward, overreaching much smaller lateral orbital spine. Rostrum narrow triangular, with interior angle of 25°; lateral margins feebly concave, with 8-9 denticles; dorsal surface somewhat concave; length slightly smaller than (0.9 ×) breadth, 0.4 × that of remaining carapace, breadth less than half carapace breadth measured at posterior carapace margin; lateral orbital spine slightly anterior to level of anterolateral spine. Pterygostomian flap anteriorly angular, produced to strong sharp spine; surface with small spines on anterior half, including row of spines along anterior part of dorsal margin.

Sternum: Excavated sternum with slightly convex anterior margin; surface smooth, without ridge and central spine, with setae along anterior margin; sternal plastron 0.8 × as long as broad (1.3 × broader than long); anterior margin of sternite 3 broadly and deeply excavated nearly in semicircular shape, with pair of small median spines basally contiguous (one of the spines occasionally missing), anterolateral margin with several small or obsolescent spines. Sternite 4 having anterolateral margin feebly convex, anteriorly produced to distinct spine and followed by a few smaller spines; posterolateral margin 0.7 × as long as anterolateral margin. Anterolateral margin of sternite 5 as long as posterior margin of sternite 4.

Abdomen. Somite 1 gently convex from anterior to posterior. Somite 2 tergite 2.5 × broader than long, pleura similar in males and females, slightly concavely divergent posteriorly, ending in rounded margin. Pleura of somites 3-5 laterally rounded. Telson slightly less than half as long as broad; posterior plate slightly emarginate or feebly convex, 1.7 × longer than anterior plate.

Eyes: Elongate, 1.9-2.1 × longer than broad, reaching or very slightly falling short of rostral tip, lateral margin convex, mesial margin concave. Cornea slightly inflated, very slightly broader than and half as long as remaining eyestalk.

Antennule and antenna: Ultimate article of antennular peduncle 2.7-2.9 × longer than high. Antennal peduncle slightly overreaching rostral tip. Article 2 with strong distolateral spine. Antennal scale articulated with article 2, ending in midlength of article 5, laterally with 1 or 2 spines; breadth 1.3-1.4 × that of article 5. Article 3 distomesially rounded or produced to small spine. Article 4 with distinct distomesial spine. Article 5 with small distomesial spine, length 1.5 × that of article 4, breadth 0.4-0.5 × height of ultimate article of antennule. Flagellum of 8-10 segments ending in midlength of P1 merus.

Mxp: Mxp1 with bases broadly separated. Mxp3 with long setae. Basis with 1 distal denticle on mesial ridge. Ischium with small spine lateral to distal end of flexor margin, crista dentata with a few to several obsolescent denticles on proximal half. Merus 2.5 × length of ischium, ridged along flexor margin, not well compressed; with 3 spines on distal third of flexor margin and 2 distolateral spines. Carpus unarmed.

P1: Slender, subcylindrical, 5.7-6.6 × (males), 4.4-6.8 × (females) longer than carapace. Ischium dorsally with basally broad, depressed, bifurcate short spine, ventromesially unarmed. Merus 1.5 × longer than carapace, dorsally bearing 2 rows of spines: 1 row along mesial margin and another row of smaller (some bifurcated) spines directly dorsolateral to it). Carpus 1.3-1.4 × longer than merus, unarmed. Palm 4.1-5.0 × (males), 6.4-6.6 (females) as long as broad, slightly (females) or distinctly (males) broadened distally, length 0.9 × that of carpus. Fingers relatively narrow distally, strongly gaping in proximal two-thirds in males, weakly so in proximal third in females; opposable margins fitting to each other in distal two-thirds (females) or in distal third (males) when closed, with row of denticles; movable finger 0.4 × length of palm, with obtuse process at midpoint of gaping portion.

P2-4: Thickly setose like P1. Meri successively shorter posteriorly (P3 merus 0.9 × length of P2 merus, P4 merus 0.8 × length of P3 merus), slightly broader on P3 and P4 than on P2, or slightly broader on P3 than on P2 and P4; length-breadth ratio, 4.1-4.2 on P2, 3.4 on P3, 2.9-3.0 on P4; dorsal margin with 7-9 small spines on P2, 6-7 on P3, unarmed on P4; P2 merus subequal to length of carapace, 1.3-1.4 × length of P2 propodus; P3 merus 1.3 × length of P3 propodus; P4 merus 1.2-1.3 × length of P4 propodus. Carpi successively shorter posteriorly; carpus-propodus length ratio, 0.6 on P2, 0.5 on P3 and P4. Propodi successively shorter posteriorly; flexor margin straight in lateral view, with pair of small terminal spines only. Dactyli subequal on P2-4; dactylus-carpus length ratio, 0.5-0.6 on P2, 0.6-0.7 on P3, 0.6-0.8 on P4; dactylus-propodus length ratio, 0.3 on P2, 0.3-0.4 on P4; truncate, bearing 7 slender spines obscured by setae on P2 and P3, 6 spines on P4, 3 or 4 of these located on terminal margin, remainder perpendicular to flexor margin, terminal spine smaller.

Eggs. Number of eggs carried, about 40; size, 0.63 × 0.68 mm - 0.61 × 0.71 mm.

REMARKS — The material registered under ZLKU 5871 contains three specimens. One is identified as *U. imparilis* and the other two, with the label "associated with a pennatulacean," are *Uroptychus scandens*. The materials reported under *U. scandens* by Baba (1981) are referable to *U. imparilis*.

This species shares truncate dactyli of P2-4 with *U. articulatus* n. sp., *U. parisus* n. sp. and *U. scandens* Benedict, 1902. Their relationships are discussed under *U. scandens* (see below).

Uroptychus inaequalis n. sp.
Figures 107, 108

Uroptychus pilosus — Ahyong & Poore 2004: 71, fig. 21 (not *U. pilosus* Baba, 1981).

TYPE MATERIAL — Holotype: **New Caledonia**, Loyalty Islands. CALSUB PL07, 20°48'S, 167°05'E, 970-489 m, 25.II.1989, ♀ 5.1 mm (MNHN-IU-2011-5948). Paratypes: **Solomon Islands**. SALOMON 2 Stn CP2197, 08°24'S, 159°23'E, 897-1057 m, 24.X.2004, 1 ♂ 4.4 mm (MNHN-IU-2011-5952). – Stn CP2273, 8°31.8'S, 157°42.8'E, 732-839 m, 5.XI.2004, 1 sp., sex indet. [female gonopores present, male gonopods distinct], 5.2 mm (MNHN-IU-2011-5949). – Stn CP2297, 9°08.8'S, 158°16.0'E, 728-777 m, 8.XI.2004, 1 ♂ 4.4 mm, 1 ♀ 4.2 mm (MNHN-IU-2011-5950 & MNHN-IU-2011-5951).

ETYMOLOGY — From the Latin *inaequalis* (unequal), alluding to the two terminal spines of different sizes on the P2-4 dactyli, the character to separate the species from *U. pilosus* Baba, 1981.

DISTRIBUTION — New South Wales, and now Solomon Islands and Loyalty Islands; in 728-1057 m.

SIZE — Males, 4.4 mm; females, 4.2-5.1 mm.

DESCRIPTION — Medium-sized species. *Carapace:* Usually shorter (0.9 ×) than, rarely as long as broad; greatest breadth 1.7 × distance between anterolateral angles. Dorsal surface smooth, slightly convex, with very faint depression between gastric and cardiac regions, sparsely or thickly with short fine setae; gastric region anteriorly abruptly descending to depressed rostrum. Lateral margins convex on branchial region, with small denticles and short setae; anterolateral cor-

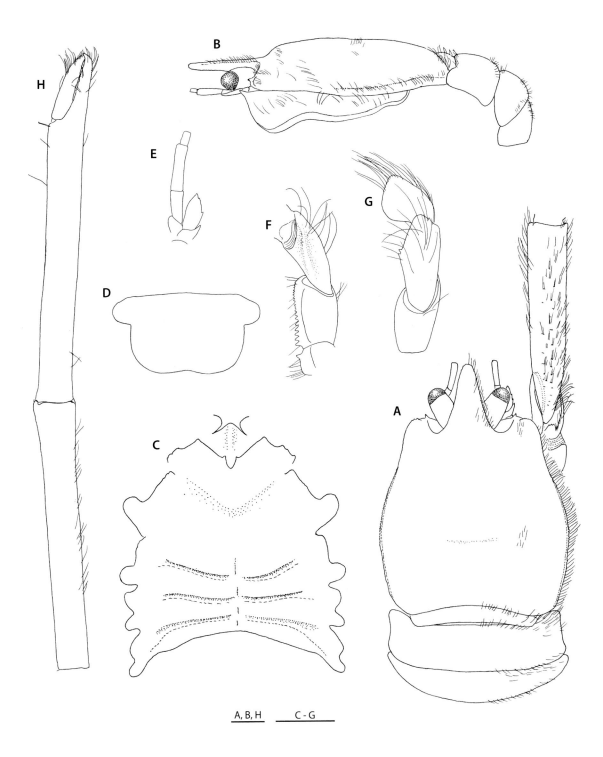

FIGURE 107

Uroptychus inaequalis n. sp., holotype, female 5.1 mm (MNHN-IU-2011-5948). **A**, carapace and anterior part of abdomen, proximal part of right P1 included, dorsal. **B**, same, lateral. **C**, sternal plastron with excavated sternum and basal parts of Mxp1. **D**, telson. **E**, left antenna, ventral. **F**, left Mxp3, ventral. **G**, same, lateral. **H**, right P1, dorsal. Scale bars: 1 mm.

ner rounded, with a few denticles or small spines (subequal to or smaller than lateral orbital spine). Rostrum straight horizontal, subtriangular with interior angle of 34-40°, distally blunt or roundish with very small denticles, dorsally flattish or slightly concave; length 0.3-0.4 × that of remaining carapace, breadth half carapace breadth measured at posterior carapace margin. Pterygostomian flap anteriorly angular, produced to sharp, anterolaterally directed spine, surface with plumose setae; posterior portion about half as high as anterior portion.

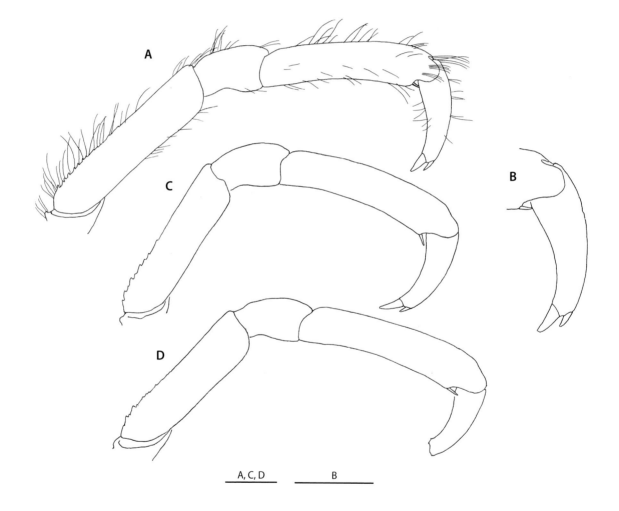

A, C, D   B

FIGURE 108

Uroptychus inaequalis n. sp., holotype, female 5.1 mm (MNHN-IU-2011-5948). **A**, right P2, lateral. **B**, same, distal part, setae omitted, lateral. **C**, right P3, lateral. **D**, right P4, lateral. Scale bars: 1 mm.

Sternum: Excavated sternum bluntly subtriangular on anterior margin; surface with rounded ridge in midline. Sternal plastron about as long as broad, slightly longer or slightly shorter, with subparallel lateral extremities. Sternite 3 well depressed, anterior margin excavated in broad V-shape, with ovate or V-shaped median notch flanked by small spine. Sternite 4 slightly narrower than sternite 5; anterolateral margin nearly straight or somewhat convex, with or without a few denticles on anterior portion, anteriorly ending in angular terminus; posterolateral margin about as long as anterolateral margin. Anterolateral margin of sternite 5 somewhat convexly divergent posteriorly, distinctly shorter than posterolateral margin of sternite 4.

Abdomen: Somite 1 moderately convex from anterior to posterior. Somite 2 tergite 2.1-2.4 × broader than long; pleuron slightly concave and slightly divergent posteriorly, posterolaterally blunt. Pleuron of somite 3 laterally blunt. Telson about half as long as broad; posterior plate 1.3-1.5 × longer than anterior plate, posterior margin feebly or distinctly emarginate.

Eye: Less than twice (1.6-1.8 x) as long as broad, slightly narrowed distally, slightly overreaching midlength of rostrum. Cornea not dilated, about half as long as remaining eyestalk.

Antennule and antenna: Ultimate article of antennule 3.0-3.3 × longer than high. Antennal peduncle reaching or slightly overreaching rostrum. Article 2 with strong distolateral spine. Antennal scale 1.5 × as broad as article 5, reaching or slightly overreaching article 4; lateral margin occasionally with a few small spines. Distal 2 articles unarmed. Article 5 1.4-1.5 × longer than article 4, breadth 0.4 × height of antennular ultimate article. Flagellum of 11-12 segments not reaching distal end of P1 merus.

Mxp: Mxp1 with bases broadly separated. Mxp3 with plumose setae on lateral surface of merus and carpus. Basis with 1 distal denticle. Ischium having crista dentata with 16-20 denticles much smaller in distal half; flexor margin not rounded distally. Merus 2.3 × longer than ischium, with small distolateral spine and 2-6 small spines distal to midlength of flexor crest. Carpus unarmed.

P1: Slender, subcylindrical, length 4.2-5.3 × (females), 5.6 × (male; missing in another male), 7.3 × (sex indet.) that of carapace, with soft fine setae dense on merus, moderate on fingers, sparse on carpus and palm. Ischium with short broad dorsal spine usually laciniate, rarely not laciniate; unarmed elsewhere. Merus dorsally with denticles and fine setae, length 1.1-1.5 × that of carapace. Carpus 1.3-1.5 × longer than merus. Palm 7.6-7.8 × (male and females), 10.0 × (sex indet.) longer than broad, 1.0-1.1 × length of carpus. Fingers not gaping (slightly gaping in proximal third in sex-indeterminate paratype), distally well incurved, crossing when closed; movable finger quarter to one-fifth length of palm, opposable margin slightly convex or with low prominence proximal to midlength in male, nearly straight in females.

P2-4: Thick mesio-laterally, sparingly with soft setae. Meri successively shorter posteriorly (P3 merus 0.8-0.9 × length of P2 merus, P4 merus 0.8-1.0 × length of P3 merus), equally broad on P2-4; dorsal margin with denticles along proximal half on P2-4; length-breadth ratio, 4.7-5.0 on P2, 4.0-4.2 on P3, 3.3-4.0 on P4; P2 merus 0.8-0.9 × length of carapace, 1.0-1.1 × length of P2 propodus; P3 merus 0.8 × length of P3 propodus; P4 merus 0.7-0.8 × length of P4 propodus. P3 carpus subequal to or very slightly shorter than P2 carpus, P4 carpus 0.9 × length of P3 carpus; length one-third that of propodus on P2-4. Propodi subequal in length on P2-4; flexor margin slightly curving, ending in pair of terminal spines (mesial one often obsolete). Dactyli subequal or slightly shorter on P2 than on P3 and P4, length 1.3-1.4 × that of carpus on P2-4, 0.4-0.5 × length of propodus on P2, 0.4 × on P3 and P4; flexor margin curving, with 2 terminal spines only, proximal much larger than distal.

REMARKS — The present material is identical with the male reported under *U. pilosus* by Ahyong & Poore (2004) from New South Wales. Ahyong & Poore listed a number of characters that do not fit the original account of *U. pilosus* and suggested the possibility of either polymorphism or the existence of another related species. Some of the differences they enumerated are actually displayed as intraspecific variations by the present material, but the broad rostrum appears to be consistent: the interior angle of the rostrum is 34-40° in *U. pilosus*, 23° in the present species. An additional difference worth noting is that the P2-4 dactyli in *U. inaequalis* are distinctly longer than instead of subequal to the carpi. In addition, the distal of the two terminal spines on the dactyli is much more slender than instead of subequal to the proximal spine.

Uroptychus inermis n. sp.

Figures 109, 110

TYPE MATERIAL — Holotype: **New Caledonia**, Norfolk Ridge. BIOCAL Stn DW36, 23°09'S, 167°11'E, 650-680 m, 29.VIII.1985, ♂ 4.7 mm (MNHN-IU-2014-16584).

ETYMOLOGY — From the Latin *inermis* (unarmed), alluding to the P1 ischium without subterminal spine, a character to separate the species from *U. granulipes* n. sp.

DISTRIBUTION — Norfolk Ridge; 650-680 m.

DESCRIPTION — *Carapace*: As long as broad; greatest breadth measured at posterior third, 1.5 × distance between anterolateral spines. Dorsal surface slightly convex from anterior to posterior, finely granulated on anterior portion and branchial region; small hepatic spine on right side only. Lateral margins slightly convex, sparsely setose, with granulate scale-like ridge at anterior end of branchial margin followed by very fine granulations, posterior quarter with low ridge; anterolateral spine well developed, extending far beyond much smaller lateral orbital spine, situated slightly posterior to level of that spine. Rostrum broadly triangular, with interior angle of 30 °, nearly horizontal, about half as long as remaining carapace, breadth half carapace breadth at posterior carapace margin; dorsal surface moderately concave. Lateral orbital spine separated from anterolateral spine by basal breadth of latter spine. Pterygostomian flap smooth on surface, anterior margin angular, produced to sharp spine.

Sternum: Excavated sternum ending in sharp spine between bases of Mxp1, surface with small spine in center. Sternal plastron slightly broader than long, successively broader posteriorly. Sternite 3 depressed well; anterior margin of broad V-shape with narrow U-shaped median sinus flanked by well-developed spine. Sternite 4 with convex anterolateral margin anteriorly blunt angular, without spine, posterolateral margin half as long as anterolateral margin. Anterolateral margin of sternite 5 strongly convex, 1.3 × longer than posterolateral margin of sternite 4

Abdomen: Somites smooth and glabrous. Somite 1 moderately convex from anterior to posterior. Somite 2 tergite 2.8 × broader than long; pleuron posterolaterally blunt angular, lateral margin shallowly concave and weakly divergent posteriorly. Pleuron of somite 3 posterolaterally blunt. Telson about half as long as broad; posterior plate twice as long as anterior plate, moderately emarginate on posterior margin.

Eye: 1.7 × longer than broad, reaching distal quarter of rostrum; mesial margin feebly concave proximal to cornea. Cornea not inflated, distinctly more than half as long as remaining eyestalk.

Antennule and antenna: Ultimate antennular article 2.3 × longer than high. Antennal peduncle slightly overreaching cornea. Article 2 distolaterally acuminate, without distinct spine. Antennal scale 1.6 × broader than article 5, slightly falling short of distal end of article 5. Distal 2 articles unarmed; article 5 1.8 × longer than article 4, breadth less than half height of antennular ultimate article. Flagellum of 19 segments terminating in distal end of P1 merus.

Mxp: Mxp1 with bases close to each other and slightly separated. Mxp3 sparsely setose. Basis with 3 denticles on mesial ridge. Ischium with about 20 denticles on crista dentata; flexor margin not rounded distally. Merus 2.2 × longer than ischium, somewhat ridged along flexor margin bearing a few very small obtuse protuberances. Carpus unarmed.

P1: Relatively massive, barely setose except for fingers; length 4.4 × that of carapace. Ischium with compressed, triangular dorsal spine, without ventromesial subterminal spine. Merus somewhat granulose dorsally; ventral and mesial surfaces with finely granulate ridges continued on to carpus; length slightly more than that of carapace. Carpus slightly longer than merus, as long as palm, granulose like merus. Palm convex on dorsal and ventral surfaces, smooth and nearly glabrous, length 2.4 × breadth. Fingers setose (setae not very long), moderately curving ventrally, distally incurved, crossing when closed; movable finger 0.6 × as long as palm, opposable margin with rounded process about at midlength with accompanying smaller process proximal to it, larger process fitting to concavity on opposable face of fixed finger when closed; opposable margin of fixed finger with 2 low prominences medially.

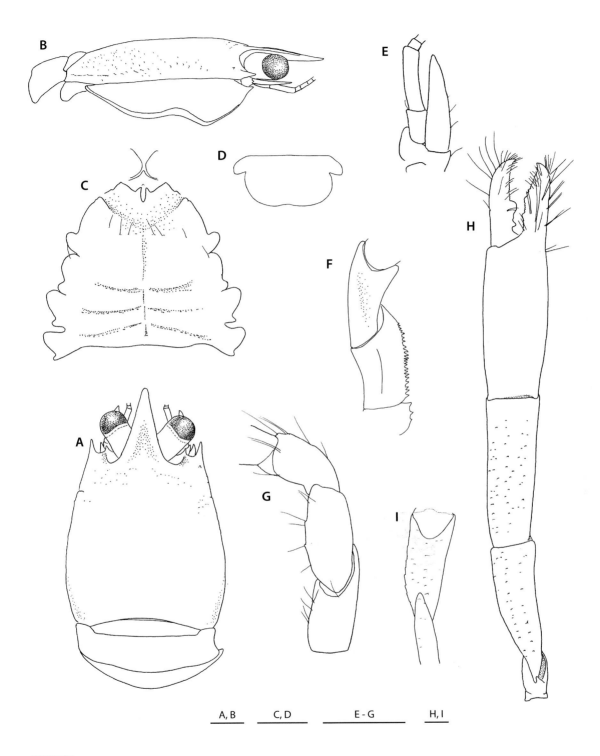

FIGURE 109

Uroptychus inermis n. sp., holotype, male 4.7 mm (MNHN-IU-2014-16584). **A**, carapace and anterior part of abdomen, dorsal. **B**, same, lateral. **C**, sternal plastron, with excavated sternum and basal parts of Mxp1. **D**, telson. **E**, left antenna, ventral. **F**, right Mxp3, setae omitted, ventral. **G**, left Mxp3, lateral. **H**, right P1, dorsal. **I**, same, proximal part, ventral. Scale bars: 1 mm.

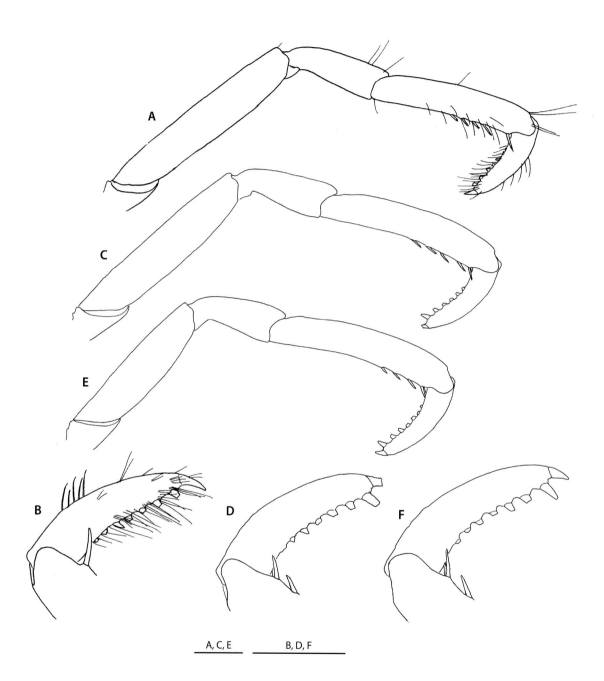

FIGURE 110

Uroptychus inermis n. sp., holotype, male 4.7 mm (MNHN-IU-2014-16584). **A**, right P2, lateral. **B**, same, distal part, lateral. **C**, right P3, setae omitted, lateral. **D**, same, distal part, lateral. **E**, right P4, lateral. **F**, same, distal part, lateral. Scale bars: 1 mm.

P2-4: Relatively slender, moderately compressed mesio-laterally, barely setose on meri, sparsely so on carpi and propodi, more setose on dactyli. Left P2 and P3 detached from body and missing. Meri successively shorter posteriorly (P3 merus 0.9 × length of P2 merus, P4 merus 0.8 × length of P3 merus), equally broad on P2-4; dorsal margin unarmed; length-breadth ratio, 5.3 on P2, 4.9 on P3, 4.0 on P4; P2 merus 0.9 × length of carapace, 1.3 × length of P2 propodus; P3 merus as long as P3 propodus; P4 merus 0.8 × length of P4 propodus. P2-3 carpi subequal, P4 carpus 0.9 × length of P3 carpus; carpus-propodus length ratio, 0.51 on P2, 0.47 on P3, 0.42 on P4. Propodi shortest on P2, subequal on P3 and P4; flexor margin nearly straight, ending in pair of spines preceded by 4 spines on P2 and P3, 3 spines on P4 at most on distal half, distalmost of these spines more remote from distal pair than from distal second. Dactyli subequal in length, shorter than carpi (dactylus-carpus length ratio, 0.9 on P2 and P3, 0.95 on P4), less than half length of propodi (dactylus-propodus length ratio, 0.4 on P2-4); flexor margin gently curving, with 8 sharp, slightly inclined, loosely arranged, proximally diminishing spines more or less obscured by fine setae; extensor margin with plumose setae on median portion.

REMARKS — The new species strongly resembles *U. maori* Borradaile, 1916 in nearly all aspects. As is suggested by the specific name, *U. inermis* is readily differentiated from *U. maori* by lack of a subterminal spine on the ventromesial margin of the P1 ischium. The combination of the following characters links the species to *U. anacaena* Baba & Lin 2009 and *U. granulipes* n. sp.: the carapace bearing a prominent anterolateral spine, the P1 merus and carpus granulose, Mxp3 spineless on the merus and carpus, and the P2-4 dactyli bearing proximally diminishing triangular spines. The most obvious differences that distinguish *U. anacaena* from the new species are: the antennal article 2 that bears a distinct distolateral spine, the carapace that is granulose on the entire surface, and the rostrum that is much less than half as broad as the the carapace measured along the posterior margin. *Uroptychus inermis* is distinguished from *U. granulipes* by the P1 ischium that is unarmed instead of bearing a pronounced subterminal spine on the ventromesial margin, the antennal article 4 is unarmed instead of bearing a small but distinct spine, and the anterolateral spine of the carapace is situated distinctly posterior to instead of directly lateral to the lateral orbital spine.

Uroptychus joloensis Van Dam, 1939
Figure 305E-G

Uroptychus joloensis Van Dam, 1939: 395, figs 2, 2a, 2b, 2c. — Ahyong & Baba 2004: 58, fig. 1. — Baba 2005: 39, 227. — Baba *et al.* 2008: 35. — Poore *et al.* 2011: 328, pl. 6, fig. I. — McCallum & Poore 2013: 164, fig. 12D.
Uroptychus kudayagi Miyake, 1961: 237, figs 1, 2. — 1982: 143, pl. 48, fig. 2. — Minemizu 2000: 165, unnumbered fig. — Kawamoto & Okuno 2003: 97, unnumbered fig.

TYPE MATERIAL — Holotype: **Philippines**, Jolo Sea, 37.8-56.7 mn male (ZMUC). [not examined].

MATERIAL EXAMINED — **Philippines**. MUSORSTOM 1 Stn CP60, 14°05'N, 120°19'E, 129-124 m, 27.III.1976, 1 ♂ 4.9 mm, 1 ov. ♀ 4.5 mm (MNHN-IU-2014-16585). MUSORSTOM 3 Stn CP134, 12°01'N, 121°57'E, 92-95 m, 5.VI.1985, 1 ♂ 3.3 mm (MNHN-IU-2014-16586). **Solomon Islands**. SALOMON 1 Stn CP1831, 10°12.1'S, 161°19.2'E, 135-325 m, 5.X.2001, 1 ♂ 2.7 mm (MNHN-IU-2014-16587). **Vanuatu**. GEMINI Stn DW48, 21°00'S, 170°04'E, 200-150 m, 04.VII.1989, 1 ♂ 2.5 mm, 1 ov. ♀ 3.1 mm (MNHN-IU-2014-16588). MUSORSTOM 8 Stn CP1131, 15°38.41'S, 167°03.52'E, 140-175 m, 11.X.1994, 1 ♂ 3.0 mm (MNHN-IU-2014-16589); 2 ov. ♀ 2.9, 3.0 mm (MNHN-IU-2014-16590). – Stn CP1133, 15°38.83'S, 167°03.06'E, 174-210 m, 11.X.1994, 1 ♀ 2.8 mm (MNHN-IU-2014-16591). – Stn CP1077, 16°04.00'S, 167°06.09'E, 180-210 m, 5.X.1994, 1 ♂ 2.7 mm (MNHN-IU-2014-16592). – Stn CP1071, 15°36.63'S, 167°16.34'E, 180-191 m, 4.X.1994, 1 ♂ 3.3 mm (MNHN-IU-2014-16593). – Stn CP961, 20°18.50'S, 169°49.90'E, 100-110 m, 21.IX.1994, 1 ov. ♀ 3.2 mm, 1 ♀ 2.0 mm (MNHN-IU-2014-16594). Santo Stn AT04, 15°32.9'S, 167°13.3'E, 97-101 m, 15.IX.2006, 1 ov. ♀ 3.0 mm (MNHN-IU-2014-16595). – Stn AT06, 15°38'S, 167°02'E, 140-167 m, 15.IX.2006, 1 ♂ 3.5 mm, 2 ov. ♀ 3.2, 3.8 mm, 1 ♀ 3.2 mm (MNHN-IU-2014-16596). – Stn AT17, 15°40'S, 167°02'E, 267-270 m, 21.IX.2006, 1 ♂ 3.2 mm (MNHN-IU-2014-16597). – Stn AT22, 15°32.3'S,

167°16.0'E, 180-227 m, 22.IX.2006, 1 ov. ♀ 3.7 mm (MNHN-IU-2014-16598). **Wallis and Futuna Islands**. MUSORSTOM 7 CP498, 14°19'S, 178°03'W 105-160 m, 10.V.1992, on *Chironephthya* sp. (Nidaliidae: Alcyonacea), 4 ov. ♀ 2.5-2.8 mm (MNHN-IU-2014-16599, MNHN-IU-2014-16600, MNHN-IU-2014-16601, MNHN-IU-2014-16602). **Fiji Islands**. MUSORSTOM 10 CP1364, 18°11.9'S, 178°34.5'E, 80-86 m, 15.VIII.1998, 1 ov. ♀ 3.2 mm (MNHN-IU-2014-16603).

DISTRIBUTION — Previously known from Jolo Sea, Kai Islands, Timor Sea and northwestern Australia in 37.8-142 m; and under the name *U. kudayagi* Miyake, 1961, from Japan (eastern Sagami Bay, west coast of Kyushu, southern Kii Peninsula, Shizuoka, and Okinawa), in 20-80 m. The present material was taken from the Philippines, Solomon Islands, Vanuatu, Wallis and Futuna Islands, and Fiji Islands, in 80-325 m. Additional material was collected from the southern coast of Papua New Guinea, in 25 m (see below under the remarks).

SIZE — Males, 2.7-4.9 mm; females 2.0-4.5 mm; ovigerous females from 2.5 mm.

DIAGNOSIS — Small species. Carapace smooth on surface, somewhat broader than long, greatest breadth about twice distance between anterolateral spines. Lateral margins feebly convexly divergent posteriorly; anterolateral spine situated closely lateral to and overreaching smaller lateral orbital spine; 2 spines about at anterior third (anterior one occasionally very small, vestigial or even obsolete, posterior one larger), followed by ridge along remaining margin. Rostrum triangular, with interior angle of 22-25°, lateral margin with small spine often obsolete, dorsal surface concave, breadth less than half carapace breadth at posterior carapace margin. Pterygostomian flap anteriorly angular, ending in sharp spine. Excavated sternum anteriorly subtriangular, with cristate ridge in midline; sternal plastron slightly longer than broad, with subparallel lateral extremities; sternite 3 having anterior margin very weakly concave, with deep median sinus usually U-shaped, flanked by small incurved spine; sternite 4 with anterolateral margin 1.5 × length of posterolateral margin; anterolateral margins of sternite 5 slightly convexly subparallel, slightly longer than posterolateral margin of sternite 4. Abdominal somite 1 without transverse ridge; somite 2 tergite 2.2-2.4 × broader than long; pleural lateral margins feebly or somewhat concave, subparallel or somewhat divergent posteriorly, posteriorly blunt; pleuron of somite 3 laterally rounded. Telson about half as long as broad; posterior plate 1.4-1.8 × longer than anterior plate, posterior margin roundish. Eyes elongate, distally narrowed, barely reaching apex of rostrum. Ultimate antennular article 2.5 × longer than high. Article 2 of antenna with sharp lateral spine; antennal scale terminating in midlength of article 5; articles 4 and 5 each with distinct distoventral spine, article 5 1.2-1.4 × longer than article 4, breadth 0.6-0.7 × height of ultimate antennular article; flagellum of 6-9 segments not reaching distal end of P1 merus. Mxp1 with bases separated from each other. Mxp3 sharply ridged along flexor margins of ischium and merus; basis with obsolescent denticles on proximal part of mesial ridge; ischium with small spine lateral to rounded distal end of flexor margin, crista dentata with obsolescent denticles; merus about twice longer than ischium, with distolateral spine and 1 or 2 small spines distal to midlength of flexor margin; carpus with distolateral spine and often with small spine on proximal lateral surface. P1 slender, subcylindrical, bearing soft fine setae; 6.2-6.4 × (males), 5.7-7.0 × (females) longer than carapace; ischium dorsally with 2 spines (distal one well-developed and sharp, proximal one small), ventromesially with short, blunt subterminal spine; merus 1.3-1.4 × longer than carapace, with a few small spines on mesio-proximal part of ventral surface; carpus 1.3-1.4 × longer than merus; palm subequal to or slightly longer than carpus; fingers slightly incurved; movable finger 0.3-0.4 × length of palm. P2-4 well compressed mesio-laterally; meri subequal in breadth on P2-4; P2 merus 0.8-0.9 × length of carapace, 1.1-1.2 × length of P2 propodus; P3 merus 0.9 × length of P2 merus, 0.9 × length of P3 propodus; P4 merus 0.9-1.0 × length of P3 merus, 0.8 × length of P4 propodus; carpi subequal, each less than half length of propodus; propodal flexor margin with pair of distal spines only; dactyli longer than carpi (dactylus-carpus length ratio, 1.0-1.1 on P2, 1.1-1.2 on P3, 1.2 on P4), flexor margin with 6 or 7 loosely arranged spines, ultimate slender, penultimate, antepenultimate and often distal quarter prominent, remaining 2 or 3 diminishing proximally, proximal-most somewhat inclined.

Eggs. Number of eggs carried up to 18; size, 0.65 mm × 0.69 mm - 0.9 mm × 1.1 mm.

COLOR — Base color translucent white, reddish on anterior part of carapace (around rostrum, including antennule and antenna) and around juncture between P1 palm and fingers (ov. ♀ 3.0 mm, MNHN-IU-2014-16595 [= Poore *et al.* 2011: pl. 6, fig. I]); additional red spot at juncture between P1 merus and carpus (specimens from Papua New Guinea, see below under the remarks); red spot absent from P1 (ov. ♀, MNHN-IU-2014-16602; McCallum & Poore, 2013: fig. 12D).

REMARKS — The species is characterized by the carapace lateral margin with two spines about at distal third (the distal of which is often obsolete), the antennal articles 4 and 5 each bearing a strong distal spine, the antennal scale barely reaching the end of the antennal article 5 (usually terminating in the midlength of this article) and the eyes distally narrowed. As noted by Baba *et al.* (2009), *Uroptychus joloensis* resembles *U. zezuensis* in morphology, coloration and habitat preference (see below under *U. zezuensis*).

The material reported under *U. kudayagi* from Japan (Miyake 1961, 1982; Minemizu 2000; Kawamoto & Okuno 2003) bears red spots at junctures between the P1 merus and carpus, and between the palm and fingers, in addition to another red spot around eyes including the rostrum. Specimens identical with this material (1 male 2.3 mm, 1 ov. female 2.1 mm, 2 eggs 0.82 × 0.86 mm, QM W25105), taken at 25 m, Bootless Bay, Port Moresby, Papua New Guinea, were made available, along with a photograph through Peter J.F. Davie (Figure 305G). On the other hand, the spots on P1 are totally absent in one of the specimens from the Wallis and Futuna Islands (MNHN-IU-2014-16602) as well as in the specimen from Western Australia (McCallum & Poore 2013: fig. 12D). The specimen from Vanuatu (MNHN-IU-2014-16595) has the spot only between the palm and fingers (Poore *et al.* 2011: pl. 6, fig. I). Inasmuch as no morphological differences are found between all of these forms of different color, these are treated as identical. However, it is not unlikely that the color differences may be validated for species discrimination by molecular data, as has been done for some species of *Galathea* (Macpherson & Robainas-Barcia 2015).

According to Minemizu (2000) and Kawamoto & Okuno (2003), this species is found among branches of alcyonacean corals, usually *Siphonogorgia dofleini*, rarely *S. dispacea* (Nidaliidae). The red spots of the animal resemble the color pattern of the host's polyps. One of the specimens from MUSORSTOM 7 Stn CP498 (MNHN-IU-2014-16601) was found on *Chironephthya* sp. (Nidaliidae).

Uroptychus kareenae n. sp.
Figures 111, 112

TYPE MATERIAL — Holotype: **Solomon Islands**. SALOMON 1 Stn DW1775, 08°13'S, 160°42'E, 498-600 m, 28.IX.2001, 1 ♂ 4.1 mm (MNHN-IU-2012-692).

DISTRIBUTION — Solomon Islands, in 498-600 m.

ETYMOLOGY — Named for Kareen Schnabel for her contributions to the knowledge of squat lobster taxonomy and distribution.

DESCRIPTION — Small species. *Carapace*: Slightly broader than long (0.9 × as long as broad); greatest breadth 1.9 × distance between anterolateral spines. Dorsal surface smooth and glabrous, feebly convex from anterior to posterior, depressed between gastric and cardiac regions. Lateral margins divergent posteriorly to point at anterior end of branchial region, with small acuminate eminence at midpoint between, then subparallel posteriorly; feebly ridged along posterior part of branchial margin; anterolateral spine overreaching lateral orbital spine, more or less close to and situated directly lateral to that spine; branchial margin with well-developed spine at anterior end, followed by 5 or 6 posteriorly diminishing spines. Rostrum narrow triangular, with interior angle of 23°, length half that of carapace, breadth slightly more than one-third carapace breadth at posterior carapace margin; dorsal surface somewhat concave. Pterygostomian flap anteriorly moderately angular, ending in distinct spine.

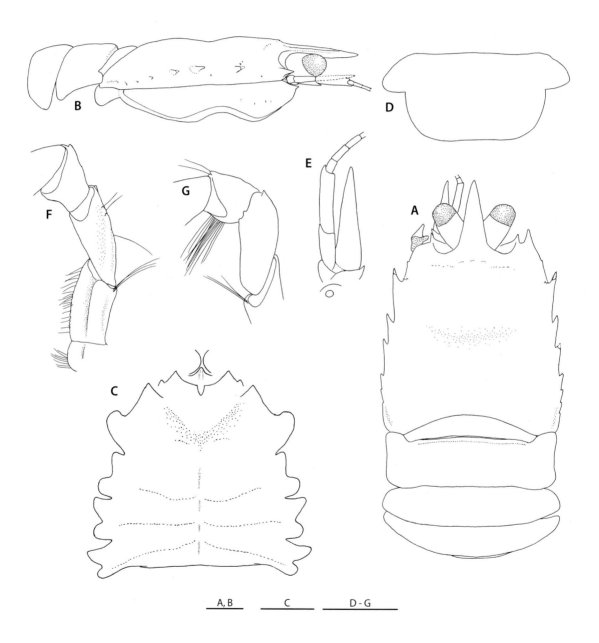

FIGURE 111

Uroptychus kareenae n. sp., holotype, male 4.1 mm (MNHN-IU-2012-692). **A**, carapace and anterior part of abdomen, dorsal. **B**, same, lateral. **C**, sternal plastron, with excavated sternum, basal parts of Mxp1 included. **D**, telson. **E**, left antenna, ventral. **F**, left Mxp3, ventral. **G**, same, lateral. Scales bars: 1 mm.

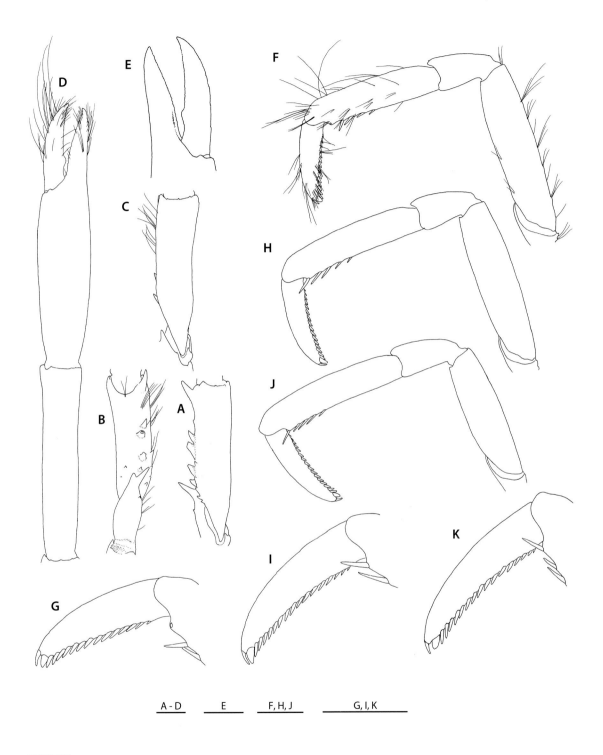

FIGURE 112

Uroptychus kareenae n. sp., holotype, male 4.1 mm (MNHN-IU-2012-692). **A**, right P1, proximal part, dorsomesial. **B**, same, ventral. **C**, same, dorsal. **D**, same, distal 3 articles, dorsal. **E**, same, fingers, ventral. **F**, left P2, lateral. **G**, same, distal part, setae omitted, lateral. **H**, left P3, setae omitted, lateral. **I**, same, distal part, lateral. **J**, left P4, setae omitted, lateral. **K**, same, distal part, lateral. Scales bars: 1 mm.

Sternum: Excavated sternum sharp triangular on anterior margin, with weak ridge in midline on surface. Sternal plastron 0.9 × as long as broad, lateral extremities slightly divergent posteriorly. Sternite 3 well depressed, anterolaterally sharp angular, anterior margin semicircular (shallowly), with V-shaped median notch. Sternite 4 having anterolateral margin slightly shorter than posterolateral margin, slightly concave, anteriorly angular, not produced to spine. Sternite 5 with anterolateral margin convex anteriorly, 0.6 × as long as posterolateral margin of sternite 4.

Abdomen: Smooth and nearly glabrous; somite 1 gently convex from anterior to posterior. Sternite 2 tergite 2.5 × broader than long; pleural lateral margins feebly concave, weakly divergent posteriorly; pleuron of somite 3 laterally blunt. Telson slightly less than half as long as broad, slightly concave on posterior margin, posterior plate 1.4 × longer than anterior plate.

Eye: Relatively long (1.7 × longer than broad), medially somewhat swollen, cornea not dilated, more than half length of remaining eyestalk.

Antennule and antenna: Ultimate article of antennular peduncle 3.5 × longer than high. Antennal article 2 with distinct lateral spine; antennal scale twice as broad as article 5 and slightly overreaching that article; article 4 with distinct distomesial spine; article 5 with small distomesial spine, length 1.7 × that of article 4, breadth 0.6 × height of ultimate article of antennule; flagellum of 12 or 13 segments far falling short of distal end of P1 merus.

Mxp: Mxp1 with bases close to each other. Mxp3 basis without denticle on mesial ridge. Ischium with distally rounded flexor margin, crista dentata with obsolescent denticles. Merus 1.8 × longer than ischium, flattish on mesial face, weakly ridged along flexor margin, distolateral spine small, flexor margin with 2 small, blunt spines distal to point two-thirds of length. Carpus with small distolateral spine.

Pereopods sparsely with soft fine setae.

P1: Slender, 5.2 × longer than carapace. Ischium dorsally with basally broad, dorso-ventrally depressed sharp spine, ventromesially with strong subterminal spine; merus 1.2 × longer than carapace, ventrally with 2 rows of spines: 1 row along mesial margin, consisting of 3 or 4 small spines, another row in line with subterminal spine of ischium, consisting of 5 large spines, distalmost largest. Carpus 1.2 × longer than merus. Palm 3.3 × longer than broad, 0.9 × length of carpus. Fingers slightly incurved distally, not gaping; movable finger half length of palm, opposable margin slightly concave, with proximal eminence; opposable margin of fixed finger convex, without prominence.

P2-4: Meri flattish on mesial face, moderately inflated on lateral face, dorsal margin rounded; subequal in length on P2 and P3, P4 merus 0.8 × length of P3 merus; successively narrower posteriorly; length-breadth ratio, 4.0 on P2 and P4, 3.4 on P4; P2 merus 0.9 × length of carapace,1.1 × longer than P2 propodus; P3 merus very slightly longer than P3 propodus; P4 merus 0.8 × length of P4 propodus. Carpi subequal, 0.4 × length of propodi on P2-4. Propodi successively longer posteriorly; flexor margin straight, ending in pair of long movable spines preceded by 4 single spines on distal two-fifths length on P2-4. Dactyli slightly longer on P3 and P4 than on P2, much longer than carpi (1.4 ×, 1.6 ×, 1.6 × longer on P2, P3, P4, respectively), distinctly more than half (0.5 on P2, 0.6 on P2 and P4) as long as propodi, well compressed mesio-laterally, proportionately broad; flexor margin nearly straight, with 16, 18, 18 spines on P2, P3, P4 respectively, all obliquely directed and very close but not contiguous to one another, ultimate slender, penultimate 2 × broader than ultimate, remaining spines as broad as ultimate; antepenultimate spine 3.9-4.8 × longer than broad.

REMARKS — *Uroptychus kareenae* n. sp. is very close to *U. adiastaltus* n. sp. (see above), *U. depressus* n. sp. (see above) and *U. levicrustus* Baba, 1988 in the shape of sternite 4 and in the spination of pereopods. *Uroptychus adiastaltus* differs from the these three related species in the spination of the branchial lateral margin (see above under *U. adiastaltus*), in having the sternite 3 anterior margin with a V-shaped emargination lacking a median notch, and in having the antennal article 4 unarmed. *Uroptychus kareenae* is differentiated from *U. adiastaltus*, *U. depressus* and *U. levicrustus* by the antennal article 5 that bears a distinct distomesial spine instead of being unarmed, and by the antepenultimate spines of the P2-4 dactyli that are long relative to breadth, with the length-breadth ratio, 3.9-4.8 instead of 2.0-2.9 (2.0-2.7 in *U. adiastaltus*, 2.7-2.9 in *U. depressus*, 2.5-2.6 in *U. levicrustus*). Moreover, the ultimate spines of the P2-4 dactyli are as broad as the antepenultimate spines in *U. kareenae* and *U. levicrustus*, while more slender in *U. adiastaltus* and *U. depressus*.

Uroptychus kareenae is distinguished from *U. depressus* by the following differences: the anterolateral spine of the carapace is closer to lateral orbital spine (these spines are separated by less than basal width of the anterolateral spine in *U. kareenae*, more than that in *U. depressus*; the eyes are more elongate in *U. kareenae* (breadth-length ratio = 1.7 in *U. kareenae*, 1.4 in *U. depressus*), reaching the distal third point of rostrum length instead of barely reaching the midlength of the rostrum; the antennal scale slightly overreaches article 5 in *U. kareenae*, reaching the first or second segment of the flagellum in *U. depressus*; the gastric region in profile is smoothly sloping down anteriorly on to the rostrum in *U. kareenae*, preceded by the depressed rostrum in *U. depressus*.

Uroptychus karubar n. sp.

Figures 113, 114

TYPE MATERIAL — Holotype: **Indonesia**, Kai Islands. KARUBAR Stn CP19, 5°15'S, 133°01'E, 605-576 m, with corals Primnoidae gen. sp., *Acanella* sp. and *Chrysogorgia* sp. (all Calcaxonia), 25.X.1991, ♂ 7.4 mm (MNHN-IU-2014-16604). Paratypes: Collected together with holotype, 4 ♂ 3.5-7.5 mm, 1 ov. ♀ 5.3 mm, 2 ♀ 4.2, 6.8 mm (MNHN-IU-2014-16605). **Solomon Islands**. SALOMON 1 Stn DW1775, 8°12.6'S, 160°41.7'E, 498-600 m, 28.IX.2001, 1 ov. ♀ 4.9 mm (MNHN-IU-2014-16606). **New Caledonia**, Norfolk Ridge. NORFOLK 2 Stn DW2075, 25°23.12'S, 168°20.07'E, 650-1000 m, 27.X.2003, 1 ♂ 4.3 mm, 3 ov. ♀ 5.1-6.3 mm (MNHN-IU-2014-16607). – Stn DW2087, 24°56.22'S, 168°21.66'E, 518-586 m, 28.X.2003, 1 ov. ♀ 6.5 mm (MNHN-IU-2014-16608).

ETYMOLOGY — Named for the cruise KARUBAR, by which the holotype was collected; used as a noun apposition.

DISTRIBUTION — Kai Islands, Solomon Islands and Norfolk Ridge; 518-1000 m.

SIZE — Males, 3.5-7.5 mm; females, 4.2-6.8 mm; ovigerous females from 4.9 mm.

DESCRIPTION — Medium-sized species. *Carapace*: broader than long (0.9 × as long as broad); greatest breadth 1.6 × distance between anterolateral spines. Dorsal surface moderately convex from side to side, slightly so from anterior to posterior, with feeble depression between gastric and cardiac regions and between anterior and posterior branchial regions; transverse row of 9-11 epigastric spines (median 3 rudimentary in ovigerous female MNHN-IU-2014-16606) preceded by depressed rostrum, followed by scattered small spines. Lateral margins somewhat convexly divergent posteriorly, with 6 spines: first anterolateral, distinctly overreaching much smaller lateral orbital spine; second smaller than first, equidistant between first and third and ventral to level of these spines; third largest, located at anterior end of anterior branchial region, with accompanying small spine dorsomesial to it; fourth to sixth on posterior branchial region, all acute, sixth followed by ridge. Rostrum straight horizontal, narrow triangular, with interior angle of 20-25°; length 0.5-0.6 × (rarely 0.7 x) postorbital carapace length, breadth less than half carapace breadth at posterior carapace margin; dorsal surface flattish; lateral margin with 4-6 spinules. Lateral orbital spine somewhat anterior to level of and moderately remote from anterolateral spine. Pterygostomian flap anteriorly angular, produced to sharp spine; surface covered with small spines.

Sternum: Excavated sternum anteriorly blunt angular, surface with weak ridge in midline. Sternal plastron slightly broader than long; lateral extremities gently divergent posteriorly. Sternite 3 well depressed, anterior margin deeply excavated, with 2 small submedian spines separated by semicircular or U-shaped sinus, laterally angular. Sternite 4 anterolaterally ending in sharp, anteriorly directed spine accompanied by small spine mesial to it; anterolateral margin somewhat concave, length subequal to that of posterolateral margin. Anterolateral margin of sternite 5 straight, slightly divergent posteriorly, anteriorly blunt angular, length 0.7-0.9 × that of posterolateral margin of sternite 4.

Abdomen: Somites smooth and glabrous. Somite 1 with distinct transverse ridge. Somite 2 tergite 2.5-2.7 × broader than long; pleuron with concavely divergent lateral margin leading to bluntly produced posterior terminus. Pleuron of

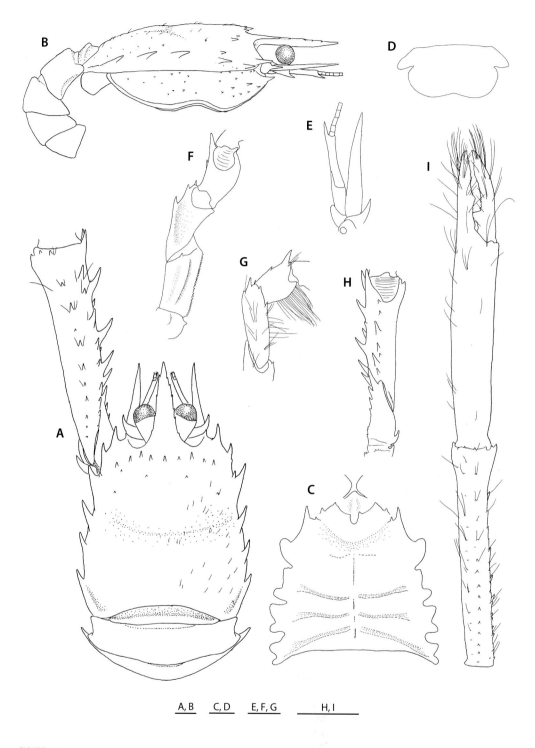

A, B   C, D   E, F, G   H, I

FIGURE 113

Uroptychus karubar n. sp., holotype, male 7.4 mm (MNHN-IU-2014-16604). **A**, carapace and anterior part of abdomen, proximal part of left P1 included, dorsal. **B**, same, lateral. **C**, sternal plastron, with excavated sternum and basal parts of Mxp1. **D**, telson. **E**, left antenna, ventral. **F**, right Mxp3, ventral. **G**, same, setae omitted, lateral. **H**, left P1, proximal part, ventral. **I**, same, distal part, dorsal. Scale bars: A-G, 1 mm; H, I, 5 mm.

somite 3 with bluntly angular lateral tip. Telson slightly less than half as long as broad; posterior plate 1.4-1.6 × longer than anterior plate, distinctly emarginate on posterior margin.

Eye: Terminating in midlength of rostrum, 1.5 × longer than broad, lateral and mesial margins subparallel. Cornea not dilated, length more than half that of remaining eyestalk.

Antennule and antenna: Ultimate article of antennule 3.5-4.1 × longer than high. Antennal peduncle extending far beyond cornea and falling short of rostral tip. Article 2 with strong lateral spine. Antennal scale twice as broad as article 5, reaching rostral tip and proximal quarter segment of flagellum. Articles 4 and 5 each with strong distomesial spine; article 5 1.4-1.9 × longer than article 4, breadth 0.5-0.6 × height of ultimate antennular article. Flagellum of 11-12 segments falling short of distal end of P1 merus.

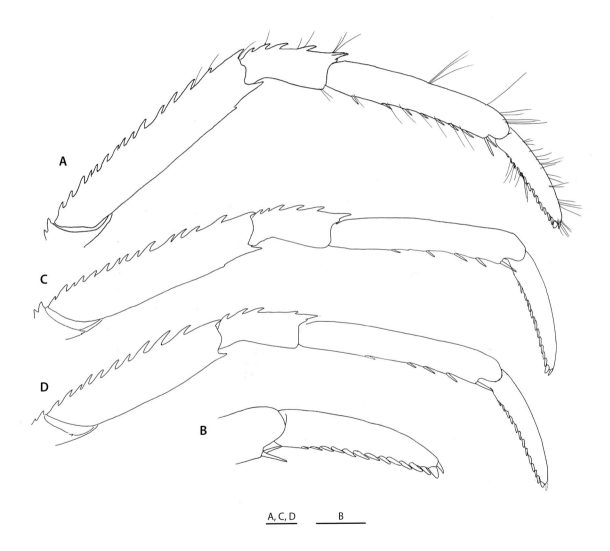

FIGURE 114

Uroptychus karubar n. sp., holotype, male 7.4 mm (MNHN-IU-2014-16604). **A**, right P2, lateral. **B**, same, distal part, setae omitted, lateral. **C**, right P3, lateral. **D**, right P4, lateral. Scale bars: 1 mm.

Mxp: Mxp1 with bases close to each other but not contiguous. Mxp3 basis with 1 denticle on mesial ridge. Ischium with small spine slightly lateral to distal end of flexor margin; crista dentata with very small denticles; flexor margin not rounded distally. Merus 1.8 × longer than ischium; flexor margin well ridged, bearing a few small spines on distal third; distolateral spine distinct. Carpus also with well-developed distolateral spine and a few small spines along extensor margin.

P1: Relatively slender, sparingly setose but fingers more setose; length 5.4-6.6 × that of carapace. Ischium dorsally with sharp, curved spine, ventromesially with well-developed subterminal spine followed by a few proximally diminishing spines. Merus 1.3-1.5 × longer than carapace, with 4 rows of spines: 2 dorsal rows of small spines, 1 mesial row of strong spines (4 or 5 in number), and ventral row of 3-5 spines. Carpus 1.3-1.5 × longer than merus, about as long as palm, with 3 (rarely 4) rows of small spines (2 dorsal, 1 mesial, 1 ventral). Palm unarmed, 3.6-4.9 × (males), 4.3-6.1 × (females) longer than broad, 0.9-1.0 × length of carpus. Fingers gaping in males, not gaping in females, distally ending in papilla-like tip, not crossing; movable finger with bluntly triangular proximal process, length 0.4-0.5 × that of palm; opposable margin of fixed finger with longitudinal groove to accommodate opposite process of movable finger when closed.

P2-4: Relatively slender, moderately compressed mesio-laterally, sparsely setose. Meri successively shorter posteriorly (P3 merus 0.9 × length of P2 merus, P4 merus 0.8-0.9 × length of P3 merus), equally broad on P2-4; length-breadth ratio, 5.1-5.7 on P2, 4.5-5.7 on P3, 3.8-5.0 on P4; dorsal margin with 14-16 spines on P2, 11-17 on P3, 11-15 on P4; ventrolateral margin with 1 or 2 spines distally; ventromesial margin with 3 equidistant spines on P2 only, proximal 2 obsolete in small specimens; P2 merus 0.8-1.0 × length of carapace, 1.1-1.2 × length of P2 propodus; P3 merus very slightly longer than or subequal to P3 propodus; P4 merus 0.8-0.9 × length of P4 propodus. Carpi subequal on P2-4, on P2-3 (shorter on P4), or on P3-4 (longer on P2); carpus-propodus length ratio, 0.33-0.42 on P2, 0.31-0.39 on P3, 0.31-0.36 on P4; extensor margin with 5-7 spines on P2, 4-7 on P3, 3-4 on P4, terminal spine with accompanying spine lateral to it. Propodi subequal on P3 and P4, shorter on P2; flexor margin nearly straight, with pair of terminal spines preceded by 5-7 slender, movable spines on P2, 2-6 spines on P3, 2-5 spines on P4. Dactyli longer than carpi (dactylus-carpus length ratio, 1.2-1.3 on P2, 1.4-1.5 on P3, 1.3-1.6 on P4), about half as long as propodi on P2-4; flexor margin feebly curving, with 14 or 15 spines, ultimate spine slender, penultimate about twice as broad as ultimate, remaining spines similar to ultimate, but distally blunt, strongly inclined and nearly contiguous to one another.

Eggs. Number of eggs carried, 9-16; size, 0.91 mm × 0.96 mm - 1.16 × 1.09 mm.

REMARKS — The combination of the following characters link the species to *U. alophus* n. sp., *U. echinatus* n. sp., *U. longior* Baba, 2005 and *U. nanophyes* McArdle, 1901: the P2-4 dactyli having the flexor margin with more than 10 closely arranged spines proximal to the pronounced penultimate spine, the P2-4 carpi with a few to several extensor marginal spines, and the carapace lateral margin with at least six spines. Among these species, *Uroptychus nanophyes* keys out first by having the antennal scale terminating in the distal end of the antennal article 5 and by having the P2 merus with a row of spines along the ventromesial margin. *Uroptychus echinatus* is distinctive in having the carapace thickly covered with setae and in having carapace lateral marginal spines much smaller. *Uroptychus karubar* is distinguished from *U. alophus* and *U. longior* by having a pair of epigastric spines and by the Mxp3 ischium bearing a small but distinct spine near the distal end of the flexor margin.

The new species also resembles *U. vegrandis* n. sp. from which it is readily distinguished by having a row of epigastric spines. More of distinguishing characters are outlined under the remarks of that species (see below).

Uroptychus lacunatus n. sp.

Figures 115, 116

TYPE MATERIAL — Holotype: **Vanuatu**. MUSORSTOM 8 Stn CP982, 19°21.80'S, 169°26.47'E, 408-410 m, 23.IX.1994, ♂ 6.1 mm (MNHN-IU-2014-16609). Paratype: Collected with holotype, 1 ov. ♀ 6.0 mm (MNHN-IU-2014-16610).

ETYMOLOGY — From the Latin *lacunatus* (hollowed out), alluding to a deeply excavated anterior margin of the sternal plastron that separates the species from *U. minor* n. sp.

DISTRIBUTION — Vanuatu; 408-410 m.

DESCRIPTION — Medium-sized species. *Carapace*: 1.2-1.3 × longer than broad; greatest breadth 1.5 × distance between anterolateral spines. Dorsal surface smooth and glabrous. Gastric and cardiac regions moderately inflated, bordered by depression; no spine on surface. Lateral margins slightly divergent posteriorly; anterolateral spine well developed, distinctly overreaching much smaller lateral orbital spine; finely denticulate, short, oblique ridge discernible under high magnification on anteriormost part of branchial region. Rostrum relatively long triangular, with interior angle of 23-25°; length 0.3-0.4 × that of remaining carapace, breadth slightly less than half carapace breadth at posterior carapace margin; dorsal surface moderately concave, ventral surface horizontal. Lateral orbital spine slightly anterior to level of anterolateral spine and separated from that spine by its basal breadth. Pterygostomian flap smooth on surface, anteriorly angular, ending in small spine.

Sternum: Excavated sternum anteriorly ending in sharp spine between bases of Mxp1, surface with small spine in center. Sternal plastron slightly longer than broad; lateral extremities divergent posteriorly. Sternite 3 strongly depressed; anterior margin strongly excavated in broad U-shape, with pair of submedian spines without notch between. Sternite 4 having anterolateral margin nearly straight, bluntly produced anteriorly; posterolateral margin somewhat more than half as long as anterolateral margin; surface with medially interrupted setiferous transverse ridge. Anterolateral margin of sternite 5 gently convexly divergent, slightly longer than posterolateral margin of sternite 4.

Abdomen: barely setose, smooth. Somite 1 convex from anterior and posterior, without distinct ridge. Somite 2 tergite 2.1 × broader than long, lateral margins slightly divergent posteriorly, posterolaterally blunt. Pleuron of somite 3 laterally blunt. Telson slightly more than half as long as broad; posterior plate 1.5 × (or slightly more) longer than anterior plate, feebly emarginate on posterior margin.

Eye: 1.6 × longer than broad, mesial and lateral margins subparallel, barely reaching apex of rostrum. Cornea not dilated, more than half as long as remaining eyestalk.

Antennule and antenna: Antennular ultimate article 1.7-1.9 × longer than high. Antennal peduncle slightly overreaching cornea but barely reaching apex of rostrum. Article 2 with small distolateral spine. Antennal scale as broad as or slightly broader than article 5, slightly overreaching midlength of article 5. Distal 2 articles unarmed; article 5 twice as long as article 4, breadth 0.4-0.5 height of antennular ultimate article. Flagellum consisting of 14 segments, barely reaching distal end of P1 merus.

Mxp: Mxp1 with bases nearly contiguous. Mxp3 glabrous on lateral surface, lacking spines on merus and carpus. Basis with 2 denticles on distal part of mesial margin. Ischium thick, crista dentata with 11-13 denticles, flexor margin distally not rounded. Merus 1.9 × longer than ischium, mesio-laterally compressed, ridged along flexor margin.

P1: Massive, 3.8-4.0 × longer than carapace, with small sharp distodorsal spine on ischium, unarmed elsewhere; ventral surface of ischium and carpus with tubercles somewhat more numerous in female paratype than in male holotype; dorsal surfaces smooth, merus and carpus barely setose. Merus slightly longer (holotype) or slightly shorter (paratype) than carapace. Carpus 1.2 × longer than merus. Palm distally setose (bearing simple setae), 2.2-2.3 × longer than broad, 0.9 × as long as carpus; lateral and mesial margins slightly convex. Fingers broad relative to length, distally incurved,

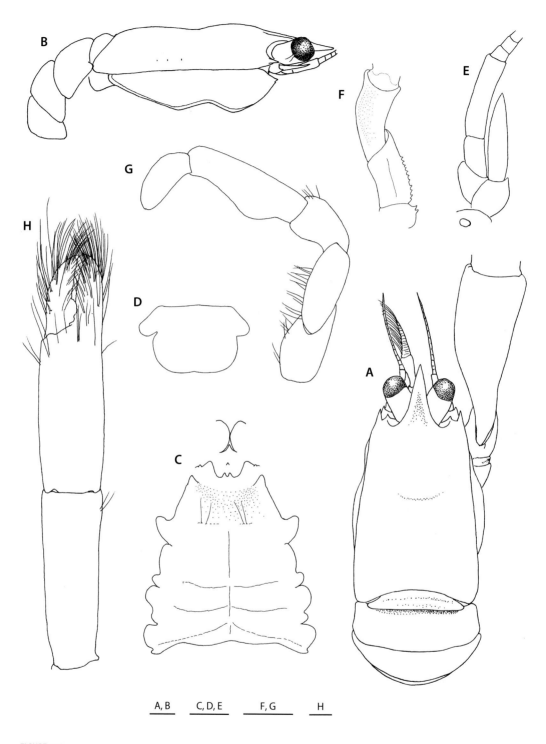

FIGURE 115

Uroptychus lacunatus n. sp., **A-C**, **E-H**, holotype, male 6.1 mm (MNHN-IU-2014-16609); **D**, paratype, ovigerous female (MNHN-IU-2014-16610). **A**, carapace and anterior part of abdomen, proximal part of right P1 included, dorsal. **B**, same, lateral. **C**, sternal plastron, with excavated sternum and basal parts of Mxp1. **D**, telson. **E**, left antenna, ventral. **F**, right Mxp3, setae omitted, ventral. **G**, left P3, setae omitted from distal articles, lateral. **H**, right P1, dorsal. Scale bars: 1 mm.

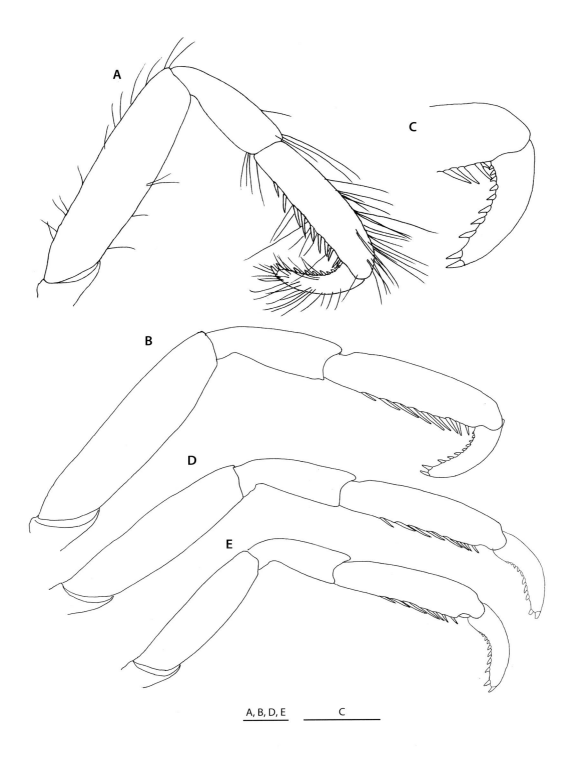

FIGURE 116

Uroptychus lacunatus n. sp., holotype, male 6.1 mm (MNHN-IU-2014-16609). **A**, right P2, lateral. **B**, same, setae omitted, lateral. **C**, same, distal part, lateral. **D**, right P3, lateral. **E**, right P4, lateral. Scale bars: 1 mm.

crossing when closed; opposable margins minutely denticulate, nearly straight along entire length of fixed finger, with prominent but low proximal process on movable finger; movable finger about half as long as palm.

P2-4: Moderately compressed mesio-laterally, broad relative to length. Meri successively shorter posteriorly (P3 merus 0.8 × length of P2 merus, P4 merus 0.7-0.8 × length of P3 merus), successively narrower posteriorly; length-breadth ratio, 4.0-4.5 on P2, 3.6-3.9 on P3, 3.4-3.7 on P4; dorsal margin not cristiform but rounded and unarmed; P2 merus 0.9 × length of carapace, 1.3-1.5 × length of P2 propodus; P3 merus 1.3 × length of P3 propodus; P4 merus 0.9 × length of P4 propodus. Carpi successively shorter posteriorly; carpus-propodus length ratio, 0.6 on P2, 0.5-0.6 on P3, 0.5 on P4. Propodi longest on P2 or subequal on P2 and P3, shortest on P4; flexor margin straight, ending in pair of spines located near juncture with dactylus, preceded by 8-10 movable spines on P2, 7-8 spines on P3, 6-7 spines on P4. Dactyli moderately curving at proximal third; dactylus-carpus length ratio, 0.7 on P2, 0.7-0.8 on P3, 1.0 on P4; dactylus-propodus length ratio, 0.4 on P2, 0.4-0.5 on P3, 0.5 on P4; flexor margin with 9-12 sharp triangular, loosely arranged, proximally diminishing and somewhat inclined spines obscured by setae, ultimate subequal to penultimate.

Eggs. 13 eggs, 1.04 mm × 1.10 mm - 1.18 mm × 1.39 mm.

REMARKS — The elongate carapace and relatively massive P1 as displayed by the species are possessed by *U. minor* n. sp. Also, *U. lacunatus* resembles *U. stenorhynchus* n. sp. in having the elongate carapace and the anteriorly angular pterygostomian flap, in the shape of sternum 3, and in the spination of the P2-4 dactyli. Their relationships are discussed under the remarks of those species (see below).

Uroptychus lanatus n. sp.
Figures 117, 118

TYPE MATERIAL — Holotype: **Indonesia**, Kai Islands. KARUBAR Stn CP17, 5°15'S, 133°01'E, 459-439 m, with coral, possibly belonging to Chrysogorgiidae (Calcaxonia), 24.X.1991, ov. ♀ 2.8 mm (MNHN-IU-2011-5958). Paratypes: Collected with holotype, same host, 3 ♂ 2.4-3.0 mm, 2 ov. ♀ 2.7, 3.0 mm (MNHN-IU-2011-5960, MNHN-IU-2011-5961, MNHN-IU-2014-16611). **Vanuatu**. MUSORSTOM 8 Stn CP975, 19°23.60'S, 169°28.93'E, 566-536 m, 22.IX.1994, 1 ♀ 3.8 mm (MNHN-IU-2011-5959).

ETYMOLOGY — From the Latin *lanatus* (soft), alluding to soft setae covering the body, characteristic of the species.

DISTRIBUTION — Kai Islands and Vanuatu; 459-536 m.

SIZE — Males, 2.4-3.0 mm; females 2.7-3.8 mm; ovigerous females from 2.7 mm.

DESCRIPTION — Small species. Body densely covered with soft fine setae. *Carapace*: 1.1-1.2 × broader than long; greatest breadth 1.7 × distance between anterolateral spines. Dorsal surface somewhat convex from anterior to posterior, without distinct groove. Lateral margins convexly divergent posteriorly, bearing several small spines; first anterolateral, largest, slightly overreaching lateral orbital spine; second rather remote from first, usually equidistant between first and third, rarely with accompanying small spine mesial to it; remaining spines situated on anterior half of posterior branchial region. Rostrum sharply triangular, with interior angle of 25-27°, deflected ventrally; length about half or slightly less than half that of remaining carapace, breadth distinctly less than half carapace breadth at posterior carapace margin; dorsal surface concave. Lateral orbital spine smaller than, moderately remote from, and slightly anterior to level of anterolateral spine. Pterygostomian flap anteriorly angular, produced to small spine, surface unarmed.

Sternum: Excavated sternum anteriorly triangular, surface with weak ridge in midline. Sternal plastron much broader than long, lateral extremities convexly divergent posteriorly. Sternite 3 weakly depressed; anterior margin nearly semicircular or

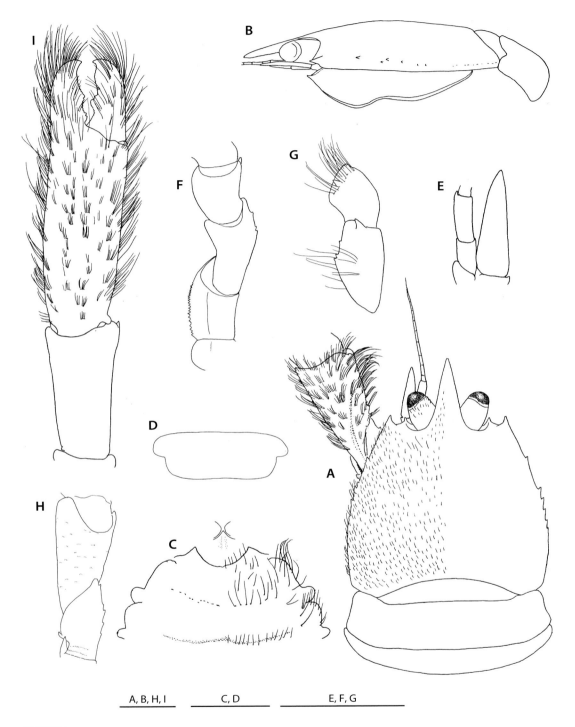

A, B, H, I   C, D   E, F, G

FIGURE 117

Uroptychus lanatus n. sp., holotype, ovigerous female 2.8 mm (MNHN-IU-2011-5958). **A**, carapace and anterior part of abdomen, proximal part of left P1 included, setae omitted from abdomen, dorsal. **B**, same, setae omitted, lateral. **C**, anterior part of sternal plastron, with excavated sternum and basal parts of Mxp1. **D**, telson. **E**, left antenna, ventral. **F**, left Mxp3, ventral. **G**, same, merus and carpus, lateral. **H**, left P1, proximal part, setae omitted, ventral. **I**, same, setae omitted from carpus, dorsal. Scale bars: 1 mm.

of very broad V-shape, without median notch and submedian spines, laterally sharply produced or bluntly angular. Sternite 4 having anterolateral margin convex, anteriorly rounded, bearing a few distinct or obsolescent denticles, posterolateral margin short, less than half as long as anterolateral margin. Anterolateral margin of sternite 5 strongly convex, much longer than (2 x) posterolateral margin of sternite 4.

Abdomen: Somite 1 slightly convex from anterior to posterior, without transverse ridge. Somite 2 tergite 2.4-2.9 × broader than long; pleuron posterolaterally blunt, lateral margin slightly concave, moderately divergent posteriorly. Pleuron of somite 3 with blunt lateral margin. Telson 0.4 × as long as broad; posterior plate nearly as long as or slightly shorter than anterior plate, posterior margin feebly or somewhat concave, not emarginate.

Eye: Relatively short (1.4-1.6 × longer than broad), terminating in midlength of rostrum, medially somewhat inflated, distally narrowed. Cornea not dilated, about half length of remaining eyestalk.

Antennule and antenna: Ultimate article of antennular peduncle 2.8-3.0 × longer than high. Antennal peduncle overreaching eyes by half length of article 5. Article 2 with small lateral spine. Antennal scale distinctly overreaching distal end of antennal peduncle, breadth 1.7 × that of article 5 or slightly more. Articles 4 and 5 each with distoventral spine very small or barely discernible. Article 5 1.3-1.5 × length of article 4, breadth 0.6-0.7 height of ultimate article of antennular peduncle. Flagellum of 10-11 segments reaching or somewhat overreaching distal end of P1 merus.

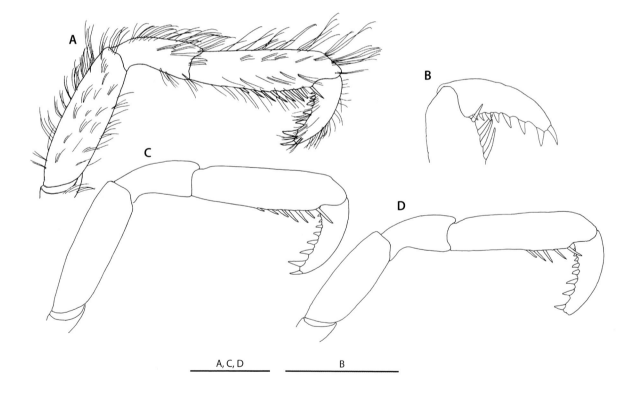

FIGURE 118

Uroptychus lanatus n. sp., holotype, ovigerous female 2.8 mm (MNHN-IU-2011-5958). **A**, right P2, lateral. **B**, same, distal part, setae omitted, lateral. **C**, right P3, setae omitted, lateral. **D**, right P4, setae omitted, lateral. Scale bars: 1 mm.

Mxp: Mxp1 with bases close to each other but not contiguous. Mxp3 with plumose setae on lateral surface of merus and carpus, other than brushes. Basis without denticles on mesial ridge. Ischium with flexor margin not rounded distally; crista dentata with numerous small distally diminishing denticles. Merus 2.1 × as long as ischium, flattish on mesial face, with distinct distolateral spine and a few small or obsolescent spines anterior to midlength of flexor margin. Carpus with small or obsolescent distolateral spine.

Pereopods thickly covered with soft, fine, plumose setae. *P1*: Massive, 4.7-4.8 × (males), 3.9-4.1 × (females) longer than carapace. Ischium dorsally with short spine, ventromesially with several small spines obscured by setae, subterminal spine very small. Merus 0.8-1.0 × length of carapace, usually with ventral distomesial spine, rarely with a few tubercular processes on proximal mesial margin. Carpus 1.1 × longer than merus. Palm 2.5-2.7 × (females), 2.5-3.0 × (males) longer than broad, 1.2-1.6 × longer than carpus. Fingers depressed, relatively broad, directed straight forward in females and small males, somewhat laterally directed in largest male, distally incurved, crossing when closed; movable finger having opposable margin with bluntly triangular median process fitting to groove between 2 low eminences on opposite face of fixed finger when closed, length usually 0.4 ×, rarely 0.5 × (non-ovigerous female) that of palm.

P2-4: Somewhat compressed mesio-laterally. Meri and carpi unarmed. Meri successively shorter posteriorly (P3 merus 0.9 × length of P2 merus, P4 merus 0.8 × length of P3 merus), subequally broad on P2 and P3, very slightly narrower on P4 than on P3; length-breadth ratio, 3.2-3.9 on P2, 3.1-3.9 on P3, 2.6-3.2 on P4; dorsal margin unarmed; P2 merus 0.7-0.8 length of carapace, 0.9 × length of P2 propodus; P3 merus 0.9 × length of P3 propodus; P4 merus 0.7-0.8 × length of P4 propodus. Carpi subequal in length on P2 and P3 or slightly longer on P2 than on P3 and shortest on P4; carpus-propodus length ratio, 0.5 on P2, 0.4-0.5 on P3, 0.4 on P4; unarmed. Propodi subequal on P2-4 or longer on P3-4 than on P2; flexor margin slightly convex (rarely somewhat more convex on P2), ending in pair of spines preceded by 6-7 slender movable spines on distal two-thirds on P2, 6 spines in distal half on P3, 4-5 spines on P4. Dactyli longer on P3 and P4 than on P2; dactylus-carpus length ratio, 0.9-1.0 on P2, 1.0-1.1 on P3, 1.2 on P4; dactylus-propodus length ratio, 0.4-0.5 on P2-4; flexor margin moderately curving, ending in slender spine preceded by 7-9 subtriangular, moderately obliquely directed spines diminishing toward juncture with propodus, ultimate somewhat more slender than penultimate.

Eggs. Number of eggs carried, 10; size, 0.83 mm × 0.83 mm - 1.0 mm × 1.25 mm (smallest eggs yolky).

REMARKS — *Uroptychus lanatus* keys out together with *U. perpendicularis* n. sp., but it is closer to *U. posticus* n. sp. Their relationships are discussed under the remarks of that species (see below).

Uroptychus latior n. sp.
Figures 119, 120

TYPE MATERIAL — Holotype: **New Caledonia**, Chesterfield Islands. CORAIL 2 Stn DE14, 21°00.69'S, 160°57.18'E, 650-660 m, 21.VII.1988, ♀ 6.1 mm (MNHN-IU-2012-682) (with rhizocephalan externa on abdomen).

ETYMOLOGY — From the Latin *latior* (broader), alluding to the carapace breadth being distinctly greater than that of the close relative *U. edisonicus* Baba & Williams, 1998.

DESCRIPTION — Small species. *Carapace*: 0.7 × as long as broad, broadest on posterior fourth; greatest breadth 2.3 × distance between anterolateral spines. Dorsal surface strongly convex from anterior to posterior and from side to side, without depression bordering gastric and cardiac regions, covered with minute pits suggesting supporting fine setae. Lateral margins convexly divergent, finely granulose, and weakly ridged along posterior fourth; anterolateral spine short, moderately remote from and overreaching lateral orbital angle, directed somewhat mesially. Rostrum horizontal, narrow triangular, with interior angle 25°; dorsal surface nearly flattish; lateral margin straight with 3 (left) or 2 (right) spinules or denticle-like small spines distally; length half that of remaining carapace, breadth one-third carapace breadth at posterior

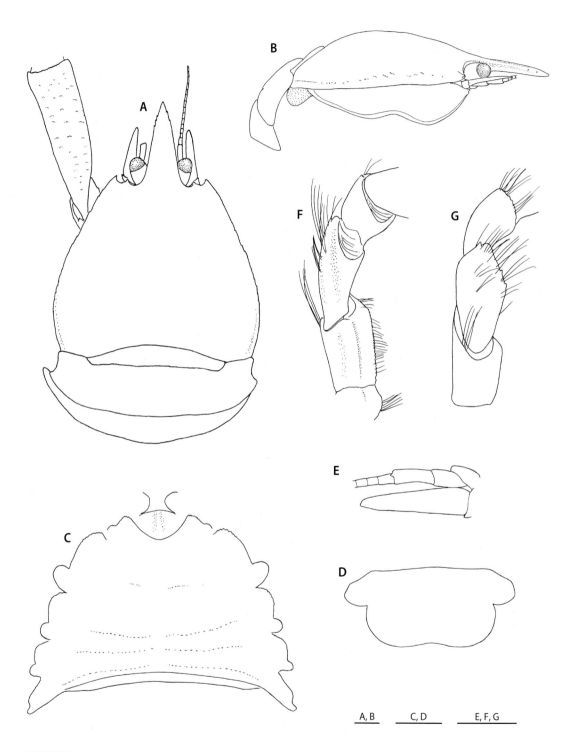

FIGURE 119

Uroptychus latior n. sp., holotype, female 6.1 mm (MNHN-IU-2012-682). **A**, carapace and anterior part of abdomen, proximal part of left P1 included, fine setae omitted, dorsal. **B**, same, lateral. **C**, sternal plastron, with excavated sternum and basal parts of Mxp1. **D**, telson. **E**, right antenna, ventral. **F**, right Mxp3, ventral. **H**, same, lateral. Scale bars: 1 mm.

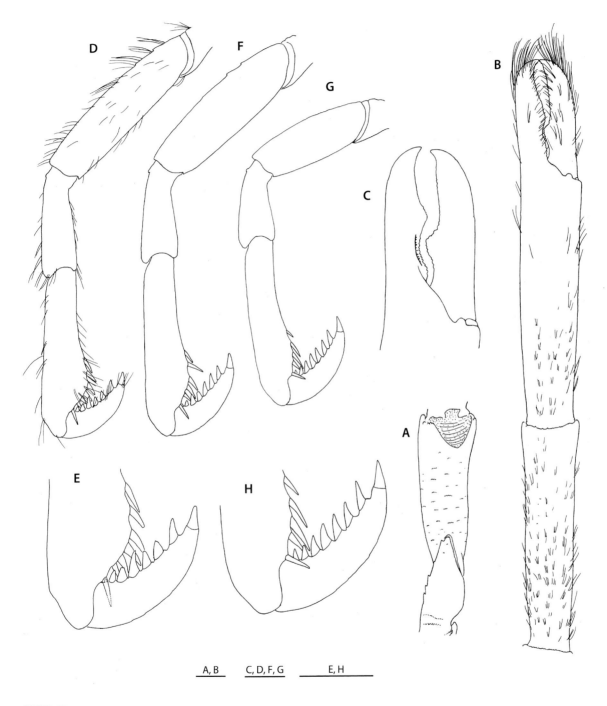

FIGURE 120

Uroptychus latior n. sp., holotype, female 6.1 mm (MNHN-IU-2012-682). **A**, left P1, proximal part, setae omitted, ventral. **B**, same, distal articles, dorsal. **C**, same, fingers, ventral. **D**, left P2, lateral. **E**, same, distal part, setae omitted, lateral. **F**, left P3, setae omitted, lateral. **G**, left P4, setae omitted, lateral, **H**, same, distal part, setae omitted, lateral. Scale bars: 1 mm.

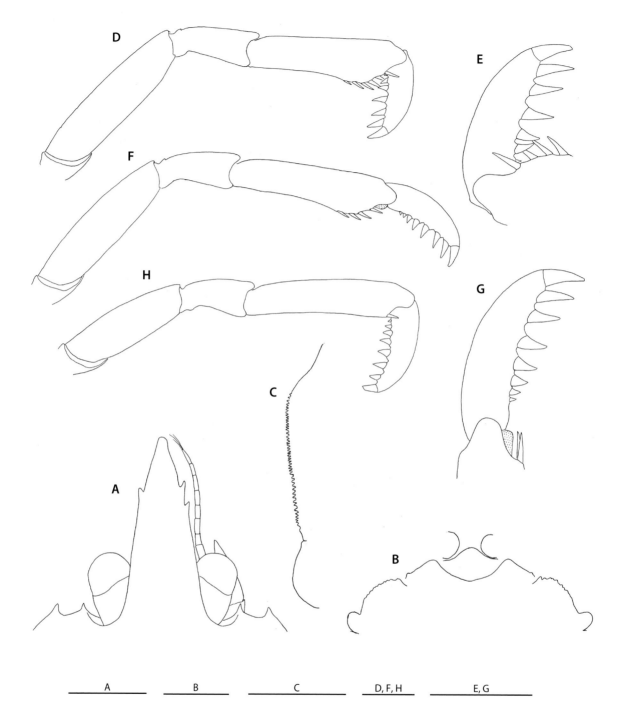

FIGURE 121

Uroptychus edisonicus Baba & Williams, 1998, holotype, ovigerous female 4.6 mm, USNM 251479. **A**, anterior part of carapace, dorsal. **B**, anterior part of sternal plastron, excavated sternum and basal parts of Mxp1 included. **C**, left Mxp3, crista dentata, ventral. **D**, right P2, setae omitted, lateral. **E**, same, distal part, setae omitted. **F**, right P3, lateral. **G**, same, distal part, lateral. **H**, right P4, lateral. Scale bars: A, B, D-H, 1 mm; C, 0.5 mm.

carapace margin. Lateral orbital angle acuminate, located at level of anterolateral spine. Pterygostomian flap smooth on surface, anterior margin angular, produced to small spine.

Sternum: Excavated sternum with slightly convex anterior margin; surface with weak ridge in midline. Sternal plastron 0.7 × as long as broad, lateral extremities divergent posteriorly. Sternite 3 shallowly depressed; anterior margin deeply excavated in semicircular shape without median notch and spines. Sternite 4 with anterolateral margin anteriorly convex and denticulate, about 2 × longer than posterolateral margin. Anterolateral margin of sternite 5 anteriorly convex and denticulate, 1.5 × longer than posterolateral margin of sternite 4.

Abdomen: Somite 1 somewhat convex from anterior to posterior, without transverse ridge. Somite 2 tergite 2.8 × broader than long; pleural lateral margins concavely divergent posteriorly, posterolateral end blunt. Pleuron of somite 3 with blunt posterolateral margin. Telson 0.4 × as long as broad; posterior plate slightly longer than anterior plate, concave on posterior margin.

Eye: 1.5 × longer than broad, distally narrowed, not reaching midlength of rostrum; cornea not dilated, slightly shorter than remaining eyestalk.

Antennule and antenna: Ultimate article of antennule 3.8 × longer than high. Antennal peduncle overreaching cornea, reaching midlength of rostrum. Article 2 with small lateral spine. Antennal scale ending in blunt tip, 2.2 × broader than article 5, extending far beyond eye and reaching end of proximal third segment of flagellum. Articles 4 and 5 each with tiny distomesial spine; article 5 1.3 × longer than article 5, breadth 0.6 × height of antennular ultimate article. Flagellum of 14 segments slightly falling short of distal end of P1 merus.

Mxp: Mxp1 with bases broadly separated. Mxp3 with relatively long plumose fine setae other than brushes on distal articles. Basis with a few obsolescent denticles on mesial ridge. Ischium with flexor margin not rounded distally; crista dentata with ca. 30 distally diminishing denticles. Merus 2.2 × longer than ischium, flexor margin roundly ridged, with 2 tiny spines or processes distal to point one-third from distal end; distolateral spine small on merus, obsolescent on carpus.

P1: 4.7 × longer than carapace, sparingly covered with relatively short fine plumose setae. Ischium with small dorsal spine, ventromesial margin with row of denticle-like spines, subterminal spine vestigial. Merus finely and sparsely granulose on surface, with small distomesial and distolateral spines on ventral surface, and row of denticles along distodorsal margin; length subequal to that of carapace. Carpus subcylindrical, 1.3 × longer than merus, distodorsal margin with row of denticles. Palm 4.1 × longer than broad, 1.1 × longer than carpus. Fingers moderately depressed and relatively broad, distally incurved, crossing when closed, not gaping; fixed finger with low eminence at midlength of opposable margin; movable finger half as long as palm, opposable margin with low proximal process fitting to longitudinal groove on opposable face of fixed finger when closed.

P2-4: Relatively broad and somewhat compressed mesio-laterally, with fine soft setae. Meri successively shorter posteriorly (P3 merus 0.9 × length of P2 merus, P4 merus 0.7 length of P3 merus); breadths greatest on P3, smaller and subequal in P2 and P4; length-breadth ratio, 4.1 on P2, 3.5 on P3, 2.7 on P4; dorsal margins rounded, not cristiform, with 2 tiny proximal spines on P2 and P3, smooth on P4. P2 merus 0.8 × as long as carapace, 1.1 × length of P2 propodus; P3 merus 0.9 × length of P3 propodus; P4 merus 0.7 × length of P4 propodus. Carpi successively shorter posteriorly; carpus-propodus length ratio, 0.6 on P2, 0.5 on P3, 0.4 on P4; extensor margins with small but distinct proximal spine on P2-4. Propodi longest on P3, shorter and subequal on P2 and P4; flexor margin strongly convex on distal portion, ending in pair of terminal spines preceded by 8, 7, 6 spines on P2, P3, P4 respectively. Dactyli shortest on P2, subequal on P3 and P4; moderately curving; dactylus-carpus length ratio, 0.8 on P2, 1.0 on P3, 1.2 on P4; dactylus-propodus length ratio, 0.4 on P2 and P3, 0.5 on P4; flexor margin with 8 strong, subtriangular spines more or less closely arranged, proximally diminishing, and obliquely directed on P2-4.

PARASITES — A rhizocephalan externa on the abdomen.

REMARKS — The combination of the following characters links the species to *U. edisonicus* Baba & Williams, 1998, *U. norfolkanus* n. sp. and *U. pedanomastigus* n. sp.: the carapace broader than long, bearing an anterolateral spine only on the lateral margin; sternite 3 with a deeply excavated, semicircular anterior margin; P1 almost spineless, with the fingers strongly incurved distally; the P2-4 propodi distally broadened, and dactyli ending in a strong spine preceded by similar but proximally diminishing spines. *Uroptychus latior* differs from *U. edisonicus* (Figure 121) in having the carapace much broader (length-breadth ratio 0.7 versus 0.8), in having the anterolateral spine of the carapace directed anteromesially instead of straight forward, in having the antennal scale extending far beyond instead of terminating at the tip of the antennal article 5 (extending far beyond instead of slightly overreaching the eyes), in having the pterygostomian flap anteriorly ending in a small spine instead of being produced to a larger spine, and in having the P2-4 dactyli with obliquely instead of perpendicularly directed flexor marginal spines. The relationships with *U. norfolkanus* and *U. pedanomastigus* are discussed under the accounts of the respective species (see below).

Uroptychus latirostris Yokoya, 1933
Figure 122

Uroptychus latirostris Yokoya, 1933: 69, figs 30. — Baba 2005: 39, figs 11-12.

TYPE MATERIAL —Neotype: **Japan**, Ashizuri-zaki, Tosa Bay, 150 m, male (ZLKU 12993). [not examined].

MATERIAL EXAMINED — **New Caledonia**, Hunter and Matthew Islands. VOLSMAR Stn DW39, 22°20′S, 168°44′E, 280-305 m, 08.VI.1989, 1 ♂ 3.3 mm (MNHN-IU-2014-16612). **New Caledonia**, Norfolk Ridge. SMIB 8 Stn DW183, 23°18′S, 168°05′E, 330-367 m, 31.I.1993, 1 ♂ 4.3 mm (MNHN-IU-2014-12791). – Stn DW190, 23°18′S, 168°05′E, 305-310 m, 31.I.1993, 3 ♂ 3.2-4.0 mm (MNHN-IU-2014-12792). **Vanuatu**. MUSORSTOM 8 Stn CP1017, 17°52.80′S, 168°26.20′E, 294-295 m, 27.IX.1994, 1 ov. ♀ 4.2 mm (MNHN-IU-2014-16613). **Solomon Islands**. SALOMON 2 Stn CP2179, 8°48.6′S, 159°43.3′E, 765-773 m, 22.X.2004, 1 ♂ 3.8 mm (MNHN-IU-2014-16614). **Indonesia**, Tanimbar Islands. KARUBAR Stn DW50, 7°59′S, 133°02′E, 184-186 m, 29.X.1991, 2 ov. ♀ 5.5, 5.6 mm, 1 ♀ 5.3 mm (MNHN-IU-2014-16615). **Philippines**. MUSORSTOM 2 Stn DR33, 13°32′N, 121°07′E, 130-137 m, with *Chrysogorgia* sp. (Calcaxonia, Chrysogorgiidae), 24.XI.1980, 3 ov. ♀ 4.2-6.0 mm, 1 ♀ 3.7 mm (MNHN-IU-2014-16616).

DISTRIBUTION — Previously known from Sagami Bay, Izu Islands, Bonin [Ogasawara] Islands, and Tosa Bay (Japan); 100-200 m. The present material is from the Philippines, Indonesia, Vanuatu, Hunter-Matthew Islands, and Norfolk Ridge, in 130-773 m.

SIZE — Males, 3.2-4.3 mm; females, 3.7-6.0 mm; ovigerous females from 4.2 mm.

DIAGNOSIS — Small to medium-sized species. Carapace as long as or slightly longer than broad; greatest breadth 1.3-1.4 × distance between anterolateral spines. Dorsal surface smooth and glabrous. Lateral margins slightly convex on posterior branchial region, bearing 3 spines: first anterolateral, reaching or slightly overreaching tip of smaller lateral orbital spine; second larger than first, placed on anterior end of anterior branchial region, followed by tiny third. Rostrum broad triangular, with interior angle of 49-51°; dorsal surface deeply concave; length about half postorbital carapace length, breadth three-quarters carapace breadth at posterior carapace margin. Lateral orbital spine directly mesial and close to anterolateral spine. Pterygostomian flap anteriorly roundish with small spine. Excavated sternum anteriorly produced to sharp subtriangular spine between close bases of Mxp1. Sternal plastron as long as broad; sternite 3 with deeply concave anterior margin without distinct submedian spines, occasionally with rudimentary spines; sternite 4 with anterolateral margin somewhat convex, about 4 × longer than posterolateral margin. Anterolateral margin of sternite 5 more than 5 × longer than posterolateral margin of sternite 4. Abdominal somites smooth, without transverse ridge on somite 1. Somite 2 tergite 2.3-2.5 × broader than long; pleuron posterolaterally strongly divergent, ending in blunt margin; pleuron of somite 3 posterolaterally blunt. Telson 0.6 × as long as broad; posterior plate semicircular or with concave posterior margin, length 1.7-1.9 × that of anterior plate. Eyes elongate, nearly reaching rostral tip; cornea half as long as remaining eyestalk. Ultimate article of antennule high relative to length, length 1.9-2.3 × height. Antennal peduncle terminating at distal end of cornea; article 2 with tiny lateral spine; antennal scale reaching or overreaching midlength of article 5, not reaching distal end of that article; article 5 with small distomesial spine, length 2.0-2.2 × that of article 4, breadth 0.3-0.5 that of antennular ultimate article; flagellum of 20-23 segments reaching or slightly overreaching distal end of P1 merus.

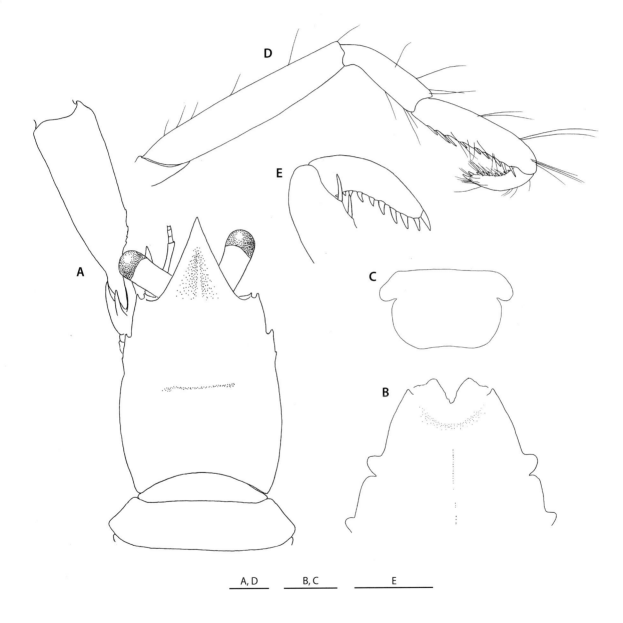

FIGURE 122

Uroptychus latirostris Yokoya, 1933, ovigerous female 4.2 mm (MNHN-IU-2014-16613). **A**, carapace and anterior part of abdomen, proximal part of left P1 included, dorsal. **B**, anterior part of sternal plastron. **C**, telson. **D**, right P2, lateral. **E**, same, distal part, setae omitted, lateral. Scale bars: 1 mm.

Mxp3 basis with a few denticles on mesial ridge, distalmost constant, others often obsolescent; ischium with 20-22 denticles on crista dentata, flexor margin distally not rounded; merus 2.0-2.3 × longer than ischium, P1 4.2-6.2 × longer than carapace, moderately massive, denticulate on proximal portion of merus, smooth elsewhere; ischium with long sharp dorsal spine, no spine elsewhere; merus 1.2-1.4 × longer than carapace; carpus 1.3-1.5 × length of merus, terminal margin with dorsal spine medially; palm 2.1-2.4 × (males), 3.2-4.9 × (females) longer than broad, varying from slightly shorter to slightly longer than carpus; fingers distally incurved, crossing when closed, with obtuse process on opposable margin of movable finger; movable finger 0.4-0.5 × length of palm. P2-4 relatively slender, with long, coarse setae on distal articles; meri successively shorter posteriorly (P3 merus 0.91-0.96 × length of P2 merus, P4 merus 0.84-0.85 × length of P3 merus), subequally broad on P2-4, length-breadth ratio, 5.6-6.0 on P2, 4.6-5.4 on P3, 4.2-4.3 on P4; P2 merus 1.0-1.3 × length of carapace, 1.6-1.7 × length of P2 propodus; P4 merus as long as or slightly longer than P4 propodus; carpi subequal; carpus-propodus length ratio, 0.6 on P2 and P3, 0.5-0.6 on P4; propodi having flexor margin terminating in pair of spines preceded by 8 spines on P2, 7 on P3, 6-7 on P4; dactyli shorter than carpi (dactylus-carpus length ratio, 0.6 on P2, 0.7 on P3, 0.7-0.8 on P4), 0.4 × as long as propodi on P2-4; flexor margin strongly curving at proximal third, with 9-10 successively diminishing subtriangular spines, ultimate largest.

Eggs. Number of eggs carried, up to 28; size, 1.28 mm × 1.06 mm - 1.24 mm × 1.04 mm.

REMARKS — The specimens examined agree well with the description of neotype (Baba 2005), except for the following: the posterior plate of the telson is slightly convex on the posterior margin in two of the three males from the Norfolk Ridge (MNHN-IU-2014-12792), whereas in the male from Vanuatu it is concave; the antennal scale is as in the neotype, terminating in the midlength of article 5 in the smallest male from the Norfolk Ridge, whereas in the other specimens it distinctly overreaches the midlength but barely reaches the distal end of that article. These differences are here considered as intraspecific variations. There still remains a problem with the systematic status of *Uroptychus cavirostris* Alcock and Anderson, 1899, which has been regarded to be different from *U. latirostris* in having the emarginate posterior margin of the telson. The material reported by Tirmizi (1964) from the western Indian Ocean, which was examined on loan from the Natural History Museum, London, is referable to an undescribed species, having a distally narrowed rostrum, in having a pair of median spines on the sternite 3 anterior margin, and in having an anterior branchial marginal spine vestigial rather than well developed. The Vanuatu material here examined occasionally has a concave posterior margin of telson just as defined for *U. cavirostris*. Access to the type material of *U. cavirostris* is now hardly possible, so discovery of topotypic material would solve the problem.

The elongate eyes, broad rostrum, spination of both the carapace lateral margin and P2-4 displayed by the present species are much like those of *U. alcocki* Ahyong & Poore, 2004 and *U. yokoyai* Ahyong & Poore, 2004. Their relationships are discussed under *U. yokoyai* (see below).

Uroptychus laurentae n. sp.

Figures 123, 124

TYPE MATERIAL — Holotype: **New Caledonia**, Chesterfield Islands. MUSORSTOM 5 Stn DW272, 24°41'S, 159°43'E, 500-540 m, 09.X.1986, 1 ♂ 2.9 mm (MNHN-IU-2014-16617).

ETYMOLOGY — The name is dedicated to the late Michéle de Saint Laurent who had shared the interest with me in squat lobsters.

DISTRIBUTION — Chesterfield Islands; 500-540 m.

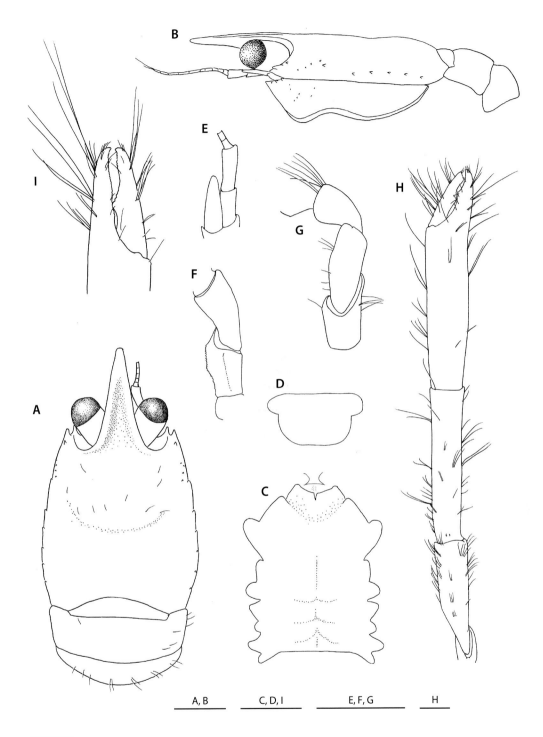

FIGURE 123

Uroptychus laurentae n. sp., holotype, male 2.9 mm (MNHN-IU-2014-16617). **A**, carapace and abdomen, dorsal. **B**, same, lateral. **C**, sternal plastron, with excavated sternum and basal parts of Mxp1. **D**, telson. **E**, right antenna, ventral. **F**, left Mxp3, setae omitted, ventral. **G**, same, lateral. **H**, right P1, dorsal. **I**, left P1, distal part, dorsal. Scale bars: 1 mm.

DESCRIPTION — Small species. *Carapace*: Slightly broader than long (0.9 × as long as broad); greatest breadth 1.5 × distance between anterolateral spines. Dorsal surface with scattered setae, moderately convex from side to side, nearly horizontal on gastric region, somewhat convex behind it, with shallow groove bordering gastric and cardiac regions. Lateral margins convex medially, with row of 7 small spines; first spine anterolateral, not reaching tip of lateral orbital spine, followed by small dorsal spine directly behind it; second spine rather remote from first, situated at anterior end of branchial margin, preceded by row of tubercle-like denticles placed somewhat dorsomesially; 5 other spines along posterior branchial region. Rostrum broad triangular, with interior angle of 20°, straight and horizontal; dorsal surface moderately concave; length 0.7 × postorbital carapace length, breadth somewhat more than half carapace breadth at posterior carapace margin. Lateral orbital spine slightly smaller than anterolateral spine, situated directly mesial and very slightly anterior to level of that spine. Pterygostomian flap with tubercles on anterior surface, anterior margin sharp angular and spiniform; dorsal margin anterior to linea anomurica with a few tubercle-like spinules.

Sternum: Excavated sternum with anterior margin nearly transverse, surface ridged in midline. Sternal plastron 1.2 × longer than broad, broadest on sternite 4, lateral extremities subparallel between sternites 5-7. Sternite 3 shallowly depressed, anterior margin shallowly excavated, with small V-shaped median notch lacking flanking spine. Sternite 4 having anterolateral margin nearly smooth, slightly convex, anteriorly blunt; posterolateral margin distinctly longer than anterolateral margin. Anterolateral margins of sternite 5 subparallel, anteriorly rounded, about half length of posterolateral margin of sterntie 4.

Abdomen: Smooth, sparsely setose. Somite 1 without transverse ridge. Somite 2 tergite 2.1 × broader than long; pleuron posterolaterally rounded, lateral margin shallowly concave and weakly divergent posteriorly. Pleuron of somite 3 posterolaterally blunt. Telson about half as long as broad, posterior plate 1.3 × longer than anterior plate, posterior margin feebly concave.

Eye: Relatively short (1.5 × longer than broad), slightly overreaching midlength of rostrum; lateral and mesial margins subparallel. Cornea not dilated, length slightly less than that of remaining eyestalk.

Antennule and antenna: Ultimate article of antennular peduncle 3.5 × longer than high. Antennal peduncle overreaching cornea by half length of article 5. Article 2 with distinct distolateral spine. Antennal scale slightly broader than article 5, ending in blunt tip, slightly falling short of midlength of article 5; lateral margin with small proximal spine on left appendage. Distal 2 articles unarmed; article 5 1.2 × longer than article 4, breadth slightly smaller than height of ultimate antennular article. Flagellum consisting of 10-11 segments, not reaching distal end of P1 merus.

Mxp: Mxp1 with bases broadly separated. Mxp3 basis with proximally rounded mesial ridge bearing no distinct denticle. Ischium with obsolescent denticles on crista dentata; flexor margin not rounded distally. Merus 2.6 × length of ischium, distolateral spine obsolescent; flexor margin not sharply cristate but moderately ridged and unarmed. Carpus spineless.

P1: Unequal in size, left smaller, presumably regenerated. Right P1 5.8 × longer than carapace. All articles sparsely setose as figured. Ischium dorsally with depressed process, ventrally with a few denticles on proximal mesial margin, without distinct subterminal spine. Merus and carpus subcylindrical and unarmed. Merus 1.3 × longer than carapace. Carpus 1.3 × longer than merus. Palm 3.5 × longer than broad, slightly longer than carpus. Fingers slightly gaping, distally somewhat incurved, ending in blunt short spine, not spooned; movable finger less than half length of palm, opposable margin with low median process fitting to narrow longitudinal groove on opposite fixed finger when closed; fixed finger directed somewhat laterally, opposable margin with low prominence slightly distal to opposite median process of movable finger.

P2-4: Sparsely setose. Meri broad relative to length, length-breadth ratio, 3.1 on P2, 2.8 on P3, 2.6 on P4, successively shorter posteriorly (P3 merus 0.9 × length of P2 merus, P4 merus 0.9 × length of P3 merus), equally broad on P2-4; dorsal margin with several eminences in proximal half. P2 merus three-quarters length of carapace, as long as P2 propodus; P3 merus 0.8 × length of P3 propodus; P4 merus 0.7 × length of P4 propodus. Carpi short relative to breadth, successively slightly shorter posteriorly, length one-third that of propodus on P2 and P3, one-fourth on P4. Propodi subequal on P2 and P3, slightly longer on P4; flexor margin with paired terminal spines only. Dactyli subequal on P2 and P3, slightly longer on P4; length 2.3-2.4 × breadth (breadth measured at base); dactylus-carpus length ratio, 1.4 on P2, 1.5 on P3, 1.7 on P4, and 0.5 × length of propodi on P2, slightly less than so on P4; relatively stout, slightly curving distally; flexor margin with

6 spines on P2, 6 or 7 on P3, 7 on P4, ultimate spine very slender and short, obscured by fine setae, very close or nearly contiguous to very broad penultimate spine, remaining spines slender, loosely arranged and obliquely directed; penultimate spine 9 × broader than ultimate, 5 × broader than antepenultimate.

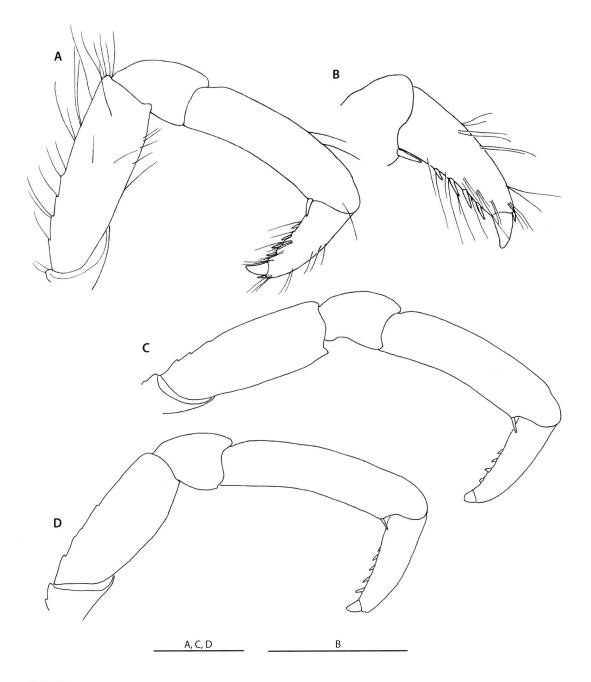

A, C, D   B

FIGURE 124

Uroptychus laurentae n. sp., holotype, male 2.9 mm (MNHN-IU-2014-16617). **A**, right P2, lateral. **B**, same, distal part, lateral. **C**, right P3, setae omitted, lateral. **D**, right P4, setae omitted, lateral. Scale bars: 1 mm.

REMARKS — The species is similar to *U. paenultimus* Baba, 2005 and *U. volsmar* n. sp. in having the anterolateral spine of the carapace very close to the lateral orbital spine, in having a short antennal scale, in having sternite 3 with the posterolateral margin relatively long, and in having the P2-4 dactyli with the penultimate spine prominent and preceded by slender spines. *Uroptychus laurentae* is differentiated from *U. paenultimus* by: the rostrum that is much longer instead of slightly shorter than broad; article 4 of the antennal peduncle that is unarmed instead of bearing a distinct distomesial spine; sternite 4 that has the posterolateral margin distinctly longer instead of slightly shorter than the anterolateral margin; the P2-4 dactyli that are shorter relative to breadth (the length-breadth ratio 2.3-2.4 versus 3.1-3.4) and that bear the penultimate spine much greater, being 9 times instead of 5 times as long as the ultimate spine. Differences between *U. laurentae* and *U. volsmar* are mentioned under the remarks of the latter (see below).

The new species also resembles *U. longicheles* Ahyong & Poore, 2004 in having the anterolateral spine of the carapace close to the lateral orbital spine and in the spination of the P2-4 dactyli. However, *U. laurentae* is distinctive in having the branchial lateral margin with 6 instead of 4 or 5 spines, in having the rostrum 1.4 times longer instead of slightly shorter than broad, in having the anterolateral spine of the carapace subequal to instead of smaller than the lateral orbital spine, in having sternite 4 with the posterolateral margin distinctly longer than instead of subequally long as the anterolateral margin, and in having the antennal article 2 unarmed instead of bearing a distolateral spine.

Uroptychus litosus Ahyong & Poore, 2004

Figure 125

Uroptychus litosus Ahyong & Poore, 2004: 52, fig. 14. — Poore 2004: 226, figs 60f, 62g (compilation). — Baba 2005: 227 (synonymies, key).

TYPE MATERIAL — Holotype: **Australia**, "Andys" Seamount, 65.5 km SSE of SE Cape, 44°10.8′S, 147°00.0′E, 800 m, male (NMV J52862). [not examined].

MATERIAL EXAMINED — **Wallis and Futuna Islands**. MUSORSTOM 7 Stn CP638, 13°37′S, 179°56′E, 820-840 m, 30.V.1992, 1 ♂ 9.3 mm (MNHN-IU-2014-16618), 1 ♂ 7.9 mm, 1 ov. ♀ 9.3 mm (MNHN-IU-2014-16619). **Solomon Islands**. SALOMON 2 Stn CP2297, 9°08.8′S, 158°16.0′E, 728-777 m, 8.XI.2004, 1 sp. (sex indet.) 7.6 mm (MNHN-IU-2014-16620).

DISTRIBUTION — Southern Tasmania, and now Wallis and Futuna Islands and Solomon Islands; 728-1120 m.

DESCRIPTION — Medium-sized species. *Carapace*: as long as broad; greatest breadth 1.7-1.8 × distance between anterolateral spines. Dorsal surface smooth and glabrous, somewhat convex from anterior to posterior, with shallow depression between gastric and cardiac regions. Epigastric region with pair of tuberculate ridges behind eyes. Lateral margins convexly divergent, bearing tubercles; low ridge along posterior fourth; anterolateral spine well developed, overreaching lateral orbital spine. Rostrum sharp triangular, with interior angle of 25-30°, horizontal, dorsal surface flattish; length 0.4 × postorbital carapace length, breadth about half carapace breadth at posterior carapace margin. Lateral orbital spine small, moderately remote from and anterior to level of anterolateral spine. Pterygostomian flap anteriorly angular, produced to distinct spine, surface smooth and glabrous.

Sternum: Excavated sternum sharply produced between bases of Mxp1, surface with spine in center. Sternal plastron 0.9 × as long as broad, sternites successively broader posteriorly. Sternite 3 depressed well, anterior margin in broad V-shape with 2 submedian spines separated by shallow, narrow notch; anterolaterally sharp angular. Sternite 4 with anterolateral margin convex, irregular and somewhat crenulated, anteriorly produced to blunt spine reaching submedian spines of sternite 3; posterolateral margin half as long as anterolateral margin. Anterolateral margin of sternite 5 convex, slightly longer than posterolateral margin of sternite 4.

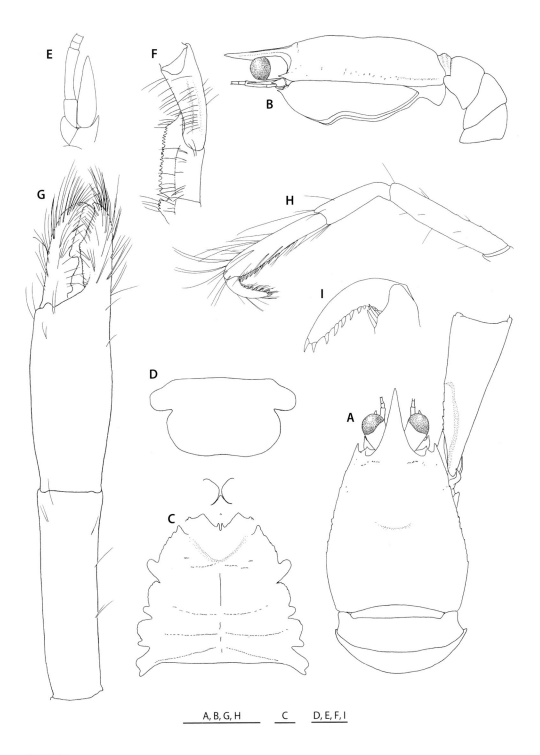

A, B, G, H   C   D, E, F, I

FIGURE 125

Uroptychus litosus Ahyong & Poore, 2004, male 9.3 mm (MNHN-IU-2014-16618). **A**, carapace and anterior part of abdomen, proximal part of right P1 included, dorsal. **B**, same, lateral. **C**, sternal plastron, with excavated sternum and basal parts of Mxp1. **D**, telson. **E**, left antenna, ventral. **F**, left Mxp3, ventral. **G**, right P1, proximal articles omitted, dorsal. **H**, left P2, lateral. **I**, same, distal part, setae omitted, lateral. Scale bars: A, B, G, H, 5 mm; C- F, I, 1 mm.

Abdomen: Tergites smooth and glabrous. Somite 1 pronouncedly convex from anterior to posterior. Somite 2 tergite 2.5-2.8 × broader than long; pleuron with concavely divergent lateral margin, bluntly produced posterolaterally. Pleuron of somite 3 with bluntly angular lateral margin. Telson 0.5-0.6 × as long as broad; posterior plate 1.3-1.5 × (males), 1.7 (female) longer than anterior plate, posterior margin somewhat concave.

Eye: Short relative to breadth (1.5 × longer than broad), overreaching midlength of, but not reaching distal end of rostrum, mesial margin concave. Cornea slightly dilated, more than half as long as remaining eyestalk.

Antennule and antenna: Ultimate article of antennular peduncle 2.3-2.8 × longer than high. Antennal peduncle overreaching cornea. Article 2 with strong lateral spine. Antennal scale slightly falling short of distal end of article 5, slightly overreaching cornea, breadth slightly more than 2 × that of article 5. Distal 2 articles unarmed. Article 5 1.8-2.3 × longer than article 4, breadth 0.4-0.5 × height of antennular ultimate article. Flagellum consisting of 18-20 segments, slightly falling short of or reaching distal end of P1 merus.

Mxp: Mxp1 with bases close to each other. Mxp3 basis with 4 denticles on mesial ridge. Ischium with about 20 denticles on crista dentata, flexor margin distally not rounded. Merus 2 × longer than ischium, unarmed, flexor margin not cristate but roundly ridged. Carpus also unarmed.

P1: Missing in female. Male P1 3.8-4.8 × longer than carapace, massive, barely setose on merus, sparsely so on carpus and palm, thickly so on fingers. Ischium with procurved dorsal spine, ventrally with a few denticle-like small spines on mesial margin, lacking subterminal spine. Merus ventrally granulose, with or without row of tubercle-like spines on mesial and ventromesial margins; length 1.0-1.2 × that of carapace. Carpus 1.2-1.3 × longer than merus. Palm rounded on mesial margin, 2.0 × longer than broad, 0.8-0.9 × length of carpus. Fingers distally incurved, crossing when closed; movable finger 0.6-0.7 × length of palm, opposable margin with 2-toothed process on proximal portion; opposable margin of fixed finger with low angular prominence distal to midlength.

P2-4: With setae sparse on meri and carpi, numerous on propodi and dactyli; setae on distal parts of carpi and propodi long, those on dactyli short. Meri successively shorter posteriorly (P3 merus 0.9 × length of P2 merus, P4 merus 0.8 × length of P3 merus); P2-3 meri subequal in breadth, P4 merus 0.9 × as broad as P3 merus; dorsal margins with obsolescent eminences on proximal half on P2, smooth on P3 and P4; length-breadth ratio, 4.8-5.5 on P2, 4.6-5.0 on P3, 4.0-4.2 on P4; P2 merus 0.8-1.0 × length of carapace, 1.1-1.2 × length of P2 propodus; P3 merus 1.1 × length of P3 propodus; P4 merus 0.9 × length of P4 propodus. Carpi subequal in length on P2 and P3 or slightly shorter on P3 than on P2; P4 carpus 0.9 × length of P3 carpus; carpus-propodus length ratio 0.6 on P2, 0.5-0.6 on P3 and P4. Propodi subequal in length on P2 and P3, slightly shorter on P4; flexor margin somewhat convex distally, with pair of terminal spines preceded by row of 7 or 8 movable slender spines on proximal half on P2, 6-7 spines on P3, 5-7 spines on P4. Dactyli successively longer posteriorly, much shorter than carpi (dactylus-carpus length ratio, 0.6 on P2, 0.6-0.7 on P3, 0.7-0.8 on P4), slightly more than one-third length of propodi (0.36-0.37 on P2, 0.35-0.38 on P3, 0.36-0.40 on P4); flexor margin with 10-12 triangular spines somewhat obliquely directed, loosely arranged and diminishing toward base of article, ultimate, penultimate and antepenultimate subequal.

Eggs. Number of eggs carried, 4 (normal number probably more); size, 1.5 mm × 1.73 mm - 1.58 mm × 1.85 mm.

REMARKS — This species has convexly divergent, spineless (other than the anterolateral spine) lateral margins of the carapace, relatively massive P1, relatively long P2-4 carpi, and proximally diminishing triangular spines on the flexor margin of the P2-4 dactyli, characters shared by *U. acostalis* Baba, 1988, *U. bardi* McCallum & Poore, 2013, and *U. orientalis* Baba & Lin, 2008. *Uroptychus litosus* differs from *U. acostalis* in having a pair of denticulate ridges on the epigastric region, in having sternite 4 with the anterolateral angle not so strongly produced as to reach the anterior margin of sternite 3, and in having the P1 ischium with a strongly procurved instead of very short triangular spine on the dorsal side. Distinctions between *U. litosus* and *U. bardi* are noted under the remarks of that species (see above). The relationships with *U. orientalis* have been discussed by Baba & Lin (2009).

The species also resembles *U. anacaena* Baba & Lin, 2008 in the spination of the P2-4 dactyli and in the carapace shape, but it differs in having the carapace dorsal surface smooth except for a pair of tuberculate epigastric ridges instead of being totally granulose, in having the anterolateral margin of sternite irregular rather than smooth, and in having the P2-4 dactyli much shorter than instead of equally long as the carpi.

Uroptychus longicarpus n. sp.

Figures 126, 127

TYPE MATERIAL — Holotype: **New Caledonia**, Norfolk Ridge. MUSORSTOM 4 Stn CP215, 22°55.7'S, 167°17.0'E, 485-520 m, 28.IX.1985, ♂ 7.5 mm (MNHN-IU-2014-16621). Paratypes: **New Caledonia**, Norfolk Ridge. MUSORSTOM 4, collected with holotype, 4 ♂ 5.4-8.0 mm, 7 ov. ♀ 7.1-7.3 mm, 1 ♀ 7.0 mm (MNHN-IU-2014-16622). – Stn CP214, 22°53.8'S, 167°13.9'E, 425-440 m, 28.IX.1985, 2 ♂ 6.8, 7.3 mm, 2 ov. ♀ 6.7, 9.0 mm (MNHN-IU-2014-16623). – Stn DW229, 22°51.5'S, 167°13.5'E, 445-460 m, 30.IX.1985, 4 ♂ 4.6-7.1 mm, 4 ov. ♀ 5.5-7.2 mm, 2 ♀ 4.3, 4.9 mm (MNHN-IU-2014-16624). NORFOLK 1 Stn DW1733, 22°56'S, 167°15'E, 427-433 m, 28.VI.2001, 3 ♂ 5.4-7.7 mm, 6 ov. ♀ 5.8-7.6 mm (MNHN-IU-2014-16625). – Stn DW1734, 22°53'S, 167°12'E, 403-429 m, 28.VI.2001, 1 ♂ 5.9 mm, 4 ov. ♀ 5.5-6.3 mm (MNHN-IU-2014-16626). NORFOLK 2 Stn CP2146, 22°50.17'S, 167°17.35'E, 518-518 m, 4.XI.2003, 2 ov. ♀ 7.0, 7.5 mm (MNHN-IU-2014-16627). – Stn DW2147, 22°49.80'S, 167°16.09'E, 496 m, 4.XI.2003, 8 ♂ 4.4-6.2 mm, 2 ov. ♀ 5.7, 6.0 mm (MNHN-IU-2014-16628). – Stn DW2148, 22°44.20'S, 167°15.97'E, 386-391 m, 4.XI.2003, 1 ov. ♀ 6.7 mm (MNHN-IU-2014-16629). – Stn DW2155, 22°52.40'S, 167°13.46'E, 453-455 m, 5.XI.2003, 2 ♂ 5.3, 5.9 mm, 1 ov. ♀ 5.7 mm (MNHN-IU-2014-16630). – Stn DW2156, 22°54.19'S, 167°15.13'E, 468-500 m, 5.XI.2003, 5 ♂ 5.4-7.3 mm, 3 ♀ 4.0-5.5 mm (MNHN-IU-2014-16631). **New Caledonia**. MUSORSTOM 4 Stn DW162, 18°35'S, 163°10'E, 535 m, 16.IX.1985, 1 ♂ 6.5 mm (MNHN-IU-2014-12793).

OTHER MATERIAL EXAMINED — Vanuatu. MUSORSTOM 8 Stn CP1026, 17°50'S, 168°39'E, 437-504 m, 28.IX.1994, 1 ♂ 6.7 mm, 1 ov. ♀ 6.8 mm, 1 ♀ 7.0 mm (MNHN-IU-2014-16632). – Stn CP1136, 15°41'S, 167°02'E, 398-400 m, 11.X.1994, 1 ♀ 7.0 mm (MNHN-IU-2014-16633). – Stn CP975, 19°23.60'S, 169°28.93'E, 566-536 m, 22.IX.1994, 1 ♂ 5.5 mm, 1 ov. ♀ 6.3 mm (MNHN-IU-2014-16634). – Stn CP975, 19°24'S, 169°29'E, 566-536 m, 22.IX.1994, 1 ♀ 5.7 mm (MNHN-IU-2014-16635). – Stn CP1088, 15°09'S, 167°15'E, 425-455 m, 6.X.1994, 1 ♀ 7.3 mm (MNHN-IU-2014-16636). New Caledonia, Loyalty Ridge. MUSORSTOM 6 Stn CP464, 21°01'S, 167°32'E, 430-420 m, 21.II.1989, 2 ov. ♀ 7.4, 7.6 mm (MNHN-IU-2014-16637). – Stn DW478, 21°08.96'S, 167°54.28'E, 400 m, 22.II.1989, 1 ov. ♀ 6.7 mm (MNHN-IU-2014-16638). New Caledonia, Isle of Pines. 500 m, 12.V.1978, coll. Intes, 1 ov. ♀ 6.9 mm (MNHN-IU-2014-16639). New Caledonia, Norfolk Ridge. BATHUS 2 Stn DW719, 22°48'S, 167°15'E, 444-445 m, 11 .V.1993, 1 ♀ 5.6 mm (MNHN-IU-2014-16640). SMIB 4 Stn DW68, 22°55'S, 167°16'E, 430-440 m, 10.III.1989, 1 ov. ♀ 6.9 mm (MNHN-IU-2014-16641). BIOCAL Stn DW44, 22°47'S, 167°14'E, 440-450 m, 30.VIII.1985, 3 ♂ 5.3-5.9 mm, 4 ov. ♀ 4.9-6.8 mm (MNHN-IU-2014-16642). – Stn CP45, 22°47'S, 167°15'E, 430-465 m, 30.VIII.1985, 3 ♂ 6.0-7.3 mm, 1 ov. ♀ 6.2 mm, 1 ♀ 5.7 mm (MNHN-IU-2014-16643). SMIB 2 Stn DW03, 22°54'S, 167°14'E, 412-428 m, 17.IX.1986, 3 ♂ 5.8-7.1 mm, 3 ov. ♀ 5.8-6.2 mm (MNHN-IU-2014-16644). – Stn DW06, 22°55'S, 167°16'E, 442-460 m, 17.IX.1986, 1 ♂ 7.8 mm, 1 ov. ♀ 7.3 mm, 2 ♀ 4.5, 4.6 mm (MNHN-IU-2014-16645). – Stn DW08, 22°53'S, 167°14'E, 435-447 m, 18.IX.1986, 2 ♂ 5.9, 6.7 mm, 1 ♀ 6.2 mm (MNHN-IU-2014-16646). – Stn DW09, 22°55'S, 167°16'E, 475-500 m, 18.IX.1986, 2 ♂ 5.5, 5.7 mm, 1 ov. ♀ 6.6 mm (MNHN-IU-2014-16647). – Stn DW10, 22°55'S, 167°16'E, 490-495 m, 18.IX.1986, 1 ov. ♀ 6.6 mm (MNHN-IU-2014-16648). – Stn DW13, 22°52'S, 167°13'E, 427-454 m, 18.IX.1986, 1 ♂ 6.3 mm (MNHN-IU-2014-16649). – Stn DW14, 22°53'S, 167°13'E, 405-444 m, 18.IX.1986, 2 ♂ 5.6, 7.5 mm, 1 ♀ 4.8 mm (MNHN-IU-2014-16650). SMIB 3 Stn DW21, 23°00'S, 167°19'E, 525-525 m, 24 .V.1987, 1 ♂ 6.2 mm, 1 ov. ♀ 7.3 mm (MNHN-IU-2014-16651). – Stn DW22, 23°03'S, 167°19.1'E, 503 m, 24.V.1987, 2 ♂ 4.7, 6.4 mm, 2 ov. ♀ 6.4, 7.0 mm (MNHN-IU-2014-16652). – Stn DW26, 22°55'S, 167°16'E, 450 m, 24.V.1987, 1 ♂ 5.7 mm (MNHN-IU-2014-16653).

ETYMOLOGY — From the Latin *longus* (long) plus *carpus* (carpus, hand), referring to the P4 carpus, which is longer than the P3 carpus, a distinctive character to separate the new species from congeners.

DISTRIBUTION — Vanuatu, Loyalty Islands, New Caledonia, Isle of Pines, and Norfolk Ridge; 386-525 m.

SIZE — Males, 4.4-8.0 mm; females, 4.0-9.0 mm; ovigerous females from 4.9 mm.

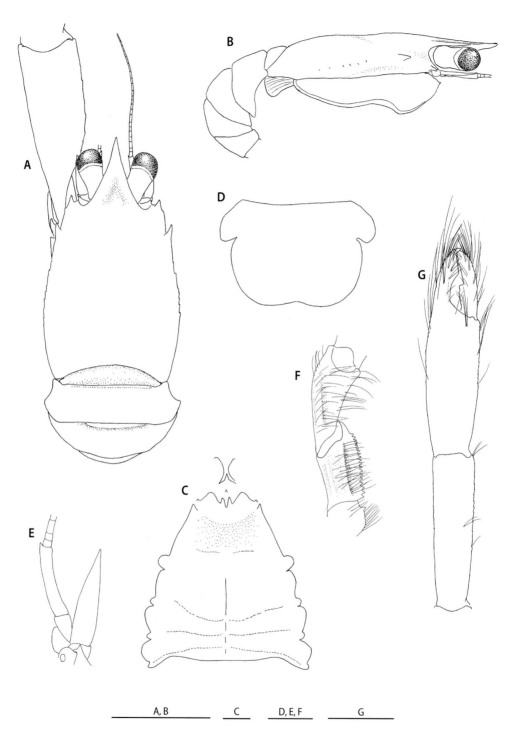

FIGURE 126

Uroptychus longicarpus n. sp., holotype, male 7.5 mm (MNHN-IU-2014-16621). **A**, carapace and anterior part of abdomen, proximal part of left P1 included, dorsal. **B**, same, lateral. **C**, sternal plastron, with excavated sternum and basal parts of Mxp1. **D**, telson. **E**, left antenna, ventral. **F**, right Mxp3, ventral. **G**, left P1, basal articles omitted, dorsal. Scale bars: A, B, G, 5 mm; C, D, E, F, 1 mm.

DESCRIPTION — Medium-sized species. *Carapace*: As long as broad; greatest breadth 1.5 × distance between anterolateral spines. Dorsal surface smooth and glabrous, slightly convex in profile, with feeble depression between gastric and cardiac regions. Lateral margins somewhat convex, bearing 2 distinct spines: first anterolateral spine reaching or slightly overreaching lateral orbital spine; second larger than first, situated at anterior end of branchial margin, followed by row of several tubercles or denticles reaching between midlength and posterior third of branchial margin. Rostrum broad triangular, with interior angle of 40-42°, nearly horizontal; dorsal surface concave; length 0.4-0.5 × postorbital carapace length, breadth slightly more than half carapace breadth at posterior carapace margin. Lateral orbital spine much smaller than anterolateral spine, situated slightly anterior to level of that spine. Pterygostomian flap anteriorly roundish, with very small spine; surface smooth.

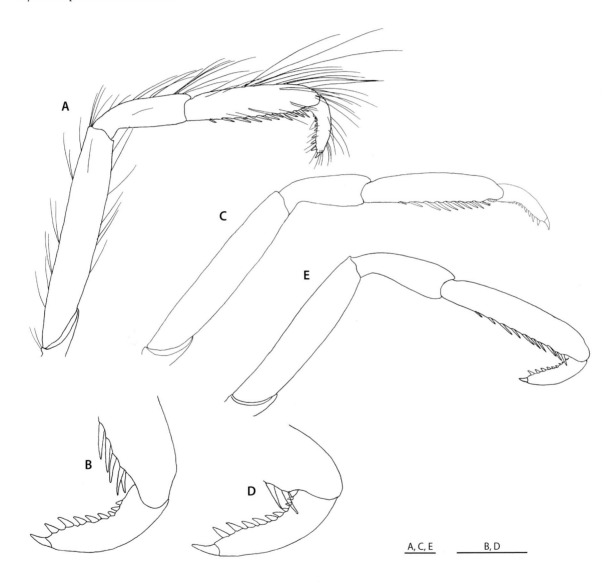

FIGURE 127

Uroptychus longicarpus n. sp., holotype, male 7.5 mm (MNHN-IU-2014-16621). **A**, right P2, lateral. **B**, same, distal part, setae omitted, lateral. **C**, right P3, setae omitted, lateral. **D**, same, distal part, lateral. **E**, right P4, lateral. Scale bars: 1 mm.

Sternum: Excavated sternum anteriorly ending in sharp spine, surface with small, laterally compressed spine or process in center. Sternal plastron as long as broad, lateral extremities gently divergent posteriorly. Sternite 3 depressed well; anterior margin deeply excavated, with 2 well-developed submedian spines nearly contiguous or separated by narrow V-shaped notch, anterolateral angle rounded. Sternite 4 having relatively long anterolateral margin slightly convex, anteriorly produced to spine of moderate-size, 3-4 × longer than posterolateral margin. Anterolateral margin of sternite 5 convexly divergent posteriorly, much longer than posterolateral margin of sternite 4.

Abdomen: Tergites smooth and glabrous. Somite 1 gently convex from anterior to posterior. Somite 2 tergite 2.4-2.7 × broader than long; pleural lateral margins strongly divergent posteriorly, with bluntly angular terminus. Pleuron of somite 3 laterally angular. Telson 0.6 × as long as broad; posterior plate 1.7-1.9 × (rarely 1.4 x) longer than anterior plate, posterior margin slightly or moderately emarginate.

Eye: Elongate, 2.0-2.4 × longer than broad, mesial and lateral margins concave, distally somewhat broadened, falling short of rostral tip. Cornea dilated, slightly more than half as long as remaining eyestalk.

Antennule and antenna: Ultimate article of antennular peduncle 2.0-2.4 × longer than high. Antennal peduncle barely reaching distal end of cornea. Article 2 with distolateral spine usually very small, occasionally barely discernible. Antennal scale reaching or slightly falling short of distal end of article 5, breadth about twice that of article 5. Article 4 unarmed. Article 5 2.4-3.2 × longer than article 4, with small distomesial spine, breadth 0.4-0.5 × height of ultimate antennular article. Flagellum of 16-23 segments reaching distal end of P1 merus.

Mxp: Mxp1 with bases close to each other. Mxp3 basis with 2 denticles on mesial ridge, distal one distinct, proximal one obsolescent and often absent. Ischium with 18-21 distally diminishing denticles on crista dentata, flexor margin distally not rounded. Merus 2 × as long as ischium, flexor margin sharply ridged, unarmed, with row of long setae; mesial face flattish. Carpus unarmed.

P1: Somewhat massive in males, nearly glabrous except for distal articles, length 4.5-5.3 × that of carapace. Ischium with strong sharp dorsal spine, unarmed elsewhere. Merus 1.1-1.2 × longer than carapace, granulose on ventral surface, lateral and mesial distoventral spines obsolete. Carpus denticulate ventrally, 1.0-1.3 × longer than merus. Palm 2.4-2.7 × (males), 2.7-2.9 × (females) longer than broad, 0.9-1.0 × length of carpus, breadth slightly more than distance between lateral orbital spines in large males; with fine tubercles on ventral surface in large males (>6.3 mm), smooth in females and small males. Fingers distally incurved, crossing when closed; movable finger 0.5-0.6 × length of palm; in males, moderately or largely gaping, fitting to each other on distal third; opposable margin with proximo-distally very broad eminence on gaping margin, opposable margin of fixed finger largely concave on proximal two-thirds; in females, fingers not gaping, movable finger with broad proximal process proximal to opposite eminence on fixed finger.

P2-4: Relatively slender, compressed mesio-laterally, with sparse long setae. Meri equally broad on P2-4; length-breadth ratio, 5.3-6.5 on P2, 5.2-6.4 on P3, 5.2-6.2 on P4; P3 merus 0.9 × length of P2 merus, P4 merus 0.9-1.0 × length of P3 merus; dorsal margins unarmed, ventrolaterally with obsolescent distal spine; P2 merus 0.90-0.95 × length of carapace, 1.5-1.6 × length of P2 propodus; P3 merus 1.3 × length of P3 propodus; P4 merus 1.1-1.2 × length of P4 propodus, Carpi unarmed, subequal in length on P2 and P3, longest on P4 (P4 carpus 1.1-1.2 × longer than P3 carpus), much longer than dactyli (carpus-dactylus length ratio, 1.7-2.0 on P2, 1.5-1.9 on P3, 1.6-2.0 on P4); carpus-propodus length ratio 0.5-0.6 on P2-4. Propodi successively longer posteriorly; flexor margin nearly straight, ending in pair of spines very close to juncture with dactylus, preceded by row of 7-10 relatively long movable spines on P2, 7-9 spines on P3, 6-9 spines on P4 (on distal 5/6 on P2, slightly more distal on P3 and P4). Dactyli 0.3-0.4 × length of propodi and 0.5-0.6 length of carpi on P2-4; flexor margin curving at proximal third, ending in strong spine preceded by 7-8 subtriangular spines more or less closely arranged, somewhat inclined and diminishing toward base of article, ultimate spine distinctly larger and broader than penultimate.

Eggs. Number of eggs carried, ca. 30, 1.27 mm × 1.32 mm - 1.40 × 1.64 mm.

REMARKS — The second lateral marginal spine of the carapace is rarely reduced to small size (in the male from Vanuatu, MNHN-IU-2014-16636) or tubercle-like process (in the specimens from Vanuatu, MNHN-IU-2014-16632). In large specimens more than 9.0 mm long, additional spine is present directly mesial to the second spine, as also are a few similar spines between the first and second.

This new species strongly resembles *U. sibogae* Van Dam, 1933 (see below) and *U. nebulosus* n. sp. in nearly all aspects. As suggested by the species name, however, the P4 carpus in *U. longicarpus* is consistently longer than the P3 carpus, instead of being 0.8-0.9 times as long in the congeners.

In the key to species, the species is closely set to *U. politus* (Henderson, 1885), but it actually is rather distant in having a spine of good size at the anterior end of the branchial margin, which is reduced to an elevated ridge, in having sternite 5 with a well convex rather than feebly convex or nearly straight anterolateral margin, and most noticeably in having the P4 carpus longer than the P3 carpus.

Uroptychus longicheles Ahyong & Poore, 2004

Figure 128

Uroptychus longicheles Ahyong & Poore, 2004: 55, fig. 15.

TYPE MATERIAL — Holotype: **Australia**, Gifford Guyot, Tasman Sea, 26°44.27'S, 159°28.93'E, 306 m, ov. ♀ (AM P65826). [not examined].

MATERIAL EXAMINED — **New Caledonia**, Chesterfield Islands. MUSORSTOM 5 Stn DW256, 25°18.0'S, 159°52.70'E, 290-300 m, 7.X.1986, 1 ♀ 1.8 mm (MNHN-IU-2014-16654).

DISTRIBUTION — Tasman Sea, and now Chesterfield Islands; 290-306 m.

DIAGNOSIS — Small species. Carapace slightly broader than long (0.9 × as long as broad); greatest breadth 1.3 × distance between anterolateral spines. Dorsal surface with weak depression bordering gastric and cardiac regions; lateral portion of posterior branchial region with low projection around last lateral spine; lateral margins weakly convex, with 5 spines: first anterolateral, distinctly posterior to level of lateral orbital spine, never overreaching that spine, subequal to second and fifth; third ventral to level of second and fourth; fifth more dorsal to this level. Rostrum half as broad as posterior carapace margin. Lateral orbital spine distinctly larger than anterolateral spine. Pterygostomian flap anteriorly angular, produced to sharp spine; surface with small spines. Excavated sternum anteriorly ending in rounded margin, surface ridged in midline. Sternal plastron as long as broad, lateral extremities subparallel between sternites 4 and 7; sternite 3 with 2 tiny submedian spines separated by narrow notch; sternite 4 with anterolateral margin about as long as posterolateral margin. Anterolateral margins of sternite 5 slightly convex, length 0.8 × that of posterolateral margin of sternite 4. Abdominal somite 1 without transverse ridge. Somite 2 tergite 2.4 × broader than long, pleuron anterolaterally and posterolaterally rounded, lateral margins somewhat concave; pleuron of somite 3 posterolaterally blunt. Telson half as long as broad, posterior plate as long as anterior plate, posterior margin barely emarginate. Eyes 1.7 × longer than broad; lateral and mesial margins subparallel, cornea not dilated, more than half length of remaining eyestalk. Ultimate article of antennule 3.2 × longer than high. Antennal article 2 with distinct lateral spine; antennal scale terminating in midlength of article 5; article 4 with distomesial spine; article 5 as long as article 4, breath about half height of antennular ultimate article. Mxp1 with bases broadly separated. Mxp3 basis without denticles on mesial ridge; ischium with flexor margin not rounded distally, crista dentata with obsolescent denticles; merus 2.8 × longer than ischium, with distinct distolateral spine; carpus unarmed. P1 slender, about 7 × longer than carapace; ischium with sharp, triangular dorsal spine, ventromesial margin unarmed; merus 1.7 × longer than carapace; carpus 1.4 × longer than merus; palm 6 × longer than broad, slightly shorter than carpus; fingers one-third length of palm, distally slightly incurved. P2 merus as long as carapace, as well as P3 merus and P2 propodus; length-breadth ratio, 4.0 on P2 and P3 (P4 missing); carpi subequal on P2 and P3, less than one-third (0.28 on P2, 0.25 on P3) length of propodus, slightly more than half length of dactylus; flexor margin of propodus with pair of terminal spines only; dactyli 3 × longer than broad (breadth measured at base), 1.6 × longer than carpi, half length of propodi, flexor margin nearly straight, bearing 7 spines, ultimate slender, penultimate pronouncedly broad and long (fully 3 × broader than ultimate and antepenultimate), remaining spines slender, obliquely directed and loosely arranged.

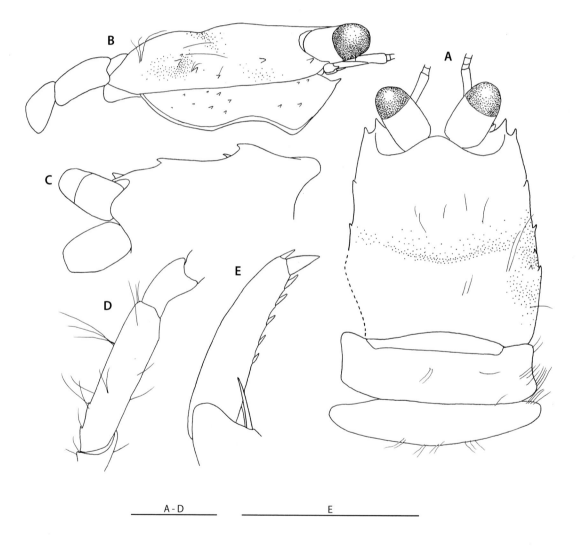

FIGURE 128

Uroptychus longicheles Ahyong & Poore, 2004, female 1.8 mm (MNHN-IU-2014-16654). **A**, carapace and anterior part of abdomen, dorsal. **B**, same, lateral. **C**, same, right half, dorsomesial. **D**, right P2, proximal part, lateral. **E**, same, distal part, setae omitted, lateral. Scale bars: 1 mm.

REMARKS — The rostrum in the present specimen is not normal, very short as displayed by the ovigerous female paratype (MNHN-IU-2014-16386) of *U. denticulisquama* n. sp. (see above). The P2-3 meri bear 2 small spines on the proximal portion of dorsal crest (P4 missing), a character not possessed by the holotype (rechecked by S. Ahyong, pers. comm). The other features agree very well with the original description.

The sternal plastron, antenna, P1 and P2 dactyli are nearly similar in *U. laurentae* n. sp., *U. longicheles*, *U. paenultimus* Baba, 2005 and *U. turgidus* n. sp. *Uroptychus longicheles* differs from *U. paenultimus* in having the carapace lateral margin with 4 or 5 distinct spines instead of 7 or 8 very small spines, in addition to the anterolateral spine of good size, and in having the posterior branchial region with a low dorsolateral projection instead of being smooth. The differences between *U. longicheles* and *U. laurentae* are outlined under the remarks of the latter (see above). Also, the reader is referred to the remarks of *U. turgidus* for its relationships with *U. longicheles* (see below).

Uroptychus longior Baba, 2005

Figures 129, 130

Uroptychus longior Baba, 2005: 43, fig. 14.

TYPE MATERIAL — Holotype: **Indonesia**, Bali Sea, 7°29'S, 114°49'E, 240 m, male (ZMUC CRU-11075). [not examined].

MATERIAL EXAMINED — **Philippines**. MUSORSTOM 2 Stn CP15, 13°55'N, 120°29'E, 326-330 m, 21.XI.980, 1 ov. ♀ 5.3 mm (MNHN-IU-2014-16655). **Indonesia**, Kai Islands. KARUBAR Stn CP05, 5°49'S, 132°18'E, 296-299 m, 22.X.1991, 1 ov. ♀ 6.4 mm (MNHN-IU-2014-16656). – Stn CP16, 05°17'S, 132°50'E, 315-349 m, 24.X.1991, 1 ♀ 5.8 mm (MNHN-IU-2014-16657), 1 ov. ♀ 6.9 mm, 1 ♀ 7.2 mm (MNHN-IU-2014-16658), 1 ♀ 3.1 mm (MNHN-IU-2014-16659). Tanimbar Islands. KARUBAR Stn CP45, 7°54'S, 132°47'E, 302-305 m, 29.X.1991, with coral Primnoidae (Calcaxonia), 1 ♂ 6.2 mm (MNHN-IU-2014-16660). **Solomon Islands**. SALOMON 1 Stn DW1788, 9°19.4'S, 160°15.4'E, 341-343 m, 30.IX.2001, 1 ♂ 4.7 mm (MNHN-IU-2014-16661). – Stn DW1826, 9°56.4'S, 161°04.0'E, 418-432 m, 4.X.2001, 1 ♀ 4.3 mm (MNHN-IU-2014-16662). **New Caledonia**, Wallis and Futuna Islands. MUSORSTOM 7 Stn DW537, 12°30'S, 176°41'W, 325-400 m, with corals of Primnoidae (Calcaxonia), 16.V.1992, 1 ♂ 4.9 mm (MNHN-IU-2014-16663). **Vanuatu**. MUSORSTOM 8 Stn DW1068, 16°16'S, 167°20'E, 536-619 m, 2.X.1994, 1 ♂ 4.6 mm (MNHN-IU-2014-16664). **New Caledonia**, Chesterfield Islands. MUSORSTOM 5 Stn DW355, 19°36'S, 158°43'E, 580 m, 18.X.1986, 1 ♀ 4.1 mm (MNHN-IU-2014-16665). **New Caledonia**. MUSORSTOM 4 Stn CP194, 18°52'S, 163°21'E, 550 m, 19.IX.1985, 1 ov. ♀ 6.1 mm (MNHN-IU-2014-16666). BATHUS 4 Stn CP930, 18°31.36'S, 163°23.63'E, 530-520 m, 7.VIII.1994, 1 ♂ 4.8 mm (MNHN-IU-2014-16667). BATHUS 2 CP766, 22°10'S, 166°01'E, 650-724 m, 17.V.1993, 1 ♂ 6.0 mm (MNHN-IU-2014-16668). **New Caledonia**, Norfolk Ridge. BIOCAL Stn DW36, 23°09'S, 167°11'E, 650-680 m, 29.VIII.1985, 1 ♀ 4.7 mm (MNHN-IU-2014-16669). CHALCAL 2 Stn CC01, 24°55'S, 168°22'E, 500-580 m, 28.X.1986, 1 ♂ 4.8 mm (MNHN-IU-2014-16670). – Stn DW72, 24°54.5'S, 168°22.3'E, 527 m, 28.X.1986, 1 ♀ 3.9 mm (MNHN-IU-2014-16671). – Stn DW73, 24°39.9'S, 168°38.1'E, 573 m, on coral Primnoidae (Calcaxonia), 29.X.1986, 1 ♂ 4.4 mm, 2 ov. ♀ 5.4, 5.5 mm (MNHN-IU-2014-16672). – Stn DW74, 24°40.36'S, 168°38.38'E, 650 m, 29.X.1986, 2 ♂ 2.9, 4.9 mm, 3 ov. ♀ 4.8-5.8 mm, 1 ♀ 5.2 mm (MNHN-IU-2014-16673). – Stn DW75, 24°39.31'S, 168°39.67'E, 600 m, 29.X.1986, 1 ov. ♀ 4.8 mm (MNHN-IU-2014-16674). NORFOLK 1 Stn DW1666, 23°42'S, 167°44'E, 469-860 m, 20.VI.2001, 1 ov. ♀ 4.4 mm (MNHN-IU-2014-16675). – Stn DW1694, 24°40'S, 168°39'E, 575-589 m, 24.VI.2001, 1 ♂ 5.5 mm (MNHN-IU-2014-16676). – Stn DW1697, 24°39'S, 168°38'E, 569-616 m, 24.VI.2001, 2 ♂ 6.2, 6.4 mm (MNHN-IU-2014-16677). – Stn DW1700, 24°40'S, 168°39'E, 605-752 m, 24.VI.2001, 1 ♂ 5.5 mm, 1 ov. ♀ 6.4 mm, 1 ♀ 3.6 mm (MNHN-IU-2014-16678). – Stn DW1707, 23°43'S, 168°16'E, 381-493 m, 25.VI.2001, 1 ov. ♀ 6.1 mm (MNHN-IU-2014-16679). NORFOLK 2 Stn DW2027, 23°26.34'S, 167°51.38'E, 465-650 m, 21.X.2003, 2 ov. ♀ 5.7, 6.2 mm (MNHN-IU-2014-16680). – Stn DW2049, 23°42.88'S, 168°15.43'E, 470-621 m, 24.X.2003, 1 ♂ 4.7 mm (MNHN-IU-2014-16681). – Stn DW2056, 24°40.32'S, 168°39.17'E, 573-600 m, 25.X.2003, 1 ov. ♀ 4.0 mm (MNHN-IU-2014-16682). – Stn DW2057, 24°40'S, 168°39'E, 555-565 m, 25.X.2003, 1 ♂ 4.7 mm, 3 ♀ 3.4-4.3 mm (MNHN-IU-2014-16683). – Stn DW2058, 24°39.76'S, 168°40.43'E, 591-1032 m, 25.X.2003, 2 ♂ 3.0, 5.1 mm, 1 ov. ♀ 4.1 mm (MNHN-IU-2014-16684). – Stn DW2060, 24°39.84'S, 168°38.50'E, 582-600 m, 25.X.2003, 1 ov. ♀ 5.3 mm (MNHN-IU-2014-16685). – Stn CP2061, 24°39.50'S, 168°40.32'E, 620-1040 m, 25.X.2003, 1 ♂ 5.5 mm, 1 ov. ♀ 4.1 mm (MNHN-IU-2014-16686). – Stn CP2062, 24°40.05'S, 168°39.70'E, 560-572 m, 25.X.2003, 1 ♂ 5.7 mm (MNHN-IU-2014-16687). – Stn DW2064, 25°17'S, 168°56'E, 609-691 m, 26.X.2003, 1 ♂ 5.3 mm, 2 ov. ♀ 4.1-4.9 mm (MNHN-IU-2014-16688). – Stn DW2075, 25°23.12'S, 168°20.07'E, 650-1000 m, 27.X.2003, 6 ♂ 4.2-5.8 mm, 6 ov. ♀ 5.3-5.8 mm, 1 ♀ 5.7 mm (MNHN-IU-2014-16689). – Stn DW2077, 25°20.63'S, 168°18.53'E, 666-1000 m, 27.X.2003, 1 ♂ 3.8 mm (MNHN-IU-2014-16690). – Stn CP2088, 24°57'S, 168°22'E, 627-1089 m, 28.X.2003, 2 ♂ 5.7, 5.8 mm, 2 ov. ♀ 5.7, 6.0 mm, 1 ♀ 4.4 mm (MNHN-IU-2014-16691).

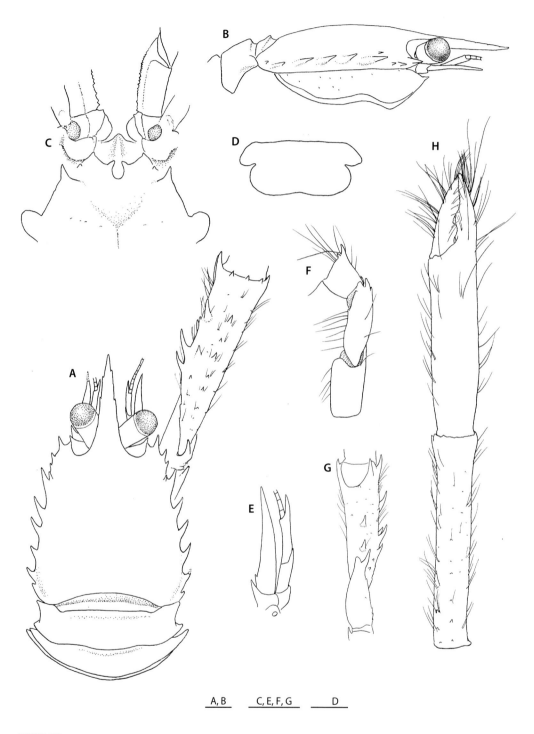

A, B   C, E, F, G   D

FIGURE 129

Uroptychus longior Baba, 2005, **A-C**, **E-H**, ovigerous female 4.8 mm (MNHN-IU-2014-16674); **D**, female 6.3 mm (MNHN-IU-2014-16692). **A**, carapace and anterior part of abdomen, proximal part of right P1 included, dorsal. **B**, same, lateral. **C**, anterior part of sternal plastron, with excavated sternum and proximal parts of Mxps, ventral. **D**, telson. **E**, right antenna, ventral. **F**, left Mxp3, lateral. **G**, right P1, proximal part, ventral. **H**, same, proximal articles omitted, dorsal. Scale bars: A-F, 1 mm; G, H, 5 mm.

– Stn CP2089, 24°44'S, 168°09'E, 227-230 m, 29.X.2003, 4 ♂ 3.7-5.3 mm, 2 ov. ♀ 5.1, 5.9 mm, 2 ♀ 4.8, 6.2 mm (MNHN-IU-2014-16692). – Stn DW2112, 23°44.44'S, 168°18.40'E, 640-1434 m, 31.X.2003, 1 ♂ 3.7 mm, 1 ov. ♀ 6.5 mm (MNHN-IU-2014-16693). – Stn CP2122, 23°21.55'S, 168°00.12'E, 560-577 m, 1.XI.2003, 1 ♂ 5.5 mm, 1 ov. ♀ 5.6 mm (MNHN-IU-2014-16694). BERYX 11 Stn DW10, 24°52.85'S, 168°21.40'E, 560-600 m, 15.X.1992, 2 ♂ 5.7, 6.2 mm, 3 ov. ♀ 4.6-6.8 mm, 1 ♀ 4.7 mm (MNHN-IU-2014-16695). **Tonga**. BORDAU 2 Stn DW1617, 23°03'S, 175°53'W, 483-531 m, 17.VI.2000, 1 ♀ 4.4 mm (MNHN-IU-2014-16696).

DISTRIBUTION — Kai Islands and Bali Sea, and now Philippines (between Lubang Islands and Luzon), Solomon Islands, Wallis-Futuna Islands, Vanuatu, Chesterfield Islands, New Caledonia, Norfolk Ridge, and Tonga; 227-1434 m.

SIZE — Males, 2.9-6.4 mm; females, 3.1-6.9 mm; ovigerous females from 4.0 mm.

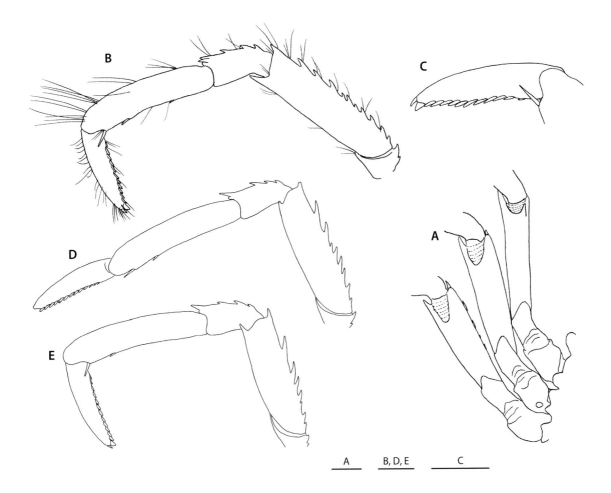

FIGURE 130

Uroptychus longior Baba, 2005, ovigerous female 4.8 mm (MNHN-IU-2014-16674). **A**, proximal parts of right P2-4, ventral. **B**, left P2, lateral. **C**, same, distal part, setae omitted, lateral. **D**, left P3, setae omitted, lateral. **E**, left P4, setae omitted, lateral. Scale bars: 1 mm.

DESCRIPTION — Medium-sized species. *Carapace*: Slightly broader than long (0.9 × as long as broad); greatest breadth 1.5 × distance between anterolateral spines. Dorsal surface nearly smooth, somewhat convex from anterior to posterior without any depression. Lateral margins somewhat divergent posteriorly, with 8 spines: first anterolateral, reaching or slightly overreaching smaller lateral orbital spine; second and third usually small, third occasionally obsolete, and both distinctly ventral to level of other spines; fourth to eighth well developed, acute, placed along branchial margin, rarely followed by additional small spine; another small spine consistently dorsomesial to fourth spine. Rostrum narrow triangular, with interior angle of 23-24° (rarely 17°); dorsal surface flattish or slightly concave, ventral surface straight horizontal; lateral margin with usually 3-5, occasionally 0-2 distal spinules; length 0.5-0.7 × that of remaining carapace, breadth slightly less than half carapace breadth at posterior carapace margin. Pterygostomian flap anteriorly angular, produced to sharp spine, surface with tiny denticle-like spines.

Sternum: Excavated sternum subtriangular between bases of Mxp1, ending in blunt tip, surface with somewhat ridged in midline. Sternal plastron about as long as broad, lateral extremities between sternites 4 and 7 somewhat divergent posteriorly. Sternite 3 shallowly depressed, anterior margin with semi-oval median sinus flanked by incurved spine on each side. Sternite 4 with smooth, nearly straight anterolateral margin anteriorly produced to pronounced spine directed anterolaterally, reaching at most base of submedian spines on sternite 3; posterolateral margin relatively long, slightly shorter than anterolateral margin. Anterolateral margin of sternite 5 slightly shorter than posterolateral margin of sternite 4.

Abdomen: Smooth and glabrous. Somite 1 with distinct transverse ridge. Somite 2 tergite 2.3-2.7 × broader than long; pleuron anterolaterally and posterolaterally blunt angular, lateral margin well concave. Pleuron of somite 3 laterally angular. Telson slightly less than half as long as broad; posterior plate moderately emarginate on posterior margin, length 1.2-1.4 × that of anterior plate.

Eye: 1.7-1.9 × longer than broad, usually overreaching, rarely terminating in midlength of rostrum; slightly broader distally. Cornea more than half as long as remaining eyestalk.

Antennule and antenna: Ultimate article of antennular peduncle 3.2-4.0 × longer than high. Antennal article 2 with strong lateral spine. Antennal scale about twice as broad as article 5, tapering, overreaching peduncle by more than half length of article 5, not reaching rostral tip, occasionally bearing 1 or 2 small spines on lateral proximal margin. Distal 2 articles each with strong distomesial spine; article 5 1.5-1.8 × longer than article 4, breadth 0.6-0.7 × height of ultimate antennular article. Flagellum consisting of 11-15 segments, not reaching distal end of P1 merus.

Mxp: Mxp1 with bases broadly separated. Mxp3 having lateral surface barely or sparsely setose. Basis with a few denticles on mesial ridge, distal one usually distinct, rarely obsolescent, others usually tiny or obsolescent. Ischium with 20-26 (rarely 30) denticles on crista dentata; flexor margin distally rounded. Merus 1.6 × as long as ischium, bearing distolateral spine and a few small spines on distal half of cristate flexor margin; mesial face flattish. Carpus with distinct distolateral spine often accompanying small spine, with or without another small proximal spine on extensor surface.

P1: 4.6-5.8 × longer than carapace, relatively slender, sparingly with soft setae, more setose on fingers. Ischium dorsally with strong spine, ventromesially with well-developed subterminal spine. Merus 1.1-1.4 × (females), 1.4 × (males) longer than carapace, with 4 rows of spines: 2 dorsal rows of distinct, often small spines, 1 dorsomesial row (distal spine and one about at midlength usually strong, and few other smaller), 1 ventral row (1 strong distal spine and 2 well-developed, occasionally small ventral spines in line with ventromesial subterminal spine of ischium). Carpus 1.2-1.4 × longer than merus, somewhat granulose, with distomesial and distolateral spines of small-size ventrally, and other small spines in 2 rows dorsally (1 in midline, another near mesial margin); all of these occasionally obsolescent. Palm 2.9-3.1 × (males), 2.8-4.9 × (females) longer than broad, 0.8-1.1 × length of carpus. Fingers fitting to each other in distal part when closed, not distinctly incurved, gaping in large males, occasionally slightly gaping in large females; movable finger about half length of palm, opposable margin with basally broad subtriangular or narrow process about at midlength in large males, low process at proximal third in females and small males; opposable face of fixed finger with proximal longitudinal groove to accommodate opposite process on movable finger when closed.

P2-4: Relatively slender and well compressed mesio-laterally; sparsely with soft setae. Meri successively shorter posteriorly (P3 merus 0.8-0.9 × length of P2 merus, P4 merus 0.9 × length of P3 merus), equally broad on P2-4; length-breadth ratio,

3.8-4.3 on P2, 3.4-3.8 on P3, 3.2-3.7 on P4; P2 merus 0.9 × length of carapace, 1.0-1.2 × length of P2 propodus; P3 merus 0.9 × length of P3 propodus; P4 merus 0.8-0.9 × length of P4 propodus; dorsal margin with row of spines; ventrolateral margin with well-developed terminal spine, rarely accompanying small spine proximal to it, ventromesial margin with terminal spine often obsolescent or absent in small specimens. Carpi subequal, shorter than dactyli, less than half length of propodi (carpus-propodus length ratio, 0.37-0.40 on P2, 0.33-0.36 on P3, 0.32-0.35 on P4), extensor margin with row of spines occasionally obsolescent on P4 (bearing distal spine only), distalmost spine accompanying additional spine lateral to it on P2. Propodi shorter on P2 than on P3 and P4; extensor margin usually unarmed; flexor margin ending in pair of spines preceded by 2 or 3 (rarely 4) spines on P2, 1 or 2 (rarely 4 or 5) on P3 and P4. Dactyli 0.5-0.6 × length of propodi and longer than carpi (dactylus-carpus length ratio, 1.1-1.4 × on P2, 1.2-1.4 on P3, 1.1-1.5 on P4); flexor margin nearly straight, bearing 12-14 spines, ultimate slender, penultimate about twice as broad as ultimate but not much longer than remainder, remaining spines also slender, obliquely directed, nearly contiguous to one another, antepenultimate slightly broader than ultimate.

Eggs. Number of eggs carried, 6-30 eggs (normal number probably more); size, 0.86 × 0.94 mm - 1.10 mm × 1.20 mm.

PARASITES — Non-ovigerous female (MNHN-IU-2014-16689) with externa of rhizocephalan parasite attached to antennular article 2.

REMARKS — The male specimen taken at NORFOLK 1 Stn DW1694 (MNHN-IU-2014-16676) has three small proximal spines on the extensor margin of P2 propodus but no other differences were found. The species is very similar to *U. alophus* n. sp. (see above for their relationships). It is also close to *Uroptychus nanophyes* McArdle, 1901. Their relationships were discussed in my earlier paper (Baba, 2005).

Uroptychus longvae Ahyong & Poore, 2004
Figure 131

Uroptychus longvae Ahyong & Poore, 2004: 58, fig. 16.

TYPE MATERIAL — Holotype: **Australia**, 342 km west of Cape Wiles, Great Australian Bight, South Australia, 34°56'S, 133°20'E, 805-816 m, ov. female (SAMC C6064). [not examined].

MATERIAL EXAMINED — **New Caledonia**, Norfolk Ridge. NORFOLK 2 Stn DW2070, 25°22.97'S, 168°57.12'E, 630-1150 m, 26.X.2003, 1 ♂ 5.0 mm (MNHN-IU-2014-16697).

DISTRIBUTION — Great Australian Bight, and now Norfolk Ridge; 630-1150 m.

DESCRIPTION — Medium-sized species. *Carapace*: Broader than long (0.8 × as long as broad), greatest breadth 1.8 × distance between anterolateral spines. Dorsal surface unarmed, covered with short fine setae, moderately convex from anterior to posterior, without groove or depression. Lateral margins convexly divergent, bearing row of short, oblique, setiferous ridges; anterolateral spine strong, overreaching acuminate lateral orbital angle. Rostrum narrow triangular, with interior angle of ca. 20°, straight horizontal on ventral surface; dorsal surface flattish; length about half that of remaining carapace, breadth one-third carapace breadth at posterior carapace margin. Lateral limit of orbit acuminate without distinct spine, situated at same level as anterolateral spine, and separated from that spine by full breadth of cornea. Pterygostomian flap with somewhat angular anterior margin without distinct spine, surface smooth.

Sternum: Excavated sternum with anterior margin slightly convex between bases of Mxp1, surface with low ridge in midline. Sternal plastron 1.4 × broader than long (three-quarters as long as broad), lateral extremities convexly divergent

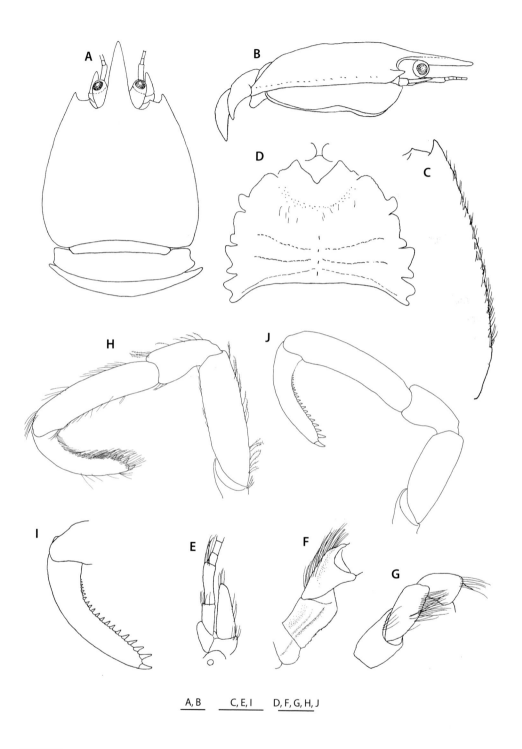

A, B   C, E, I   D, F, G, H, J

FIGURE 131

Uroptychus longvae Ahyong & Poore, 2004, male 5.0 mm (MNHN-IU-2014-16697). **A**, carapace and anterior part of abdomen, dorsal. **B**, same, lateral. **C**, lateral margin of carapace, right, dorsal. **D**, sternal plastron, with excavated sternum and basal parts of Mxp1. **E**, left antenna, ventral. **F**, right Mxp3, ventral. **G**, same, lateral. **H**, left P3, lateral. **I**, same, distal part, setae omitted, lateral. **J**, left P4, setae omitted, lateral. Scale bars: 1 mm.

posteriorly. Anterior margin of sternite 3 excavated in broad V-shape, without spines but obsolescent median notch, lateral ends bluntly angular. Sternite 4 with anterolateral margin anteriorly convex with obsolescent denticles, twice as long as posterolateral margin. Anterolateral margin of sternite 5 anteriorly rounded, 1.5 × longer than posterolateral margin of sternite 4.

Abdomen: Smooth, with fine setae. Somite 1 without transverse ridge. Somite 2 tergite 3.3 × broader than long; pleuron rounded on anterolateral end, bluntly angular on posterolateral end, lateral margin concavely divergent posteriorly. Pleuron of somite 3 tapering. Telson slightly less than half as long as broad; posterior plate 1.2 × longer than anterior plate, posterior margin distinctly emarginate.

Eye: Short (1.6 × longer than broad), barely reaching midlength of rostrum, slightly narrowed distally. Cornea more than half as long as remaining eyestalk.

Antennule and antenna: Ultimate article of antennule 2.9 × longer than high. Antennal peduncle overreaching cornea. Article 2 without distinct lateral spine. Antennal scale more than 2 × as broad as article 5, ending in midlength of that article. Article 5 1.3 × longer than article 5, breadth 0.6 × height of antennular basal article; no spine on articles 4 and 5 but tuft of setae on each mesio-distoventral margin; flagellum of 11-12 segments.

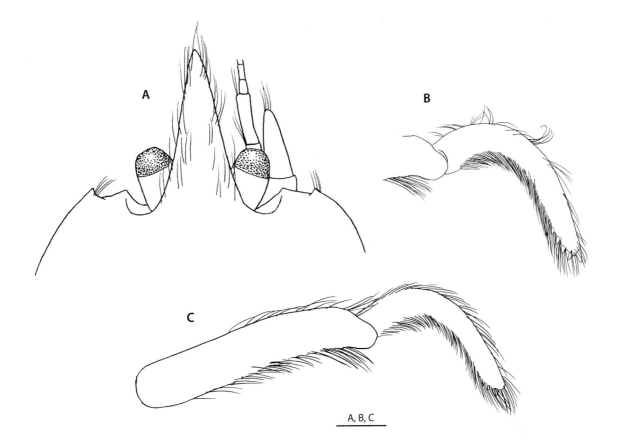

FIGURE 132

Uroptychus onychodactylus Tirmizi, 1964, holotype, ovigerous female 7.5 mm (BMNH 1966.2.3.41). **A**, anterior part of carapace, dorsal. **B**, right P2, distal part, lateral. **C**, right P3, distal part, lateral. Scale bar: 1 mm.

Mxp: Mxp1 with bases broadly separated. Mxp3 basis with obsolescent denticle. Ischium with flexor margin not rounded distally, crista dentata with numerous, very small denticles. Merus 2 × longer than ischium, unarmed. Carpus also unarmed.

P1: Missing.

P3-4: P2 missing; left P3, left P4, and right P4 available, relatively thick, not strongly compressed mesio-laterally, sparingly bearing soft setae. P3 merus 0.9 × length of P3 propodus; P3 merus 0.9 × length of P3 propodus; P4 merus 0.7 × length of P3 merus, 0.7 × length of P4 propodus; length-breadth ratio, 3.3 on P3, 2.6 on P4. P4 carpus 0.9 length of P3 carpus, carpus-propodus length ratio, 0.4 on P3 and P4. Propodi without flexor marginal spine, slightly longer on P3 than on P4. Dactyli subequal on P3 and P4, strongly curving at proximal quarter of length; dactylus-carpus length ratio, 1.7 on P3, 1.5 on P4; dactylus-propodus length ratio, 0.6 on P3, 0.7 on P4; flexor margin thickly setose, with row of 20 subtriangular spines successively diminishing toward proximal end of article, more or less close to one another, somewhat obliquely directed (proximal spines perpendicular to margin), all obscured by setae; ultimate spine slightly larger than penultimate.

REMARKS — The specimen examined agrees well with the description of the holotype, except that the P3 carpus is relatively short.

The species is very similar to *U. setosidigitalis* Baba, 1977 from off Midway Island and *U. onychodactylus* Tirmizi, 1964 from the Maldives. In addition to the differences noted by Ahyong & Poore (2004), *U. longvae* is distinguished from these two species by the anterolateral spine of the carapace that is strong rather than small, extending far beyond instead of never overreaching the acuminate lateral orbital angle. Examination of the type material of *Uroptychus onychodactylus* shows that the P2-4 dactyli are more strongly curved (Figure 132), a distinctive difference from *U. longvae* and *U. setosidigitalis*.

Uroptychus longvae also resembles *U. foulisi* Kensley, 1977 in having no flexor marginal spine on the P2-4 propodi, in having proximally diminishing flexor marginal spines on the dactyli, and in having the anterolateral spine of the carapace distinctly larger than the lateral orbital spine. *Uroptychus foulisi* is known only from the brief description of the type material from the western Indian Ocean, but is clearly different from *U. longvae* in having distinct carapace lateral spines in addition to the anterolateral one and in having fewer (8 versus 18-20), more obliquely directed flexor marginal spines on the P2-4 dactyli.

Uroptychus lumarius n. sp.
Figures 133, 134

TYPE MATERIAL — Holotype: **New Caledonia**, Norfolk Ridge. BATHUS 3 Stn CP811, 23°41'S, 168°15'E, 383-408 m, 28.XI.1993, ♂ 3.6 mm (MNHN-IU-2013-8516). Paratypes: **Solomon Islands**. SALOMON 1 Stn DW1852, 09°47'S, 160°53'E, 236-250 m, 07.X.2001, 1 ov. ♀ 3.6 mm (MNHN-IU-2014-16698). **New Caledonia**, Norfolk Ridge. BATHUS 2 CP736 23°03'S, 166°58'E, 452-464 m, 13.V.1993, 1 ♂ 3.4 mm, 2 ov. ♀ 3.2, 3.3 mm, 1 ♀ 3.2 mm (MNHN-IU-2014-16699). BATHUS 3, station data as for the holotype, 1 ♂ 3.5 mm, 4 ov. ♀ 3.3-3.5 mm (MNHN-IU-2014-16700). – Stn DW818, 23°44'S, 168°16'E, 394-401 m, 28.XI.1993, 1 ♂ 3.0 mm, 1 ov. ♀ 2.8 mm (MNHN-IU-2014-16701). LITHIST Stn DW13, 23°45.0'S, 168°16.7'E, 400 m, 12.VIII.1999, 1 ♂ 3.1 mm (MNHN-IU-2014-16702). – Stn CP15, 23°40'S, 168°15'E, 389-404 m, 12.VIII.1999, 1 ♀ 3.5 mm (MNHN-IU-2014-16703). NORFOLK 1 Stn CP1706, 23°44'S, 168°17'E, 383-394 m, 25.VI.2001, 1 ♀ 3.3 mm (MNHN-IU-2014-16704). – Stn DW1707, 23°43'S, 168°16'E, 381-493 m, 25.VI.2001, 1 ov. ♀ 3.5 mm (MNHN-IU-2014-16705). NORFOLK 2 Stn CP2048, 23°43.82'S,168°16.24'E, 380-389 m, with Primnoidae gen. sp. (Calcaxonia), 24.X.2003, 10 ♂ 2.9-3.5 mm, 7 ov. ♀ 3.2-3.5 mm (MNHN-IU-2014-16706). – Stn CP2050, 23°42.17'S, 168°15.72'E, 377-377 m, 24.X.2003, 1 ♂ 3.0 mm, 1 ov. ♀ 2.9 mm (MNHN-IU-2014-16707). – Stn DW2052, 23°42.29'S, 168°15.27'E, 473-525 m, 24.X.2003, 1 ov. ♀ 3.2 mm (MNHN-IU-2014-16708). – Stn DW2108, 23°46.52'S, 168°17.12'E, 403-440 m, 31.X.2003, 1 ♂ 2.8 mm, 1 ♀ 3.3 mm (MNHN-IU-2014-16709). – Stn DW2109, 23°47.46'S, 168°17.04'E, 422-495 m, 31.X.2003, 1 ♂ 3.0 mm, 1 ov. ♀ 3.2 mm (MNHN-IU-2014-16710). BERYX 11 Stn CP51, 23°44'S, 168°17'E,

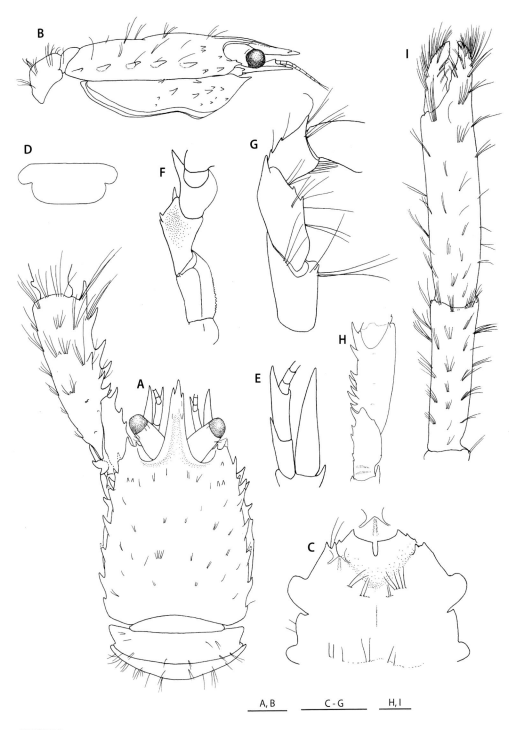

FIGURE 133

Uroptychus lumarius n. sp., holotype, male 3.6 mm (MNHN-IU-2013-8516). **A**, carapace and anterior part of abdomen, proximal part of left P1 included, dorsal. **B**, same, lateral. **C**, anterior part of sternal plastron, with excavated sternum and basal parts of Mxp1. **D**, telson. **E**, left antenna, ventral. **F**, right Mxp3, setae omitted, ventral. **G**, same, lateral. **H**, right P1, proximal part, setae omitted, ventral. **I**, same, proximal articles omitted, dorsal. Scale bars: 1 mm.

390-400 m, 21.X.1992, 5 ♂ 3.2-3.7 mm, 4 ov. ♀ 3.2-3.6 mm, 2 ♀ 3.5, 3.9 mm (MNHN-IU-2014-16711). – Stn CP52, 23°47.45'S, 168°17.05'E, 430-530 m, 21.X.1992, 1 ♂ 2.8 mm (MNHN-IU-2014-16712). **Philippines.** MUSORSTOM 2 Stn CP51, 14°00'N, 120°17'E, 170-187 m, with corals of Antipatharia (Hexacorallia), 27.XI.1980, 1 ♂ 3.9 mm (MNHN-IU-2014-16713). MUSORSTOM 3 Stn CP134, 12°01'N, 121°57'E, 92-95 m, with possibly *Dendronephthya* sp. (Alcyoniina: Nephtheidae), 5.VI.1985, 1 ov. ♀ 3.5 mm (MNHN-IU-2014-16714).

ETYMOLOGY — From the Latin *lumarius* (of thorns), alluding to strong spines on the flexor margin of P2-4 dactyli, by which the species is easily distinguished from the other three congeners, *U. floccus*, *U. quinarius* and *U. dualis*.

DISTRIBUTION — Philippines, Solomon Islands, and Norfolk Ridge; 92-530 m.

SIZE — Males, 2.8-3.9 mm; females, 2.8-3.9 mm; ovigerous females from 2.8 mm.

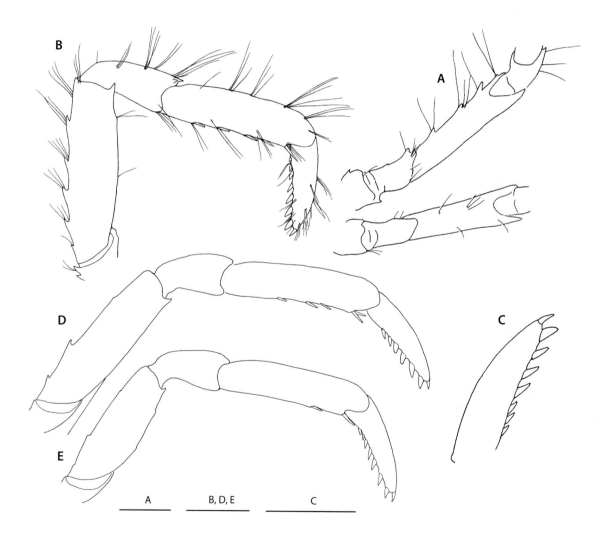

FIGURE 134

Uroptychus lumarius n. sp., holotype, male 3.6 mm (MNHN-IU-2013-8516). **A**, left P2 and P3, distal part omitted, ventral. **B**, right P2, lateral. **C**, same, distal part, setae omitted. **D**, right P3, setae omitted, lateral. **E**, right P4, setae omitted, lateral. Scale bars: 1 mm.

DESCRIPTION — Small species. *Carapace*: As long as broad; greatest breadth 1.5 × distance between anterolateral spines. Dorsal surface slightly convex from anterior to posterior, bearing shallow depression between gastric and cardiac regions; with sparse short fine setae and a few small spines on hepatic region: usually 2 placed side by side mesial to between third and fourth lateral spines, rarely 3, 1 or 0. Lateral margins slightly divergent posteriorly, bearing 8 spines along entire length, rarely with additional obsolescent spine behind anterior third spine; first anterolateral, moderately developed, overreaching much smaller lateral orbital spine; second and third small; fourth to seventh strong and subequal, last spine small or obsolete. Rostrum narrow triangular, with interior angle of about 20°, laterally with small subapical spine (rarely 2) on each side, dorsally excavated longitudinally; length 0.5-0.7 × that of remaining carapace, breadth half carapace breadth at posterior carapace margin. Lateral orbital spine small, situated at same level as anterolateral spine. Pterygostomian flap anteriorly angular, produced to strong spine, surface with small spines on anterior half, one consistent directly below linea anomurica between third and fourth lateral spines.

Sternum: Excavated sternum with longitudinal ridge in midline on anterior half, anterior margin broadly triangular or convex between bases of Mxp1. Sternal plastron about as long as broad or slightly longer, lateral extremities subparallel between sternites 4 and 7. Sternite 3 moderately depressed; anterior margin shallowly excavated with deep median notch usually of narrow, rarely moderately broad U-shape or narrow V-shape flanked by small or obsolescent spine; anterolateral angle sharply produced. Sternite 4 with nearly straight anterolateral margin anteriorly ending in 1 or 2 (placed side by side) small spines; posterolateral margin relatively long, slightly shorter than anterolateral margin. Anterolateral margin of sternite 5 as long as posterolateral margin of sternite 4.

Abdomen: Tergites smooth, sparsely setose. Somite 1 moderately convex from anterior to posterior. Somite 2 tergite 3.1-3.5 × broader than long; pleuron anterolaterally and posterolaterally bluntly angular, lateral margin strongly concave, not divergent posteriorly. Pleuron of somite 3 tapering, laterally angular. Telson less than half (0.33-0.45 x) as long as broad; posterior plate 1.0-1.3 × longer than anterior plate, posterior margin nearly transverse or feebly concave.

Eye: Elongate (1.9 × longer than broad), slightly overreaching midlength of rostrum, distally narrowed, bearing a few setae dorsally. Cornea not dilated, about half as long as remaining eyestalk.

Antennule and antenna: Ultimate article of antennular peduncle 2.5-2.9 × longer than high. Antennal peduncle overreaching cornea. Article 2 with sharp, strong lateral spine. Antennal scale overreaching article 5 by length of proximal 2 segments of flagellum, slightly falling short of rostral tip, breadth 2.0-2.5 × that of article 5. Distal 2 articles with strong distomesial spine. Article 5 1.3 × longer than article 4, breadth two-thirds height of ultimate antennular article. Flagellum of 9-14 segments far falling short of distal end of P1 merus (rarely 19 segments reaching distal end of P1 merus (male, MNHN-IU-2014-16713)).

Mxp: Mxp1 with bases distantly separated. Mxp3 with sparse long setae on ischium, merus and carpus. Basis with or without obsolescent denticles. Ischium with distally diminishing denticles on crista dentata, distolaterally bearing small spine somewhat lateral to rounded distal end of flexor margin. Merus 2 × longer than ischium, mesio-laterally compressed, mesial face smooth, slightly concave; flexor margin with 1 or 2 small spines distal to point one-third from distal end; distolateral spine well-developed. Carpus with 1 well-developed distolateral and 1 or 2 small extensor proximal spines.

P1: Sparingly setose, 4.1-5.5 × longer than carapace. Ischium dorsally with 2 spines (distal larger), ventromesially with strong subterminal spine proximally followed by small spines. Merus with field of 3 spines in oblique row on proximomesial surface (from dorsal to ventral) and 4 strong ventromesial spines; length 1.0-1.2 × that of carapace. Carpus 1.2-1.4 × (rarely subequal to) length of merus, ventrally bearing distomesial and distolateral spines. Palm 3.0-3.5 × (males), 2.7-3.7 × (females) longer than broad, 1.2-1.4 × (rarely 1.1 x) longer than carpus. Fingers somewhat incurved distally; opposable margins not gaping in both sexes, that of fixed finger with low eminence distal to position of opposite median low process on movable finger; movable finger 0.4-0.5 × length of palm.

P2-4: Moderately compressed mesio-laterally, relatively broad, setose like P1. Meri successively shorter posteriorly (P3 merus 0.9 × length of P2 merus, P4 merus 0.8-0.9 × length of P3 merus), equally broad on P2-4; length-breadth ratio, 3.6-4.5 on P2, 3.3-3.9 on P3, 3.1-3.4 on P4; P2 merus 0.7-0.9 × length of carapace, 1.1-1.2 × length of P2 propodus; P3 merus subequal to length of P3 propodus; P4 merus 0.8 × length of P4 propodus; dorsal margin with row of 3-5 small spines distinct on P2, 2 spines occasionally obsolescent on P3, no spine but a few setiferous eminences on P4, unarmed on distal end; ventrolaterally with distal spine, ventromesially with 4-5 spines on P2, no spine on P3 and P4. Carpi successively very slightly shorter, length 0.4 that of propodus. Propodi shorter on P2 than on P3 and P4; flexor margin nearly straight, ending in pair of spines preceded by 4 or 5 spines on P2, 3 or 4 spines on P3, 1 or 2 spines on P4. Dactyli subequal in length on P3 and P4, slightly shorter on P2; slightly longer than carpi (dactylus-carpus length ratio, 1.1-1.3 on P2, 1.2-1.3 on P3, 1.3-1.5 on P4), about half as long as propodi (dactylus-propodus length ratio, 0.4-0.5 on P2 and P3, 0.5-0.6 on P4); similar in spination on P2-4; flexor margin straight, ending in slender spine preceded by 6 or 7 (usually 7) spines relatively large, subtriangular, obliquely directed, loosely arranged, and diminishing toward base of article, penultimate somewhat closer to ultimate than to antepenultimate, antepenultimate broader than ultimate.

Eggs. Number of eggs carried, 5-15; size, 0.9 mm × 1.0 mm - 1.0 mm × 1.3 mm.

REMARKS — The male collected from the Philippines at MUSORSTOM 3 Stn CP134 (MNHN-IU-2014-16714) is somewhat different from the others in having weaker spination of P2 merus, with the dorsal margin bearing a few obsolescent spines on the proximal part and the ventromesial margin unarmed but a very small proximal spine on the left side only. This specimen was collected together with the alcyonacean, possibly *Dendronephthya* sp. (Nephtheidae), whereas the lot MNHN-IU-2014-16713 from the Philippines was with antipatharian corals, and the specimens MNHN-IU-2014-16706 from the Norfolk Ridge was with alcyonacean coral Primnoidae. Given the different hosts, it is not unlikely that these may represent cryptic species but all are provisionally placed in *U. lumarius*.

The carapace bearing strong lateral spines and the P1 merus bearing a field of three spines in an oblique row on the proximal portion of the mesial surface link the species to *U. dualis* n. sp., *U. floccus* N. sp., and *U. quinarius* n. sp. These three congeners are characterized by sternite 4 with the posterolateral margin long relative to the anterolateral margin, and the P2-4 dactyli with numerous obliquely directed flexor marginal spines among which the distal second is pronouncedly broad and the remainder are slender and close to one another. In this new species, these spines are well developed, subtriangular, successively smaller toward the base of article, and loosely arranged. In addition, the carapace lateral margin bears 7-8 spines in *U. lumarius*, 6 spines in *U. dualis* and *U. floccus*, and 5 spines in *U. quinarius*.

Uropotychus macrolepis n. sp.

Figures 135, 136

TYPE MATERIAL — Holotype: **New Caledonia**. MUSORSTOM 4 Stn CP213, 22°51.30'S, 167°12.0'E, 405-430 m, on corals of Antipatharia (Hexacorallia), 28.IX.1985, ♂ 3.1 mm (MNHN-IU-2013-8517). Paratypes: MUSORSTOM 4, station data as for the holotype, 1 ov. ♀ 3.2 mm (MNHN-IU-2014-16715). New Caledonia, Norfolk Ridge. BATHUS 3 Stn DW817, 23°42'S, 168°16'E, 405-410 m, 28.XI.1993, 1 ♂ 3.2 mm (MNHN-IU-2014-16716). NORFOLK 2 Stn CP2048, CP2048, 23°43.82S, 168°16.24'E, 380-389 m, 24.X.2003, 1 ♂ 2.6 mm (MNHN-IU-2014-16717).

ETYMOLOGY — From the Greek *macros* (long) and *lepis* (scale), alluding a long antennal scale displayed by the species.

DISTRIBUTION — New Caledonia and Norfolk Ridge; 380-430 m.

SIZE — Males, 2.6-3.2 mm; ovigerous female, 3.2 mm.

DESCRIPTION — Small species. *Carapace*: Strongly broadened posteriorly, 1.3-1.4 × broader than long (0.7 × as long as broad); greatest breadth 1.8-1.9 × distance between anterolateral spines. Dorsal surface moderately convex from anterior to posterior, smooth, with very sparse soft setae, lacking distinct depression. Lateral margins divided into anterior and posterior parts by concavity or constriction at point one-third from anterior end or directly anterior to posterior bran-

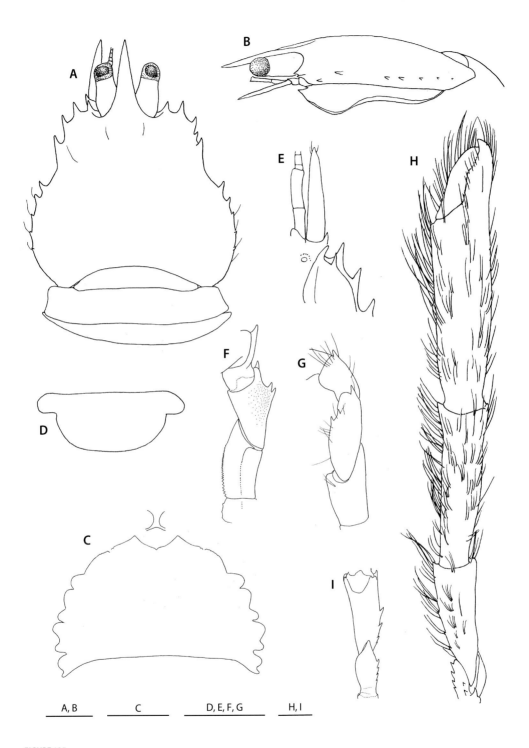

A, B　　　　C　　　　D, E, F, G　　　H, I

FIGURE 135

Uroptychus macrolepis n. sp., holotype, male 3.1 mm (MNHN-IU-2013-8517). **A**, carapace and anterior part of abdomen, dorsal. **B**, same, lateral. **C**, sternal plastron, with excavated sternum and basal parts of Mxp1. **D**, telson. **E**, antenna and anterior part of pterygostomian flap, ventral. **F**, left Mxp3, setae omitted, ventral. **G**, same, lateral. **H**, right P1, dorsal. **I**, same, proximal part, setae omitted, ventral. Scale bars: 1 mm.

chial margin, anterior part more or less convexly divergent, with 4 spines; first spine anterolateral, well developed, reaching or slightly overreaching lateral orbital spine; second and third somewhat ventral in position (third absent in 2 specimens); fourth as large as first; posterior part (posterior branchial margin) strongly convex, with 5 relatively small, posteriorly diminishing spines, anteriormost of these remotely separated from preceding spine, last spine situated at posterior third of posterior branchial margin. Rostrum narrowly triangular, with interior angle of about 20°, deflected ventrally, dorsally concave; length about half that of remaining carapace, breadth one-third carapace breadth at posterior carapace margin. Lateral orbital spine slightly smaller than or subequal to anterolateral spine, situated very slightly anterior to but well separated from that spine. Pterygostomian flap anteriorly angular, produced to acute spine followed by 1 or 2 distinct spines anterior to anterior linea anomurica; no spines on surface; height of posterior half 0.3 × that of anterior half.

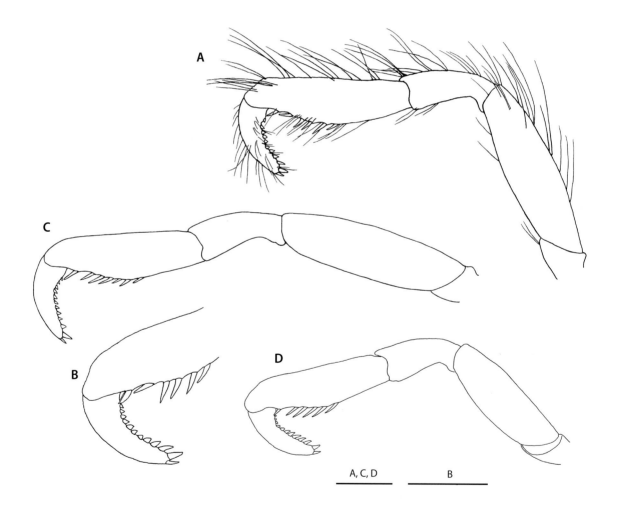

FIGURE 136

Uroptychus macrolepis n. sp., holotype, male 3.1 mm (MNHN-IU-2013-8517). **A**, left P2, lateral. **B**, same, distal part, setae omitted, lateral. **C**, left P3, setae omitted, lateral. **D**, left P4, setae omitted, lateral. Scale bars: 1 mm.

Sternum: Excavated sternum with broad subtriangular anterior margin ending in blunt tip, bearing weak ridge in midline. Sternal plastron much broader than long (0.6 × as long as broad), lateral extremities convexly divergent posteriorly, sternite 6 broadest. Sternite 3 very shallowly depressed, anterior margin shallowly or moderately concave, with obsolescent or small median notch, lacking submedian spines; anterolateral end sharply angular. Sternite 4 with convex anterolateral margin not produced anteriorly, posterolateral margin very short. Anterolateral margin of sternite 5 strongly convex, about twice as long as posterolateral margin of sternite 4.

Abdomen: Tergites smooth, polished, sparsely with short setae, each somite short relative to width. Somite 1 without transverse ridge. Somite 2 tergite 3.2-3.3 × broader than long; pleuron slightly concavely divergent posteriorly, blunt at end. Pleuron of somite 3 with blunt lateral end. Telson slightly less than half as long as broad; posterior plate feebly concave or feebly convex on posterior margin, length 1.3-1.8 × that of anterior plate.

Eye: Elongate (2.6-3.2 × longer than broad), reaching at most distal third of rostrum, narrowed distally (in smallest male, eyes 2.2 × longer than broad, distally not narrowed). Cornea distinctly less than half length of remaining eyestalk.

Antennule and antenna: Ultimate article of antennular peduncle 2.8-3.0 × longer than high. Antennal peduncle not overreaching cornea. Article 2 with acute, short lateral spine. Antennal scale overreaching distal end of peduncle by half length of article 5, reaching or slightly overreaching rostral tip, breadth 1.5 × that of article 5. Distal 2 articles unarmed; article 5 1.3-1.5 × longer than article 4, breadth three-fifths to four-fifths height of ultimate article of antennule. Flagellum of 10-12 segments reaching or falling short of distal end of P1 merus.

Mxp: Mxp1 with bases close to each other. Mxp3 with relatively short setae. Basis with a few obsolescent denticles on mesial ridge. Ischium with distally diminishing small denticles on crista dentata, flexor margin distally rounded. Merus 2 × as long as ischium, with prominent distolateral spine and 2 smaller spines distal to point two-thirds of sharply ridged flexor margin. Carpus with distinct distolateral spine.

P1: 5.3-5.6 × longer than carapace, thickly covered with fine, plumose, long setae. Ischium dorsally with strong spine, ventromesially with 3-6 small spines, subterminal spine absent. Merus 1.0-1.2 × length of carapace, with 3-5 mesial, 1 ventral distomesial, 1 very reduced ventral distolateral spine. Carpus 1.2-1.3 × longer than merus, mesially with 2 small blunt spines (distodorsal and distoventral). Palm 2.7-3.3 × longer than broad, 1.3 × longer than carpus, unarmed. Fingers not gaping in both sexes, strongly incurved distally, crossing when closed; movable finger slightly shorter than fixed finger, length 0.4-0.5 × that of palm, opposable margin with low proximal process; fixed finger directed feebly laterally, opposable margin with median eminence.

P2-4: Unarmed except for 2 distal articles, with long fine plumose setae particularly thick along dorsal or extensor margins. Meri relatively broad, distally somewhat narrowed; length-breadth ratio, 3.5-3.8 on P2, 3.3-3.5 on P3, 2.7-3.0 on P4; dorsal margin with or without obsolescent eminences; P2 merus as long as or slightly shorter than (0.9 × length of) carapace, slightly longer than P2 propodus; P3 merus slightly broader and slightly longer than P2 merus, 1.1 × length of P3 propodus; P4 merus very slightly narrower than or as broad as P2 merus, 0.7-0.8 × length of P3 merus, 0.8-0.9 × length of P4 propodus. Carpi about as long as dactyli on P2-4; P3 carpus as long as P2 carpus, P4 carpus 0.9-1.0 × length of P2 carpus, 0.8-1.0 × length of P3 carpus; carpus-propodus length ratio, 0.5 on P2 and P3, 0.4-0.5 on P4. Propodi subequal on P2-4 or slightly longer on P2 or P3 (not consistent), relatively broad; flexor margin convex (broadest portion measured at midlength), ending in pair of movable spines preceded by 5-7 slender spines in zigzag arrangement on distal half, distalmost of these unpaired spines somewhat more remote from distal pair than from their distal second. Dactyli relatively stout, strongly curved at proximal third; length 0.4-0.5 × that of propodus on P2-4; flexor margin with 11-13 short, triangular, obliquely directed spines, ultimate slender, penultimate broader than antepenultimate, antepenultimate broader than ultimate, ultimate and penultimate nearly contiguous at base.

Eggs. Eggs carried 4 in number, measuring 1.10 mm in diameter.

REMARKS — The combination of the following characters links the species to *U. duplex* n. sp. and *U. zigzag* n. sp.: the posteriorly broadened carapace bearing lateral spines, the antennal scale overreaching the antennal article 5, the P2 merus subequal to or slightly shorter than the P3 merus, and the P2-4 propodi inflated on the distal part of flexor margin.

Uroptychus duplex and *U. macrolepis* share a well-depressed cephalothorax, the posterior half in particular, with the pterygostomian flap much lower in the posterior half than in the anterior half. *Uroptychus duplex* is distinguished from *U. macrolepis* by sternite 3 that has the anterior margin transverse along the median third length, not concavely excavated; the sternal plastron is strongly broadened posteriorly and broadest on sternite 7; and the anterolateral spine of the carapace is much larger than instead of slightly larger or subequal to the lateral orbital spine.

The relationships with *U. zigzag* are discussed under the remarks of that species (see below).

Uroptychus magnipedalis n. sp.

Figures 137, 138

TYPE MATERIAL — Holotype: **Vanuatu**. MUSORSTOM 8 Stn CP1088, 15°09'S, 167°15'E, 425-455 m, 6.X.1994, ♂ 6.3 mm (MNHN-IU-2014-16718). Paratype: Collected with holotype, 1 ♂ 7.2 mm (MNHN-IU-2014-16719).

ETYMOLOGY — From the Latin *magnus* (large) and *pedalis* (of the foot), referring to a massive P1 of the species.

DISTRIBUTION — Vanuatu; 425-455 m.

DESCRIPTION — Medium-sized species. *Carapace*: Slightly broader than long (0.9 × as long as broad); greatest breadth 1.8 × distance between anterolateral spines. Dorsal surface almost glabrous, slightly granulate sparsely, somewhat convex from anterior to posterior, without depression between gastric and cardiac regions. Lateral margins convexly divergent, ridged along posterior half; anterolateral spine slightly overreaching much smaller lateral orbital spine, followed by 3 very small spines anterior to 1 relatively large (smaller than anterolateral spine) and a few very small spines on posterior branchial region. Rostrum moderately broad triangular, with interior angle of 26-28°; length 0.4-0.5 × that of remaining carapace, breadth less than half carapace breadth at posterior carapace margin; dorsal surface concave; lateral margin with 1 (2 on left side in holotype) subterminal spinule. Lateral orbital spine situated slightly anterior to level of anterolateral spine. Pterygostomian flap anteriorly angular, produced to acute spine; surface with very small sparse tubercles on anterior portion.

Sternum: Excavated sternum with weak ridge in midline, anterior margin strongly convex. Sternal plastron slightly broader than long, lateral extremities weakly divergent posteriorly. Sternite 3 moderately depressed; anterior margin shallowly concave, with deep, narrow V-shaped or U-shaped median notch without flanking spine. Sternite 4 with anterolateral margin convex and anteriorly lobe-like with a few denticles; posterolateral margin about half as long as anterolateral margin. Anterolateral margin of sternite 5 anteriorly strongly convex, 1.2 × longer than posterolateral margin of sternite 4.

Abdomen: Smooth, polished and glabrous. Somite 1 convex from anterior to posterior, without transverse ridge. Somite 2 tergite 2.7-3.0 × broader than long; pleuron anterolaterally and posterolaterally blunt or bluntly angular, lateral margin slightly concave and slightly divergent posteriorly. Pleura of somites 3 and 4 tapering to angular or blunt tip. Telson about half as long as broad; posterior plate convex on posterior margin, length 1.2 × that of anterior plate.

Eye: Twice as long as broad, slightly narrowed distally, reaching anterior third of rostrum. Cornea not dilated, half as long as remaining eyestalk.

Antennule and antenna: Ultimate article of antennule 2.6-3.0 × longer than high. Antennal peduncle overreaching cornea. Article 2 with strong distolateral spine. Antennal scale overreaching article 5, about twice as broad as that article. Distal 2 articles each with well-developed distomesial spine, that of article 4 with small accompanying spine in holotype. Article 5 1.4 × longer than article 4, breadth 0.6-0.7 × height of antennular ultimate article. Flagellum consisting of 15-16 segments, slightly falling short of distal end of P1 merus.

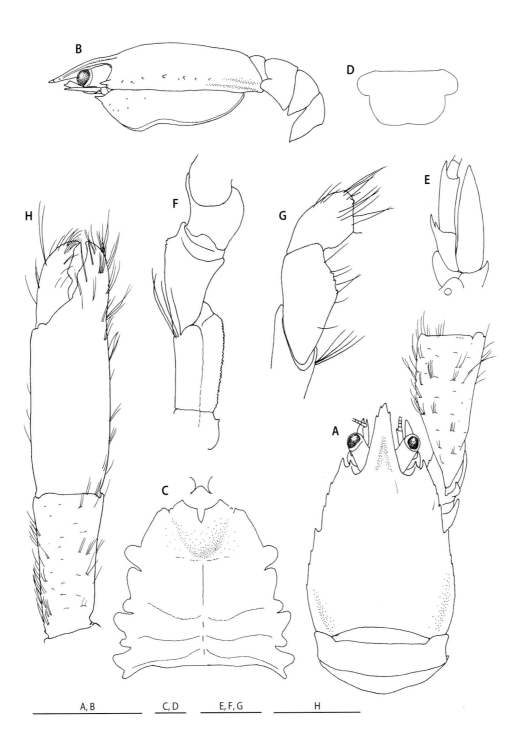

A, B   C, D   E, F, G   H

FIGURE 137

Uroptychus magnipedalis n. sp., holotype, male 6.3 mm (MNHN-IU-2014-16718). **A**, carapace and anterior part of abdomen, proximal part of right P1 included, dorsal. **B**, same, lateral. **C**, sternal plastron, with excavated sternum and basal parts of Mxp1. **D**, telson. **E**, left antenna, ventral. **F**, right Mxp3, ventral. **G**, same, lateral. **H**, right P1, proximal part omitted, dorsal. Scale bars: A, B, H, 5 mm; C-G, 1 mm.

Mxp: Mxp1 with bases broadly separated. Mxp3 basis with 1 or 2 denticles on mesial ridge. Ischium bearing long setae lateral to rounded distal end of flexor margin; crista dentata with about 30 denticles. Merus 1.6 × longer than ischium; flexor margin convex, sharply ridged, bearing a few denticles on distal third. Carpus with obsolescent distolateral process.

P1: Massive, 4.2-4.5 × longer than carapace, sparingly bearing short, fine setae. Ischium dorsally with strong spine, ventromesially with row of tubercular processes, without subterminal spine. Merus slightly longer than carapace, ventrally bearing blunt distomesial and distolateral spines. Carpus 1.1-1.2 × longer than merus, with 2 distoventral spines. Palm unarmed, 1.9-2.1 × longer than broad, 1.2 × longer than carpus, breadth at midlength slightly more than distance between anterolateral spines of carapace. Fingers short relative to breadth, distally incurved, crossing when closed, not gaping; movable finger having opposable margin with blunt median process fitting to opposite concavity on fixed finger when closed, length half that of palm.

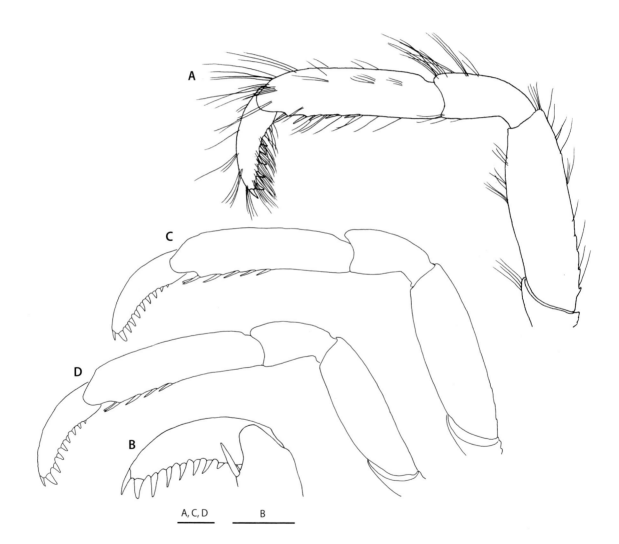

FIGURE 138

Uroptychus magnipedalis n. sp., holotype, male 6.3 mm (MNHN-IU-2014-16718). **A**, left P2, lateral. **B**, same, distal part, setae omitted, lateral. **C**, left P3, setae omitted, lateral. **D**, left P4, setae omitted, lateral. Scale bars: 1 mm.

P2-4: With short setae sparse on proximal articles, numerous on distal articles, moderately compressed mesiolaterally. Meri relatively flattened, successively shorter posteriorly (P3 merus 0.9 × length of P2 merus, P4 merus 0.8 × length of P3 merus), breadths subequal on P2 and P3, P4 merus 0.9 × as broad as P3 merus; dorsal crest with several denticles or eminences distinct on P2, obsolescent on P3, absent on P4; length-breadth ratio, 3.3-3.5 on P2, 3.0-3.4 on P3, 2.9-3.1 on P4; P2 merus 0.8 × length of carapace, 1.0-1.2 × length of P2 propodus; P3 merus 0.9 × length of P3 propodus; P4 merus 0.8 × length of P4 propodus. P3 carpus 0.9 × length of P2 carpus, P4 carpus 0.9-1.0 × length of P3 carpus, less than half (0.42-0.45 on P2, 0.38-0.40 on P3, 0.36-0.38 on P4) length of propodi. Propodi subequal in length on P2 and P3, longer on P4 in holotype, shortest on P2 and longer on P3 than on P4 in paratype; flexor margin straight, ending in pair of spines preceded by 6-7 spines on P2, 4 spines on P3, 3 spines on P4. Dactyli slightly shorter on P2 than on P3 and P4, slightly shorter than carpi on P2, subequal to that article on P3 and P4, slightly less than half length of propodi on P2-4; flexor margin gently curving, ending in slender spine preceded by 7-8 relatively long, sharp spines loosely arranged, slightly obliquely directed, and successively diminishing proximally, ultimate more slender than penultimate, penultimate slightly larger than antepenultimate.

REMARKS — *Uroptychus magnipedalis* resembles *U. baeomma* n. sp. in having a subapical spine on each side of the rostrum, in having elongate eyes, and in the shapes of the P2-4, sternal plastron and Mxp3. However, this new species may be differentiated from that species by the carapace bearing only one instead of 5 well-developed spines on the branchial margin, the antennal peduncle bearing much stronger instead of obsolescent spine on each of the distal two articles, and the antennal scale overreaching rather than falling short of the distal end of article 5.

Uroptychus maori Borradaile, 1916

Figures 139, 140

Uroptychus maori Borradaile, 1916: 92, fig. 6. — Schnabel 2009: 555, figs 8, 9.

TYPE MATERIAL — Holotype: **New Zealand**, off Three Kings Islands, 34°15.60'S, 174° 6.00'E, 183 m, male (BMNH 1917.1.29.116). [not examined].

MATERIAL EXAMINED — **New Caledonia**, Norfolk Ridge. BIOCAL Stn CP52, 23°06'S, 167°47'E, 540-600 m, 31.VIII.1985, 1 ov. ♀ 11.3 mm (MNHN-IU-2014-16720), 1 ♂ 8.2 mm (MNHN-IU-2014-16721). **New Caledonia**, Loyalty Ridge. BERYX 2 Stn CH16, 23°35.60'S, 169°36.52'E, 660-675 m, 29.X.1991, 1 ov. ♀ 12.9 mm (MNHN-IU-2014-16722). HALIPRO 1 Stn CH872, 23°02'S, 166°52'E, 620-700 m, 30.III.1994, 1 ♂ 8.1 mm (MNHN-IU-2014-16723). **New Caledonia**, Hunter and Matthew Islands. VOLSMAR Stn DW05, 22°26'S, 171°46'E, 620-700 m, 01.VI.1989, 1 ♂ 7.6 mm, 1 ov. ♀ 11.2 mm (MNHN-IU-2014-16724).

DISTRIBUTION — New Zealand (Three Kings Islands, West Norfolk Ridge, and Bay of Plenty), 183-700 m; and now Loyalty Ridge, Hunter-Matthew and Norfolk Ridge; 540-700 m.

SIZE — Males 7.6-8.2 mm; females, 11.2-12.9 mm; ovigerous females from 11.2 mm.

DESCRIPTION — Large species. *Carapace*: As long as broad; greatest breadth 1.7 × distance between anterolateral spines. Dorsal surface granulose, somewhat convex from anterior to posterior, with distinct depression between gastric and cardiac regions. Lateral margins somewhat divergent posteriorly, slightly convex on both anterior and posterior branchial regions, with elevated ridge at anterior end of branchial region, ridged near posterior end; anterolateral spine well developed, overreaching much smaller lateral orbital spine, situated slightly posterior to and separated from that spine

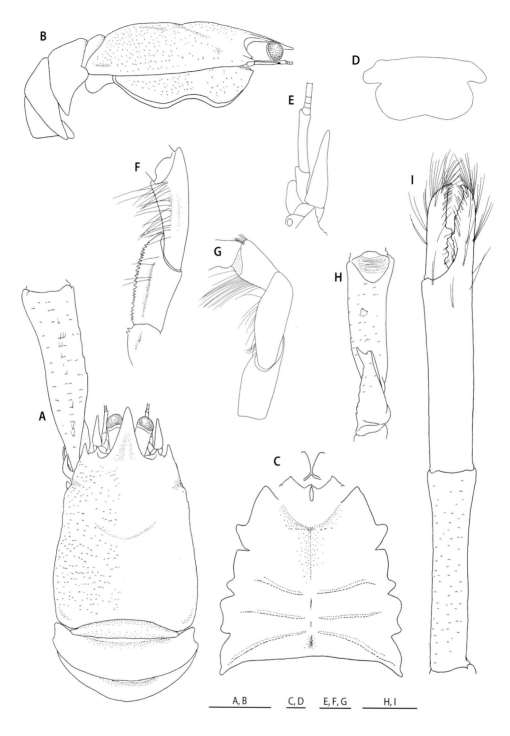

FIGURE 139

Uroptychus maori Borradaile, 1916, ovigerous female 11.3 mm (MNHN-IU-2014-16720). **A**, carapace and anterior part of abdomen, proximal part of left P1 included, dorsal. **B**, same, lateral. **C** sternal plastron, with excavated sternum and basal parts of Mxp1. **D**, telson. **E**, left antenna, ventral. **F**, left Mxp3, ventral. **G**, same, lateral. **H**, left P1, proximal part, ventral. **I**, same, proximal part omitted, dorsal. Scale bars: A, B, H, I, 5 mm; C-G, 1 mm.

by basal breadth. Rostrum relatively broad triangular, with interior angle of 30-35°; about as long as broad; length 0.3-0.4 × postorbital carapace length, breadth less than half carapace breadth at posterior carapace margin; dorsal surface concave. Pterygostomian flap granulose, anteriorly angular, acuminate at tip or ending in tiny spine.

Sternum: Excavated sternum anteriorly produced to small sharp spine, surface with weak ridge in midline. Sternal plastron as long as or slightly shorter than broad; lateral extremities gently divergent posteriorly. Sternite 3 well depressed; anterior margin of broad V-shape with 2 incurved submedian spines separated by narrow U-shaped sinus, laterally angular. Sternite 4 with anterolateral margin smooth, slightly convex, anteriorly subtriangular, 1.5 × longer than posterolateral margin. Anterolateral margin of sternite 5 1.2-1.3 × longer than posterolateral margin of sternite 4.

Abdomen: Somite 1 moderately convex from anterior to posterior. Somite 2 tergite 2.5-2.8 × broader than long; pleuron posterolaterally tapering, lateral margin concavely divergent. Pleuron of somite 3 tapering. Telson about half as long as broad; posterior plate 1.3-1.9 × longer than anterior plate, emarginate on posterior margin.

Eye: 2.0-2.2 × longer than broad, reaching or overreaching anterior quarter of rostrum; lateral and mesial margins subparallel or slightly concave. Cornea more than half as long as remaining eyestalk.

Antennule and antenna: Antennular ultimate article 2.4-3.0 × longer than high. Antennal peduncle slightly overreaching cornea, barely reaching rostral tip. Article 2 with distolateral spine usually small, occasionally tiny and discernible only under high magnification. Antennal scale 1.6-1.9 × broader than article 5, overreaching midlength of but barely reaching distal end of article 5. Article 4 with short blunt distomesial spine. Article 5 unarmed, length 1.9-2.1 × that of article 4, breadth 0.4-0.5 × height of antennular ultimate article. Flagellum of 22-24 segments varying from barely reaching to overreaching distal end of P1 merus.

Mxp: Mxp1 with bases close to each other. Mxp3 basis with a few denticles on mesial ridge. Ischium half as long as merus, crista dentata with 28-40 denticles, flexor margin not rounded distally. Merus 2.0 × longer than ischium, sharply ridged along flexor margin without distinct spines but a number of very small tubercles or granules; with no distolateral spine. Carpus unarmed.

P1: Length 4.2-5.0 × that of carapace; ventrally granulated on merus and carpus, tuberculate on merus in large specimens, setose on fingers. Ischium with strong dorsal spine, ventromesially with well-developed subterminal spine proximally followed by tubercles. Merus varying from slightly shorter to slightly longer than carapace, mesially with tubercles or small spines, ventrally with a few spines in line with subterminal spine of ischium, median one usually larger. Carpus 1.2-1.3 × length of merus. Palm 2.4-3.5 × longer than broad, as long as or slightly shorter than carpus; lateral and mesial margins not carinated but rounded. Fingers not gaping, distally incurved, crossing when closed; movable finger with 2-toothed broad process on proximal half of opposable margin, proximal one smaller; length 0.4-0.5 × that of palm; opposable margin of fixed finger with low prominence at distal third and low processes at midlength.

P2-4: Relatively thick mesio-laterally, very sparsely setose except for distal 2 articles. Meri successively shorter posteriorly (P3 merus 0.9 × length of P2 merus, P4 merus 0.8 × length of P3 merus), subequally broad on P2-4; length-breadth ratio, 5.0-5.7 on P2, 4.1-5.3 on P3, 3.4-4.3 on P4; dorsal margin with row of denticle-like very small spines distinct at most on proximal half on P2, obsolescent on P3, absent on P4; P2 merus 0.8-0.9 × length of carapace, 1.3-1.4 × length of P2 propodus; P3 merus 1.1 × length of P3 propodus; P4 merus 0.9 × length of P4 propodus. Carpi successively slightly shorter posteriorly or P3 and P4 carpi subequal; carpus-propodus length ratio, 0.5-0.6 on P2, 0.4-0.5 on P3 and P4. Propodi subequal on P3 and P4, shortest on 2; flexor margin straight, ending in pair of spines preceded by 9-12 spines on P2, 6-10 on P3, 6-8 on P4. Dactyli slightly longer on P3 and P4 than on P2, slightly shorter than carpi (dactylus-carpus length ratio, 0.7-0.8 on P2, 0.8-0.9 on P3, 0.9-1.0 on P4), dactylus-propodus length ratio, 0.4 on P2 and P3, 0.4-0.5 on P4; flexor margin curving at proximal third, with 11 or 12 sharp triangular, loosely arranged, slightly obliquely directed, proximally diminishing spines on P2 and P3, 10-13 spines on P4, distal 3 subequal; extensor margin with plumose setae at least on median third.

Eggs. Number of eggs carried, 12-20; size, 1.16 mm × 1.26 mm - 2.05 mm × 1.84 mm.

REMARKS — The present specimens agree well with the species account of Schnabel (2009), except that the rostrum is short relative to breadth (length-breadth ratio, 1.1 versus 1.4), and the pterygostomian flap is anteriorly acuminate or ends in a tiny spine rather than being produced to a distinct spine.

This species resembles *U. brachydactylus* Tirmizi, 1964 , *U. brucei* Baba, 1986a and *U. granulipes* n. sp. in the carapace shape and the spination of P2-4. *Uroptychus maori* is readily distinguished from *U. brachydactylus* by the P1 ischium that bears a strong subterminal spine instead of being unarmed; the P1 palm is not carinated along the mesial margin; the P2-4 dactyli are longer relative to the propodi, the dactylus-propodus length ratio being 0.40-0.44 on P2, 0.43-0.45 on P3, 0.40-0.47 on P4, instead of 0.27-0.30 on P2-4.

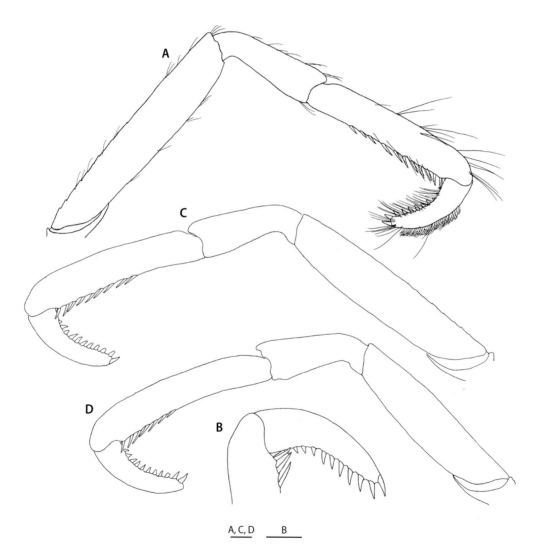

A, C, D   B

FIGURE 140

Uroptychus maori Borradaile, 1916, ovigerous female 11.3 mm (MNHN-IU-2014-16720). **A**, right P2, lateral. **B**, same, distal part, setae omitted, lateral. **C**, left P3, setae omitted, lateral. **D**, left P4, setae omitted, lateral. Scale bars: 1 mm.

Uroptychus maori looks much closer to *U. brucei* and *U. granulipes* than to *U. brachydactylus*. Their differences are very slight but this new species may be differentiated from *U. brucei* by the following: the anterolateral spine of the carapace is closer to the lateral orbital spine than in *U. brucei* in which it is separated by twice its basal breadth when viewed from dorsal side; the P1 merus and carpus are granulose instead of smooth on the dorsal surface in large specimens; the palm has the mesial margin rounded instead of sharply ridged; the antennal article 4 bears a small blunt distoventral process rather than a distinct spine, and article 5 is unarmed instead of bearing a distinct distomesial spine; the antennal article 2 bears a tiny instead of a distinct lateral spine; and the P2-4 dactyli are relatively long, the dactylus-carpus length ratio being 0.7-0.8 on P2, 0.8-0.9 on P3, 0.9-1.0 on P4, instead of 0.6 on P2 and P3, 0.7 on P4. Differences between *U. maori* and *U. granulipes* are discussed under the latter species (see above).

Uroptychus marcosi n. sp.
Figures 141, 142

TYPE MATERIAL — Holotype: **New Caledonia**, Chesterfield Islands. MUSORSTOM 5 Stn CC365, 19°43'S, 158°48'E, 710 m, 19.X.1986, with corals of Isididae (Calcaxonia), ♀ 7.2 mm (MNHN-IU-2014-16725).

ETYMOLOGY — The species is named for Marcos S. Tavares for his long-lasting friendship.

DISTRIBUTION — Chesterfield Islands; 710 m.

DESCRIPTION — Medium-sized species. *Carapace*: 1.2 × longer than broad; greatest breadth 1.6 × distance between anterolateral spines. Dorsal surface smooth and glabrous, slightly convex from anterior to posterior, with feeble depression between gastric and cardiac regions. Lateral margins posteriorly divergent to point one-quarter from posterior end, then gently convex; ridged along posterior two-fifths of length; anterolateral spine small, ending in somewhat acuminate lateral limit of orbit. Rostrum narrowly triangular, with interior angle of 20°, nearly straight and horizontal, slightly overreaching cornea; length 0.4 × that of remaining carapace, breadth less than half carapace breadth at posterior carapace margin; dorsal surface flattish. Lateral limit of orbit without distinct spine, situated distinctly anterior to and well separated from anterolateral spine. Pterygostomian flap anteriorly angular, ending in very small spine; surface smooth.

Sternum: Excavated sternum subtriangular on anterior margin, surface with small spine in center. Sternal plastron slightly shorter than broad, lateral extremities divergent posteriorly. Sternite 3 shallowly depressed, anterior margin broadly and deeply excavated, with 2 small submedian spines (additional small spine lateral to right spine); anterolateral angle rounded. Sternite 4 with slightly convex anterolateral margin anteriorly ending in small spine followed by posteriorly diminishing eminences; posterolateral margin half as long as anterolateral margin. Anterolateral margin of sternite 5 convexly divergent posteriorly, about as long as posterolateral margin of sternite 4.

Abdomen: Smooth and glabrous. Somite 1 smooth on surface, moderately convex from anterior to posterior. Somite 2 tergite 2.3 × broader than long; pleuron posterolaterally blunt, lateral margins somewhat concave and strongly divergent posteriorly. Pleuron of somite 3 posterolaterally rounded. Telson slightly more than half as long as broad; posterior plate distinctly emarginate on posterior margin, length 1.5 × that of anterior plate.

Eye: Relatively broad (1.6 × longer than broad), falling short of apex of rostrum, slightly broadened distally, mesial margin somewhat concave, lateral margin convex. Cornea somewhat inflated, nearly as long as remaining eyestalk.

Antennule and antenna: Antennular ultimate article twice as long as high. Antennal peduncle terminating in corneal distal margin, relatively slender. Article 2 with small but distinct distolateral spine. Antennal scale ending in midlength of article 5, breadth about 2 × that of article 5. Distal 2 articles unarmed; article 5 2.3 × longer than article 4, breadth less than half height of antennular ultimate article. Flagellum of 15 segments barely reaching distal end of P1 merus.

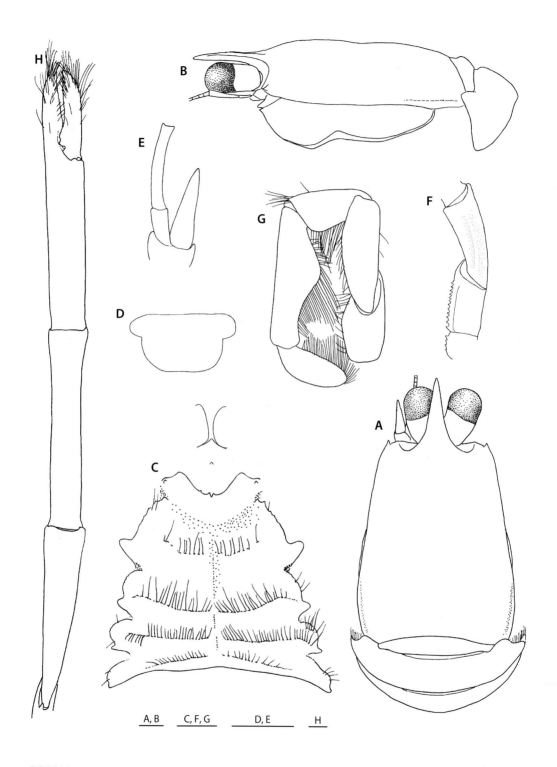

FIGURE 141

Uroptychus marcosi n. sp., holotype, female 7.2 mm (MNHN-IU-2014-16725). **A**, carapace and anterior part of abdomen, dorsal. **B**, same, lateral. **C**, sternal plastron, with excavated sternum and basal parts of Mxp1. **D**, telson. **E**, left antenna, ventral. **F**, left Mxp3, setae omitted, ventral. **G**, same, lateral. **H**, left P1, dorsal. Scale bars: 1 mm.

Mxp: Mxp1 with bases close to each other. Mxp3 basis with 4 denticles on mesial ridge. Ischium with 12-14 denticles on proximal three-quarters of crista dentata, flexor margin distally not rounded. Merus elongate, 2.5 × longer than ischium; flexor margin ridged and unarmed; distolateral spine absent. Carpus also unarmed.

P1: Slender, 4.6 × longer than carapace. Ischium dorsally with moderate-sized, flattish spine, ventromesially with a few proximal spines and no subterminal spine. Merus with blunt low distomesial process ventrally, length slightly more than that of carapace. Carpus unarmed, 1.2 × longer than merus. Palm with subparallel mesial and lateral margins, 4.2 × longer than broad, finely granulate ventrally, length 0.8 × that of carpus. Fingers with relatively short setae as illustrated,

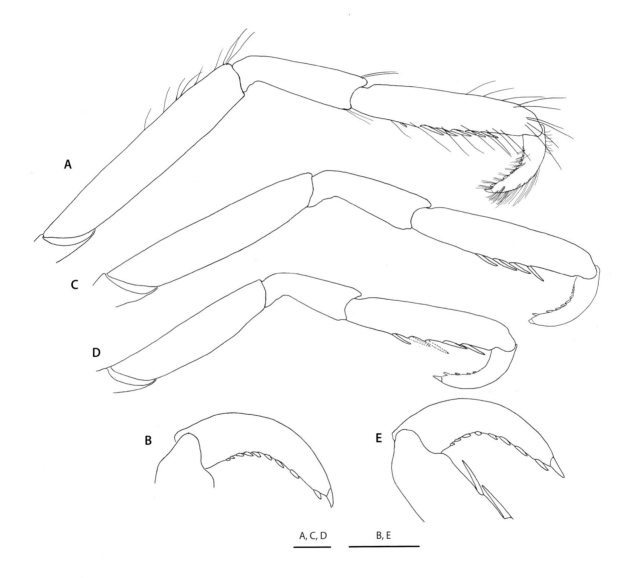

FIGURE 142

Uroptychus marcosi n. sp., holotype, female 7.2 mm (MNHN-IU-2014-16725). **A**, right P2, lateral. **B**, same, distal part, setae omitted, lateral. **C**, right P3, setae omitted, lateral. **D**, right P4, lateral. **E**, same, distal part, lateral. Scale bars: 1 mm.

not distinctly crossing when closed, each distally ending in papilla-like small spine; movable finger half as long as palm, opposable margin with low proximal process fitting into longitudinal groove on opposite face of fixed finger when closed.

P2-4: Relatively slender, with sparse long setae. Meri well compressed mesio-laterally, unarmed, successively shorter posteriorly (P3 merus 0.9 × length of P2 merus, P4 merus 0.7-0.8 × length of P3 merus), equally broad on P2-3, P4 merus 0.9 × as broad as P3 merus; length-breadth ratio, 5.4 on P2, 4.8 on P3, 4.1 on P4; P2 merus 0.9 × length of carapace, 1.2 × length of P2 propodus; P3 merus 1.1 × length of P3 propodus; P4 merus subequal to length of P4 propodus. Carpi slightly more than half length of propodi, much longer than dactyli (carpus-dactylus length ratio, 1.5 on P2, 1.4 on P3, 1.2 on P4); P3 carpus 0.9 × length of P2 carpus, P4 carpus 0.8 × length of P3 carpus. Propodi subequal in length on P2 and P3, shorter on P4; flexor margin straight, with row of 6 or 7 slender spines on P2, 4 or 5 spines on P3 and P4, distalmost single, located at point one-fourth from juncture with dactylus on P2, more remote on P3 and P4. Dactyli strongly curving at proximal third, shorter than carpi (dactylus-carpus length ratio, 0.7 on P2 and P3, 0.8 on P4), 0.4 × as long as propodi on P2-4; flexor margin with 10 spines, ultimate spine largest, penultimate more remote from antepenultimate than from ultimate on P2 and P3, remaining spines oriented parallel to flexor margin and diminishing toward base of article.

REMARKS — The new species resembles *U. australis* (Henderson, 1885) in the shapes of the carapace and sternal plastron, and in the spination of P2-4 dactyli. However, they are readily distinguished by the following differences. In *U. marcosi*, the P4 merus is longer, being 0.7-0.8 instead of 0.5-0.6 times the length of P3 merus; the P2-4 propodi bear the distalmost of the flexor marginal spines single and noticeably remote from the juncture with the dactylus whereas the distalmost is paired with another spine mesial to it and very close to the juncture in *U. australis*; the lateral limit of the orbit is angular, without spine instead of ending in a distinct spine, and the antennal scale terminates in the midlength of the antennal article 5, instead of reaching the distal end of that article.

Uroptychus marcosi also shares the following characters with *U. bispinatus* Baba, 1988 and *U. vandamae* Baba, 1988: the carapace lateral margin bearing an anterolateral spine only; the P2-4 dactyli with the flexor spines oriented parallel to the margin; the P2-4 propodi with the terminal one of the flexor spines single, not paired; and the antennal scale short, terminating at most in the midlength of article 5. *Uroptychus marcosi* differs from *U. bispinatus* in having the posterolateral margin of sternite 4 distinctly shorter than instead of as long as the anterolateral margin, in having the pterygostomian flap anteriorly produced to a distinct spine instead of being roundish with a very small spine, in having the P2-4 propodi with 2 or 3 instead of 4 or 5 flexor marginal spines distant from the juncture with the dactyli, and in having the carapace distinctly longer than instead of as long as broad.

Uriptychus vandamae has a rostrum that extends far beyond the eyes; the anterolateral spine of the carapace is relatively strong; the terminal one of the flexor spines of P2-4 propodi is closer to the juncture with the dactyli and remote from the distal second spine; and the antepenultimate spine of P2-4 dactyli is noticeably distant from the penultimate (Figure 275), all to mention the differences from *U. marcosi*.

Uroptychus megistos n. sp.
Figures 143, 144

TYPE MATERIAL — Holotype: **Vanuatu**. MUSORSTOM 8 Stn CP983, 19°21.61'S, 169°27.76'E, 480-475 m, 23.IX.1994, ov. ♀ 5.2 mm (MNHN-IU-2014-16726). Paratypes: Collected with holotype, 1 ♂ 4.3 mm (MNHN-IU-2014-16727). **Solomon Islands**. SALOMON 1 Stn DW1826, 9°56.4'S, 161°04.0'E, 418-432 m, 4.X.2001, 1 ♂ 6.0 mm, 1 ov. ♀ 6.6 mm (MNHN-IU-2014-16728).

ETYMOLOGY — From the Greek *megistos* (largest, greatest), alluding to the ultimate of the flexor marginal spines of P2-4 dactyli, that is the largest among the others, by which the species is distinguished from *U. duplex* n. sp. and *U. zigzag* n. sp.

DISTRIBUTION — Vanuatu and Solomon Islands; 418-480 m.

SIZE — Males, 4.3, 6.0 mm; ovigerous females, 5.2, 6.6 mm.

DESCRIPTION — Medium-sized species. *Carapace*: Posteriorly broadened, 1.3-1.4 × broader than long (0.7-0.8 × as long as broad); greatest breadth 2.1 × distance between anterolateral spines. Dorsal surface covered with short fine setae, smoothly convex from anterior to posterior, without any depression; small scattered spines on hepatic region and transverse row of very tiny spines on anterior gastric region (both barely discernible in holotype). Lateral margins convexly divergent posteriorly, with row of spines of irregular sizes: first anterolateral spine strong, overreaching lateral orbital spine; second and third very small (third obsolete on right side in holotype) placed on hepatic region; fourth situated at anterior end of anterior branchial margin, larger than second and third but much smaller than first, followed by a few small spines; about 10 short spines on posterior branchial margin, posteriorly diminishing and tending to be laciniate. Rostrum triangular, with interior angle of 28°, nearly horizontal; dorsal surface flattish, with fine setae; length less than half that of remaining carapace, breadth one-third carapace breadth at posterior carapace margin. Lateral orbital spine very small, slightly anterior to level of and well separated from anterolateral spine. Pterygostomian flap anteriorly somewhat angular, produced to small spine, anterior surface with several spinules.

Sternum: Excavated sternum blunt subtriangular or nearly transverse on anterior margin, surface with longitudinal ridge in midline. Sternal plastron about twice broader than long, lateral extremities convexly divergent posteriorly. Sternite 3 weakly depressed, anterior margin deeply excavated, with semicircular or narrow median notch, lacking submedian spines. Sternite 4 with smooth anterolateral margin anteriorly blunt angular, twice as long as posterolateral margin. Anterolateral margin of sternite 5 strongly convex, 1.5 × longer than posterolateral margin of sternite 4.

Abdomen: Smooth, sparingly with short fine setae. Somite 2 tergite 2.6-2.9 × broader than long; pleural lateral margins concavely divergent, rounded on anterolateral corner, angular at posterolateral terminus. Pleuron of somite 3 with bluntly angular lateral end. Telson more than one-third as long as broad, less than half as long; posterior plate slightly longer than anterior plate, posterior margin nearly transverse, slightly or deeply concave.

Eye: Elongate (1.6 × longer than broad), slightly overreaching midlength of rostrum, proximally somewhat broader. Cornea not dilated, half as long as remaining eyestalk.

Antennule and antenna: Ultimate article of antennule 3.0-3.3 × longer than high. Antennal peduncle overreaching cornea. Article 2 distolaterally angular, without distinct spine. Antennal scale 1.7-1.8 × broader than article 5, bearing long setae distally, varying from slightly falling short of to slightly overreaching tip of article 5. Distal 2 articles each with distomesial spine accompanying plumose setae; article 5 1.3-1.4 × longer than article 4, breadth 0.6-0.7 × height of ultimate article of antennule. Flagellum consisting of 12-18 segments, fully or barely reaching distal end of P1 merus.

Mxp: Mxp1 with bases broadly separated. Mxp3 basis lacking denticles on mesial ridge. Ischium rather thick, with tuft of plumose setae lateral to slightly rounded distal end of flexor margin; crista dentata with numerous (more than 40) very small denticles. Merus 2.0 × as long as ischium, setose (relatively long plumose setae along flexor margin); distolateral spine distinct; flexor margin sharply ridged, with 2-4 denticles on distal third; mesial face flattish. Distolateral spine of carpus obsolete.

P1: 4.8-5.9 × (males), 4.0-4.9 × (females) longer than carapace, covered with soft, fine plumose setae. Ischium dorsally with dorsoventrally flattened triangular spine, ventromesially with row of 4-7 spinules including very small subterminal spine. Merus and carpus with short denticulate ridges supporting setae. Merus 1.0-1.2 × length of carapace; mesial face with several small denticle-like processes or spines on proximal half, ventral surface with distomesial and distolateral processes, both blunt; distal margin with papilla-like, very small spines. Carpus 1.3-1.4 × longer than merus, dorsally with a few tubercles near proximal end and a few obsolescent papilla-like tubercles on distal margin, ventrally with distomesial and distolateral processes, both blunt and short. Palm 3.3-3.4 × longer than broad, slightly longer (1.1 x) than carpus; surface smooth but a few small tubercles near proximal part of dorsal side. Fingers directed somewhat laterally, proportionately broad, distally incurved, crossing when closed; opposable margin with median eminence on fixed finger, rounded process on movable finger; length 0.4 × that of palm.

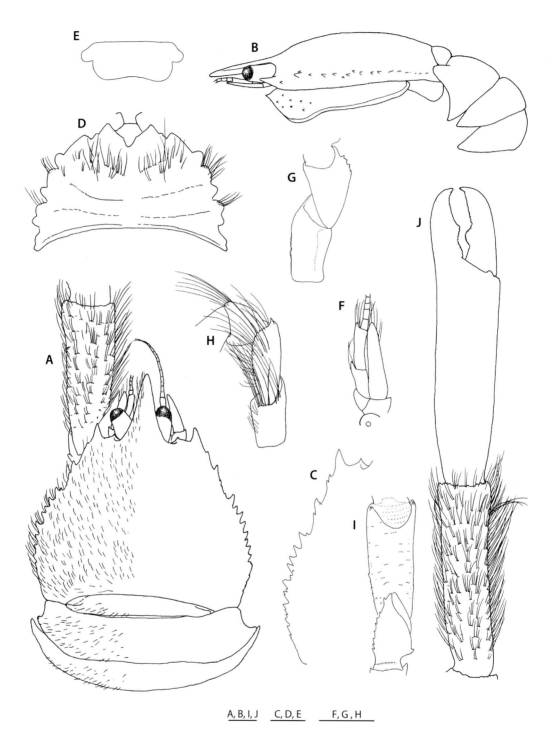

A, B, I, J   C, D, E   F, G , H

FIGURE 143

Uroptychus megistos n. sp., holotype, ovigerous female 5.2 mm (MNHN-IU-2014-16726), **A**, carapace and anterior part of abdomen (setae omitted on right half), with proximal part of left P1, dorsal. **B**, same, setae omitted, lateral. **C**, lateral margin of carapace, dorsal. **D**, sternal plastron, with excavated sternum and basal parts of Mxp1. **E**, telson. **F**, left antenna, ventral. **G**, left Mxp3, setae omitted, ventral. **H**, same, lateral. **I**, left P1, proximal part, setae omitted, ventral. **J**, same, setae omitted on distal articles, dorsal. Scale bars: 1 mm.

P2-4: setose like P1, broad relative to length, moderately depressed. Meri successively shorter posteriorly (P3 merus 0.9 × length of P2 merus, P4 merus 0.7-0.8 × length of P3 merus), breadths subequal on P2 and P3, P4 merus 0.9 × breadth of P3; length-breadth ratio, 3.4-3.6 on P2, 2.7-2.9 on P3, 2.1-2.5 on P4; P2 merus 0.8 × length of carapace, about as long as P2 propodus; P3 merus 0.9 × length of P3 propodus; P4 merus 0.7-0.8 × length of P4 propodus; dorsal margin with obsolescent eminences on proximal portion. Carpi successively shorter posteriorly (P3 carpus 0.9 × length of P2 carpus, P4 carpus 0.8-0.9 × length of P3 carpus); carpus-propodus length ratio, 0.5-0.6 on P2, 0.5 on P3, 0.4-0.5 on P4. Propodi longer on P3 than on P2 and P4, slightly longer on P2 than on P4; flexor margin distinctly convex on distal half, ending in pair of spines preceded by row of 5-7 movable spines along distal third on P2, 4-6 spines on P3, 3-5 spines on P4. Dactyli subequal on P2-4 or shorter on P2 than on P3 and P4; dactylus-carpus length ratio, 0.7 on P2, 0.8 on P3, 0.9-1.0 on P4; dactylus-propodus length ratio 0.3-0.4 on P2, 0.4 on P3 and P4; flexor margin gently curving, with 6 or 7 strong triangular spines somewhat obliquely directed, loosely arranged and proximally diminishing, ultimate strongest.

Eggs. About 30 eggs carried; size, 1.07 mm × 1.24 mm - 1.08 mm × 1.24 mm.

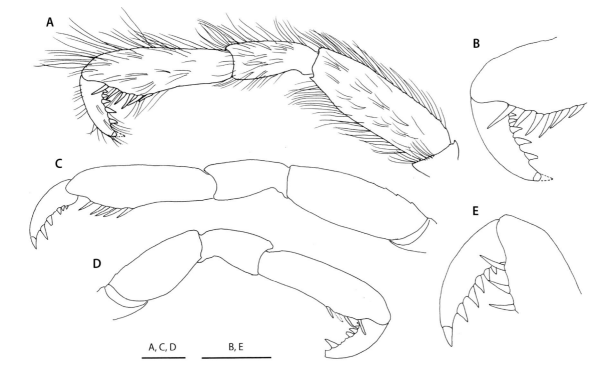

FIGURE 144

Uroptychus megistos n. sp., holotype, ovigerous female 5.2 mm (MNHN-IU-2014-16726). **A**, left P2, lateral. **B**, same, distal part, setae omitted, lateral. **C**, left P3, setae omitted, lateral. **D**, left P4, setae omitted, lateral. **E**, same, distal part, setae omitted, lateral. Scale bars: 1 mm.

REMARKS — Slight morphological differences are observed between Vanuatu and Solomon Islands specimens. The P4 merus is shorter in the Vanuatu specimens (P4 merus-P3 merus length ratio, 0.7 versus 0.8). Also, the P2-4 propodi in the Vanuatu specimens are broader, with the length-breadth ratio 3.4-3.7 versus 4.2-4.3 on P2, 3.9-4.0 versus 4.3-4.4 on P3, 3.6-4.3 versus 4.5-4.7 on P4.

The new species resembles *U. paracrassior* Ahyong & Poore, 2004 from off southern Queensland in having the spinous carapace lateral margin and in having the distally broadened P2-4 propodi and in the spination of the P2-4 dactyli, especially the ultimate being subequal to or slightly larger than the penultimate. *Uroptychus megistos* can be separated from that species by: sternite 3 without submedian spines, the P2-4 dactyli with 5 instead of 9-11 flexor marginal spines, the pterygostomian flap that is unarmed instead of bearing 2 spines between the anterior terminal spine and the anterior end of the linea anomurica, and the antennal article 2 with a very short instead of strong distolateral spine. In addition, the lateral orbital spine is tiny in *U. megistos*, whereas relatively large in *U. paracrassior*.

Uroptychus megistos also resembles *U. duplex* n. sp., *U. macrolepis* n. sp. and *U. zigzag* n. sp. in having the spinous carapace lateral margin and in having the distally convex flexor margins of P2-4 propodi. However, these three species have a different spination of the P2-4 dactyli, the ultimate spine being smaller than instead of equally strong as the penultimate as in *U. megistos*. *Uroptychus megistos* is differentiated from *U. duplex* by the anterior margin of sternite 3 that is deeply excavated with a median notch, instead of being transverse on the median third, without median notch; the P2-4 dactyli bear fewer flexor spines (7 or 8 instead of 12 or 13), and the antennal articles 4 and 5 each bear a distoventral spine instead of being unarmed. *Uroptychus macrolepis* has elongate eyes, sternite 3 without a distinct median notch on the anterior margin, the flexor spines of the P2-4 dactyli more numerous (11-13), the ultimate of these being much more slender than the penultimate, and the distal two articles of the antennal peduncle unarmed, all features that separate the species from *U. megistos*.

The relationships with *U. zigzag* are discussed under the remarks of that species (see below).

Uroptychus mesodme n. sp.

Figures 145, 146

TYPE MATERIAL — Holotype: **Fiji Islands**. BORDAU 1 Stn CP1461, 18°09'S, 178°48'W, 560 m, 6.III.1999, 1 ♂ 3.9 mm (MNHN-IU-2014-16729).

ETYMOLOGY — From the Greek *mesodme* (something between), alluding to a pair of small spines between two larger epigastric spines.

DISTRIBUTION — Fiji Islands, 560 m.

DESCRIPTION — *Carapace*: Broader than long (0.8 × as long as broad); greatest breadth 1.6 × distance between antero-lateral spines. Dorsal surface smooth, gastric region smoothly sloping down to rostrum, posterior portion (cardiac and intestinal regions) somewhat convex; somewhat ridged along posterior fifth of lateral margin; epigastric region with 2 pairs of spines, mesial pair small, situated behind eyes, lateral pair well developed, somewhat lateral to midpoint of upper orbital margin. Lateral margins constricted at anterior third; anterior part (anterior to constrictions) gently divergent posteriorly, with 3 spines, first anterolateral, strong, directed straight forward, reaching proximal part of cornea, second small, situated at level of first, directed anteriorly, third strong, somewhat dorsal to level of second, directed anterolaterally; posterior part (posterior to constrictions) well convex, with 5 short, posteriorly diminishing spines (anteriormost preceded by small spine on left side; left second small). Rostrum narrow triangular, distally curving dorsally, with interior angle of 21°; length 1.2 × breadth, 0.4 × that of remaining carapace; dorsal surface moderately concave; breadth less than half carapace breadth at posterior carapace margin. Lateral orbital spine small, situated somewhat anterior to level of

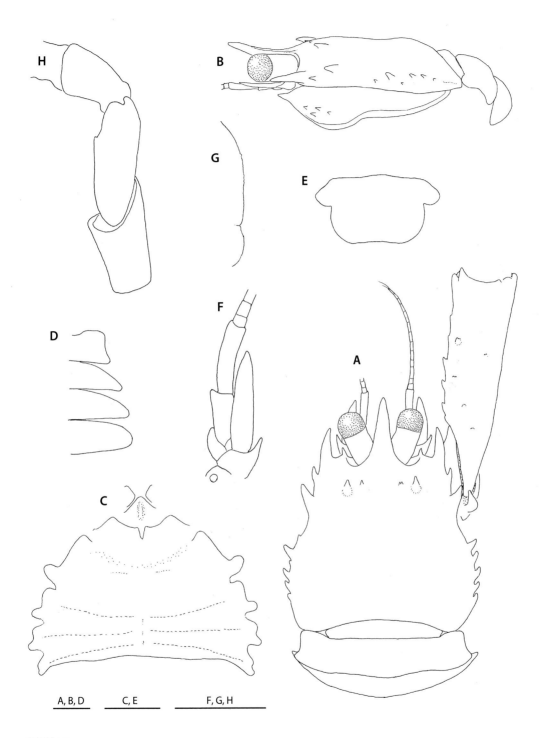

A, B, D   C, E   F, G, H

FIGURE 145

Uroptychus mesodme n. sp., holotype, male 3.9 mm (MNHN-IU-2014-16729). **A**, carapace and anterior part of abdomen, proximal part of right P1 included, dorsal. **B**, same, lateral. **C**, sternal plastron, with excavated sternum and basal parts of Mxp1. **D**, right pleura of abdominal somites 2-5, dorsolateral. **E**, telson. **F**, left antenna, ventral. **G**, right Mxp3, crista dentata, ventral. **H**, left Mxp3, setae omitted, lateral. Scale bars: 1 mm.

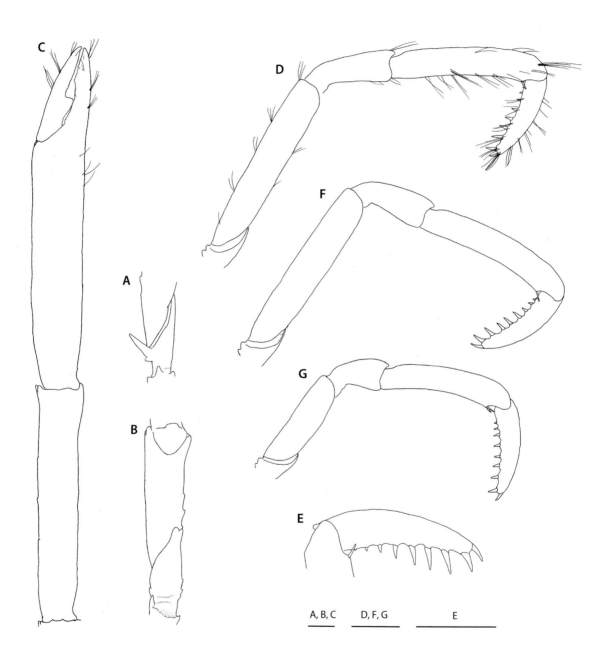

FIGURE 146

Uroptychus mesodme n. sp., holotype, male holotype 3.9 mm (MNHN-IU-2014-16729). **A**, right P1, proximal part, lateral. **B**, same, ventral. **C**, same, distal three articles, dorsal. **D**, right P2, lateral. **E**, same, distal part, setae omitted, lateral. **F**, right P3, setae omitted, lateral. **G**, right P4, setae omitted, lateral. Scale bars: 1 mm.

anterolateral spine. Pterygostomian flap anteriorly angular, produced to strong spine, surface with 2 (right) or 4 (left) spines on anterior part; height of posterior portion 0.5 × that of anterior portion.

Sternum: Excavated sternum with sharp ridge in midline leading onto anterior margin produced triangularly between bases of Mxp1. Sternal plastron 1.5 × broader than long; lateral extremities convexly divergent behind sternite 4, sternite 7 slightly narrower than sternite 6. Sternite 3 shallowly depressed; anterior margin moderately concave, with V-shaped median notch without flanking spine. Sternite 4 short relative to breadth; anterolateral margin slightly convex, anteriorly rounded, length 1.6 × that of posterolateral margin. Sternite 5 with convex anterolateral margin as long as posterolateral margin of sternite 4.

Abdomen: Somites short relative to breadth. Somite 1 well convex from anterior to posterior. Somite 2 tergite 3.0 × broader than long; pleuron moderately produced posterolaterally, ending in blunt margin. Pleuron of somite 3 laterally tapering to blunt end, those of somites 4-5 laterally rounded. Telson 0.5 × as long as broad; posterior plate 1.3 × longer anterior plate, posterior margin slightly concave.

Eye: Elongate, 2.0 × longer than broad, barely reaching apex of rostrum, slightly broadened proximally. Cornea 0.6 × length of remaining eyestalk.

Antennule and antenna: Ultimate article of antennular peduncle 2.9 × longer than high. Antennal article 2 with strong distolateral spine. Antennal scale proportionately broad, 1.3 × broader than article 5, ending in point two-thirds of length of article 5. Article 4 with bluntly produced distomesial spine. Article 5 unarmed, length 1.3 × that of article 4, breadth 0.5 × height of ultimate article of antennule. Flagellum of 14 segments reaching distal end of P1 merus.

Mxp: Mxp1 with bases broadly separated. Mxp3 basis without denticle on mesial ridge. Ischium having flexor margin distally rounded, crista dentata with denticles nearly obsolete, only visible under high magnification. Merus twice as long as ischium, mesial face flattish; flexor margin roundly ridged on proximal half, sharply ridged on distal half, bearing 1 blunt very short spine at distal third; lateral face with short blunt distal spine. Carpus unarmed.

P1: 7.2 × longer than carapace, smooth and barely setose except for fingers. Ischium with strong dorsal spine accompanying small spine proximally; ventromesial margin with subterminal spine very short and blunt, not reaching distal end of ischium, proximally with a few obsolescent protuberances. Merus with 2 rows each of short blunt spines (dorsomesial row of 2 spines, ventromesial row of 5 spines), ventral distomesial spine blunt and short; length 1.4 × that of carapace. Carpus 1.5 × length of merus, mesial margin with terminal spine followed proximally by 3 rudimentary spines. Palm unarmed, depressed, two-thirds as high as broad, 4.5 × longer than broad, lateral and mesial margins subparallel. Fingers relatively slender distally, sparingly setose, moderately gaping in proximal two-thirds, not incurved distally; fixed finger directed slightly laterally, with low eminence at distal third; movable finger 0.4 × as long as palm, opposable margin with prominent median process (distal margin of process perpendicular to opposable margin); no longitudinal groove on ventromesial face of fixed finger.

P2-4: Relatively slender, well compressed mesio-laterally, with sparse simple setae, more setose on propodi and dactyli. Ischia with 2 short dorsal spines. Meri subequal in length on P2 and P3, shortest on P4 (P4 merus 0.6 × length of P3 merus); subequally broad on P2 and P3, slightly narrower on P4; dorsal margins unarmed; length-breadth ratio, 5.0 on P2 and P3, 3.2 on P4; P2 merus as long as carapace, 1.1 × longer than P2 propodus; P3 merus 1.1 × length of P3 propodus, P4 merus 0.6 × length of P4 propodus. Carpi successively shorter posteriorly, 0.5 × length of propodus on P2, 0.4 × on P3 and P4. Propodi subequal in length on P2 and P3, shortest on P4; flexor margin slightly concave in lateral view, with pair of slender terminal spines only. Dactyli subequal in length on P2-4; dactylus-carpus length ratio, 1.0 on P2, 1.2 on P3, 1.7 on P4; dactylus-propodus length ratio, 0.5 on P2 and P3, 0.6 on P4; flexor margin slightly curving in lateral view, with 9 sharp spines proximally diminishing, loosely arranged and nearly perpendicular to margin, ultimate spine more slender than penultimate.

REMARKS — As mentioned above under the account of *U. clarki* n. sp., *U. mesodme* resembles *U. paraplesius* n. sp. and *U. trispinatus* n. sp., sharing the carapace lateral margin with a group of 3 spines on the anterior part remotely separated from a group of 3-5 spines on the posterior part, the antennal scale overreaching the midlength of article 5, and article 4 bearing a distomesial spine. Characters distinguishing *U. mesodme* from *U. paraplesius* and *U. trispinatus* are discussed under the latter two species (see below).

Uroptychus micrommatus n. sp.

Figures 147-149

TYPE MATERIAL — Holotype: **Indonesia**, Tanimbar Islands. KARUBAR Stn CP69, 8°42'S, 131°53'E, 356-368 m, 2.XI.1991, ♂ 8.2 mm (MNHN-IU-2012-676). Paratypes: **Indonesia**, Tanimbar Islands. KARUBAR Stn CC42, 07°53'S, 132°42'E, 354-350 m, 28.X.1991, 1 ♂ 7.3 mm (MNHN-IU-2012-677). – Stn CP69, 8°42'S, 131°53'E, 356-368 m, 2.XI.1991, 6 ♂ 3.4-9.0 mm, 5 ov. ♀ 8.1-9.8 mm, 8 ♀ 3.7-10.2 mm (MNHN-IU-2012-675). Kai Islands. KARUBAR Stn CP09, 5°23'S, 132°29'E, 368-389 m, unidentified host, 23.X.1991, 1 ♂ 10.3 mm, 1 ov. ♀ 11.4 mm (MNHN-IU-2012-678). **Solomon Islands**. SALOMON 2 Stn CP2262, 7°56.4'S, 156°51.2'E, 460-487 m, 3.XI.2004, 2 ♂ 4.3, 8.7 mm (MNHN-IU-2013-12295, MNHN-IU-2013-8576). – Stn CP2263, 7°54.8'S, 156°51.3'E, 485-520 m, 3.XI.2004, 1 ♂ 9.2 mm (MNHN-IU-2013-8577). – Stn CP2264, 7°52.4'S, 156°51.0'E, 515-520 m, 3.XI.2004, 1 ♂ 9.0 mm, 1 ov. ♀ 9.7 mm, 1 ♀ 6.9 mm (MNHN-IU-2013-8578).

ETYMOLOGY — From the Greek *mikrommatos* (small-eyed), alluding to the eye cornea much smaller than the remaining eyestalk, a character to separate the species from *U. occultispinatus* Baba, 1988.

DISTRIBUTION — Indonesia (Kai and Tanimbar Islands) and Solomon Islands; 354-520 m.

SIZE — Males, 3.4-10.3 mm; females, 3.7-11.4 mm; ovigerous females from 8.1 mm.

DESCRIPTION — Large species. Body and appendages covered with setae. *Carapace*: Slightly broader than long (0.9 × as long as broad). Dorsal surface slightly convex from anterior to posterior, bearing distinct groove between gastric and cardiac regions, and anterior and posterior branchial regions; covered with scale-like short ridges (somewhat elevated in large specimens, very weak in small specimens) supporting setae, bearing tiny tubercles or denticles, usually on anterior gastric region; greatest breadth 1.5-1.9 × (usually 1.7 x) distance between anterolateral spines. Lateral margins convex, with distinct spines: first anterolateral spine well developed, overreaching lateral orbital spine, followed by 1 or 2 spinules often obsolete on hepatic margin; second spine situated at anterior end of anterior branchial margin, as large as anterolateral spine, followed by 1 or 2 spines usually very small, occasionally of good size or obsolete; third spine located at anterior end of posterior branchial margin, stronger than second spine, followed by 3 distinct spines, posteriormost usually small, rarely obsolete, followed by ridge along lateral margin. Rostrum triangular, with interior angle of 27-30 °; dorsal surface somewhat concave; lateral margin rarely with very small subapical spine; length 0.4-0.5 × that of carapace, breadth about half carapace breadth at posterior carapace margin. Lateral orbital spine very small, moderately remote from and slightly anterior to level of anterolateral spine. Pterygostomian flap anteriorly angular, produced to small spine, surface granulose.

Sternum: Excavated sternum having anterior margin strongly convex between bases of Mxp1, surface with low ridge in midline. Sternal plastron slightly broader than long. Sternite 3 depressed well, anterolateral angle sharply produced; anterior margin representing broad V-shape, with U-shaped or semicircular median sinus flanked by very small, obsolescent spine. Sternite 4 having straight or slightly convex, smooth anterolateral margin anteriorly blunt angular, often bearing a few denticles, 1.3-1.6 × longer than posterolateral margin. Sternite 5 slightly broader than sternite 4, anterolateral margins strongly convex anteriorly, 1.0-1.3 × length of posterolateral margin of sternite 4. Sternite 6 broadest; sternite 7 narrower than sternite 6.

Abdomen: Somite 1 antero-posteriorly strongly convex. Somite 2 tergite 2.6-2.9 × broader than long; pleural lateral margin weakly concave and weakly divergent posteriorly, posterolateral end blunt. Pleuron of somite 3 with blunt lateral margin. Telson half as long as broad; posterior plate concave or emarginate on posterior margin, usually 1.2-1.3 × longer than (rarely subequal to) anterior plate.

Eye: Elongate (1.7 × longer than broad), markedly narrowed distally, terminating in midlength of rostrum. Cornea small, less than half length of remaining eyestalk.

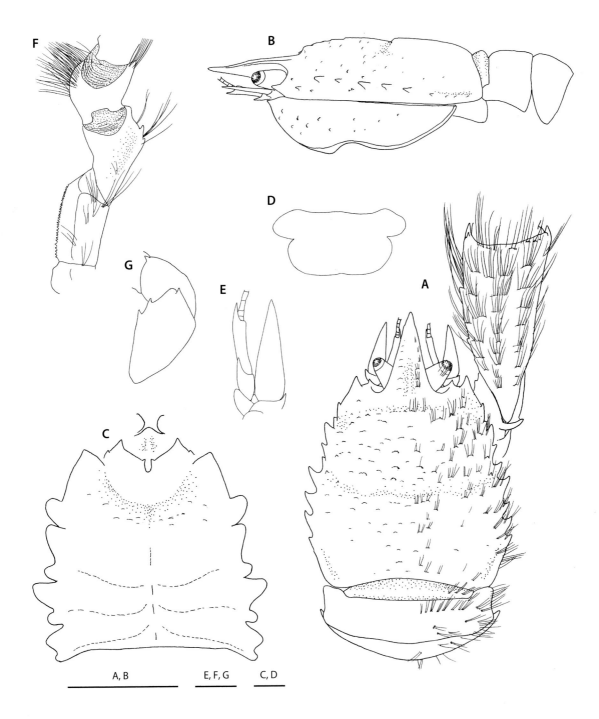

FIGURE 147

Uroptychus micrommatus n. sp., holotype, male 8.2 mm (MNHN-IU-2012-676). **A**, carapace and abdomen, setae omitted from left half, proximal part of P1 included, dorsal. **B**, same, lateral. **C**, sternal plastron, with excavated sternum and basal parts of Mxp1. **D**, telson. **E**, left antenna, ventral. **F**, left Mxp3, distal articles omitted, ventral. **G**, same, merus and carpus, setae omitted, lateral. Scale bars: A, B; 5 mm; C-G, 1 mm.

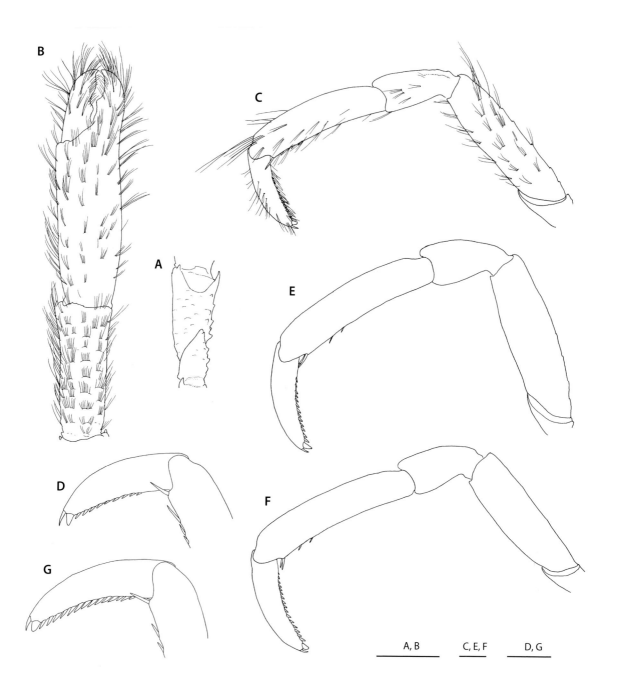

FIGURE 148

Uroptychus micrommatus n. sp., holotype, male 8.2 mm (MNHN-IU-2012-676). **A**, right P1, proximal part, ventral. **B**, right P1, dorsal. **C**, left P2, lateral. **D**, same, distal part, setae omitted. **E**, left P3, setae omitted, lateral. **F**, left P4, setae omitted, lateral. **G**, same, distal part, setae omitted. Scale bars: A, B, 5 mm; C-G, 1 mm.

Antennule and antenna: Ultimate article of antennular peduncle relatively slender, 3.1-3.5 × (2.6 × in small specimens) longer than high. Antennal peduncle extending far beyond cornea, barely reaching apex of rostrum. Article 2 with small but distinct distolateral spine. Antennal scale tapering, relatively broad basally, more than 2 × broader than article 5, overreaching article 5. Article 4 with strong distomesial spine. Article 5 about twice as long as article 4, breadth 0.6 × height of ultimate article of antennule; with strong distomesial spine. Flagellum of 18-20 segments reaching distal end of P1 merus (15 segments overreaching P1 merus by distal 7 segments in smallest specimen).

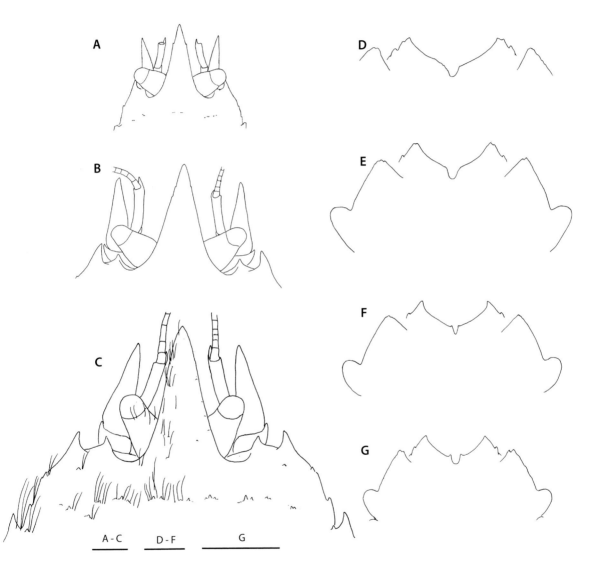

FIGURE 149

Uroptychus micrommatus n. sp., **A-C**, carapace, dorsal. **D-G**, anterior part of sternal plastron. **A**, paratype, female 3.7 mm (MNHN-IU-2012-675). **B**, paratype, female 8.7 mm (MNHN-IU-2012-675). **C**, paratype, female 10.3 mm (MNHN-IU-2012-675). **D**, paratype, male 10.3 mm (MNHN-IU-2012-675). **E**, paratype, female 9.0 mm (MNHN-IU-2012-675). **F**, paratype, female 8.0 mm (MNHN-IU-2012-675). **G**, paratype, female 3.7 mm (MNHN-IU-2012-675). Scale bars: 1 mm.

Mxp: Mxp1 with bases broadly separated. Mxp3 basis with 1 distal denticle usually obsolescent, often absent on mesial ridge. Ischium with distally rounded flexor margin, crista dentata with numerous (34-40) denticles. Merus broad relative to length, 1.6 × longer than ischium, flattish on mesial face; distolateral spine small; flexor margin ridged, with a few small, distinct spines distal to point one-third from distal end. Distolateral spine of carpus small or obsolescent.

P1: Massive, setose, length 3.4-3.7 × that of carapace. Ischium dorsally with relatively short, basally broad, depressed spine, ventrally with tubercular processes along mesial margin, with subterminal spine almost vestigial or obsolete. Merus and carpus somewhat tuberculose, each with distomesial and distolateral spines ventrally; additional short spines along mesial margin of merus. Merus 0.8-0.9 × as long as carapace. Carpus 1.0 × (small specimens)-1.2 × length of merus. Palm 2.2-2.7 × longer than broad, 1.2-1.4 × longer than carpus. Fingers directed somewhat laterally in large specimens, not gaping, distally crossing. Movable finger 0.4 × as long as palm (slightly more than half as long in small specimens), opposable margin with low median process proximal to position of opposite low eminence on fixed finger.

P2-4: Relatively setose. Meri with setiferous short rugae, successively shorter posteriorly (P3 merus 0.9-1.0 × length of P2 merus, P4 merus 0.9 × length of P3 merus), P2 and P3 meri subequally broad, P4 merus slightly narrower than P3 merus; dorsal margin with several eminences. P2 merus 0.7-0.8 length of carapace, subequal to length of P2 propodus; P3 merus 0.9 × length of P3 propodus; P4 merus 0.7-0.8 × length of P4 propodus; length-breadth ratio, 3.5-3.7 on P2, 3.1-3.3 on P3, 3.0-3.4 on P4. Carpi successively slightly shorter posteriorly (P3 carpus 0.9-1.0 × length of P2 carpus, P4 carpus 0.9-1.0 × length of P3 carpus); carpus-propodus length ratio, 0.4-0.5 on P2, 0.4 on P3, 0.3-0.4 on P4. Propodi successively longer posteriorly or shorter on P2 than on P3 and P4 and subequal on P3 and P4; flexor margin slightly concave in lateral view, ending in pair of spines preceded by 2-3 movable spines on P2, 1-3 spines on P3 and P4. Dactyli proportionately broad, about half as long as propodi (slightly more than half as long in small specimens); dactylus-carpus length ratio, 1.0-1.1 (1.3-1.4 in small specimens) on P2, 1.1-1.2 (1.4-1.5 in small specimens) on P3, 1.4 (1.5-1.7 in small specimens) on P4; flexor margin slightly curving, with 10-16 spines obscured by thick setae, ultimate slender, usually slightly longer than penultimate and distinctly longer than antepenultimate; penultimate spine slightly more than 2 × as broad as antepenultimate, remaining proximal spines slender and obliquely directed, close to one another but not contiguous.

Eggs. More than 100 eggs carried; diameter, 1.2 mm.

REMARKS — The granules and setiferous scale-like ridges on the carapace dorsal surface are pronounced in large specimens, less so or not discernible in small specimens. The hepatic marginal spines are obsolescent in small specimens. The sternal plastron has the anterior margin with a median notch flanked by a spine usually very small, occasionally obsolescent or totally absent (Figure 149).

Small populational differences are observed. The Indonesian specimens (poc, 4.2-8.9 mm) differ from the Solomon Islands specimens (poc, 3.4-11.0 mm) in the following particulars: the antennular ultimate article is higher relative to length (the length-height ratio 2.6-3.4 versus 3.5-4.6); the P1 merus is broader, with the length-breadth ratio 0.8-0.9 versus 1.1-1.3 (the length measured in ventral midline, the breadth measured at midlength), the P1 merus-carapace length ratio being 0.8-0.9 versus 1.0. These differences are regarded here as intraspecific variations for the time being, awaiting molecular data (analyses using specimens of MNHN-IU-2012-677 and MNHN-IU-2012-12295 have failed (L. Corbari, personal. comm.).

Uroptychus micrommatus resembles *U. occultispinatus* Baba, 1988 (replacement name for *U. granulatus* var. *japonicus* Balss, 1913a) from Japan and the Philippines in nearly all aspects, but they key out in remotely different couplets due to the presence or absence of median notch and/or submedian spines on the anterior margin of sternite 3. *Uroptychus micrommatus* is characterized by eyes that are strongly narrowed distally rather than uniformly broad as in *U. occultispinatus* (Figure 150B; Balss 1913b: fig. 18). The P1 ischium in *U. micrommatus* has no distinct subterminal spine on the ventromesial margin, whereas the spine is short and distinct in *U. occultispinatus* (Figure 150E).

Uroptychus micrommatus resembles *U. dentatus* Balss, 1913 from the western Indian Ocean and *U. crassipes* Van Dam, 1939 (see above) in the carapace ornamentation, and in having relatively massive P1 and in the spination of the P2-4 dactyli. The shape of the sternal plastron, especially the median excavation of the anterior margin in *U. dentatus*

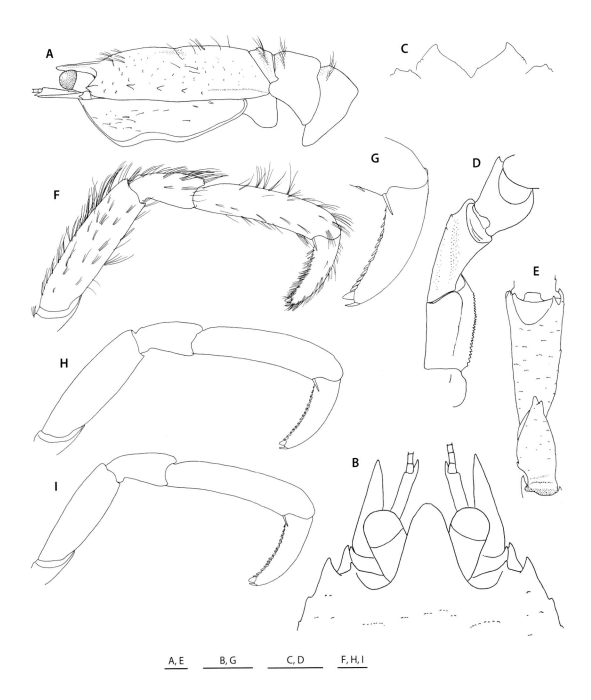

FIGURE 150

Uroptychus occultispinatus Baba, 1988, ovigerous female 6.9 mm from *Albatross* Stn 5529 (USNM 150308). **A**, carapace and anterior part of abdomen, lateral. **B**, anterior part of carapace, dorsal. **C**, anterior part of sternal plastron. **D**, right Mxp3, ventral. **E**, right P1, proximal part, ventral. **F**, right P2, lateral. **G**, same, distal part, setae omitted, lateral. **H**, right P3, lateral. **I**, right P4, lateral. Scale bars: 1 mm.

is variable as illustrated (Figure 151) for a female (6.1 mm) of the syntypes of *U. dentatus* from Valdivia Station 264 in 1079 m (Musée Zoologique, Strasbourg, MZS 349) and for the material reported by Baba (1990) from Madagascar. This variability is similar to that in the present material of *U. micrommatus* n. sp. (Figure 150). However, *U. dentatus* differs from *U. micrommatus* in having the Mxp3 merus unarmed instead of bearing a few distinct spines as in *U. micrommatus*. In addition, the eyes of *U. dentatus* are not narrowed distally, and the anterolateral margin of sternite 4 is slightly longer than the posterolateral margin, instead of being much longer (1.3-1.6 times longer) as in *U. micrommatus*.

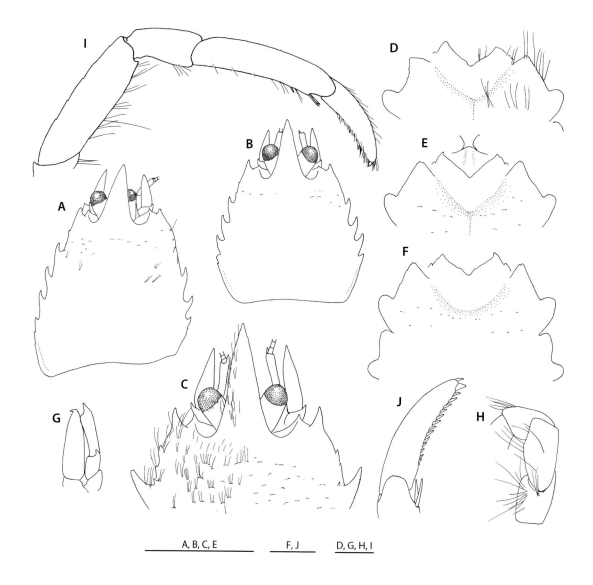

FIGURE 151

Uroptychus dentatus Balss, 1913. **A**, **D**, **G-J**, syntype, female 6.1 mm (MZS 349); **B**, female 5.2 mm from Madagascar, Vauban Stn CH 104 (MNHN-IU-2014-12825); **C**, male 9.0 mm, Vauban Stn CH 104 (MNHN-IU-2014-12825); **E**, same; **F**, female 5.2 (MNHN-IU-2014-12825). **A-C**, carapace. **D-F**, anterior part of sternum. **G**, right antenna. **H**, left Mxp3, lateral. **I**, right P2, lateral. **J**, same, distal part. Scale bars: A-C, E, 5 mm; D, F-I, J, 1 mm.

Uroptychus micrommatus can be separated from *U. crassipes* by the following differences: the rostral lateral margin is smooth instead of bearing a distinct subapical spine on each side; the eyes are distally narrowed even in small specimens rather than having the lateral and mesial margins subparallel; the P1 ischium has a reduced instead of well-developed subterminal spine on the ventromesial margin; and the P1 merus and carpus each bear almost vestigial instead of distinct spines on the dorsoterminal margin.

Uroptychus minor n. sp.

Figures 152, 153

TYPE MATERIAL — Holotype: **New Caledonia**. BIOCAL Stn DW51, 23°05'S, 167°45'E, 680-700 m, 31.VIII.1985, ♂ 2.2 mm (MNHN-IU-2013-8518). Paratypes: **New Caledonia**. Collected with holotype, 2 ♂ 1.6, 2.3 mm (MNHN-IU-2014-16730). CALSUB Dive PL20, 22°53,7'S, 167°23'E, 616-555 m, 10.III.1989, 2 ♀ 1.6, 2.1 mm (MNHN-IU-2014-16731). MUSORS-TOM 4 Stn DW151, 19°07.0'S, 163°22.0'E, 200 m, 14.IX.1985, 1 ♀ 2.0 mm (MNHN-IU-2014-16732). – Stn DW197, 18°51'S, 163°21'E, 560 m, 20.IX.1985, 1 ♀ 2.4 mm (MNHN-IU-2014-16733). SMIB 6 Stn DW123, 18°56.6'S, 163°25.0'E, 330-360 m, 3.III.1990, 1 ov. ♀ 2.6 mm (MNHN-IU-2014-16734). BATHUS 4 Stn DW924, 18°54.85'S, 163°24.34'E, 344-360 m, 7.VIII.1994, 1 ♀ 2.4 mm (MNHN-IU-2014-16735). **New Caledonia**, Norfolk Ridge. SMIB 4 Stn DW55, 23°21'S, 168°05'E, 215-260 m, 09.III.1989, 1 ♂ 2.0 mm, 2 ov. ♀ 1.5, 1.6 mm (MNHN-IU-2014-16736). – Stn DW57, 23°21'S, 168°04'E, 210-260 m, 09.III.1989, 1 ♂ 1.7 mm (MNHN-IU-2014-16737). **New Caledonia**, Chesterfield Islands. MU-SORSTOM 5 Stn DW308, 22°09'S, 159°23'E, 450-635 m, 12.X.1986, 1 ♀ 2.4 mm (MNHN-IU-2014-16738). EBISCO Stn DW2606, 19°36.0'S, 158°42.0'E, 442-443 m, 18.X.2005, 1 ♀ 3.0 mm (MNHN-IU-2014-16739).

ETYMOLOGY — From the Latin *minor* (smaller), alluding to the small size of the species.

DISTRIBUTION — New Caledonia, Norfolk Ridge and Chesterfield Islands; 200-700 m.

SIZE — Males, 1.6-2.3 mm; females, 1.5-3.0 mm; ovigerous females from 1.5 mm.

DESCRIPTION — Small species. Body elongate and subcylindrical. *Carapace*: Much longer than broad (1.3-1.4 × longer); greatest breadth 1.4 × distance between anterolateral spines. Dorsal surface strongly convex from side to side, nearly straight from anterior to posterior, feebly depressed on gastric-cardiac and gastric-branchial boundaries; sparsely or moderately with short fine setae. Lateral margins slightly divergent posteriorly or slightly convex medially, with row of very fine, short, oblique ridges often barely discernible; anterolateral spine slightly overreaching smaller lateral orbital spine. Rostrum very short, broad triangular, with interior angle of 46-48°, slightly deflected ventrally, varying from barely reaching to slightly overreaching midlength of eye, never reaching cornea, length 0.12-0.27 × that of remaining carapace, breadth half carapace breadth at posterior carapace margin. Lateral orbital spine small, situated directly mesial to ante-rolateral spine. Pterygostomian flap anteriorly angular, produced to distinct spine, smooth on surface.

Sternum: Excavated sternum with sharp ridge in ventral midline, anterior margin subtriangular. Sternal plastron 1.2 × longer than broad. Sternite 3 strongly depressed, anterolateral angle sharply produced anteriorly, anterior margin of broad V-shape with or without small median notch. Sternite 4 as broad as sternite 6, anterolateral margin slightly convex, anteriorly angular; posterolateral margin 0.7 × length of anterolateral margin. Sternite 5 narrower than sternite 4, anterolateral margin roundly produced anteriorly, about as long as posterolateral margin of sternite 4. Sternite 7 broader than sternite 6.

Abdomen: Somites long relative to breadth, covered with short fine setae. Somite 1 without transverse ridge. Somite 2 tergite 1.8-2.2 × broader than long; pleuron posterolaterally rounded, lateral margins feebly concave and nearly subparallel. Pleura of somites 3 and 4 with rounded lateral margin. Telson slightly more than half as long as broad; posterior plate semicircular, not emarginate on posterior margin, length 1.2-1.7 × that of anterior plate.

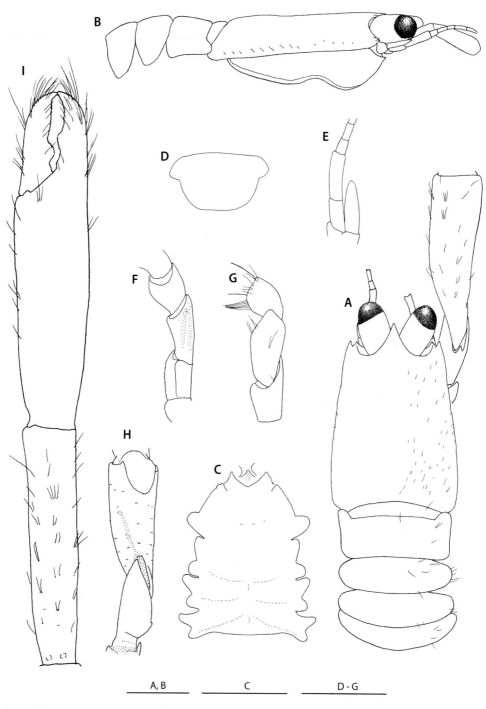

FIGURE 152

Uroptychus minor n. sp., holotype, male 2.2 mm (MNHN-IU-2013-8518). **A**, carapace and abdomen, proximal part of right P1 included, dorsal. **B**, same, lateral. **C**, sternal plastron, with excavated sternum and basal parts of Mxp1. **D**, telson. **E**, left antenna, ventral. **F**, left Mxp3, setae omitted, ventral. **G**, same, lateral. **H**, right P1, proximal part, setae omitted, ventral. **I**, same, dorsal. Scale bars: 1 mm.

Eye: Elongate (1.7 × longer than broad), overreaching rostrum, broad relative to carapace breadth, one-third as broad as distance between anterolateral spines of carapace; lateral and mesial margins somewhat convex. Cornea not dilated, length about half that of remaining eyestalk.

Antennule and antenna: Ultimate article of antennular peduncle 2.3-2.7 × longer than high. Antennal peduncle extending far beyond eyes. Article 2 with very small distolateral spine. Antennal scale slightly broader than article 5, proportionately broad distally, terminating in midlength of article 5. Distal 2 articles unarmed; article 5 1.5 × longer than article 4, breadth slightly less than half height of ultimate article of antennule. Flagellum of 7-11 segments barely reaching distal end of P1 merus.

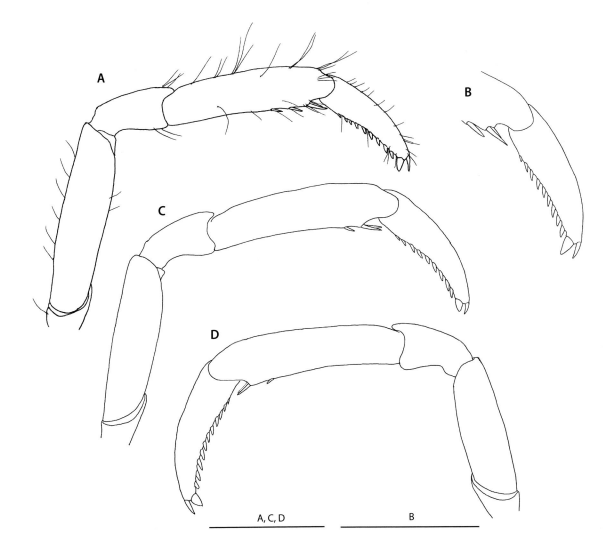

FIGURE 153

Uroptychus minor n. sp., holotype, male 2.2 mm (MNHN-IU-2013-8518). **A**, right P2, lateral. **B**, same, distal part, setae omitted, lateral. **C**, right P3, setae omitted, lateral. **D**, left P4, setae omitted, lateral. Scale bars: 1 mm.

Mxp: Mxp1 with bases broadly separated. Mxp3 sparsely setose on lateral surfaces of merus and carpus. Basis without denticles on mesial ridge. Ischium with very small, rather obsolescent denticles on crista dentata; flexor margin rounded distally. Merus 2.3 × longer than ischium, moderately compressed mesio-laterally; flexor margin with a few very small denticles distal to point one-third from distal end. Carpus unarmed.

P1: Massive especially on palm, almost unarmed, 3.7-5.4 × longer than carapace, sparsely with short soft setae. Ischium dorsally bearing basally broad, short, blunt spine, ventrally unarmed but a few obsolescent tubercles on proximal mesial margin. Merus and carpus with weak short ridges supporting setae. Merus as long as or slightly longer than carapace. Carpus 1.1-1.5 × longer than merus. Palm 2.1-3.4 × longer than broad, 1.0-1.4 × length of carpus. Fingers more setose than proximal articles, distally incurved, crossing when closed; movable finger 0.4-0.6 × length of palm, opposable margin with low obtuse proximal process somewhat proximal to position of opposite low eminence on fixed finger.

P2-4: Slender, sparsely setose, unarmed on meri and carpi. Meri successively shorter posteriorly (P3 merus 0.7-0.8 × length of P2 merus, P4 merus 0.9 × length of P3 merus); breadths subequal on P2-4; length-breadth ratio, 3.9-4.0 on P2, 3.3 on P3, 2.8-2.9 on P4; P2 merus 0.7-0.8 × length of carapace, about as long as or slightly shorter than P2 propodus; P3 merus 0.8 × length of P3 propodus; P4 merus 0.7 × length of P4 propodus. Carpi subequal, much shorter than dactyli (carpus-dactylus length ratio, 0.7 on P2, 0.5-0.6 on P3 and P4); carpus-propodus length ratio, 0.4 on P2, 0.3-0.4 on P3, 0.3 on P4. Propodi shorter on P2 than on P3 and P4 and subequal on P3 and P4 or longer on P2 than on P3 and P4; propodus-dactylus length ratio, 1.5-1.6 on P2-4; flexor margin nearly straight, ending in pair of spines preceded by 2 or 3 spines on P2, 1 spine on P3 and P4. Dactyli gently narrowed distally; dactylus-carpus length ratio, 1.4-1.5 on P2, 1.7-1.8 on P3 and P4; more than half length of propodi on P2-4; flexor margin slightly curving, ending in slender spine preceded by 8-12 spines, penultimate spine pronounced, more than twice broader than ultimate and antepenultimate spines, remaining spines slender and much shorter than ultimate, obliquely directed, closely arranged but not contiguous to one another, successively diminishing toward base of article.

Eggs. Number of eggs carried, 2-5; size; 0.7 mm × 0.8 mm - 0.9 mm × 1.0 mm.

REMARKS — The elongate, subcylindrical body with the short rostrum in the shape of an equilateral triangle is unique among the species of *Uroptychus*. It is interesting to note that the body size is small relative to the size of eggs. Two eggs carried by the smallest ovigerous female (1.5 mm) each measure two-thirds the greatest breadth of the carapace.

The elongate carapace is also possessed by *U. lacunatus* n. sp. (see above). However, *U. lacunatus* is characterized by the narrowly triangular rostrum overreaching the eyes, the anterolateral spine of the carapace well developed, extending far beyond the lateral orbital spine, the P2-4 carpi longer than the dactyli, and the dactyli having proximally diminishing spines on the flexor margin (the ultimate and penultimate subequal); all distinctive differences from *U. minor*.

Uroptychus modicus n. sp.
Figures 154, 155

TYPE MATERIAL — Holotype: **New Caledonia**, Isle of Pines. SMIB 8 DW198, 22°51.6'S, 167°12.4'E, 414-430 m, 1.II.1993, ♀ 6.8 mm (MNHN-IU-2014-16740). Paratypes: **New Caledonia**, Isle of Pines. SMIB 2 Stn DW05, 22°56'S, 167°15'E, 398-410 m, 17.IX.1986, 1 ov. ♀ 6.0 mm, 1 ♀ 5.8 mm (MNHN-IU-2014-16741). **New Caledonia**, Norfolk Ridge. BATHUS 3 Stn CP833, 23°03'S, 166°58'E, 441-444 m, 30.XI.1993, 1 ov. ♀ 6.6 mm (MNHN-IU-2014-16742).

ETYMOLOGY — From the Latin *modicus* (medium, moderate), alluding to the moderate size of the species.

DISTRIBUTION — Isle of Pines and Norfolk Ridge; 398-444 m.

SIZE — Females, 5.8-6.8 mm; ovigerous females from 6.0 mm; males have not been collected.

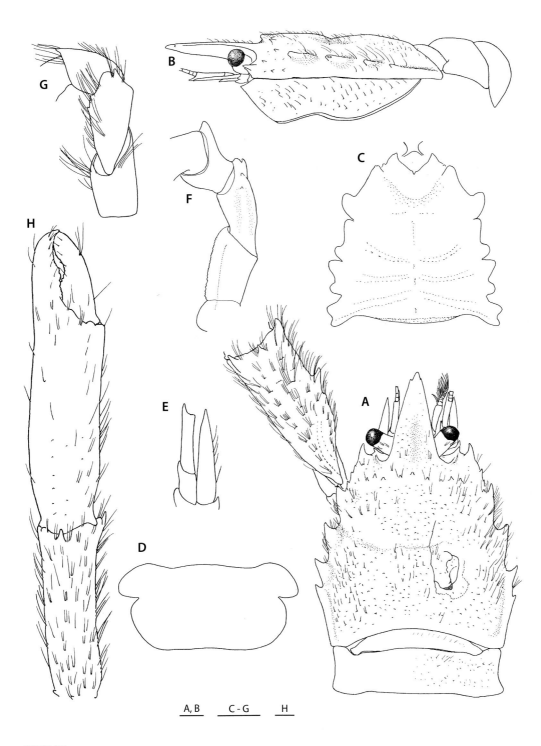

FIGURE 154

Uroptychus modicus n. sp., holotype, female 6.8 mm (MNHN-IU-2014-16740). **A**, carapace and anterior part of abdomen, proximal part of left P1 included, dorsal. **B**, same, abdomen denuded, lateral. **C**, sternal plastron, with excavated sternum and basal parts of Mxp1. **D**, telson. **E**, left antenna, ventral. **F**, left Mxp3, setae omitted, ventral. **G**, same, lateral. **H**, left P1, proximal part omitted, dorsal. Scale bars: A, B, G, 5 mm; C-F, 1 mm.

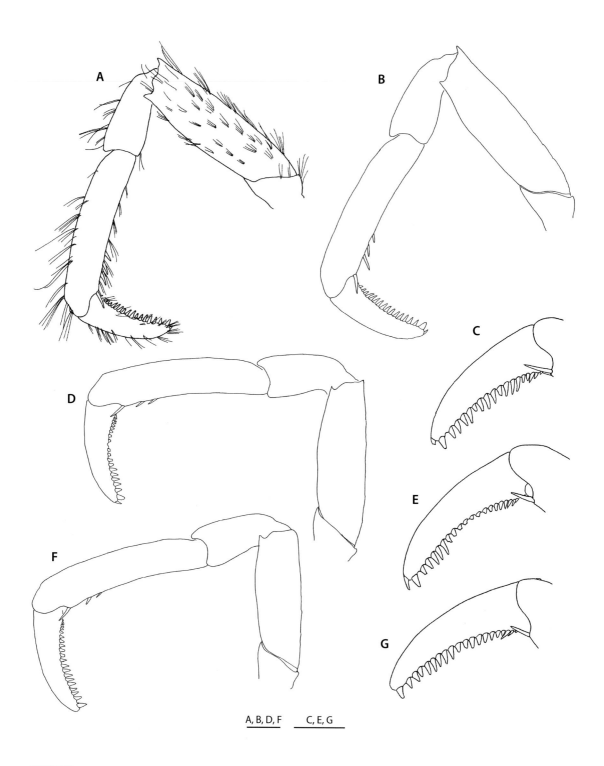

A, B, D, F   C, E, G

FIGURE 155

Uroptychus modicus n. sp., holotype, female 6.8 mm (MNHN-IU-2014-16740). **A**, left P2, lateral. **B**, same, setae omitted, lateral. **C**, same, distal part, setae omitted, lateral. **D**, left P3, setae omitted, lateral. **E**, same, distal part, setae omitted, lateral. **F**, left P4, lateral. **G**, same, distal part, lateral. Scale bars: 1 mm.

DESCRIPTION — Medium-sized species. *Carapace*: 1.1 × broader than long (0.9 × as long as broad); greatest breadth 1.7 × distance between anterolateral spines. Dorsal surface weakly convex from anterior to posterior, with shallow depression between indistinct gastric and cardiac regions, covered with short fine setae; epigastric region with transverse row of 7 or 9 small spines (1 spine in midline flanked by 3 or 4 spines) preceded by depressed rostrum. Lateral margins convexly divergent posteriorly; anterolateral spine well developed, overreaching lateral orbital spine, followed by 2 small hepatic marginal spines in line with anterolateral spine and distinctly ventral to following lateral spines; branchial margin with 4 spines, anterior 3 strong and subequal, last smaller or tiny, followed by ridge leading to posterior end of margin; first spine on branchial margin with accompanying small, often obsolete spine mesial to it. Rostrum broad triangular, with interior angle of about 30°; dorsal surface concave, strongly depressed from level of gastric region; lateral margin with a few obsolescent spinules distally; length 0.5-0.6 × that of remaining carapace, breadth less than carapace breadth at posterior carapace margin. Lateral orbital spine small, slightly anterior to level of anterolateral spine. Pterygostomian flap anteriorly angular, produced to small spine; surface with short soft setae.

Sternum: Excavated sternum with strongly convex anterior margin, surface bearing low, relatively broad ridge in midline. Sternal plastron as long as broad, lateral extremities between sternites 4 and 7 gently divergent posteriorly. Sternite 3 strongly depressed, anterior margin in broad V-shape bearing small or obsolescent median notch without submedian spines; anterolateral angle well produced. Sternite 4 having anterolateral margin nearly straight or slightly concave, anteriorly roundish with a few tubercles or denticles; posterolateral margin relatively short, slightly more than half length of anterolateral margin. Anterolateral margin of sternite 5 anteriorly rounded, about as long as posterolateral margin of sternite 4.

Abdomen: Smooth, sparingly with short, soft setae. Somite 1 without transverse ridge. Somite 2 tergite 2.6-2.8 × broader than long; pleuron anterolaterally bluntly rectangular, posterolaterally blunt angular; lateral margin concavely divergent posteriorly. Pleuron of somite 3 laterally tapering. Telson half as long as broad; posterior plate somewhat concave on posterior margin, length 1.5-1.6 × that of anterior plate.

Eye: Relatively small, elongate (1.9-2.1 × longer than broad), slightly narrowed distally, and reaching midlength of rostrum. Cornea more than half as long as remaining eyestalk.

Antennule and antenna: Ultimate article of antennular peduncle 3.3-3.8 × longer than high. Antennal peduncle extending far beyond cornea. Article 2 with short, distinct distolateral spine. Antennal scale slightly falling short of to slightly overreaching distal end of article 5, breadth 1.5-1.6 × that of article 5. Distal 2 articles each armed with strong distomesial spine; article 5 1.7-1.9 × longer than article 4, breadth 0.6-0.7 × height of ultimate article of antennule. Flagellum consisting of 10-13 segments, reaching distal end of P1 merus.

Mxp: Mxp1 with bases broadly separated. Mxp3 basis without denticles on mesial ridge. Ischium with obsolescent denticles on crista dentata; flexor margin distally not rounded. Blunt distolateral spines on each of merus and carpus. Merus 2 × longer than ischium, flexor margin roundly ridged, bearing 2 small spines distal to point two-thirds of length.

P1: Moderately massive, 4.6 × longer than carapace, covered with soft, fine setae. Ischium dorsally with basally broad, short spine, ventromesially with tubercle-like spines. Granulose short ridges supporting setae on merus and carpus. Merus as long as carapace, with 2 stout mesial, 2 distodorsal, and 2 distoventral spines. Carpus 1.2 × length of merus, with 3 terminal spines: dorsal, ventromesial, ventrolateral. Palm 2.8 × longer than broad, 1.2 × longer than carpus, unarmed. Fingers short relative to breadth, curving ventrally, distally crossing when closed; opposable margins not gaping, that of movable finger with low or moderate-sized proximal process proximal to position of low eminence on opposite margin of fixed finger; movable finger half as long as palm.

P2-4: Setose like P1. Meri mesio-laterally compressed, successively shorter posteriorly (P3 merus 0.9 × length of P2 merus, P4 merus 0.9 × length of P3 merus), breadths subequal on P2-4; dorsal and ventrolateral margins smooth, bearing terminal spine only, but ventrolateral terminal spine obsolete on P4; length-breadth ratio, 3.5-3.6 on P2, 3.0-3.4 on P3, 3.1-3.9 on P4; P2 merus 0.7 × length of carapace, 1.0 × length of P2 propodus; P3 merus 0.9 × length of P3 propodus; P4 merus 0.7-0.8 × length of P4 propodus. Carpi subequal, unarmed, slightly shorter than dactyli (carpus-dactylus length ratio, 0.9 on P2, 0.8 on P3 and P4); carpus-propodus length ratio, 0.46-0.49 on P2, 0.42-0.46 on P3, 0.41-0.42 on P4; slightly shorter than dactyli, unarmed. Propodi subequal in length on P2-4; flexor margin feebly curving, ending in pair of spines preceded by 3 or 4 single movable spines on distal third on P2, 2 or 3 spines on P3, 1 or 2 spines on P4, distalmost of these single spines much more distant from distal pair than from distal second. Dactyli relatively stout; dactylus-propodus length ratio, 0.5-0.6 on P2-4; dactylus-carpus length ratio, 1.1 on P2, 1.2 on P3, 1.3 on P4; flexor margin gently curving, with 18-21 acute spines subperpendicular to margin on distal portion, somewhat inclined on proximal portion, ultimate spine more slender than penultimate, remaining spines relatively close to one another but not contiguous, diminishing toward proximal end of article.

Eggs. Number of eggs carried, 19-30; size, 1.42 mm × 1.58 mm - 1.42 mm × 1.83 mm.

REMARKS — The spination of the carapace lateral margin and the shape of the rostrum and eyes displayed by *U. modicus* n.s p. are very similar to those of *U. baeomma* n. sp. (see above). *Uroptychus modicus* is distinguished from that species by the following: the transverse row of epigastric spines is distinct in *U. modicus*, absent in *U. baeomma*; the flexor spines of the P2-4 dactyli are more numerous on *U. modicus* (18-21 versus 8-10); sternite 3 has the anterior margin deeply excavated in V-shape with a small or obsolescent median notch in *U. modicus*, whereas it is shallowly excavated, with a deep median notch in *U. baeomma*; and the antennal articles 4 and 5 each bear a strong instead of tiny or obsolescent distomesial spine.

The spination of P2-4, especially the dactyli ending in a small ultimate spine proximally preceded by triangular spines, and sternite 3 having a V-shaped anterior margin without submedian spines suggest that *U. modicus* resembles *U. posticus* n. sp. Characters distinguishing these species are outlined under *U. posticus* (see below).

Uroptychus multispinosus Ahyong & Poore, 2004

Figures 156, 157

Uroptychus multispinosus Ahyong & Poore, 2004: 60, fig. 17.

TYPE MATERIAL — Holotype: **Australia**, E of Southport, Queensland, 318 m, female (AM P31415). [not examined].

MATERIAL EXAMINED — **New Caledonia**, Norfolk Ridge. CHALCAL 2 Stn DW74, 24°49.36'S, 168°38.38'E, 650 m, 29.X.1986, 1 ♂ 3.2 mm (MNHN-IU-2014-16743).

DISTRIBUTION — Southern Queensland, and now New Caledonia; 318-650 m.

DESCRIPTION — Small species. *Carapace*: As long as broad; greatest breadth 1.6 × distance between anterolateral spines. Dorsal surface convex from anterior to posterior, with feeble depression between gastric and cardiac regions, bearing scattered short setae. Lateral margins posteriorly divergent but slightly convex on posterior branchial region; anterolateral spine small, barely reaching tip of lateral orbital spine, followed by 5 or 6 very small, obsolescent spines along entire margin; no ridge along posterior lateral margin. Rostrum narrow triangular, with interior angle of 15°, horizontal on ventral surface, dorsally concave, laterally with small subterminal spine on each side; length about half that of remaining carapace, breadth about half carapace breadth at posterior carapace margin. Lateral orbital spine as large as anterolateral spine, situated slightly anterior to level of, and separated from that spine by its basal breadth. Pterygostomian flap with spinules on surface, anteriorly angular, produced to small spine.

Sternum: Excavated sternum moderately convex on anterior margin, surface with distinct ridge in midline. Sternal plastron somewhat broader than long. Sternite 3 shallowly depressed; anterior margin shallowly excavated, with broad median notch flanked by very small spine, lateral end with a few very small spines. Sternite 4 anterolaterally blunt angular, anterolateral margin smooth, more than twice as long as posterolateral margin. Sternite 5 slightly broader than sternite 4, anterolateral margins anteriorly convex, subparallel, much longer than posterolateral margin of sternite 4. Sternites 6 and 7 subequal in breadth, broader than sternite 5.

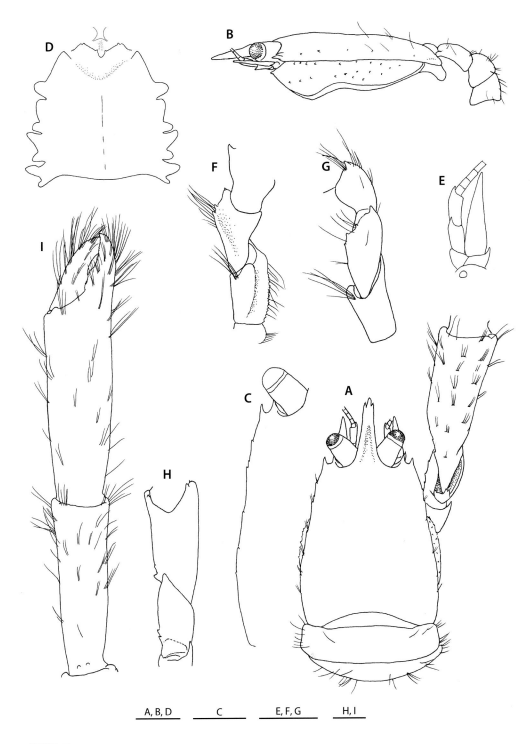

A, B, D   C   E, F, G   H, I

FIGURE 156

Uroptychus multispinosus Ahyong & Poore, 2004, male 3.2 mm (MNHN-IU-2014-16743). **A**, carapace and anterior part of abdomen, proximal part of P1 included, dorsal. **B**, same, lateral. **C**, lateral margin of carapace, dorsal. **D**, sternal plastron, with excavated sternum and basal parts of Mxp1. **E**, left antenna, ventral. **F**, right Mxp3, ventral. **G**, left Mxp3, lateral. **H**, left P1, proximal part, ventral. **I**, right P1, dorsal. Scale bars: 1 mm.

Abdomen: More setose than carapace. Somite 1 moderately convex from anterior to posterior. Somite 2 tergite 2.4 × broader than long; pleuron posterolaterally blunt; lateral margin slightly concave and somewhat divergent posteriorly. Pleuron of somite 3 with blunt lateral terminus. Telson slightly less than half as long as broad; posterior plate somewhat concave on posterior margin, length 0.8 × that of anterior plate.

Eye: 1.8 × longer than broad, slightly overreaching midlength of rostrum, somewhat narrowed distally. Cornea half as long as remaining eyestalk.

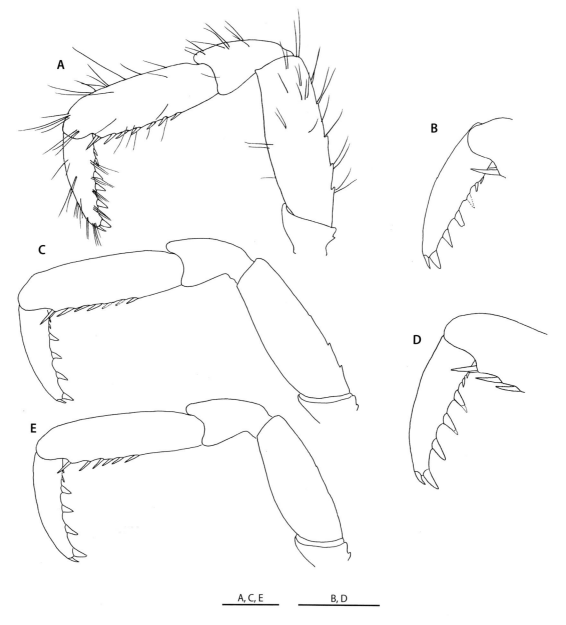

FIGURE 157

Uroptychus multispinosus Ahyong & Poore, 2004, male 3.2 mm (MNHN-IU-2014-16743). **A**, left P2, lateral. **B**, same, distal part, setae omitted, lateral. **C**, left P3, setae omitted, lateral. **D**, same, distal part, setae omitted, lateral. **E**, left P4, setae omitted, lateral. Scale bars: 1 mm.

Antennule and antenna: Ultimate article of antennular peduncle 2.9 × longer than high. Antennal peduncle relatively short, terminating in distal end of cornea. Article 2 with sharp lateral spine. Antennal scale overreaching peduncle by full length of article 5, breadth 1.7 × that of article 5. Distal 2 articles short relative to breadth, each with distinct distomesial spine; article 5 as long as article 4; breadth 0.7 × height of ultimate antennular article. Flagellum of 9 segments far falling short of distal end of P1 merus.

Mxp: Mxp1 with bases broadly separated. Mxp3 basis without distinct denticles on mesial ridge. Ischium with very small denticles on crista dentata, bearing tuft of setae lateral to rounded distal end of flexor margin. Merus twice as long as ischium, flattish on mesial face; flexor margin ridged, with 2 or 3 small spines on distal third; distolateral spine short and small. Carpus also with distolateral spine.

P1: 5.7 × longer than carapace, setose, relatively massive. Ischium with well-developed distodorsal spine; ventromesial margin with 2 or 3 small denticle-like processes on proximal portion, lacking subterminal spine. Merus with granulate scale-like ridges supporting setae on dorsal surface and tubercles on proximal mesial portion; length 1.1 × that of carapace. Carpus 1.1 × longer than merus, ventrally bearing small distolateral spine. Palm unarmed, about 3 × longer than broad, distally somewhat broader, 1.2 × longer than carpus. Fingers stout, distally crossing when closed. Movable finger falling short of distal end of fixed finger when closed, opposable margin with median process fitting to between 2 low processes on opposite fixed finger; length slightly less than half (0.45) that of palm.

P2-4: Broad relative to length, moderately setose. Meri successively shorter posteriorly (P3 merus 0.9 × length of P2 merus, P4 merus 0.9 × length of P3 merus), equally broad on P2 and P3, slightly narrower on P4 (P4 merus 0.9 × as broad as P3); length-breadth ratio, 2.8 on P2, 2.9 on P3, 2.5 on P4; dorsal margin with a few small proximal spines distinct on P2 and P3, obsolescent on P4. P2 merus 0.7 × length of carapace, subequal to length of P2 propodus; P3 merus 0.9 × length of P3 propodus; P4 merus 0.8 length of P4 propodus. Carpi successively slightly shorter posteriorly; distinctly shorter than dactyli, carpus-dactylus length ratio, 0.9 on P2, 0.7 on P3 and P4; carpus-propodus length ratio, 0.4 on P3 and P4. Propodi subequal on P3 and P4, shorter on P2 (P2 propodus 0.8 × length of P3 propodus); flexor margin nearly straight, ending in pair of spines preceded by 6 spines on P2 and P3, 5 spines on P4. Dactyli slender relative to propodi, length more than half (0.52 on P2, 0.53 on P3, 0.56 on P4) that of propodus; dactylus-carpus length ratio, 1.2 on P2, 1.4 on P3, 1.5 on P4; flexor margin nearly straight on P2, slightly curving on P3, more distinctly so on P4, bearing 7 spines, ultimate spine slender and very close to penultimate, 2 proximal spines small and slender and strongly inclined, 4 other spines strong, somewhat inclined, and widely spaced.

REMARKS — The present specimen agrees with the original description of *U. multispinosus*, with only small differences: the dorsal margins of P2 and P3 meri bear small spines instead of being unarmed and the lateral marginal spines of the carapace are very small and obsolescent instead of small and distinct. These difference are here regarded as individual variations.

This species strongly resembles *U. vicinus* n. sp. Their relationships are discussed under the remarks of that species (see below).

Uroptychus nanophyes McArdle, 1901

Figures 158, 159, 305H

Uroptychus nanophyes McArdle, 1901: 525. — Alcock & McArdle 1902: pl. 57, figs 1, 1a. — Van Dam 1940: 96, fig. 1. — Baba 2005: 48, 228, fig. 16. — Poore *et al.* 2011: 329, pl. 7, fig. A.
Not *Uroptychus nanophyes* — Baba 1981: 117, fig. 5 (nr. *U. sexspinosus* Balss, 1913a).

TYPE MATERIAL — Syntypes: **India**, NE coast of Sri Lanka, 926 m (ZSIC). [not examined].

MATERIAL EXAMINED — **Vanuatu**. MUSORSTOM 8 Stn CP982, 19°21.80'S, 169°26.47'E, 408-410 m, 23.IX.1994, 1 ov. ♀ 5.7 mm (MNHN-IU-2014-16744). – Stn CP1026, 17°50.35'S, 168°39.33'E, 437-504 m, 28.IX.1994, 1 ♂ 4.2 mm (MNHN-IU-2014-16745). **New Caledonia**, Loyalty Ridge. MUSORSTOM 6 Stn DW420, 20°29.27'S, 166°43.35'E, 600 m, on a species of Primoidae (Alcyonacea), 16.II.1989, 3 ♂ 6.9-7.9 mm, 2 ♀ 5.5, 7.2 mm, 3 spec. (sex indet.) 2.7-3.1 mm (MNHN-IU-2014-16746). **New Caledonia**, Hunter and Matthew Islands. VOLSMAR Stn DW05, 22°26'S, 171°46'E, 620-700 m, 01.VI.1989, 2 ♂ 8.8, 9.1 mm, 2 ♀ 5.1 mm, carapace broken (MNHN-IU-2014-16747). **New Caledonia**, Norfolk Ridge. CHALCAL 2 Stn CC01, 24°55'S, 168°22'E, 500-580 m, 28.X.1986, 2 ov. ♀ 4.3, 6.5 mm (MNHN-IU-2014-16748). – Stn CP25, 23°38.6'S, 167°43.12'E, 418 m, 30.X.1986, 1 ♂ 5.3 mm (MNHN-IU-2014-16749). – Stn CP26, 23°18.15'S, 168°03.58'E, 296 m, 3.X.1986, 2 ♂ 8.7, 11.7 mm (MNHN-IU-2014-16750). – Stn DW76, 23°40.5'S, 167°45.2'E, 470 m, on Primoidae gen. sp. (Alcyonacea), 30.X.1986, 1 ♂ 11.3 mm (MNHN-IU-2014-16751), 1 ov. ♀ 10.3 mm (MNHN-IU-2014-16752). AZTÈQUE Stn CH06, 23°37.9'S, 167°42.5'E, 425-470 m, 14.II.1990, 5 ♂ 9.2-11.8 mm, 4 ov. ♀ 10.3-11.1 mm (MNHN-IU-2014-16753). SMIB 2 Stn

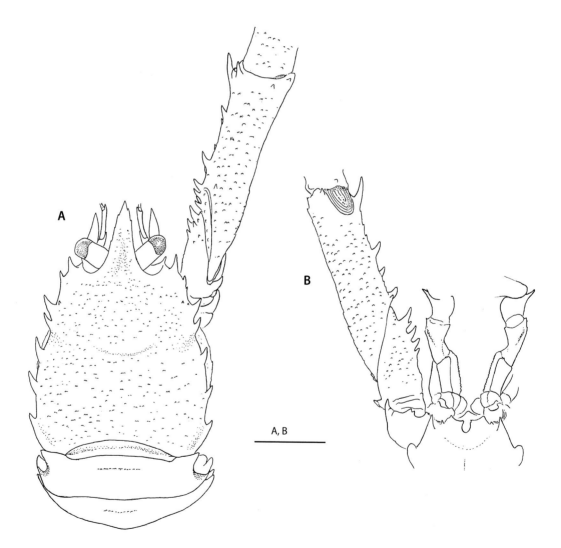

FIGURE 158

Uroptychus nanophyes McArdle, 1901, male 11.2 mm (MNHN-IU-2014-16753). **A**, carapace and abdomen, left P1 included, dorsal. **B**, anterior part of sternal plastron with excavated sternum and proximal parts of Mxp1, Mxps 3 right P1, ventral. Scale bar: 5 mm.

DW13, 22°52'S, 167°13'E, 427-454 m, 18.IX.1986, 1 ♂ 7.7 mm (MNHN-IU-2014-16754). SMIB 4 Stn DW58, 22°59'S, 167°23'E, 480-560 m, 10.III.1989, 1 ♀ 6.7 mm (MNHN-IU-2014-16755). – Stn DW61, 23°01'S, 167°22'E, 520-550 m, 10.III.1989, 1 ♂ 8.9 mm, 1 ov. ♀ 8.0 mm (MNHN-IU-2014-16756), 1 ♂ 10.7 mm (MNHN-IU-2014-16757). SMIB 8 Stn 167, 23°38'S, 168°43'E, 430-452 m, 29.I.1993, 1 ♂ 10.1 mm (MNHN-IU-2014-16758). – Stn DW201, 22°58.6'S, 167°20.3'E, 500-504 m, 2.II.1993, 2 ov. ♀ 9.4, 9.8 mm (MNHN-IU-2014-16759). BERYX 11 Stn CH30, 23°37'S, 167°42'E, 420-470 m, 18.X.1992, 2 ♂ 7.1, 12.2 mm (MNHN-IU-2014-16760). – Stn CP31, 23°39.12'S, 167°43.65'E, 430-440 m, 18.X.1992, 4 ♂ 8.0-11.1 mm, 4 ov. ♀ 8.5-10.1 mm, 3 ♀ 5.3-8.8 mm (MNHN-IU-2014-16761). – Stn CP32, 23°37.70'S, 167°43.45'E, 420-460 m, 18.X.1992, 3 ♂ 7.7-9.7 mm, 1 ov. ♀ 11.2 mm (MNHN-IU-2014-16762), 1 ♀ 9.4 mm (MNHN-IU-2014-16763). – Stn DW38, 23°37.53'S, 167°59.42'E, 550-690 m, 19.X.1992, 1 ov. ♀ 7.2 mm, 1 ♀ 3.0 mm (MNHN-IU-2014-16764); 6 ♂ 5.3-11.8 mm, 2 ov. ♀ 6.8, 8.0 mm (MNHN-IU-2014-16765). – Stn CP49, 23°45.22'S, 168°17.06'E, 400-460 m, 21.X.1992, 3 ♂ 5.3-6.0 mm (MNHN-IU-2014-16766). BATHUS 3 Stn CP814, 23°47.60'E, 168°17.10'E, 444-530 m, 28.XI.1993, 1 ♂ 5.6 mm (MNHN-IU-2014-16767); 1 ♂ 5.9 mm, 1 ov. ♀ 3.3 mm (MNHN-IU-2014-16768). – Stn DW818, 23°44'S, 168°16'E, 394-401 m, 28.XI.1993, 1 ♂ 6.6 mm (MNHN-IU-2014-16769). – Stn CH820, 23°43'S, 168°16'E, 405-411 m, 28.XI.1993, 1 ov. ♀ 5.4 mm (MNHN-IU-2014-16770). LITHIST Stn CC06, 23°37.5'S, 167°42.1'E, 440-579 m, 10.VIII.1999, 2 ov. ♀ 9.8, 10.6 mm (MNHN-IU-2014-16771). – Stn DW05, 23°38.2'S, 167°42.9'E, 433-500 m, 10.VIII.1999, 2 ♂ 5.9, 9.0 mm, 1 ov. ♀ 10.0 mm (MNHN-IU-2014-16772). – Stn CP02, 23°37.1'S, 167°41.1'E, 442 m, 10.VIII.1999, 1 ov. ♀ 9.6 mm (MNHN-IU-2014-16773). HALIPRO 2 Stn BT 94, 23°33'S, 167°42'E, 448-880 m, 24.XI.1996, 1 ov. ♀ 9.9 mm (MNHN-IU-2014-16774). – Stn BT 95, 24°00'S, 162°08'E, 1224-1233 m, 25.XI.1996, 1 ♀ 8.6 mm (MNHN-IU-2014-16775). NORFOLK 1, Stn DW1666, 23°42'S, 167°44'E, 469-860 m, 20.VI.2001, 2 ♂ 4.9, 9.7 mm (MNHN-IU-2014-16776). – Stn DW1707, 23°43'S, 168°16'E, 381-493 m, 25.VI.2001, 2 ♂ 5.1, 5.7 mm (MNHN-IU-2014-16777). NORFOLK 2 Stn CP2029, 23°38.88'S, 167°44.05'E, 438-445 m, 22.X.2003, 17 ♂ 3.7-11.4 mm, 5 ov. ♀ 7.8-10.4 mm, 12 ♀ 3.2-7.2 mm, 2 sp. (sex indet.) 3.3, 3.7 mm (MNHN-IU-2014-16778). – Stn CP2030, 23°39.01'S, 167°44.04'E, 440-440 m, 22.X.2003, 1 ov. ♀ 10.9 mm, 1 ♀ 6.7 mm (MNHN-IU-2014-16779). – Stn DW2031, 23°38.83'S, 167°44.01'E, 440-440 m, 22.X.2003, 1 sp. (sex indet.) 5.0 mm (MNHN-IU-2014-16780). – Stn DW2032, 23°39.14'S, 167°43.39'E, 420-450 m, 22.X.2003, 1 ♂ 3.0 mm, 2 ♀ 5.3, 5.5 mm (MNHN-IU-2014-16781), 2 ♂ 11.3, 12.4 mm, 3 ov. ♀ 9.2-11.0 mm (MNHN-IU-2014-16782). – Stn DW2036, 23°37.81'S, 167°38.78'E, 571-610 m, 22.X.2003, 8 ♂ 4.8-11.1 mm, 13 ov. ♀ 7.8-9.8 mm (MNHN-IU-2014-16783). – Stn DW2041, 23°40.93'S, 168°01.29'E, 400-400 m, 23.X.2003, 1 ov. ♀ 5.3 mm (MNHN-IU-2014-16784). – Stn DW2049, 23°42.88'S, 168°15.43'E, 470-621 m, 23.X.2003, 1 ♂ 2.9 mm (MNHN-IU-2014-16785). – Stn DW2052, 23°42.29'S, 168°15.27'E, 473-525 m, 24.X.2003, 3 ♂ 4.7-5.9 mm (MNHN-IU-2014-16786). – Stn DW2056, 24°40.32'S, 168°39,17'E, 573-600 m, 25.X.2003, 2 ♂ 5.1, 9.6 mm, 1 sp. (sex indet.) 4.5 mm (MNHN-IU-2014-16787). – Stn CP2083, 24°53.23'S, 168°21.86'E, 530-540 m, sponge, 28.X.2003, 1 ♂ 5.5 mm, 1 ov. ♀ 5.4 mm, 1 ♀ 6.0 mm (MNHN-IU-2014-16788). – Stn DW2109, 23°47.46'S, 168°17.04'E, 422-495 m, 31.X.2003, 1 ♂ 3.2 mm (MNHN-IU-2014-16789). **Solomon Islands**. SALOMON 1 Stn DW1788, 9°19.4'S, 160°15.4'E, 341-343 m, 30.IX.2001, 1 ♂ 3.8 mm (MNHN-IU-2014-16790). – Stn CP1790, 9°19.2'S, 160°10.8'E, 357 m, 30.IX.2001, 1 ♀ 5.2 mm (MNHN-IU-2014-16791). – Stn CP1831, 10°12.1'S, 161°19.2'E, 135-325 m, 5.X.2001, 1 ov. ♀ 6.4 mm (MNHN-IU-2014-16792).

SIZE — Males, 2.9-12.4 mm; females, 3.6-11.2 mm; ovigerous females from 3.3 mm.

DISTRIBUTION — Northeast coast of Sri Lanka, Java Sea and the Kai Islands, in 66-926 m; now Solomon Islands, Vanuatu, Loyalty Ridge, Hunter-Matthew Islands, and Norfolk Ridge, in 296-1233 m.

DESCRIPTION — *Carapace*: slightly broader than long; greatest breadth 1.5-1.6 × distance between anterolateral spines. Dorsal surface feebly convex, anteriorly descended to depressed rostrum, feebly depressed between gastric and cardiac-anterior branchial regions, nearly smooth or slightly granulose in small specimens, distinctly granulose in large specimens; epigastric region smooth but a few small spines laterally, occasionally with a few tiny or obsolescent spines medially; small spine mesial to third lateral spine. Lateral margins convexly divergent (less convexly in small specimens), with 6 spines; first anterolateral, distinctly overreaching small lateral orbital spine; second small, situated on hepatic median margin and ventral to level of first and third spines, occasionally followed by 1 (rarely 2) very small spine; 4 strong spines (third to sixth) along entire length of branchial margin (last one situated near posterior end), widely spaced, occasionally intervened by small or moderate-sized spine between third and fourth, rarely small spine between fourth and fifth, and fifth and sixth, last spine rarely followed by very small spine; ridged along posterior third. Rostrum elongate triangular, with interior angle of 24-27°, length 0.4-0.6 × that of remaining carapace (greater in small specimens), breadth less than half carapace breadth at posterior carapace margin; lateral margin with at most 6 lateral spinules in distal two-thirds. Pterygostomian flap granulose (in large specimens) or with tiny spines (in small specimens) on surface, anteriorly sharp angular, produced to strong spine.

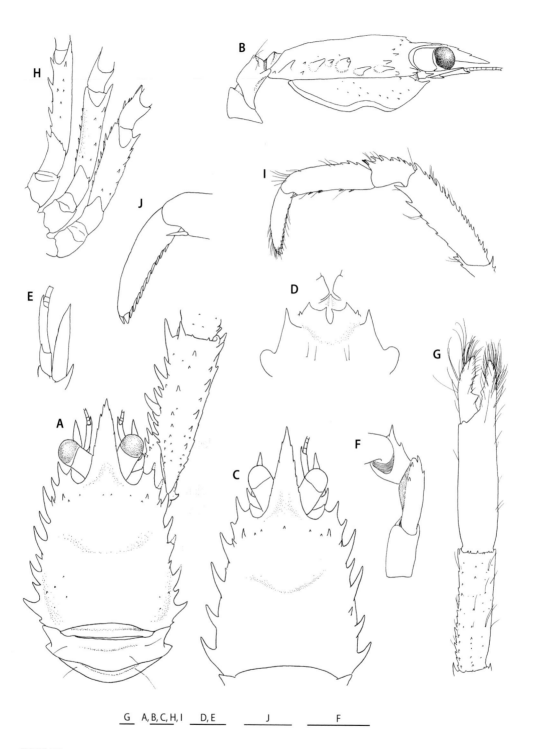

G   A, B, C, H, I   D, E   J   F

FIGURE 159

Uroptychus nanophyes McArdle, 1901. **A**, **B**, **D-I**, male 5.3 mm; **C**, male 5.5 mm (MNHN-IU-2014-16766). **A**, carapace and anterior part of abdomen, proximal part of right P1 included, dorsal. **B**, same, lateral. **C**, carapace, dorsal. **D**, anterior part of sternal plastron, with excavated sternum, proximal part of Mxp1 included, ventral. **E**, left antenna, ventral. **F**, left Mxp3, ventral. **G**, right P1, proximal articles omitted, dorsal. **H**, proximal parts of P2-4, ventral. **I**, left P2, lateral. **J**, same, distal part, setae omitted, lateral. Scale bars: 1 mm.

Sternum: Excavated sternum with anterior margin sharply or bluntly subtriangular, surface with ridge in midline. Sternal plastron about as long as or slightly longer than broad; lateral extremities gently divergent posteriorly (in small specimens) or sternites 6 and 7 subequal in breadth and broader than sternites 4 and 5. Sternite 3 well depressed, anterolaterally sharply angular, anterior margin shallowly excavated, medially bearing U- or V-shaped or oval sinus, flanked by small incurved spine. Sternite 4 narrow relative to length, anterolateral margin nearly straight, anteriorly produced to very strong processes directed straight forward, reaching anterior end of sternite 3; posterolateral margin 0.7-0.8 × as long as anterolateral margin. Anterolateral margin of sternite 5 two-thirds length of posterior margin of sternite 4.

Abdomen: Tergites smooth and glabrous. Somite 1 with sharp transverse ridge. Somites 2 and 3 transversely ridged anteriorly; somite 2 tergite 2.3-2.6 × broader than long; pleural lateral margin concavely divergent posteriorly, sharply edged, anterolaterally and posterolaterally sharp angular. Pleura of somites 3 and 4 tapering to sharp tip. Telson 0.4-0.5 × as long as broad; posterior plate 1.1-1.4 × longer than anterior plate, posterior margin distinctly or moderately emarginate or somewhat concave.

Eye: 1.6-1.8 × longer than broad, slightly overreaching midlength of rostrum; lateral and mesial margins subparallel. Cornea slightly inflated, slightly more than half as long as remaining eyestalk.

Antennule and antenna: Ultimate article of antennular peduncle slender, 3.0-3.7 × longer than high. Antennal peduncle slender, extending far beyond cornea. Article 2 with strong distolateral spine. Antennal scale varying from barely reaching to slightly overreaching article 5, breadth 1.4-1.7 × that of antennal article 5, lateral margin rarely bearing a few small proximal spines. Distal 2 articles each with well-developed distomesial spine; article 5 2.1-2.4 × longer than article 4, breadth 0.5-0.7 × height of ultimate antennular article. Flagellum consisting of 14-22 segments (numerous in large specimens), not reaching distal end of P1 merus.

Mxp: Mxp1 with bases close to each other. Mxp3 basis with 3-5 denticles on mesial ridge, distalmost larger. Ischium with 16-25 (in small specimens), 25-30 (in large specimens) denticles on crista dentata; with 1 or 2 small spines (rarely hardly visible in small specimens) lateral to somewhat rounded distal end of flexor margin. Merus 1.8-2.2 × (greater in large specimens) longer than ischium, flattish on mesial face, flexor margin somewhat cristate with small spines on distal third; distolateral spine well developed, occasionally with a few small accompanying spines in large specimens. Carpus with distolateral spine proximally followed by a few small spines along extensor margin.

Pereopods: Scarcely setose except for distal articles.

P1: Relatively slender, 4.2-6.0 × longer than carapace, granulose on surface in large specimens, tuberculose except for palm and fingers in small specimens. Ischium dorsally bearing strong spine with small accompanying spine proximally, ventromesially with prominent subterminal spine proximally followed by successively diminishing spines. Merus with 5 rows of spines (2 dorsal, 2 mesial, 1 ventral) in small specimens; dorsal and ventral spines reduced in large specimens; length 1.1-1.3 × that of carapace. Carpus with 1 mesial and 1 dorsal row of small spines often reduced to small size or obsolete; length 1.0-1.5 × that of merus. Palm nearly smooth, 3.0-4.4 × longer than broad, with no sexual difference in proportion, subequal to or slightly longer than carpus. Fingers feebly incurved distally, not distinctly crossing, somewhat gaping in males and large females, occasionally not gaping in males; movable finger 0.4-0.5 × as long as palm, opposable margin with subtriangular proximal process fitting to longitudinal groove on opposite fixed finger when closed.

P2-4: Meri successively shorter posteriorly (P3 merus 0.8-0.9 × length of P2 merus, P4 merus 0.9 × length of P3 merus), equally broad on P2-4; length-breadth ratio 3.5-4.1 on P2, 3.1-3.7 on P3, 2.9-3.2 on P4; dorsal margin with 13-17 spines on P2, 12-16 spines on P3, 9-14 spines on P4; ventrolateral margin with a few to several spines on P2, distalmost well developed, larger than distalmost dorsal spine, other spines occasionally obsolescent on P3 and P4; another row of ventromesial spines usually on P2, occasionally absent on P3 and P4 but distal spine consistent. P2 merus 0.8-0.9 × length of carapace, 1.2 × length of P2 propodus; P3 merus 0.9-1.0 × length of P3 propodus; P4 merus 0.8-0.9 × length of P4 propodus. Carpi subequal in length or successively very slightly shorter posteriorly, relatively short; carpus-propodus length ratio, 0.36-0.44, 0.34-0.43, 0.33-0.38 on P2, P3, P4 respectively; extensor margin with row of spines distinct on P2 and P3, occasionally obsolescent on P4. Propodi successively longer posteriorly or shorter on P2 than on P3 and P4; extensor margin with or without proximal spines; flexor margin ending in pair of spines preceded by 2 or 3 (occasionally

1 or 4) in small specimens, 3-6 spines in large specimens. Dactyli longer than carpi (dactylus-carpus length ratio, 1.1-1.4 on P2, 1.2-1.4 on P3, 1.2-1.6 on P4; larger in small specimens), 0.5-0.6 × length of propodi on P2-4; flexor margin nearly straight, with row of spines strongly inclined and contiguous to one another, ultimate slender, penultimate pronouncedly broad, fully twice as broad as antepenultimate, remaining spines 9-10 in number, slender and uniform in breadth; antepenultimate subequal to ultimate in size.

Eggs. Up to 40 eggs carried; size, 0.95 × 1.01 mm - 1.07 × 1.17 mm.

Color. Male from Smib 4 Stn DW61 (MNHN-IU-2014-16757): Body and appendages pale red, abdomen translucent.

The photograph of the specimen from Vanuatu [SANTO Stn AT28] in Poore *et al.* (2011: pl. 7, fig. A) shows a different color pattern: body and P2-4 paler; P1 pale red with deep red bands, P2-4 with reddish bands, 2 on merus and propodus, 1 on carpus and dactylus.

REMARKS — At first glance, there seem to be two different forms: one is granulose on the carapace and P1 (Figure 158), and the other is non-granulose (Figure 159). However, the non-granulose form is small, with the maximum carapace length of 6.0 mm in males, 6.5 mm in females, whereas the granulose form measures 4.9-12.4 mm in males, 5.1-11.2 mm in females. These two forms share a row of spines along the ventromesial margin of P2 merus. In the granulose form the P1 carpus are less spinose than in the non-ovigerous form. No clear differences are found in the other characters. In addition, these forms are collected together in some stations (MNHN-IU-2014-16778; MNHN-IU-2014-16781; MNHN-IU-2014-16782). The female illustrated in Alcock & McArdle (1902: pl. 57: fig. 1a) appears to represent the non-granulose form, but the male carapace dorsum (pl. 57: fig. 1) is covered with small spines, with a distinct row of epigastric spines. Such a male is not included in the present collection. P3 and P4 in the female (pl. 57, fig. 1a) are illustrated to bear extra spines in addition to the terminal one on the ventrolateral margin (P2 is viewed from dorsolateral side so this spination is not visible), the feature constant on P2 merus in the present material. The presence of a row of spines along the ventromesial margin of the P2 merus that is also consistent in the present material is not mentioned. Examination of the type now in the collection of the Zoological Survey of India would elaborate on its specific status.

The specimens reported under *U. nanophyes* from Japan by Baba (1981) are now removed from the synonymy of the species because of the following characters that are not in agreement with the present material: the carapace sizes (poc 5.1-9.0 mm) are about the same as those of the granulose specimens but the carapace dorsal surface is smooth, not granulose, bearing scattered tiny spines on the posterior half; the epigastric region bears an uninterrupted transverse row of spines; the P2-4 meri bear only a distal spine, lacking a row of spines, on each of the ventrolateral and ventromesial margins; and the ultimate of the flexor marginal spines of P2-4 dactyli is larger than instead of subequal to the antepenultimate. This material resembles *U. sexspinosus* Balss, 1913a in having scattered small spines on the carapace (Balss described that the carapace dorsum is smooth and devoid of hairs and spines, but the illustration (Balss 1913b: fig. 21) shows scattered spines). According to the description by Balss (1913b), the P2-4 propodi in *U. sexspinosus* are entire along the flexor margin. To the best of my knowledge, a pair of terminal spines are consistent in those including *U. nanophyes* that have the P2-4 dactyli with a prominent penultimate spine proximally preceded by obliquely directed, closely arranged spines. It is not unlikely that the spines may have been overlooked. Unfortunately, the type could not be located (Baba *et al.* 2008). *Uroptychus nanophyes* differs from *U. sexspinosus* in having the carapace dorsum with spines restricted to the anterior portion mostly lateral to the epigastric region rather than scattered over the surface including posterior half. In addition, the P2-4 propodi in *U. sexspinosus* are entire along the flexor margin, whereas *U. nanophyes* has a pair of terminal spines preceded by 2 or 3 (occasionally 1 or 4) spines.

Uroptychus nanophyes also resembles *U. alophus* n. sp. and *U. vegrandis* n. sp., sharing the rostral lateral margin with a few small lateral spines, the spinose carapace lateral margin, and the P2-4 meri with a row of dorsal spines and the dactylar flexor margin with obliquely directed, closely arranged spines proximal to the pronounced penultimate spine. Characters distinguishing *U. nanophyes* from *U. alophus* and *U. vegrandis* are discussed under the accounts of those species.

Uroptychus naso (Van Dam, 1933)

Uroptychus naso Van Dam, 1933: 23, figs 35-37. — Van Dam 1939: 402 (part); 1940: 97. — Baba 1969: 42 (part), fig. 2a; 1988: 39. — Wu *et al.* 1998: 81, figs 5, 12B. — Baba 2005: 49. — Baba *et al.* 2008: 37, fig. 1F. — Baba *et al.* 2009: 47 (part), figs 38, 40. — Poore & Andreakis 2011: 158, figs 4a, 5a, 6, 7.

TYPE MATERIAL — Syntypes: **Indonesia**, Kur Island and Taam Island, Kai Islands, 204-304 m, 2 males from SIBOGA Stn 253, 304 m (ZMA De. 101.692) ; 1 male and 1 female syntype from SIBOGA Stn 251, 204 m (ZMA De. 101.667). [not examined].

MATERIAL EXAMINED — Indonesia, Kai Islands. KARUBAR Stn CP05, 05°49'S, 132°18'E, 296-299 m, 22.X.1991, 2 ♂ 7.3, 7.3 mm (MNHN-IU-2014-16793). – Stn CP16, 5°17'S, 132°50'E, 315-349 m, 24.X.1991, 1 ♂ 7.2 mm (MNHN-IU-2014-16794). **Indonesia**, Tanimbar Islands. KARUBAR Stn CP82, 9°32'S, 131°02'E, 219-215 m, 4.XI.1991, 2 ♂ 6.9, 7.2 mm, 2 ov. ♀ 7.1, 7.7 mm, 1 ♀ 5.2 mm (MNHN-IU-2014-16795). – Stn CP86, 9°26'S, 131°13'E, 225-223 m, 4.XI.1991, 3 ♂ 7.8-9.1 mm, 2 ov. ♀ 6.6, 8.5 mm, 2 ♀ 8.0, 8.3 mm (MNHN-IU-2014-16796). **Philippines**. MUSORSTOM 1 Stn CP03, 14°01'N, 120°15'E, 183-185 m, 19.III.1976, 19.III.1976, 1 ov. ♀ 8.2 mm (MNHN-IU-2014-16797). – Stn CP35, 14°08'N, 120°17'E, 186-187 m, 23.III.1976, 1 ♂ 11.8 mm, 2 ov. ♀ 7.7, 8.3 mm (MNHN-IU-2014-16798). – Stn CP36, 14°00'N, 120°17'E, 210-187 m, 23.III.1976, 1 ♂ 3.2 mm (MNHN-IU-2014-16799). MUSORSTOM 2 Stn CP01, 14°00'N, 120°18'E, 198-188 m, 20.XI.1980, 1 ♂ 11.3 mm, 1 ♀ 6.5 mm (MNHN-IU-2014-16800). – Stn CP19, 14°00.5'N, 120°16.5'E, 189-192 m, 22.XI.1980, 1 ov. ♀ 10.9 mm (MNHN-IU-2014-16801). – Stn CP53, 14°01'N, 120°17'E, 215-216 m, 27.XI.1980, 3 ♂ 6.4-8.7 mm (MNHN-IU-2014-16802). – Stn CP54, 14°00'N, 120°10'E, 170-174 m, 27.XI.1980, 1 ov. ♀ 11.0 mm (MNHN-IU-2014-16803).

DISTRIBUTION — Southern Japan, East China Sea, Taiwan, Philippines, Indonesia, and northern Western Australia; 128-440 m. Poore & Andreakis (2011) believed that the material reported by Van Dam (1940) from Java Sea can be referred to this species. It was taken in 68-71 m.

SIZE — Males 3.2-11.8 mm; females, 5.2-11.0 mm; ovigerous females from 6.6 mm.

DIAGNOSIS — Medium-sized to large species. Carapace 0.9-1.2 × as long as broad; greatest breadth 1.8-2.0 × distance between anterolateral spines. Dorsal surface covered with tubercles, bearing a few to several small spines on hepatic region; deep cervical groove bordering gastric and cardiac regions, and anterior and posterior branchial regions, anterior cervical groove indistinct. Lateral margins divergent posteriorly; anterolateral spine small but distinctly over-reaching lateral orbital spine, followed by a few small spines on hepatic region and relatively large spines on branchial region (anterior branchial margin with 2 spines, posterior smaller than anterior, often followed by much smaller spine; posterior branchial margin with 7 or 8 spines relatively widely spaced anteriorly). Rostrum broad, long triangular, with interior angle of 30-33°, somewhat deflected ventrally or nearly horizontal; dorsally covered with tubercles and depressed in midline, laterally with 6-8 small spines on anterior half, proximalmost spine located slightly to greatly proximal to midlength (at most to point one-third from proximal end); length 0.7-0.8 × (longer in young specimens) that of carapace, breadth slightly more than half carapace breadth at posterior carapace margin. Pterygostomian flap covered with small spines, anteriorly angular, produced to strong spine. Excavated sternum anteriorly ending in convex margin, surface with longitudinal ridge in midline. Sternal plastron 1.1-1.2 × longer than broad, lateral extremities somewhat divergent posteriorly; sternite 3 moderately depressed; anterior margin shallowly excavated, bearing deep U-shaped median sinus flanked by small or obsolescent spine; anterolateral margin of sternite 4 denticulate, occasionally with distinct terminal spine, nearly straight, 1.1-1.3 × longer than posterolateral margin. Anterolateral margin of sternite 5 subequal to or shorter than posterolateral margin of sternite 4. Abdominal somite 1 with transverse ridge; somite 2 tergite 2.2-2.4 × broader than long, pleural lateral margin moderately concave, slightly divergent posteriorly; pleuron of somite 3 laterally blunt. Telson 0.4 × as long as broad; posterior plate distinctly emarginate on posterior margin, length 1.3 × that of anterior plate. Eyes short, ending in proximal third of rostrum. Ultimate antennular article 3 × longer than high. Antennal peduncle having distal 2 articles each mesially with distoventral spine; article 5 1.4-1.6 × as long as article 4, breadth 0.6 × height of antennule ultimate article; antennal scale barely reaching apex of article 5, often with 1 or 2 small lateral spines proximally; flagellum of 9-10 segments very short, about as long as distal 2 articles combined. Mxp1 with bases strongly produced mesially, hence close to each other but somewhat separated. Mxp3 basis with 4 or 5 obsolescent denticles; ischium having flexor margin distally ending in spine laterally accompanying a few small spines, crista dentata with very small, obsolescent denticles; merus 1.7-1.9 × longer than ischium, flattish on mesial face, with distinct distolateral spine, flexor margin with a few distinct spines distal to mid-length; carpus with distinct distolateral spine and small spines on lateral face. P1 depressed distally, especially palm and fingers, covered with spinules and scattered short setae, bearing spines along mesial and lateral margins, mesial spines larger; ischium with strong dorsal spine laterally and dorsally accompanying small spines, ventromesial margin with strong subterminal spine often followed by small spines; merus 1.1-1.3 × longer than carapace; carpus 1.0-1.2 × as long as merus; palm as long as carpus; fingers broad relative to length, slightly incurved distally, movable finger with 1 blunt tooth on opposable margin, occasionally with 2 teeth in males, length 0.4-0.5 (longer in small specimens) that of palm. P2-4 broad relative to length, bearing fine setae much longer than those on P1; meri successively shorter posteriorly (P3 merus 0.9 × length of P2 merus, P4 merus 0.8 × length of P3 merus), equally broad on P2-4; length-breadth ratio, 3.1-3.5 on P2, 2.8-3.0 on P3, 2.3-2.5 on P4; P2 merus 0.8-0.9 × length of carapace, 1.4-1.6 × length of P2 propodus; P3 merus 1.3 × length of P3 propodus; P4 merus 0.9-1.0 × length of P4 propodus; meri covered with denticles on lateral surface, mesial face flattish, dorsal margin sharply ridged bearing row of sharp spines continued on to carpus, distal portion of ventrolateral margin with several spines as equally sharp as but larger than those on extensor margin, terminal spine strongest; carpi subequal on P2-3, shorter on P4, each less than half (0.4) length of propodus on P2-4; propodi successively longer posteriorly or slightly shorter on P2 than on P3-4; flexor margin straight, with pair of distal spines preceded by smaller spines (4-10 on P2, 3-8 on P3, 3-6 on P4; fewer in small specimens), some of these on P2 and P3 in zigzag arrangement along midline; lateral and mesial faces with no dense tufts of short setae on P2-4; dactyli short relative to breadth, less than half as long as propodi, dactylus-carpus length ratio, 0.7-0.8 on P2, 0.8-0.9 on P3, 0.9-1.0 on P4; flexor margin slightly convex distally, with row of 9-11 somewhat inclined spines, penultimate 2 × broader than antepenultimate, remaining spines slender, ultimate slightly broader and distinctly longer than antepenultimate.

Eggs. Number of eggs carried up to 60; size, 1.46 × 1.32 mm (largest) in diameter.

Color. Illustrated by Poore & Andreakis (2011: fig. 4a).

PARASITES — The smaller female from MUSORSTOM 2 Stn CP51 (MNHN-IU-2014-16860) bears a rhizocephalan externa.

REMARKS — Poore & Andreakis (2011) reviewed the *naso* complex based on morphological and molecular data, describing two new species: *U. cyrano* from northern Western Australia and *U. pinocchio* from the Philippines, Taiwan and Japan. These two species have dense tufts of short setae on P2 and two longitudinal bands on the carapace, the clear differences from *U. naso*. Some of the features that they believed to characterize *U. naso* as different from these congeners are not exactly applicable: the rostrum that is said to be deflected ventrally is often horizontal in the present material. An additional character distinguishing *U. naso* from the two congeners is that the sternal plastron is shorter relative to breadth, the length-breadth ratio, 1.1-1.2 versus 1.4. The coloration is also clearly different between *U. naso* and *U. pinocchio*, as shown by Poore & Andreakis (2011).

Uroptychus nebulosus n. sp.
Figures 160, 161, 305I

Uroptychus sibogae — Poore *et al.* 2011: 330, pl. 8, fig. A (not *U. sibogae* Van Dam, 1933).

TYPE MATERIAL — Holotype: **Tonga**. BORDAU 2 Stn CH1621, 24°19'S, 176°23'W, 570-573 m, 18.VI.2000, ♂ 9.5 mm (MNHN-IU-2014-16804). Paratypes: **Tonga**. BORDAU 2, station data as for the holotype, 1 ♂ 8.7 mm, 1 ov. ♀ 9.0 mm (MNHN-IU-2014-16805). – Stn CP1620, 24°18'S, 176°20'W, 572 m, 18.VI.2000, 1 ♂ 9.6 mm, 1 ov. ♀ 11.1 mm (MNHN-IU-2014-16806). **Solomon Islands**. SALOMON 1 Stn DW1788, 09°19'S, 160°15'E, 341-343 m, 30.IX.2001, 2 ♂ 4.8, 5.1 mm (MNHN-IU-2014-16807). SALOMON 2 Stn CP2262, 7°56.4'S, 156°51.2'E, 460-487 m, 3.XI.2004, 1 ♂ 4.9 mm (MNHN-IU-2014-16808). – Stn CP2263, 7°54.8'S, 156°51.3'E, 485-520 m, 3.XI.2004, 1 ov. ♀ 6.8 mm (MNHN-IU-2014-16809). **Vanuatu**. SANTO Stn AT09, 15°42'S, 167°01'E, 481 m, 17.IX.2006, 2 ♂ 6.6, 7.6 mm (MNHN-IU-2014-16810). – Stn AT10, 15°41.1'S, 167°00.5'E, 509-559 m, 17.IX.2006, 1 ♂ 6.5 mm (MNHN-IU-2014-16811). – Stn AT73, 15°40.8'S, 167°00.5'E, 514-636 m, 07.X.2006, 1 ♂ 8.1 mm (MNHN-IU-2014-16812), 3 ♂ 5.6-7.9 mm, 1 ov. ♀ 6.0 mm, 1 ♀ 5.7 mm (MNHN-IU-2014-16813). **New Caledonia**, Hunter and Matthew Islands. VOLSMAR Stn DW05, 22°26'S, 171°46'E, 620-700 m, 01.VI.1989, 4 ♂ 4.4-6.2 mm, 1 ov. ♀ 5.7 mm (MNHN-IU-2014-16814). **New Caledonia**. MUSORSTOM 4 Stn CP216, 22°59.5'S, 167°22.0'E, 490-515 m, with Chrysogorgiidae gen. sp. (Calcaxonia), 29.IX.1985, 1 ♂ 5.7 mm (MNHN-IU-2014-16815). **French Polynesia**, Austral Islands. BENTHAUS Stn DW2001, 22°26.6'S, 151°20.1'W, 200-550 m, 23.XI.2002, 1 ov. ♀ 6.1 mm (MNHN-IU-2014-16816).

ETYMOLOGY — From the Latin *nebulosus* (indefinite, obscure), alluding to the systematic status of this species which is very similar to *U. sibogae* Van Dam, 1933 and *U. longicarpus* n. sp., but the species is distinctive in subtle detail.

DISTRIBUTION — Solomon Islands, Vanuatu, Hunter-Matthew Islands, New Caledonia, Tonga and French Polynesia; 200-700 m.

SIZE — Males, 4.4-9.6 mm; females, 5.7-11.1 mm; ovigerous females from 5.7 mm.

DESCRIPTION — Medium-sized to large species. *Carapace*: As long as or slightly longer than broad; greatest breadth 1.5 × distance between anterolateral spines. Dorsal surface smooth, with very fine, short setae discernible under high magnification, slightly convex from anterior to posterior, feebly concave between gastric and cardiac regions. Lateral margins slightly convex and slightly divergent posteriorly; anterolateral spine small, located posterior to level of smaller lateral orbital spine, and never overreaching its tip; another spine located at anterior end of branchial region, followed by row of feebly denticulate short ridges; ridged along posterior sixth of length. Rostrum elongate, sharp triangular, with interior angle 27-35°; dorsal surface moderately concave; length 0.4-0.5 × that of remaining carapace, breadth slightly more than half carapace breadth at posterior carapace margin. Pterygostomian flap anteriorly angular, produced to very small spine, surface smooth.

Sternum: Excavated sternum with sharp anterior spine between bases of Mxp1, surface with spine in center. Sternal plastron as long as broad or slightly shorter; lateral extremities divergent posteriorly. Sternite 3 strongly depressed, anterior margin strongly excavated, with 2 well-developed submedian spines nearly contiguous or separated by narrow notch, anterolateral angle rounded or angular. Sternite 4 long relative to breadth, anterolaterally sharp angular; anterolateral margin smoothly straight, length 2.6-3.2 × that of posterolateral margin. Anterolateral margin of sternite 5 anteriorly rounded, 1.5-1.6 × longer than posterolateral margin of sternite 4.

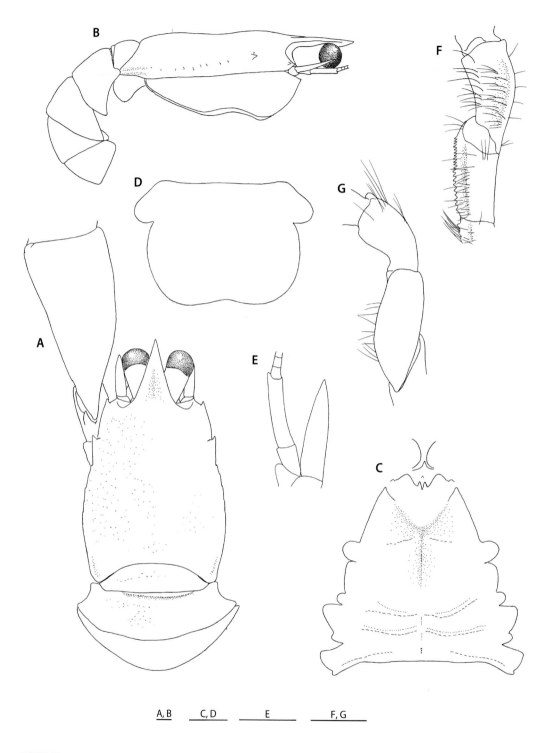

FIGURE 160

Uroptychus nebulosus n. sp., holotype, male 9.5 mm (MNHN-IU-2014-16804). **A**, carapace and abdomen, proximal part of left P1 included, dorsal. **B**, same, lateral. **C**, sternal plastron, with excavated sternum and basal parts of Mxp1. **D**, telson. **E**, left antenna, ventral. **F**, left Mxp3, ventral. **G**, same, lateral. Scale bars: 1 mm.

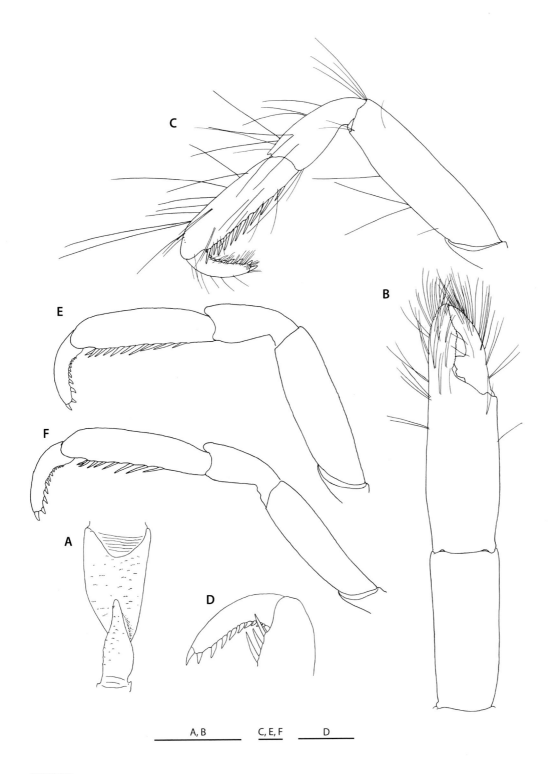

FIGURE 161

Uroptychus nebulosus n. sp., holotype, male 9.5 mm (MNHN-IU-2014-16804). **A**, right P1, proximal part, setae omitted, ventral. **B**, same, proximal part omitted, dorsal. **C**, left P2, lateral. **D**, same, distal part, setae omitted, lateral. **E**, left P3, setae omitted, lateral. **F**, right P4, setae omitted, lateral. Scale bars: 1 mm.

Abdomen: Smooth, with very fine setae. Somite 1 antero-posteriorly convex, not ridged. Somite 2 tergite 2.2-2.4 × broader than long; pleuron with anterolaterally rounded, posterolaterally angular, lateral margin concavely strongly divergent. Pleuron of somite 3 with angular lateral terminus. Telson 0.6-0.7 × as long as broad; posterior plate 1.5-2.1 × longer than anterior plate, emarginate on posterior margin.

Eye: 1.6-2.0 × longer than broad, overreaching midlength of rostrum but not reaching rostral tip; lateral and mesial margins subparallel. Cornea slightly inflated, more than half as long as remaining eyestalk.

Antennule and antenna: Ultimate article of antennule 2.0-2.5 × longer than high. Antennal peduncle not reaching distal end of cornea. Article 2 with very small or obsolescent lateral spine. Antennal scale slightly falling short of or nearly reaching distal end of article 5, 1.6 × broader than article 5. Article 4 unarmed. Article 5 with small distomesial spine, length 2.5-2.6 × that of article 4, breadth 0.4 height of ultimate antennular article. Flagellum of 22-24 segments slightly falling short of distal end of P1 merus.

Mxp: Mxp1 with bases close to each other. Mxp3 basis with 3 or 4 denticles on mesial ridge, distalmost consistent, others often obsolescent. Ischium with 20-25 denticles on crista dentata, flexor margin distally not rounded. Merus about twice as long as ischium, unarmed, moderately ridged along flexor margin. Carpus unarmed.

P1: 4.7-5.0 × longer than carapace, massive, setose on fingers and distal part of palm, ventrally granulose except for palm and fingers; triangular dorsal spine on ischium, unarmed elsewhere. Merus 1.1-1.2 × longer than carapace. Carpus 1.1-1.3 × longer than merus. Palm 2.1-2.6 × (males), 2.5-2.7 × (females) longer than broad, 0.9-1.0 × longer than carpus. Fingers gaping in males, not gaping in females and small males, distally incurved, crossing when closed; movable finger 0.6-0.7 × length of palm, opposable margin with medially incised proximal process.

P2-4: Broad relative to length, dactyli narrow. Meri moderately compressed mesio-laterally, sparingly with long setae, successively shorter posteriorly (P3 merus 0.9 × length of P2 merus, P4 merus 0.8-0.9 × length of P3 merus); successively narrower posteriorly (P4 merus 0.9 × breadth of P2 merus, 0.8-0.9 × breadth of P3 merus); length-breadth ratio, 3.5-3.7 on P2, 3.3-3.9 on P3, 3.6-4.0 on P4; P2 merus 0.8 × length of carapace, 1.2-1.4 × length of P2 propodus; P3 merus subequal to length of P3 propodus; P4 merus 0.9 × length of P4 propodus. Carpi successively shorter posteriorly (P3 carpus 0.9 × length of P2 carpus, P4 carpus 0.9 × length of P3 carpus); carpus-dactylus length ratio, 1.3-1.7 on P2, 1.2-1.4 on P3, 1.0-1.3 on P4; carpus-propodus length ratio, 0.6 on P2, 0.5 on P3, 0.4-0.5 on P4. Propodi subequal in length on P3 and P4 (shortest on P2) or on P2 and P3 (longest on P4); flexor margin straight, terminating in pair of spines preceded by 9-11 spines on P2, 8-9 spines on P3, 6-8 spines on P4. Dactyli subequal, 0.6-0.8 × length of carpi on P2, 0.7-0. 8 × on P3, 0.8-1.0 × on P4, 0.4 × length of propodi on P2-4; flexor margin strongly curving at proximal quarter, with 9-10 subtriangular, somewhat inclined, proximally decreasing spines on P2, 10-11 spines on P3, 10-12 spines on P4, ultimate largest, penultimate and antepenultimate subequal.

Eggs. Number of eggs carried, about 60; size, 1.28 mm × 1.36 mm - 1.37 mm × 1.40 mm.

Color. Pale orange overall, posterior half of carapace translucent white (male 8.1 mm, MNHN-IU-2014-16812; male 6.5 mm, MNHN-IU-2014-16811). The color illustration given under *U. sibogae* by Poore *et al.* (2011) is from one of the photos (MNHN-IU-2014-16811) available to me.

REMARKS — The species is very closely related to *U. sibogae* Van Dam, 1933 (see below) and *U. longicarpus* n. sp. in nearly all aspects. *Uroptychus nebulosus* is distinguished from *U. sibogae* by subtle morphological differences: the P2 merus is longer and broader than instead of subequal to or slightly narrower than P3 merus; the P2-4 meri are broader relative to length (the length-breadth ratio being 3.7-3.9, 3.4-4.1, 3.6-3.8 versus 5.0-5.6, 5.0-5.3, 4.2-4.6 on P2, P3, P4 respectively; the P2 merus is much shorter than instead of subequal to the carapace (0.8 versus 1.0); and the cornea is not so well inflated as in *U. sibogae*. *Uroptychus longicarpus* n. sp. is unique in the genus in having the P4 carpus distinctly longer than the P3 carpus.

Uroptychus nigricapillis Alcock, 1901

Figures 162-173

Uroptychus nigricapillis Alcock, 1901: 283, pl. 3, fig. 3. — Alcock & McArdle 1902: pl. 56, fig. 3. — Van Dam 1933: 26; 1940: 98, fig. 2. — Baba 1981: 116, fig. 4; 1988: 40; 1990: 947 (part). — Baba 2005: 50 (part). — Baba *et al.* 2008: 37. — Baba *et al.* 2009: 50 (part), figs 41, 43. — Poore *et al.* 2011: 329, pl. 7, fig. E.

Uroptychus gracilimanus — Ahyong & Poore 2004: 40, fig. 10 (not *U. gracilimanus* Alcock, 1901).

Not *Uroptychus nigricapillis* — Laurie 1926: 123 (= *U. longioculus* Baba, 1990). — Tirmizi 1964: 390, figs 4, 5. — Baba 2005: 50 (part) (new species). — Ahyong & Baba 2004: 60, fig. 2. — Baba *et al.* 2009: 50 (part), fig. 42. — Poore *et al.* 2011: 329, pl. 7, fig. F (new species).

TYPE MATERIAL — Holotype: **Andaman Sea**, 669 fms (1224 m), female, ZSI 3443/10). [not examined].

MATERIAL EXAMINED — **Solomon Islands**. SALOMON 1 Stn CP1807, 09°42'S, 160°53'E, 1077-1135 m, 02.X.2001, 1 ♂ 10.8 mm (MNHN-IU-2014-16817). SALOMON 2 Stn CP2182, 8°47.0'S, 159°37.9'E, 762-1060 m, 22.X.2004, 1 ♂ 8.3 mm (MNHN-IU-2014-16818). – Stn CP2230, 6°27.8'S, 156°24.3'E, 837-945 m, 29.X.2004, 1 ♀ 9.2 mm (MNHN-IU-2013-12298). – Stn CP2260, 8°03.5'S, 156°54.5'E, 399-427 m, 3.XI.2004, 1 ♂ 7.5 mm, 1 ♀ 7.8 mm (MNHN-IU-2014-16819). – Stn CP2261, 8°01.9'S, 156°54.1'E, 433-470 m, 3.XI.2004, 1 ♀ 9.6 mm (MNHN-IU-2014-16820). **Wallis and Futuna Islands**. MUSORSTOM 7 Stn CP564, 11°46'S, 178°27'W, 1015-1020 m, with *Chrysogorgia* sp. (Calcaxonia, Chrysogorgiidae), 20.V.1992, 3 ♂ 7.9-8.3 mm, 2 ov. ♀ 8.7-9.1 mm, 4 ♀ 7.1-8.8 mm (MNHN-IU-2014-16821). – Stn CP567, 11°47'S, 178°27'W, 1010-1020 m, 20.V.1992, 1 ♂ 6.7 mm, 1 ov. ♀ 7.8 mm, 4 ♀ 7.0-9.4 mm (MNHN-IU-2014-16822), 1 ov. 9.0 mm (MNHN-IU-2013-12297). **Vanuatu**. MUSORSTOM 8 Stn CP990, 18°51.63'S, 168°50.98'E, 980-990 m, 24.IX.1994, 1 ov. ♀ 8.0 mm (MNHN-IU-2014-16823). – Stn CP1008, 18°53.29'S, 168°52.65'E, 919-1000 m, 25.IX.1994, 1 ♂ 8.8 mm (MNHN-IU-2014-16824). – Stn CP1037, 18°03.70'S, 168°54.40'E, 1058-1086 m, 29.IX.1994, 1 ov. ♀ 8.6 mm, 1 ♀ 9.8 mm (MNHN-IU-2014-16825). – Stn CP1125, 15°57.63'S, 166°38.43'E, 1160-1220 m, 10.X.1994, 2 ♂ 6.4, 7.3 mm (MNHN-IU-2014-16826). – Stn CP1129, 16°00.73'S, 166°39.94'E, 1014-1050 m, 10.X.1994, 6 ♂ 5.0-8.9 mm, 3 ov. ♀ 7.1-8.9 mm, 1 ♀ 6.6 mm (MNHN-IU-2010-5420, MNHN-IU-2013-12296, MNHN-IU-2014-16827). **New Caledonia, Chesterfield Islands**. MUSORSTOM 5 Stn CP324, 21°15.01'S, 157°51.33'E, 970 m, 14.X.1986, 1 ♂ 8.3 mm, 1 ov. ♀ 8.2 mm (MNHN-IU-2010-5421). **New Caledonia**. BIOCAL Stn CP55, 23°20'S, 167°30'E, 1160-1175 m, 1.IX.1985, 1 ov. ♀ 7.7 mm (MNHN-IU-2014-16828). – Stn CP61, 24°11'S, 167°32'E, 1070 m, 2.IX.1985, 2 ov. ♀ 9.2, 11.0 mm (MNHN-IU-2014-16829).

DISTRIBUTION — Western Indian Ocean (Mozambique Channel, Zanzibar, off Kenya, South Arabian coast, Madagascar and Maldives), Andaman Sea, west of Makassar, Java Sea, Flores Sea off southern Sulawesi, between Siquijor and Bohol, South China Sea, Taiwan and Japan (southeastern Kyushu), in [66] 450-1939 m, and now Solomon Islands, Wallis and Futuna Islands, Vanuatu, Chesterfield Islands and New Caledonia, in 399-1220 m.

SIZE — Males, 5.0-10.8 mm; females, 6.6-11.0 mm; ovigerous females from 7.1 mm.

DESCRIPTION — Medium-sized species. *Carapace*: 1.1-1.2 × longer than broad; greatest breadth 1.5 × distance between anterolateral spines. Dorsal surface smooth, with shallow depression between gastric and cardiac regions; pair of epigastric spines varying from small to good size. Lateral margins somewhat convex, anterolateral spine usually small but larger than lateral orbital spine, located posterior to level of that spine, usually not reaching, occasionally reaching or slightly overreaching it; anterior end of branchial region with somewhat elevated ridge or very small spine, followed by denticle-like or obsolescent very small spines; ridged along posterior half or posterior third. Rostrum narrowly triangular, with interior angle of 21-25°, straight horizontal or directed slightly ventrally; dorsal surface flattish; length 0.4-0.5 × that of remaining carapace, breadth half carapace breadth at posterior carapace margin. Pterygostomian flap anteriorly not sharply produced but somewhat roundish, ending in small spine; surface glabrous and spineless.

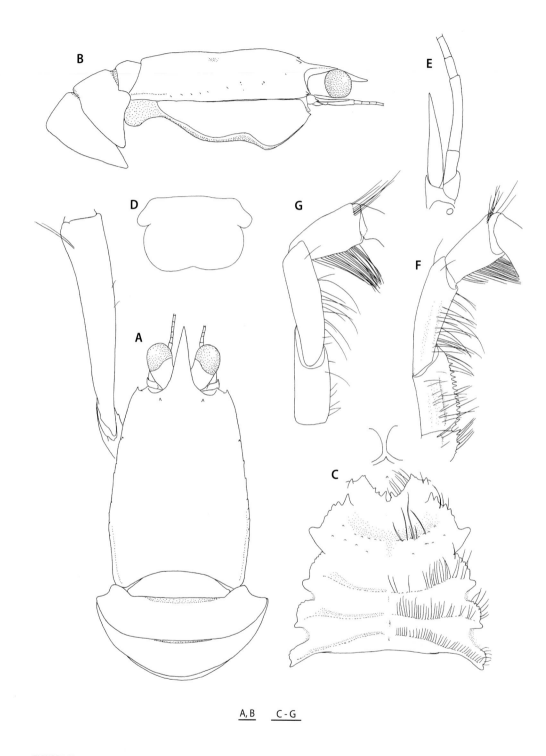

A, B   C - G

FIGURE 162

Uroptychus nigricapillis Alcock, 1901, ovigerous male 9.0 mm (MNHN-IU-2013-12297). **A**, carapace and anterior part of abdomen, proximal part of left P1 included, dorsal. **B**, same, lateral. **C**, sternal plastron, with excavated sternum and basal parts of Mxp1. **D**, telson. **E**, right antenna, ventral. **F**, right Mxp3, ventral. **G**, same, lateral. Scale bars: 1 mm.

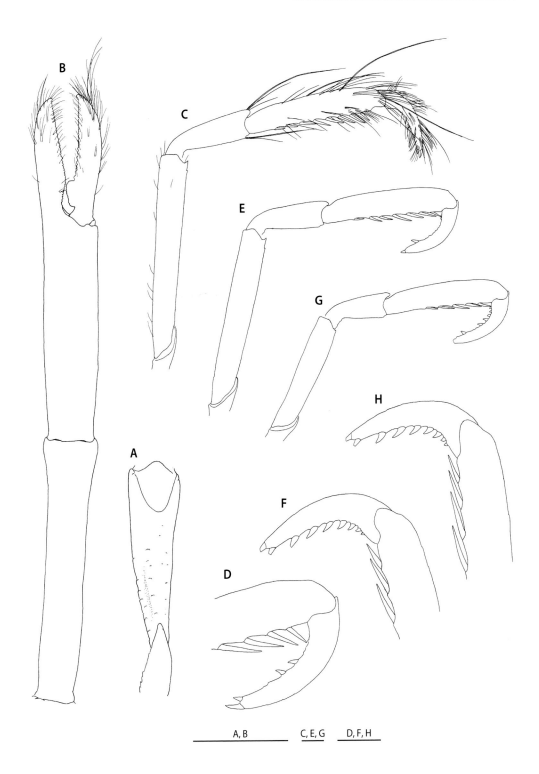

FIGURE 163

Uroptychus nigricapillis Alcock, 1901, ovigerous male 9.0 mm (MNHN-IU-2013-12297). **A**, left P1, proximal part, ventral. **B**, same, proximal part omitted, dorsal. **C**, right P2, lateral. **D**, same, distal part, setae omitted, lateral. **E**, right P3, setae omitted, lateral. **F**, same, distal part, lateral. **G**, right P4, lateral. **H**, same, distal part, lateral. Scale bars: A, B, 5 mm; C-H, 1 mm.

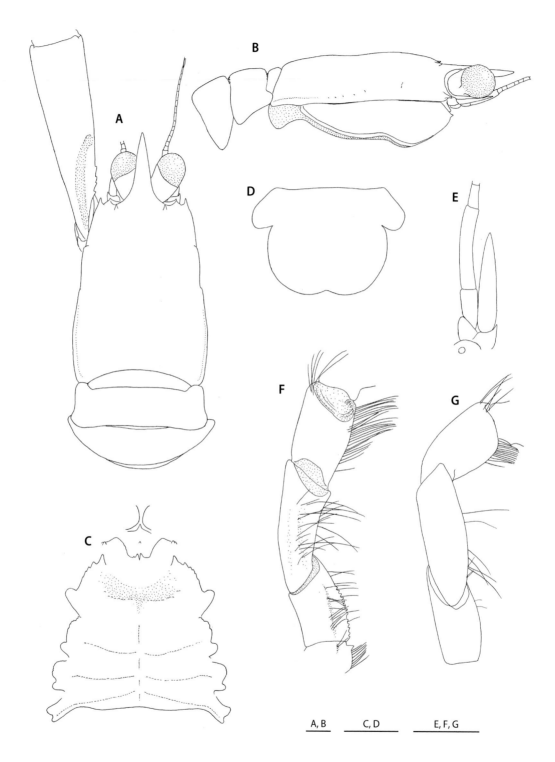

FIGURE 164

Uroptychus nigricapillis Alcock, 1901, male 6.2 mm (MNHN-IU-2013-12296). **A**, carapace and anterior part of abdomen, proximal part of left P1 included, dorsal. **B**, same, lateral. **C**, sternal plastron, with excavated sternum and basal parts of Mxp1. **D**, telson. **E**, left antenna, ventral. **F**, right Mxp3, ventral. **G**, same, lateral. Scar bars: 1 mm.

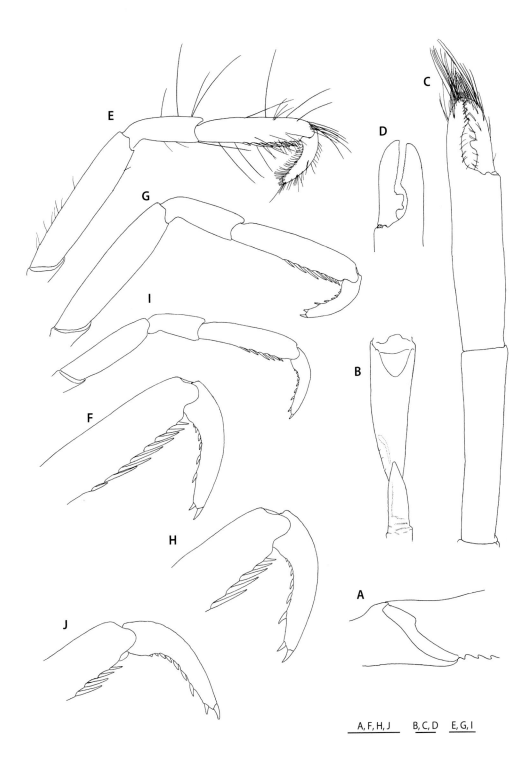

FIGURE 165

Uroptychus nigricapillis Alcock, 1901, male 6.2 mm (MNHN-IU-2013-12296). **A**, left P1, proximal part, setae omitted, lateral. **B**, same, ventral. **C**, same, proximal part omitted, dorsal. **D**, right P1, fingers, dorsal. **E**, left P2, lateral. **F**, same, distal part, setae omitted, lateral. **G**, right P3, lateral. **H**, same, distal part, lateral. **I**, right P4, lateral. **J**, same, distal part, lateral. Scale bars: 1 mm.

A, B ___ C, D ___ E, F, G

FIGURE 166

Uroptychus nigricapillis Alcock, 1901, female 10.0 mm (USNM 151644) from Indonesia. **A**, carapace and anterior part of abdomen, dorsal. **B**, same, lateral. **C**, sternal plastron, with excavated sternum and basal parts of Mxp1. **D**, telson. **E**, left antenna, ventral. **F**, right Mxp3, ventral. **G**, same, lateral. Scale bars: A, B, 5 mm; C-G, 1 mm.

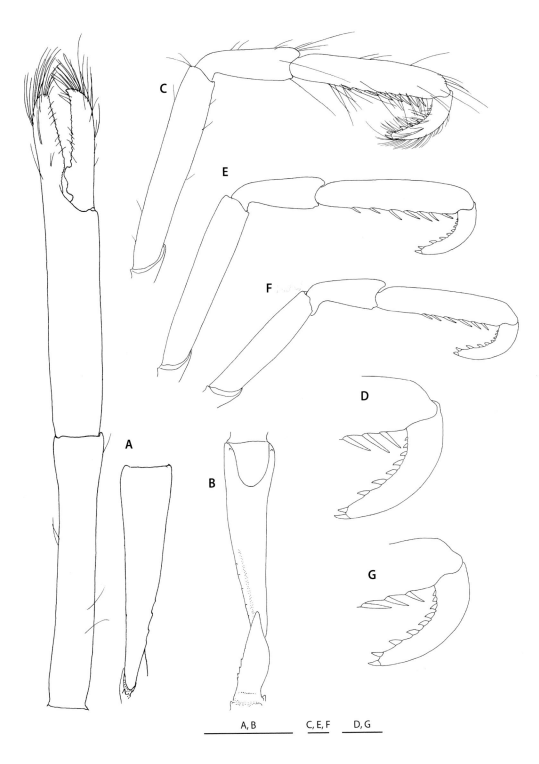

FIGURE 167

Uroptychus nigricapillis Alcock, 1901, female 10.0 mm (USNM 151644) from Indonesia. **A**, left P1, dorsal. **B**, same, proximal part, ventral. **C**, right P2, lateral. **D**, same, distal part, setae omitted, lateral. **E**, right P3, setae omitted, lateral. **F**, right P4, lateral. **G**, same, distal part, lateral. Scale bars: A, B, 5 mm; C-G, 1 mm.

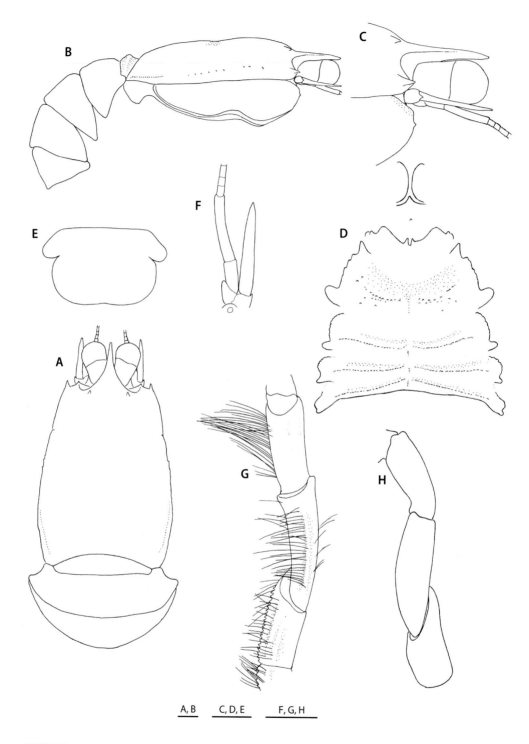

A, B   C, D, E   F, G, H

FIGURE 168

Uroptychus nigricapillis Alcock, 1901, ovigerous female 7.3 mm (USNM 151639) from the Philippines. **A**, carapace and anterior part of abdomen, dorsal. **B**, same, lateral. **C**, carapace, anterior part, lateral. **D**, sternal plastron, with excavated sternum and basal parts of Mxp1. **E**, telson. **F**, left antenna, ventral. **G**, left Mxp3, ventral. **H**, same, lateral. Scale bars: 1 mm.

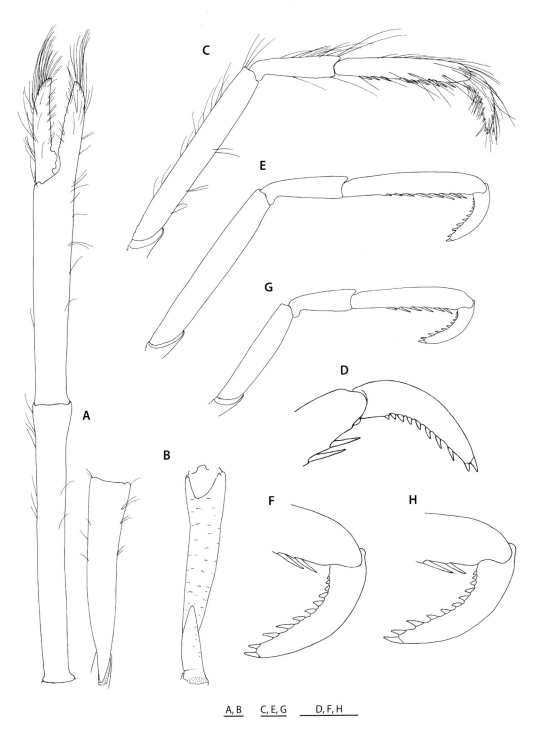

FIGURE 169

Uroptychus nigricapillis Alcock, 1901, ovigerous female 7.3 mm (USNM 151639) from the Philippines. **A**, right P1, dorsal. **B**, same, proximal part, ventral. **C**, right P2, lateral. **D**, same, distal part, setae omitted, lateral. **E**, right P3, setae omitted, lateral. **F**, same, distal part, lateral. **G**, right P4, lateral. **H**, same, distal part, lateral. Scale bars: 1 mm.

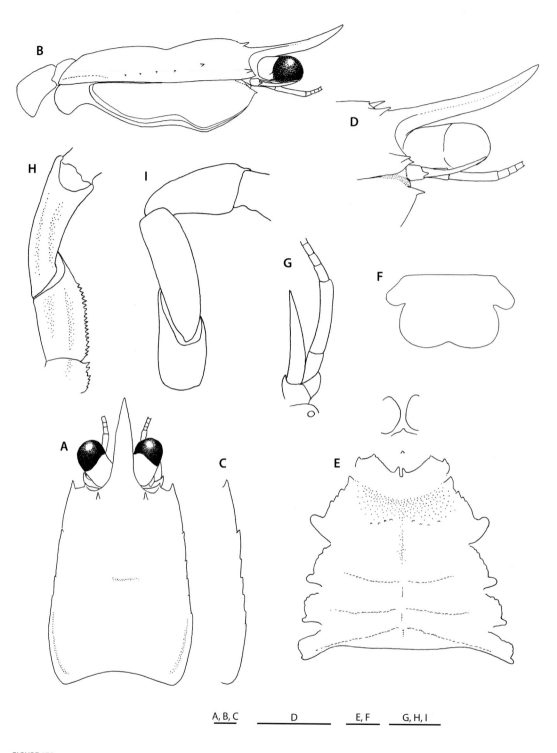

FIGURE 170

Uroptychus nigricapillis Alcock, 1901, male 7.9 mm (USNM 151638) from the Philippines. **A**, carapace, dorsal. **B**, same, anterior part of abdomen included, lateral. **C**, carapace lateral margin, slightly dorsomesial. **D**, carapace, anterior part, lateral. **E**, sternal plastron, with excavated sternum and basal parts of Mxp1. **F**, telson. **G**, right antenna, ventral. **H**, right Mxp3, ventral. **I**, same, lateral. Scale bars: 1 mm.

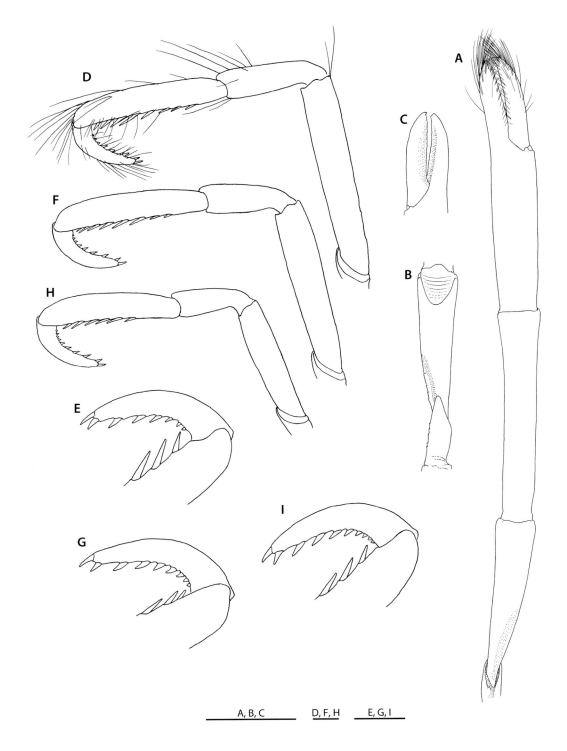

A, B, C   D, F, H   E, G, I

FIGURE 171

Uroptychus nigricapillis Alcock, 1901, male 7.9 mm (USNM 151638) from the Philippines. **A**, right P1, dorsal. **B**, same, proximal part, ventral. **C**, same, fingers, ventral. **D**, left P2, lateral. **E**, same, distal part, setae omitted, lateral. **F**, left P3, setae omitted, lateral. **G**, same, distal part, lateral. **H**, left P4, lateral. **I**, same, distal part, lateral. Scale bars: A-C, 5 mm; B-I, 1 mm.

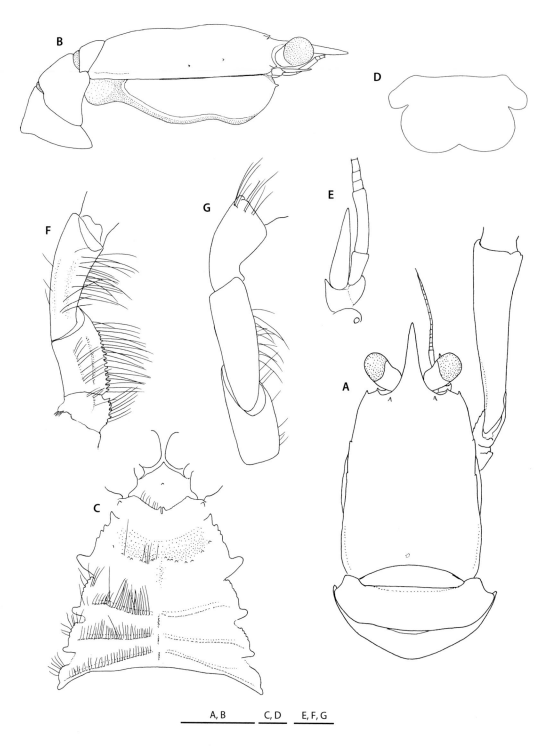

FIGURE 172

Uroptychus nigricapillis Alcock, 1901, female 10.6 mm from Mozambique Channel (MNHN-IU-2010-5450). **A**, carapace and anterior part of abdomen, proximal part of right P1 included, dorsal. **B**, same, lateral. **C**, sternal plastron, with excavated sternum and basal parts of Mxp1. **D**, telson. **E**, right antenna, ventral. **F**, right Mxp3, ventral. **G**, same, lateral. Scale bars: A, B, 5 mm; C-G, 1 mm.

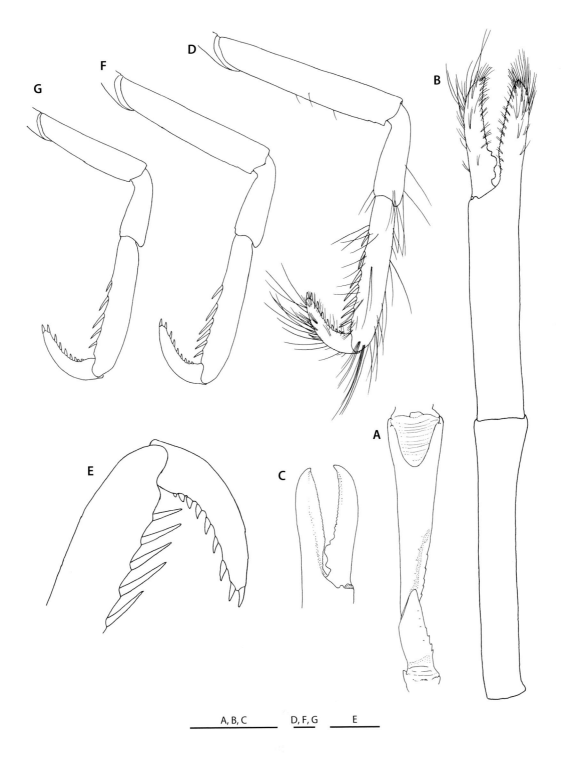

FIGURE 173

Uroptychus nigricapillis Alcock, 1901, female 10.6 mm from Mozambique Channel (MNHN-IU-2010-5450). **A**, right P1, proximal, ventral. **B**, same, proximal articles omitted, dorsal. **C**, same, fingers, ventral. **D**, right P2, lateral. **E**, same, distal part, setae omitted, lateral. **F**, right P3, setae omitted, lateral. **G**, right P4, lateral. Scale bars: A-C, 5 mm; B-G, 1 mm.

Sternum: Excavated sternum with anterior margin broadly triangular with small median spine, surface with small spine in center. Sternal plastron 0.9 × as long as broad, lateral extremities gently divergent posteriorly. Sternite 3 well depressed; anterior margin strongly excavated, with subovate or narrow median notch flanked by spine often accompanying a few small spines lateral to each, anterolaterally sharp angular. Sternite 4 with transverse row of denticles or tubercles on surface; anterolateral margin with posteriorly diminishing spines, length 1.5 × that of posterolateral margin.

Abdomen: Smooth and glabrous. Somite 1 with antero-posteriorly convex transverse ridge. Somite 2 tergite 2.2-2.7 × broader than long, pleural lateral margin strongly divergent posteriorly, posterolaterally blunt. Pleura of somites 3 and 4 also bluntly angular laterally. Telson 0.55-0.68 × as long as broad, posterior plate 1.5-2.0 × length of anterior plate, distinctly emarginate on posterior margin.

Eye: 1.6-2.0 × longer than broad, overreaching midlength of and not reaching apex of rostrum; mesial margin concave. Cornea slightly broader than and about as long as remaining eyestalk.

Antennule and antenna: Ultimate article of antennular peduncle 1.9-2.0 × longer than high. Antennal peduncle slender, not overreaching cornea. Article 2 with distinct distolateral spine. Antennal scale terminating in or overreaching midlength of and barely reaching distal end of article 5; breadth 1.5 × that of article 5. Articles 4 and 5 unarmed; article 5 2.2-2.6 × longer than article 4, breadth less than half height of ultimate article of antennule. Flagellum of 12-15 segments barely reaching distal end of P1 merus.

Mxp: Mxp1 with bases very close to each other. Mxp3 relatively slender, barely setose on lateral surfaces of ischium and merus. Basis with 3-5 proximally diminishing denticles on mesial ridge. Ischium with flexor margin not rounded distally, crista dentata with 14-15 denticles. Merus and carpus unarmed; merus twice as long as ischium, with weak ridge along flexor margin bearing sparse long setae, mesial and lateral faces convex.

P1: 4.5-5.0 × longer than carapace, slender, subcylindrical but somewhat depressed on fingers and palm, with setae relatively short on fingers, almost glabrous elsewhere; small tubercles occasionally present on ventral surfaces of palm, carpus and merus in large specimens, obsolete or absent in small specimens. Ischium with basally broad, depressed, short triangular dorsal spine, ventromesial margin with tubercular processes on proximal part, lacking subterminal spine. Merus 1.1-1.3 × longer than carapace. Carpus 1.1-1.3 × longer than merus. Palm 2.9-3.4 × (males), 3.9-4.5 ×, rarely 5.8 × (females) longer than broad, subequal to or slightly shorter than carpus. Fingers not gaping in females (opposable margin of movable finger with low proximal process); gaping in males (movable finger with obtuse process occasionally bidentate at midlength of gaping portion); fingers fitting in distal third, ending in small incurved spine, slightly crossing when closed; movable finger 0.5-0.6 × length of palm.

P2-4: Slender and well compressed mesio-laterally, bearing long setae on distal articles. Meri successively shorter posteriorly (P3 merus 0.8-0.9 × length of P2 merus, P4 merus 0.7-0.8 × length of P3 merus), subequally broad on P2-3, slightly narrower on P4; length-breadth ratio, 5.2-6.5 on P2, 4.7-6.0 on P3, 3.9-4.8 on P4; dorsally unarmed, ventrolaterally bearing terminal spine; P2 merus 0.8-0.9 × length of carapace, 1.2-1.3 × (1.1 × in small specimens) longer than P2 propodus; P4 merus 0.8-0.9 × length of P4 propodus. Carpi successively shorter posteriorly; carpus-propodus length ratio, 0.54-0.56 on P2, 0.51-0.54 on P3, 0.43-0.50 on P4; carpus-dactylus length ratio, 1.2-1.4 on P2, 1.0-1.2 on P3, 0.9-1.0 on P4. Propodi successively shorter posteriorly; flexor margin with long, slender spines usually along entire length at least on P2: 7-8 spines on P2, 5-6 spines on P3, 4-5 spines on P4; terminal spine single, relatively remote from juncture with dactylus, equidistant between distal second spine and juncture or closer to second. Dactyli more slender than propodi, strongly curving at proximal third; dactylus-propodus length ratio, 0.41-0.46 on P2, 0.41-0.47 on P3, 0.46-0.48 on P4; flexor margin with relatively large, loosely spaced, obliquely directed, triangular spines 8-10 (mostly 9) in number, ultimate larger than penultimate, antepenultimate more remote from penultimate than from distal quarter.

Eggs. Number of eggs carried 5-20; size, 1.93 mm × 2.07 mm - 2.10 mm × 2.50 mm.

REMARKS — According to S.T. Ahyong (pers. comm.), all the material reported by Ahyong & Poore (2004) under *U. gracilimanus* is referable to this species: the anterior margin of sternite 3 is usually moderately excavated on the

Australian specimens of Ahyong & Poore (2004), sometimes shallowly excavated as illustrated by Ahyong & Poore (2004: fig 10E), but usually deeper being more like Figure 162C.

Most of the present material generally fits the description and illustration of the type material from the Andaman Sea (Alcock 1901: 283, pl. 3, fig. 3; Alcock & McArdle 1902: pl. 56, fig. 3) but it is not in complete agreement in having the carapace lateral marginal spines usually tiny or obsolescent, occasionally small and not so distinct as illustrated for the type and in having the epigastric spines small, not so clearly large as in the type. If the figure of the Investigator material is correctly illustrated, the P4 merus is as long as the P4 propodus whereas it is 0.8-0.9 times as long in the MNHN specimens (Figures 162, 163).

The specimens taken at MUSORSTOM 8 Stn CP1129, Vanuatu (MNHN-IU-2014-16827) have more closely arranged spines along the propodal flexor margins of P2-4, compared with the rest of the material examined. In one of the males (Figures 164, 165), sternite 3 is more strongly excavated.

The Albatross material reported earlier from the Philippines and Indonesia (Baba 1988) seems to contain two or three different species: 1) one ovigerous female from Albatross Station 5660 (USNM 151644) (Figures 166, 167) looks identical with the present material; 2) one male and two females from Albatross Station 5527 (USNM 151639) are very similar to the above but differ in having the pterygostomian flap more roundish anteriorly with a tiny spine and in having the P2 merus slightly longer than the carapace (Figures 168, 169); 3) one male from Albatross Station 5274 (USNM 151638) has a upturned rostrum, the carapace lateral spines more distinct and much like that of the type, the pterygostomian flap anteriorly more or less angular with a distinct spine and the P2 merus 0.9 times as long as the carapace (Figures 170, 171).

The material from Madagascar (Baba 1990) contains two species. One is represented by a female (6.1 mm) from Vauban Station CH138 (MNHN-IU-2013-7805). It is similar to the western Australian specimen reported by Ahyong & Baba (2004; see below), having relatively long P2-4 carpi (carpus-dactylus length ratio, 2.0 on P2, 1.9 on P3, 1.7 on P4), but it differs from that specimen in having obsolescent carapace lateral spines, in having the anterolateral spines overreaching the lateral orbital spine, and in having the pterygostomian flap anteriorly more angular and produced to a distinct spine; this species is apparently new. The other species is represented by the rest of the material from Madagascar (see Baba 1990), which is identical with a specimen collected by Mainbaza Station CP 3139 in the Mozambique Channel [23°33.51'S, 36°7.27'E, 1092-1195 m, 11 Apr 2009, 1 female 10.6 mm, MNHN-2010-5450 (see Figures 172, 173)]. These specimens are morphologically hardly discriminated from the material around New Caledonia and vicinity.

The specimens from three different localities in the western Indian Ocean collected by the John Murray Expedition (Tirmizi 1964) are identical with those taken at Galathea Station 241 off Kenya (Baba 2005; ZMUC CRU-11279). These specimens are different from the above-mentioned western Indian Ocean material, having a small spine in midline slightly posterior to the position of a pair of epigastric spines. They appear to be an undescribed species.

Ahyong & Baba (2004) described a western Australian specimen that has well-developed carapace lateral spines and deep cervical groove, noting that the P2 merus is longer than the carapace. In addition, the specimen has longer carpi on P2-4 (the P2 carpus is nearly twice the length of the P2 dactylus), whereas this article in the holotype is shorter, 1.2 times longer than the dactylus if the figure (Alcock & McArdle 1902: pl. 56, fig. 3) is correctly depicted; and the P3 merus is subequally long as the P2 merus rather than shorter (0.9 times as long). These features are consistent in additional material examined (Soela Stn NWS-29, 17°55.5' S, 118°19.5' E, 450-454 m, 27.I.1984, 1 ovigerous female 11.5 mm (NTM Cr. 000649); Soela Stn NWS-47, 18°34.3' S, 117°30.0' E, 404 m, 01.II.1984, 1 ovigerous female 10.0 mm (NTM Cr. 000650); Soela Stn NWS-52, 18°05.8' S, 118°10.0' E, 408-396 m, 02.II.1984, 1 male 10.8 mm (NTM Cr. 000566)). Baba *et al.* (2009) noted that the three specimens collected off Taiwan and identified as *U. nigricapillis* are identical with the western Australian specimen of Ahyong & Baba (2004). In the MNHN material, the P2 merus is 0.8-0.9 times as long as the carapace and the P3 merus is 0.8-0.9 times the length of P2 merus. The western Australian and Taiwanese specimens have been described as *U. michaeli* by Ahyong & Baba (unpublished).

All of the above-mentioned specimens may be referable to more than four species constituting a species complex, which should be solved by examination of the type material of *U. nigricapillis*. However, access to the type is hardly possible, because repeated inquiries to the Zoological Survey of India, the repository of the type, have been ignored. All of these are provisionally placed in *U. nigricapillis* sensu lato until the systematic status of *U. nigricapillis* is established by examination of the type material or by a discovery of topotypic material.

The material from Saya de Malha Bank in the western Indian Ocean (Laurie 1926) will in all probability be referable to *U. longioculus* Baba, 1990, according to the inconsistencies listed by Laurie.

Uroptychus norfolkanus n. sp.

Figures 174, 175

TYPE MATERIAL — Holotype: **New Caledonia**, Norfolk Ridge. BERYX 11 Stn DW38, 23°37.53'S, 167°39.42'E, 550-690 m, 19.X.1992, ov. ♀ 4.8 mm (MNHN-IU-2012-681).

ETYMOLOGY — Named for the type locality of the species.

DESCRIPTION — Small species. *Carapace*: 1.4 × broader than long (0.7 × as long as broad), broadest on posterior third; greatest breadth 2.1 × distance between anterolateral spines. Dorsal surface convex smoothly from anterior to posterior and from side to side, without depression bordering gastric and cardiac regions, sparsely covered with small pits suggesting to support short fine setae. Lateral margins convex on posterior third, finely granulose, and weakly ridged along posterior third; anterolateral spine short, distinctly larger than lateral orbital spine, directed forward. Rostrum directed somewhat ventrally, narrow triangular, with interior angle of 22°, distally proportionately broad; dorsal surface somewhat concave; lateral margin more or less convex, with 3 obsolescent spinules distally; length half that of remaining carapace, breadth one-third that of carapace measured along posterior margin. Lateral orbital angle with small, distinct spine separated from anterolateral spine by more than basal breadth of latter spine, and located at level of that spine. Pterygostomian flap smooth on surface, anterior margin angular, produced to distinct spine.

Sternum: Excavated sternum with convex anterior margin; surface with weak low ridge in midline. Sternal plastron 0.7 × as long as broad, broadest on sternite 7; lateral extremities convexly divergent posteriorly. Sternite 3 shallowly depressed; anterior margin broadly excavated in subsemicircular shape with 2 obsolescent submedian spines separated by small shallow concavity, laterally blunt angular. Sternite 4 with anterolateral margin convex and denticulate, anteriorly roundish, posterior margin half as long as anterolateral margin. Anterolateral margin of sternite 5 anteriorly feebly angular, 1.5 × longer than posterolateral margin of sternite 4.

Abdomen: Somite 1 moderately convex from anterior to posterior, without transverse ridge. Somite 2 tergite 3.0 × broader than long; pleural lateral margin concavely divergent posteriorly, posterolaterally blunt. Pleuron of somite 3 with blunt posterolateral terminus. Telson 0.4 × as long as broad; posterior plate equally long as anterior plate, concave on posterior margin.

Eye: 1.7 × longer than broad, barely reaching midlength of rostrum, mesial and lateral margins subparallel. Cornea not inflated, length more than half length of remaining eyestalk.

Antennule and antenna: Ultimate article of antennule 2.9 × longer than high. Antennal peduncle overreaching cornea, slightly overreaching midlength of rostrum. Article 2 with small lateral spine. Antennal scale tapering distally, 2.2 × broader than article 5, extending far beyond eye and slightly beyond proximal second segment of flagellum. Articles 4 and 5 each with small distomesial spine, article 5 1.4 × longer than article 4, breadth two-thirds height of antennular ultimate article. Flagellum of 9 segments slightly overreaching rostral tip, far falling short of distal margin of P1 merus.

Mxp: Mxp1 with bases broadly separated. Mxp3 with fine plumose setae other than brushes on distal articles. Basis having mesial ridge lobe-like and nearly smooth. Ischium with flexor margin not rounded distally; crista dentata with denticles obsolescent on proximal half, very tiny on distal half. Merus 2.1 × longer than ischium; with small distolateral spine; flexor margin roundly ridged, with 6 tubercle-like processes on distal third. Carpus with small distolateral spine.

FIGURE 174

Uroptychus norfolkanus n. sp., holotype, ovigerous female 4.8 mm (MNHN-IU-2012-681). **A**, carapace and anterior part of abdomen, proximal part of left P1 included, dorsal. **B**, same, lateral. **C**, sternal plastron, with excavated sternum and basal parts of Mxp1. **D**, anterior part of sternal plastron. **E**, telson. **F**, left antenna, ventral. **G**, right Mxp3, ventral. **H**, same, setae omitted, lateral. Scale bars: 1 mm.

P1: 4.3 × longer than carapace, sparingly covered with short fine plumose setae. Ischium with small dorsal spine, ventromesial margin with row of small spines, with no distinct subterminal spine. Merus with small distomesial and distolateral spines on ventral surface, row of denticles along distodorsal margin, and a few small spines on proximal mesial margin; length 0.9 × that of carapace. Carpus subcylindrical, 1.3 × longer than merus, distodorsal margin with row of denticles. Palm 2.7 × longer than broad, 1.3 × longer than carpus. Fingers moderately depressed and relatively broad, distally incurved, crossing when closed, not gaping; fixed finger with low eminence at midlength of opposable margin; movable finger 0.5 × as long as palm, opposable margin with denticulate eminence (higher than opposite eminence on fixed finger) proximal to midlength.

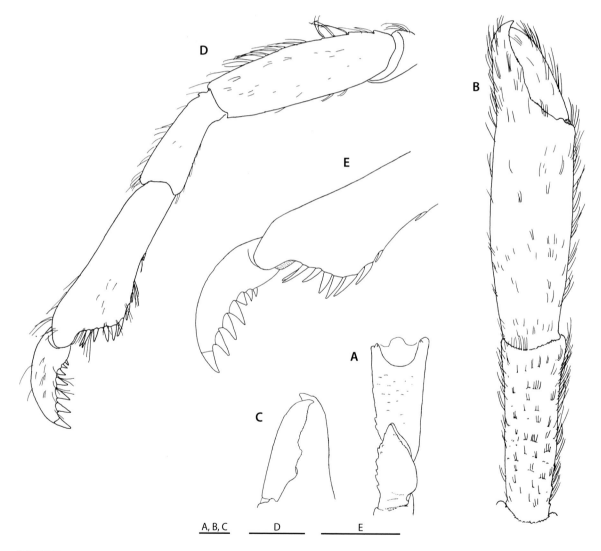

FIGURE 175

Uroptychus norfolkanus n. sp., holotype, ovigerous female 4.8 mm (MNHN-IU-2012-681). **A**, left P1, proximal part, setae omitted, ventral. **B**, same, proximal part omitted, dorsal. **C**, same, fingers, ventral. **D**, left P3, lateral. **E**, same, distal part, setae omitted, lateral. Scale bars: 1 mm.

P2-4: P2 and P4 missing, right P3 broken, only left P3 available. P3 relatively thick mesio-laterally, with fine soft setae. Merus 3.4 × broader than long, about as long as propodus; dorsal margin rounded, not cristate, with 2 small denticle-like spines proximally; ventrolateral margin distally ending in small spine. Carpus 1.3 × longer than dactylus; dorsal margin with tiny proximal spine. Propodus 2.0 × longer than carpus; flexor margin strongly convex on distal third, with pair of terminal spines preceded by 7 spines, 6 of these situated on convex margin, last somewhat more proximal in position. Dactylus 0.8 × length of carpus, 0.4 × length of propodus; tapering distally and well curved; flexor margin with 7 triangular spines proximally diminishing, more or less close to one another and somewhat inclined but proximal spines nearly perpendicular to margin.

Eggs. Number of eggs carried, 15; size, 0.79 × 1.00 - 0.79 × 1.04 mm.

REMARKS — The new species resembles *U. edisonicus* Baba & Williams, 1998, *U. latior* n. sp. and *U. pedanomastigus* n. sp. in the carapace shape, in having a deeply excavated, semicircular anterior margin of sternite 3, in having the distally broadened P2-4 propodi, in the arrangement of flexor marginal spines of P2-4 dactyli, and in having nearly spineless P1 with fingers distally strongly incurved. *Uroptychus norfolkanus* is distinguished from *U. edisonicus* (Figure 121) by the antenna that has articles 4 and 5 each with a distomesial spine instead of being unarmed, and the antennal scale reaching the proximal third segment of the antennal flagellum instead of terminating in the distal end of article 5; the P3 propodus has an additional spine remotely proximal to the distal group of flexor marginal spines, which spine is absent in *U. edisonicus* (P2 and P4 missing in *U. norfolkanus*); and the P3 dactylus is shorter than instead of subequal to the P3 carpus (0.8 versus 1.0). *Uroptychus norfolkanus* differs from *U. latior* in having the anterolateral spine directed straight forward instead of anteromesially, in having the pterygostomian flap anteriorly produced to a distinct spine instead of bearing a tiny spine, in having the P3 propodi with an additional spine remotely proximal to the distal group of flexor marginal spines, in having the eyes subcylindrical instead of distally narrowed, and in having the anterior margin of sternite 3 with obsolescent submedian spines, not smoothly semicircular. The relationships between *U. norfolkanus* and *U. pedanomastigus* are discussed under the account of the latter species (see below).

Uroptychus numerosus n. sp.

Figures 176, 177

TYPE MATERIAL — Holotype: **New Caledonia**. BIOGEOCAL Stn DW307, 20°35.38'S, 166°55.25'E, 470-480 m, 1.V.1987, ♂ 4.1 mm (MNHN-IU-2014-16830). Paratype: Station data as for the holotype, 1 ov. ♀ 4.3 mm (MNHN-IU-2014-16831).

ETYMOLOGY — From the Latin *numerosus* (many), referring to numerous spines on the rostral lateral margin, the character to separate the species from related species.

DISTRIBUTION — New Caledonia; 470-480 m.

DESCRIPTION — Small species. *Carapace*: As long as broad; greatest breadth 1.6 × distance between anterolateral spines. Dorsal surface covered with numerous, sharp, slender spines, slightly convex in profile, with depression bordering gastric and cardiac regions. Lateral margins convex along branchial regions; anterolateral spine overreaching lateral orbital spine. Rostrum straight, directed slightly dorsally, long and narrow but proportionately broad distally, with interior angle of ca. 10°, bearing 9 dorso-anteriorly directed spines along entire lateral margin, and pair of small spines on basal part of dorsal surface; length 0.8 × that of remaining carapace, breadth much less than half carapace breadth at posterior carapace margin. Pterygostomian flap covered with small spines, anteriorly somewhat roundish, produced to strong spine.

Sternum: Excavated sternum sharply cristate in anterior midline, anterior margin convex. Sternal plastron slightly longer than broad; sternites 4-6 successively slightly broader posteriorly, sternite 7 slightly narrower than sternite 6.

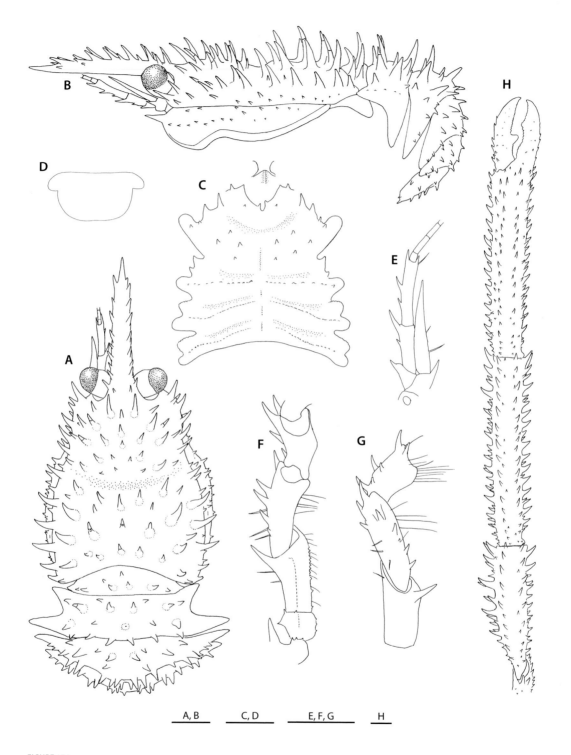

FIGURE 176

Uroptychus numerosus n. sp., holotype, male 4.1 mm (MNHN-IU-2014-16830). **A**, carapace and anterior part of abdomen, dorsal. **B**, same, lateral. **C**, sternal plastron, with excavated sternum and basal parts of Mxp1. **D**, telson. **E**, left antenna, ventral. **F**, right Mxp3, ventral. **G**, same, lateral. **H**, right P1, dorsal. Scale bars: 1 mm.

Sternite 3 depressed well, anterior margin deeply excavated in semicircular shape, with 2 submedian spines separated by U-shaped sinus and flanked by small spine on surface. Sternite 4 with small spines arranged in posteriorly concentric arcs on posterior surface; anterolateral margin short, 0.7 × as long as posterolateral margin, with 2 strong spines, one at anterior terminus, another at midlength. Sternite 5 with 1 or 2 anterolateral spines, length about half that of posterolateral margin of sternite 4.

Abdomen: Somite 1 with anterior row of 3 spines (right one missing in holotype) and posterior row of 6 spines. Somite 2 tergite 2.2 × broader than long; pleuron anterolaterally sharply produced, posterolaterally strongly produced and tapering to sharp point; pleural lateral margins strongly concave and strongly divergent posteriorly; tergite with spines arranged

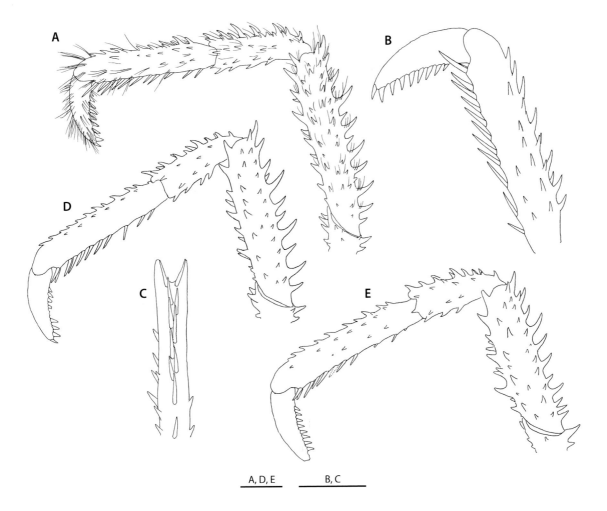

A, D, E   B, C

FIGURE 177

Uroptychus numerosus n. sp., holotype, male 4.1 mm (MNHN-IU-2014-16830). **A**, left P2, lateral. **B**, same, distal part, setae omitted, lateral. **C**, same, propodus, flexor. **D**, left P3, lateral. **E**, left P4, lateral. Scale bars: 1 mm.

roughly in 3 rows: anterior row of 8 spines, median row of 4 or 5 spines, posterior row of 6-7 spines. Somite 3 with spines roughly arranged in 3 rows: anterior row of 6-8 spines, median row of 8 spines, posterior row of 4-10 spines. Somite 4 with 3 rows of spines (4-7 anterior, 6-10 median, 8-16 posterior); pleura of somites 3 and 4 tapering laterally to sharp point. Somite 5 with 16-25 spines roughly in 3 rows. Somite 6 with 26-30 spines more or less transversely arranged on surface and 7 spines along posterior margin. Telson 0.5 × as long as broad; posterior plate 1.6-1.7 × longer than anterior plate, not emarginate on posterior margin. Uropod having protopod with small, elongate, distally blunt projection on mesial margin; endopod 2.2 × longer than broad.

Eye: Globular, short relative to breadth, 1.3-1.4 × longer than broad, convex on mesial margin, terminating at most in proximal quarter of rostrum. Cornea narrower than greatest breadth of eyestalk, slightly shorter than remaining eyestalk.

Antennule and antenna: Ultimate article of antennular peduncle 3.5-3.6 × longer than high. Antennal peduncle extending far beyond cornea. Article 2 with sharp lateral spine. Article 3 with distinct mesial spine. Antennal scale as broad as article 5, overreaching midlength of, but not reaching distal end of article 5. Article 4 with strong distomesial spine and additional small spine at midlength of mesial margin. Article 5 1.4 × longer than article 4, with distomesial and distolateral spines (both much smaller than distomesial spine of article 4) plus 2 mesial marginal spines; breadth 0.5 × height of ultimate antennal article. Flagellum consisting of 6 segments, terminating in apex of rostrum, not reaching distal end of P1 merus.

Mxp: Mxp1 with bases broadly separated. Mxp3 coxa with strong ventrolateral spine. Basis with a few obsolescent denticles on mesial ridge. Ischium with strong distolateral spine (accompanying small spine directly mesial to it in paratype); flexor margin distally not rounded; crista dentata with 7-11 distally diminishing denticles, proximal denticles loosely arranged. Merus twice as long as ischium, with 5-6 flexor marginal, 2 lateral (including strong distal spine) and 2 extensor marginal spines. Carpus with 3 extensor marginal and 3 lateral spines.

P1: 5.5-6.5 × length of carapace, sparsely setose. Ischium with strong dorsal and strong ventromesial subterminal spine. Merus 1.2-1.4 × carapace length, surface covered with spines in 8 rows (2 dorsal, 2 lateral, 2 ventral, 2 mesial) continued on to carpus and palm. Carpus 1.3-1.4 × length of merus. Palm 5.3-5.6 × longer than broad, as long as carpus. Fingers distinctly gaping in male, slightly gaping in female, ending in incurved spine, crossing when closed; fixed finger with low eminence distal to midlength of opposable margin, bearing row of small spines on proximal half of lateral margin, and small spines on proximal part of dorsal surface; movable finger 0.4 × length of palm; opposable margin with prominent median process fitting to longitudinal groove on opposite face of fixed finger when closed; lateral margin with row of spines.

P2-4: Meri successively shorter posteriorly (P3 merus 0.9 × length of P2 merus, P4 merus 0.9 × length of P3 merus), broader on P4 than on P2 and P3; length-breadth ratio, 5.4-5.5 on P2, 4.7-4.8 on P3, 3.8-4.0 on P4; dorsal row of strong spines (ca. 8 in number) interspaced by smaller spines, 3 rows of lateral spines small on P2-3, relatively large on P4, 3 rows of ventral spines (ventrolateral larger), all these rows continued on to carpus and propodus (not on flexor margin); P2 merus subequal to carapace length, 1.2 × length of P2 propodus; P3 merus 1.1 × length of P3 propodus; P4 merus 0.9 × length of P4 propodus. Carpi subequal; carpus-propodus length ratio, 0.6 on P2 and P3, 0.5 on P4; carpus-dactylus length ratio, 1.2-1.4 on P2, 1.0-1.4 on P3, 1.3 on P4. Propodi subequal in length on P2-3, longer on P4; flexor margin straight, with pair of terminal movable spines preceded by row of spines, distal 4-7 spines in zigzag arrangement and others in straight line. Dactyli shorter than carpi (dactylus-carpus length ratio, 0.7 on P2, 0.7 on P2, 0.8 on P4), 0.4 × as long as propodi on P2-4; flexor margin slightly curving, with 8 or 9 somewhat oblique, proximally diminishing spines on P2 and P3, 9 or 10 spines on P4, ultimate somewhat narrower than penultimate.

Eggs. Number of eggs carried, 6; size, 1.00 mm × 1.40 mm - 1.06 mm × 1.42 mm; eggs postembryonic.

REMARKS — The new species is grouped together with *U. ciliatus* (Van Dam, 1933), *U. quartanus* n. sp. *U. senarius* n. sp., *U. spinimanus* Tirmizi, 1964 and *U. spinirostris* (Ahyong & Poore, 2004). Their relationships are discussed under *U. senarius* (see below). From all of these relatives *U. numerosus* differs in the following particulars: the rostrum is proportionately broad distally, not triangular, with a row of 9 lateral spines along the entire length; the abdominal somite 1 bears a transverse row of 6 spines, and the somites 2-6 each bear more numerous spines; and sternite 4 bears 2 rows of spines arranged in concentric arcs.

Uroptychus obtusus n. sp.

Figures 178, 179

TYPE MATERIAL — Holotype: **New Caledonia**, Norfolk Ridge. NORFOLK 2 Stn DW2049, 23°42.88'S, 168°15.43'E, 470-621 m, 24.X.2003, ♂ 2.6 mm (MNHN-IU-2014-16832).

ETYMOLOGY — From the Latin *obtusus* (blunt, dull), alluding to distomesial spines of antennal articles 4-5 that are distally blunt, not sharp as in other species, characteristic of the new species.

DISTRIBUTION — Norfolk Ridge; 470-621 m.

DESCRIPTION — Carapace 1.4 × as broad as long; greatest breadth 1.4 × distance between anterolateral spines. Dorsal surface nearly horizontal in profile on gastric region; cardiac and branchial regions somewhat inflated; hepatic and branchial regions with very tiny spines laterally (1 slightly larger spine directly behind anterolateral spine on right side in holotype). Lateral margins convex on hepatic region, and again well convex on branchial region; anterolateral spine strong, directed straight forward, extending far beyond antennal article 2, reaching midlength of cornea. Rostrum horizontal, with interior angle of 30°, dorsally excavated, laterally slightly convex, with denticle-like small spines distally; length about three-quarters that of remaining carapace, breadth half carapace breadth at posterior carapace margin. Lateral orbital spine very small, situated at same level as anterolateral spine. Pterygostomian flap anteriorly produced to strong spine reaching distal end of cornea; surface with tubercle-like spines.

Sternum: Excavated sternum with sharp ridge in midline, anteriorly subtriangular between bases of Mxp1. Sternal plastron slightly shorter than broad, sternites successively broader posteriorly. Sternite 3 well depressed, anterolaterally angular; anterior margin very weakly excavated, with deep narrow median notch. Sternite 4 having anterolateral margin very weakly divergent posteriorly, about as long as posterolateral margin, anterior end rounded or bluntly angular. Sternite 5 with anterior lobe smaller than posterior lobe.

Abdomen: Glabrous. Somite 1 antero-posteriorly convex. Somite 2 tergite 2.7 × broader than long; pleuron with slightly concave lateral margin very weakly divergent posteriorly, ending in blunt tip. Pleuron of somite 3 laterally blunt. Telson half as long as broad; posterior plate moderately emarginate, length 1.4 × that of anterior plate.

Eye: Short relative to breadth (1.4-1.5 × longer than broad), ending in midlength of rostrum, slightly broadened at midlength. Cornea not dilated, slightly more than half as long as remaining eyestalk.

Antennule and antenna: Ultimate article of antennular peduncle 3 × longer than high. Antennal peduncle overreaching cornea by length of article 5. Article 2 fused with antennal scale, with blunt spine at ordinary place of distolateral margin. Antennal scale with 2 small, blunt lateral spines, slightly overreaching midlength of article 5, breadth 1.5 × that of article 5. Articles 4 with blunt distomesial spine ventrally. Article 5 produced ventrodistally into bluntly truncate mesial spine reaching second segment of antennal flagellum, bearing a few small spines in ventral midline; length 1.3 × that of article 4, breadth three-quarters height of antennular ultimate article. Flagellum consisting of 8 segments, falling short of distal end of P1 merus.

Mxp: Mxp1 with bases broadly separated. Mxp3 basis with 4 or 5 obsolescent denticles on mesial ridge. Ischium with 7 or 8 obsolescent denticles on crista dentata, flexor margin distally rounded. Merus about 3 × longer than ischium, flexor margin not cristate but roundly ridged, bearing 2 blunt spines in distal third; distolateral spine distinct. Carpus with distolateral spine and small extensor proximal spine, and another small spine between.

P1: Subcylindrical, scarcely setose, length 8.6 × that of carapace. Ischium dorsally with strong spine, ventrally with a few tubercle-like proximal spines only. Merus 1.7 × longer than carapace; dorsally with spines arranged roughly in 2 longitudinal rows; ventrally with distolateral and distomesial spines, and a few proximal spines; mesially with a few strong spines. Carpus narrower than merus, covered with very small spines, distally lacking ventrolateral and ventromesial spines; length 1.8 × that of merus. Palm 4.7 × longer than broad, 0.9 × length of carpus, bearing tubercle-like spines on proximal

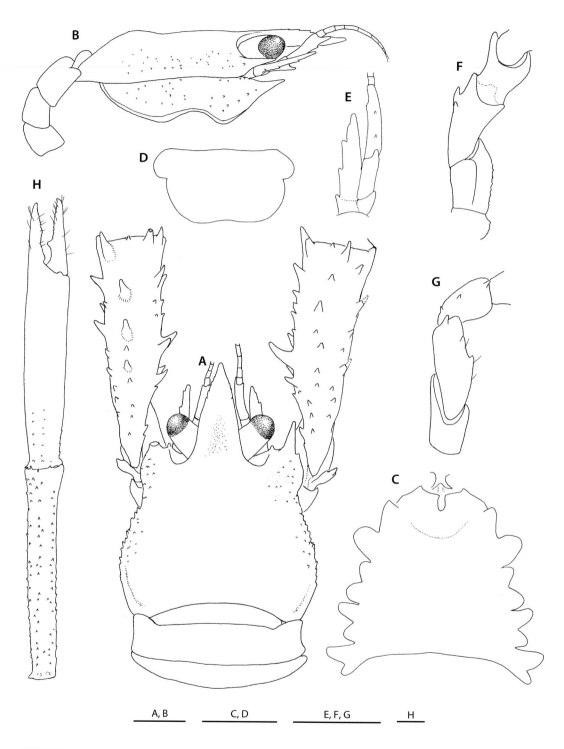

A, B   C, D   E, F, G   H

FIGURE 178

Uroptychus obtusus n. sp., holotype, male 2.6 mm (MNHN-IU-2014-16832). **A**, carapace and anterior part of abdomen, proximal parts of P1 included, dorsal. **B**, same, lateral. **C**, sternal plastron, with excavated sternum and basal parts of Mxp1. **D**, telson. **E**, right antenna, ventral. **F**, right Mxp3, setae omitted, ventral. **G**, same, lateral. **H**, left P1, dorsal. Scale bars: 1 mm.

part of dorsal surface, smooth elsewhere. Fingers with relatively short, sparse setae, tips slightly incurved, crossing when closed; opposable margin of fixed finger sinuous; that of movable finger with lobe-like process at proximal third fitting to longitudinal groove on opposite face of fixed finger when closed; length 0.4 × that of palm.

P2-4: Right P2 and P3 missing. Relatively thick, barely setose except for dactyli. Meri equally broad on P2-4, subequal in length on P2 and P3, shortest on P4 (P4 merus 0.8 × length of P3 merus); length-breadth ratio, 3.2 on P2 and P3, 2.6 on P4; dorsal margin dully crested, with row of short spines (10 on P2 and P3, 7 on P4) continued onto carpi; ventral surface more or less flattish; ventrolateral margin with terminal spine distinct on P2 and P3, obsolete on P4; P2 merus 0.9 × length of carapace, slightly shorter than P2 propodus; P3 merus 0.8 × length of P3 propodus; P4 merus 0.7 × length

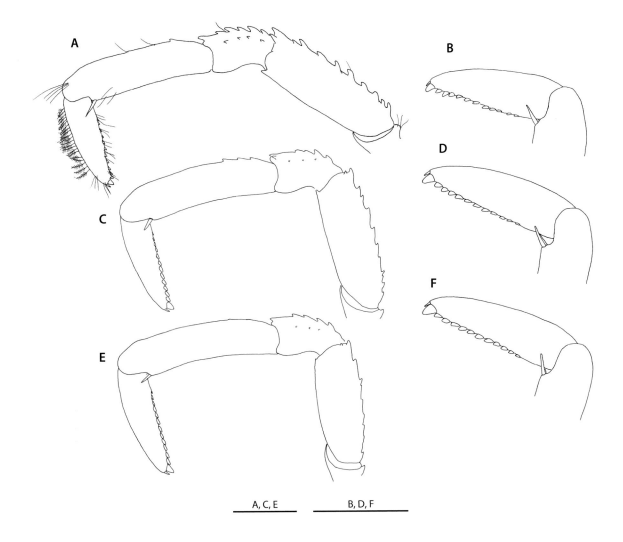

FIGURE 179

Uroptychus obtusus n. sp., holotype, male 2.6 mm (MNHN-IU-2014-16832). **A**, left P2, lateral. **B**, same, distal part, setae omitted, lateral. **C**, left P3, setae omitted, lateral. **D**, same, distal part, lateral. **E**, left P4, lateral. **F**, same, distal part, lateral. Scale bars: 1 mm.

of P4 propodus. Carpi subequal on P2-4, length less than half that of propodi (carpus-propodus length ratio, 0.4 on P2 and P3, 0.3 on P4); extensor margin with 6 spines subparalleling 4 or 5 smaller spines on lateral surface. Propodi shorter on P2 than on P3 and P4; extensor margin with 1 or 2 small spines proximally; flexor margin straight on P2 and P3, slightly concave on P4, bearing pair of terminal spines only. Dactyli shorter on P2 than on P3 and P4, subequal on P3 and P4; length much more than that of carpi (dactylus-carpus length ratio, 1.5 on P2 and P3, 1.7 on P4) and 0.6 that of propodi; flexor margin nearly straight, with short, broad penultimate spine preceded by much smaller, obliquely directed, blunt, very closely arranged spines, ultimate spine very slender and shorter than penultimate; penultimate spine 1.5 × broader than antepenultimate; extensor margin fringed with plumose setae.

REMARKS — The spination of pereopods in the new species is very similar to that of *U. vulcanus* n. sp. Their relationships are discussed under the account of that species (see below).

Uroptychus paenultimus Baba, 2005
Figures 180, 181

Uroptychus paenultimus Baba, 2005: 49, fig. 17.

TYPE MATERIAL — Holotype: **Indonesia**, Kai Islands, 320 m, ov. female (ZMUC CRU-11318). [not examined].

MATERIAL EXAMINED — **Tonga**. BORDAU 2 Stn DW1615, 23°03'S, 175°53'W, 482-504 m, 17.VI.2000, 1 ♀ 2.3 mm (MNHN-IU-2014-16833).

DISTRIBUTION — Kai Islands, and now Tonga; 320-504 m.

DESCRIPTION — Small species. *Carapace*: Slightly broader than long (0.9 × as long as broad); greatest breadth 1.4 × distance between anterolateral spines. Dorsal surface moderately convex from anterior to posterior, without distinct groove, barely setose. Lateral margins medially convex, with anterolateral spine of good size followed by 7 or 8 very small spines; anterolateral spine close to and slightly posterior to level of lateral orbital spine, barely reaching tip of that spine; ridged along posterior third. Rostrum triangular, slightly shorter than broad, with interior angle of ca. 35°, straight horizontal; lateral margins smooth; dorsal surface concave; length about half that of remaining carapace, breadth slightly more than half carapace breadth at posterior carapace margin. Lateral orbital spine subequal in size to anterolateral spine. Pterygostomian flap anteriorly angular, produced to small spine; surface with very small spines.

Sternum: Excavated sternum with slightly convex anterior margin, surface ridged in midline. Sternal plastron 1.1 × longer than broad, lateral extremities subparallel. Sternite 3 moderately depressed, anterior margin of wide V-shape with small median notch. Sternite 4 having anterolateral margin slightly convex, anteriorly angular; posterolateral margin slightly shorter than anterolateral margin. Anterolateral margins of sternite 5 subparallel, anteriorly rounded, slightly shorter than posterolateral margin of sternite 4.

Abdomen: With sparse soft setae. Somite 1 convex from anterior to posterior. Somite 2 tergite 2.3 × broader than long; lateral margins slightly concavely subparallel, posterolaterally blunt. Pleuron of somite 3 laterally blunt. Telson 0.6 × as long as broad, posterior plate slightly longer than anterior plate, posterior margin feebly convex.

Eye: Relatively broad (1.4 × longer than broad), overreaching midlength and falling short of apex of rostrum; lateral and mesial margins somewhat convex. Cornea not dilated, about half length of remaining eyestalk.

Antennule and antenna: Ultimate article of antennule 2.9 × longer than high. Antennal peduncle overreaching eyes. Article 2 with acute distolateral spine. Antennal scale reaching midlength of article 5, slightly broader than that article; lateral margin with small spine at midlength. Article 4 with distinct distomesial spine. Article 5 1.4 × longer than article 4, breadth 0.6 × height of antennular ultimate article. Flagellum consisting of 9 segments, not reaching distal end of P1 merus.

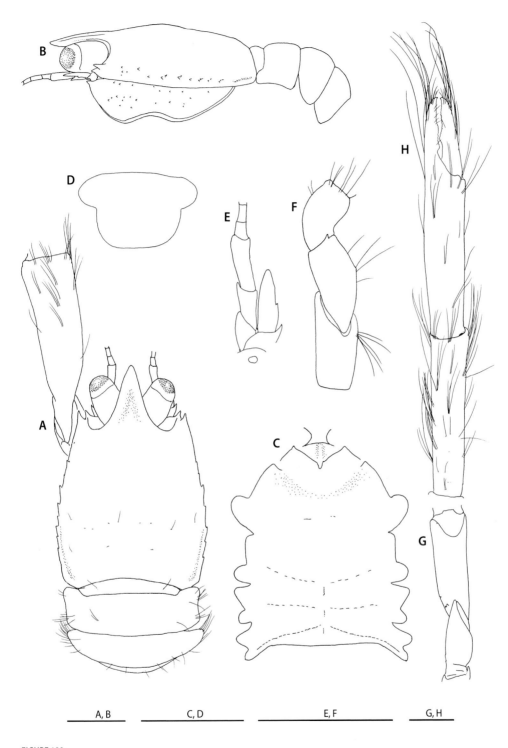

FIGURE 180

Uroptychus paenultimus Baba, 2005, female 2.3 mm (MNHN-IU-2014-16833). **A**, carapace and anterior part of abdomen, proximal part of left P1 included, dorsal. **B**, same, lateral. **C**, sternal plastron with excavated sternum and basal parts of Mxp1. **D**, telson. **E**, left antenna, ventral. **F**, right Mxp3, lateral. **G**, left P1, proximal part, setae omitted, ventral. **H**, same, proximal part omitted, dorsal. Scale bars: 1 mm.

Mxp: Mxp1 with bases broadly separated. Mxp3 basis without denticles on mesial ridge. Ischium distolaterally with long setae, crista dentata with very small, obsolescent denticles; flexor margin distally rounded. Merus 1.8 × longer than ischium, flexor margin moderately ridged, somewhat angular at distal third; distolateral spine small. Carpus unarmed.

P1: 5 × longer than carapace, sparingly bearing long soft setae. Ischium dorsally with broad, depressed, lobe-like distal process, ventromesially unarmed. Merus 1.2 × longer than carapace; dorsally smooth, ventrally with a few tubercle-like spines mesial to distal end of ischium. Carpus 1.2 × longer than merus. Palm 3.6 × longer than broad, subequal to length of carpus. Fingers not gaping, somewhat incurved distally, crossing when closed; opposable margin of fixed finger somewhat sinuous, with low eminence medially; movable finger about half as long as palm, opposable margin with low proximal process.

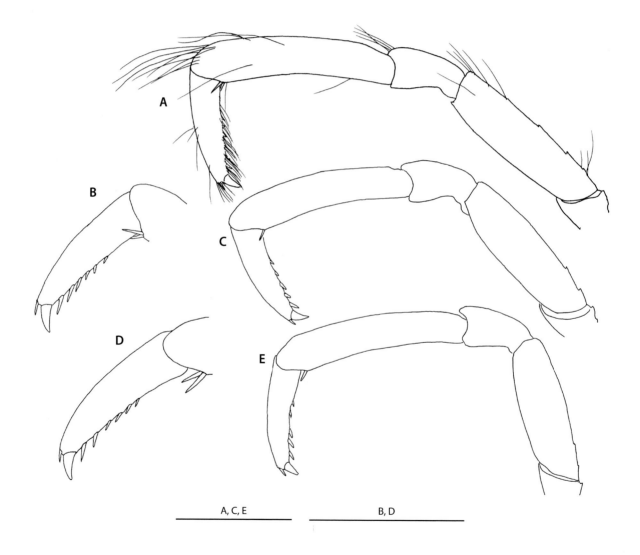

FIGURE 181

Uroptychus paenultimus Baba, 2005, female 2.3 mm (MNHN-IU-2014-16833). **A**, left P2, lateral. **B**, same, distal part, setae omitted, lateral. **C**, left P3, setae omitted, lateral. **D**, same, distal part, setae omitted, lateral. **E**, left P4, setae omitted, lateral. Scale bars: 1 mm.

P2-4: Moderately compressed mesio-laterally, sparingly with long setae. Meri successively shorter posteriorly (P3 merus 0.9 × length of P2 merus, P4 merus 0.9 × length of P3 merus), equally broad on P2-4; length-breadth ratio, 3.8-3.9 on P2, 3.5-3.7 on P3, 3.3-3.4 on P4; extensor margin with a few very small spines distinct on proximal half on P2 and P3, obsolete on P4. P2 merus 0.8 × length of carapace, 0.8 × length of P2 propodus; P3 merus 0.8 × length of P3 propodus; P4 merus 0.7 × length of P4 propodus; Carpi successively slightly shorter posteriorly, about one-third length of propodi and 0.6 × length of dactyli on P2-4. Propodi successively slightly shorter posteriorly; flexor margin slightly concave in lateral view, bearing pair of distal spines only. Dactyli slightly shorter on P2 than on P3 and P4, proportionately broad, 3.1-3.4 × longer than broad (breadth measured at base), 1.6 × longer than dactyli and about half as long as propodi on P2-4; flexor margin nearly straight, with 7 or 8 spines obscured by setae, penultimate spine pronouncedly broad and long (5 × broader than ultimate spine), preceded by 5 or 6 slender spines, loosely arranged, moderately inclined and successively diminishing toward base of article, ultimate spine very slender, shorter than antepenultimate.

REMARKS — The present material differs from the holotype (ovigerous female 3.1 mm) in the following: the carapace bears small lateral spines instead of denticle-like spines; the flexor margin of Mxp3 merus is unarmed instead of bearing one or two small spines; and the P2 merus is 0.8 instead of 1.0 times the length of P2 propodus. Molecular analyses would help determine if these are different species.

This species is very similar to *U. laurentae* n. sp. and *U. longicheles* Ahyong & Poore, 2004. Their relationships are discussed under the respective species (see above).

Uroptychus paku Schnabel, 2009

Figure 182

Uroptychus paku Schnabel, 2009: 562, fig. 12.

TYPE MATERIAL — Holotype: **New Zealand**, Esperance Rock, Kermadec Ridge, 32°11.10'S, 179°05.20'W, 122-307 m, female (NIWA 9805). [not examined].

MATERIAL EXAMINED — New Caledonia, Norfolk Ridge. BERYX 11 Stn CP52, 23°47.45'S, 168°17.05'E, 430-530 m, 21.X.1992: 1 ♀ 3.4 mm (MNHN-IU-2010-5477).

DISTRIBUTION — Kermadec Ridge, and now Norfolk Ridge; 122-530 m.

DIAGNOSIS — Carapace very slightly shorter than broad, greatest breadth 1.3 × distance between anterolateral spines. Dorsal surface with row of 7 small epigastric spines, pair of small submedian spines on anterior cardiac region and 2 longer spines on posterior branchial region; lateral margins slightly divergent posteriorly, bearing 6 spines, first anterolateral, strong, slightly posterior to level of but overreaching much smaller lateral orbital spine; second small, placed on hepatic margin ventral to level of first and third, third to sixth strong, last (sixth) strongest, situated near posterior end. Rostrum relatively broad triangular, with interior angle of 23°, dorsal surface somewhat concave, length four-fifths that of remaining carapace, breadth about half carapace breadth at posterior carapace margin. Pterygostomian flap anteriorly angular, produced to strong spine, surface with several small spines. Excavated sternum bluntly produced anteriorly, surface strongly ridged in midline; sternal plastron 1.2 × longer than broad; lateral extremities somewhat divergent posteriorly; sternite 3 depressed well, anterior margin moderately excavated, with 2 obsolescent submedian spines flanking small median notch; sternite 4 having anterolateral margins relatively short, about as long as posterolateral margin; anterolateral margin of sternite 5 0.7 × as long as anterolateral margin of sternite 4. Abdominal somite 1 with distinct transverse ridge; somite 2 tergite 2.1 × broader than long, pleural lateral margin strongly concave,

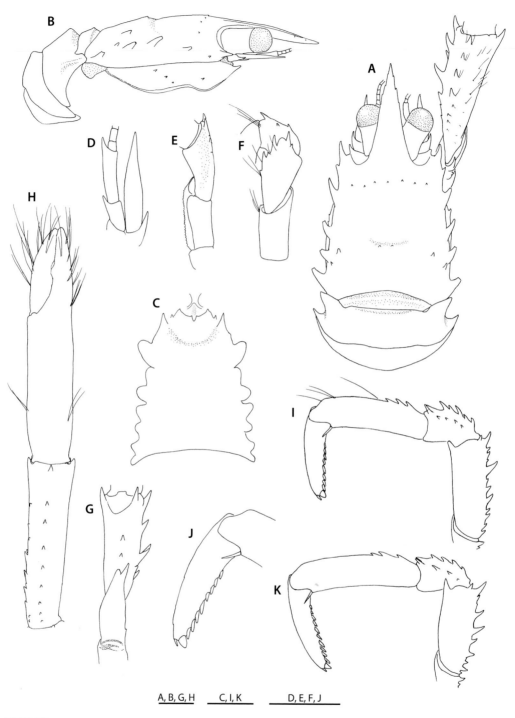

A, B, G, H   C, I, K   D, E, F, J

FIGURE 182

Uroptychus paku Schnabel, 2009, female 3.4 mm (MNHN-IU-2010-5477). **A**, carapace and anterior part of abdomen, proximal part of right P1 included, dorsal. **B**, same, lateral. **C**, sternal plastron, with excavated sternum and basal parts of Mxp1. **D**, left antenna, ventral. **E**, left Mxp3, setae omitted, ventral. **F**, same, lateral. **G**, right P1, proximal part, ventral. **H**, same, proximal part omitted, dorsal. **I**, left P2, lateral. **J**, same, distal part, setae omitted, lateral. **K**, left P4, setae omitted, lateral. Scale bars: 1 mm.

anteriorly and posteriorly produced; pleural lateral margin of somite 3 tapering. Telson slightly less than half as long as broad; posterior plate 1.3 × longer than anterior plate, posterior margin distinctly emarginate. Eyes 2.0 × longer than broad, somewhat overreaching midlength of rostrum; lateral margin concave, cornea more than half as long as remaining eyestalk. Ultimate article of antennular peduncle 2.7 × longer than high. Antennal peduncle slightly overreaching eye; article 2 with strong lateral spine; antennal scale overreaching article 5; distal 2 articles each with strong distomesial spine; breadth of article 5 0.8 × height of ultimate article of antennule; flagellum of 10 or 11 segments not reaching distal end of P1 merus. Mxp1 with bases broadly separated. Mxp3 basis with 1 obsolescent denticle on distal part of mesial ridge; ischium with very small denticle-like spine directly lateral to rounded distal end of flexor margin, crista dentata with small, distally obsolescent denticles; merus 1.7 × longer than ischium, with well-developed distolateral spine accompanying small spine near base, flexor margin sharply ridged with a few distinct spines; carpus with distinct distolateral spine and a few small spines on extensor and lateral faces. P1 5 × as long as carapace, slender, with spines on merus and carpus; ischium dorsally with procurved spine, ventromesially with well-developed subterminal spine; merus as long as carapace; carpus 1.3 × longer than merus; palm 3.0 × longer than broad, about as long as carpus; fingers feebly crossing distally, length 0.6 × that of palm. P2-4 short relative to breadth; meri successively shorter posteriorly (P3 merus 0.9 × length of P2 merus, P4 merus 0.95 × length of P3 merus), subequally broad and about 2.5 × as long as broad on P2-4; P2 merus 0.6 × length of carapace, 0.8 × length of P2 propodus; P4 merus as long as or very slightly shorter than P4 propodus; dorsal margins with row of spines, distalmost with accompanying small spine mesial to it; carpi successively slightly shorter posteriorly; carpus-propodus length ratio, 0.4 on P2-4; extensor margin with row of 6 spines (5 on left P4) paralleling row of 4 or 5 smaller spines on lateral surface; propodi successively slightly longer posteriorly; flexor margin with pair of relatively long movable terminal spines only; extensor margin with 3 or 4 spines proximally; dactyli proportionately broad, much longer than carpi (dactylus-carpus length ratio, 1.5 on P2, 1.7 on P3 and P4), 0.6 × length of propodi on P2-4, flexor margin nearly straight, with row of 10 (P2)-12 (P3, P4) obliquely directed spines nearly contiguous to one another, ultimate more slender than antepenultimate, penultimate pronounced, more than 2 × broader than antepenultimate, remainder similar but proximal spines diminishing toward base of article.

REMARKS — The material differs from the holotype in having sternite 4 more produced anteriorly, the pterygostomian flap bearing small spines, the P2-4 propodi proximally bearing extensor marginal spines and carpi bearing additional row of small spines paralleling the row of extensor marginal spines, and 7 epigastric spines. However, they share nearly all essential features including the long rostrum, a pair of spines on the posterior branchial region and a pair of small spines on the cardiac region. The above-mentioned differences will in all probability be size-related; the type material is small, the carapace length being about half that of the present material.

The relationships with *U. sexspinosus* Balss, 1913a are discussed by Schnabel (2009). The spinose carapace lateral margin, spinose P1-4 and the arrangement of spines on P2-4 dactyli link the species to *U. nanophyes* McArdle, 1901. *Uroptychus paku* is readily distinguished from that species by the longer rostrum that is four-fifths as long instead of at most half as long as the remaining carapace; the branchial and cardiac regions each bear a pair of dorsal spines; the P2-4 propodi bear on the flexor margin a pair of terminal spines only instead of bearing additional row of spines proximal to the distal pair. In addition, the P2-4 meri in *U. paku* bear a terminal spine only on the ventrolateral margin, whereas in *U. nanophyes* of more than 5.6 mm (poc) a row of spines are usually present on each of the ventrolateral and ventromesial margins of the P2 merus, occasionally obsolescent on P3 and P4.

Uroptychus palmaris n. sp.

Figures 183, 184

TYPE MATERIAL — Holotype: **New Caledonia**, Norfolk Ridge. NORFOLK 1 Stn CP1669, 23°41'S, 168°01'E, 302-325 m, 21.VI.2001, ov. ♀ 2.2 mm (MNHN-IU-2011-5975). Paratypes: **New Caledonia**, Norfolk Ridge. CHALCAL 2 Stn CH03, 24°48'S, 168°09'E, 257 m, 27.X.1986, with antipatharian coral (Hexacorallia), 2 ♂ 1.7, 2.0 mm, 2 ov. ♀ 2.0, 2.3 mm (MNHN-IU-2014-16834). – Stn CH04, 24°44'S, 168°10'E, 253 m, 27.X.1986, antipatharian coral (Hexacorallia), 3 ♂ 1.7-1.8 mm (MNHN-IU-2011-5977, MNHN-IU-2014-16835), 1 ov. ♀ 2.1 mm (MNHN-IU-2011-5976). – Stn DW70, 24°46.0'S, 168°09.0'E, 232 m, 27.X.1986, 1 ♂ 1.7 mm (MNHN-IU-2014-16836).

ETYMOLOGY — From the Latin *palmaris* (pertaining to the palm of the hand) for the P1 palm of the species which is massive and large relative to the carapace.

DISTRIBUTION — Norfolk Ridge; 232-325 m.

SIZE — Males, 1.7-2.0 mm; females, 2.0-2.3 mm; ovigerous females from 2.0 mm.

DESCRIPTION — Small species. *Carapace*: 1.1-1.2 × broader than long (0.8-0.9 × as long as broad); greatest breadth 1.8 × distance between anterolateral spines. Dorsal surface smooth, sparsely setose, feebly convex from anterior to posterior, without distinct groove. Lateral margins convexly divergent posteriorly, bearing 6 spines; first anterolateral, well developed, directed anterolaterally, overreaching much smaller lateral orbital spine; second small, ventral in position, third to sixth located on branchial region, fourth more remote from third than from fifth, sixth (last) situated about at midlength of posterior branchial margin. Rostrum narrow triangular, with interior angle of 20-23°, somewhat deflected ventrally; dorsal surface concave; lateral margin with a few (usually 2) small spines near apex; length 0.6-0.7 × that of remaining carapace, breadth half carapace breadth at posterior carapace margin. Lateral orbital spine closely mesial to anterolateral spine. Pterygostomian flap anteriorly angular, produced to acute spine, surface smooth and unarmed.

Sternum: Excavated sternum smooth, without spine and ridge on surface, anterior margin broad triangular. Sternal plastron 1.3-1.4 × broader than long, broadened posteriorly. Sternite 3 shallowly depressed; anterior margin shallowly excavated, with 2 very small median spines contiguous at base, lacking median notch between, lateral limits sharply angular. Sternite 4 having anterolateral margin short relative to breadth, anteriorly with 1 or 2 small spines placed side by side, posterolateral margin half length of anterolateral margin. Anterolateral margin of sternite 5 anteriorly rounded, much longer than posterolateral margin of sternite 4.

Abdomen: Smooth and sparsely setose. Somite 1 without transverse ridge. Somite 2 tergite 2.4-2.7 × broader than long; lateral margin moderately concave and divergent posteriorly, posterolaterally blunt. Pleuron of somite 3 laterally rounded. Telson about half as long as broad; posterior plate slightly longer to slightly shorter than anterior plate, posterior margin nearly transverse, slightly concave or slightly convex.

Eye: 2 × longer than broad, overreaching midlength of, barely reaching distal third point of rostrum, slightly narrowed distally. Cornea not dilated, about half as long as remaining eyestalk.

Antennule and antenna: Ultimate article of antennular peduncle 3.0-3.4 × longer than high. Antennal peduncle overreaching cornea but never reaching rostral tip. Article 2 with very small, occasionally obsolescent lateral spine. Antennal scale 1.8 × broader than article 5, varying from barely reaching to somewhat overreaching distal end of article 5. Articles 4 and 5 subequal in length or article 5 slightly longer, each with distinct distomesial spine; breadth of article 5 two-thirds height of ultimate antennular article. Flagellum of 10-11 segments barely reaching distal end of P1 merus.

Mxp: Mxp1 with bases close to each other. Mxp3 sparsely setose except for distal 2 articles. Basis without denticles on mesial ridge. Ischium with more than 18 distally diminishing denticles on crista dentata, flexor margin not rounded

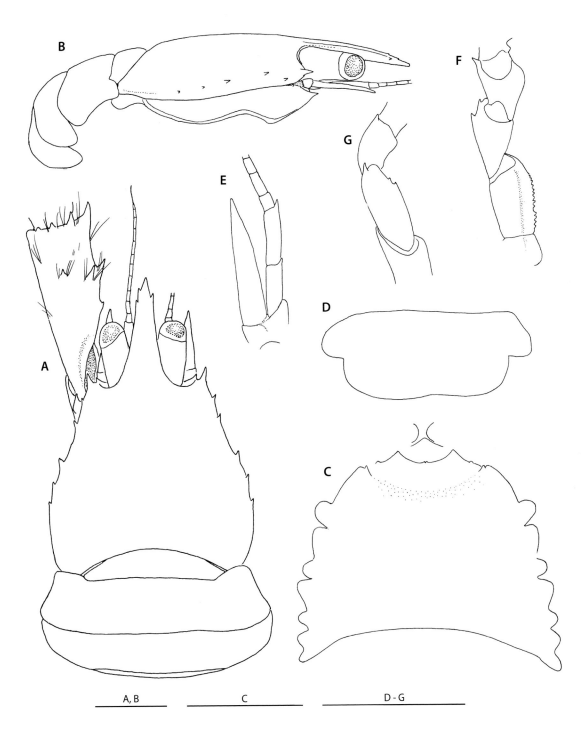

FIGURE 183

Uroptychus palmaris n. sp., holotype, ovigerous female 2.2 mm (MNHN-IU-2011-5975). **A**, carapace and anterior part of abdomen, proximal part of left P1 included, dorsal. **B**, same, lateral. **C**, sternal plastron, with excavated sternum and basal parts of Mxp1. **D**, telson. **E**, right antenna, ventral. **F**, right Mxp3, setae omitted, ventral. **G**, same, setae omitted, lateral. Scale bars: 1 mm.

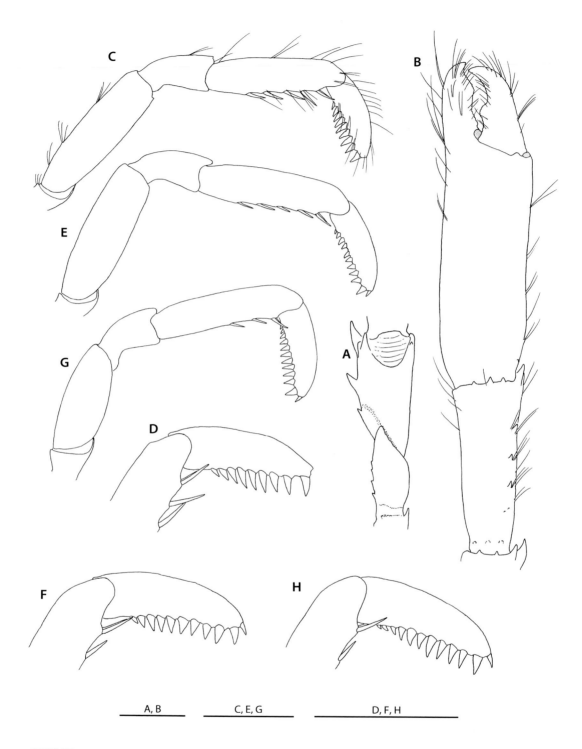

FIGURE 184

Uroptychus palmaris n. sp., holotype, ovigerous female 2.2 mm (MNHN-IU-2011-5975). **A**, left P1, proximal part, setae omitted, ventral. **B**, same, proximal part omitted, dorsal. **C**, right P2, lateral. **D**, same, distal part, setae omitted, lateral. **E**, right P3, setae omitted, lateral. **F**, same, distal part, setae omitted, lateral. **G**, right P4, setae omitted, lateral. **H**, same, distal part, setae omitted, lateral. Scale bars: 1 mm.

distally. Merus about twice as long as ischium, with well-developed distolateral spine and 1 or 2 small spines distal to point two-thirds of cristate flexor margin; mesial face flattish. Carpus with distinct distolateral spine.

P1: Sparsely with fine setae, 5.2-7.1 × longer than carapace, massive on palm, greatest breadth often more than distance between anterolateral spines of carapace. Ischium with sharp, strong dorsal spine, ventromesial margin with small subterminal spine proximally followed by 4 or 5 spinules. Merus 1.1 × length of carapace, bearing a few to several spines on each of mesial and ventromesial margins and dorsal surface, distomesial and mid-mesial spines well developed. Carpus 1.0-1.2 × longer than merus, distally broadened, terminally with 4 or 5 spines (2 or 3 dorsal, 1 mesial, 1 ventromesial; mesial one stronger), dorsomesially with a few to several small, often well-developed spines, dorsally 3 tubercle-like spines on proximal portion. Palm massive, 2.4-2.8 × longer than broad, smooth, without spines; length 1.0-1.3 × that of carpus. Fingers broad relative to length, distally incurved, crossing when closed; movable finger not reaching tip of fixed finger, opposable margin with low, blunt median process; length 0.4-0.6 × that of palm; fixed finger having opposable margin with prominence distal to position of opposite process of movable finger.

P2-4: Relatively short, moderately depressed, sparsely setose. Meri unarmed but dorsal margin with a few weak eminences supporting long setae on P2; length-breadth ratio, 3.3-3.5 on P2, 3.1-3.5 on P3, 2.5-2.8 on P4; breadths subequal on P2-4; P2 merus three-quarters length of carapace, as long as P2 propodus. P3 merus subequal to, rarely slightly shorter than P2 merus, 0.9 × length of P3 propodus; P4 merus 0.8 × length of P3 merus, 0.7-0.8 × length of P4 propodus. Carpi subequal in length on P2 and P3 or slightly shorter on P3 than on P2, shortest on P4 (P4 carpus 0.9 × length of P3 carpus), slightly less than half as long as propodi (carpus-propodus length ratio, 0.40-0.44, 0.36-0.44, 0.34-0.42 on P2, P3, P4 respectively). Propodi shortest on P2, longer on P4 than on P3; flexor margin nearly straight, ending in pair of movable spines preceded by 4 or 5 similar spines in distal two-thirds of length on P2 and P3, 1 or 2 spines in distal third on P4. Dactyli proportionately broad, longer than carpi (dactylus-carpus length ratio, 1.2-1.3 on P2, 1.4-1.5 on P3 and P4), dactylus-propodus length ratio, 0.5-0.6 on P2 and P3, 0.6 on P4; flexor margin nearly straight, ending in slender terminal spine preceded by 9 or 10 well-developed triangular spines perpendicular to margin and successively diminishing toward base of article.

Eggs. Number of eggs carried 2-5; size, 0.92 mm × 0.88 mm - 1.10 mm × 1.25 mm.

REMARKS — As suggested by the species name, the P1 palm of the species is pronouncedly broad, often more than the distance between the anterolateral spines of the carapace.

Uroptychus palmaris somewhat resembles *U. yaldwyni* Schnabel, 2009 from the Kermadec Ridge, in having the anterolateral spine that is close to the smaller lateral orbital spine, the elongate eyes, an antennal article 5 that is slightly longer than article 4, and the P2-4 dactyli that are longer than the carpi, with the ultimate spine much more slender than the penultimate spine. However, the new species is distinguished from that species by the following: the last carapace lateral spine is located at the midpoint instead of near the posterior end of the branchial margin; the anterior margin of sternite 3 is weakly instead of strongly excavated, with 2 very small median spines closely placed side by site instead of acute submedian spines separated by a distinct notch; the anterolateral margin of sternite 4 is twice instead of subequal to the length of the posterolateral margin, and the surface of the pterygostomian flap is more sharply angular, the surface being smooth instead of covered with small spines.

Uroptychus palmaris also resembles *U. triangularis* Miyake & Baba, 1967 in having two small, closely placed median spines on the concave anterior margin of sternite 3, but it differs from that species in having the branchial marginal spines rather small instead of short but broad at base, in having the P2-4 meri unarmed instead of bearing a row of dorsal spines. At first glance, the carapace in *U. palmaris* is not so strongly broadened posteriorly as in *U. triangularis* (the greatest breadth is 1.8 versus 2.2 times the distance between left and right anterolateral spines).

Uroptychus paraplesius n. sp.

Figures 185-188

TYPE MATERIAL — Holotype: **Indonesia**, Kai Islands. KARUBAR Stn CP16, 5°17'S, 132°50'E, 315-349 m, 24.X.1991, ♂ 5.3 mm (MNHN-IU-2014-16837). Paratype: Collected with holotype, 1 ov. ♀ 6.0 mm (MNHN-IU-2014-16838). **New Caledonia**, Isle of Pines. BIOCAL Stn CP52, 23°06'S, 167°47'E, 600-540 m, 31.VIII.1985, with rhizocephalan parasite, 1 ♂ 3.2 mm (MNHN-IU-2013-8574).

DISTRIBUTION — Kai Islands (Indonesia) and Isle of Pines; 315-600 m.

ETYMOLOGY — From the Greek *paraplesios* (somewhat similar), alluding to the morphological characteristics of the species, which are somewhat similar to those of *U. trispinatus* n. sp. described in this paper.

DESCRIPTION — Small species. *Carapace:* Broader than long (0.8-0.9 × as long as broad); greatest breadth 1.8 × distance between anterolateral spines. Dorsal surface smooth, moderately convex from anterior to posterior, with shallow depression between gastric and cardiac regions; epigastric region with pair of small spines. Lateral margins with constriction at anterior third; anterior part (anterior to constrictions) divergent posteriorly, with 3 prominent spines, all directed anteriorly and somewhat laterally, first anterolateral, overreaching base of antennal scale, second slightly smaller than first, situated at same level as first, third strongest, distinctly dorsal to level of first and second; posterior part (posterior to constrictions) gently convex, with 4 or 5 sharp, posteriorly diminishing spines, all directed anterolaterally like those on anterior part, last spine followed by feeble ridge leading to posterior end. Rostrum narrow triangular, with interior angle of 21-25°, straight horizontal, dorsal surface concave; length 0.4 × that of remaining carapace, breadth much less than half carapace breadth at posterior carapace margin. Lateral orbital spine small, slightly anterior to level of antero-lateral spine. Pterygostomian flap anteriorly angular, produced to strong spine, followed by a few well-developed spines on anterior part; height of posterior portion 0.5 × that of anterior portion.

Sternum: Excavated sternum sharply ridged in midline on surface, anterior margin produced triangularly between bases of Mxp1. Sternal plastron 1.3-1.5 × broader than long, lateral extremities convexly divergent behind sternite 4, sternites 6 and 7 subequal in breadth. Sternite 3 shallowly depressed; anterior margin excavated in shallow V-shape, with V-shaped median notch. Sternite 4 having anterolateral margin feebly convex, anteriorly blunt angular, 1.7-1.9 × longer than posterolateral margin. Sternite 5 with medially convex anterolateral margin longer than posterolateral margin of sternite 4.

Abdomen: Somite 1 convex from anterior to posterior. Somite 2 tergite 3.3 × longer than broad, pleuron posterolaterally moderately produced, somewhat more produced in female. Pleuron of somite 3 blunt on lateral terminus, those of somites 4 and 5 laterally rounded. Telson about half as long as broad; posterior plate 1.7 × longer than anterior plate, posterior margin feebly emarginate.

Eye: Elongate, 2.0-2.1 × longer than broad, overreaching midlength of, not overreaching point distal third of rostrum, distally narrowed. Cornea feebly dilated, about half as long as remaining eyestalk.

Antennule and antenna: Ultimate article of antennular peduncle 2.4-2.9 × longer than high. Antennal peduncle reaching apex of rostrum. Article 2 with strong distolateral spine. Antennal scale 1.5 × as broad as article 5, reaching end of first segment of flagellum. Article 4 with strong distomesial spine accompanying smaller spine at its ventromesial base. Article 5 with short distomesial spine, length 1.4-1.5 × that of article 4, breadth 0.5-0.6 × height of ultimate article of antennule. Flagellum of 9 segments not reaching distal end of P1 merus.

Mxp: Mxp1 with bases moderately separated. Mxp3 basis with 1 distal denticle on mesial ridge. Ischium having flexor margin distally rounded, crista dentata with 20-23 distally diminishing denticles. Merus twice as long as ischium; mesial face flattish; flexor margin roundly ridged on proximal half, sharply ridged on distal half, produced around distal third bearing 2 or 3 small spines; lateral face with distinct distal spine. Carpus unarmed.

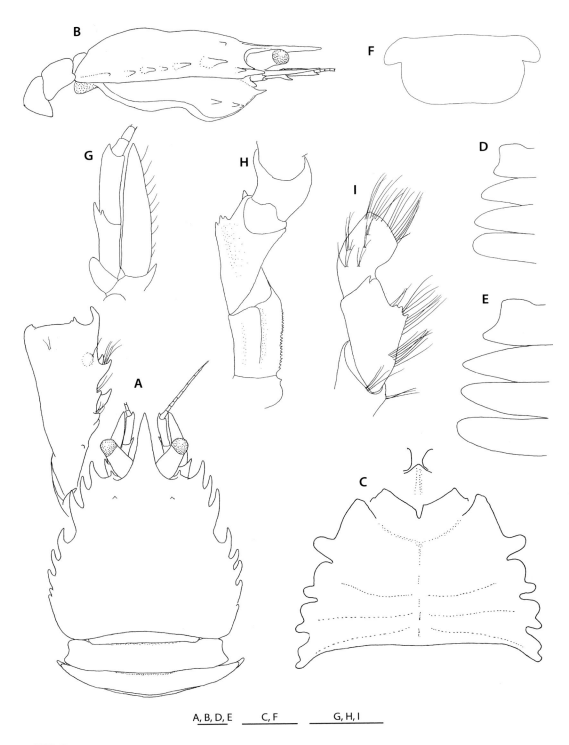

FIGURE 185

Uroptychus paraplesius n. sp., **A-D**, **F-I**, holotype, male 5.3 mm (MNHN-IU-2014-16837); **E**, ovigerous female paratype 6.0 mm (MNHN-IU-2014-16838). **A**, carapace and anterior part of abdomen, proximal part of left P1 included. **B**, same, lateral. **C**, sternal plastron, with excavated sternum and basal parts of Mxp1. **D**, left pleura of abdominal somites 2-5, dorsolateral. **E**, same. **F**, telson. **G**, left antenna. **H**, right Mxp3, setae omitted, ventral. **I**, same, lateral. Scale bars: 1 mm.

P1: 5.2-5.4 × longer than carapace, smooth and sparsely setose, more setose on fingers. Ischium with strong dorsal spine; ventromesially with small blunt subterminal spine not overreaching distal end of ischium, and several low protuberances on proximal half of length. Merus mesially with relatively stout spines roughly in 3 rows (mesial, ventromesial, dorsomesial; dorsal spines smaller), and a few protuberances on proximal ventral surface; length slightly more (1.04-1.15 x) than that of carapace. Carpus 1.4 × length of merus, with several small spines in 2 rows along mesial margin. Palm unarmed, somewhat

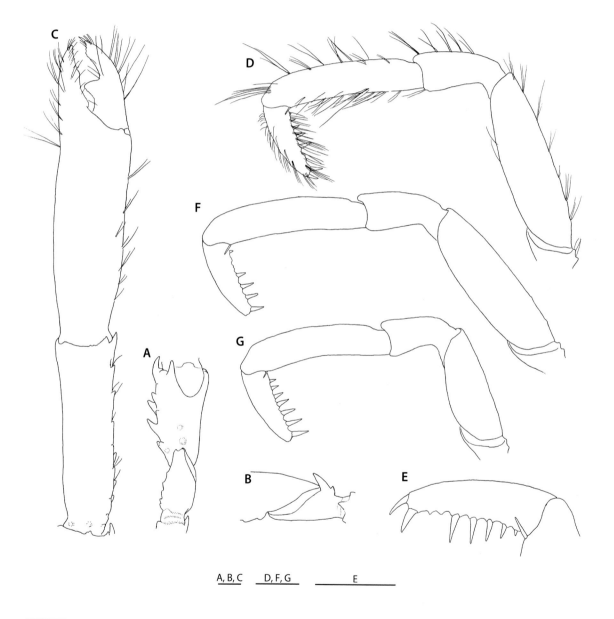

FIGURE 186

Uroptychus paraplesius n. sp., holotype, male 5.3 mm (MNHN-IU-2014-16837). **A**, left P1, proximal part, ventral. **B**, same, proximal part, lateral. **C**, same, distal three articles, dorsal. **D**, left P2, lateral. **E**, same, distal part, setae omitted, lateral. **F**, left P3, setae omitted, lateral. **G**, left P4, setae omitted, lateral. Scale bars: 1 mm.

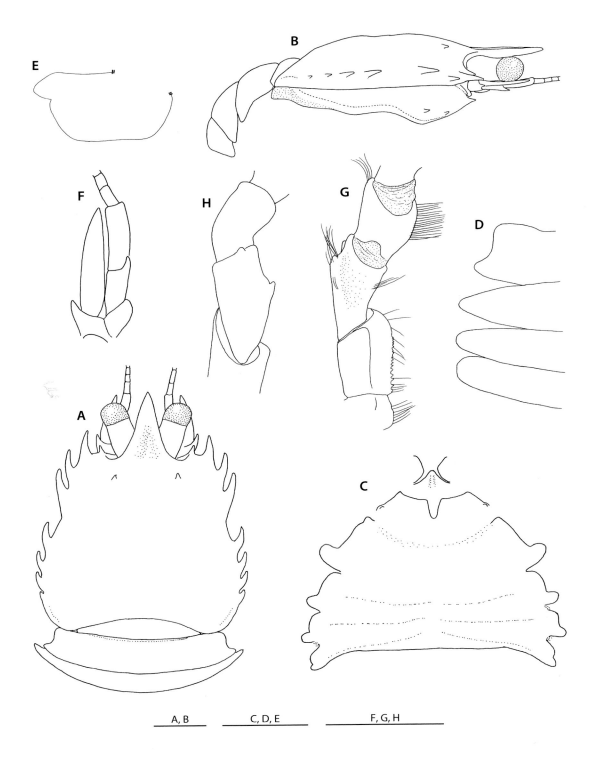

FIGURE 187

Uroptychus paraplesius n. sp., paratype, male 3.2 mm (MNHN-IU-2013-8574). **A**, carapace and anterior part of abdomen, dorsal. **B**, same, lateral. **C**, sternal plastron, with excavated sternum and basal parts of Mxp1. **D**, left pleura of abdominal somites 2-5, dorsolateral. **E**, telson, right part distorted. **F**, right antenna, ventral. **G**, right Mxp3, ventral. **H**, same, lateral. Scale bars: 1 mm.

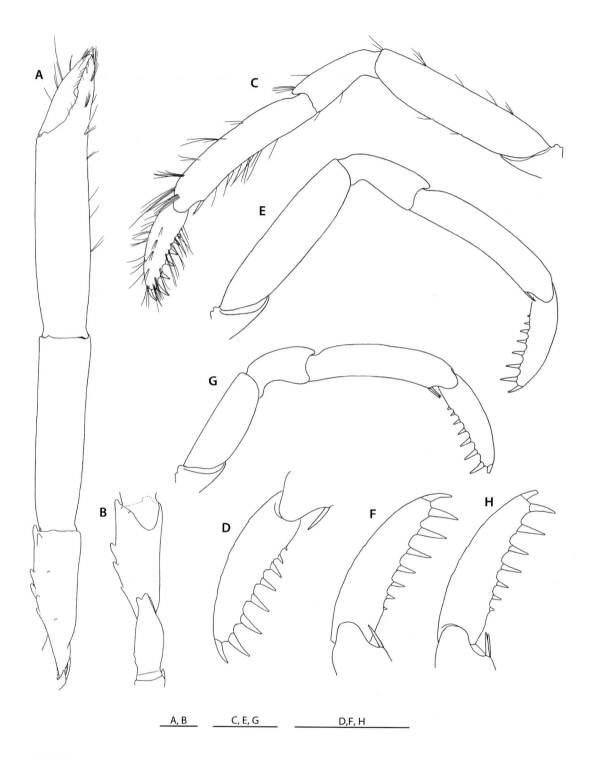

FIGURE 188

Uroptychus paraplesius n. sp., paratype, male 3.2 mm (MNHN-IU-2013-8574). **A**, right P1, dorsal. **B**, same, proximal part, ventral. **C**, left P2, lateral. **D**, same, distal part, setae omitted, lateral. **E**, right P3, setae omitted, lateral. **F**, same, distal part, lateral. **G**, right P4, lateral. **H**, same, distal part, lateral. Scale bars: 1 mm.

depressed (height-breadth ratio, 0.8), with lateral and mesial margins subparallel or slightly convex; length 3.0-3.2 × breadth, 1.2 × that of carpus. Fingers relatively stout, moderately gaping, feebly incurved distally, slightly crossing when closed; not spooned along opposable margins; fixed finger inclined somewhat laterally, opposable margin sinuous; movable finger 0.4 × as long as palm, opposable margin with prominent median process (distal margin of process perpendicular to opposable margin) fitting into narrow longitudinal groove on opposite ventromesial face of fixed finger when closed.

P2-4: Relatively slender, well compressed mesio-laterally, with sparse simple setae, more setose on dactyli. Ischia with 2 short dorsal spines. Meri subequal in length on P2 and P3 (female) or very slightly shorter on P3 than on P2 (male), shortest on P4 (P4 merus 0.7 × length of P3 merus); subequally broad on P2 and P3, narrower on P4; dorsal margins unarmed; length-breadth ratio, 3.5-4.0 on P2, 3.3-3.7 on P3, 2.5 on P4; P2 merus 0.7-0.8 × length of carapace, subequal to length of P2 propodus; P3 merus very slightly shorter than P3 propodus. P4 merus 0.7 × length of P4 propodus. Carpi subequal in length on P2 and P3 and shorter on P4; 0.4 × as long as propodus on P2-4. Propodi subequal in length on P2 and P3, shorter on P4; flexor margin slightly concave in lateral view, with pair of slender terminal spines only. Dactyli subequal in length on P2-4, slightly less than half as long as propodus; dactylus-carpus length ratio, 1.0 on P2, 1.1 on P3, 1.3-1.5 on P4; flexor margin slightly curving, with 7 or 8 sharp spines loosely arranged, proximally diminishing and nearly perpendicular to margin, ultimate spine slightly more slender than penultimate.

Eggs. Number of eggs carried, 30; size, 1.08 × 1.29 mm - 1.13-1.29 mm.

REMARKS — As mentioned under the remarks of *U. clarki* n. sp., *U. paraplesius* resembles *U. mesodme* n. sp. and *U. trispinatus* n. sp., sharing the carapace lateral margin having a group of 3 spines on the anterior part remotely separated from another group of 3-5 spines on the posterior part, the antennal scale overreaching the midlength of article 5, and article 4 bearing a distomesial spine. *Uroptychus paraplesius* is distinguished from *U. mesodme* by the anterolateral spine of the carapace that is directed anterolaterally instead of straight forward; a pair of epigastric spines are small instead of well-developed with an additional pair of smaller spines between; the eyes are more narrowed distally; the antennal articles 4 and 5 each bear a distinct distomesial spine, instead of being unarmed (although article 5 is distomesially bluntly produced). The relationships between *U. paraplesius* and *U. trispinatus* are discussed under the account of the latter species (see below).

The male paratype (poc, 3.2 mm) from the Isle of Pines (MNHN-IU-2013-8574) differs from the other two types (poc, 5.3, 6.0 mm) from the Kai Islands in having the antennal article 5 unarmed and article 4 lacking a small spine at the base of the well-developed distomesial spine, in having the anterolateral spine of the carapace directed straight forward, in having the P1 ischium and merus longer relative to breadth, and the carpus bearing low eminences along the mesial margin (Figures 187, 188). These differences appear to be size-related. Discovery of additional material would allow genetic analyses to clarify its systematic status.

Uroptychus parisus n. sp.

Figures 189, 190

Uroptychus scandens — McCallum & Poore 2013: 165, fig. 12F.

TYPE MATERIAL — Holotype: **New Caledonia**. HALIPRO 1 Stn CP877, 23°03'S, 166°59'E, 464-480 m, 31.III.1994, ov. ♀ 3.3 mm (MNHN-IU-2013-8563). Paratypes: **New Caledonia**. HALIPRO 1 Stn CP877, 23°03'S, 166°59'E, 464-480 m, 31.III.1994, collected together with holotype, 1 ov. ♀ 3.4 mm (MNHN-IU-2014-16839). BATHUS 2 Stn CP738, 23°02'S, 166°56'E, 558-647 m, 13.V.1993, 1 ov. ♀ 3.3 mm (MNHN-IU-2014-16840). MUSORSTOM 4 Stn CP169, 18°55'S, 163°12'E, 600 m, 17.IX.1985, 1 ov. ♀ 3.6 mm (MNHN-IU-2014-16841). – Stn CP170, 18°57'S, 163°13'E, 485 m, 17.IX.1985, on *Pennatula* sp. (Pennatulacea, Pennatulidae), 1 ♂ 2.9 mm, 2 ov. ♀ 3.2, 3.3 mm (MNHN-IU-2014-16842); 1 ♂ 2.7 mm, 1 ov. ♀ 3.6 mm, 1 ♀ 3.7 mm (MNHN-IU-2013-8562). – Stn CP195, 18°54'S, 163°23'E, 470 m, 19.IX.1985, 1 ov. ♀ 3.1 mm

(MNHN-IU-2014-16843). – Stn CP236, 22°11.3'S, 167°15.0'E, 495-550 m, 2.X.1985, 1 ♂ 3.5 mm, 6 ov. ♀ 3.1-3.5 mm, 3 ♀ 2.8-3.3 mm (MNHN-IU-2014-16844). – Stn CP238, 22°13.0'S, 167°14.0'E, 500-510 m, 2.X.1985, 1 ov. ♀ 3.2 mm (MNHN-IU-2013-8570). **New Caledonia**, Chesterfield Islands. MUSORSTOM 5 Stn CP389, 20°44.95'S, 160°53.67'E, 500 m, 22.X.1986, 1 ov. ♀ 4.1 mm (MNHN-IU-2014-16845). New Caledonia, Loyalty Ridge. MUSORSTOM 6 CP467, 21°06'S, 167°32'E, 530-575 m, 21.II.1989, 1 ov. ♀ 3.6 mm (MNHN-IU-2014-16850). **New Caledonia**, Norfolk Ridge. NORFOLK 2 Stn CP2121, 23°23'S, 168°00'E, 486-514 m, 1.XI.2003, 1 ♂ 4.2 mm (MNHN-IU-2014-12794). **Solomon Islands**. SALOMON 1 Stn CP1786, 09°21'S, 160°25'E, 387 m, 30.IX.2001, 1 ♂ 3.1 mm, 4 ov. ♀ 3.4-4.1 mm, 2 ♀ 3.1, 3.4 mm (MNHN-IU-2013-8568, MNHN-IU-2013-12276). – Stn CP1804, 9°32.0'S, 160°37.4'E, 309-328 m, 2.X.2001, 1 ♀ 4.2 mm (MNHN-IU-2014-16851). – Stn CP1851, 10°27.6'S, 162°00.6'E, 297-350 m, 6.X.2001, 2 ♂ 2.4, 3.5 mm, 4 ov. ♀ 3.2-3.5 mm, 4 ♀ 3.7-4.1 mm (MNHN-IU-2013-8564). SALOMON 2 Stn CP2227, 6°37.2'S, 156°12.7'E, 508-522 m, 28.X.2004, 1 ♂ 3.4 mm, 1 ov. ♀ 3.0 mm (MNHN-IU-2014-16852). – Stn CP2287, 8°40.8'S, 157°24.6'E, 253-255 m, 6.XI.2004, 1 ov. ♀ 3.9 mm, 1 ♀ 3.3 mm (MNHN-IU-2014-16853). **Vanuatu**. MUSORSTOM 8 Stn CP984, 19°19.62'S, 169°26.43'E, 480-544 m, 23.IX.1994, 1 ♂ 3.8 mm (MNHN-IU-2014-16846). – Stn CP1026, 17°50.35'S, 168°39.33'E, 437-504 m, 28.IX.1994, 1 ♂ 3.3 mm, 1 ov. ♀ 3.9 mm (MNHN-IU-2014-16847). – Stn CP1136, 15°40.62'S, 167°01.60'E, 398-400 m, 11.X.1994, 2 ov. ♀ 4.5, 4.5 mm (MNHN-IU-2013-12277 & MNHN-IU-2013-12278). – Stn CP1049, 16°39'S, 168°03'E, 469-525 m, 1.X.1994, 2 ov. ♀ 4.0, 4.5 mm, 1 ♀ 4.7 mm (MNHN-IU-2014-16848). SANTO Stn AT120, 15°40.4'S, 167°01.0'E, 431-445 m, 19.X.2006, 1 ♂ 4.2 mm, 1 ov. ♀ 3.8 mm (MNHN-IU-2014-16849). **Philippines**. MUSORSTOM 1 Stn CP42, 13°54'N, 120°29'E, 379-407 m, 24.III.1976, 1 ♀ 4.3 mm (MNHN-IU-2013-12279).

ETYMOLOGY — From the Greek *parisos* (evenly balanced), alluding to the eyestalk being evenly broad from proximal to distal, not broadened proximally as in the related species *Uroptychus scandens* Benedict, 1902.

DISTRIBUTION — Chesterfield Islands, Solomon Islands, Vanuatu, Loyalty Ridge, New Caledonia, Norfolk Ridge, Philippines; 253-647 m.

SIZE — Males, 2.4-4.2 mm; females, 2.8-4.7 mm; ovigerous females from 3.0 mm.

DESCRIPTION — Small species. *Carapace*: Broader than long (0.8 × as long as broad); greatest breadth 1.5-1.8 × distance between anterolateral spines. Dorsal surface setose laterally, with very small spines on epigastric, hepatic, anterior branchial regions, and often on anterior part of posterior branchial region, those on hepatic region somewhat larger. Lateral margins often slightly constricted between hepatic and branchial regions, convex along branchial region, with row of small spines; anterolateral spine larger than others, directed somewhat anterolaterally, overreaching much smaller lateral orbital spine. Rostrum narrow triangular, directed straight forward on dorsal surface, directed anterodorsally on ventral surface, with interior angle of 23-25°; lateral margins somewhat concave, more strongly convergent near tip, with several obsolescent denticles; dorsal surface somewhat concave; length slightly smaller than (0.8-0.9 x) breadth, 0.4 × that of remaining carapace, breadth slightly less than half carapace breadth at posterior carapace margin. Lateral orbital spine situated slightly anterior to level of anterolateral spine. Pterygostomian flap anteriorly somewhat roundish, produced to strong sharp spine; surface with small spines on anterior half, including row of spines along dorsal margin anterior to linea anomurica.

Sternum: Excavated sternum with slightly convex anterior margin; surface smooth, without ridge and central spine, with setae along anterior margin. Sternal plastron 0.8-0.9 × as long as broad, lateral extremities between sternites 4 and 7 convex, sternite 6 broadest; anterior margin of sternite 3 broadly and deeply excavated nearly in semicircular shape, with pair of small median spines, often with single spine instead or with no spine. Sternite 4 having anterolateral margin feebly convex, anteriorly produced to distinct spine followed by a few smaller spines proximally, length 1.4 × as long as posterolateral margin. Anterolateral margin of sternite 5 as long as posterior margin of sternite 4.

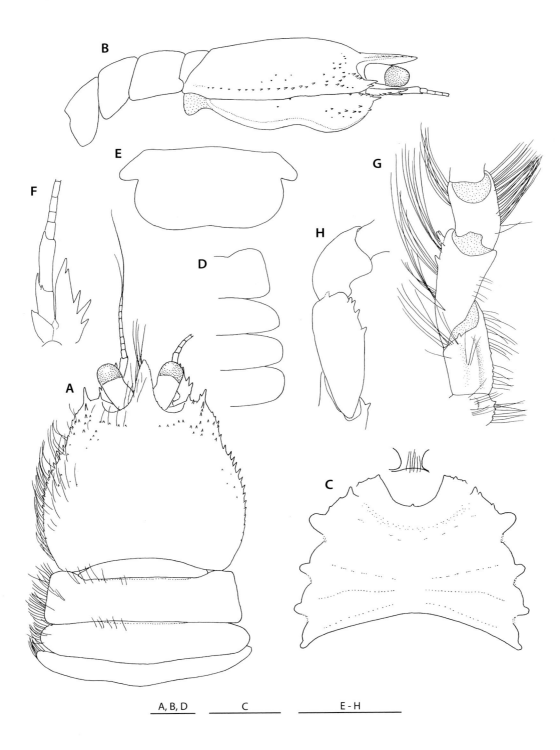

FIGURE 189

Uroptychus parisus n. sp., holotype, ovigerous female 3.3 mm (MNHN-IU-2013-8563). **A**, carapace and anterior part of abdomen, dorsal. **B**, same, setae omitted, lateral. **C**, sternal plastron, with excavated sternum and basal parts of Mxp1. **D**, right pleura of abdominal somites 2-5, dorsolateral. **E**, telson. **F**, left antenna, ventral. **G**, right Mxp3, ventral. **H**, same, setae omitted, lateral. Scale bars: 1 mm.

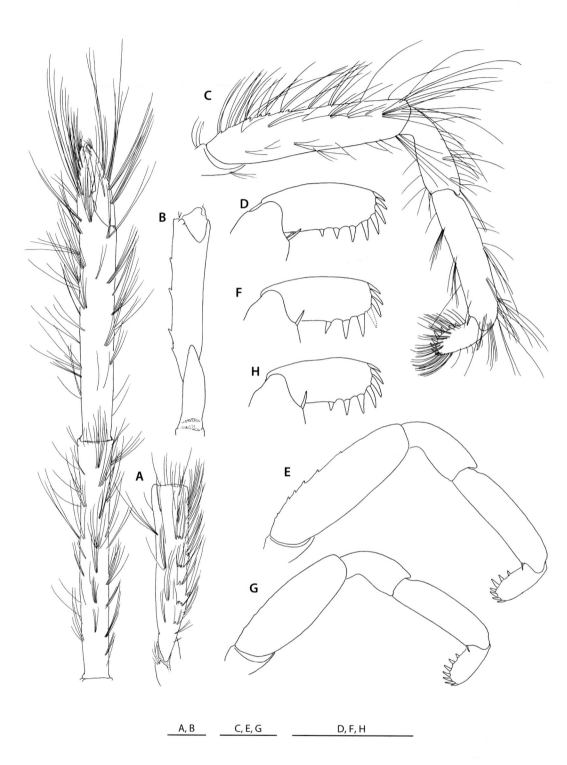

A, B   C, E, G   D, F, H

FIGURE 190

Uroptychus parisus n. sp., holotype, ovigerous female 3.3 mm (MNHN-IU-2013-8563). **A**, left P1, dorsal. **B**, same, proximal part, setae omitted, ventral. **C**, right P2, lateral. **D**, same, distal part, setae omitted. **E**, right P3, setae omitted, lateral. **F**, same, distal part, lateral. **G**, right P4, setae omitted, lateral. **H**, same, distal part, lateral. Scale bars: 1 mm.

Abdomen. Pleura similar in both sexes. Somite 1 gently convex from anterior to posterior. Somite 2 tergite 2.5-2.9 × broader than long, pleural lateral margin feebly or barely concavely divergent posteriorly, ending in rounded margin. Pleura of somites 3-5 laterally rounded. Telson 0.4-0.5 × as long as broad; posterior plate slightly or moderately emarginate, 1.5-2.0 × longer than anterior plate.

Eyes: Elongate, 2.0-2.2 × longer than broad, very slightly overreaching apex of rostrum, equally broad proximally and distally; lateral margin convex, mesial margin concave. Cornea 0.5 × length of remaining eyestalk and slightly inflated, breadth slightly larger than greatest breadth of remaining eyestalk.

Antennule and antenna: Ultimate article of antennular peduncle 2.6-3.3 × longer than high. Antennal peduncle reaching or slightly overreaching apex of rostrum. Antennal scale fused with article 2, 1.7-1.8 × broader than article 5, ending in point half to four-fifths length of article 5, laterally with 1-3 sharp spines including spine at ordinary site of article 2, proximalmost usually strong, rarely small. Articles 4 with distomesial spine (with accompanying smaller spine ventrolateral to it in holotype). Article 5 with small distomesial spine, length 1.5 × that of article 4, breadth 0.4-0.5 × height of ultimate article of antennule. Flagellum of 8-10 segments ending in midlength of P1 merus, apical seta somewhat longer than flagellum.

Mxp: Mxp1 with bases broadly separated. Mxp3 with long setae. Basis with 3 or 4 proximally diminishing denticles on mesial ridge. Ischium with small but distinct spine lateral to distal end of flexor margin, crista dentata with several obsolescent denticles on proximal half. Merus more than twice (2.5 x) length of ischium, ridged along flexor margin; with 4 spines on distal third of flexor margin and 1 or 2 distolateral spines. Carpus unarmed.

P1: Slender, subcylindrical, 6.0-7.1 × (males), 5.8-6.6 (females) × longer than carapace. Ischium dorsally with basally broad, depressed, bifurcate short spine, ventromesially unarmed. Merus 1.3-1.6 × longer than carapace, dorsally bearing 2 rows of single, bifurcate or trifurcate spines (1 row along mesial margin and another row directly mesiodorsal to it). Carpus 1.3-1.4 × longer than merus, unarmed. Palm 4.3-4.9 × (males), 3.9-6.7 × (females) longer than broad, 0.9-1.0 × as long as carpus, very slightly broadened distally. Fingers relatively narrow distally, somewhat gaping in proximal third; opposable margins fitting to each other in distal two-thirds with row of very small spines or denticles; movable finger 0.4-0.5 × length of palm, with obtuse process at midpoint of gaping portion.

P2-4: Thickly setose like P1. Meri successively shorter posteriorly (P3 merus 0.8-0.9 × length of P2 merus, P4 merus 0.8-0.9 × length of P3 merus), slightly broader on P3 and P4 than on P2; length-breadth ratio, 4.1-4.7 on P2, 3.1-3.7 on P3, 2.7-3.1 on P4; dorsal margins with 5-7 small spines on proximal half on P2 and P3, unarmed on P4; P2 merus as long as carapace, rarely slightly shorter; 1.3-1.4 × longer than P2 propodus; P3 merus 1.2-1.3 × length of P3 propodus; P4 merus 1.1-1.2 × length of P4 propodus. Carpi successively shorter posteriorly; carpus-propodus length ratio, 0.5 on P2-4. Propodi successively shorter posteriorly; flexor margin straight in lateral view, with pair of small distal spines only. Dactyli successively shorter posteriorly, shorter than carpus; dactylus-carpus length ratio, 0.6-0.7 on P2, 0.7-0.8 on P3 and P4; dactylus-propodus length ratio, 0.3 on P2, 0.3-0.4 on P3, 0.4 on P4; truncate, bearing 7 or 8 slender spines obscured by setae, 3-4 of these located on terminal margin, remainder on flexor margin, distalmost (terminal and directly adjacent to extensor margin) slightly smaller than distal second.

Eggs. Number of eggs carried, 7-14; size, 0.71 × 0.79 mm - 0.92 × 0.71 mm.

REMARKS — The new species is grouped together with *U. articulatus* n. sp., *U. imparilis* n. sp. and *U. scandens* Benedict, 1902 by having truncate dactyli of P2-4. Their relationships are discussed under *U. scandens* (see below).

Molecular data provided by L. Corbari (personal comm.) suggest that the female from the Philippines (MNHN-IU-2013-12279), the two females from Vanuatu (MNHN-IU-2013-12277 and 12278) and one of seven specimens from the Solomon Islands (MNHN-IU-2013-12276) are different from one another at a specific level. However, I am at a loss to discover any morphological differences, for which extensive studies are required.

The material reported under *U. scandens* from western Australia by McCallum & Poore (2013) may be referable to this species (A. McCallum, person. comm.).

Uroptychus pectoralis n. sp.

Figure 191

TYPE MATERIAL — Holotype: **New Caledonia**, Chesterfield Islands. MUSORSTOM 5 Stn CC366, 19°45'S, 158°46'E, 650 m, 19.X.1986, with ?*Chironephthya* sp. (Alcyoniina: Nidaliidae), ♀ 9.1 mm (MNHN-IU-2014-16854).

ETYMOLOGY — From the Latin *pectoralis* (pertaining to the breast), alluding to the elongate sternal plastron with the anterior margin deeply excavated, a character to separate the species from related species.

DISTRIBUTION — Chesterfield Islands; 650 m.

DESCRIPTION — *Carapace*: As long as broad; greatest breadth 1.6 × distance between anterolateral spines. Dorsal surface glabrous, slightly convex from anterior to posterior, with very shallow depression between gastric and cardiac regions. Lateral margins posteriorly convex; anterolateral spine very small, well separated from, situated distinctly posterior to, and falling short of lateral orbital angle, followed by posteriorly diminishing granules. Rostrum relatively broad triangular, with interior angle of 30°, slightly deflected ventrally; dorsal surface feebly concave; length 0.4 × that of remaining carapace, breadth half carapace breadth at posterior carapace margin. Lateral limit of orbit angular or acuminate, without distinct spine. Pterygostomian flap anteriorly roundish with very tiny spine; surface smooth.

Sternum: Excavated sternum broad triangular and anteriorly blunt, surface with small spine in center. Sternal plastron relatively narrow, about as long as broad; lateral extremities posteriorly divergent. Sternite 3 strongly depressed; anterior margin deeply excavated, representing V-shape, with ill-defined median notch, laterally roundish. Sternite 4 with anterolateral margin long relative to breadth, slightly convex and moderately divergent posteriorly, anteriorly angular, with posteriorly diminishing crenulations; posterolateral margin short, one-third length of anterolateral margin. Sternite 5 with anterolateral margin anteriorly convex, twice as long as posterolateral margin of sternite 4.

Abdomen: Smooth and glabrous. Somite 1 without transverse ridge. Somite 2 tergite 2.3 × broader than long; pleuron with moderately concave, strongly divergent lateral margin, posterolaterally tapering to blunt point. Pleuron of somite 3 with blunt lateral tip. Telson slightly more than half as long as broad; posterior plate twice as long as anterior plate, posterior margin slightly concave.

Eye: Elongate (1.9 × longer than broad), overreaching point two-thirds of rostrum, not reaching apex of rostrum, lateral and mesial margins feebly concave. Cornea slightly dilated, about half as long as remaining eyestalk.

Antennule and antenna: Ultimate article of antennular peduncle 2.3 × longer than high. Antennal peduncle terminating in corneal margin. Article 2 unarmed. Antennal scale 1.7 × broader than article 5, slightly falling short of distal end of article 5. Article 4 unarmed. Article 5 2.5 × longer than article 4, distomesially bearing short blunt spine; breadth less than half height of ultimate antennular peduncle. Flagellum of 21 segments nearly reaching distal end of P1 merus.

Mxp: Mxp1 with bases close to each other. Mxp3 barely setose laterally. Basis with 1 or 2 denticles on mesial ridge, distal one larger (proximal one missing on right side). Ischium having crista dentata with 22 denticles diminishing toward proximal and distal ends of article, flexor margin distally not rounded. Merus about twice as long as ischium, moderately thick mesio-laterally, unarmed, flexor margin sharply ridged. Carpus unarmed.

P1: Massive, 4.9 × longer than carapace, unarmed, setose distally, glabrous elsewhere. Ischium dorsally with distal spine of moderate size, ventrally feebly granulose, ventromesially unarmed. Merus with scattered granulate ridges on ventral surface, length slightly more than that of carapace. Carpus moderately depressed on dorsal surface, 1.2 × longer than merus, slightly longer than palm. Palm about 3 × longer than broad, slightly less than twice as long as movable finger. Fingers setose, ending in incurved spine; fixed finger with low eminence at midlength of opposable margin; movable finger with disto-proximally broad proximal process fitting into longitudinal groove on opposite face of fixed finger when closed.

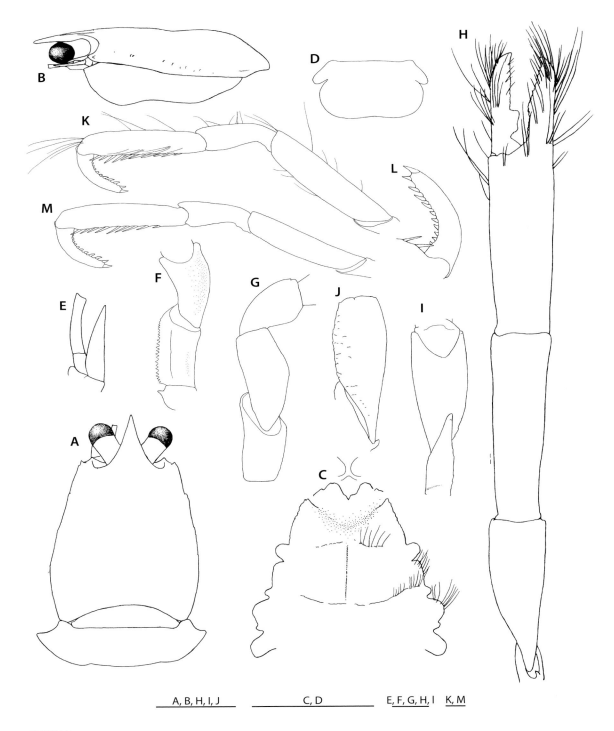

FIGURE 191

Uroptychus pectoralis n. sp., holotype, female 9.1 mm (MNHN-IU-2014-16854). **A**, carapace and anterior part of abdomen, dorsal. **B**, carapace, lateral. **C**, sternal plastron, with excavated sternum and basal parts of Mxp1. **D**, telson. **E**, left antenna, ventral. **F**, left Mxp3, setae omitted, ventral. **G**, right Mxp3, lateral. **H**, right P1, dorsal. **I**, same, proximal part, ventral. **J**, same, proximal part, mesial. **K**, left P3, lateral. **L**, same, distal part, setae omitted, lateral. **M**, left P4, setae omitted, lateral. Scale bars: A, B, H, I, J, 5 mm; C, D, E, F, G, K, L, M, 1 mm.

P2-4: Left P2, right P2 and right P4 detached from body and missing. Remaining legs with scattered long setae. Meri moderately compressed, unarmed; length-breadth ratio, 4.5 on P3, 4.6 on P4; P3 merus very slightly longer than P3 propodus; P4 merus 0.9 × as long as and 0.9 × as broad as P3 merus, 0.9 × as long as P4 propodus. P4 carpus 0.9 × length of P3 carpus; carpus-propodus length ratio, 0.5 on P3, 0.4 on P4; carpus-dactylus length ratio, 1.2 on P3, 1.1 on P4. Propodi longer on P4 than on P3, slightly more than twice as long as dactyli; flexor margin nearly straight, with pair of distal spines slightly distant from juncture with dactylus, preceded by 10 (on P3) or 8-9 (on P4) spines along distal three-fifths of length. Dactyli much more slender than propodi and strongly curving; flexor margin with 11 or 12 subtriangular spines, ultimate longest, remaining spines somewhat inclined and proximally diminishing; length 0.4 × that of propodi on P3 and P4, and 0.8 × that of carpi on P3, 0.9 × on P4.

REMARKS — *Uroptychus pectoralis* resembles *U. gordonae* Tirmizi, 1964 from the Maldives, *U. laperousazi* Ahyong & Poore, 2004 and *U. latus* Ahyong & Poore, 2004, both from southern Australia, sharing the carapace with the small anterolateral spine only, the P2-4 dactyli with proximally diminishing triangular spines and sternite 3 without submedian spines. The new species is distinguished from these three species by the following differences: the carapace is as long as instead of much broader than long, as also is the sternal plastron; sternite 3 is more deeply excavated on the anterior margin; the anterolateral spine of the carapace is distinctly posterior rather than directly lateral to the lateral orbital spine. The type material of *U. gordonae* (male holotype and 1 female paratype, BMNH 1966.2.3.17-18) examined on loan shows that the flexor margins of the P2-4 propodi each bear 2 or 3 single spines proximal to the distal pair (8-10 spines in *U. pectoralis*); also, the flexor marginal spines of the dactyli are less numerous (6-8 versus 11 or 12).

Uroptychus pedanomastigus n. sp.

Figures 192, 193

TYPE MATERIAL — Holotype: **New Caledonia**, Norfolk Ridge. BIOCAL Stn CP30, 23°09'S, 166°41'E, 1140 m, 29.VIII.1985, 1 ov. ♀ 3.5 mm (MNHNIU-2012-679).

ETYMOLOGY — From the Greek *pedanos* (short) plus *mastigos* (whip), alluding to a short antennal flagellum possessed by the new species.

DESCRIPTION — Small species. *Carapace*: Much broader than long (0.7 × as long as broad), broadest on posterior third; greatest breadth 2.1 × distance between anterolateral spines. Dorsal surface strongly convex from anterior to posterior, moderately from side to side, without depression, sparsely covered with short fine setae. Lateral margins convexly divergent posteriorly, granulate, and weakly ridged along posterior half; anterolateral spine small, slightly larger than lateral orbital spine, directed anteromesially. Rostrum deflected ventrally in proximal two-thirds, horizontal in distal third, narrow triangular, with interior angle 20°; dorsal surface feebly concave; lateral margin with 2 spines distally, proximal one obsolescent; length two-thirds that of remaining carapace, breadth about half carapace breadth at posterior carapace margin. Lateral orbital angle with small, distinct spine separated from anterolateral spine by basal breadth of latter spine, and located somewhat anterior to level of that spine. Pterygostomian flap smooth on surface, anterior margin angular, produced to distinct spine; anterior half much higher (about 2 x) than posterior half.

Sternum: Excavated sternum with triangular anterior margin; surface smooth, with weak low ridge in midline. Sternal plastron 0.7 × as long as broad, lateral extremities convexly divergent posteriorly; broadest on sternite 6. Sternite 3 shallowly depressed; anterior margin in broad V-shape without median notch and submedian spines. Sternite 4 with anterolateral margin convex and denticulate, 2 × longer than posterolateral margin. Anterolateral margin of sternite 5 anteriorly convex, 1.3 × longer than posterolateral margin of sternite 4.

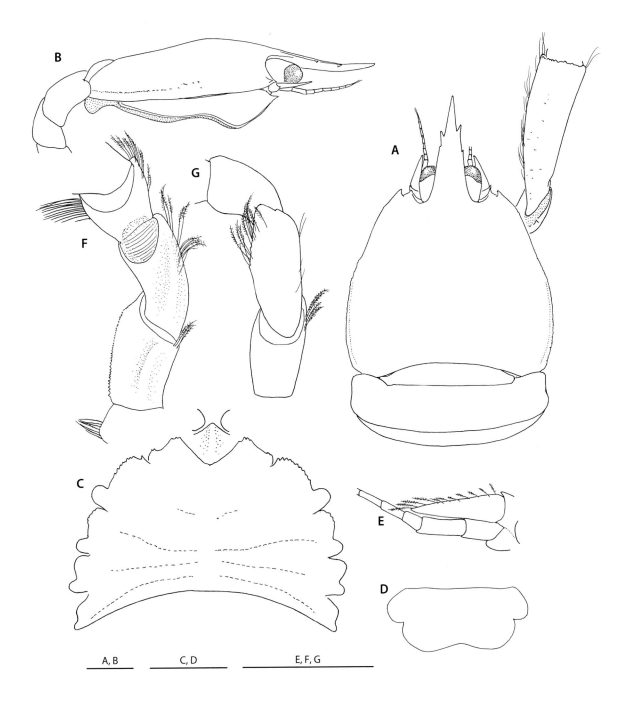

FIGURE 192

Uroptychus pedanomastigus n. sp., holotype, ovigerous female 3.5 mm (MNHN-IU-2012-679). **A**, carapace and anterior part of abdomen, proximal part of right P1 included, setae on carapace omitted, dorsal. **B**, same, setae omitted, lateral. **C**, sternal plastron, with excavated sternum and basal parts of Mxp1. **D**, telson. **E**, left antenna, ventral. **F**, left Mxp3, ventral. **H**, same, lateral. Scale bars: 1 mm.

Abdomen: Somite 1 weakly convex from anterior to posterior, without transverse ridge. Somite 2 tergite 2.7 × broader than long; pleural lateral margin concave, slightly divergent posteriorly, posterolateral end blunt. Pleuron of somite 3 with blunt posterolateral margin. Telson 0.4 × as long as broad; posterior plate as long as anterior plate, posterior margin medially deeply concave.

Eye: 1.4 × longer than broad, distally narrowed. Cornea not inflated, half as long as remaining eyestalk.

Antennule and antenna: Ultimate article of antennule 3.2 × longer than high. Antennal peduncle overreaching cornea, barely reaching midlength of rostrum. Article 2 with small distolateral spine. Antennal scale tapering distally, 1.9 × broader than article 5, overreaching distal end of article 5, slightly overreaching eye but not reaching apex of rostrum. Articles 4 and 5 each with small distomesial spine; article 5 1.2 × longer than article 4, breadth about half height of antennular ultimate article. Flagellum of 6 or 7 segments barely reaching rostral tip, far falling short of distal margin of P1 merus.

Mxp: Mxp1 with bases broadly separated. Mxp3 with fine plumose setae other than brushes on distal articles. Basis having mesial ridge proximally lobe-like and nearly smooth. Ischium with flexor margin not rounded distally; crista dentata with more than 35 denticles distally diminishing. Merus 2.4 × longer than ischium, with obsolescent distolateral spine; flexor margin roundly ridged, with a few tubercles on distal half. Carpus unarmed.

P1: 4.3 × longer than carapace, sparklingly with short fine soft setae along mesial and lateral margins. Ischium with small dorsal spine, ventromesial margin with row of very small spines, with or without obsolescent subterminal spine. Merus with small distomesial and distolateral spines on ventral surface and tubercle-like small spines on distodorsal margin, unarmed elsewhere, length subequal to that of carapace. Carpus subcylindrical, 1.2 × longer than merus, distodorsal margin with row of small, denticle-like spines. Palm 4.0 × longer than broad, 1.2 × longer than carpus. Fingers moderately depressed and relatively broad, distally incurved, crossing when closed, feebly gaping, denticulate on opposable margins; fixed finger with low eminence at midlength of opposable margin; movable finger 0.4 × as long as palm, opposable margin with marginally denticulate process proximal to midlength, fitting to longitudinal groove on opposite face of fixed finger when closed.

P2-4: Relatively thick mesio-laterally, with fine soft setae. Meri successively shorter posteriorly (P3 merus 0.8 × length of P2 merus, P4 merus 0.8 × length of P3 merus); breadths slightly greater on P3 and P4 than on P2; length-breadth ratio, 4.0 on P2, 3.7 on P3, 3.2 on P4; dorsal margins rounded, not carinated, unarmed; P2 merus 0.7 × as long as carapace, subequal to length of P2 propodus; P3 merus 0.9 × length of P3 propodus; P4 merus 0.7 × length of P4 propodus. Carpi subequal on P2 and P3, slightly shorter on P4; carpus-propodus length ratio, 0.5 on P2 and P3, 0.3 on P4; carpus-dactylus length ratio, 1.2 or 1.3 on P2, 1.0 on P3, 0.6 on P4; extensor margin with obsolescent proximal process. Propodi subequal on P2 and P3 and longer on P4; flexor margins moderately convex on distal portion on P2, less so on P3, feebly so on P4; ending in pair of terminal spines preceded by 6 (right) or 8 (left) spines on P2, 8 on P3, 5 on P4, all along entire length, those on proximal half much smaller and broadly spaced from one another. Dactyli 0.4 × as long as propodi on P2, 0.5 × on P3 and P4; dactylus-carpus length ratio, 0.8 on P2, 1.0 on P3, 1.6 on P4; flexor margins somewhat curving, with 7, 7, 8 well-developed, triangular spines on P2, P3, P4 respectively, ultimate strongest, remaining spines proximally diminishing and perpendicular to margin.

Eggs. Number of eggs carried, 3; size, 1.03 × 1.00 mm - 1.13 × 0.96 mm.

REMARKS — This species resembles *U. edisonicus* Baba & Williams, 1998 from the hydrothermal vent site at Edison Seamount, Bismarck Archipelago, *U. latior* n. sp. and *U. norfolkanus* n. sp. (for their similarity, see above under the account of *U. latior*). *Uroptychus pedanomastigus* is distinguished from *U. edisonicus* (Figure 121) by the anterolateral spine that is directed anteromesially instead of straight forward; the excavated sternum has a triangular instead of convex anterior margin; the P2-4 propodi bear flexor marginal spines placed along the entire margin (although proximal spines are widely spaced) instead of restricted to the distal portion. *Uroptychus pedanomastigus* and *U. latior* n. sp. share the anteromesially directed anterolateral spine of the carapace, the P2-4 carpi bearing a small proximal process on the extensor margin, and the Mxp3 crista dentata bearing a row of regularly arranged small denticles. However, *U. pedanomastigus* is distinctive in the anterior margin of the excavated sternum that is triangular instead of gently convex; the antennal scale

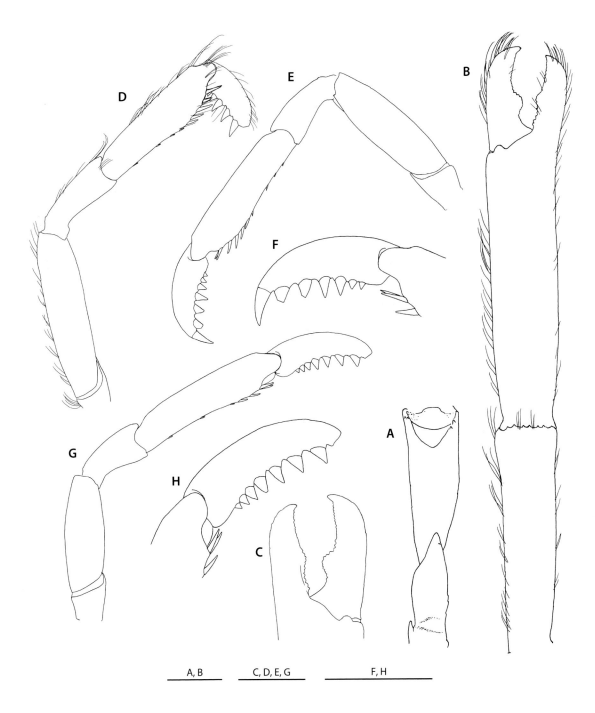

FIGURE 193

Uroptychus pedanomastigus n. sp., holotype, ovigerous female 3.5 mm (MNHN-IU-2012-679). **A**, right P1, proximal part, setae omitted, ventral. **B**, same, distal three articles, dorsal. **C**, same, fingers, setae omitted, ventral. **D**, right P2, lateral. **E**, left P3, setae omitted, lateral. **F**, same, distal part, lateral. **G**, right P4, lateral. **H**, same, distal part, lateral. Scale bars: 1 mm.

slightly overreaching instead of extending far beyond the eye; the antennal flagellum barely reaching instead of extending far beyond the apex of the rostrum; the P2-4 propodi bearing additional loosely arranged spines proximal to the closely arranged distal group of flexor marginal spines; and the P2-4 dactyli bearing flexor marginal spines perpendicularly instead of obliquely directed.

Uroptychus pedanomastigus is distinguished from *U. norfolkanus* by the excavated sternum that is triangular instead of convex on the anterior margin; the anterolateral spine of the carapace is directed anteromesially rather than straight forward and closer to instead of more distant from the lateral orbital spine (this is clearly recognizable when viewed from dorsal side); the flexor marginal spines of the P2-4 dactyli are perpendicularly instead of obliquely directed; and the flexor margin of the P3 propodus (P2 and P4 are missing in *U. norfolkanus*) bears three instead of one spine remotely proximal to the distal group of closely arranged spines. In addition, the P3 propodus is more strongly inflated on the distal flexor margin in *U. norfolkanus*. Generally this inflation becomes weaker in smaller specimens (Ahyong *et al.* 2015, for *U. insignis* (Henderson, 1885)), but the ovigerous female of *U. norfolkanus* is smaller than the ovigerous female holotype of *U. pedanomastigus*.

Uroptychus perpendicularis n. sp.

Figure 194

TYPE MATERIAL — Holotype: **New Caledonia**, Norfolk Ridge. NORFOLK 2, Stn DW2052, 23°42.29'S, 168°15.27'E, 473-525 m, 24.X.2003, ♂ 3.0 mm (MNHN-IU-2014-16855).

ETYMOLOGY — From the Latin *perpendicularis* (upright, perpendicular), referring to spines on the P2-4 dactyli that are perpendicular to the flexor margin in the new species.

DISTRIBUTION — Norfolk Ridge; 473-525 m.

DESCRIPTION — *Carapace*: Slightly broader than long (0.93 × as long as broad); greatest breadth 1.6 × distance between anterolateral spines. Dorsal surface feebly convex from anterior to posterior, bearing very sparse short setae. Lateral margins convex on posterior branchial regions, bearing 7 or 8 spines; anterolateral (first) spine well developed, distinctly larger than and slightly overreaching lateral orbital spine, other 6 or 7 spines small; second spine on hepatic margin, ventral to level of first and third; third spine located on anterior end of branchial margin, accompanying small spine dorsomesial to it; remaining 5 spines on posterior branchial margin; 1 or 2 small hepatic spines mesial to between first and second spines (or directly behind anterolateral spine). Rostrum narrow triangular, with interior angle of 20°, nearly horizontal, length more than half that of remaining carapace, breadth half carapace breadth at posterior carapace margin; lateral margin with small subterminal spine; dorsal surface concave. Pterygostomian flap anteriorly angular, produced to sharp spine, surface with row of spines directly below anterior part of linea anomurica.

Sternum: Excavated sternum anteriorly produced between bases of Mxp1, bearing weak ridge in midline. Sternal plastron slightly broader than long, lateral extremities between sternites 4-7 gently divergent posteriorly. Sternite 3 weakly depressed; anterior margin concavely excavated, with small median notch without flanking spine, laterally rounded. Sternite 4 having anterolateral margin anteriorly angular (right) or ending in small blunt spine (left), twice as long as posterolateral margin. Sternite 5 with anterolateral margin anteriorly convex, 1.3 × longer than posterolateral margin of sternite 4.

Abdomen: Smooth, nearly glabrous. Somite 1 without transverse ridge. Somite 2 tergite 3.0 × broader than long; pleural lateral margin somewhat concavely divergent posteriorly, anterior and posterior ends rounded. Pleuron of somite 3 laterally blunt. Telson half as long as broad; posterior plate slightly convex on posterior margin, length 1.4 × that of anterior plate.

Eye: Elongate (twice as long as broad), slightly overreaching midlength of rostrum; lateral and mesial margins subparallel. Cornea not dilated, about half length of remaining eyestalk.

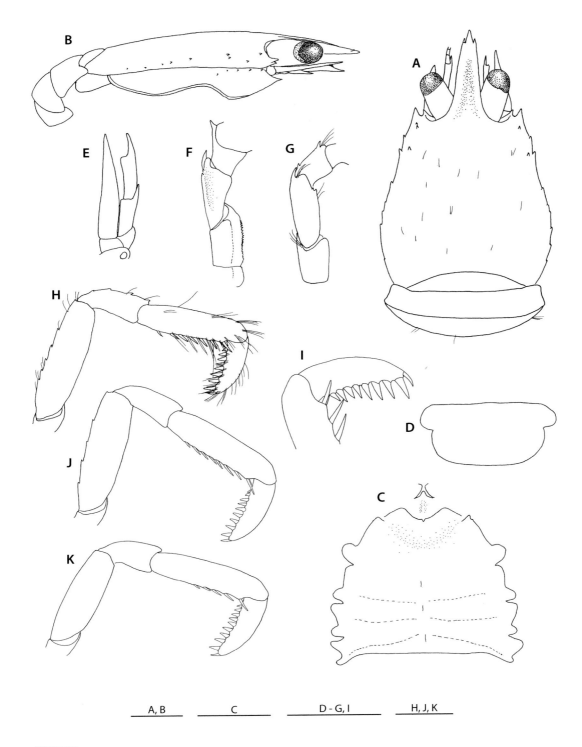

FIGURE 194

Uroptychus perpendicularis n. sp., holotype, male 3.0 mm (MNHN-IU-2014-16855). **A**, carapace and anterior part of abdomen, dorsal. **B**, same, lateral. **C**, sternal plastron, with excavated sternum and basal parts of Mxp1. **D**, telson. **E**, right antenna, ventral. **F**, right Mxp3, ventral. **G**, same, lateral. **H**, right P2, lateral. **I**, same, distal part, setae omitted, lateral. **J**, right P3, setae omitted, lateral. **K**, right P4, setae omitted, lateral. Scale bars: 1 mm.

Antennule and antenna: Ultimate article of antennular peduncle 3.1 × longer than high. Antennal peduncle somewhat overreaching cornea. Article 2 unarmed. Antennal scale overreaching peduncle, terminating at tip of strong distomesial spine of article 5, breadth 1.5 × that of article 5. Distal 2 articles each with strong distomesial spine; article 4 very slightly longer than article 4, breadth 0.6 × height of antennular ultimate article. Flagellum consisting of 11-12 segments, not reaching distal end of P1 merus.

Mxp: Mxp1 close to each other. Mxp3 with sparse short setae other than brushes on distal articles. Basis lacking denticles on mesial ridge. Ischium with distally rounded flexor margin; crista dentata with more than 20 denticles. Merus 2.5 × longer than ischium, flattish on mesial face; flexor margin sharply ridged, with 2 small spines about at distal quarter; distolateral spine strong. Carpus with distolateral spine and 1 distinct spine on proximal part of extensor margin.

P1: Missing.

P2-4: Compressed mesio-laterally, relatively broad. Meri subequal in length on P2 and P3, P4 merus 0.9 × length of P3 merus; length-breadth ratio, 3.6 on P2, 3.1 on P3 and P4; dorsal margin with row of very small spines distinct on P2 and P3, obsolete on P4; P2 merus 0.8 × length of carapace, 1.2 × length of P2 propodus; P3 merus as long as P3 propodus; P4 merus 0.9 × length of P4 propodus. Carpi subequal in length on P2 and P3, P4 carpus 0.9 × length of P3 carpus; carpus-propodus length ratio, 0.50 on P2, 0.46 on P3, 0.43 on P4; extensor margin with row of small spines on P2 only. Propodi subequal in length on P3 and P4, shorter on P2, flexor margin with pair of terminal spines preceded by 6 spines on P2, 5 spines on P3, 3 or 4 spines on P4. Dactyli subequal on P3 and P4, shorter on P2; as long as carpi on P2, slightly longer on P3 and P4; half as long as propodi on P2, slightly more than so on P3 and P4; flexor margin feebly curving, ending in slender spine preceded by 9 strong, long triangular spines perpendicular to margin, close to one another and successively diminishing proximally.

REMARKS — *Uroptychus perpendicularis* resembles *U. multispinosus* Ahyong & Poore, 2009 (see above) in having the carapace lateral margin with small spines, the rostrum with a subapical spine on each side, elongate eyes, the antennal peduncle with a strong spine on each of articles 4 and 5, and the antennal scale overreaching the peduncle. In *U. perpendicularis*, however, the anterolateral spine of the carapace is distinctly larger than instead of subequal to the lateral orbital spine; the antennal article 2 is unarmed instead of bearing a distinct distolateral spine; and the flexor marginal spines of the P2-4 dactyli are perpendicular to the margin rather than obliquely directed, and more numerous (10 versus 7) and closer to one another.

Uroptychus perpendicularis also resembles *U. vicinus* n. sp. Their relationships are discussed under the remarks of that species (see below).

The combination of the following characters links the species to *U. lanatus* n. sp.: the carapace with small lateral spines, the P2-4 propodi with a row of spines proximal to the pair of terminal spines, and the P2-4 dactyli with the ultimate of the flexor marginal spines more slender than the antepenultimate. These species can be distinguished by the following differences: the flexor marginal spines of the P2-4 dactyli in *U. perpendicularis* are subperpendicularly directed, whereas these are obliquely directed in *U. lanatus*; the rostrum in *U. perpendicularis* bears a subapical spine on each side, which spine is absent in *U. lanatus*; the antennal articles 4 and 5 in *U. perpendicularis* each bear a strong distomesial spine instead of being unarmed; and the P2-3 meri bear a row of dorsal spines instead of being spineless.

Uroptychus philippei n. sp.
Figures 195, 196

TYPE MATERIAL — Holotype: **New Caledonia**, Chesterfield Islands. CORAIL 2 Stn DE16, 20°47.75'S, 160°55.87'E, 500 m, 21.VII.1988, ov. ♀ 4.0 mm (MNHN-IU-2013-8529). Paratypes: **New Caledonia**, Chesterfield Islands. MUSORSTOM 5 Stn CP389, 20°44.95'S, 160°53.67'E, 500 m, 22.X.1986, 1 ov. ♀ 4.3 mm (MNHN-IU-2013-8527). CORAIL 2 Stn DE16, station data as for holotype, 2 ov. ♀ 3.9, 4.0 mm, 1 ♀ 3.6 mm (MNHN-IU-2013-8528); 5 ♂ 3.3-4.3 mm, 4 ov. ♀ 3.6-4.1 mm, 3 ♀ 3.5-4.1 mm (MNHN-IU-2014-16856).

ETYMOLOGY — Named for Philippe Bouchet of MNHN for his support and friendship.

DISTRIBUTION — Chesterfield Islands; in 500 m.

SIZE — Males, 3.3-4.3 mm; females, 3.5-4.3 mm; ovigerous females from 3.6 mm.

DESCRIPTION — Small species. *Carapace*: 0.8-0.9 × as long as broad (1.1-1.2 × broader than long); greatest breadth 1.7-1.9 × distance between anterolateral spines. Dorsal surface well convex from side to side and from anterior to posterior, with or without feeble groove between gastric and cervical regions, covered with short fine setae, smooth, occasionally bearing denticles sparse on hepatic and anterior branchial regions. Lateral margins moderately convexly divergent posteriorly with denticles arranged on short oblique ridges; anterolateral spine small, somewhat smaller than lateral orbital spine. Rostrum triangular, with interior angle of 25-30°, nearly horizontal; length 1.1-1.3 × breadth, slightly less than half that of remaining carapace, breadth less than half carapace breadth at posterior carapace margin; dorsal surface flattish, feebly concave at base; lateral margin with or without obsolescent spine near tip. Lateral orbital spine distinctly anterior to level of anterolateral spine. Pterygostomian flap covered with tiny denticles, anteriorly angular, produced to distinct spine.

Sternum: Excavated sternum with anterior margin convex between bases of Mxp1, surface with weak ridge in midline on anterior half. Sternal plastron 0.9-1.0 × as long as broad, lateral extremities slightly divergent or subparallel between sternites 4 and 7. Sternite 3 shallowly depressed, with anterior margin gently concave with U-shaped median sinus flanked by tiny spine, and laterally blunt or with a few denticles. Sternite 4 with relatively short anterolateral margin anteriorly ending in angular corner; posterolateral margin about as long as or slightly shorter than anterolateral margin. Anterolateral margin of sternite 5 anteriorly somewhat convex, slightly shorter than posterolateral margin of sternite 4.

Abdomen: Setose like carapace. Somite 1 convex from anterior to posterior. Somite 2 tergite 2.4 × broader than long; pleuron posterolaterally blunt, lateral margin weakly concave and slightly divergent posteriorly. Pleura of somites 3 and 4 laterally rounded. Telson slightly less than half as long as broad; posterior plate 1.1-1.3 × longer than anterior plate, posterior margin feebly concave or moderately emarginate.

Eye: Long relative to breadth (1.7-2.0 × longer than broad), ending in distal third of rostrum, slightly broader proximally or with subparallel lateral and mesial margins. Cornea not dilated, length 0.6-0.8 × that of remaining eyestalk.

Antennule and antenna: Ultimate article of antennule 2.6-2.8 × longer than high. Antennal peduncle relatively slender, not reaching rostral tip. Article 2 with distinct distolateral spine. Antennal scale sharply tapering, 1.6-1.8 × broader than article 5, slightly overreaching article 4 and reaching (usually slightly falling short of) at most midlength of article 5, with no lateral spine. Article 4 with small distomesial spine. Article 5 unarmed, 1.3-1.4 × longer than article 4, breadth slightly less than half height of antennular ultimate article. Flagellum of 11-13 segments slightly falling short of distal end of P1 merus.

Mxp: Mxp1 with bases broadly separated. Mxp3 basis without denticles on convex mesial ridge. Ischium having flexor margin not rounded distally, crista dentata with 28 small denticles. Merus 1.9 × longer than ischium, lateral surface moderately setose, flexor margin not cristate, with or without denticle-like small spine distal to midlength.

P1: Slender, with fine setae, lacking spines; length 5.0-6.2 × (males), 4.2-5.2 × (females) that of carapace. Ischium with short, flattish, laciniate dorsal process, unarmed elsewhere. Merus 1.2-1.3 × longer than carapace. Carpus subcylindrical, 1.2-1.4 × longer than merus. Palm somewhat depressed (height-breadth ratio, 0.7-0.8), 0.8-0.9 × length of carpus, 3.1-4.0 × (males), 3.4-4.5 × (females) longer than broad. Fingers slightly gaping, distally slightly incurved, not spooned; movable finger half as long as palm or slightly less than so; opposable margin with small subtriangular process proximal to opposing low prominence located at midlength of fixed finger.

P2-4: Moderately compressed mesio-laterally, with soft fine setae, along extensor margin in particular. Meri successively shorter posteriorly (P3 merus 0.9 × length of P2 merus, P4 merus 0.8-0.9 × length of P3 merus), subequally broad on P2 and P3, very slightly narrower on P4 than on P3; length-breadth ratio, 3.8-4.6 on P2, 3.6-4.1 on P3, 3.1-3.8 on 4; P2 merus 0.8-0.9 × length of carapace, 1.1-1.2 × length of P2 propodus; P3 merus 1.0-1.1 × length of P3 propodus; P4 merus 0.8-0.9 × length of P4 propodus; extensor margin smooth or with several crenulations or eminences on P2 and P3, smooth

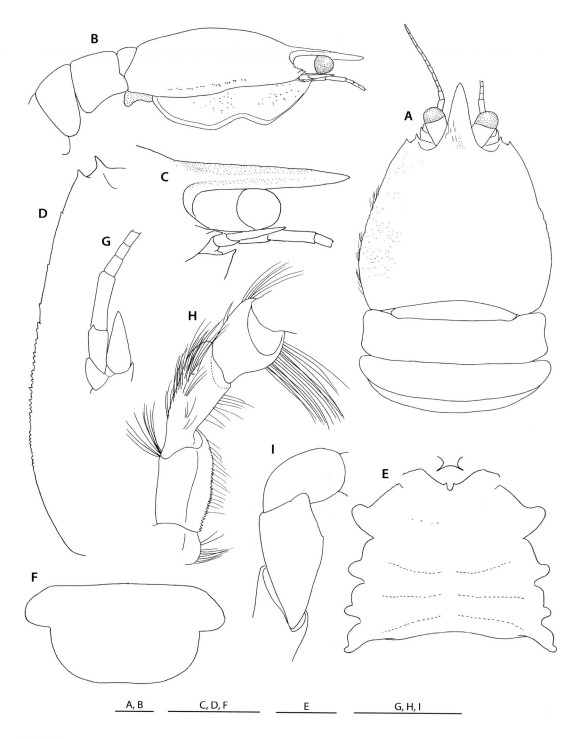

FIGURE 195

Uroptychus philippei n. sp., holotype, ovigerous female 4.0 mm (MNHN-IU-2013-8529). **A**, carapace and anterior part of abdomen, dorsal. **B**, same, lateral. **C**, same, anterior part, lateral. **D**, same, lateral margin, dorsal. **E**, sternal plastron, with excavated sternum and basal parts of Mxp1. **F**, telson. **G**, left antenna, ventral. **H**, right Mxp3, ventral. **I**, same, setae omitted, lateral. Scale bars: 1 mm.

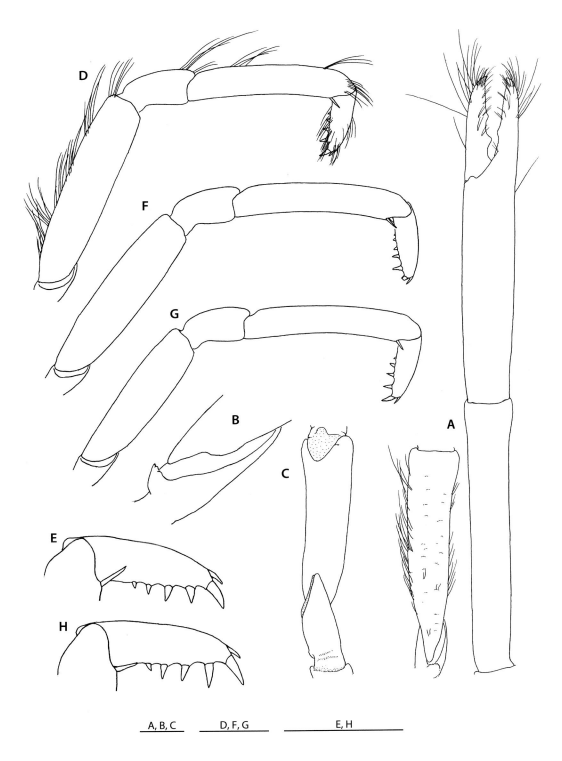

FIGURE 196

Uroptychus philippei n. sp., holotype, ovigerous female 4.0 mm (MNHN-IU-2013-8529). **A**, right P1, dorsal. **B**, same, proximal part, lateral. **C**, same, proximal part, ventral. **D**, right P2, lateral. **E**, same, distal part, setae omitted, lateral. **F**, right P3, lateral. **G**, right P4, lateral. **H**, same, distal part, lateral. Scale bars: 1 mm.

on P4. Carpi subequal, unarmed, carpus-propodus length ratio, 0.4 on P2, 0.3 on P3 and P4. Propodi shorter on P2 than on subequal P3 and P4; flexor margin with pair of movable slender terminal spines only. Dactyli proportionately broad in lateral view, subequal to length of carpi (dactylus-carpus length ratio, 1.0 × on P2, 1.1 × on P3 and P4), less than half that of propodi (dactylus-propodus length ratio, 0.4 on P2, 0.3-0.4 on P3, 0.3 on P4); flexor margin nearly straight, with 6 or 7 spines often obscured by setae, ultimate small, slender and very close to strongest penultimate spine preceded by 4 or 5 spines successively diminishing, loosely arranged and nearly perpendicular to flexor margin, but proximal-most slightly inclined; antepenultimate spine slightly narrower than (at most three-quarters as broad as) penultimate.

Eggs. Number of eggs carried, 7; size, 0.80 × 0.90 mm - 0.90 × 1.04 mm; 2 eggs, 0.96-1.13 mm.

REMARKS — *Uroptychus philippei* n. sp., *U. bertrandi* n. sp., *U. sarahae* n. sp., *U. rutua* Schnabel, 2009 and *U. toka* Schnabel, 2009 are grouped together, sharing the small anterolateral spine of the carapace, the short antennal scale, the P2-4 dactyli with loosely arranged, perpendicularly directed flexor marginal spines, and the pterygostomian flap covered with denticle-like small spines. *Uroptychus philippei* is different from all of these related species in having the P2-4 dactyli subequal to instead of longer than carpi (the carpus-dactylus length ratio, 1.0 (P2), 1.1 (P3-4) in *U. philippei*; 1.1-1.3 (P2-3), 1.2-1.4 (P4) in *U. bertrandi*; 1.5 (P2-3), 1.4 (P4) in *U. rutua*; 1.3 (P2-3), 1.4 (P4) in *U. toka*; 1.1-1.3 (P2-3), 1.2-1.3 (P4) in *U. sarahae*) and in having the dorsal surface of the rostrum concave in the proximal half and flattish in the distal half instead of distinctly hollowed out along the entire length. In addition, the dorsal surface of the carapace in *U. philippei* is usually smooth, occasionally bearing sparse denticles instead of bearing numerous denticle-like small spines around the hepatic region. *Uroptychus philippei* is distinguished from *U. sarahae* and *U. toka* by the P2-4 dactyli bearing the penultimate spine less than 1.3 instead of fully 2 times broader than the antepenultimate spine. *Uroptychus bertrandi* is distinctive in having proximally strongly inflated eyes, and also is *U. rutua* in having broad prominences on the gastric region.

Uroptychus pilosus Baba, 1981

Uroptychus pilosus Baba, 1981: 126, figs 10, 11. — Baba 2005: 53, 230 (Makassar Strait, 1600 m).
Not *Uroptychus pilosus* — Ahyong & Poore 2004: 71, fig. 21 (= *U. inaequalis* n. sp.). — Poore 2004: 226, fig. 62b.

TYPE MATERIAL — Holotype: **Japan**, Kumanonada off E coast of Kii Peninsula, 1120-1160 m, male (NSMT-Cr. 6172). [not examined].

MATERIAL EXAMINED — Indonesia, Kai Islands. KARUBAR Stn CP20, 5°15'S, 132°59'E, 769-809 m, with Chrysogorgiidae gen. sp. (Calcaxonia), 25.X.1991, 2 ♂ 6.0, 6.3 mm, 1 ♀ 5.5 mm (MNHN-IU-2014-16857).

DISTRIBUTION — Known from Japan (Kumanonada off east coast of Kii Peninsula and off southeastern Kyushu) and Makassar Strait, in 1120-1600 m, and now from Indonesia (Kai Islands), in 769-809 m.

DIAGNOSIS — Moderate-sized species. Body covered with short soft fine setae. Carapace slightly broader than long; greatest breadth 1.7 × distance between anterolateral spines. Dorsal surface markedly convex on anterior gastric region preceded by depressed rostrum, well depressed between gastric and cardiac regions. Lateral margins with no spines, convex on branchial region; anterolateral corner rounded, with a few denticle-like or obsolescent small spines. Rostrum short triangular, with interior angle of 23°, directed slightly upwards, breadth much more than half carapace breadth at posterior carapace margin. Pterygostomian flap smooth on surface, anteriorly sharp angular, produced to distinct spine. Excavated sternum with bluntly triangular anterior margin, surface ridged in midline. Sternal plastron slightly longer than broad, lateral extremities subparallel between sternites 4-7; sternite 3 having anterior margin deeply excavated, with broad U-shaped median notch devoid of flanking spine; sternite 4 with posterolateral margin as long as anterolateral margin; anterolateral margin of sternite 5 0.7 × as long as posterolateral margin of sternite 4. Abdominal somite 2 tergite 2.2-2.3 × broader than long; pleuron posterolaterally blunt angular, lateral margin concavely divergent posteriorly; pleuron of somite 3 with blunt posterolateral terminus. Telson slightly more than half (0.53-0.57) as long as broad, posterior plate 1.5 × longer than anterior plate, distinctly emarginate on posterior margin. Eyes elongate, overreaching midlength of rostrum, lateral and mesial margins subparallel; cornea less than half length of remaining eyestalk. Ultimate article of antennule slightly more than 3 × longer than high. Antennal peduncle overreaching eye by full length of article 5; article 2 with lateral spine; antennal scale terminating in or slightly overreaching distal end of antennal article 4; article 5 relatively long (three-quarters length of article 4); flagellum of 14-15 segments not reaching distal end of P1 merus; Mxp1 with bases broadly separated. Mxp3 basis having mesial ridge proximally lobe-like, distally with 1 denticle; ischium not rounded on distal end of flexor margin, crista dentata with about 20 denticles diminishing distally; merus 2.3 × longer than ischium, relatively thick mesio-laterally, flexor margin with small spines or denticles distal to midlength. P1 slender, unarmed but ischium with short, basally broad dorsal spine; merus 1.3-1.5 × longer than carapace; carpus 1.5-1.7 × length of merus; palm slightly longer than carpus, about 8 × as long as broad; fingers ending in small incurved spine, crossing when closed; movable finger one-fourth to one-fifth length of palm, opposable margin with subtriangular proximal process. P2-4 subcylindrical; meri successively shorter posteriorly (P3 merus 0.8 × length of P2 merus, P4 merus 0.7-0.8 × length of P3 merus), equally broad on P2-4; dorsal margin with small spines on proximal third at least on P2 and P3; P2 merus about as long as or slightly shorter than carapace, slightly longer than P2 propodus; P4 merus 0.8-0.9 × length of P4 propodus; carpi successively slightly shorter posteriorly, length about one-third (0.3-0.4) that of propodus; propodi successively slightly shorter posteriorly, flexor margin concave in lateral view, with pair of terminal spines only; dactyli nearly as long as carpi on P2-4, each with 2 terminal spines of subequal size.

REMARKS — This species shares with *U. inaequalis* n. sp. and *U. plautus* n. sp. the P2-4 dactyli bearing only two terminal spines, a character rather unusual among the species of *Uroptychus*. Characters distinguishing *U. pilosus* from *U. inaequalis* and from *U. plautus* are outlined under the remarks of the latter two species (see below).

Uroptychus pinocchio Poore & Andreakis, 2011

Uroptychus naso — Van Dam 1939: 402 (part). — Baba 1969: 42 (part), figs 1, 2b. — Baba *et al.* 2008: 37 (part), fig. 1F. — Baba *et al.* 2009: 47 (part), fig. 39.
Uroptychus pinocchio Poore & Andreakis, 2011: 164, figs 4c, 4d, 5c, 10, 11.

TYPE MATERIAL — Holotype: **Philippines**, 14°01'N, 120°17'E, 184-186 m, male (MNHN-IU-2014-16859 [=MNHN-Ga 6232]). [examined].

OTHER MATERIAL EXAMINED — **Philippines**. MUSORSTOM 1 Stn CP61, 14°00'N, 120°17'E, 202-184 m, 27.III.1976, 1 ♂ 7.1 mm (MNHN-IU-2014-16858 [= MNHN-Ga 6230]). MUSORSTOM 2 Stn CP02, 14°00'N, 120°17'E, 186-184 m, 20.XI.1980, 1 ♂ 7.1 mm (holotype, MNHN-IU-2014-16859 [=MNHN-Ga 6232]). – Stn CP51, 14°00'N, 120°17'E, 170-187 m, 27.XI.1980, 1 ov. ♀ 8.7 mm, 1 ♀ 6.2 mm (MNHN-IU-2014-16860 [=MNHN-Ga 6234]).

DISTRIBUTION — Philippines, NW Kyushu (Japan) and Taiwan; in 153-225 m.

DIAGNOSIS — Carapace as long as broad; greatest breadth 1.6-1.7 × distance between anterolateral spines. Dorsal surface covered with setiferous tubercles, and with conical tubercles or small spines laterally and around cervical groove, small spines on hepatic region and behind lateral part of rostrum; with deep cervical groove bordering gastric and cardiac regions and anterior and posterior branchial regions, anterior cervical groove indistinct. Lateral margins divergent posteriorly; anterolateral spine relatively small, distinctly overreaching small or obsolete lateral orbital spine, followed by a few small spines on hepatic region and relatively large spines on branchial region; anterior branchial margin with 2 spines, anterior larger than anterolateral spine, posterior smaller than anterior, often followed by much smaller spine; posterior branchial margin with 7-9 large spines relatively widely spaced anteriorly, closely so posteriorly, usually with intermediate small spines, followed by a few small close spines. Rostrum broad, long triangular, with interior angle of 30°, nearly horizontal; dorsal surface covered with tubercles and depressed in midline, laterally with 6-8 small spines on anterior half; length 0.8-0.9 × that of carapace, breadth about half carapace breadth at posterior carapace margin. Pterygostomian flap covered with conical tubercles, anteriorly angular and produced to strong spine. Excavated sternum anteriorly ending in convex margin, surface with longitudinal ridge in midline. Sternal plastron 1.4 × longer than broad, lateral extremities subparallel; sternite 3 moderately depressed; anterior margin shallowly excavated, bearing deep U-shaped median notch without flanking spine, anterolaterally rounded or somewhat angular with obsolescent denticles; anterolateral margin of sternite 4 denticulate, nearly straight, about as long as posterolateral margin. Anterolateral margin of sternite 5 shorter (0.7-0.8 x) than posterolateral margin of sternite 4. Abdominal somite 1 with well-elevated transverse ridge; somite 2 tergite 2.0-2.2 × broader than long, pleural lateral margin moderately concave, slightly divergent posteriorly; pleuron of somite 3 laterally blunt. Telson 0.4 × as long as broad; posterior plate distinctly emarginate, length 1.3 × that of anterior plate. Eyes short, barely reaching proximal third of rostrum. Ultimate antennular article 3 × longer than high. Antennal peduncle having distal 2 articles each mesially with distoventral spine; article 5 1.4 × as long as article 4, breadth 0.6 × height of antennule ultimate article; antennal scale barely reaching apex of article 5; flagellum of 9-10 segments very short, about as long as distal 2 articles combined. Mxp1 with bases close to each other. Mxp3 basis with 3-5 obsolescent denticles; ischium with 3-5 small spines lateral to distal end of flexor margin, crista dentata with 10-12 distally diminishing denticles; merus 1.7-2.0 × longer than ischium, flattish on mesial face, with distinct distolateral spine, flexor margin with a few distinct spines distal to midlength; carpus with distinct distolateral spine and small spines on lateral face. P1 somewhat depressed distally, covered dorsally and ventrally with elevated short ridges bearing a few to several small spines or tubercle-like spines (median one larger in small female) and short setae, laterally and mesially with distinct spines usually of small size; ischium with strong dorsal spine with laterally and dorsally accompanying small spines, ventromesial margin with strong subterminal spine followed by row of small spines; merus 1.0-1.1 × longer than carapace; carpus 1.3-1.4 × as long as merus; palm as long as or slightly longer than carpus, height 0.8 × breadth in midlength; fingers depressed, slightly incurved distally, movable finger with 2 processes on opposable margin in both sexes, length 0.4 × that of palm. P2-4 broad relative to length, bearing fine setae much longer than those on P1; meri successively shorter posteriorly (P3 merus 0.8 × length of P2 merus, P4 merus 0.8 × length of P3 merus), equally broad on P2-4; length-breadth ratio, 3.6-3.8 (3.3) on P2, 2.8-3.0 (2.6) on P3, 2.3-2.5 on P4; P2 merus 0.7-0.8 × length of carapace, 1.4-1.6 × length of P2 propodus; P3 merus subequal to length of P3 propodus; P4 merus 0.8-0.9 × length of P4 propodus; meri covered with denticles on lateral surface, mesial face flattish, dorsal margin sharply ridged bearing row of sharp spines with subparalleling small spines directly lateral to it, distal portion of ventrolateral margin with several spines as equally sharp as but larger than those on extensor margin, terminal spine strongest; carpi subequal on P2-3, shorter on P4, or successively shorter posteriorly; each less than half (0.4) length of propodus on P2-4, extensor margin with 7 spines (rarely 9 on P4); propodi successively longer posteriorly; flexor margin straight, with pair of distal spines preceded by smaller, proximally diminishing spines or robust setae (5-8 on P2, 4-7 on P3, 2-4 on P4); lateral and mesial faces with dense tufts of short setae arranged in 2 rows; dactyli short relative to breadth, less than half as long as propodi, dactylus-carpus length ratio, 0.8-1.0 on P2, 1.0 on P3-4; flexor margin with row of somewhat inclined spines or robust setae (14-16 on P2, 10 on P3 and P4), penultimate much broader, fully 2 × broader than antepenultimate, ultimate somewhat broader and distinctly longer than antepenultimate, remaining spines slender and more like setae.

PARASITES — The smaller female from MUSORSTOM 2 Stn CP51 (MNHN-IU-2014-16860) bears a rhizocephalan externa.

REMARKS — This species is grouped together with *U. naso* Van Dam, 1933 and *U. cyrano* Poore & Andreakis, 2011. The relationships between *U. naso* and *U. pinocchio* are discussed under *U. naso* (see above).

Uroptychus plautus n. sp.

Figures 197, 198

TYPE MATERIAL — Holotype: **Indonesia**, Kai Islands. KARUBAR Stn CC10, 5°21'S, 132°30'E, 329-389 m, with coral Primnoidae (Calcaxonia), 23.X.1991, ov. ♀ 6.8 mm (MNHN-IU-2014-16861).

ETYMOLOGY — From the Latin *plautus* (broad), alluding to the broad rostrum and broad sternal plastron of the species.

DISTRIBUTION — Kai Islands, Indonesia; in 329-389 m.

DESCRIPTION — Medium-sized species. *Carapace*: 1.2 × broader than long (0.85 × as long as broad); greatest breadth 2.0 × distance between anterolateral spines. Dorsal surface finely granulate (discernible under high magnification), sparingly with short soft setae, anteriorly elevated from level of rostrum in profile, bearing depression between gastric and cardiac regions. Lateral margins convexly divergent, somewhat ridged along posterior portion; anterolateral corner angular, ending in very tiny spine not reaching lateral orbital angle. Rostrum broader than long, distally blunt triangular, with interior angle of 42°; dorsal surface horizontal, somewhat concave; length 0.3 × that of remaining carapace, breadth half carapace breadth at posterior carapace margin. Lateral limit of orbit acuminate, slightly anterior and relatively close to anterolateral spine. Pterygostomian flap anteriorly roundish, without distinct spine.

Sternum: Excavated sternum with anterior margin broad, blunt triangular, surface ridged in midline. Sternal plastron 1.5 × as broad as long, lateral extremities convexly divergent posteriorly. Sternite 3 shallowly depressed, anterior margin moderately concave with small median notch lacking flanking spine, laterally rounded. Sternite 4 short relative to breadth; anterolateral margin anteriorly rounded, length 1.3 × that of posterolateral margin. Anterolateral margin of sternite 5 1.2 × longer than posterolateral margin of sternite 4.

Abdomen: Sparsely setose; Somites long relative to breadth. Somite 1 dorsally convex from anterior to posterior. Somite 2 tergite 2.1 × broader than long; pleural lateral margin somewhat concavely divergent posteriorly, with blunt posterolateral terminus. Pleuron of somite 3 laterally blunt. Telson half as long as broad; posterior plate slightly longer than anterior plate, posterior margin distinctly emarginate.

Eye: Relatively small, elongate (2.2 × longer than broad), slightly broadened proximally, slightly overreaching midlength of rostrum. Cornea slightly dilated, about half as long as remaining eyestalk.

Antennule and antenna: Ultimate article of antennule 3.8 × longer than high. Antennal peduncle overreaching cornea, not reaching apex of rostrum. Article 2 distolaterally angular and acuminate. Antennal scale 2.0 × broader than article 5, slightly falling short of midlength of article 5. Distal 2 articles unarmed. Article 5 1.5 × longer than article 4, breadth 0.8 × height of antennular ultimate article. Flagellum consisting of 10 segments, slightly falling short of distal end of P1 merus.

Mxp: Mxp1 with bases broadly separated. Mxp3 setose. Basis without denticles on mesial ridge. Ischium with about 30 distally diminishing denticles on crista dentata, flexor margin distally not rounded. Merus 2.2 × longer than ischium, relatively thick, unarmed, flexor margin roundly ridged. Carpus unarmed.

P1: Left P1 missing. Right P1 short, not massive, 3.5 × longer than carapace; sparingly with soft fine setae, spineless except for ischium. Ischium with short, basally broad, depressed dorsal spine, ventrally unarmed. Merus 0.85 × length of carapace. Carpus 1.1 × length of merus. Palm 4.0 × longer than broad, as long as carpus. Fingers not gaping, each ending in small incurved spine; fixed finger feebly sinuous; movable finger slightly less than half length of palm; opposable margin with 2 low processes, proximal one fitting to longitudinal groove on opposite face of fixed finger.

P2-4: Left P2 and right P3 missing. Relatively thick, not compressed mesio-laterally, setose like P1. Meri successively shorter posteriorly (P3 merus 0.96 × length of P2 merus, P4 merus 0.90 × length of P3 merus), equally broad on P2-4; length-breadth ratio, 4.5 on P2, 4.3 on P3, 3.8-3.9 on P4; unarmed; P2 merus 0.8 × length of carapace, slightly (1.06 x) longer than P2 propodus; P3 merus as long as P3 propodus; P4 merus 0.9 × length of P4 propodus. Carpi successively slightly shorter posteriorly, 1.3 × longer than dactyli on P2-4; carpus-propodus length ratio, 0.40 on P2, 0.36 on P3, 0.31

FIGURE 197

Uroptychus plautus n. sp., holotype, ovigerous female 6.8 mm (MNHN-IU-2014-16861). **A**, carapace and anterior part of abdomen, dorsal. **B**, same, lateral. **C**, anterior part of carapace, dorsal. **D**, excavated sternum and basal parts of Mxp1. **E**, anterior part of sternal plastron. **F**, telson. **G**, left antenna, ventral. **H**, left Mxp3, ventral. **I**, same, lateral. **J**, right P1, dorsal. **K**, same, fingers, setae omitted, ventral. Scale bars: 1 mm.

on P4. Propodi slightly shorter on P2 than on P3 and P4; flexor margin nearly straight, with pair of very tiny, vestigial terminal spines only. Dactyli subequal, slightly curving, 0.8 × as long as carpi and 0.3 × as long as propodi on P2-4, bearing 2 distal spines, terminal one shorter and more slender than subterminal spine.

Eggs. Number of eggs carried, 4; size, 1.5 × 1.7 mm.

REMARKS — The P2-4 dactyli bearing only two terminal spines are very unusual in the species of *Uroptychus* but possessed by *U. pilosus* Baba, 1981, *U. inaequalis* n. sp. (see above) and *U. plautus* n. sp. *Uroptychus plautus* is distinguished from *U. pilosus* by the rostrum that is much broader, with the interior angle of 42° instead of 23°; the anterolateral corner of the carapace is distinctly angular instead of roundish; the dorsal margins of P2-4 meri are unarmed instead of bearing a row of small spines at least on the proximal portion; the pterygostomian flap is anteriorly roundish instead of angular bearing a sharp spine; the antennal article 2 is distolaterally angular and acuminate instead of bearing a distinct spine; and the ultimate spine on the P2-4 dactyli is smaller than instead of subequal to the penultimate spine. The new species is differentiated from *U. inaequalis* by the following particulars: the P2-4 dactyli are shorter instead of longer than carpi; the P4 merus is 0.90 instead of 0.75 times length of P4 propodus; the Mxp3 merus is unarmed instead of bearing distolateral and flexor marginal spines; the pterygostomian flap is anteriorly roundish instead of produced to a distinct spine; the antennal article 2 is distolaterally acuminate instead of bearing a strong spine; the anterolateral corner of carapace is angular instead of roundish; and the sternal plastron is 1.5 times broader than instead of as broad as long.

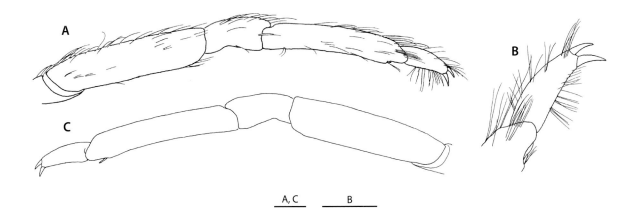

FIGURE 198

Uroptychus plautus n. sp., holotype, ovigerous female 6.8 mm (MNHN-IU-2014-16861). **A**, right P2, lateral. **B**, same, distal part, lateral. **C**, left P3, setae omitted, lateral. Scale bars: 1 mm.

Uroptychus plumella n. sp.

Figures 199, 200

TYPE MATERIAL — Holotype: **New Caledonia**, Loyalty Ridge. BATHUS 3 Stn DW778, 24°43'S, 170°07'E, 750-760 m, 24.XI.1993, 1 ov. ♀ 6.6 mm (MNHN-IU-2012-683).

ETYMOLOGY — The specific name is a noun in apposition from the Latin *plumella*, diminutive of *pluma* (= soft feather), alluding to short soft plumose setae covering carapace and pereopods.

DISTRIBUTION — Loyalty Ridge, 750-760 m.

DESCRIPTION — Small species. *Carapace*: 1.3 × broader than long (0.8 × as long as broad), broadest on posterior fourth; greatest breadth 2.1 × distance between anterolateral spines. Dorsal surface convex smoothly from anterior to posterior and from side to side, without depression, thickly covered with short fine plumose setae. Lateral margins convexly divergent, finely granulose, and weakly ridged along posterior fourth; anterolateral spine distinctly larger than lateral orbital spine, directed forward. Rostrum horizontal in proximal three-quarters, upcurved in distal quarter, narrow triangular, with interior angle 22°; dorsal surface somewhat concave; lateral margin straight with 1 (right) or 3 (left) vestigial denticles distally; length half that of remaining carapace, breadth one-third carapace breadth at posterior carapace margin. Lateral orbital angle acuminate, situated directly mesial to and separated from anterolateral spine by basal breadth of latter spine. Pterygostomian flap smooth on surface, anteriorly angular, produced to small spine; anterior half 1.3 × higher than posterior half.

Sternum: Excavated sternum with convex anterior margin, surface with weak low ridge in midline. Sternal plastron 0.7 × as long as broad; lateral extremities divergent posteriorly. Sternite 3 shallowly depressed; anterior margin deeply emarginate, medially V-shaped, without submedian spines. Sternite 4 with anterolateral margin convex and denticulate, about 2 × longer than posterolateral margin. Anterolateral margin of sternite 5 anteriorly convex and denticulate, 1.3 × longer than posterolateral margin of sternite 4.

Abdomen: Somite 1 somewhat convex from anterior to posterior, without transverse ridge. Somite 2 tergite 2.6 × broader than long; pleural lateral margin concavely divergent posteriorly, posterolateral end blunt. Pleuron of somite 3 laterally blunt. Telson 0.4 × as long as broad; posterior plate slightly longer than anterior plate, emarginate on posterior margin.

Eye: 1.4 × longer than broad, not reaching midlength of rostrum, mesial and lateral margins subparallel. Cornea not inflated, slightly shorter than remaining eyestalk.

Antennule and antenna: Ultimate article of antennule 4.2 × longer than high. Antennal peduncle overreaching cornea, overreaching midlength of rostrum. Article 2 distolaterally acuminate, lacking distinct spine. Antennal scale 2.2 × broader than article 5, extending slightly beyond proximal second segment of flagellum and greatly overreaching eye. Articles 4 and 5 each with tiny distomesial spine (that of article 4 tubercle-like), article 5 1.4 × longer than article 5, breadth 0.6 × height of antennular ultimate article. Flagellum of 16 segments reaching distal margin of P1 merus.

Mxp: Mxp1 with bases moderately separated. Mxp3 with relatively long plumose setae other than brushes on distal articles. Basis having mesial ridge nearly smooth. Ischium with flexor margin not rounded distally; crista dentata with denticles widely spaced along most of length, minute and closely set on distal part. Merus 2.2 × longer than ischium, flexor margin roundly ridged, with 4 tiny spines or processes along distal third. Distodorsal spine on each of merus and carpus.

P1: 4.5 × longer than carapace, sparingly covered with relatively short fine plumose setae. Ischium with small dorsal spine, ventromesial margin with row of denticle-like spines, lacking distinct subterminal spine. Merus with small distomesial and distolateral spines, row of denticles along distodorsal margin, granulate along proximal mesial margin; length 0.9 × that of carapace. Carpus subcylindrical, 1.5 × longer than merus, distodorsal margin with small but distinct median spine flanked by denticles. Palm 3.7 × longer than broad, 1.1 × longer than carpus. Fingers moderately

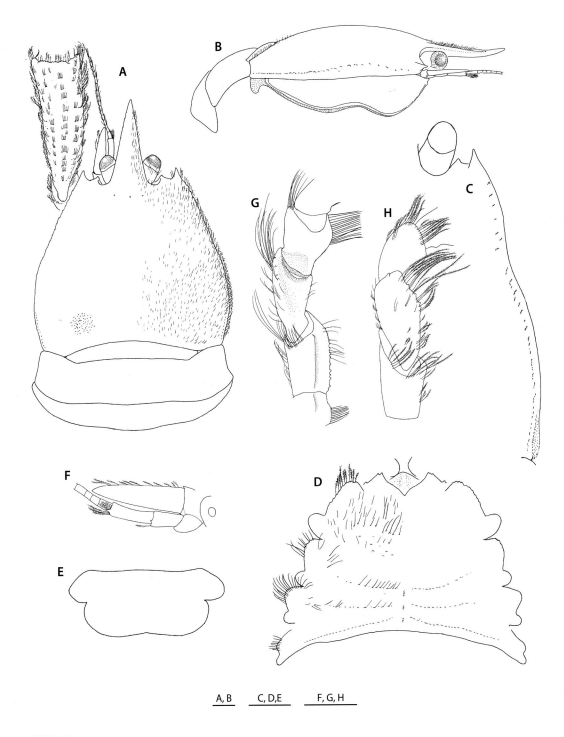

FIGURE 199

Uroptychus plumella n. sp., holotype, ovigerous female 6.6 mm (MNHN-IU-2012-683). **A**, carapace and anterior part of abdomen, proximal part of left P1 included, dorsal. **B**, same, setae on carapace omitted, lateral. **C**, lateral margin of carapace, dorsolateral. **D**, sternal plastron, with excavated sternum and basal parts of Mxp1. **E**, telson. **F**, left antenna, ventral. **G**, right Mxp3, ventral. **H**, same, lateral. Scale bars: 1 mm.

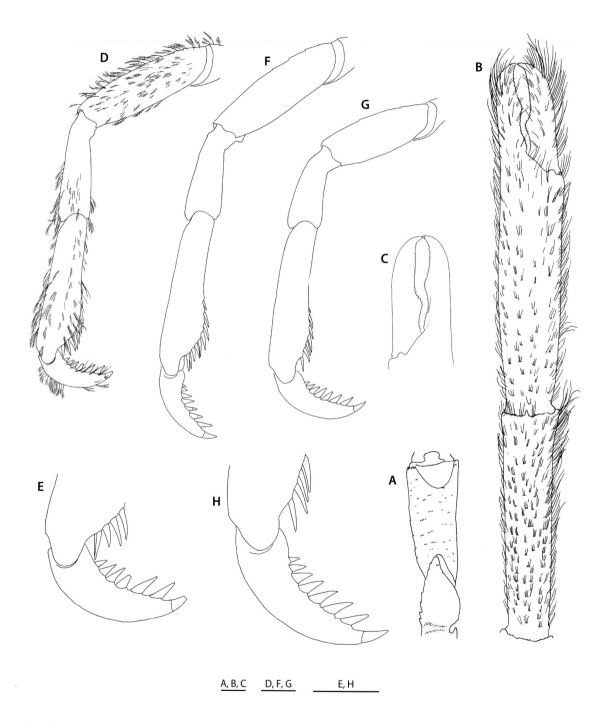

FIGURE 200

Uroptychus plumella n. sp., holotype, ovigerous female 6.6 mm (MNHN-IU-2012-683). **A**, left P1, proximal part, setae omitted, ventral. **B**, same, distal three articles, dorsal. **C**, same, fingers, setae omitted, ventral. **D**, left P2, lateral. **E**, same, distal part, setae omitted, lateral. **F**, left P3, setae omitted, lateral. **G**, left P4, setae omitted, lateral. **H**, same, distal part, setae omitted, lateral. Scale bars: 1 mm.

depressed and relatively broad, distally incurved, crossing when closed, not gaping; fixed finger with low process at midlength of opposable margin; movable finger half as long as palm, opposable margin with larger process proximal to position of opposite process on fixed finger.

P2-4: Relatively broad and somewhat compressed mesio-laterally, with fine soft setae. Meri slightly shorter on P3 than on P2, much shorter on P4 than on P3 (P4 merus 0.8 length of P3 merus); breadths slightly smaller on P2 than on P3 and P4; length-breadth ratio, 3.6 on P2, 3.2 on P3, 2.5 on P4; dorsal margins rounded, not cristiform, with a few spines very tiny on proximal half on P2 and P3, obsolescent on P4. P2 merus 0.7 × as long as carapace, 1.0 × (left) or 0.9 × (right) length of P2 propodus; P3 merus 0.8 × (left) or 1.0 × (right) length of P3 propodus; P4 merus 0.6 × length of P4 propodus. Carpi successively shorter posteriorly; carpus-propodus length ratio, 0.6 on P2, 0.4 (left) or 0.5 (right) on P3, 0.5 on P4. Propodi shortest on P2, longer and subequal on P3 and P4; flexor margin moderately convex on distal portion, ending in pair of terminal spines preceded by 8, 7, 6 spines on P2, P3, P4 respectively. Dactyli shortest on P2, subequal on P3 and P4; moderately curving; dactylus-carpus length ratio, 0.7, 0.7-0.8, 1.1 on P2, P3, P4 respectively; flexor margins with 7, 8, 8 strong, proximally diminishing, obliquely directed spines on P2, P3, P4 respectively.

Eggs. Five eggs carried; size, 1.29 × 1.13 mm - 1.33 × 1.25 mm.

REMARKS — *Uroptychus plumella* n. sp. is characterized, along with *U. senticarpus* n. sp. and *U. shanei* n. sp., by the combination of the following characters: the carapace longer than broad, laterally unarmed other than the anterolateral spine; sternite 3 with the anterior margin emarginate in a broad V-shape; P1 nearly spineless, with the fingers distally strongly incurved; the P2-4 dactyli ending in a strong spine preceded by similar, proximally diminishing spines. However, the plumose setae on the carapace and P1 as displayed by this species are not possessed by the two related species. Characters to distinguish these species are outlined under the accounts of the related species (see below).

Uroptychus politus (Henderson, 1885)

Figures 201, 202

Diptychus politus Henderson, 1885: 420.
Uroptychus politus — Henderson 1888: 178, pl. 6: figs 2, 2a, 2b. — Thomson 1899: 196 (list). — Baba 1974: 387, fig. 5; 2005: 219 (key), 230 (list). — Schnabel 2009: 564 (examination of female syntype).

TYPE MATERIAL — Syntype: **New Zealand**, north of the Kermadec Islands, 600 fms (1098 m), female (BMNH 1888: 33). [examined].

MATERIAL EXAMINED — **New Zealand**, Kermadec Islands. CHALLENGER Stn 171, 28°33'S, 177°50'W, 600 fms (1098 m), 1 ov. ♀ 5.4 mm, syntype (BMNH 1888: 33). **Solomon Islands**. SALOMON 2 Stn CP2197, 8°24.2'S, 159°22.5'E, 897-1057 m, 24.X.2004, 2 ♂ 5.3, 7.5 mm, 1 ov. ♀ 7.2 mm (MNHN-IU-2014-16862). **New Caledonia**, Loyalty Islands. BIOGEOCAL Stn CP297, 20°37'S, 167°11'E, 1230-1240 m, 28.IV.1987, 1 ♂ 6.3 mm (MNHN-IU-2014-16863).

DISTRIBUTION — Kermadec Islands, and now Solomon Islands and Loyalty Islands; in 897-1240 m.

SIZE — Males, 5.3-7.5 mm; ovigerous females, 5.4-7.2 mm.

DESCRIPTION — Medium-sized species. *Carapace*: Smooth, without setae and spines, 1.1 × longer than broad; greatest breadth 1.7 × distance between anterolateral spines. Dorsal surface somewhat convex, with or without depression between gastric and cardiac regions. Lateral margins convexly divergent posteriorly (divergent to point one-fifth from posterior end, then convergent), ridged along posterior half; short, elevated, granulated ridge bearing small spine

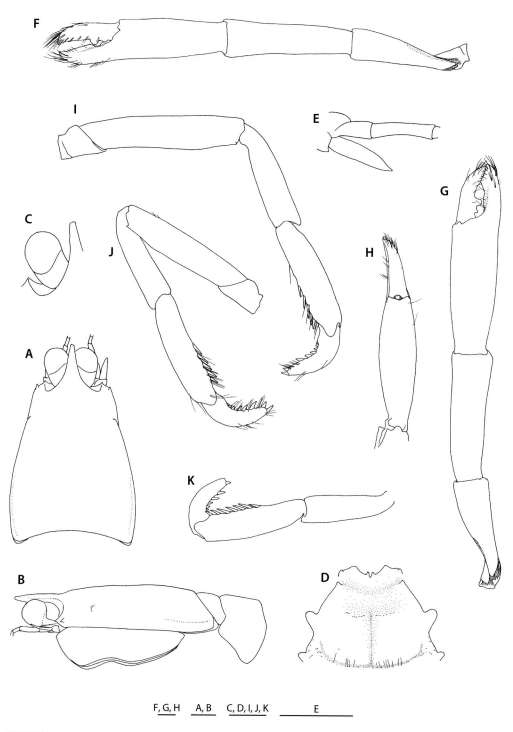

F, G, H   A, B   C, D, I, J, K   E

FIGURE 201

Uroptychus politus (Henderson, 1885), **A-F**, syntype, ovigerous female 5.4 mm (BMNH 1888: 33); **G, H**, possibly male syntype; **I-K**, pereopods detached from body. **A**, carapace, dorsal. **B**, same, anterior part of abdomen included, lateral. **C**, anterior part of carapace, showing left eye, dorsolateral. **D**, anterior part of sternal plastron. **E**, left antenna, ventral. **F**, left P1, dorsal. **G**, right P1, dorsal. **H**, same, distal part, mesial. **I**, right P2, lateral. **J**, left P3, lateral. **K**, right P3, distal 3 articles, lateral. Scale bars: 1 mm.

(obsolete in holotype) at one-fifth length from anterior end (anterior end of branchial region); anterolateral spine small, distinctly posterior to level of lateral orbital spine, barely or fully reaching tip of that spine. Rostrum narrow triangular, with interior angle of 18-21°, horizontal; dorsal surface somewhat concave; length about half that of remaining carapace, breadth slightly less than half carapace breadth at posterior carapace margin. Lateral orbital spine somewhat smaller than anterolateral spine. Pterygostomian flap anteriorly roundish with or without very small spine.

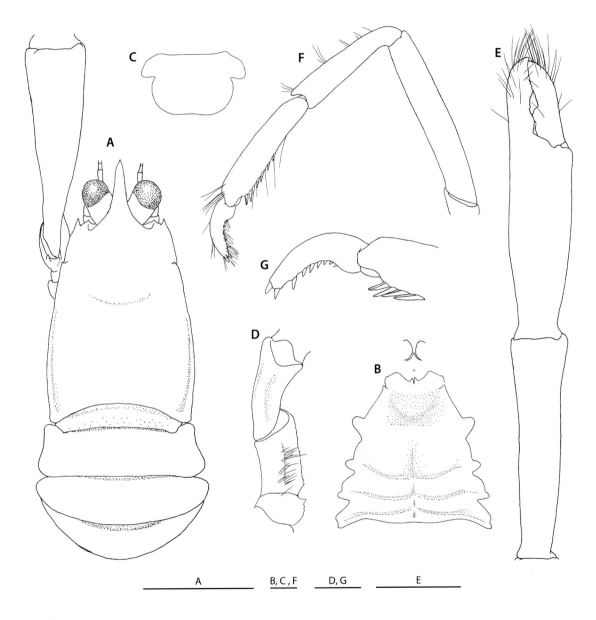

FIGURE 202

Uroptychus politus (Henderson, 1885), male 7.5 mm (MNHN-IU-2014-16862). **A**, carapace and anterior part of abdomen, proximal part of left P1 included, dorsal. **B**, sternal plastron, with excavated sternum and basal parts of Mxp1. **C**, telson. **D**, right Mxp3, ventral. **E**, left P1, dorsal. **F**, left P2, lateral. **G**, same, distal part, setae omitted, lateral. Scales bars: A, E, 5 mm; B, C, D, F, G, 1 mm.

Sternum: Excavated sternum strongly produced anteriorly between bases of Mxp1, bearing small spine in center of surface. Sternal plastron slightly broader than long, lateral margins somewhat divergent posteriorly. Sternite 3 moderately depressed; anterior margin moderately concave, with submedian spines nearly contiguous at base or separated by small notch. Sternite 4 long relative to width, surface with transverse ridge preceded by depression; anterolateral margin nearly straight, anteriorly ending in short process, length 1.6-1.8 × that of posterolateral margin. Sternite 5 two-thirds as long as sternite 4, anterolateral margins feebly convex and moderately divergent posteriorly, about as long as posterolateral margin of sternite 4.

Abdomen: Smooth and glabrous, lacking setae. Somite 1 somewhat convex from anterior to posterior. Somite 2 tergite 2.3-2.5 × broader than long; pleuron posterolaterally blunt, lateral margin feebly concave and strongly divergent posteriorly. Pleuron of somite 3 posterolaterally bluntly angular. Telson slightly more than half (0.56-0.66) as long as broad; posterior plate 1.7-1.8 × longer than anterior plate, posterior margin moderately emarginate or slightly concave.

Eye: Short relative to breadth, reaching at most distal quarter of rostrum, mesial margin somewhat concave. Cornea more than half that of remaining eyestalk.

Antennule and antenna: Ultimate article of antennule 2.7-3.2 × longer than high. Antennal peduncle relatively slender, overreaching cornea. Article 2 distolaterally angular, without distinct spine. Antennal scale 1.4 × as broad as article 5, overreaching distal end of article 4, falling short of midlength of article 5. Distal 2 articles unarmed; article 5 1.7-1.9 × length of article 4, breadth 0.4 × height of antennular ultimate article. Flagellum consisting of 14-18 segments, slightly falling short of distal end of P1 merus.

Mxp: Mxp1 with bases close to each other. Endopod of Mxp3 barely setose laterally, slender. Basis with a few denticles on mesial ridge, distalmost distinct, others obsolescent. Ischium having crista dentata with 5-10 (7 in holotype) small denticles rather distant from one another; flexor margin distally not rounded. Merus 1.9 × longer than ischium, unarmed; flexor margin roundly ridged. Carpus also unarmed.

P1: Smooth, nearly glabrous except for fingers, length 4.3-5.6 × that of carapace. Ischium with short dorsal spine, ventromesially unarmed. Merus 1.2 × longer than carapace. Carpus 2.6-4.0 × longer than broad, 1.2-1.3 × longer than merus. Palm 2.7-3.0 × longer than broad, about as long as carpus. Fingers relatively broad, slightly incurved distally, gaping in both sexes (not gaping in female syntype), opposable margins straight in distal quarter when gaping, in distal half when not gaping; movable finger about half as long as palm, opposable margin with 2 distinct proximal processes opposed to low process on fixed finger.

P2-4: Relatively thick mesio-laterally, sparsely setose but barely so on meri. Meri successively shorter posteriorly (P3 merus 0.9 × length of P2 merus, P4 merus 0.9 × length of P3 merus), equally broad on P2-4; unarmed; length-breadth ratio, 6.1-6.9 (5.4 in syntype) breadth on P2, 5.7-6.6 (4.3 in syntype) on P3, 5.6-6.0 × on P4; P2 merus 0.87-0.98 × as long as carapace, 1.4 × longer than P2 propodus; P3 merus 1.1 × length of P3 propodus; P4 merus 1.0-1.1 × length of P4 propodus. Carpi successively shorter posteriorly (P3 carpus 0.9 × length of P2 carpus, P4 carpus subequal to or slightly shorter than P3 carpus); much longer than dactyli (carpus-dactylus length ratio, 2.1-2.3 on P2, 1.9-2.0 on P3, 1.6-1.7 on P4); carpus-propodus length ratio, 0.8-0.9 on P2, 0.7-0.8 on P3, 0.6-0.7 on P4. Propodi subequal on P2-4; flexor margin with pair of terminal spines slightly distant from juncture with dactylus, preceded by 6-7 spines on P2, 5-6 spines on P3, 3-5 spines on P4 at most on distal half. Dactyli strongly curving at proximal third, ending in strong spine preceded by 8-10 successively diminishing, somewhat inclined spines on flexor margin on P2, 9 or 10 spines on P3, 10 or 11 spines on P4, ultimate slightly longer than penultimate, antepenultimate closer to distal quarter than to penultimate; length 0.4 × that of propodus on P2-4, and 0.5 × that of carpi on P2 and P3, 0.6 × on P4.

Eggs. Number of eggs carried, 16 (2 in syntype); size, 2.18 mm × 2.22 mm - 2.22 mm × 2.43 mm (1.50 × 1.92 mm in syntype).

REMARKS — Henderson (1885) reported one male from north of the Kermadec Islands, and in his extensive description of the Challenger material, he provided an additional female. In an earlier paper (Baba, 1974), the illustration of the female syntype drawn by R. W. Ingle was provided under "holotype." At that time Ingle did not find the male syntype so that it was called the holotype. In the lot now available of the syntypes (BM 1888:33), a pair of P1 (possibly of the male, due to its massiveness) and a left P1 (possibly of the female) are present. Left two and right two walking legs, and left Mxp3 removed from the ovigerous female syntype are available (Figure 201).

The syntype has relatively broad P2 and P3, the length-breadth ratios of meri and propodi being somewhat greater than those of the other specimens examined. In addition, the anterolateral spine is as small as the lateral orbital spine and does not reach the tip of that spine, whereas in the present material the spine is larger than and fully reaches the tip of the lateral orbital spine. These difference are here considered as intraspecific variations. The species is very characteristic in the shape of the sternite 5, the anterolateral margin of which is very weakly convex instead of well convex as in most of the other known species.

The species is close to *U. salomon* n. sp. described below. Their relationships are discussed under that species.

Uroptychus pollostadelphus n. sp.

Figures 203, 204

TYPE MATERIAL — Holotype: **New Caledonia**, Norfolk Ridge. NORFOLK 2 Stn DW2103, 23°56.96'S, 167°43.70'E, 717-737 m, 30.X.2003, ♂ 6.8 mm (MNHN-IU-2014-16864).

ETYMOLOGY — From the Greek *pollostos* (smallest) plus *adelphos* (twin), alluding to a pair of very small epigastric spines.

DISTRIBUTION — Norfolk Ridge; 717-737 m.

DESCRIPTION — *Carapace*: Nearly as long as broad; greatest breadth 1.7 × distance between anterolateral spines. Dorsal surface smooth and glabrous, slightly convex from anterior to posterior; epigastric region with pair of very small spines. Lateral margins convexly divergent posteriorly; anterolateral spine small, reaching tip of lateral orbital spine; denticulate short ridge on anterior end of branchial region, followed by row of obsolescent denticles. Rostrum straight horizontal, narrow triangular, with interior angle of 20°; dorsal surface flattish; length about half that of remaining carapace, breadth slightly less than half carapace breadth at posterior carapace margin. Lateral limit of orbit angular, ending in very small spine, distinctly anterior to level of anterolateral spine. Pterygostomian flap nearly smooth on surface, anteriorly roundish, with angular end.

Sternum: Excavated sternum anteriorly sharp triangular, surface with small spine in center. Sternal plastron slightly shorter than broad, lateral extremities gently divergent posteriorly. Sternite 3 moderately depressed, with broad V-shaped anterior margin bearing 2 submedian spines separated by narrow notch. Sternite 4 having anterolateral margin slightly convex, anteriorly ending in blunt but distinct process, slightly more than 1.5 × as long as posterolateral margin. Sternite 5 having anterolateral margin moderately convex, as long as posterolateral margin of sternite 4.

Abdomen: Smooth and glabrous. Somite 1 well convex from anterior to posterior. Somite 2 tergite 2.6 × broader than long; pleuron posterolaterally bluntly angular, lateral margin concave and moderately divergent posteriorly. Pleuron of somite 3 tapering to angular tip. Telson 0.6 × as long as broad; posterior plate emarginate on posterior margin, length 1.7 × that of anterior plate.

Eye: 1.7 × longer than broad, overreaching midlength of rostrum, mesial margin concave, lateral margin convex. Cornea slightly inflated, length subequal to that of remaining eyestalk.

Antennule and antenna: Ultimate article of antennule 2 × longer than high. Antennal peduncle slightly overreaching cornea. Article 2 with small but distinct distolateral spine. Antennal scale reaching distal end of article 5, 1.8 × broader than article 5. Distal 2 articles unarmed. Article 5 twice as long as article 4, breadth about one-third height of antennular ultimate article. Flagellum consisting of 16 segments, barely reaching distal end of P1 merus.

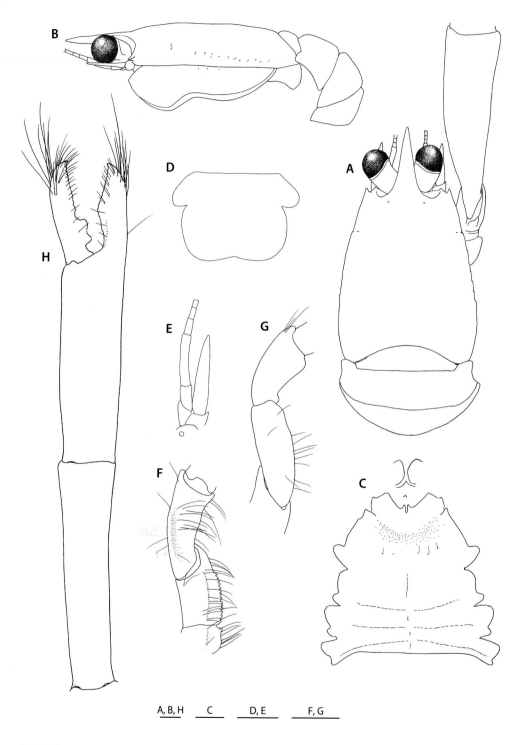

FIGURE 203

Uroptychus pollostadelphus n. sp., holotype, male 6.8 mm (MNHN-IU-2014-16864). **A**, carapace and anterior part of abdomen, proximal part of right P1 included, dorsal. **B**, same, lateral. **C**, sternal plastron, with excavated sternum and basal parts of Mxp1. **D**, telson. **E**, left antenna, ventral. **F**, right Mxp3, ventral. **G**, same, lateral. **H**, right P1, dorsal. Scale bars: 1 mm.

Mxp: Mxp1 with bases close to each other. Mxp3 sparsely setose on lateral surface. Basis with 1 denticle on mesial ridge. Ischium with 14 denticles on crista dentata, flexor margin not rounded at distal end. Merus 2.2 × longer than ischium, moderately ridged along flexor margin, and unarmed. Carpus also unarmed.

P1: 5.3 × longer than carapace, smooth and barely setose except for fingers. Ischium with short, subtriangular dorsal spine; ventromesial margin bearing a few tubercles, lacking subterminal spine. Merus 1.25 × longer than carapace, with low blunt distomesial and distolateral processes ventrally. Carpus unarmed, length 1.3 × that of merus. Palm 3.0 × longer than broad, slightly shorter than carpus. Fingers relatively broad distally, ending in small, somewhat incurved spine, slightly crossing when closed; movable finger half as long as palm, opposable margin with basally broad (disto-proximally), distally bilobed process; fixed finger with median prominence on opposable margin.

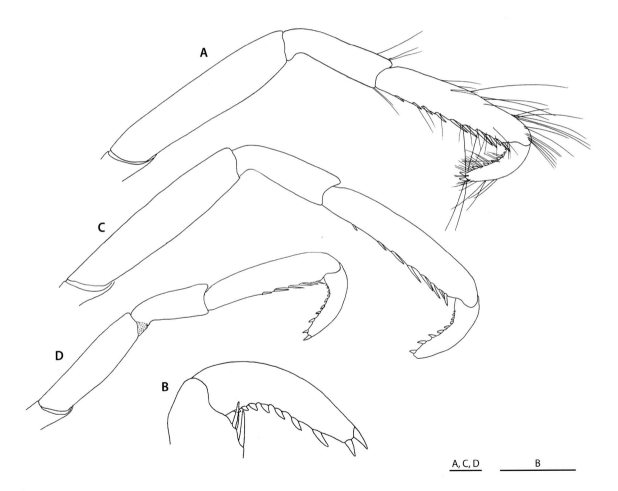

A, C, D _____   B _____

FIGURE 204

Uroptychus pollostadelphus n. sp., holotype, male 6.8 mm (MNHN-IU-2014-16864). **A**, right P2, lateral. **B**, same, distal part, setae omitted, lateral. **C**, right P3, setae omitted, lateral. **D**, right P4, lateral. Scale bars: 1 mm.

P2-4: Slender. Meri mesio-laterally compressed, unarmed; P2 merus slightly shorter than carapace, subequal to P3 merus in length and breadth, 1.2 × longer than P2 propodus; P3 merus subequal to length of P3 propodus; P4 merus much smaller than P3 merus (0.6 × length and 0.7 × breadth of P3 merus), 0.8-0.9 × length of P4 propodus; length-breadth ratio, 4.9 on P2, 4.5-4.7 on P3, 3.6-3.8 on P4. Carpi subequal in length on P2 and P3, shorter on P4; carpus-propodus length ratio, 0.5 on P2 and P3, 0.4-0.5 on P4; carpus-dactylus length ratio, 1.5 on P2, 1.3 on P3, 1.0 on P4. Propodi longest on P3, shortest and narrowest on P4; flexor margin straight, with pair of terminal spines preceded by 9 spines along entire length on P2, 7 spines on P3, 4 spines on P4. Dactyli longest on P3, shortest on P4, shorter than carpi on P2 and P3, about as long on P4; dactylus-propodus length ratio, 0.40 on P2 and P3, 0.46-0.47 on P4; flexor margin strongly curving at proximal third, with 9 or 10 triangular, moderately inclined spines diminishing toward base of article, ultimate strongest, penultimate close to ultimate, antepenultimate more proximal to midlength between penultimate and distal quarter.

REMARKS — This species strongly resembles *U. sagamiae* Baba, 2005 in nearly all details, from which it is differentiated by sternite 4, which is smooth instead of granulose on the posterior surface; the P1 palm is ventrally smooth instead of granulose; and the extensor margins of P2-4 dactyli have no plumose setae as in *U. sagamiae*. Both species are medium-sized, so the above mentioned differences may not be considered as age related variations.

The new species also resembles *U. benthaus* n. sp. (see above) in the carapace ornamentation especially bearing a relatively small anterolateral spine and a pair of small epigastric spines, and in the spination of P2-4 propodi and dactyli. *Uroptychus pollostadelphus* is differentiated from that species by the P2-4 that are broader, *e.g.* the meri having the length-breadth ratio, 4.5-4.9 on P2-3, 3.6-3.8 on P4 instead of 6.1-7.0 on P2-3, 4.8-5.1 on P4; the P4 merus is much shorter, 0.6 instead of 0.7 times the length of P3 merus; and the pterygostomian flap bears the anterior margin almost roundish instead of sharply produced to a distinct spine.

The species also resembles *U. septimus* n. sp. Their relationships are discussed under the account of that species (see below).

Uroptychus poorei n. sp.
Figures 205, 206

TYPE MATERIAL — Holotype: **New Caledonia**, Norfolk Ridge. NORFOLK 1 Stn CP1714, 23°22'S, 168°03'E, 257-269 m, 26.VI.2001, ♂ 3.1 mm (MNHN-IU-2011-5967). Paratypes: **New Caledonia**, Norfolk Ridge. CHALCAL 2 Stn CH03, 24°48'S, 168°09'E, 257 m, 27.X.1986, with *Chironephthya* sp. (Alcyoniina: Nidaliidae), 1 ov. ♀ 2.0 mm (MNHN-IU-2011-5966). SMIB 4 Stn DW45, 24°45'S, 168°08'E, 245-260 m, 08.III.1989, 1 ♂ 2.5 mm (MNHN-IU-2011-5963). Beryx 11 Stn CP17, 24°48.00'S, 168°08.89'E, 250-270 m, 16.X.1992, 1 ov. ♀ 2.9 mm (MNHN-IU-2011-5965).

ETYMOLOGY — Named for Gary C.B. Poore for his friendship and for his leadership to organize projects of squat lobster studies.

DISTRIBUTION — Norfolk Ridge; 245-270 m.

SIZE — Males, 2.5-3.1 mm; ovigerous females, 2.0-2.9 mm.

DESCRIPTION — Small species. *Carapace*: Broader than long (0.8-0.9 × as long as broad); greatest breadth 1.5-1.6 × distance between anterolateral spines. Dorsal surface unarmed, moderately convex from side to side, slightly convex on gastric and cardiac regions in profile, with weak depression between; with distinct ridge along branchial margin. Lateral margins divergent posteriorly, with strong anterolateral spine directed straight forward, overreaching lateral orbital spine, unarmed elsewhere. Rostrum directed slightly ventrally, relatively narrow, sharp triangular, with interior angle of

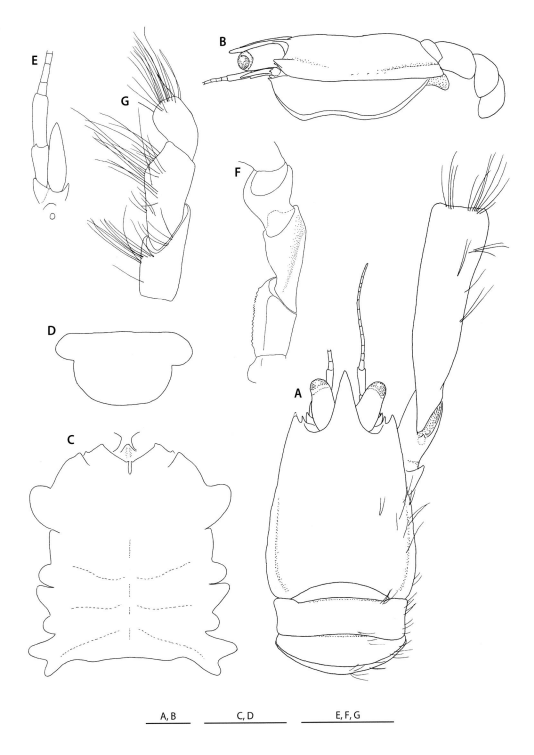

FIGURE 205

Uroptychus poorei n. sp., holotype, male 3.1 mm (MNHN-IU-2011-5967). **A**, carapace and anterior part of abdomen, proximal part of right P1 included, dorsal. **B**, same, setae omitted, lateral. **C**, sternal plastron, with excavated sternum and basal parts of Mxp1. **D**, telson. **E**, left antenna, ventral. **F**, left Mxp3, setae omitted, ventral. **G**, same, lateral. Scale bars: 1 mm.

A, B, C   D, F   E, G   H, I, J

FIGURE 206

Uroptychus poorei n. sp., **A-G**, holotype, male 3.1 mm (MNHN-IU-2011-5967); **H-J**, ovigerous female paratype (MNHN-IU-2011-5966).
A, right P1, proximal part, lateral. **B**, same, proximal part, setae omitted, ventral. **C**, same, distal three articles, dorsal. **D**, right P2,
lateral. **E**, same, distal part, setae omitted, lateral. **F**, right P3, lateral. **G**, same, distal part, lateral. **H**, left P2, lateral. **I**, right P3, lateral.
J, right P4, lateral. Scale bars: 1 mm.

30°, ending in sharp tip; dorsal surface concave; lateral margin smooth; length 0.4 × that of remaining carapace, breadth about half carapace breadth at posterior carapace margin. Lateral orbital spine small, situated directly mesial to antero-lateral spine. Pterygostomian flap anteriorly angular, produced to distinct spine, smooth on surface.

Sternum: Excavated sternum convex or subtriangular on anterior margin, surface with weak ridge in midline. Sternal plastron about as long as broad, lateral extremities subparallel between sternites 4-7. Sternite 3 slightly depressed from level of sternite 4 in ventral view; anterior margin moderately concave in broad V-shape with narrow or broad deep median notch separating small submedian spines. Sternite 4 having anterolateral margin somewhat convex and anteriorly blunt angular; posterolateral margin as long as anterolateral margin. Anterolateral margins of sternite 5 subparallel, anteriorly rounded, 1.5-1.8 × longer than posterolateral margin of sternite 4.

Abdomen: With sparse long setae. Somite 1 convex from anterior to posterior. Somite 2 tergite 2.0-2.5 × broader than long; pleural lateral margin slightly concave, nearly subparallel in dorsal view, posterolaterally rounded. Pleuron of somite 3 with broadly rounded lateral margin. Telson half as long as broad; posterior plate 1.2-1.4 × length of anterior plate, posterior margin gently rounded or convex, not emarginate.

Eye: Long relative to breadth (length twice breadth), distally narrowed, lateral margin convex, slightly falling short of rostral tip. Cornea relatively small, length less than one-third that of remaining eyestalk.

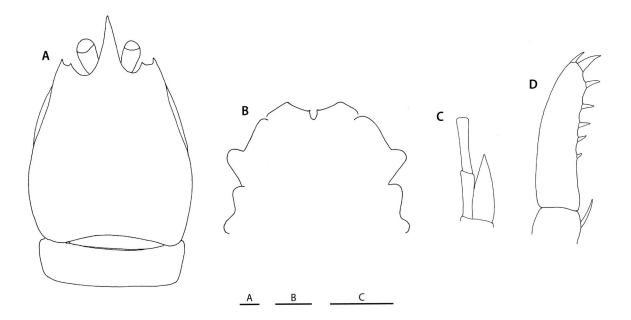

FIGURE 207

Uroptychus bacillimanus Alcock & Anderson, 1899, female syntype. **A**, carapace and anterior part of abdomen, dorsal. **B**, anterior part of sternal plastron. **C**, left antenna, ventral. **D**, right P3, distal part, setae omitted. Scale bars: 1 mm (Illustrated by K.K. Tiwari, 1977).

Antennule and antenna: Ultimate article of antennular peduncle 2.9-3.3 × longer than high. Antennal peduncle overreaching cornea, reaching rostral tip. Article 2 with strong lateral spine. Antennal scale 1.2-1.3 × broader than article 5, fully or barely reaching midlength of article 5 (varying from slightly overreaching article 4 to terminating in midlength of article 5). Article 4 with tiny ventral distomesial spine. Article 5 unarmed, 1.4 × longer than article 4, breadth 0.5-0.6 × height of ultimate article of antennule. Flagellum of 9-14 segments not reaching distal end of P1 merus.

Mxp: Mxp1 with bases broadly separated. Mxp3 basis without denticle on mesial ridge. Ischium with flexor margin rounded distally, crista dentata with about 20 distally diminishing denticles. Merus 2.3 × longer than ischium, without distolateral spine; flexor margin well ridged along distal third of length, without spine; mesial face concave. Carpus unarmed.

P1: 7.2-7.3 × (males), 6.5-6.7 × (females) longer than carapace, slender, with sparse setae of moderate length, smooth without spine. Ischium with small dorsal spine, unarmed on ventral surface and ventromesial margin. Merus 1.6 × longer than carapace. Carpus 1.4 × longer than merus. Palm distally broadened in males, medially somewhat broader in females, 3.7-4.0 × (males), 4.9-6.7 × (females) longer than broad, 1.0 × (males), 0.8-0.9 × (females) length of carpus, somewhat depressed. Fingers slightly crossing when closed, gaping in males, not gaping in females, sparingly with setae of moderate length; movable finger 0.3-0.4 × as long as palm, opposable margin with prominent, proximally lowering process at midpoint of gaping portion; no longitudinal groove on opposite side of fixed finger.

P2-4: Setose (with long setae), relatively thick mesio-laterally; P2 merus 0.8 × length of carapace, 1.0-1.1 × length of P2 propodus; P3 merus 0.9 × length of P2 merus, 0.9 × length of P3 propodus; P4 merus 0.9 × length of P3 merus, 0.9-1.0 × (males), 0.7-0.8 (females) length of P4 propodus; length-breadth ratio, 3.7-3.8 on P2, 3.4 on P3, 3.3-3.5 on P4; dorsal margin not cristate but rounded, unarmed. Carpi subequal, 0.3-0.4 × (P2) or 0.3 × (P3 and P4) as long as propodi, 0.7-0.8 × as long as dactyli. Propodi shorter on P2 than on P3-4; flexor margin nearly straight, with pair of slender terminal spines only. Dactyli relatively stout, 1.2-1.4 × longer than carpi on P2-4, 0.4 × as long as propodi on P2-4; flexor margin nearly straight, with 6 relatively loosely arranged spines, ultimate slender, penultimate prominent, about 2 × broader than antepenultimate, preceded by proximally diminishing spines perpendicular to flexor margin, proximal-most obliquely directed.

Eggs. Number of eggs carried, 2-7; size, 0.82 × 0.98 mm - 0.92 × 1.04 mm.

REMARKS — The carapace shape and the distally narrowed eyes in the new species are very much like those of *U. amabilis* Baba, 1977 from off Noumea, New Caledonia. *Uroptychus poorei* is readily distinguished from that species by the following differences: sternite 3 has the anterior margin deeply excavated representing a broad V-shape instead of being shallowly excavated and medially transverse; sternite 4 has the anterolateral margin as long as instead of much longer than the posterolateral margin; the antennal peduncle has a small distomesial spine on article 4 only instead of a distinct spine on each of articles 4 and 5; the antennal scale terminates at most in the midlength, instead of slightly falling short of the end, of article 5; the Mxp3 merus is unarmed instead of bearing a distinct distolateral spine and a few flexor marginal spines; and the penultimate of the flexor marginal spines of P2-4 dactyli is much stronger than instead of subequal to the antepenultimate spine. In addtion, *U. amabilis* is a shallow water species taken in 30 m, whereas *U. poorei* ranges between 250 and 270 m.

The original description of *U. bacillimanus* Alcock & Anderson, 1899 is very brief, but the illustrations (Alcock & Anderson, 1899: pl. 45, figs 3, 3a) show that the species is somewhat similar to this new species, although the distal parts of the P2-4 dactyli are not clearly depicted. At my request in 1977, K. K. Tiwari, then the director of the Zoological Survey of India, Calcutta, provided me with illustrations of one of the syntypes (Figure 207). In fact, its P2-4 spination is nearly the same as that of the new species, but the eyes are not distally narrowed, the anterolateral spine of the carapace is separated from and not closer at base to the lateral orbital spine in dorsal view, and the antennal article 2 has no distolateral spine, all to mention obvious differences from *U. poorei*.

Uroptychus posticus n. sp.

Figures 208, 209

TYPE MATERIAL — Holotype: **Vanuatu**. MUSORSTOM 8 Stn CP974, 19°21.51'S, 169°28.26'E, 492-520 m, 22.IX.1994, ov. ♀ 5.1 mm (MNHN-IU-2014-16865). Paratype: **Vanuatu**. MUSORSTOM 8 Stn CP1089, 15°08.82'S, 167°17.23'E, 494-516 m, 6.X.1994, 1 ♀ 4.6 mm (MNHN-IU-2014-16866).

ETYMOLOGY — From the Latin *posticus* (that which is behind), alluding to a small spine that is located behind the rostrum.

DISTRIBUTION — Vanuatu; 492-520 m.

DESCRIPTION — Medium-sized species. *Carapace*: 1.2 × broader than long; greatest breadth 1.7 × distance between anterolateral spines. Dorsal surface convex from anterior to posterior and from side to side, sparingly covered with short fine setae; small spine in midline directly behind rostrum (flanked by small spine behind each eye in paratype). Lateral margins somewhat convexly divergent posteriorly, with 8 or 9 spines; first anterolateral, well developed, slightly over-reaching lateral orbital spine (with accompanying spine directly behind it in holotype); second situated at anterior end of branchial margin, rather remote from first, somewhat dorsal in position, accompanying 1 small spine dorsomesial to it (and another small spine ventral to it in holotype); remaining spines placed at same level in profile; third as large as first, fourth larger than others, with 0-3 accompanying small spines; remaining spines posteriorly diminishing; posterior quarter of lateral margin ridged. Rostrum narrow triangular, with interior angle of 17°, somewhat deflected dorsally, dorsally flattish, laterally with small denticle distally; length half that of remaining carapace, breadth much more than half carapace breadth at posterior carapace margin. Lateral orbital spine relatively large but smaller than anterolateral spine, situated directly mesial to and moderately remote from that spine. Pterygostomian flap anteriorly angular, ending in small spine followed by 1 or 2 very small spines along upper margin (discernible under high magnification); smooth on surface.

Sternum: Excavated sternum anteriorly broad triangular, surface with weak longitudinal ridge in midline. Sternal plastron 1.2 × broader than long; lateral extremities slightly convexly divergent. Sternite 3 well depressed, anterior margin of broad V-shape without submedian spines (in holotype) or with very small median notch flanked by obsolescent spine (in paratype), laterally angular. Sternite 4 with scattered tubercles on surface, relatively short; anterolateral margin convex and irregular, length slightly less than twice (1.8 x) that of posterolateral margin. Sternite 5 as long as sternite 4, anterolateral margin strongly convex, subequal to or slightly longer than posterolateral margin of sternite 4.

Abdomen: Nearly glabrous on tergites, bearing fine setae on pleura. Somite 1 gently convex from anterior to posterior, without transverse ridge. Somite 2 tergite 2.4-2.5 × broader than long; pleuron anterolaterally rounded, posterolaterally angular, lateral margin concavely strongly divergent. Pleuron of somite 3 tapering to blunt tip. Telson half as long as broad; posterior plate 1.5 × longer than anterior plate, posterior margin moderately concave or distinctly emarginate.

Eye: Overreaching midlength of rostrum, 2 × longer than broad, feebly narrowed distally. Cornea about half as long as remaining eyestalk.

Antennule and antenna: Ultimate article of antennular peduncle 3 × longer than high. Antennal peduncle overreaching cornea. Article 2 with distinct lateral spine. Antennal scale 1.8 × as broad as article 5, somewhat overreaching article 5, laterally with small spine in holotype, smooth in paratype. Distal 2 articles each with very small distomesial spine; article 5 1.3 × longer than article 4, breadth 0.8 × height of ultimate article of antennule. Flagellum of 13-16 segments nearly reaching distal end of P1 merus.

Mxp: Mxp1 with bases nearly contiguous. Mxp3 basis without distinct denticles on mesial ridge. Ischium with more than 30 small denticles on crista dentata; flexor margin distally not rounded. Merus 1.9 × longer than ischium, moderately thick mesiolaterally, bearing small distolateral spine; flexor margin proximally with rounded ridge, distally with keeled ridge, bearing 2 or 3 small spines on distal third. Carpus with small distolateral spine.

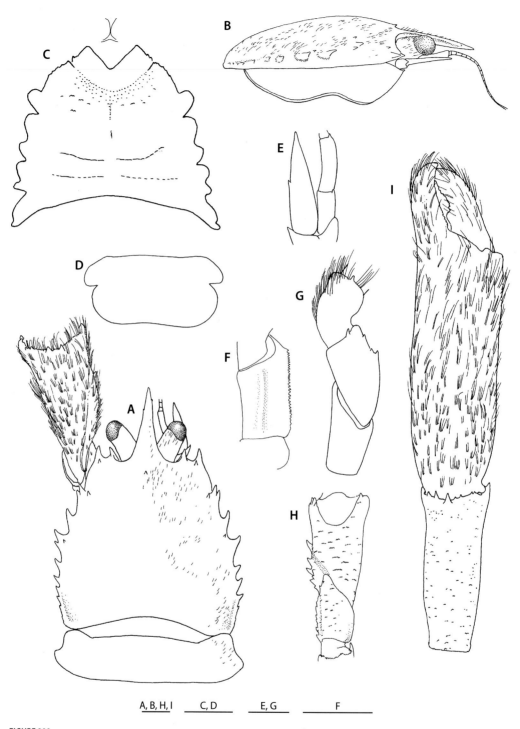

FIGURE 208

Uroptychus posticus n. sp., holotype, ovigerous female 5.1 mm (MNHN-IU-2014-16865). **A**, carapace (setae on left half omitted) and abdomen, proximal part of left P1 included, dorsal. **B**, same, abdomen omitted, lateral. **C**, sternal plastron, with excavated sternum and basal parts of Mxp1. **D**, telson. **E**, right antenna, ventral. **F**, right Mxp3, basis and ischium, setae omitted, ventral. **G**, same, ischium, merus and carpus, lateral. **H**, left P1, proximal part, setae omitted, ventral. **I**, same, distal articles, setae on carpus omitted, dorsal. Scale bars: 1 mm.

P1: Massive, with soft fine setae; length 4.4-4.8 × that of carapace. Ischium with short, basally broad, depressed spine; ventromesial margin with row of small spines, lacking distinct subterminal spine. Merus slightly shorter than carapace; mesial face denticulate, with several spines distinct in 2 rows (paratype) or obsolescent (holotype) on proximal half; dorsal surface with row of small spines distinct in paratype, obsolete in holotype; distal margin with short blunt spines; greatest breadth equal to distance between anterolateral spines of carapace. Carpus 1.3 × longer than merus. Palm slightly depressed, 2.6-2.8 × longer than broad, 1.2-1.5 × longer than carpus. Fingers broad relative to length, slightly gaping, inclined somewhat laterally, strongly incurved distally, crossing when closed; movable finger with low process on proximal third, length 0.4-0.5 × that of palm; fixed finger straight or somewhat concave on opposable margin, without distinct concavity to accommodate opposite process of movable finger when closed.

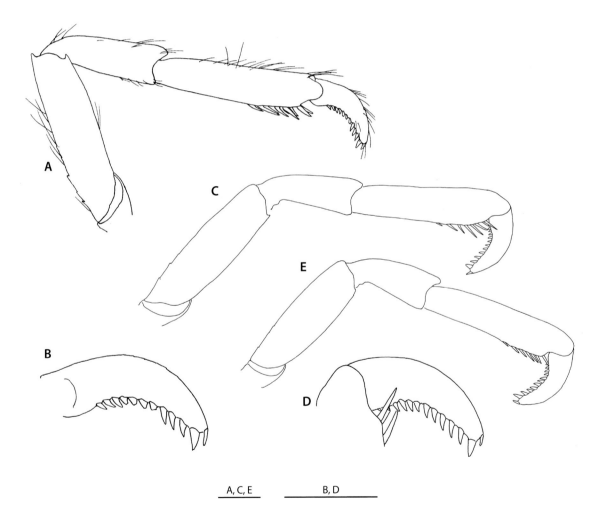

FIGURE 209

Uroptychus posticus n. sp., holotype, ovigerous female 5.1 mm (MNHN-IU-2014-16865). **A**, right P2, lateral. **B**, same, distal part, setae omitted, lateral. **C**, right P3, setae omitted, lateral. **D**, same, distal part, lateral. **E**, right P4, lateral. Scale bars: 1 mm.

P2-4: Relatively short. Meri successively shorter posteriorly (P3 merus 0.9 × length of P2 merus, P4 merus 0.8 × length of P3 merus), equally broad on P2-4; length-breadth ratio, 3.4-3.7 on P2, 3.1-3.3 on P3, 2.8-3.2 on P4; slightly compressed mesio-laterally, slightly granulate on surface, flexor face rounded; dorsal margin with 4 or 5 very small spines on P2 and P3, few in paratype and obsolete in holotype on P4, other than very small distal spine; ventrolaterally with terminal spine present on P2 and P3, absent on P4; P2 merus 0.8 × length of carapace, subequal to or very slightly longer than P2 propodus; P3 merus 0.9 × length of P3 propodus; P4 merus 0.8 × length of P4 propodus. Carpi successively shorter posteriorly (P3 carpus 0.9 × length of P2 carpus, P4 carpus 0.9 × length of P3 carpus), distinctly longer than dactyli (carpus-dactylus length ratio, 1.5-1.6 on P2, 1.4-1.5 on P3, 1.2-1.5 on P4) ; carpus-propodus length ratio, 0.56-0.59 on P2, 0.51-0.55 on P3, 0.46-0.49 on P4. Propodi subequal on P3 and P4, longest on P2; flexor margin somewhat convex around distal fifth, bearing pair of distal spines preceded by 6 spines roughly in zigzag arrangement (mesial and lateral to midline of flexor face) and a few others in midline, unarmed on proximal half. Dactyli shorter than carpi (dactylus-carpus length ratio, 0.6-0.7 on P2 and P3, 0.7-0.9 on P4), 0.4 × length of propodi on P2-4, strongly curving; flexor margin with row of 12-14 relatively large spines, ultimate spine slender, penultimate largest among others, remaining spines somewhat inclined and diminishing toward proximal end of article.

Eggs. About 20 eggs carried, measuring 1.04 mm × 1.28 mm.

PARASITES — The female paratype bears a rhizocephalan externa on the abdomen.

REMARKS — The species is characterized by the elongate eyes, the presence of a spine directly behind the rostrum, the spinose lateral margins of the carapace, and the zigzag arrangement of flexor spines on the P2-4 propodi.

Uroptychus posticus mostly closely resembles *U. exilis* n. sp. (see above, for their similarities). However, *U. exilis* has no spine directly behind the rostrum and anterolateral spine, the anterior margin of sternite 3 is shallowly V-shaped, and the P2-4 dactyli are longer than the carpi.

The spination of the carapace including the relatively large lateral orbital spine, the shapes of the antenna, Mxp3 and P1, and the spination of the distal two articles of P2-4 in this new species are similar to those of *U. paracrassior* Ahyong & Poore, 2004 from southern Queensland. *Uroptychus posticus* is distinguished from that species by having a spine directly behind the rostrum, the ventromesial subterminal spine of the P1 ischium vestigial instead of well developed, the ultimate spine of the P2-4 dactyli much more slender than instead of subequal to the penultimate spine, and the pterygostomian flap with an anteriorly produced spine only, not successively followed by two additional spines as in *U. paracrassior*.

This species resembles *U. lanatus* n. sp. (see above) in the Mxp3 ischium that is not rounded on the distal end of the flexor margin, and in having the massive P1 and the long antennal scale. However, *U. lanatus* is largely different in having no spine behind the rostrum, in having the lateral spines of the carapace much smaller instead of relatively large, in having the lateral orbital spine small instead of relative large compared to the anterolateral spine, and in having the P1 merus mesially unarmed instead of bearing a row of spines.

As mentioned under the remarks of *U. modicus* n. sp. (see above), *U. posticus* also resembles that species, from which it is distinguished by having more numerous, smaller spines (7-8 instead of 4 spines) on the branchial margin, by having a median spine only instead of a transverse row of spines on the epigastric region, and by having less numerous flexor marginal spines (11-13 instead of 18-22 spines, proximal to the slender ultimate spine) on the P2-4 dactyli.

Uroptychus poupini n. sp.
Figures 210, 211, 214E-H

TYPE MATERIAL — Holotype: **Fiji Islands**. BORDAU 1 Stn CP1444, 17°11'S, 178°41'W, 398-409 m, 3.III.1999, 1 ♂ 6.7 mm (MNHN-IU-2014-16867). Paratypes: **Fiji Islands**. MUSORSTOM 10 Stn CP1317, 17°11.99'S, 178°14.14'E, 471-475 m, 5.VIII.1998, 1 ♂ 9.2 mm (MNHN-IU-2014-16868). – Stn CP1327, 17°13,26'S, 177°51,62'E, 370-389 m, 7.VIII.1998, 2 ♂ 8.5, 8.7 mm, 7 ov. ♀ 6.5-8.8 mm (MNHN-IU-2014-10145, MNHN-IU-2014-10146, MNHN-IU-2014-16869). – Stn

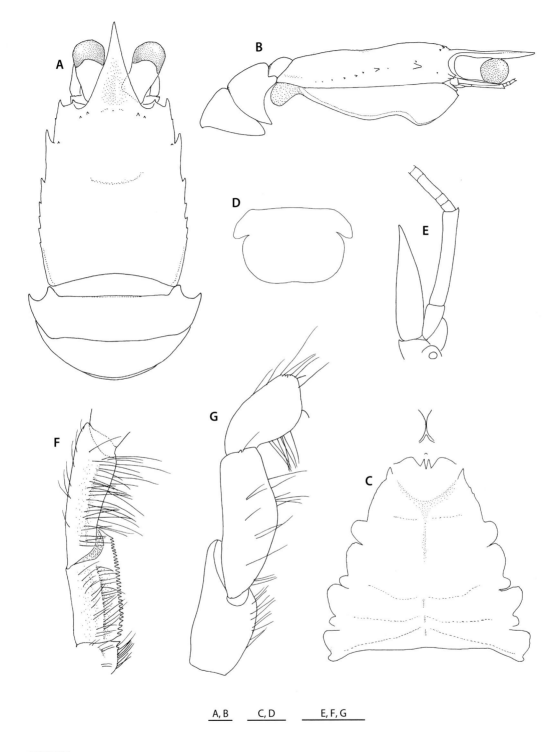

A, B   C, D   E, F, G

FIGURE 210

Uroptychus poupini n. sp., holotype, male 6.7 mm (MNHN-IU-2014-16867). **A**, carapace and anterior part of abdomen, dorsal. **B**, same, lateral. **C**, sternal plastron, with excavated sternum and basal parts of Mxp1. **D**, telson. **E**, right antenna, ventral. **F**, right Mxp3, ventral. **G**, same, lateral. Scale bars: 1 mm.

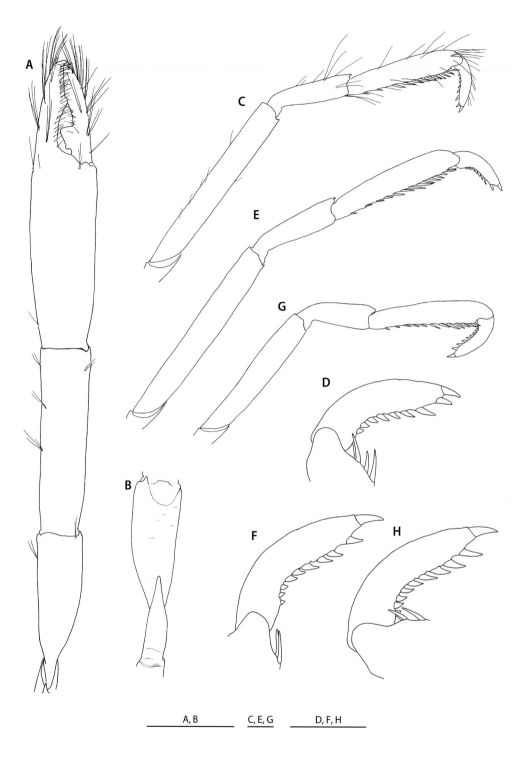

FIGURE 211

Uroptychus poupini n. sp., holotype, male 6.7 mm (MNHN-IU-2014-16867). **A**, left P1, dorsal. **B**, same, proximal part, ventral. **C**, right P2, lateral. **D**, same, distal part, setae omitted, lateral. **E**, right P3, setae omitted, lateral. **F**, same, distal part, lateral. **G**, right P4, lateral. **H**, same, distal part, lateral. Scale bars: A, B, 5 mm; C-H, 1 mm.

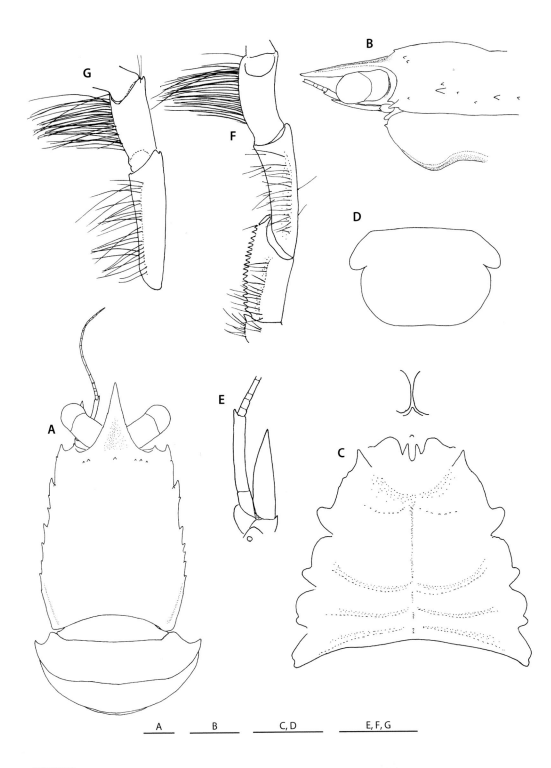

FIGURE 212

Uroptychus longioculus Baba, 1990, paratype, ovigerous female 4.9 mm (MNHN-IU-2014-12822 [=Ga1536]). **A**, carapace and anterior part of abdomen, dorsal. **B**, same, anterior part, lateral. **C**, sternal plastron, with excavated sternum and basal parts of Mxp1. **D**, telson. **E**, left Mxp3, ventral. **F**, same, merus and carpus, ventrolateral. Scale bars: 1 mm.

CP1360, 17°59,57'S, 178°48,20'E, 402-444 m, 13.VIII.1998, 1 ♂ 7.1 mm, 1 ♀ 5.0 mm (MNHN-IU-2014-16870). BORDAU 1 Stn CP1411, 16°05'S, 179°28'W, 390-403 m, 26.II.1999, 3 ♂ 4.8-7.3 mm, 3 ov. ♀ 5.8-6.9 mm (MNHN-IU-2014-16871). – Stn CP1412, 16°06'S, 179°28'W, 400-407 m, 26.II.1999, 1 ♂ 5.2 mm, 1 ov. ♀ 7.7 mm, 1 ♀ 6.1 mm (MNHN-IU-2014-16872). – Stn CP1434, 17°11'S, 178°41'W, 400-401 m, 2.III.1999, 1 ♂ 6.9 mm (MNHN-IU-2014-16873). – Stn CP1467, 18°12'S, 178°36'W, 417-427 m, 6.III.1999, 1 ♂ 7.1 mm, 1 ov. ♀ 7.3 mm (MNHN-IU-2014-16874). – Stn CP1468, 18°16'S, 178°41'W, 478-500 m, 7.III.1999, 2 ♂ 4.9, 6.0 mm, 1 ov. ♀ 6.8 mm, 2 ♀ 5.8, 7.0 mm (MNHN-IU-2014-16875). – Stn DW1421, 17°08'S, 178°59'W, 403-406 m, 28.II.1999, 1 ♂ 3.8 mm, 1 ov. ♀ 6.0 mm, 1 ♀ 4.9 mm (MNHN-IU-2014-16876). – Stn DW1469, 19°40'S, 178°10'W, 314-377 m, 8.III.1999, 1 ♂ 6.7 mm (MNHN-IU-2014-16877). – Stn CP1493, 18°43'S, 178°24'W, 429-440 m, 11.III.1999, 1 ♂ 8.7 mm, 1 ♀ 5.0 mm (MNHN-IU-2014-16878). **Tonga**. BORDAU 2 Stn CP1592, 19°08'S, 174°17'W, 391-426 m, 14.VI.2000, 1 ov. ♀ 6.8 mm (MNHN-IU-2014-16879). – Stn CH1596, 19°06'S, 174°18'N, 371-437 m, 14.VI.2000, 1 ♂ 7.9 mm, 2 ov. ♀ 7.4, 8.7 mm (MNHN-IU-2014-16880). – Stn DW1633, 22°59'S, 175°35'W, 442-453 m, 20.VI.2000, 1 ♂ 6.3 mm, 1 ♀ 6.7 mm (MNHN-IU-2014-16881). **New Caledonia**. BATHUS 1 CP658, 21°13'S, 165°55'E, 515-580 m, 12.III.1993, 1 ♂ 4.3 mm, 1 ov. ♀ 5.8 mm, 1 ♀ 6.3 mm (MNHN-IU-2014-16882). MUSORSTOM 4 Stn CP236, 22°11.3'S, 167°15.0'E, 495-550 m, 2.X.1985, 1 ♀ 6.5 mm (MNHN-IU-2014-16883).

ETYMOLOGY — Named for Joseph Poupin for his friendship and for the inspiration he provided to me with specimens he collected.

DISTRIBUTION — Fiji Islands, Tonga, and New Caledonia, in 314-580 m.

SIZE — Males 3.8-9.2 mm; females, 4.9-8.8 mm; ovigerous females from 5.8 mm.

DESCRIPTION — Medium-sized species. *Carapace*: 1.0-1.1 × longer than broad; greatest breadth 1.3-1.4 × distance between anterolateral spines. Dorsal surface somewhat depressed between indistinct gastric and cardiac regions; epigastric region with small median spine directly behind rostrum and 2 small spines behind each eye (lateral of these occasionally obsolescent). Lateral margins with row of spines: anterolateral spine well-developed, slightly overreaching smaller lateral orbital spine, occasionally followed by small spine on hepatic margin; spine at anterior end of anterior branchial region largest, with accompanying small spine mesial to it; spine at anterior end of posterior branchial region usually smaller than preceding pronounced spine, rarely about as large as anterolateral spine or reduced to small size, occasionally preceded by 1 or 2 small spines, followed by 3 or 4 posteriorly diminishing spines. Rostrum sharp triangular, with interior angle of 32-35°; dorsal surface concave; lateral margin with a few very tiny denticle-like spines; length half or slightly less than half that of remaining carapace; breadth about half carapace breadth at posterior carapace margin. Pterygostomian flap anteriorly roundish with tiny spine.

Sternum: Excavated sternum anteriorly produced to sharp spine, surface with small central process. Sternal plastron as long as or slightly shorter than broad, lateral extremities divergent posteriorly. Sternite 3 strongly depressed, anterior margin deeply excavated, with 2 well-developed submedian spines separated usually by deep narrow notch, occasionally by shallow notch; laterally angular or rounded. Sternite 4 with left and right anterolateral ends each produced to acute spine directed somewhat laterally or straight forward, anterolateral margin relatively long, about twice length of posterolateral margin, 0.63-0.74 × distance between anteriorly produced left and right spines. Anterolateral margin of sternite 5 1.4-1.5 × longer than posterolateral margin of sternite 4.

Abdomen: Somite 1 without transverse ridge; somite 2 tergite 2.1-2.5 × broader than long; pleuron tapering to sharp or bluntly angular tip, lateral margin strongly concave and strongly divergent posteriorly. Pleuron of somite 3 posterolaterally moderately angular. Telson 0.6-0.7 × as long as broad; posterior plate slightly or distinctly emarginate on posterior margin, length 1.3-2.0 × that of anterior plate (greater in small specimens).

Eye elongate, 2 × longer than broad, barely or fully reaching apex of rostrum, mesial margin concave, cornea moderately dilated, half as long as remaining eyestalk.

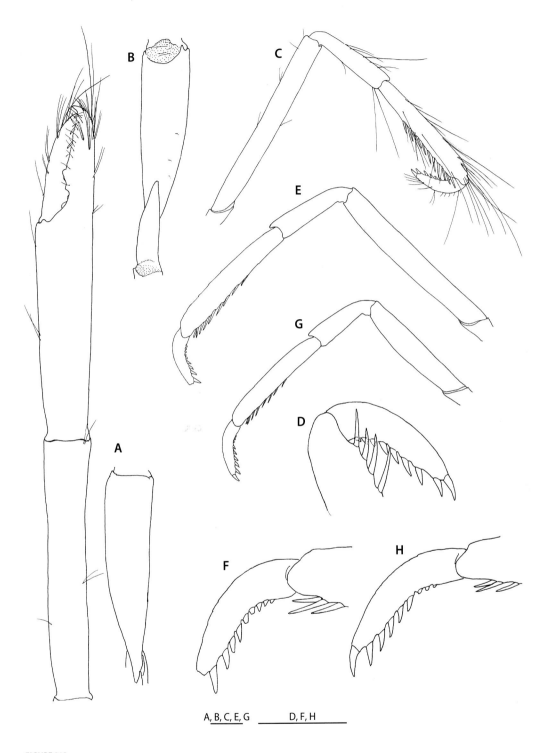

FIGURE 213

Uroptychus longioculus Baba, 1990, paratype, ovigerous female 4.9 mm (MNHN-IU-2014-12822 [=Ga1536]). **A**, right P1, dorsal. **B**, same, proximal part, ventral. **C**, right P2, lateral. **D**, same, distal part, setae omitted, lateral. **E**, left P3, setae omitted, lateral. **F**, same, distal part, lateral. **G**, left P4, lateral. **H**, same, distal part, lateral. Scale bars: 1 mm.

FIGURE 214

Anterior part of sternal plastron. **A-D**, *Uroptychus longioculus* Baba, 1990, paratypes; **E-H**, *U. poupini* n. sp., paratypes. **A**, ovigerous female 4.0 mm (MNHN-IU-2014-12822 [=Ga1536]). **B**, ovigerous female 4.2 mm (MNHN-IU-2014-12823 [=Ga1457]). **C**, ovigerous female 4.9 mm (MNHN-IU-2014-12822 [=Ga1536]). **D**, female 5.2 mm (MNHN-IU-2014-12824 [=Ga1465]). **E**, male 4.1 mm (MNHN-IU-2014-16872). **F**, female 7.4 mm (MNHN-IU-2014-16880). **G**, male 7.5 mm (MNHN-IU-2014-16880). **H**, female 8.7 mm (MNHN-IU-2014-16880).

Antennule and antenna: Ultimate article of antennular peduncle 1.9-2.3 × longer than high. Antennal peduncle slender, not reaching distal end of cornea; article 2 with very small lateral spine; antennal scale barely or fully reaching distal end of article 5, breadth 1.9-2.0 × that of article 5; article 4 without spine; article 5 with ventral distomesial spine of very small to good size, about 3 × longer than article 4, breadth 0.4 × height of ultimate article of antennule; flagellum consisting of 14-22 segments (fewer in small specimens), barely reaching distal end of P1 merus.

Mxps: Mxp1 with bases nearly contiguous. Mxp3 basis with 1 or 2 denticles on mesial ridge. Ischium with flexor margin not rounded distally, crista dentata with 19-28 denticles. Merus 2.0-2.2 × longer than ischium, well ridged along flexor margin; merus and carpus each with very small distolateral spine.

P1: Moderately massive in large males, slender in females and small males, length 5.2-5.5 × that of carapace length in both sexes; granulate on ventral surface of merus and carpus in large males, less so in large females. Ischium with strong, curved dorsal spine, ventromesially with no submedian spine. Merus and carpus with small lateral and mesial distoventral spines; merus 1.2-1.4 × longer than carapace. Carpus 1.2-1.3 × longer than merus. Palm 2.3-3.3 × (males), 2.6-4.3 × (females) longer than broad; usually smooth on ventrolateral margin but a row of tubercles or very small spines continued on to fixed finger in large males and occasionally in large females; length 0.8-0.9 × that of carpus. Fingers gaping somewhat in proximal half, moderately in proximal third or strongly in proximal two-thirds, distally incurved, crossing when closed; movable finger about half or slightly more than half as long as palm, opposable margin with pronounced proximal process.

P2-4: Very slender, compressed mesio-laterally, with long setae especially on carpi and propodi. P3 merus subequally long as and slightly broader than P3 merus, P4 merus 0.7-0.8 × length of P3 merus, slightly narrower than P3 merus; P2 merus 1.0-1.2 × length of carapace, 1.6-1.7 × length of P2 propodus; P3 merus 1.4 × length of P3 propodus; P4 merus 1.0-1.1 × length of P4 propodus; length-breadth ratio, 7.3-9.7 on P2, 6.8-8.9 on P3, 5.7-6.3 on P4. Carpi much longer than dactyli (carpus-dactylus length ratios successively smaller posteriorly, 1.9-2.3 on P2, 1.7-2.0 on P3, 1.4-1.7 on P4); P3 carpus slightly longer than or subequal to P2 carpus, P4 carpus shortest. Propodi with flexor margin ending in pair of terminal spines preceded by row of 10-15 spines. Dactyli about one-third length of propodi on P2-4; dactylus-carpus length ratio, 0.4-0.5 on P2, 0.5-0.6 on P3, 0.6-0.7 on P4, flexor margin strongly curving at proximal third, with 9-10 spines diminishing toward base of article, ultimate spine larger than others, penultimate closer to ultimate than to antepenultimate on P2, somewhat closer to antepenultimate on P3 and P4.

Eggs. Number of eggs carried, 4-30; size, 1.08 mm × 1.20 mm - 1.30 mm × 1.60 mm.

REMARKS — All morphological features displayed by this species from the Western Pacific are very similar to those of *U. longioculus* Baba, 1990 from Madagascar (see Figures 212, 213, 214A-D) but the sizes are clearly different. The largest in the present material measures cl 12.8 mm (poc 9.2 mm) in males and 12.7 mm (poc 8.8 mm) in females, whereas in *U. longioculus* it is cl 8.9 mm in both sexes; the smallest ovigerous female is also larger in the present material (cl 8.5 mm [poc 5.8 mm] versus cl 5.8 mm).

DNA analyses employed at my request showed a high genetic divergence between them (Corbari *et al.* personal comm.), so that a closer examination has been made to detect possible morphological differences. The only difference I have found resides on sternite 4, which has a longer anterolateral margin relative to the breadth measured at the anterior end in *U. poupini*: the length ratio of the anterolateral margin to the distance between left and right anteriorly produced anterolateral spines is 0.63-0.73 in *U. poupini*, 0.53-0.56 in *U. longioculus*. Also, the ratio of the breadth at the anterior end (between anterolateral spines) to the greatest breadth measured between the left and right posterolateral lobes is 0.45-0.49 in *U. poupini*, whereas it is 0.55 in *U. longioculus* (Figure 214).

Uroptychus psilus n. sp.

Figures 215, 216, 308A

TYPE MATERIAL — Holotype: **New Caledonia**, Norfolk Ridge. SMIB 4 Stn DW63, 22°58.7'S, 167°21.1'E, 520 m, 10.III.1989, ov. ♀ 13.0 mm (MNHN-IU-2014-16884). Paratypes: **New Caledonia**, Norfolk Ridge. SMIB 2 Stn DW10, 22°50'S, 167°16'E, 490-495 m, 18.IX.1986, 2 ♂ 8.0, 8.1 mm, 1 ov. ♀ 10.0 mm (MNHN-IU-2014-16885). – Stn DW13, 22°52'S, 167°13'E, 427-454 m, 18.IX.1986, 1 ♂ (carapace missing), 1 ov. ♀ 12.3 mm (MNHN-IU-2014-16886). – Stn DC26, 22°59'S, 167°23'E, 500-535 m, 21.IX.1986, 1 ♂ 8.3 mm, 1 ov. ♀ 10.6 mm (MNHN-IU-2014-16887). SMIB 4 Stn DW64, 22°55'S, 167°15'E, 455-460 m, 10.III.1989, 1 ♂ 9.1 mm (MNHN-IU-2014-16888). NORFOLK 2 Stn CP2146, 22°50.17'S, 167°17.35'E, 518-518 m, 4.XI.2003, 1 ♂ 8.4 mm (MNHN-IU-2014-16889). MUSORSTOM 4 Stn CP214, 22°53.8'S, 167°13.9'E, 425-440 m, 28.IX.1985, 1 ov. ♀ 8.2 mm (MNHN-IU-2014-16890). – Stn CP215, 22°55.7'S, 167°17.0'E, 485-520 m, 28.IX.1985, 1 ♂ 4.6 mm, 1 ov. ♀ 10.3 mm (MNHN-IU-2014-16891). – Stn CP216, 22°59.5'S, 167°22.0'E, 490-515 m, with coral Chrysogorgiidae (Calcaxonia), 29.IX.1985, 9 ♂ 8.0-12.9 mm, 4 ov. ♀ 10.8-12.2 mm, 2 ♀ 8.1, 11.5 mm (MNHN-IU-2014-16892), 1 ♂ 5.9 mm (MNHN-IU-2014-16893). HALIPRO 2 Stn BT94, 23°33'S, 167°42'E, 448-880 m, 24.XI.1996, 1 ♂ 13.0 mm (MNHN-IU-2014-16894). SMIB 8 Stn DW200, 22°59.21'S, 168°20.23'E, 514-525 m, 2.II.1993, 1 ov. ♀ 10.1 mm (MNHN-IU-2014-16895). **New Caledonia**, Loyalty Ridge. BERYX 2 Stn CH16, 23°35.60'S, 169°36.52'E, 660-675 m, 29.X.1991, 1 ov. ♀ 13.7 mm (MNHN-IU-2014-16896). **Tonga**. BORDAU 2 Stn CH1621, 24°19'S, 176°23'W, 570-573 m, 18.VI.2000, 1 ♀ 6.0 mm (MNHN-IU-2014-16897).

ETYMOLOGY — From the Greek *psilos* (bare, smooth), alluding to the smooth carapace surface of the species.

DISTRIBUTION — Norfolk Ridge, Loyalty Ridge and Tonga; 425-880 m.

SIZE — Males, 4.6-13.0 mm; females, 6.0-13.7 mm; ovigerous females from 8.2 mm.

DESCRIPTION — Large species. *Carapace:* As long as broad; greatest breadth 1.6-1.7 × distance between anterolateral spines. Dorsal surface smooth and barely setose, well convex from anterior to posterior, usually with no depression, occasionally with faint depression between gastric and cardiac regions. Lateral margins convex posteriorly, with somewhat elevated small ridge at anterior quarter, followed by a few minute tubercles or granules; anterolateral spine small, reaching but not overreaching tip of lateral orbital spine. Rostrum broad triangular, as long as broad, with interior angle of 35-37°, length 0.3-0.4 that of remaining carapace, breadth half carapace breadth at posterior carapace margin; dorsal surface moderately concave; lateral margins slightly concave. Lateral orbital spine somewhat smaller than anterolateral spine, slightly anterior to level of, and moderately remote from lateral orbital spine. Pterygostomian flap smooth on surface, anterior margin somewhat roundish, ending in small spine.

Sternum: Excavated sternum anteriorly triangular, ending in acute spine, surface somewhat ridged in midline, with small spine in center. Sternal plastron gradually broadened posteriorly, slightly shorter than broad. Sternite 3 strongly depressed, anterior margin relatively narrow and excavated in broad V-shape with narrow median notch flanked by small spine; anterolateral angle sharply produced. Sternite 4 long relative to breadth; anterolateral margin convex, moderately divergent posteriorly, anteriorly produced to distinct spine, followed by posteriorly diminishing serrations; posterolateral margin slightly less than half as long as anterolateral margin. Sternite 5 more than half length of sternite 4, anterolateral margin strongly convex anteriorly, 1.7 × longer than posterolateral margin of sternite 4.

Abdomen: Smooth and glabrous. Somite 1 gently convex from anterior to posterior. Somite 2 tergite 2.4-2.8 × broader than long; pleuron posterolaterally blunt angular, lateral margins concave, moderately divergent posteriorly. Pleuron of somite 3 with bluntly or more or less sharply angular lateral tip. Telson 0.4-0.5 × as long as broad; posterior plate slightly or distinctly concave on posterior margin, length 1.3-1.7 × that of anterior plate.

FIGURE 215

Uroptychus psilus n. sp., holotype, ovigerous female 13.0 mm (MNHN-IU-2014-16884). **A**, carapace and anterior part of abdomen, proximal part of left P1 included, dorsal. **B**, same, lateral. **C**, sternal plastron, with excavated sternum and basal parts of Mxp1. **D**, telson. **E**, right antenna, ventral. **F**, right Mxp3, setae omitted, ventral. **G**, same, merus and carpus, lateral. **H**, left P1, dorsal. **I**, same, proximal part, ventral. Scale bars: A, B, G, H, 5 mm; C-F, 1 mm.

Eye: More or less elongate (1.5-1.7 × longer than broad), overreaching distal third of rostrum, falling short of rostral tip; mesial and lateral margins concave. Cornea somewhat dilated, length more than half that of remaining eyestalk.

Antennule and antenna: Ultimate article of antennule 2.3-2.7 × longer than high. Antennal peduncle terminating in distal end of cornea. Article 2 with small lateral spine. Antennal scale twice broader than article 5, varying from slightly overreaching midlength of to almost reaching distal end of antennal article 5. Article 4 unarmed. Article 5 with small or obsolescent distomesial spine, length 2.0-2.3 × that of article 4; breadth 0.5 × height of antennular ultimate article. Flagellum of 30-34 segments distinctly overreaching distal end of P1 merus (right flagellum in holotype with 24 segments, barely reaching distal end of P1 merus, presumably regenerated).

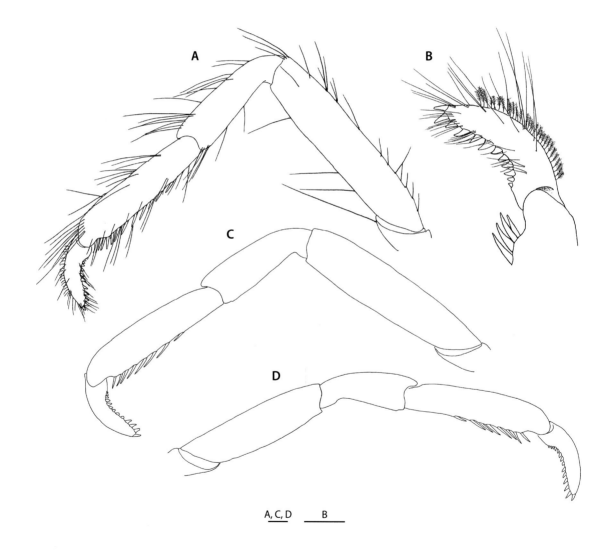

A, C, D   B

FIGURE 216

Uroptychus psilus n. sp., holotype, ovigerous female 13.0 mm (MNHN-IU-2014-16884). **A**, left P2, lateral. **B**, same, distal part, lateral. **C**, left P3, setae omitted, lateral. **D**, right P4, setae omitted, lateral. Scale bars: 1 mm.

Mxp: Mxp1 with bases nearly contiguous to each other. Mxp3 barely setose on lateral surface. Basis with 4-5 denticles on mesial ridge. Ischium with 18-24 denticles on crista dentata, flexor margin distally not rounded. Merus 1.5 × longer than ischium, moderately thick, flexor margin with sharp ridge bearing 0-3 obsolescent denticles at distal third; mesial face setose. Carpus unarmed.

P1: Massive, 4.0-4.9 × longer than carapace. Ischium dorsally with dorsoventrally depressed, short triangular spine, ventrally with relatively short subterminal spine followed proximally by row of tubercle-like spines. Merus dorsally smooth, ventrally with small spines or tubercle-like spines; length subequal to that of carapace. Carpus 1.1-1.2 × longer than merus, smooth and barely setose dorsally, with distinct or tubercle-like spines on ventral surface. Palm 1.9-2.0 × (males), 2.3-2.6 × (females) longer than broad, length varying from slightly shorter to slightly longer than carpus, distally sparsely setose. Fingers nearly straight in lateral view, setose, relatively broad, ending in incurved small spine, crossing when closed; movable finger 0.6-0.8 × as long as palm in both sexes; opposable margin with 2 blunt proximal processes arising from broad base, distal one stronger in males, subequal to proximal process in females, proximal process in females fitting to opposite groove of fixed finger when closed; fixed finger somewhat sinuous on opposable margin.

P2-4: With sparse long setae, slender, unarmed except for distal 2 articles. Meri well compressed mesio-laterally; P3 merus subequal to P2 merus or slightly shorter (0.91-1.00 × length of P2 merus), P4 merus 0.71-0.78 × length of P3 merus; breadths successively slightly smaller posteriorly or slightly larger on P3 than on P2, P4 merus 0.8-0.9 × as broad as P3 merus; length-breadth ratio, 4.1-5.2 on P2, 3.9-4.7 on P3, 3.5-4.0 on P4; P2 merus 0.8-0.9 × length of carapace, 1.4 × length of P2 propodus; P3 merus 1.2 × length of P3 propodus; P4 merus 0.9-1.0 × length of P4 propodus, 0.9 × as broad as P3 merus. Carpi successively shorter posteriorly or P2-3 carpi subequal (P3 carpus 0.9-1.0 × length of P2 carpus, P4 carpus 0.8-0.9 × length of P3 carpus), carpus-propodus length ratio, 0.7 on P2, 0.6-0.7 on P3, 0.5-0.6 on P4. Propodus shorter on P2 than on P3 and P4, or subequal on P2-4; length-breadth ratio, 4.4-5.1 on P2, 4.8-5.2 on P3, 4.8-5.2 on 4; flexor margin straight or feebly concave, ending in pair of spines preceded by 7-11 slender movable spines on distal four-fifths of length on P2, 8-10 spines on distal three-quarters on P3, 5-9 spines in distal two-thirds on P4. Dactyli shorter than carpi (dactylus-carpus length ratio, 0.5-0.6 on P2, 0.6-0.7 on P3, 0.7-0.8 on P4), less than half as long as propodi (dactylus-propodus length ratio, 0.4 on P2 and P3, 0.3-0.4 on P4), curving at proximal third; flexor margin with 10-13 triangular, somewhat obliquely directed, proximally diminishing spines, arranged as illustrated; ultimate spine very slightly smaller than or subequal to penultimate, subequal to antepenultimate; extensor margin with row of plumose setae.

Eggs. Eggs carried, 6-40 in number; eggs ready to hatch, 1.5-1.8 mm; those presumably shortly after spawning, 0.94 mm × 0.97 mm - 1.00 mm × 1.05 mm.

Color. Male 9.1 mm from Norfolk Ridge (MNHN-IU-2014-16888): Pale pink-orange all over the surface, abdomen more or less translucent.

REMARKS — The combination of the following characters links the new species to *U. brucei* Baba, 1986a, *U. maori* Borradaile, 1916 and *U. granulipes* n. sp.: the carapace lateral margin with an anterolateral spine only, sternite 3 with a pair of submedian spines, the P1 ischium with a well-developed subterminal spine on the ventromesial margin, the P2-4 propodi with a pair of terminal spines proximally preceded by a row of single slender spines, and the dactyli with the penultimate spine subequal to or smaller than the ultimate and subequal to the antepenultimate. However, *U. psilus* is readily distinguished from these congeners by the anterolateral spine rather small, not overreaching instead of distinctly overreaching the lateral orbital spine, by the P1 merus covered with small, distinct spines instead of being granulose on the ventral surface, and by the antennal article 4 spineless instead of bearing a distomesial spine.

Uroptychus psilus also resembles *U. septimus* n. sp. in the carapace shape, especially in having the anterolateral spine small, not overreaching the lateral orbital spine. Characters distinguishing *U. psilus* from *U. septimus* are discussed under the remarks of the latter species (see below).

The species account of *U. indicus* Alcock, 1901 (Alcock 1901: 284) suggests that this new species may be close to that species. However, the original description is not detailed enough to warrant its systematic status, as also not are subsequent descriptions (Van Dam 1933, 1937; Tirmizi 1964). In fact, four lots of the SIBOGA material of *U. australis* var. *indicus* (see Van Dam, 1933) are referable to *U. vandamae* (see below). Examination of the John Murray material of *U. australis* var. *indicus* (see Tirmizi 1964) discloses that the specimens from Stations 108 and 122 in Zanzibar area are identical with the Madagascar material reported under *U. gracilimanus* by Baba (1990) (= undescribed species, see Baba 2005; and under the remarks of *U. gracilimanus*). In addition, the material reported under *U. nitidus* from off Durban (Kensley 1977) will in all probability be referred to this species. A female specimen collected by the R/V Meiring Naude at Station 121 south of Durban, 900-625 m and identified by B. Kensley but not included in his 1977 paper, now in the collection of the Smithsonian Institution (USNM 1101919), is also identical with this undescribed species.

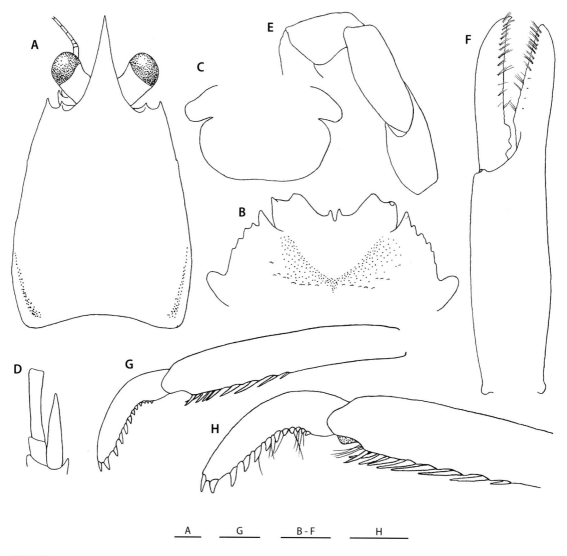

FIGURE 217

Uroptychus indicus Alcock, 1901, syntype, ovigerous female 7.2 mm (ZSI 9328/9). **A**, carapace, dorsal. **B**, anterior part of sternal plastron. **C**, telson. **D**, left antenna, ventral. **E**, left Mxp3, setae omitted, lateral. **F**, right P1, distal part, dorsal. **G**, left P2, dactylus and propodus, setae omitted, lateral. **H**, same, distal part, lateral. Scale bars: 1 mm.

Fortunately, one of the syntypes (ovigerous female) of *U. indicus* was examined on loan through the courtesy of K. K. Tiwari, then the director of the Zoological Survey of India, in 1974. The morphological features of *U. indicus* (Figure 217) obtained by the examination of the syntype are not well detailed, but *U. psilus* n. sp. can be distinguished from that species by the following: the carapace is as long as instead of longer than broad, bearing no ridge along the posterior part of the branchial lateral margin; the antennal article 2 bears a small instead of pronounced distolateral spine; the P2 propodus bears flexor marginal spines along the entire length instead of restricted to the distal half. The flexor marginal spines of the P2 dactylus are closer to each other in the distal portion (the antepenultimate spine is equidistant between the penultimate and the distal quarter spines in *U. psilus*, much more remote from the penultimate than from the distal quarter in *U. indicus*), with the ultimate spine subequal to instead of distinctly more slender than the penultimate.

Uroptychus quartanus n. sp.
Figures 218, 219

TYPE MATERIAL — Holotype: **Indonesia**, Kai Islands. KARUBAR Stn CP16, 5°17′S, 122°50′E, 315-349 m, 24.X.1991, ♂ 4.2 mm (MNHN-IU-2014-16898).

ETYMOLOGY — From the Latin *quartanus* (of the fourth), referring to the fourth thoracic sternite that is much broader than the other sternites in the new species.

DISTRIBUTION — Kai Islands; 315-349 m.

DESCRIPTION — Body and appendages very spinose, sparsely with relatively short setae.

Carapace: 0.9 × as long as broad; greatest breadth 1.3 × distance between anterolateral spines. Dorsal surface moderately convex from anterior to posterior, posterior cervical groove distinct, cardiac region well circumscribed. Strong spines as described below and scattered small spines: 3 equally strong spines in midline, first spine directly behind rostrum, flanked by smaller spine; second spine located at posterior gastric region, flanked by slightly anteriorly located strong spine; third spine on posterior part of cardiac region, flanked by equally strong spine situated on posterior branchial region. Lateral margins convexly divergent posteriorly with weak constrictions at anterior third (between anterior and posterior branchial regions), with 3 spines along hepatic region and 5 strong spines along branchial region; anterolateral spine well developed, directed anterolaterally, slightly overreaching lateral orbital spine; second and third small, ventral to level of remainder, last spine near posterior end. Rostrum directed somewhat dorsally, narrow triangular, with interior angle of 20°, laterally with 1 (left) or 2 (right) anterodorsally directed submedian spines; length half that of remaining carapace, breadth somewhat more than half carapace breadth at posterior carapace margin. Lateral orbital spine small, situated directly lateral to and rather remote from anterolateral spine. Pterygostomian flap anteriorly roundish, ending in sharp spine, surface covered with small spines.

Sternum: Excavated sternum with subtriangular anterior margin, anterior surface sharply cristate in midline. Sternal plastron slightly broader than long, broadest on sternite 4, lateral extremities subparallel between sternites 5-7. Sternite 3 having anterior margin strongly excavated, representing V-shape, with 2 submedian spines separated by narrow notch; anterolaterally angular, ending in 2 small spines. Sternite 4 with 2 strong spines on anterolateral margin, posterior stronger; surface with denticle-like very small spines arranged in concentric arc and fine setae; posterolateral margin about as long as anterolateral margin. Sternite 5 with anterolateral spine, anterolateral margins subparallel, shorter than posterolateral margin of sternite 4.

Abdomen: With sparse setae and sharp spines. Somite 1 with 3 (1 median and 2 lateral) spines. Somite 2 tergite 2.4 × broader than long; with anterior row of 4 well-developed spines and 2 posterior spines each placed laterally; pleuron anterolaterally angular, posterolaterally produced and tapering to sharp point; lateral margin strongly concave. Somite 3 with anterior row of 6 large spines followed by 2 small submedian spines; pleuron tapering to sharp point. Somite 4 with

FIGURE 218

Uroptychus quartanus n. sp., holotype, male 4.2 mm (MNHN-IU-2014-16898). **A**, carapace and anterior part of abdomen, dorsal. **B**, same, lateral. **C**, sternal plastron, with excavated sternum and basal parts of Mxp1. **D**, telson. **E**, left antenna, ventral. **F**, left Mxp3, ventral. **G**, same, lateral. Scale bars: 1 mm.

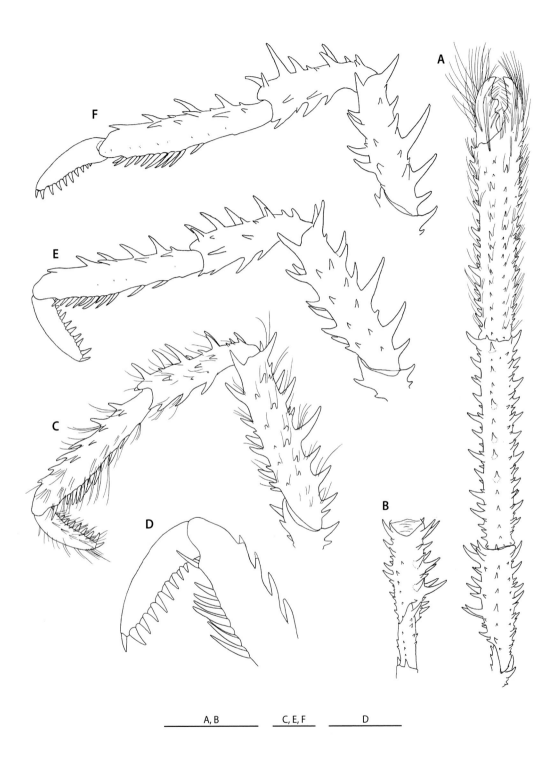

FIGURE 219

Uroptychus quartanus n. sp., holotype, male 4.2 mm (MNHN-IU-2014-16898). **A**, right P1, dorsal. **B**, same, proximal part, ventral. **C**, left P2, lateral. **D**, same, distal part, setae omitted, lateral. **E**, left P3, lateral. **F**, left P4, lateral. Scale bars: A, B, 5 mm; C-F, 1 mm.

anterior row of 6 spines, with no posterior spines. Somite 5 with 2 anterior spines. Somite 6 unarmed. Telson 0.65 × as long as broad; posterior plate 1.7 × longer than anterior plate, subsemicircular. Uropodal protopod with small protuberance on posterior margin near endopod; endopod 1.8 × longer than broad.

Eye: 2 × longer than broad, distally narrowed, somewhat constricted proximal to cornea, slightly overreaching midlength of rostrum. Cornea half as long as remaining eyestalk.

Antennule and antenna: Ultimate article of antennule 2.7 × longer than high. Antennal peduncle reaching rostral tip. Article 2 with sharp distolateral spine. Antennal scale narrower than peduncle, reaching distal end of article 5. Article 3 with small distomesial spine. Article 4 with ventral distomesial spine. Article 5 1.7 × longer than article 4, with 1 distodorsal and 3 ventral spines (distoventral largest); breadth 0.6 × height of antennular ultimate article. Flagellum of 9 segments falling short of distal end of P1 merus.

Mxp: Mxp1 with bases broadly separated. Mxp3 with strong distolateral spine on coxa, ischium, merus and carpus. Basis with a few obsolescent denticles on mesial ridge. Ischium with flexor margin not rounded distally, crista dentata with 5-6 denticles irregularly distant from one another. Merus 2.1 × longer than ischium, with 1 distolateral, 3 flexor marginal spines. Carpus with 1 strong distolateral and 2 smaller extensor marginal spines.

P1: 7.7 × longer than carapace, moderately setose. Spines in 8 rows on merus, carpus, propodus: 2 dorsal, 2 mesial (dorsal and ventral), 2 lateral (dorsal and ventral), and 2 ventral. Ischium with strong ventromesial subterminal spine and strong distodorsal spine, other than small spines. Merus 1.5 × longer than carapace. Carpus 1.7 × length of merus. Palm 4.8 × longer than broad, subequal to carpus. Fingers broad relative to length, slightly incurved distally, somewhat gaping; movable finger with small spine at proximal quarter of mesial margin, opposable margin with 2 submedian processes; fixed finger with low process or prominence opposing to proximal process of movable finger.

P2-4: Moderately setose. Meri successively shorter posteriorly (P3 merus 0.9 × length of P2 merus, P4 merus 0.9 × length of P3 merus), equally broad on P2-4; P2 merus as long as carapace, subequal to length of P2 propodus; P3 merus 0.9 × length of P3 propodus; P4 merus 0.7 × length of P4 propodus; length-breadth ratio, 4.7 on P2, 4.1 on P3, 3.7 on P4; 6 rows of spines: 2 lateral, 1 ventrolateral, 1 ventromesial, 1 dorsal, 1 dorsomesial; median one of dorsal spines largest, equally long as breadth of article. Carpi with rows of prominent spines like those on meri; P3 carpus 0.9 × length of P2 carpus, P4 carpus subequal to P3 carpus; carpus-propodus length ratio, 0.6 on P2-4; carpus-dactylus length ratio, 1.8 on P2, 1.5 on P3 and P4. Propodi slightly longer on P2 than on P3, subequal on P3 and P4; extensor margin somewhat convex, with row of 6 spines paralleled laterally by smaller spines and mesially by strong spines; flexor margin with pair of terminal spines preceded by 9 spines (distal 7 in zigzag arrangement, distalmost much more remote from distal pair than from distal second on P2). Dactyli subequal on P2-4, 0.6 × length of carpi on P2 and P3, 0.7 × on P4, and 0.4 × length of propodi on P2-4; flexor margin very slightly curving at proximal third, with row of 10 sharp, perpendicularly directed spines, ultimate smaller than penultimate, subequal to antepenultimate.

REMARKS — This specimen was collected together with a male specimen of *U. ciliatus*. The antennae and eyes are so similar between these specimens that they seemed identical. However, *Uroptychus quartanus* is distinguished from *U. ciliatus* by the following particulars: the carapace and pereopods bear less numerous, relatively more pronounced spines; the excavated sternum is sharply cristate in the midline, with an triangular anterior margin, whereas it is moderately ridged and anteriorly rounded in *U. ciliatus*; sternite 4 is distinctly broader than sternites 5-7 that are equally broad, whereas sternites 3-7 are successively broader posteriorly in *U. ciliatus*; the abdominal somite 5 bears two anterior spines and the somite 6 is spineless, whereas both of the somites bear anterior and posterior rows of spines in *U. ciliatus*; the pair of terminal spines on the P2 propodus are more remotely separated from the next proximal spine; the P2 merus is as long as instead of 1.4 times longer than the carapace; the P4 merus is 1.1-1.2 instead of 0.8 times as long as the P4 propodus; and most obviously, the ultimate of the flexor marginal spines of the P2-4 dactyli is more slender than instead of larger than the penultimate.

Uroptychus quartanus also resembles *U. senarius* n. sp. in having a relatively broad and short rostrum and in having spineless abdominal somite 6. Their relationships are discussed under the account of *U. senarius* (see below).

Uroptychus quinarius n. sp.
Figures 220, 221

TYPE MATERIAL — Holotype: **New Caledonia**, Norfolk Ridge. BATHUS 3 Stn CP811, 23°41'S, 168°15'E, 383-408 m, 28.XI.1993, ov. ♀ 2.7 mm (MNHN-IU-2013-8519). Paratypes: **New Caledonia**, Norfolk Ridge. BATHUS 3, collected with holotype, 5 ♂ 3.1-3.3 mm, 9 ov. ♀ 2.6-3.5 mm, 1 ♀ 2.2 mm (MNHN-IU-2014-16899). – Stn CP812, 23°43.38'S, 168°15.98'E, 391-440 m, 28.XI.1993, 3 ♂ 2.9-3.5 mm, 9 ov. ♀ 3.1-3.6 mm, 1 ♀ 3.3 mm (MNHN-IU-2014-16900). – Stn CP813, 23°45'S, 168°17'E, 410-415 m, 28.XI.1993, 9 ♂ 2.3-3.5 mm, 1 ov. ♀ 4.1 mm, 12 ♀ 3.1-4.0 mm (MNHN-IU-2014-16901). – Stn CP814, 23°48'S, 168°17'E, 444-530 m, 28.XI.1993, 3 ♂ 2.8, 3.3 mm, 5 ov. ♀ 2.9-3.6 mm (MNHN-IU-2014-16902). – Stn DW817, 23°42'S, 168°16'E, 405-410 m, 28.XI.1993, 1 ♂ 3.3 mm, 9 ov. ♀ 3.0-3.7 mm, 1 ♀ 3.0 mm (MNHN-IU-2014-16903), 2 ♂ 2.7, 3.3 mm (MNHN-IU-2014-16904). – Stn DW818, 23°44'S, 168°16'E, 394-401 m, 28.XI.1993, 12 ♂ 2.3-3.8 mm, 24 ov. ♀ 2.4-3.5 mm, 5 ♀ 2.1-3.8 mm (MNHN-IU-2014-16905). NORFOLK 2 Stn CP2048, 23°43.82'S, 168°16.24'E, 380-389 m, with Primnoidae gen. sp. (Calcaxonia), 24.X.2003, 16 ♂ 2.2-3.4 mm, 24 ov. ♀ 2.5-3.6 mm, 7 ♀ 2.2-3.3 mm (MNHN-IU-2014-16906). – Stn DW2049, 23°42.88'S, 168°15.43'E, 470-621 m, 24.X.2003, 1 ♂ 3.2 mm, 4 ov. ♀ 3.0-3.4 mm (MNHN-IU-2014-16907). – Stn CP2050, 23°42.17'S, 168°15.72'E, 377-377 m, 24.X.2003, 4 ♂ 3.1-3.7 mm, 9 ov. ♀ 2.9-3.8 mm, 1 ♀ 3.5 mm (MNHN-IU-2014-16908). – Stn DW2052, 23°42.29'S, 168°15.27'E, 473-525 m, 24.X.2003, 1 ♂ 2.9 mm, 2 ov. ♀ 3.2, 3.6 mm, 1 ♀ 3.0 mm (MNHN-IU-2014-16909). – Stn DW2108, 23°46.52'S, 168°17.12'E, 403-440 m, 31.X.2003, 7 ♂ 2.3-3.3 mm, 18 ov. ♀ 3.4-4.2 mm, 3 ♀ 2.9-3.3 mm (MNHN-IU-2014-16910). – Stn DW2109, 23°47.46'S, 168°17.04'E, 422-495 m, 31.X.2003, 9 ♂ 2.6-3.8 mm, 19 ov. ♀ 2.7-3.8 mm, 2 ♀ 2.8, 3.5 mm (MNHN-IU-2014-16911). – Stn DW2110, 23°48.34'S, 168°16.81'E, 493-850 m, 31.X.2003, 2 ov. ♀ 3.4, 3.6 mm (MNHN-IU-2014-16912). – Stn CP2111, 23°48.56'S, 168°16.78'E, 500-1074 m, 31.X.2003, 2 ov. ♀ 3.7, 3.9 mm, 1 ♀ 3.3 mm (MNHN-IU-2014-16913). – Stn DW2112, 23°44.44'S, 168°18.40'E, 640-1434 m, 31.X.2003, 1 ov. ♀ 3.8 mm (MNHN-IU-2014-16914). – Stn DW2142, 23°00.51'S, 168°16.90'E, 550-550 m, 3.XI.2003, 1 ov. ♀ 3.2 mm, 1 ♀ (carapace broken) (MNHN-IU-2014-16915).

OTHER MATERIAL EXAMINED — New Caledonia, Norfolk Ridge. BERYX 11 Stn CP51, 23°45'S, 168°17'E, 390-400 m, 21.X.1992, 1 ov. ♀ 3.1 mm (MNHN-IU-2014-16916), 1 ♀ 3.5 mm (MNHN-IU-2014-16917). – Stn CP52, 23°47.45'S, 168°17.05'E, 430-530 m, 21.X.1992, 4 ♂ 2.6-3.3 mm, 3 ov. ♀ 2.8-3.6 mm, 1 ♀ 3.2 mm (MNHN-IU-2014-16918). SMIB 8 Stn DW179, 23°45.87', 168°16.95', 400-405 m, 30.I.1993, 1 ♂ 3.3 mm, 1 ♀ 3.0 mm (MNHN-IU-2014-16919). BATHUS 2 CP736, 23°03'S, 166°58'E, 452-464 m, 13.V.1993, 2 ov. ♀ 3.0, 3.1 mm, 1 ♀ 3.1 mm (MNHN-IU-2014-16920). SMIB 5 Stn DW98, 23°02'S, 168°16'E, 320-335 m, 14.IX.1989, 1 ov. ♀ 3.2 mm (MNHN-IU-2013-8625). NORFOLK 1 Stn DW1704, 23°45'S, 168°16'E, 400-420 m, 25.VI.200, 9 ♂ 2.8-3.7 mm, 16 ov. ♀ 2.9-3.9 mm, 1 ♀ 3.0 mm (MNHN-IU-2013-8626). – Stn CP1706, 23°44'S, 168°17'E, 383-394 m, 25.VI.2001, 4 ♂ 2.7-3.1 mm, 3 ov. ♀ 2.7-4.0 mm (MNHN-IU-2014-16921). – Stn DW1707, 23°43'S, 168°16'E, 381-493 m, 25.VI.2001, 1 ov. ♀ 3.1 mm (MNHN-IU-2014-16922). – Stn CP1708, 23°43'S, 168°16'E, 381-384 m, 25.VI.2001, 1 ♂ 3.8 mm (MNHN-IU-2014-16923). LITHIST Stn DW13, 23°45.0'S, 168°16.7'E, 400 m, 12.VIII.1999, 10 ♂ 2.3-3.9 mm, 15 ov. ♀ 2.9-3.9 mm, 4 ♀ 2.4-3.6 mm (MNHN-IU-2014-16924). – Stn CP15, 23°40.4'S, 168°15.0'E, 389-404 m, 12.VIII.1999, 1 ov. ♀ 3.1 mm (MNHN-IU-2014-16925). – Stn CP16, 23°43.2'S, 168°16.2'E, 379-391 m, 12.VIII.1999, 3 ♂ 2.0-3.6 mm, 3 ov. ♀ 3.1-3.5 mm (MNHN-IU-2014-16926). **Philippines.** MUSORSTOM 3 Stn CP133, 11°58'N, 121°52'E, 334-390 m, with Primnoidae gen. sp. (Calcaxonia), 5.VI.1985, 1 ♂ 2.3 mm (MNHN-IU-2014-16927). **Indonesia**, Kai Islands. KARUBAR Stn CP16, 5°17'S, 132°50'E, 315-349 m, 24.X.1991, 1 ov. ♀ 3.7 mm (MNHN-IU-2014-16928).

ETYMOLOGY — From the Latin *quinarius* (= consisting of five), referring to five spines on the carapace lateral margin, one of the characters to separate the species from *U. dualis* n. sp., *U. floccus* n. sp. and *U. lumarius* n. sp.

SIZE — Males, 2.3-3.9 mm; females, 2.1-4.2 mm; ovigerous females from 2.4 mm.

DISTRIBUTION — Philippines, Kai Islands (Indonesia), southern New Caledonia and Norfolk Ridge; 377-1074 m.

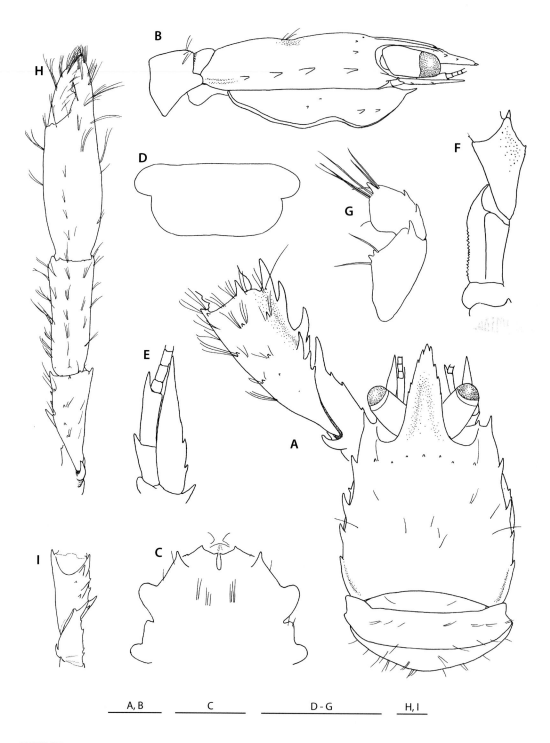

FIGURE 220

Uroptychus quinarius n. sp., holotype, ovigerous female 2.7 mm (MNHN-IU-2013-8519). **A**, carapace and anterior part of abdomen, proximal part of left P1 included, dorsal. **B**, same, lateral. **C**, anterior part of sternal plastron, with excavated sternum and basal parts of Mxp1. **D**, telson. **E**, left antenna, ventral. **F**, left Mxp3, proximal part, setae omitted, ventral. **G**, same, merus and carpus, setae omitted, lateral. **H**, right P1, dorsal. **I**, same, proximal part, setae omitted, ventral. Scale bars: 1 mm.

DESCRIPTION — Small species. *Carapace*: Slightly broader than long (0.9 × as long as broad); greatest breadth 1.4 × distance between anterolateral spines. Dorsal surface somewhat convex in profile, with shallow depression bordering gastric and cardiac regions, bearing sparse long setae; epigastric region with transverse row of spines usually small, often obsolescent, lateral-most often larger than others. Lateral margins convex on branchial region, bearing 5 spines; first anterolateral, directed slightly anterolaterally, overreaching much smaller lateral orbital spine, subequal to second and third; third situated about at midlength, second equidistant between first and third; last (fifth) situated at point quarter from posterior end, followed by ridge. Rostrum proportionately broad distally, with interior angle of 25-30°; length 0.4-0.8 × that of remaining carapace (longer in small specimens), breadth slightly more than half carapace breadth at posterior carapace margin; dorsal surface concave; lateral margin somewhat convex, with 2-4 small spines on distal half. Lateral orbital spine situated very slightly anterior to level of anterolateral spine. Pterygostomian flap anteriorly angular, produced to strong sharp spine, surface with 2 spines on anterior half and another few smaller spines on posterior half.

Sternum: Excavated sternum anteriorly rounded or broad blunt subtriangular, surface with ridge in midline. Sternal plastron nearly as long as broad, somewhat broader posteriorly. Sternite 3 moderately depressed, anterior margin weakly concave, bearing pair of small, often obsolescent submedian spines separated by deep sinus. Sternite 4 with nearly straight anterolateral margin anteriorly produced to spine often with accompanying spine lateral and/or mesial to it, posterolateral margin as long as anterolateral margin. Anterolateral margin of sternite 5 anteriorly rounded, length two-thirds that of posterolateral margin of sternite 4.

Abdomen: Smooth, sparsely setose (with short setae). Somite 1 not ridged transversely. Somite 2 tergite 2.2-2.3 × broader than long; pleuron posterolaterally bluntly angular, lateral margin concave and moderately divergent posteriorly. Pleuron of somite 3 with blunt lateral terminus. Telson slightly less than half as long as broad; posterior plate as long as or slightly longer than anterior plate, posterior margin feebly concave or nearly transverse.

Eye: Elongate, twice as long as broad, overreaching midlength of rostrum; lateral and mesial margins somewhat convex proximally. Cornea not dilated, half as long as remaining eyestalk.

Antennule and antenna: Ultimate article of antennular peduncle slender, 3.8-4.0 × longer than high. Antennal peduncle overreaching cornea. Article 2 with strong distolateral spine. Antennal scale 1.7-1.8 × broader than article 5, distally sharp, overreaching article 5 including distal spine; lateral margin with 1 or 2 small, often obsolete proximal spines. Distal 2 articles each with ventral distomesial spine; article 5 1.4-1.5 × length of article 4, breadth 0.7 × height of antennular ultimate article. Flagellum of 7-12 segments not reaching distal end of P1 merus.

Mxp: Mxp1 with bases distinctly separated. Mxp3 basis with 1 distal denticle (often obsolete) on proximally convex mesial ridge. Ischium with tuft of a few long setae often accompanied by small spine lateral to rounded distal end of flexor margin; crista dentata with 15-17 denticles. Merus 1.5 × longer than ischium, flattish on mesial face; flexor margin sharply ridged, bearing a few small spines at angular midpoint, distolateral spine well developed. Carpus also with well-developed distolateral spine and a few extensor marginal spines.

P1: More than 4 × longer than carapace, sparingly setose. Ischium dorsally with strong, curved spine, ventromesially with strong subterminal spine proximally followed by a few small spines. Merus 1.1-1.3 × longer than carapace, ventrally with row of 3 equidistant spines other than large distomesial and distolateral spines, mesially with field of 3 proximal spines, dorsally with a few small spines. Carpus nearly smooth, ventrally with distolateral and distomesial spines; occasionally with a few tubercular processes or small spines on proximal mesial margin and on proximal part of dorsal surface; length 1.2-1.4 × that of merus. Palm subequal to or slightly longer than carpus, massive with convex lateral and mesial margins, length 2.2-2.5 × (males), 2.4-3.2 × (females) breadth. Fingers with feebly incurved tips, not distinctly crossing when closed, gaping in both sexes but weakly so in females; movable finger 0.4-0.6 × as long as palm, opposable margin with subtriangular median process fitting into opposite longitudinal groove on fixed finger when closed.

P2-4: Sparsely setose, moderately compressed mesio-laterally. Meri equally broad on P2-4; length-breadth ratio, 3.9-5.1 on P2, 3.3-4.7 on P3, 2.8-3.9 on P4; dorsal margin with row of spines distinct on P2 and P3, often obsolescent on P4, without spine at distal end; ventrolateral margin ending in distal spine. P2 merus subequal to or slightly longer than P3 merus, 1.2-1.3 × longer than P2 propodus, about as long as carapace; P3 merus subequal to length of P3 propodus; P4 merus 0.8-

0.9 × length of P3 merus, 0.8-0.9 × length of P4 propodus. Carpi subequal, shorter than dactyli (carpus-dactylus length ratio, 0.7 on P2 and P3, 0.6 on P4); carpus-propodus length ratio, 0.4 on P2 and P3, 0.3-0.4 on P4; extensor margin with 2 distal spines, and often small proximal spine. Propodi subequal on P3 and P4, slightly shorter on P2; flexor margin nearly straight, terminating in pair of spines preceded by 5-6 (rarely 4) basally articulated, slender spines, distalmost of these more remote from terminal pair than from distal second. Dactyli proportionately broad, sparsely setose; dactylus-carpus length ratio, 1.4-1.5 on P2, 1.5-1.7 on P3, 1.6-1.8 on P4; dactylus-propodus length ratio, 0.6 on P2-4; flexor margin nearly straight, with 13-15 spines closely arranged (not contiguous) and obliquely directed, ultimate slender, as broad as antepenultimate, penultimate more than 2 × broader than ultimate.

Eggs. Number of eggs carried, 2-11; size, 0.67 mm × 0.75 mm.

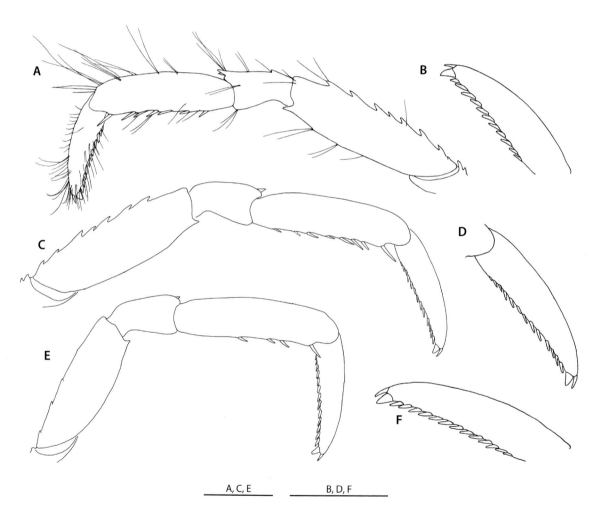

A, C, E   B, D, F

FIGURE 221

Uroptychus quinarius n. sp., holotype, ovigerous female 2.7 mm (MNHN-IU-2013-8519). **A**, left P2, lateral. **B**, same, distal part, setae omitted, lateral. **C**, right P3, setae omitted, lateral. **D**, same, distal part, setae omitted, lateral. **E**, right P4, setae omitted, lateral. **F**, same, distal part, setae omitted, lateral. Scale bars: 1 mm.

REMARKS — The spination of P1, especially in having three closely arranged spines on the proximal mesial surface of the merus, is rather characteristic among the species of the genus, being shared by *U. quinarius*, *U. dualis* n. sp., *U. floccus* n. sp. and *U. lumarius* n. sp. *Uroptychus quinarius* is readily distinguished from the congeners by having 5 instead of 6-8 spines on the carapace lateral margin. In addition, the spination of the P2-4 dactyli is similar in *U. quinarius*, *U. dualis* and *U. floccus*, but the row of dorsal spines on the P2-3 meri is present in *U. quinarius*, absent in *U. dualis* and *U. floccus*. The spination of the P2-4 dactyli is different between *U. quinarius* and *U. lumarius*: in *U. quinarius*, the ultimate flexor spine is slender and the penultimate spine is pronouncedly broad, preceded by slender, obliquely directed, closely arranged spines, whereas in *U. lumarius*, the slender ultimate spine is preceded by relatively large, subtriangular, loosely arranged and proximally diminishing spines.

The combination of the following characters links *U. quinarius* to *U. vegrandis* n. sp.: the carapace lateral margin with 5 well-developed, widely spaced spines, the rostrum with small lateral spines, the antennal articles 4 and 5 each with a distinct distal spine, the antennal scale overreaching the peduncle, and the spination of P2-4. Discriminating characters between these species are given under the remarks of that species (see below).

Uroptychus raymondi Baba, 2000

Uroptychus raymondi Baba, 2000: 250, fig. 3. — Ahyong & Poore 2004: 73, fig. 22.

TYPE MATERIAL — Holotype: **Australia**, off St. Helens, Tasmania, 645 m, ov. female (TM G3517). [not examined].

MATERIAL EXAMINED — **New Caledonia**, Norfolk Ridge. NORFOLK 2 Stn DW2047, 23°43.04'S, 168°01.92'E, 759-807 m, 23.X.2003, 1 ♂ 6.7 mm (MNHN-IU-2014-16929).

DISTRIBUTION — Victoria, Tasmania, and now Norfolk Ridge; 644-807 m.

DIAGNOSIS — Large species. Carapace broader than long (0.8 × as long as broad), broadest at midlength; greatest breadth 1.7 × distance between anterolateral spines. Dorsal surface granulose and setose, with distinct depression between gastric and cardiac regions and between gastric and anterior branchial regions. Lateral margins with strong spine at midlength, followed by distinct ridge bearing a few low, posteriorly diminishing crenulations; anterolateral spine strong, overreaching lateral orbital angle. Rostrum sharp triangular, with interior angle of 25°, more than half as long as carapace, breadth more than half carapace breadth at posterior carapace margin, surface slightly excavated longitudinally. Lateral limit of orbit angular, without distinct spine. Pterygostomian flap anteriorly angular, produced to small spine. Excavated sternum with convex anterior margin; sternal plastron slightly shorter than broad, lateral extremities gently divergent posteriorly; sternite 3 well depressed, anterior margin representing broad V-shape, with U-shaped median sinus separating obsolescent submedian spines; sternite 4 with anterolateral margin anteriorly angular, about as long as posterolateral margin; anterolateral margin of sternite 5 slightly convexly divergent posteriorly, length subequal to that of posterolateral margin of sternite 4. Abdominal somite 1 antero-posteriorly well convex but not sharply ridged; somite 2 tergite 2.6 × broader than long, pleuron anterolaterally bluntly angular, posterolaterally blunt; pleuron of somite 3 with bluntly angular lateral terminus. Telson half as long as broad; posterior plate 1.5 × longer than anterior plate, posterior margin moderately emarginate. Ultimate article of antennule 3.0 × longer than high. Antennal article 2 with small lateral spine; antennal scale terminating in midlength of article 5; distal 2 articles each with obsolescent distal spine; article 5 1.5 × length of article 4, breadth 0.6 × height of ultimate article of antennule, flagellum of 16 segments slightly falling short of distal end of P1 merus. Mxp1 with bases broadly separated. Mxp3 basis with obsolescent denticles on mesial ridge; ischium with distally diminishing denticles on crista dentata, flexor margin distally rounded; merus twice as long as ischium, relatively thick, flexor margin with a few tubercle-like or obsolescent spines on distal third, distolateral spine short and blunt. Pereopods sparingly with soft setae. P1 with tubercles on dorsal surface of merus and on proximal dorsal part of carpus; ischium dorsally with basally broad, short spine, ventromesially unarmed; merus slightly shorter than carapace; carpus 1.2 × longer than merus; palm as long as carpus; fingers distally incurved, crossing when closed, opposable margins sinuous. P2-4 relatively short and thick; meri having length-breadth ratio, 3.3 on P2, 2.9 on P3, 2.6 on P4, dorsal margins with row of small widely spaced spines; P2 merus 0.7 × length of carapace, as long as P2 propodus; P3 merus 0.9 × length of P2 merus, P4 merus 0.9 × length of P3 merus, 0.7 × length of P4 propodus; carpi subequal, relatively short, 0.7 × length of dactyli on P2, 0.6 × on P3 and P4, one-third length of propodi on P2-4; propodi having flexor margin slightly concave, with pair of terminal spines only; dactyli longer than carpi (dactylus-carpus length ratio, 1.5 on P2 and P3, 1.6 on P4), about half as long as propodi, proportionately broad distally, flexor margin nearly straight, with 10-11 spines, ultimate spine slender, penultimate spine pronouncedly broad, preceded by 8-9 slender spines somewhat inclined, closely arranged but not contiguous to one another.

REMARKS — The specimen agrees well with the previous species accounts (Baba 2000; Ahyong & Poore 2004). No additional characters of significance are noted.

Uroptychus remotispinatus Baba & Tirmizi, 1979
Figure 222

Uroptychus remotispinatus Baba & Tirmizi, 1979: 52, figs 1, 2. — Baba 2005: 55. — Baba *et al.* 2009: 57, figs 47-48. — Poore *et al.* 2011: 329, pl. 7, fig. H.

TYPE MATERIAL — Holotype: **Japan**, Bungo Strait between Kyushu and Shikoku, 1320 m, ov. female (USNM 150318). [not examined].

MATERIAL EXAMINED — New Caledonia. BIOGEOCAL Stn CP216, 22°49'S, 166°22'E, 2175-2250 m, 10.IV.1987, 2 ♂ 4.9, 6.9 mm, 1 ♀ 6.3 mm (MNHN-IU-2014-16930). **New Caledonia**, Norfolk Ridge. BIOCAL Stn CP57, 23°44'S, 166°58'E, 1490-1620 m, 1.IX.1985, 1 ov. ♀ 6.4 mm (MNHN-IU-2014-16931). – Stn CP60, 23°59'S, 167°08'E, 1530-1480 m, 02.IX.1985, with *Acanella* sp. (Alcyonacea: Isididae), 10 ♂ 7.1-8.7 mm, 8 ov. ♀ 5.8-8.8 mm, 4 ♀ 6.1-7.4 mm (MNHN-IU-2014-16932). – Stn CP61, 24°11'S, 167°34'E, 1070-1070 m, 02.IX.1985, 2 ♂ 4.2, 7.3 mm, 3 ov. ♀ 7.3-8.5 mm (MNHN-IU-2014-16933). – Stn CP62, 24°19'S, 167°49'E, 1395-1410 m, 02.IX.1985, 1 ♂ 6.7 mm (MNHN-IU-2014-16934). **Wallis and Futuna Islands**. MUSORSTOM 7 Stn DW620, 12°34'S, 178°11'W, 1280 m, 28.V.1992, with *Acanella* sp. (Alcyonacea: Isididae), 4 ♂ 5.5-7.8 mm, 4 ov. ♀ 7.2-8.8 mm, 4 ♀ 5.4-7.8 mm (MNHN-IU-2014-16935). – Stn CP621, 12°35'S, 178°11'W, 1280-1300 m, with *Acanella* sp. (Alcyonacea: Isididae), 28.V.1992, 22 ♂ 5.5-10.1 mm, 23 ov. ♀ 7.0-10.0 mm, 20 ♀ 5.2-9.0 mm (MNHN-IU-2014-16936). – Stn CP622, 12°34'S, 178°11'W, 1280-1300 m, with *Acanella* sp. (Alcyonacea: Isididae), 28.V.1992, 17 ♂ 5.5-10.1 mm, 16 ov. ♀ 7.5-9.7 mm, 13 ♀ 4.1-9.3 mm (MNHN-IU-2014-16937). – Stn CP623, 12°34'S, 178°15'W, 1280-1300 m, with *Acanella* sp. (Alcyonacea: Isididae), 28.V.1992, 17 ♂ 5.3-9.4 mm, 16 ov. ♀ 6.9-9.5 mm, 10 ♀ 4.0-9.3 mm (MNHN-IU-2014-16938).

DISTRIBUTION — Off Durban, off Mozambique, Madagascar Strait, Taiwan, Japan, and now New Caledonia, Norfolk Ridge and Wallis and Futuna Islands (SW Pacific); 850-2250 m.

SIZE — Males, 4.2-10.1 mm; females, 4.0-10.0 mm; ovigerous females from 5.8 mm.

DIAGNOSIS — Medium-sized species. *Carapace*: Slightly longer than broad, greatest breadth 1.5-1.6 × distance between anterolateral spines. Dorsal surface smooth; lateral margin with anterolateral spine reaching or overreaching lateral orbital angle, and small tubercle-like spines often obsolescent on anterior branchial margin, ridged along posterior portion. Rostrum short, varying from barely to fully reaching, rarely overreaching distal margin of cornea, with interior angle

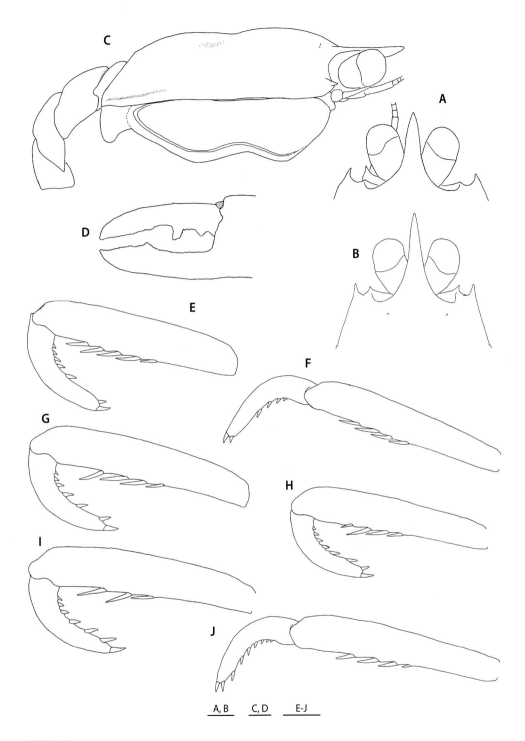

FIGURE 222

Uroptychus remotispinatus Baba & Tirmizi, 1979, **A**, **C**, **D**, **E**, **G**, **I**, male 9.0 mm (MNHN-IU-2014-16936); **B**, **F**, **H**, **J**, ovigerous female 7.3 mm, Madagascar, *Vauban* Stn CH138 (MNHN-Ga1523). **A**, **B**, anterior part of carapace, dorsal. **C**, carapace and anterior part of abdomen, lateral. **D**, distal part of left P1, dorsal. **E**, **F**, left P2, distal articles, setae omitted, lateral. **G**, **H**, left P3, distal articles, setae omitted, lateral. **I**, **J**, left P4, distal articles, setae omitted, lateral. Scale bars: 1 mm.

of 16-22°, dorsally flattish, straight horizontal or somewhat curving dorsally, length one-third that of carapace, breadth half carapace breadth at posterior carapace margin. Lateral limit of orbit angular usually without spine, rarely with small spine. Pterygostomian flap anteriorly roundish, bearing very small, often obsolescent spine. Excavated sternum inflated in center, with low longitudinal ridge in midline, anterior margin bluntly produced, often broad triangular; sternal plastron slightly broader than long, lateral extremities divergent posteriorly; sternite 3 shallowly depressed, anterior margin broadly concave with pair of median spines separated by small notch; anterolateral margin of sternite 4 with short anterior spine or process followed by a number of diminishing spines or tubercles, length nearly equal to that of posterolateral margin. Sternite 5 having anterolateral margin 0.6-0.7 × as long as posterolateral margin of sternite 4. Somite 2 tergite 2.2-2.6 × broader than long; pleuron posterolaterally blunt, lateral margin feebly or weakly concave and moderately divergent posteriorly; pleuron of somite 3 with blunt posterolateral terminus. Telson slightly more than half as long as broad; posterior plate 1.5-1.9 × longer than anterior plate, posterior margin emarginate. Eye short relative to breadth, 1.5 × longer than broad, slightly narrowed proximally, usually overreaching cornea; cornea 0.7 × as long as remaining eyestalk. Ultimate antennular article twice as long as high. Article 2 of antenna with short lateral spine. Antennal peduncle overreaching eye; antennal scale barely reaching, fully reaching or slightly overreaching distal end of article 4, occasionally slightly falling short of midlength of article 5; article 5 1.5 × longer than article 4, breadth 0.4 × height of antennular ultimate article; no spine on articles 4 and 5; flagellum of 14-16 segment not reaching distal end of P1 merus. Mxp1 with bases close to each other. Mxp3 basis with 3-5 denticles on mesial ridge; ischium having flexor margin distally not rounded, crista dentata with 16-22 denticles; merus 2.4 × longer than ischium, rather thick mesio-laterally, flexor margin distinctly ridged, not cristate, without spine; carpus unarmed. P1 4.4-4.9 × longer than carapace; slender, often denticulate on ventral surface of merus, with small ventro-distomesial and obsolescent ventro-distolateral spines on merus and carpus and short dorsal spine on ischium, unarmed elsewhere; merus 1.1-1.3 × longer than carapace; carpus 1.2 × (males), 1.2-1.4 × (females) longer than merus; palm shorter than (0.9 × in males, 0.8 × in females) carpus. Fingers proportionately broad distally, not crossing distally when closed; gaping in large males; movable finger 0.6 × (0.7 × in small females) length of palm, opposable margin with prominent proximal process disto-proximally broad in females, bidentate in males. P2-4 relatively long and slender; meri successively shorter posteriorly (P3 merus 0.9 × length of P3 merus, P4 merus 0.8-0.9 × length of P3 merus), equally broad on P2-4, dorsal margin moderately rounded, not sharply ridged; length-breadth ratio, 5.7-6.6 on P2, 5.3-5.9 on P3, 4.6-5.1 on P4; P2 merus 0.9 × (rarely 0.8 x) length of carapace, 1.2-1.3 × length of P2 propodus; P3 merus 1.1 × length of P3 propodus; P4 merus 0.9 × length of P4 propodus; carpi successively slightly shorter posteriorly, or subequal on P2 and P3 and slightly shorter on P4, longer than dactyli; carpus-propodus length ratio, 0.6-0.7 on P2 and P3, 0.5-0.7 on P4; propodi successively slightly longer posteriorly; flexor margin straight, with row of slender, movable spines (4-5 on P2, 4 on P3, 3 on P4), terminal spine single, considerably remote from juncture with dactylus; dactyli subequal on P2-4, 0.4 × length of propodi and 0.7-0.8 × length of carpi; slender, strongly curving; flexor margin with 2 groups of spines usually on P2, occasionally on P3 (additional spine between on P3 and P4), 2 distal spines remotely separated from proximal group, proximal group consisting of 4-6 spines, all spines strongly oblique but not oriented parallel to flexor margin, ultimate spine longer than penultimate.

Eggs. Number of eggs carried, 7-25; size, 1.6 mm × 1.5 mm - 2.30 mm × 2.50 mm.

Color. A specimen from Taiwan was illustrated by Baba *et al.* (2009) and Poore *et al.* (2011).

PARASITES — Rhizocephalan externa on one male and one female from MUSORSTOM 7 Stn CP622 (MNHN-IU-2014-16937) and on one male and two females from Stn CP623 (MNHN-IU-2014-16938).

REMARKS — The unique spination of the dactylus for which the species is named, is consistent without exception on P2; on P3 and P4, although, there is an additional spine that is present between the proximal and distal groups of spines and that is variable in location, being more or less close to the proximal group, more or less distant from that group or rarely equidistant between. Reexamination of the Madagascar specimens reported earlier (Baba 1990: 947) disclosed that the spination is nearly the same as in the present material.

The rostrum also varies from barely reaching to fully reaching the cornea (rarely somewhat overreaching), but not extending as far forward as in the holotype (Baba & Tirmizi 1979: fig. 1). In the Madagascar specimens, however, it usually considerably overreaches the eye, and in some specimens it is relatively short, only slightly overreaching the cornea as in the present material; P1 bears tubercles on mesial and ventral surfaces of the merus in large specimens of both sexes, and even on the carpus in large males.

The distalmost flexor marginal spine of the P2-4 propodi that is distantly remote from the juncture with the dactylus is one of the recognition characters of the species. This spination is shared with *U. bispinatus* Baba, 1988 (see above; Figure 38).

Uroptychus salomon n. sp.

Figures 223, 224

TYPE MATERIAL — Holotype: **Solomon Islands**. SALOMON 2 Stn CP2197, 8°24.2'S, 159°22.5'E, 897-1057 m, 24.X.2004, ov. ♀ 8.4 mm (MNHN-IU-2014-16939). Paratypes: station data as for the holotype, 1 ♂ 5.2 mm, 1 ov. ♀, 7.8 mm (MNHN-IU-2014-16940).

ETYMOLOGY — Named for the cruise SALOMON 2, by which the type material of this species was collected; used as a noun in apposition.

DISTRIBUTION — Solomon Islands; 897-1057 m.

DESCRIPTION — Medium-sized species. *Carapace*: As long as broad; greatest breadth 1.9 × distance between anterolateral spines. Dorsal surface smooth and glabrous, slightly convex from anterior to posterior, anteriorly smoothly sloping down to rostrum, with feeble depression between gastric and cardiac regions. Lateral margins moderately convex posteriorly, with 2 spines; first anterolateral spine small, reaching or slightly overreaching much smaller lateral orbital spine; second spine smaller than anterolateral spine, situated at anterior end of branchial region, followed by row of several short oblique ridges; distinctly ridged along posterior half of branchial margin. Rostrum long triangular, with interior angle of 21-25°; dorsal surface flattish, straight horizontal; ventral surface directed somewhat dorsally in profile; lateral margin entire; length slightly less than half (0.45-0.47 x) that of remaining carapace, breadth less than half carapace breadth at posterior carapace margin. Lateral orbital spine slightly anterior to level of and moderately remote from anterolateral spine. Pterygostomian flap smooth on surface, anterior margin roundish but slightly angular at anterior end bearing very small spine.

Sternum: Excavated sternum anteriorly sharp triangular, surface with longitudinal ridge in midline around center. Sternal plastron as long as or slightly shorter than broad, lateral extremities gently divergent posteriorly. Sternite 3 strongly depressed; anterior margin well concave, with small median notch flanked by small spine (with 1 or 2 small accompanying spines lateral to each in holotype). Sternite 4 with granulate transverse ridge on surface; anterolateral margin as long as posterolateral margin, anteriorly ending in small spine, somewhat convex medially bearing a few to several denticles. Sternite 5 short, length less than half (0.4) that of sternite 4; anterolateral margins somewhat convex and irregular, posteriorly divergent, length 0.4 × that of posterolateral margin of sternite 4.

Abdomen: Tergites smooth and glabrous. Somite 1 well convex from anterior to posterior. Somite 2 tergite 2.1-2.5 × broader than long; pleuron anterolaterally rounded, posterolaterally bluntly angular; lateral margin concave and strongly divergent posteriorly. Pleuron of somite 3 with bluntly angular lateral terminus. Telson 0.6 × as long as broad; posterior plate 1.6 × longer than anterior plate, emarginate on posterior margin.

Eye: Short relative to breadth, 1.3-1.4 × longer than broad, proximally narrowed, slightly or distinctly overreaching midlength of rostrum. Cornea somewhat swollen, about as long as remaining eyestalk.

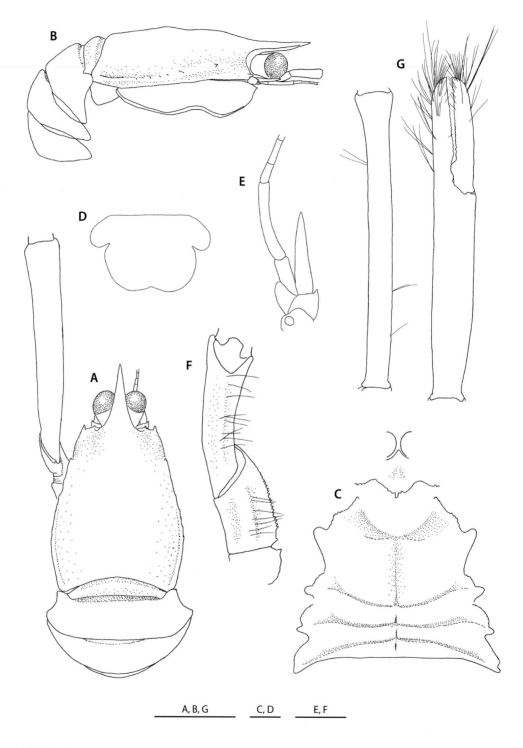

FIGURE 223

Uroptychus salomon n. sp., holotype, ovigerous female 8.4 mm (MNHN-IU-2014-16939). **A**, carapace and anterior part of abdomen, proximal part of left P1 included, dorsal. **B**, same, lateral. **C**, sternal plastron, with excavated sternum and basal parts of Mxp1. **D**, telson. **E**, left antenna, ventral. **F**, right Mxp3, ventral. **G**, left P1, dorsal. Scale bars: A, B, G, 5 mm; C-F, 1 mm.

Antennule and antenna: Ultimate article of antennule 2.9-3.7 × longer than high. Antennal peduncle terminating in corneal margin. Article 2 with small lateral spine. Antennal scale slightly less than twice as broad as article 5, ending in midlength of that article. Distal 2 articles unarmed. Article 5 1.8-2.3 × longer than article 4, breadth 0.3-0.4 × height of antennular ultimate article. Flagellum of 12-16 segments not reaching distal end of P1 merus.

Mxp: Mxp1 with bases close to each other. Mxp3 basis with 1 distal denticle on mesial ridge. Ischium with very small denticles on crista dentata; flexor margin distally not rounded. Merus 2.7 × longer than ischium, unarmed, moderately ridged along flexor margin. Carpus also unarmed.

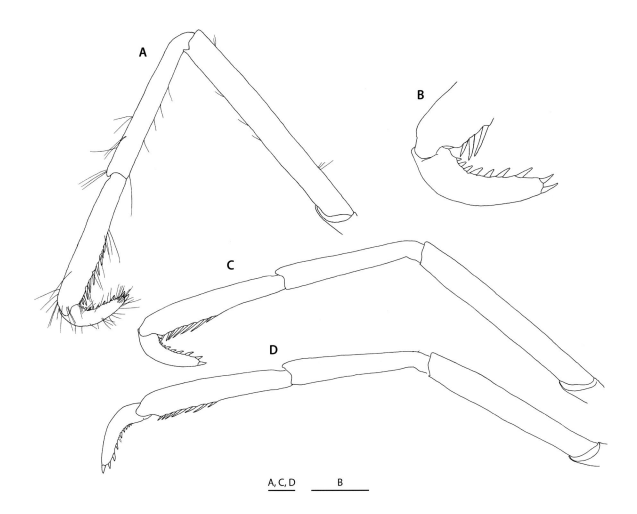

FIGURE 224

Uroptychus salomon n. sp., holotype, ovigerous female 8.4 mm (MNHN-IU-2014-16939). **A**, left P2, lateral. **B**, same, distal part, setae omitted, lateral. **C**, left P3, setae omitted, lateral. **D**, left P4, setae omitted, lateral. Scale bars: 1 mm.

P1: Slender, especially merus and carpus, setose on fingers, sparsely so on palm, carpus and merus; length 4.6-5.9 × that of carapace. Ischium with short dorsal spine, ventromesially unarmed. Merus 1.2-1.5 × longer than carapace, unarmed. Carpus 12 × longer than broad, 1.3-1.4 × length of merus. Palm 0.7 × length of carpus, 3.7-4.8 × longer than broad. Fingers slightly gaping in proximal two-fifths in male, not gaping in females, distally ending in small incurved spine; movable finger 0.6-0.7 × that of palm, opposable margin with obtuse proximal process.

P2-4 slender, relatively thick mesio-laterally, sparsely setose except for distal part. Meri successively shorter posteriorly (P3 merus 0.9 × length of P2 merus, P4 merus 0.9 × length of P3 merus); equally broad on P2-4; length-breadth ratio, 8.7-9.3 (females) or 7.3 (male) on P2, 7.5-8.3 (females) or 6.4 (male) on P3, 6.6-7.2 (females) or 5.8 (male) on P4; dorsal margin unarmed; ventrolateral margin with terminal spine; P2 merus about as long as (females) or slightly shorter (male) than carapace, 1.6 × longer than P2 propodus; P3 merus 1.4 × length of P3 propodus; P4 merus 1.0-1.2 × length of P4 propodus. Carpi unarmed, P3 carpus 0.95-0.96 × length of P2 carpus, P4 carpus 0.93-0.99 × length of P3 carpus; much longer than dactyli (carpus-dactylus length ratio, 2.2-2.6 on P2, 2.0-2.2 on P3, 2.0-2.2 on P4); carpus-propodus length ratio, 0.9-1.0 on P2, 0.8-0.9 on P3 and P4. Propodi subequal in length on P3 and P4, shorter on P2 (female holotype and male), or subequal on P2 and P3, longer on P4 (female paratype); flexor margin straight, ending in pair of spines preceded by 7 or 8 slender movable spines in distal half on P2, 6 or 7 spines on P3, 6 spines on P4. Dactyli 0.4 × length of carpi on P2, 0.5 × on P3 and P4; strongly curving at proximal quarter, with 9 or 10 proximally diminishing, obliquely directed flexor marginal spines, ultimate subequal to or slightly larger than penultimate, antepenultimate closer to fourth than to penultimate on P2, equidistant between on P3 and P4.

Eggs. Number of eggs carried, 8-16; size, 1.27 mm × 1.31 mm - 1.38 mm × 1.68 mm.

REMARKS — The new species resembles *U. politus* (Henderson, 1885) in the carapace shape and the spination of P2-4, and in having the P2-4 carpi much longer than the dactyli. However, *U. salomon* is distinguished from that species by the much more slender pereopods, especially P1 that has the carpus 12 times instead of 2.6-4.0 times longer than broad; sternite 4 has the anterolateral margin as long as instead of 1.6-1.8 times longer than the posterolateral margin; sternite 5 is very short, 0.4 instead of 0.7-0.8 times as long as sternite 4, with the anterolateral margin less than half (0.4) instead of subequal to the length of the posterolateral margin of sternite 4; and the crista dentata of Mxp3 ischium bears distinct instead of obsolescent denticles.

Uroptychus sarahae n. sp.

Figures 225, 226

TYPE MATERIAL — Holotype: **New Caledonia**, Loyalty Ridge. CALSUB Stn PL08, 20°48,3'S, 167°05'E, 880-516 m, 26.II.1989, ov. ♀ 3.2 mm (MNHN-IU-2013-8525). Paratypes: **New Caledonia**, Norfolk Ridge. BATHUS 3 Stn CP844, 23°06'S, 166°46'E, 908 m, 10.XII.1993, 1 ov. ♀ 4.1 mm (MNHN-IU-2013-8531). NORFOLK 2 Stn DW2056, 24°40.32'S, 168°39.17'E, 573-600 m, 25.X.2003, 1 ♀ 4.2 mm (MNHN-IU-2013-8532). – Stn DW2065, 25°15.65'S, 168°55.62'E, 750-800 m, 26.X.2003, 1 ♂ 4.5 mm (MNHN-IU-2013-8533). **Vanuatu**. MUSORSTOM 8 Stn DW1128, 16°02.14'S, 166°38.39'E, 778-811 m, 10.X.1994, 1 ♂ 3.6 mm, 1 ov. ♀ 3.2 mm (MNHN-IU-2013-8526).

ETYMOLOGY — Named for Sarah Samadi of MNHN for her help with DNA analyses.

DISTRIBUTION — Vanuatu, Loyalty Islands and Norfolk Ridge; 516-908 m.

SIZE — Males, 3.6-4.5 mm; females, 3.2-4.2 mm; ovigerous females from 3.2 mm.

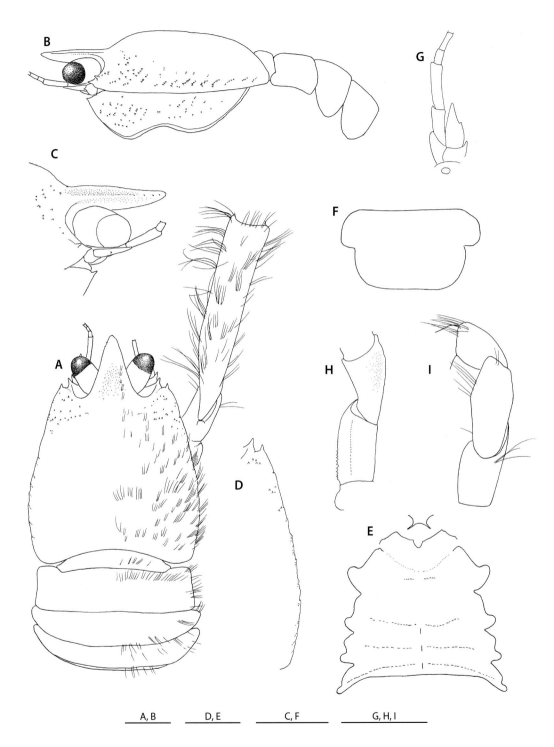

FIGURE 225

Uroptychus sarahae n. sp., holotype, ovigerous female 3.2 mm (MNHN-IU-2013-8525). **A**, carapace and anterior part of abdomen, proximal part of right P1 included, dorsal. **B**, same, lateral. **C**, anterior part of carapace, lateral. **D**, lateral part of carapace, dorsal. **E**, sternal plastron, with excavated sternum and basal parts of Mxp1. **F**, telson. **G**, left antenna, ventral. **H**, left Mxp3, ventral. **I**, same, lateral. Scale bars: 1 mm.

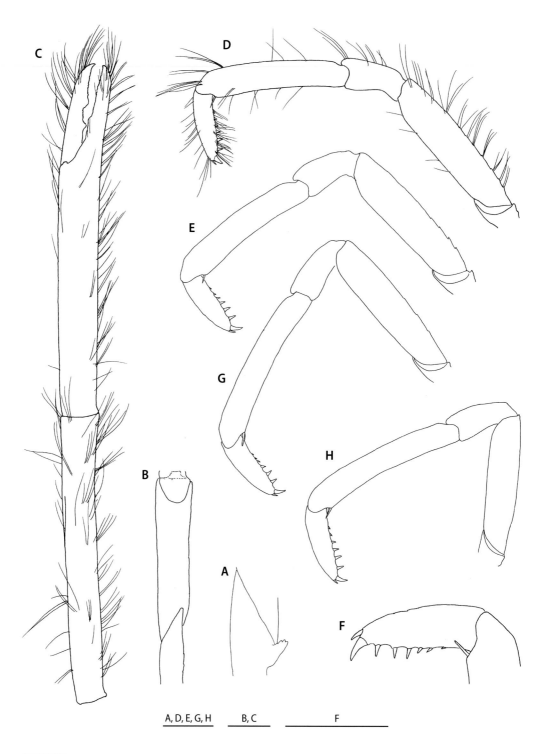

A, D, E, G, H   B, C   F

FIGURE 226

Uroptychus sarahae n. sp., holotype, ovigerous female 3.2 mm (MNHN-IU-2013-8525). **A**, right P1, proximal part (ischium), lateral. **B**, same, proximal part (ischium and merus), setae omitted, ventral. **C**, same, proximal part omitted, dorsal. **D**, left P2, lateral. **E**, same, setae omitted, lateral. **F**, same, distal part, setae omitted, lateral. **G**, left P3, lateral. **H**, left P4, lateral. Scale bars: 1 mm.

DESCRIPTION — Small species. *Carapace*: 0.8-0.9 × as long as broad; greatest breadth 1.6-1.7 × distance between ante-rolateral spines. Dorsal surface well convex from side to side and from anterior to posterior, without distinct groove or depression, covered with short fine setae, sparingly bearing denticles on hepatic, anterior branchial and lateral epigastric regions. Lateral margins somewhat convexly divergent posteriorly with denticles arranged on short oblique ridges; ante-rolateral spine small, located slightly posterior to level of lateral orbital spine (in dorsal view, well separated from that spine at base). Rostrum broad triangular, with interior angle of 30-35°, nearly horizontal; length 0.9-1.0 × breadth, less than half that of remaining carapace, breadth less than half carapace breadth at posterior carapace margin; dorsal surface distinctly depressed along midline; lateral margin with obsolescent spine near tip. Lateral orbital spine subequal to or slightly larger than anterolateral spine. Pterygostomian flap high relative to length, covered with denticles or tubercle-like small spines on surface, anteriorly angular, ending in small spine.

Sternum: Excavated sternum with anterior margin convex between bases of Mxp1, surface with weak ridge in midline on anterior half. Sternal plastron nearly as long as broad, lateral extremities slightly divergent between sternites 4 and 7. Sternite 3 shallowly depressed, with anterior margin gently concave with U-shaped median sinus flanked by very tiny or obsolescent spine; anterolateral end rounded or somewhat angular. Sternite 4 with relatively short anterolateral margin anteriorly rounded; posterolateral margin relatively long, slightly shorter than anterolateral margin. Anterolateral margin of sternite 5 anteriorly somewhat convex, about as long as posterolateral margin of sternite 4.

Abdomen: Setose like carapace. Somite 1 convex from anterior to posterior. Somite 2 tergite 2.4-2.6 × broader than long; pleuron posterolaterally blunt, lateral margin weakly concave and slightly divergent posteriorly. Pleura of somites 3 and 4 laterally rounded. Telson about half as long as broad; posterior plate 1.1-1.3 × longer than anterior plate, feebly concave on posterior margin.

Eye: Broad relative to length (1.4-1.6 × longer than broad), reaching at most distal third of rostrum, medially somewhat inflated. Cornea not dilated, length more than half that of remaining eyestalk.

Antennule and antenna: Ultimate article of antennule 3.3-3.5 × longer than high. Antennal peduncle relatively slender, barely or nearly reaching rostral tip. Article 2 with distinct distolateral spine. Antennal scale 1.6-1.8 × broader than article 5, slightly overreaching article 4, barely reaching midlength of article 5, laterally bearing 1 or 2 very small spines. Article 4 with small distomesial spine. Article 5 unarmed, 1.2-1.3 × longer than article 4, breadth slightly less than half height of antennular ultimate article. Flagellum of 11 segments slightly falling short of distal end of P1 merus.

Mxp: Mxp1 with bases broadly separated. Mxp3 with sparse setae on lateral surface. Basis without denticles on convex mesial ridge. Ischium having flexor margin not rounded distally; crista dentata with 10-15 distally diminishing denticles. Merus 2 × longer than ischium, unarmed, flexor margin not cristate but rounded, occasionally with obsolescent denticle distal to midlength.

P1: Slender, subcylindrical, with fine setae, lacking spines; length 5.1-5.9 × (males), 4.7-5.6 × (females) that of carapace. Ischium with short, flattish, laciniate dorsal process, unarmed elsewhere. Merus 1.1-1.3 × longer than carapace. Carpus 1.3-1.5 × longer than merus. Palm 0.9-1.0 × length of carpus, 4.6-4.7 × (males), 5.4-5.9 × (females) longer than broad. Fingers slightly gaping, ending in incurved spine, distally not spooned; movable finger 0.4 × length of palm, opposable margin with small subtriangular process proximal to opposing low prominence located at midlength of fixed finger.

P2-4: Moderately compressed, with soft fine setae. Meri successively shorter posteriorly (P3 merus 0.9 (rarely 1.0) × length of P2 merus, P4 merus 0.9 × length of P3 merus), equally broad on P2 and P3, very slightly narrower on P4 than on P3; length-breadth ratio, 4.2-4.4 on P2, 3.8-4.3 on P3, 3.7-4.0 on 4; P2 merus 0.8-0.9 × length of carapace, subequal to length of P2 propodus; P3 merus 0.8-0.9 × length of P3 propodus; P4 merus 0.7-0.8 × length of P4 propodus; extensor margin with a few eminences, denticles or very tiny spines on proximal half on P2, obsolescent on P3 and P4. Carpi subequal, unarmed, length less than half that of propodi (carpus-propodus length ratio, 0.34-0.36 on P2, 0.32-0.34 on P3, 0.29-0.32 on P4). Propodi successively slightly longer posteriorly, or shorter on P2 than on subequal P3 and P4; flexor margin with pair of movable slender terminal spines only. Dactyli proportionately broad in lateral view, length 1.1-1.3 × (P2 and P3), 1.2-1.3 × (P4) that of carpi, 0.4 × that of propodi; flexor margin nearly straight, with 6-9 loosely arranged spines often obscured by setae, ultimate small, slender, and very close to strongest penultimate spine preceded by 4-7 successively diminishing spines nearly perpendicular to flexor margin but proximal 1 or 2 usually obliquely directed; ultimate smaller than antepenultimate, penultimate 1.5-2.0 × broader than antepenultimate.

Eggs. Number of eggs carried, 2-6; size, 0.88 × 1.04 mm - 0.98 × 1.08 mm.

REMARKS — This new species resembles *U. bertrandi* n. sp., *U. philippei* n. sp., *U. rutua* Schnabel, 2009, *U. toka* Schnabel, 2009 and *U. volsmar* n. sp. in the shapes of the carapace, sternal plastron, antenna, third maxilliped and pereopods. *Uroptychus sarahae* shares with *U. toka* and *U. volsmar* the penultimate spine of the P2-4 dactyli that is prominent, 2 times as broad as the antepenultimate, whereas in the other three species that spine is less pronounced (less than 1.3 times). *Uroptychus sarahae* differs from *U. toka* and *U. volsmar* in having the anterolateral spine of the carapace remote from instead of close to the lateral orbital spine, separated by a U-shape instead of V-shape when viewed dorsally. *Uroptychus bertrandi* is unique in having proximally strongly inflated eyes. *Uroptychus philippei* differs from all the others in having the relatively long rostrum and smoother dorsal surface of the carapace (see under *U. philippei*). *Uroptychus rutua* has two broad prominences on the gastric region, which are absent in the other species.

Uroptychus scandens Benedict, 1902
Figures 227, 228

Uroptychus scandens Benedict, 1902: 298, fig. 42. — Miyake & Baba 1967: 227 (not fig. 2).
Not *Uroptychus scandens* – Baba 1981: 132 (= *U. imparilis* n. sp.).
Identity questioned:
Uroptychus scandens – Balss 1913b: 27, fig. 20. — Van Dam l933: 27, fig. 38. — Yokoya 1933: 68. — Van Dam 1937: 102; 1940: 97. — Miyake *in* Miyake & Nakazawa 1947: 734, fig. 2123. — Miyake 1960: 97, pl. 48, fig. 7; 1965: 634, fig. 1040. — Baba 1969: 47. — Kim & Choe 1976: 43, fig. 1 — Takeda 1982: 50, fig. 148. — Baba 2005: 58.

TYPE MATERIAL — Holotype: **Japan**, off Honshu, Suruga Bay, Ose Zaki, S 56° W, 1.6 miles, 124-119 m, ov. female (USNM 26166). [examined].

OTHER MATERIAL EXAMINED — **Japan**. Off Honshu Island, ALBATROSS Stn 3715 [Suruga Bay, Ose Zaki, S 56° W, 1.6 miles], 119-124 m, 11.V.1900, holotype, ov. ♀ 3.8 mm (USNM 26166). Off Daio-zaki, 26.IV.1936, I. Kubo coll., 26.IV.1936, 1 ♂ 3.4 mm, 3 ov. ♀ 3.2-3.8 mm, 3 ♀ 2.7, 3.3 mm (1 ♀, carapace missing) (ZLKU 4868). Off Nojima Zaki, Chiba Pref., SOYO-MARU Stn B1, 34°58.4'N, 140°04.3'E, 120-130 m, 26.V.1963, 1 ♂ 3.2 mm, 1 ♀ 3.1 mm (NSMT). Tosa Bay, 150 m, on pennatulacean, 31.I.1959, K. Kurohara coll., 1 ♂ 3.2 mm, 1 ov. ♀ 3.3 mm (ZLKU 5871). **East China Sea**. 31°29.8'N, 128°01.5'E, 145 m, on pennatulacean, 23.VI.1963, H. Yamashita coll., 1 ♂ 3.1 mm, 1 ov. ♀ 3.1 mm, 3 ♀ 2.7-3.8 mm (ZLKU 9443). **Solomon Islands**. SALOMON 2 Stn CP2210, 7°34.2'S, 157°41.8'E, 240-305 m, 26.X.2004, 1 ♂ 3.6 mm (MNHN-IU-2014-16941).

DISTRIBUTION — Japan, East China Sea and Solomon Islands; 119-305 m.

SIZE — Males, 3.1-3.6 mm; females 2.7-3.8 mm; ovigerous females from 3.1 mm.

DESCRIPTION — Small species. *Carapace*: Broader than long (0.8 × as long as broad); greatest breadth 1.7 × distance between anterolateral spines. Dorsal surface moderately convex from anterior to posterior, with very small spines on anterior part around epigastric, hepatic and anterior lateral branchial regions, those on hepatic region larger. Lateral margins convexly divergent posteriorly, with row of small spines; anterolateral spine larger than others, overreaching much smaller lateral orbital spine, directed somewhat anterolaterally. Rostrum narrow triangular, with interior angle of 21°, straight horizontal on ventral surface; lateral margin somewhat concave, with several obsolescent denticles; dorsal

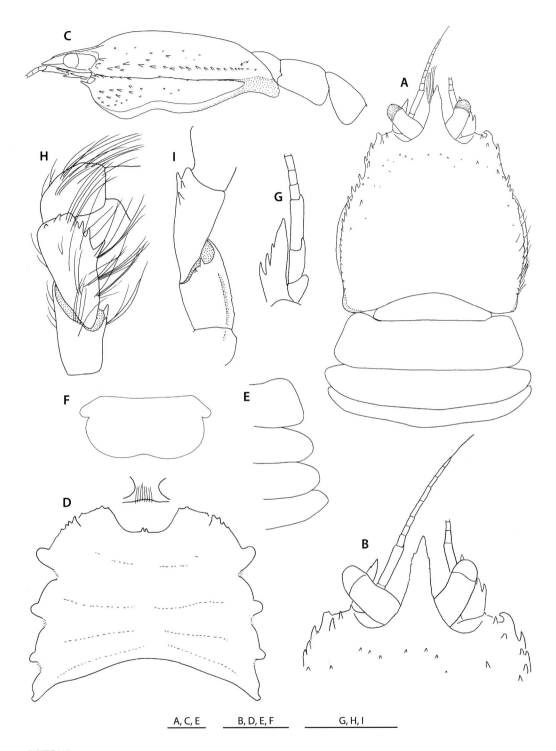

FIGURE 227

Uroptychus scandens Benedict, 1902, holotype, ovigerous female 3.8 mm (USNM 26166). **A**, carapace and anterior part of abdomen, dorsal. **B**, same, anterior part, dorsal. **C**, same, lateral. **D**, sternal plastron, with excavated sternum and basal parts of Mxp1. **E**, pleura of abdominal somites 2-5, dorsolateral. **F**, telson. **G**, right antenna, ventral. **H**, right Mxp3, lateral. **I**, same, setae omitted, ventral. Scale bars: 1 mm.

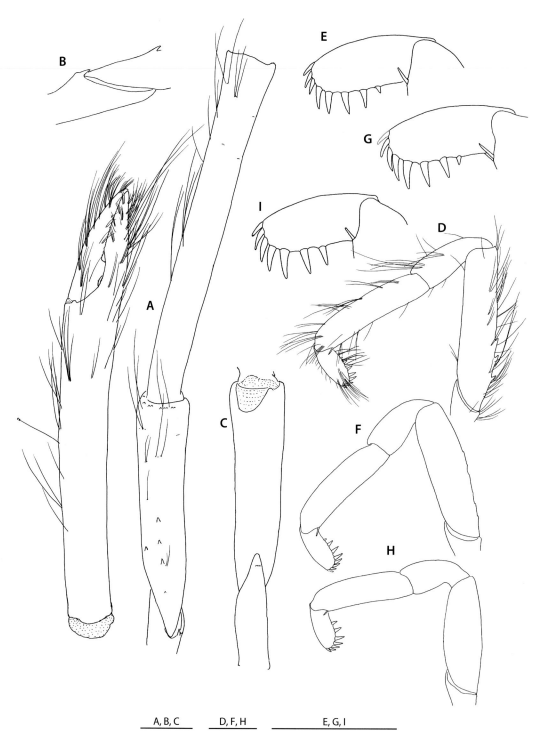

FIGURE 228

Uroptychus scandens Benedict, 1902, holotype, ovigerous female 3.8 mm (USNM 26166). **A**, right P1, dorsal. **B**, same, proximal part, lateral. **C**, same proximal part, ventral. **D**, left P2, lateral. **E**, same, distal part, setae omitted, lateral. **F**, left P3, setae omitted, lateral. **G**, same, distal part, lateral. **H**, left P4, lateral. **I**, same, distal part, lateral. Scale bars: 1 mm.

surface somewhat concave; length slightly smaller than breadth, 0.4 × that of remaining carapace, breadth less than half carapace breadth at posterior carapace margin; lateral orbital spine slightly anterior to level of anterolateral spine. Pterygostomian flap anteriorly somewhat rounded, produced to strong sharp spine; surface with small spines on anterior half, including row of spines along linea anomurica.

Sternum: Excavated sternum with slightly convex anterior margin; surface smooth, without ridge and central spine, with setae along anterior margin. Sternal plastron 0.8 × as long as broad, lateral extremities subparallel between sternites 5 and 7. Sternite 3 with anterior margin broadly and deeply excavated in subsemicircular shape with pair of small median spines basally contiguous. Sternite 4 with anterolateral margin feebly convex with small, posteriorly diminishing spines on anterior half, length 1.3 × as long as posterolateral margin. Sternite 5 with anterolateral margin somewhat convex, as long as posterolateral margin of sternite 4.

Abdomen. Somite 1 moderately convex from anterior to posterior. Somite 2 tergite 2.2 × broader than long, pleuron slightly concavely divergent posteriorly, ending in rounded margin. Pleura of somites 3-5 laterally rounded. Telson half as long as broad; posterior plate distinctly emarginate, 2.0-2.4 × longer than anterior plate.

Eyes: Elongate, 2.2-2.4 × longer than broad, slightly falling short of apex of rostrum, swollen around proximal third, lateral margin convex, mesial margin concave around distal third. Cornea short, less than half length (0.3-0.4) of remaining eyestalk.

Antennule and antenna: Ultimate article of antennular peduncle relatively slender, 3.7 × longer than high. Antennal peduncle slightly falling short of apex of rostrum. Antennal scale fused with article 2, 1.5 × broader than article 5, terminating in midlength of article 5, laterally with 2 or 3 spines, proximalmost situated at ordinary site of article 2. Articles 4 with distinct distomesial spine. Article 5 with small or obsolescent distomesial spine, length 1.9-2.0 × that of article 4, breadth 0.6 × height of ultimate article of antennule. Flagellum of 7-10 segments falling short of distal end of P1 merus, apical seta half length of flagellum.

Mxp: Mxp1 with bases broadly separated. Mxp3 with long setae. Basis with 1-4 denticles on mesial ridge. Ischium with small spine (and much smaller accompanying spine in holotype) lateral to distal end of flexor margin, crista dentata with a few obsolescent denticles. Merus more than twice (2.7 ×) length of ischium, ridged along flexor margin, not well compressed; with 3 o 4 spines on distal third of flexor margin and 1 or 2 spines on distolateral margin. Carpus unarmed.

P1: Slender, subcylindrical, 6.9 × (male), 5.4 × (female) longer than carapace. Ischium dorsally with basally broad, depressed, short spine, ventromesially unarmed. Merus 1.2-1.4 × longer than carapace, dorsal surface with several small spines along proximal half of mesial margin and denticles along distal margin. Carpus 1.3-1.5 × longer than merus, unarmed. Palm 5.8 × (male), 4.9 × (female) longer than broad, 0.9-1.0 × as long as carpus, slightly broadened distally. Fingers relatively narrow distally, somewhat gaping in proximal half; opposable margins fitting to each other in distal half, with row of very small spines or denticles when closed; movable finger 0.4 × length of palm, with obtuse process at midpoint of gaping portion.

P2-4: Thickly setose like P1. Meri successively shorter posteriorly (P3 merus 0.8-0.9 × length of P2 merus, P4 merus 0.8 × length of P3 merus), slightly broader on P3 and P4 than on P2; length-breadth ratio (holotype), 4.3 on P2, 3.3 on P3, 2.8 on P4; dorsal margins with small spines, 3 in number on P2, obsolescent on P3, absent on P4; P2 merus 0.9 × length of carapace; 1.2 × longer than P2 propodus; P3 merus 1.2 × length of P3 propodus; P4 merus subequal to P4 propodus. Carpi successively shorter posteriorly; carpus-propodus length ratio, 0.5 on P2-P4. Propodi successively shorter posteriorly; flexor margin straight in lateral view, with pair of slender terminal spines only. Dactyli subequal in length on P2-4, shorter than carpus; dactylus-carpus length ratio, 0.6 on P2, 0.7 on P3 and P4; dactylus-propodus length ratio, 0.3 on P2, 0.3-0.4 on P3 and P4; truncate, bearing 7 or 8 slender spines obscured by setae, 3-4 of these located on terminal margin, remainder on flexor margin, distalmost smaller than distal second.

Eggs. Number of eggs (yolky) carried, 28; size, 0.60 × 0.57 mm - 0.65 × 0.59 mm.

REMARKS — In an earlier stage of the present study, *Uroptychus articulatus* n. sp., *U. imparilis* n. sp., and *U. parisus* n. sp., had been placed together under *U. scandens* Benedict, 1902 on the basis of the uniquely truncate dactyli of P2-4. In addition, they share nearly all of essential characters such as the anterior carapace with denticles or small spines, all pereopods slender and very setose with the same spination, and also have Mxp3 and sternal plastron almost similar to one another. Careful examination of the material at hand and the type material of *U. scandens* made available on loan showed that there are two different shapes of eyes. One is proximally swollen and distally narrowed, and the other is nearly uniform in breadth, with the cornea broader than the greatest breadth of the remaining eyestalk. Also the antennal scale is either articulated or fused with article 5. Whether these represented intraspecific or interspecific differences was determined by DNA analyses (COI gene sequences) conducted at my request by S. Samadi, M.-C. Boisselier and L. Corbari. They showed genetic divergences between the four different forms are sufficiently high to secure their specific value.

The antennal scale is articulated in *U. articulatus* and *U. imparilis*, whereas it is fused with article 2 in *U. parisus* and *U. scandens*. These couples each are different in the shape of the eyestalk. *Uroptychus articulatus* and *U. scandens* have an elongate, proximally broadened and distally narrowed eyestalk, with the cornea distinctly narrower than the maximum breadth of the eyestalk, whereas in *U. imparilis* and *U. parisus* the eyestalk is equally broad proximally and distally, with the cornea slightly broader than the eyestalk at its maximum. Other features are extremely similar in the four species.

Examination of the material reported by Miyake & Baba (1967) from the East China Sea showed that it is referable to *U. scandens* sensu stricto but the antenna shown in their illustration that is articulated with antennal article 2 is apparently from another specimen of a different species, probably *U. imparilis* n. sp.

The previously reported specimens need to be reexamined for correct identification (see under synonymy, for identification questioned). The specimens reported by Balss (1913b: fig. 19) from Japan and by Kim & Choe (1976: fig. 1) from Korea may be referable to *U. scandens* because of the elongate eyestalks although there is no mention or illustration of the antennal scale. The other possible species is *U. articulatus* that shares with *U. scandens* the elongate eyestalks but it is common in New Caledonia and vicinity.

The distributional range of the species appears to be restricted to Japan and the East China Sea, but exceptional is the male from the Solomon Islands (MNHN-IU-2014-16941) that could not be discriminated from the type material. This specimen is provisionally placed in this species until more material and molecular data become available.

Uroptychus seductus n. sp.
Figures 229, 230

TYPE MATERIAL — Holotype: **Indonesia**, Tanimbar Islands. KARUBAR Stn CP82, 9°32'S, 131°02'E, 219-215 m, with *Thouarella* sp. (Calcaxonia, Primnoidae), 4.XI.1991, ov. ♀ 2.8 mm (MNHN-IU-2014-16942). Paratypes: Collected with holotype, 4 ov. ♀ 2.3-2.8 mm (MNHN-IU-2014-16943).

ETYMOLOGY — From the Latin *seductus* (remote, apart), referring to the flexor marginal spines of P2-4 dactyli, which are remote from one another, a character to separate the species from *U. annae* n. sp.

DISTRIBUTION — Tanimbar Islands (Indonesia); 215-219 m.

SIZE — Ovigerous females, 2.3-2.8 mm; no males were collected.

DESCRIPTION — Small species. *Carapace*: 1.1 × broader than long (0.9 × as long as broad); greatest breadth 1.7 × distance between anterolateral spines. Dorsal surface smooth, with scattered short setae on surface, slightly convex from anterior to posterior, without depression. Lateral margins slightly convexly divergent posteriorly, with 7 spines, 3 on hepatic margin and 4 on branchial margin: first anterolateral, strong, overreaching lateral orbital spine; second and third small and ventral to level of remainder; fourth to seventh equidistant, fourth and fifth as large as or somewhat smaller than first; sixth small; seventh small, obsolete or absent, situated at midlength of posterior branchial margin. Rostrum narrow triangular, with interior angle of 20-30°; dorsal surface deeply concave; ventral surface straight horizontal; lateral margin straight or somewhat concave, with subapical spine (absent in holotype); length 0.4-0.5 × that of remaining carapace, breadth less than half carapace breadth at posterior carapace margin. Lateral orbital spine much smaller than, directly mesial and close to anterolateral spine. Pterygostomian flap anteriorly angular, produced to sharp spine; surface with a few to several small spines, one of these consistent directly below linea anomurica (below third or between third and fourth of lateral marginal spines of carapace).

Sternum: Excavated sternum having anterior margin convex, surface with longitudinal ridge in midline. Sternal plastron 0.8-0.9 × as long as broad; lateral extremities slightly divergent posteriorly. Sternite 3 shallowly depressed, anterior margin shallowly excavated, with 2 small submedian spines separated by small notch, laterally angular or with 2 small spines place side by side. Sternite 4 about as broad as sternite 5; anterolateral margin nearly straight and smooth, anteriorly bearing a few small spines, length twice that of posterolateral margin. Sternite 5 about as long as sternite 4, anterolateral margin 1.3 × longer than posterolateral margin of sternite 4.

Abdomen: Tergites smooth and sparsely setose. Somite 1 smooth, not transversely ridged. Somite 2 tergite 2.0-2.4 × broader than long, pleuron posterolaterally blunt, lateral margin feebly concave, weakly divergent posteriorly. Pleuron of somite 3 with rounded lateral margin. Telson 0.3-0.4 × as long as broad, posterior plate slightly shorter than anterior plate, broadly concave on posterior margin.

Eye: 2 × longer than broad, somewhat narrowed distally, overreaching midlength of, not reaching apex of rostrum. Cornea less than half as long as remaining eyestalk.

Antennule and antenna: Ultimate article of antennular peduncle 2.6-2.9 × longer than high. Antennal peduncle terminating in distal margin of cornea. Article 2 with strong lateral spine. Antennal scale 1.4 × broader than article 5, overreaching antennal peduncle by full length of article 5. Distal 2 articles each with strong distomesial spine; article 5 1.1-1.2 × longer than article 4, breadth 0.7-0.8 × height of ultimate article of antennule. Flagellum consisting of 6-8 segments, falling far short of distal end of P1 merus.

Mxp: Mxp1 with bases broadly separated. Mxp3 with sparse long setae on lateral face. Basis without denticles on mesial ridge. Ischium with small spine and long setae lateral to rounded distal end of flexor margin; crista dentata with about 20 distally and proximally diminishing denticles. Merus 1.5 × longer than ischium, strongly compressed mesio-laterally, flexor margin sharply ridged, bearing 2 small close spines on angularly produced portion at distal third. Strong distolateral spine on merus and carpus.

P1: Slender, 4.5-4.8 × length of carapace, with distally softened long setae. Ischium with 2 dorsal spines, distal one strong, proximal one small, ventrally with well-developed subterminal spine on mesial margin. Merus 1.1-1.2 × longer than carapace, with 2 distoventral (1 large mesial, 1 lateral), 1 distodorsal (mesial), 3 proximomesial, and 1 or 2 small or obsolescent ventral spines. Carpus ventrally with distomesial and distolateral spine, length 1.1-1.2 × that of merus. Palm 3.1-3.3 × longer than broad, 1.1-1.2 × longer than carpus. Fingers distally incurved, crossing when closed; movable finger about half as long as palm, opposable margin with low median process fitting to groove proximal to low median eminence on opposable margin of fixed finger.

P2-4: Relatively short and broad, somewhat compressed mesio-laterally, setose like P1. Meri subequal in length on P2 and P3, shortest on P4 (P4 merus 0.8 × length of P3 merus), subequally broad on P2-4; length-breadth ratio, 3.5 on P2, 3.2-3.4 on P3, 2.7-2.9 on P4; dorsal margins with 2-5 eminences supporting setae, proximal 2 or 3 of these replaced by very small spines usually on P2, occasionally on P3; ventromesial margin with 1 or 2 very small spines each accompanying long setae on P2 only, distal one terminal, proximal one near proximal end of article; ventrolateral margin distally ending in spine; P2 merus 0.8 × length of carapace, 1.1-1.2 × length of P2 propodus; P3 merus subequal to length of P3 propodus; P4 merus 0.8 × length of P4 propodus. Carpi subequal, shorter than dactyli (carpus-dactylus length ratio, 0.8 on P2 and P3, 0.7 on P4), slightly less than half length (0.47-0.48 on P2, 0.41-0.44 on P3 and P4) of propodi. Propodi subequal on

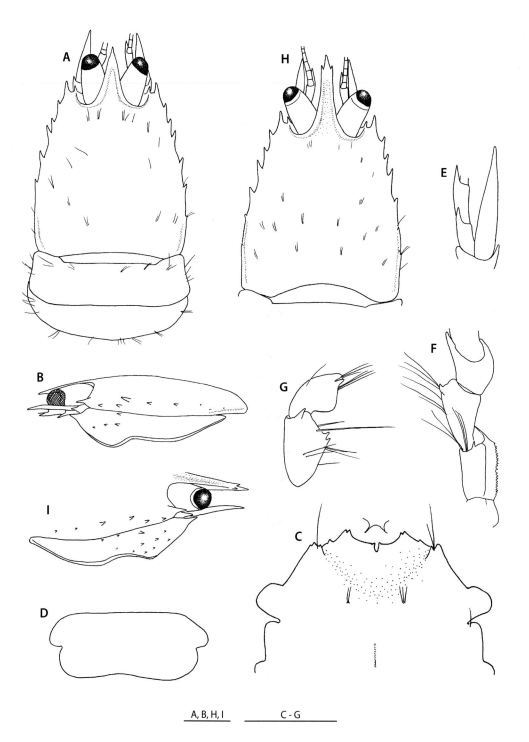

FIGURE 229

Uroptychus seductus n. sp., **A-G**, holotype, ovigerous female 2.8 mm (MNHN-IU-2014-16942); **H**, **I**, paratype, ovigerous female 2.8 mm (MNHN-IU-2014-16943). **A**, carapace and anterior part of abdomen, dorsal. **B**, same, abdomen and setae omitted, lateral. **C**, anterior part of sternal plastron, with excavated sternum and basal parts of Mxp1. **D**, telson. **E**, left antenna, ventral. **F**, right Mxp3, ventral. **G**, same, merus and carpus, lateral. **H**, carapace, dorsal. **I**, carapace, showing arrangement of spines along lateral margin and on pterygostomian flap, lateral. Scale bars: 1 mm.

P3 and 4, slightly shorter on P2; flexor margin slightly convex, ending in pair of spines preceded by 6-7 slender spines on P2, 5 or 6 spines on P3 and P4. Dactyli 0.6 × length of propodi on P2-4, longer than carpi (dactylus-carpus length ratio, 1.2 on P2, 1.3 on P3, 1.4 on P4), markedly narrowed distally; flexor margin slightly curving, with 6-8 spines loosely arranged (penultimate, antepenultimate and distal quarter spines spaced by twice their basal breadth), ultimate slender, penultimate to distal quarter equally strong and somewhat inclined, remaining spines short, slender and more obliquely directed.

Eggs. Number of eggs carried, 6-8; size, 1.70 mm × 1.90 mm.

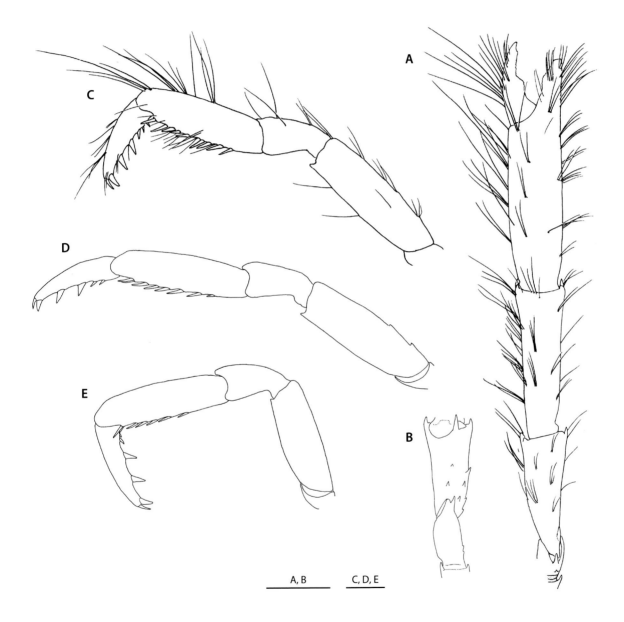

FIGURE 230

Uroptychus seductus n. sp., holotype, ovigerous female 2.8 mm (MNHN-IU-2014-16942). **A**, right P1, dorsal. **B**, same, proximal part, setae omitted, ventral. **C**, left P2, lateral. **D**, left P3, setae omitted, lateral. **E**, left P4, setae omitted, lateral. Scale bars: 1 mm.

REMARKS — The species strongly resembles *U. oxymerus* Ahyong & Baba, 2004 from northwestern Australia, *U. tridentatus* (Henderson, 1885) as redefined in this paper (see below), and *U. annae* n. sp. in the spination of the carapace lateral margin, in having a small spine lateral to the distal end of flexor margin of Mxp3 ischium, and in having the antennal articles 4 and 5 each bearing a strong distomesial spine. *Uroptychus seductus* is readily distinguished from these congeners by the P2-4 dactyli that are more strongly tapering, with more widely spaced flexor marginal spines.

The P2 and P3 propodi each bear a row of flexor marginal spines along the entire length, in addition to the distal pair, in *U. seductus* and *U. tridentatus* (P4 is unknown in *U. tridentatus*) whereas only a few spines proximal to the distal pair in *U. annae* and *U. oxymerus*. *Uroptychus seductus* differs from *U. tridentatus*, in addition to the difference in the spination of P2-3 dactyli (see above), in having the antennal scale overreaching the antennal peduncle by the full length of article 5 instead of reaching at most the second segment of flagellum.

The new species is also close to *U. vicinus* n. sp. in the spination of the carapace and P2-4, and the shapes of the antenna and sternum. Their relationships are discussed under the remarks of that species (see below).

Uroptychus senarius n. sp.
Figures 231, 232, 306B

TYPE MATERIAL — Holotype: **New Caledonia**, Norfolk Ridge. SMIB 5 Stn DW103, 23°16'S, 168°04'E, 300-315 m, 14.IX.1989, ov. ♀ 4.5 mm (MNHN-IU-2014-16944). Paratypes: **Solomon Islands.** SALOMON 2 Stn DW2301, 9°06.9'S, 158°20.6'E, 267-329 m, 8.XI.2004, 1 ov. ♀ 6.2 mm (MNHN-IU-2014-16945). **Vanuatu.** MUSORSTOM 8 Stn CP1024, 17°48.21'S, 168°38.77'E, 335-370 m, 28.IX.1994, 1 ov. ♀ 4.9 mm (MNHN-IU-2014-16946).

ETYMOLOGY — From the Latin *senarius* (consisting of six), referring to six spines on the rostrum (three spines on each lateral margin).

DISTRIBUTION — Solomon Islands, Vanuatu and Norfolk Ridge; 267-370 m.

SIZE — Ovigerous females, 4.5-6.2 mm; no males were collected.

DESCRIPTION — Medium-sized species. *Carapace*: 1.2 × broader than long (0.8 × as long as broad); greatest breadth 1.6 × distance between anterolateral spines. Dorsal surface somewhat convex from anterior to posterior, with distinct depression between gastric and cardiac regions, and shallower depression between cardiac and branchial regions; with large, slender upright spines and scattered small spines. Lateral margins slightly convexly divergent posteriorly, with row of strong sharp spines; anterolateral spine overreaching lateral orbital spine. Rostrum relatively broad triangular, distally curving dorsally, with 3 pairs of antero-dorsally directed lateral spines; dorsal surface flattish; length 1.4 × breadth, slightly more than half length of carapace, breadth slightly less than half carapace breadth at posterior carapace margin. Lateral orbital spine much smaller than and distinctly anterior to level of anterolateral spine. Pterygostomian flap covered with small spines, anteriorly roundish and produced to sharp spine.

Sternum: Excavated sternum strongly cristate in midline, narrowly produced anteriorly. Sternal plastron 0.9 × as long as broad, lateral extremities between sternites 4-7 straight divergent. Sternite 3 well depressed, anterior margin in broad V-shape with 2 submedian spines separated by V-shaped notch, anterolaterally sharp angular, flanked by small spine. Sternite 4 with 2 well-developed, anteroventrally directed spines on anterolateral margin, distal spine smaller, situated at anterior end; posterior surface with scattered granules supporting relatively thick setae; posterolateral margin as long as anterolateral margin. Sternite 5 with well-developed anterolateral spine, lateral margin posteriorly divergent, distinctly shorter than posterolateral margin of sternite 4.

A, B   C - G

FIGURE 231

Uroptychus senarius n. sp., holotype, ovigerous female 4.5 mm (MNHN-IU-2014-16944). **A**, carapace and anterior part of abdomen, dorsal. **B**, same, setae omitted lateral. **C**, sternal plastron, with excavated sternum and basal parts of Mxp1. **D**, telson. **E**, left antenna, ventral. **F**, right Mxp3, setae omitted, ventral. **G**, same, lateral. Scale bars: 1 mm.

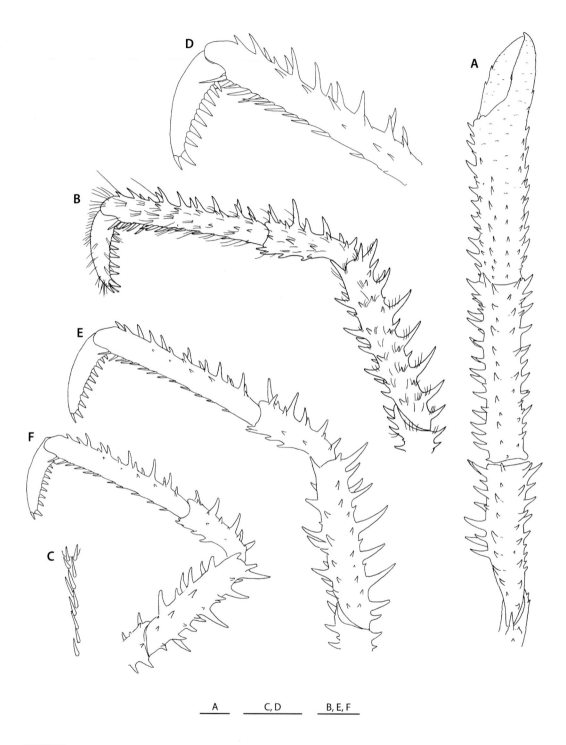

FIGURE 232

Uroptychus senarius n. sp., holotype, ovigerous female 4.5 mm (MNHN-IU-2014-16944). **A**, right P1, dorsal. **B**, left P2, lateral. **C**, same, showing arrangement of spines on propodus flexor margin, flexor. **D**, same, propodus and dactylus, setae omitted, lateral. **E**, left P3, lateral. **F**, left P4, lateral. Scale bars: 1 mm.

Abdomen: Somite 1 with 4 strong spines in transverse row and a few additional small spines. Somite 2 tergite 2.2-2.5 × broader than long; pleuron anterolaterally produced and angular, posterolaterally strongly produced and tapering to sharp point; lateral margin strongly convex and strongly divergent posteriorly. Pleuron of somite 3 also tapering to sharp point. Somites 2-4 with 2 transverse rows of well-developed spines: 6 anterior and 5-8 posterior spines on each of somites 2 and 3, 4 anterior and 2-4 posterior spines on somite 4. Somites 5 and 6 with fine granules supporting setae. Somite 5 with 2-4 anterior and 0-2 posterior spines. Somite 6 unarmed. Uropod having protopod smooth along margins; endopod 1.4 × longer than broad. Telson 0.5-0.6 × as long as broad; posterior plate 1.1-1.3 × longer than anterior plate, distinctly emarginate on posterior margin.

Eye: 2 × as long as broad, slightly overreaching midlength of rostrum, proximally broadened, constricted proximal to cornea. Cornea narrower than and more than half as long as remaining eyestalk.

Antennule and antenna: Ultimate article of antennule 3.2-3.5 × longer than high. Antennal peduncle extending far beyond eye, falling short of apex of rostrum. Article 2 with strong lateral spine. Antennal scale 1.6 × broader than article 5, terminating in or slightly overreaching midlength of article 5; lateral margin with several broad setae. Article 3 with distinct distomesial spine and small ventral spine near juncture with article 4. Article 4 with strong distomesial spine and smaller ventromesial spine at midlength. Article 5 slightly shorter than article 4, with well-developed distomesial spine and 1-2 smaller ventromesial spines; breadth 0.6 × height of antennular ultimate article. Flagellum of 7-9 segments overreaching rostrum by proximal 3-4 segments, far falling short of distal end of P1 merus.

Mxp: Mxp1 with bases close to each other but not contiguous. Mxp3 with strong distolateral spine on each of coxa, ischium, merus and carpus. Basis without denticles on mesial ridge. Ischium with 1 or 2 spines close to distolateral spine; crista dentata with distally diminishing denticles. Merus 1.7 × longer than ischium, with 3-4 spines along flexor margin and strong distolateral spine; extensor margin with very small or moderate-sized distal spine. Carpus with 2 spines on extensor surface.

P1: 4.5 × longer than carapace, with sparse short setae; 9 rows of spines on merus continued on to palm (3 dorsal, 2 lateral, 2 mesial, 2 ventral). Ischium with well-developed dorsal spine and prominent subterminal spine on ventromesial margin. Merus 1.1-1.2 × length of carapace. Carpus 1.1 × length of merus. Palm 3.0-3.3 × longer than broad, slightly shorter than carpus. Fingers broad relative to length, distally crossing when closed. Movable finger 0.5-0.6 × length of palm; opposable margin with low median process; mesial margin with small spines on proximal half; fixed finger with obsolescent spines on proximal lateral margin, sinuous on opposable margin.

P2-4: Subcylindrical, with rows of spines. Meri successively shorter posteriorly (P3 merus 0.94-0.95 × length of P2 merus, P4 merus 0.8-0.9 × length of P3 merus), equally broad on P2-4; P2 merus subequal to or slightly shorter than carapace, subequal to or slightly longer than P2 propodus; P3 merus 0.9-1.0 × [0.8 × on female paratype MNHN-IU-2014-16457] length of P3 propodus; P4 merus 0.8-0.9 × length of P4 propodus; length-breadth ratio, 5.8-6.1 on P2, 5.0-5.7 on P3, 4.6-4.7 on P4; 5 rows of spines (dorsal and ventrolateral rows of strong spines, 2 lateral and 1 ventromesial row of small spines) continued on to carpus. Carpi subequal, longer than dactyli (carpus-dactylus length ratio, 1.3-1.5 on P2, 1.2-1.4 on P3, 1.1-1.3 on P4); carpus-propodus length ratio, 0.5 on P2, 0.4-0.5 on P3, 0.4 on P4. Propodi subequal on P3 and P4, shortest on P2; extensor margin with row of spines accompanying subparalleling row of fewer spines laterally and mesially; flexor margin straight, with pair of terminal spines preceded by row of spines (distal 5 in zigzag arrangement). Dactyli shorter than carpi (dactylus-carpus length ratio, 0.7-0.8 on P2, 0.7-0.9 on P3, 0.8-1.0 on P4); dactylus-propodus length ratio, 0.4 on P2-3, 0.3-0.4 on P4; flexor margin slightly curving, with row of 9-10 sharp spines subperpendicular to margin, ultimate smaller than penultimate and subequal to antepenultimate and remainder.

Eggs. Number of eggs carried, 3-17; size, 1.06 mm × 1.02 mm - 1.12 mm × 1.23 mm.

Color. Ovigerous female from Vanuatu (MNHN-IU-2014-16946): Base color translucent bluish gray to pale pink, pereopods with dark orange bands on meri, carpi and propodi.

REMARKS — The following six species are grouped together by the noticeably spinose carapace and pereopods, and the arrangement of spines on the P2-4 dactyli (the ultimate spine preceded by subperpendicularly or obliquely directed, proximally diminishing spines), and the sternite 3 anterior margin bearing two submedian spines separated by a notch or sinus: *Uroptychus ciliatus* (Van Dam, 1933), *U. numerosus* n. sp., *U. quartanus* n. sp., *U. senarius* n. sp., *U. spinirostris* (Ahyong & Poore, 2004), and *U. spinimanus* Tirmizi, 1964. *Uroptychus chacei* (Baba, 1986b) looks very close to this group but is different in having no submedian spines and no median notch on the anterior margin of sternite 3. Within this group, *Uroptychus spinimanus* keys out first by having no spine on the abdomen and no strong spine on the lateral margin of sternite 4. *Uroptychus numerosus* differs from the other species in having an elongate, distally broad (not tapering) rostrum with more numerous spines (9 versus 1-5). *Uroptychus ciliatus* differs from *U. spinirostris*, *U. quartanus* and *U. senarius* in having the ultimate of the flexor marginal spines of P2-4 dactyli, which is subequal to or slightly larger than instead of more slender than the penultimate. *Uroptychus spinirostris* differs from *U. quartanus* and *U. senarius* in having a relatively long rostrum (the length being three-quarters instead of at most slightly more than half that of the carapace) and more spinose abdominal somites, especially somites 6 that bears numerous spines instead of no spine.

Uroptychus senarius is distinguished from *U. quartanus* by the sternal plastron that is successively broader posteriorly instead of broadest on sternite 4 and subequally broad on sternites 5-7; the ventral surface of sternite 4 has scattered granules and relatively thick setae instead of a row of denticles arranged in a concentric arc and fine setae; the abdominal somite 1 bears a row of four strong spines instead of three spines; the abdominal somite 3-5 bear two transverse (anterior and posterior) rows of spines instead of an anterior row only; the antennal scale terminates in the midlength of article 5 rather than reaching the end of that article; the crista dentata of Mxp3 bears a row of distally diminishing, numerous, closely arranged spines instead of loosely arranged, less numerous spines; the posterior plate of the telson is distinctly emarginate instead of subsemicircular on the posterior margin.

The three pairs of lateral rostral spines as possessed by *U. senarius* are also known in the larger male of *U. spinirostris* reported by McCallum & Poore (2013) from western Australia, which usually bears two pairs of spines (see above). However, the rostrum in *U. senarius* is broader relative to length (length-breadth ratio, about 1.5 versus more than 2.0). Also, sternite 4 in this new species is spineless on the ventral surface instead of bearing one or two pairs of spines.

The ovigerous female from MUSORSTOM 8 Stn CP1024 (MNHN-IU-2014-16946) is somewhat different from the others: the rostrum bears two lateral spines on each side and an additional small spine proximal to the right proximal spine; the P2-4 meri are relatively broad compared to those of the other specimens, with the length-breadth ratio, 5.0 on P2, 4.8 on P3 and 4.0 on P4, versus 5.8-6.1 on P2, 5.4-5.7 on P3, 4.6-4.7 on P4 in the other specimens; and P3 merus is relatively short (it is as long as P3 propodus, 0.8 times as long in the other specimens). This specimen is provisionally placed in *U. senarius* until more material becomes available.

Uroptychus senticarpus n. sp.

Figures 233, 234

TYPE MATERIAL — Holotype: **New Caledonia**, Norfolk Ridge. NORFOLK 2 Stn DW2072, 25°21.25'S, 168°57.16'E, 1000-1005 m, 26.X.2003, ♀ 3.7 mm (MNHN-IU-2013-8512). Paratype: Collected with holotype, 1 ♂ 3.7 mm (MNHN-IU-2013-8513).

ETYMOLOGY — The specific name is a noun in apposition from the Latin *sentis* (= thorn) plus *carpus* (= wrist joint), alluding to a pronounced spine on the dorsodistal margin of P1 carpus.

DISTRIBUTION — Norfolk Ridge, 1000-1005 m.

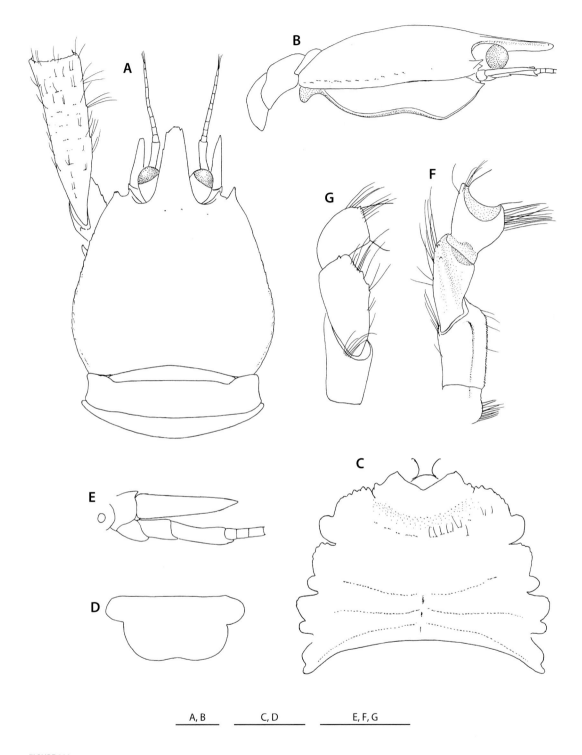

FIGURE 233

Uroptychus senticarpus n. sp., holotype, female 3.7 mm (MNHN-IU-2013-8512). **A**, carapace and anterior part of abdomen, proximal part of left P1 included, setae on carapace omitted, dorsal. **B**, same, setae omitted, lateral. **C**, sternal plastron, with excavated sternum and basal parts of Mxp1. **D**, telson. **E**, right antenna, ventral. **F**, right Mxp3, ventral. **G**, same, lateral. Scale bars: 1 mm.

DESCRIPTION — *Carapace*: 1.3 × broader than long (0.8 × as long as broad), broadest on posterior third; greatest breadth 1.8 × distance between anterolateral spines. Dorsal surface convex from anterior to posterior, smoothly sloping down to rostrum, thickly covered with small pits suggesting supporting fine setae. Lateral margins convex, with row of finely granulate ridges; anterolateral spine distinctly larger than lateral orbital spine, directed anterolaterally. Rostrum distally broken, horizontal, narrow triangular, with interior angle 20°; dorsal surface somewhat concave; lateral margin straight with a few very small spines distally; length presumably more than half that of remaining carapace, breadth 0.4 × that of carapace measured along posterior margin. Lateral orbital angle produced to small spine, separated from anterolateral spine by basal breadth of latter spine, and located slightly anterior to that spine. Pterygostomian flap smooth on surface, anterior margin angular, produced to small spine.

Sternum: Excavated sternum with convex anterior margin; surface with low ridge in midline. Sternal plastron 0.8 × as long as broad; lateral extremities convexly divergent posteriorly. Sternite 3 shallowly depressed; anterior margin excavated in broad V-shape, without median notch and submedian spines (with very small median notch in paratype). Sternite 4 with anterolateral margin convex and denticulate, about 2 × longer than posterolateral margin. Anterolateral margin of sternite 5 anteriorly convex and denticulate, 1.5 × longer than posterolateral margin of sternite 4.

Abdomen: Somite 1 convex from anterior to posterior, without transverse ridge. Somite 2 tergite 3.1 × broader than long; pleural lateral margin concavely divergent posteriorly, posterolateral end blunt. Pleuron of somite 3 with blunt posterolateral terminus. Telson 0.5 × as long as broad; posterior plate 1.5 × longer than anterior plate, deeply concave on posterior margin.

Eye: 1.3 × longer than broad, distally narrowed, lateral and mesial margins somewhat convex; cornea two-thirds as long as remaining eyestalk.

Antennule and antenna: Ultimate article of antennule 4.2 × longer than high. Antennal peduncle extending far beyond eye. Article 2 with small lateral spine. Antennal scale 2.0 × broader than article 5, extending slightly beyond proximal first segment of flagellum, greatly overreaching eye. Articles 4 and 5 each with small distomesial spine, article 5 1.7 × longer than article 5, breadth 0.5 × height of antennular ultimate article. Flagellum of 7 segments slightly falling short of distal margin of P1 merus.

Mxp: Mxp1 with bases broadly separated. Mxp3 with relatively long simple (not plumose) setae other than brushes on distal articles. Basis smooth on mesial ridge. Ischium with flexor margin not rounded distally; crista dentata with about 35 small denticles of subequal size. Merus 2.2 × longer than ischium, flexor margin roundly ridged, with 3 tiny spines on distal third. Small distodorsal spine on each of merus and carpus.

P1: 4.7 × longer than carapace, sparingly covered with relatively short fine setae. Ischium with small triangular dorsal spine, ventromesial margin with a few denticles proximally, subterminal spine vestigial. Merus with minutely denticulate short ridges supporting setae on surface; with small distomesial and distolateral spines on ventral surface, row of denticles along distodorsal margin, and very small spines along mesial margin; length subequal to that of carapace. Carpus 1.2 × longer than merus, with a few small dorsal tubercles and a few small mesial spines on proximal portion, distodorsal margin with well-developed median spine flanked by a few denticles. Palm 4.1 × longer than broad, 1.1 × longer than carpus. Fingers moderately depressed and relatively narrow, not gaping, distally incurved, crossing when closed; fixed finger sinuous on opposable margin; movable finger half as long as palm, opposable margin with low proximal process.

P2-4: Relatively broad and somewhat compressed mesio-laterally, with fine soft setae. Meri successively shorter posteriorly (P3 merus 0.9 × length of P2 merus, P4 merus 0.7 length of P3 merus); breadths subequal on P2-4; length-breadth ratio, 4.4 on P2, 3.8 on P3, 2.5 on P4; dorsal margin rounded, not cristiform, smooth and unarmed on P2-4. P2 merus 0.7 × as long as carapace, 1.1 × length of P2 propodus; P3 merus 0.9 × length of P3 propodus; P4 merus 0.7 × length of P4 propodus. Carpi successively shorter posteriorly; carpus-propodus length ratio, 0.5 on P2 and P3, 0.4 on P4. Propodi longer on P3 than on P2 and P4, subequal on P2 and P4; flexor margin somewhat convex on distal portion, ending in pair of terminal spines preceded by 5, 3, 2 spines on P2, P3, P4 respectively at most on distal third. Dactyli successively longer posteriorly, moderately curving; flexor margin with 7, 7, 8 strong spines loosely arranged, proximally diminishing and somewhat inclined on P2, P3, P4 respectively; dactylus-carpus length ratio, 0.8, 0.9, 1.2 on P2, P3, P4 respectively.

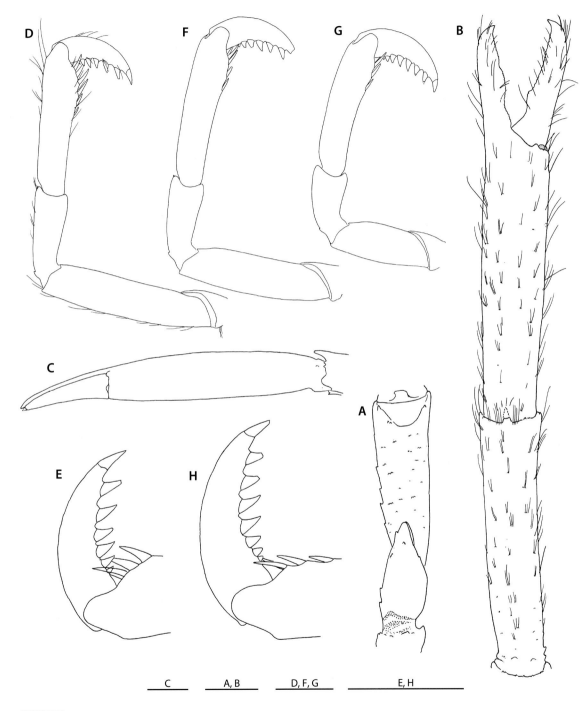

FIGURE 234

Uroptychus senticarpus n. sp., holotype, female 3.7 mm (MNHN-IU-2013-8512). **A**, left P1, proximal part, setae omitted, ventral. **B**, same, distal three articles, dorsal. **C**, right P1, distal part, setae omitted, mesial. **D**, right P2, lateral. **E**, same, distal part, setae omitted, lateral. **F**, right P3, setae omitted, lateral. **G**, right P4, setae omitted, lateral. **H**, same, distal part, setae omitted, lateral. Scale bars: 1 mm.

REMARKS — The anterior margin of sternite 3 is somewhat different between the available two type specimens: a broad V-shape in the holotype, additional very small median notch in the paratype. *Uroptychus senticarpus* n. sp., *U. plumella* n. sp. and *U. shanei* n. sp. belong to a group of species characterized by the carapace lateral margin bearing an anterolateral spine only, sternite 3 with the anterior margin emarginate in a broad V-shape, P1 nearly spineless, with fingers distally strongly incurved, and the P2-4 dactyli ending in a strong spine preceded by similar, proximally diminishing spines.

Uroptychus senticarpus is distinguished from *U. plumella* by the anterolateral spine of the carapace that is directed anterolaterally instead of straight forward; the antennal article 2 is produced to a small spine instead of being acuminate without a distinct spine, and the articles 4 and 5 each bearing a distinct spine instead of a tiny one (that of article 4 tubercle-like); the Mxp3 crista dentata bears numerous minute denticles instead of loosely arranged, distally diminishing denticles; and the pterygostomian flap is produced to a sharp spine instead of bearing small spine. Distinguishing characters between *U. senticarpus* and *U. shanei* are discussed under the accounts of the latter species (see below).

Uroptychus septimus n. sp.
Figures 235, 236

TYPE MATERIAL — Holotype: **Wallis and Futuna Islands**. MUSORSTOM 7 Stn DW523, 13°12'S, 176°16'W, 455-515 m, with corals of Chrysogorgiidae (Calcaxonia), 13.V.1992, ov. ♀ 7.1 mm (MNHN-IU-2014-16947).

ETYMOLOGY — From the Latin *septimus* (seventh), alluding to the seventh cruise of MUSORSTOM, by which the species was taken.

DISTRIBUTION — Wallis and Futuna Islands (SW Pacific); 455-515 m.

DESCRIPTION — Medium-sized species. *Carapace*: Nearly as long as broad; greatest breadth 1.5 × distance between anterolateral spines. Dorsal surface smooth and glabrous, nearly horizontal in profile, smoothly sloping down anteriorly to rostrum. Lateral margins slightly convex; anterolateral spine small, located somewhat posterior to level of lateral orbital angle, and not overreaching that spine; tubercle-like small spine at point one-fifth from anterior end (anterior end of branchial margin), followed by granulation; no ridge along posterior portion. Rostrum narrow triangular, with interior angle of 23°, slightly overreaching eyes; dorsal surface somewhat concave; ventral surface straight horizontal; lateral margins somewhat concave; length less than half that of remaining carapace, breadth half carapace breadth at posterior carapace margin. Lateral limit of orbit acuminate, lacking distinct spine. Pterygostomian flap with somewhat roundish anterior margin lacking distinct spine; surface smooth.

Sternum: Excavated sternum with anterior margin with small median spine between bases of Mxp1; surface with small spine in center. Sternal plastron 1.1 × broader than long, lateral extremities straight divergent posteriorly. Sternite 3 depressed well, anterior margin strongly excavated in broad V-shape, with pair of sharp submedian spines separated by V-shaped notch, anterolateral angle rounded. Sternite 4 having convex anterolateral margin anteriorly ending in spine nearly as large as submedian spines of sternite 3, followed by posteriorly diminishing denticles; posterolateral margin about half as long as anterolateral margin. Sternite 5 having anterolateral margin strongly convex, 1.3 × longer than posterolateral margin of sternite 4.

Abdomen: Smooth and glabrous, with relatively long somites. Somite 1 feebly convex from anterior to posterior. Somite 2 tergite 2.1 × broader than long; pleuron posterolaterally blunt; lateral margins concavely strongly divergent posteriorly. Pleuron of somite 3 with blunt lateral terminus. Telson two-thirds as long as broad; posterior plate distinctly emarginate, length 1.8 × that of anterior plate.

Eye: Slightly falling short of apex of rostrum, about twice as long as broad, lateral and mesial margins somewhat concave. Cornea somewhat inflated, length much more than half that of remaining eyestalk.

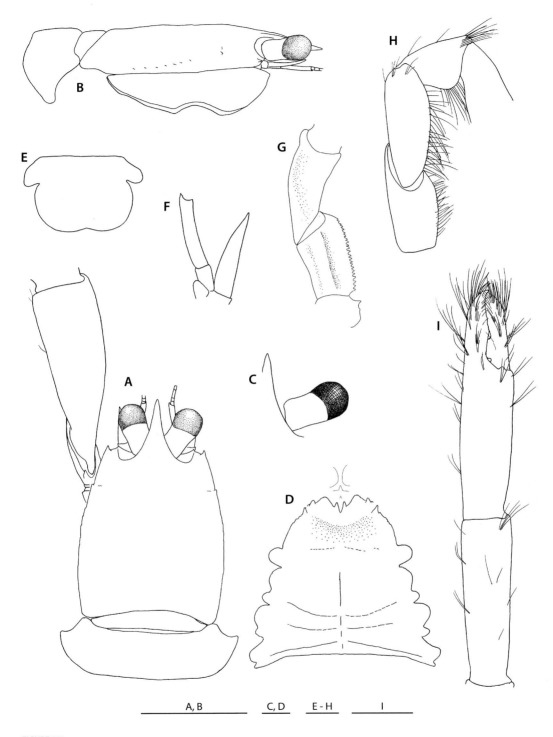

FIGURE 235

Uroptychus septimus n. sp., holotype, ovigerous female 7.1 mm (MNHN-IU-2014-16947). **A**, carapace and anterior part of abdomen, proximal part of left P1 included, dorsal. **B**, same, lateral. **C**, right eye and part of rostrum, dorsal. **D**, sternal plastron, with excavated sternum and basal parts of Mxp1. **E**, telson. **F**, left antenna, ventral. **G**, right Mxp3, setae omitted, ventral. **H**, same, lateral. **I**, left P1, dorsal. Scale bars: A, B, I, 5 mm; C-H, 1 mm.

Antennule and antenna: Ultimate article of antennular peduncle 2.5 × longer than high. Antennal peduncle relatively slender, reaching distal end of cornea. Article 2 distolaterally acuminate, lacking distinct spine. Antennal scale 1.5 × broader than article 5, tapering, slightly falling short of distal end of article 5 and distal end of eye. Article 4 unarmed. Article 5 with small distomesial spine; length 2.7 × that of article 4, breadth half height of ultimate article of antennule. Flagellum consisting of 17 or 20 segments, nearly reaching distal margin of P1 merus.

Mxp: Mxp1 with bases very close to each other. Mxp3 scarcely setose laterally. Basis with 1 distal denticle proximally followed by 1 or 2 obsolescent denticles. Ischium with 25-26 denticles on crista dentata, flexor margin distally not rounded. Merus 1.9 × longer than ischium, unarmed, not well compressed mesio-laterally, ridged along smoothly convex flexor margin. Carpus with obsolescent distolateral spine.

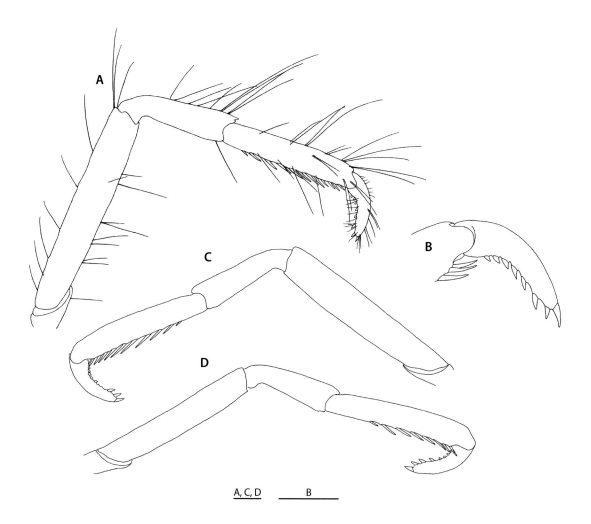

FIGURE 236

Uroptychus septimus n. sp., holotype, ovigerous female 7.1 mm (MNHN-IU-2014-16947). **A**, right P2, lateral. **B**, same, distal part, setae omitted, lateral. **C**, left P3, setae omitted, lateral. **D**, right P4, setae omitted, lateral. Scale bars: 1 mm.

P1: Moderately massive, 5.2 × longer than carapace; sparsely setose except for fingers, setae relatively short. Ischium dorsally with broad short spine, ventromesially feebly denticulate, without subterminal spine. Merus 1.3 × longer than carapace, unarmed except for very small distomesial and distolateral spines on ventral surface, ventromesially feebly granular. Carpus 1.2 × longer than merus. Palm smooth dorsally, rounded on mesial margin, 2.8 × longer than broad, 0.9 × length of carpus. Fingers slightly gaping, distally crossing when closed; fixed finger with low median eminence on opposable margin; movable finger about half as long as palm, opposable margin with disto-proximally broad, bluntly bidentate proximal process.

P2-4: Slender, with sparse long setae. Meri moderately compressed mesio-laterally, dorsally unarmed, ventrolaterally ending in small spine; successively diminishing posteriorly (P3 merus 0.9 × length of P2 merus, P4 merus 0.9 × length of P3 merus), subequally broad on P2 and P3, somewhat narrower on P4 (P4 merus 0.9 × breadth of P3 merus); length-breadth ratio, 6.2 on P2, 5.6 on P3 and P4; P2 merus as long as carapace, 1.5 × longer than P2 propodus; P3 merus 1.2 × length of P3 propodus; P4 merus as long as P4 propodus. Carpi subequal in length on P2 and P3, shortest on P4; much longer than dactyli (carpus-dactylus length ratio, 2.0 on P2, 1.7 on P3, 1.5 on P4); carpus-propodus length ratio, 0.7 on P2, 0.6 on P3, 0.5 on P4; extensor margin with small distal spine at most on P2. Propodi longest on P3, shortest on P2; flexor margin nearly straight, with pair of distal spines slightly distant from juncture with dactylus and preceded by row of spines (8 or 11 on P2, 8 on P3, 6 or 7 on P4) nearly along entire length on P2, slightly more distally on P3 and P4, distalmost of these unpaired spines equidistant between distal second and terminal pair. Dactyli relatively slender, strongly curving at proximal third, ending in prominent spine preceded by 7-8 smaller flexor marginal spines moderately inclined, loosely arranged and diminishing toward base of article; dactylus-carpus length ratio, 0.5 on P2, 0.6 on P3, 0.7 on P4; dactylus-propodus length ratio, 0.35-0.36 on P2, 0.35 on P3, 0.36 on P4.

Eggs. Number of eggs carried, 7; diameter, 1.4 mm.

REMARKS — The new species resembles *U. comptus* Baba, 1988, *U. empheres* Ahyong & Poore, 2004, *U. pollostadelphus* n. sp. and *U. sagamiae* Baba, 2005 in the carapace shape, in having the pterygostomian flap roundish on the anterior margin, and in having the P2-4 dactyli with the ultimate of the flexor spines strongest. *Uroptychus comptus* and *U. empheres* can be separated from the other species including *U. septimus* by having no epigastric spines. They also differ from *U. septimus* in having no spine on the antennal article 5 and in having a ventrally granulose P1 merus. *Uroptychus septimus* is differentiated from *U. pollostadelphus* by the P2-4 meri that are successively shorter posteriorly, with the P4 merus 0.9 times as long as the P3 merus (in *U. pollostadelphus*, the P2 merus is as long as the P3 merus and the P4 merus is much shorter, 0.6 times the length of the P3 merus); the antennal article 2 is distolaterally acuminate instead of bearing a distinct spine, and the article 5 bears a distinct distomesial spine instead of being unarmed. *Uroptychus septimus* differs from *U. sagamiae* Baba, 2005 in having the antennal article 2 distolaterally acuminate instead of bearing a distolateral spine; the antennal scale is barely reaching instead of overreaching the distal end of the article 5; the P1 palm is smooth instead of denticulate on the ventral surface; the P2 merus is as long as instead of much shorter than the carapace and 1.5 times longer than instead of as long as the P2 propodus; and the P2 carpus is longer, being 0.7 instead of 0.5 times the length of the P2 propodus.

Uroptychus setifer n. sp.
Figures 237, 238, 306C

Uroptychus sp. — Kawamoto & Okuno 2003: 98, unnumbered fig.; 2006: 98, unnumbered fig.

TYPE MATERIAL — Holotype: **Vanuatu**. SANTO Stn NB12, 15°33.1'S, 167°09.6'E, 20 m, 19.IX.2006, ov. ♀ 2.2 mm (MNHN-IU-2010-5436). Paratypes: **Japan**. Kume-jima, Ryukyu Islands, 35 m, with Scuba, associated with coral, 27. VI.2001, T. Kawamoto coll., 1 ♂ 3.2 mm (CMNH-ZC 00620), 1 ♂ 3.2 mm (CMNH-ZC 02157).

ETYMOLOGY — From the Latin *seta* (bristle, seta) and *fer* (the suffix meaning bear), referring to long setae on the body and pereopods. When alive, the setae are whitish.

DISTRIBUTION — Vanuatu and Japan (Ryukyu Islands); 12-35 m.

DESCRIPTION — Small species. *Carapace*: About as long as broad; greatest breadth 1.6 × distance between anterolateral spines. Dorsal surface sparingly with long setae, moderately convex from anterior to posterior, cardiac region more convex; with weak (holotype) or distinct (paratypes) depression between gastric and cardiac regions; 6 dorsal spines: 3 in transverse row on epigastric region, followed by 2 placed side by side on mesogastric region, and 1 on cardiac region; 2 small spines each on mesial part of posterior branchial region distinct (holotype and larger male paratype) or absent (smaller male paratype). Lateral margins convexly (holotype) or nearly straight (paratypes) divergent, with 6 spines; first anterolateral, moderate in size, overreaching lateral orbital spine; second relatively remote from first (with 1 or 2 much smaller spines between, in paratypes), situated on anterior end of anterior branchial margin, distinctly posterior to level of epigastric spines; third to sixth situated on posterior branchial region and slightly dorsomesial to margin; last situated at posterior quarter of posterior branchial margin, followed by ridge. Rostrum narrow triangular, with interior angle of 22-23,° nearly horizontal; dorsal surface moderately concave, length 0.47 (paratypes)-0.53 × (holotype) that of carapace, breadth half carapace breadth at posterior carapace margin. Lateral limit of orbit with tiny spine, directly mesial to anterolateral margin. Pterygostomian flap smooth on surface, with small spine on roundish anterior margin.

Sternum: Excavated sternum with convex anterior margin, surface smooth, without ridge. Sternal plastron 0.8 × as long as broad, lateral extremities divergent posteriorly. Sternite 3 well depressed, anterior margin shallowly excavated, with V-shaped median notch separating obsolescent submedian spines; anterolaterally angular. Sternite 4 with anterolateral margin relatively short, bearing 2 spines, anterior spine at anterior end and smaller than posterior spine situated slightly anterior to midlength, posterolateral margin as long as anterolateral margin. Sternite 5 slightly shorter than sternite 4; anterolateral margin feebly convex, moderately divergent posteriorly, length 1.4-1.5 × that of posterolateral margin of sternite 4.

Abdomen: Smooth, sparsely with tufts of a few to several long setae. Somite 1 gently convex from anterior to posterior, without distinct ridge. Somite 2 tergite 2.1 × broader than long; pleuron tapering, with lateral margin concavely strongly divergent posteriorly. Pleuron of somite 3 bluntly angular on posterolateral terminus. Telson half as long as broad; posterior plate 1.2 × longer than anterior plate, weakly concave on posterior margin.

Eye: Elongate, twice as long as broad, distally narrowed, slightly falling short of rostral tip. Cornea 0.3 × as long as remaining eyestalk.

Antennule and antenna: Ultimate article of antennule 4.5 × longer than high. Antennal peduncle reaching apex of rostrum. Article 2 unarmed. Antennal scale slightly broader than article 5, ending in midlength of that article. Articles 4 and 5 each with distomesial spine; article 5 1.5 × (holotype)-1.9 × (paratypes) longer than article 4, breadth 0.7 × height of ultimate article of antennule. Flagellum consisting of 7 or 8 (holotype)-10 (paratypes) segments, far falling short of distal end of P1 merus.

Mxp: Mxp1 with bases separated from each other. Mxp3 slender, with sparse long setae laterally. Basis without denticles on mesial ridge. Ischium with distinct spine lateral to distal end of flexor margin; crista dentata with obsolescent denticles; flexor margin not rounded distally. Merus 1.5 × longer than ischium, narrow relative to length, with distolateral and median flexor marginal spine. Carpus also with distolateral spine.

P1: 6.6 × longer than carapace, slender, with tufts of long setae. Ischium with sharp dorsal spine, ventromesially with subterminal spine. Merus 1.5 (holotype)-1.6 × (paratypes) longer than carapace; 4 rows of spines (more numerous in male paratypes) each continued on to carpus. Carpus 1.3 × longer than merus, as high as broad. Palm 5.0 × (holotype) or 3.1-3.4 × (paratypes) longer than broad; mesial margin smooth (holotype) or cristate with 8 or 9 spines (paratypes), lateral margin smooth (holotype) or with obsolescent spines (paratypes); length 1.1-1.2 × that of carpus. Fingers not gaping, distally incurved, crossing when closed; movable finger 0.3 × length of palm, opposable margin with subtriangular blunt median process proximal to opposite eminence on fixed finger.

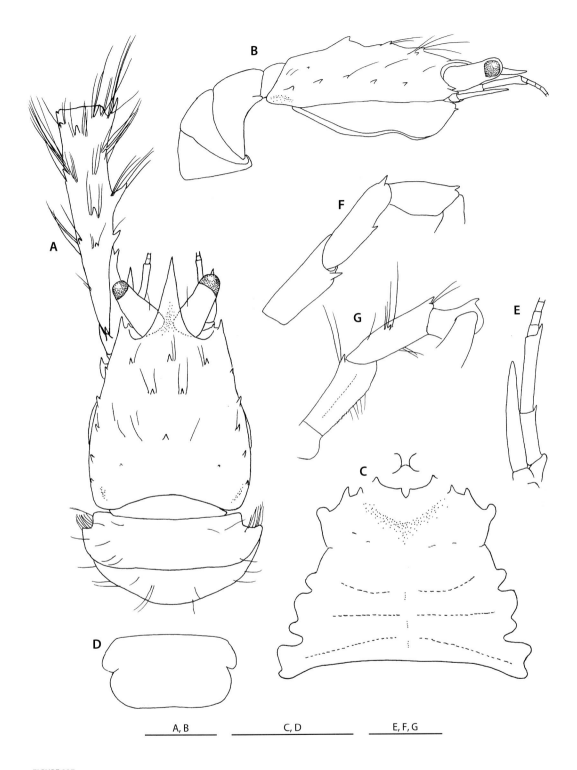

FIGURE 237

Uroptychus setifer n. sp., holotype, ovigerous female 2.2 mm (MNHN-IU-2010-5436). **A**, carapace and anterior part of abdomen, proximal part of left P1 included, dorsal. **B**, same, lateral. **C**, sternal plastron, with excavated sternum and basal parts of Mxp1. **D**, telson. **E**, right antenna, ventral. **F**, right Mxp3, lateral. **G**, same, ventral. Scale bars: A-D, 1 mm; E-G, 0.5 mm.

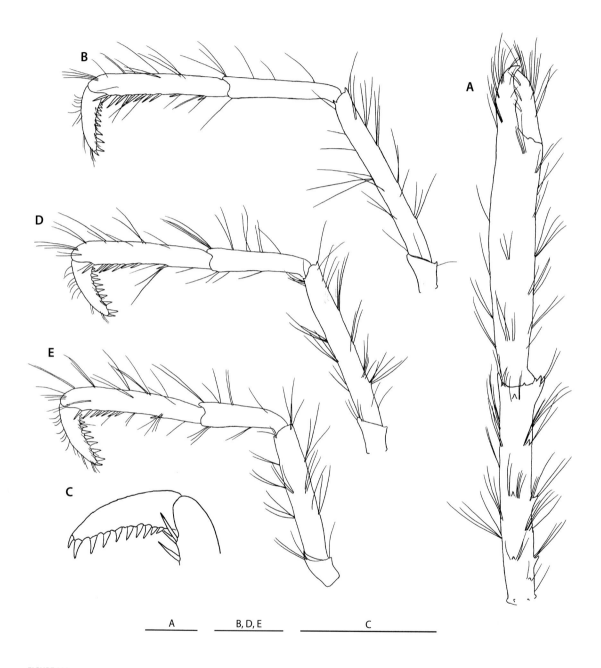

FIGURE 238

Uroptychus setifer n. sp., holotype, ovigerous female 2.2 mm (MNHN-IU-2010-5436). **A**, left P1, dorsal. **B**, left P2, lateral. **C**, same, distal part, setae omitted, lateral. **D**, left P3, lateral. **E**, left P4, lateral. Scale bars: 1 mm.

P2-4: Slender, somewhat compressed mesio-laterally, setose like P1. Ischium with small dorsal spine. Meri successively shorter posteriorly (P3 merus 0.9 × length of P2 merus, P4 merus 0.8 or 0.9 × length of P3 merus); breadths subequal on P2-4; small distal spine on dorsal and ventrolateral margins on P2-3, distodorsal spine obsolete on P4; small denticle-like spine at proximal quarter of dorsal margin on P2 and P3, obsolete on P4. P2 merus 1.2 × (holotype) or 1.6 × (paratypes) longer than carapace, 1.3 (holotype) or 1.5 × (paratypes) longer than P2 propodus; P3 merus 1.2 × (holotype) or 1.3 × (paratypes) longer than P3 propodus, P4 merus 0.9 × (holotype) or 1.3 × (paratypes) length of P4 propodus. Carpi unarmed, successively shorter posteriorly; carpus-propodus length ratio, 0.8 (holotype) or 1.0 (paratypes) on P2, 0.7 (holotype) or 0.8 (paratypes) on P3, 0.6 (holotype) or 0.8 (paratypes) on P4. Propodi subequal (holotype) or successively shorter posteriorly (paratypes); flexor margin straight, ending in pair of terminal spines preceded by 8 (holotype) or 11 (paratypes) spines on P2, 7 or 8 spines on P3, 7 spines on P4, proximal-most distinctly remote from proximal second. Dactyli subequal, 0.4 × as long as propodi, much shorter than carpi (dactylus-carpus length ratio, 0.5 on P2, 0.6 on P3 and P4); flexor margin gently curving, ending in slender spine preceded by row of 9 or 10 sharp triangular, somewhat obliquely directed, proximally diminishing spines.

Color. Holotype, ovigerous female (MNHN-IU-2010-5436): Generally translucent light purplish blue, with scattered brown spots, setae whitish. One of the Japanese specimens is illustrated by Kawamoto & Okuno (2003: 98).

REMARKS — The morphological differences between the holotype from Vanuatu and the paratypes from the Ryukyu Islands, Japan (see above under the description) leave some doubt about their identity, given that these specimens are from disjunct localities. On the other hand, the elongate eyestalks, tufts of long and whitish setae on the body and appendages, and the general colorations are consistent. These differences are here considered to be size-related, the Japanese specimens being much larger than the Vanuatu specimen. It is worth noting that this species appears to be a shallow water inhabitant.

The carapace spination is very similar to that of *U. angustus* n. sp. (see above) but not exactly the same. The metagastric spines are two in number, placed side by side in *U. setifer*, instead of a single as in *U. angustus*. *Uroptychus setifer* is readily distinguished from *U. angustus* by the following differences: the rostrum is laterally unarmed instead of bearing a subterminal spine on each side; the abdominal somite 1 is convex from anterior to posterior instead of bearing a distinct transverse ridge; the eyes are elongate and distally narrowed instead of uniformly broad proximally and distally; the antennal article 2 is unarmed instead of bearing a small but distinct distolateral spine; P2-4 are much more slender and unarmed instead of bearing spines on the dorsal or extensor margins of meri, carpi and propodi; the P2-4 dactyli are much shorter than instead of at most subequal to the carpi; and the pterygostomian flap is anteriorly roundish with a very tiny spine instead of being produced to a strong spine. In addition, *U. setifer* is distinctive in having longer setae in tufts on pereopods. It is worth noting that the breadth of the antennal article 5 is greater than the height of the antennular ultimate article in *U. setifer* and *U. angustus*, the unusual character also shared by *U. buantennatus* n. sp.

The spinose carapace and P1, the unarmed abdomen, and the spineless P2-4 meri and carpi link the species to *U. fusimanus* Alcock & Anderson, 1899. This congener is known only by the type material with the brief description. However, *U. setifer* is readily differentiated from that species by the less spinose carapace and P1 and by having the antennal scale barely reaching instead of fully reaching the tip of the peduncle.

According to Kawamoto & Okuno (2003, 2006), the species is rare in the Ryukyu Islands and found associated with antipatharians (black corals) growing on overhanging rocks of drop-offs in the open sea with good tidal current.

Uroptychus shanei n. sp.

Figures 239, 240

TYPE MATERIAL — Holotype: **Vanuatu**. Boa 0 Stn CP2307, 16°38.2'S,167°58.2'E, 586-646 m, 14.XI.2004, 1 ♀ 5.3 mm (MNHN-IU-2010-5481).

ETYMOLOGY — Named for Shane T. Ahyong for his contributions to the knowledge of squat lobsters and for his willingness to read a draft of the manuscript of this paper.

DISTRIBUTION — Vanuatu; 586-646 m.

DESCRIPTION — Small species. *Carapace*: 1.2 × broader than long (0.8 × as long as broad), broadest on posterior third; greatest breadth 1.8 × distance between anterolateral spines. Dorsal surface moderately convex from anterior to posterior, unarmed and sparingly covered with short fine setae. Lateral margins convex, without ridge on posterior portion; anterolateral spine small, slightly overreaching smaller lateral orbital spine; branchial margin with row of short setiferous ridges as to be seen as small crenulations in dorsal view. Rostrum straight horizontal, relatively broad triangular, with interior angle of 30°; dorsal surface flattish, lateral margin smooth; length half that of remaining carapace, breadth somewhat less than half carapace breadth at posterior carapace margin. Lateral orbital spine moderately remote from and located directly mesial to anterolateral spine. Pterygostomian flap smooth with sparse short fine setae on surface, anterior margin angular, ending in small spine.

Sternum: Excavated sternum with broadly convex anterior margin; surface smooth, without ridge and central spine. Sternal plastron 0.8 × as long as broad, lateral extremities convexly divergent posteriorly on sternites 4-6, subparallel on sternites 6-7. Sternite 3 shallowly depressed; anterior margin in broad V-shape without submedian spines, medially roundly excavated. Sternite 4 with anterolateral margin 1.5 × longer than posterolateral margin, anteriorly convex with obsolescent denticles. Anterolateral margin of sternite 5 anteriorly convex, 1.2 × longer than posterolateral margin of sternite 4.

Abdomen: Somite 1 without transverse ridge. Somite 2 tergite 3.4 × broader than long; pleural lateral margins concavely weakly divergent posteriorly, posterolateral end blunt. Pleuron of somite 3 with blunt lateral end. Telson half as long as broad; posterior plate 1.3 × longer than anterior plate, distinctly emarginate on posterior margin.

Eye: Short, 1.5 × longer than broad, not reaching midlength of rostrum; mesial and lateral margins subparallel. Cornea not dilated, half as long as remaining eyestalk.

Antennule and antenna: Ultimate article of antennule 2.9 × longer than high. Antennal peduncle extending far beyond cornea, falling short of apex of rostrum. Article 2 distolaterally acuminate, without distinct spine. Antennal scale twice as broad as article 5, proportionately broad distally, distally blunt, reaching distal end of article 5, greatly overreaching eye. Articles 4 and 5 unarmed; article 5 1.4 × longer than article 5, breadth slightly more than half height of antennular ultimate article. Flagellum of 12 segments ending in distal margin of P1 merus.

Mxp: Mxp1 with bases broadly separated. Mxp3 with relatively long fine setae other than brushes on distal articles. Basis with mesial ridge lobe-like, without denticles. Ischium with flexor margin not rounded distally; crista dentata with about 30 minute denticles. Merus 2 × longer than ischium, distally broader in lateral view, distolateral spine small but distinct; flexor margin moderately ridged, with 2 small spines on distal quarter. Carpus unarmed.

P1: 4.4 × longer than carapace, covered with long, fine, soft setae arising from short ridges. Ischium with short blunt dorsal spine, ventromesially with rudimentary denticles, without distinct subterminal spine. Merus with small blunt spines along mesial margin, slightly shorter than carapace. Carpus 1.1 × longer than merus, unarmed. Palm 2.6 × longer than broad, 1.2 × longer than carpus. Fingers moderately depressed, distally incurved, crossing when closed; fixed finger with low process at midlength of opposable margin; movable finger half as long as palm, opposable margin with obtuse proximal process.

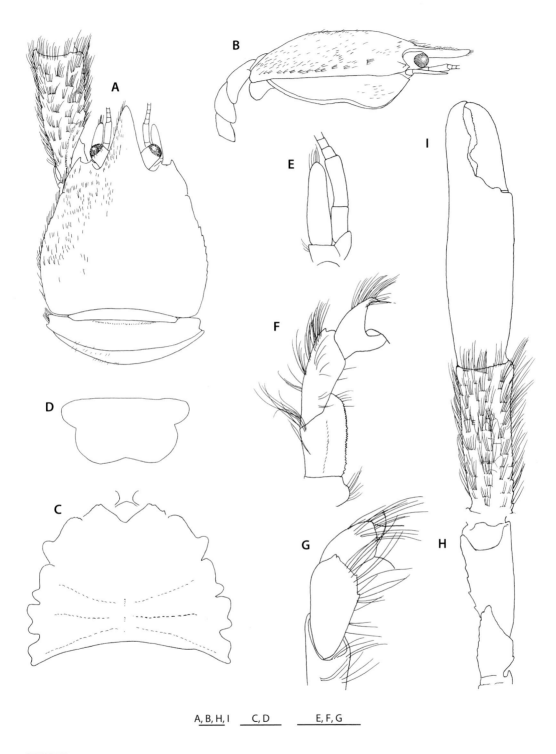

FIGURE 239

Uroptychus shanei n. sp., holotype, female 5.3 mm (MNHN-IU-2010-5481). **A**, carapace and anterior part of abdomen, proximal part of left P1 included, dorsal. **B**, same, lateral. **C**, sternal plastron, with excavated sternum and basal parts of Mxp1. **D**, telson. **E**, right antenna, ventral. **F**, right Mxp3, ventral. **G**, same, ventral. **H**, left P1, setae on fingers and palm omitted, dorsal. **I**, same, proximal part, setae omitted, ventral. Scale bars: 1 mm.

P2-4: Relatively broad and moderately compressed mesio-laterally, with fine soft setae. Meri slightly shorter on P3 than on P2, shorter on P4 than on P3 (P4 merus 0.8 × length of P3 merus); breadths slightly smaller on P4 than on P2 and P3; length-breadth ratio, 4.0 on P2, 3.8 on P3, 3.2 on P4; dorsal margins proximally irregular with obsolescent eminences, distally unarmed. P2 merus 0.7 × as long as carapace, as long as P2 propodus; P3 merus 0.9 × length of P3 propodus; P4 merus 0.8 × length of P4 propodus. Carpi subequal on P2 and P3, slightly shorter on P4; carpus-propodus length ratio, 0.6 on P2, 0.5 on P3 and P4. Propodi longest on P3, shortest on P2; flexor margin somewhat convex on distal portion, ending in pair of terminal spines preceded by 6 spines on P2 (proximal-most remote from proximal second and situated somewhat proximal to midlength), 3 spines on P3 and P4 (situated on distal third). Dactyli 0.4 × as long as propodi on P2-4, dactylus-carpus length ratio, 0.7 on P2, 0.8 on P3, 1.0 on P4; flexor margin moderately curving, distally ending in strong spine preceded by 5-6 elongate triangular, loosely arranged and proximally diminishing spines, penultimate spine more remotely separated from ultimate than from antepenultimate by twice distance between penultimate and antepenultimate.

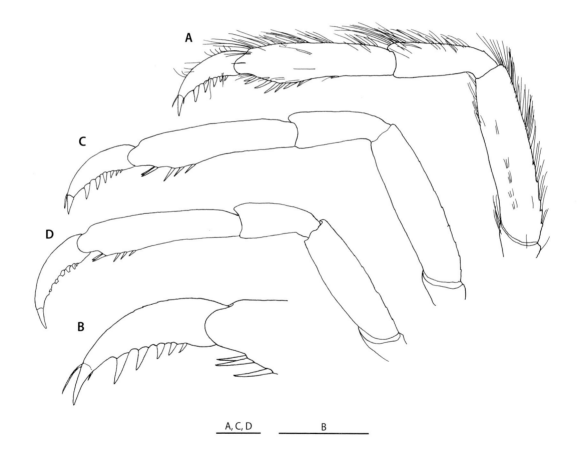

A, C, D   B

FIGURE 240

Uroptychus shanei n. sp., holotype, female 5.3 mm (MNHN-IU-2010-5481). **A**, left P2, lateral. **B**, same, distal part, setae omitted, lateral. **C**, left P3, setae omitted, lateral. **D**, left P4, setae omitted, lateral. Scale bars: 1 mm.

REMARKS — The new species resembles *U. plumella* n. sp. and *U. senticarpus* n. sp. (see under *U. plumella* for their similarities). *Uroptychus shanei* is distinctive in the shape and spination of the P2-4 dactyli and in the spination of the P1 carpus; the dactyli are more sharply tapering distally, with the penultimate of the flexor marginal spines more remote from the ultimate than from the antepenultimate, separated by twice instead of equal distance between the penultimate and antepenultimate spines; the P1 carpus is unarmed instead of bearing a spine at midpoint of the distodorsal margin. In addition, the antenna is different from one another; the antennal article 2 is distolaterally acuminate without distinct spine in *U. shanei* and *U. plumella*, bearing a small spine in *U. senticarpus*; the antennal articles 4 and 5 are unarmed in *U. shanei*, whereas these articles bear a tiny spine in *U. plumella*, a distinct spine in *U. senticarpus*; the antennal scale is distally blunt or rounded, terminating at the distal end of article 5 in *U. shanei*, whereas it tapers to a sharp point, extending far beyond article 5 in *U. plumella* and *U. senticarpus*. The anterolateral spine of the carapace is different in direction: directed straight forward in *U. plumella*, anteromesially in *U. shanei*, and anterolaterally in *U. senticarpus*.

Uroptychus sibogae Van Dam, 1933

Figure 241

Uroptychus sibogae Van Dam, 1933: 28, figs 39-41. — Baba 1981: 119, fig. 6; 1988: 45, fig. 17.
Not *Uroptychus sibogae* Poore *et al.* 2011: 330, pl. 8, fig. A (= *U. nebulosus* n. sp.).

TYPE MATERIAL — Holotype: **Indonesia**, West of Manado, Celebes, 1901 m, male, (ZMA De.101.665). [not examined].

MATERIAL EXAMINED — **Philippines**. MUSORSTOM 2 Stn CP15, 13°55'N, 120°29'E, 326-330 m, 21.XI.1980, 2 ♂ 5.8, 7.2 mm, 2 ov. ♀ 6.2, 6.7 mm (MNHN-IU-2014-16948). MUSORSTOM 3 Stn CP97, 14°00'N, 120°18'E, 189-194 m, 1.VI.1985, 1 ♀ 3.8 mm (MNHN-IU-2014-16949). **Indonesia**, Kai Islands. KARUBAR Stn CP16, 5°17'S, 132°50'E, 315-349 m, 24.X.1991, 5 ♂ 7.0-8.5 mm, 5 ov. ♀ 6.7-7.9 mm (MNHN-IU-2014-16950). – Stn CP27, 5°33'S, 132°51'E, 304-314 m, 26.X.1991, 1 ♂ 6.9 mm (MNHN-IU-2014-16951). – Stn DW32, 5°47'S, 132°51'E, 170-206 m, 26.X.1991, 1 ♂ 5.2 mm, 1 ov. ♀ 5.9 mm (MNHN-IU-2014-16952). Tanimbar Islands. KARUBAR Stn CP46, 8°01'S, 132°51'E, 271-273 m, 29.X.1991, 3 ♂ 4.7-6.8 mm, 1 ov. ♀ 6.4 mm (MNHN-IU-2014-16953). **Wallis and Futuna**. MUSORSTOM 7 Stn CP606, 13°21,4'S, 176°08,3'W, 420-430 m, 26 May 1992, 1 ♀ 4.2 mm (MNHN-IU-2010-5426).

DISTRIBUTION — West of Manado (Celebes [Sulawesi]), Molucca Sea off west coast of Halmahera, and Japan (Izu Shoto) in 170-1901 m, and now from the Philippines (off west coast of Luzon) and Indonesia (Kai and Tanimbar Islands), in 170-349 m.

SIZE — Males, 3.8-8.5 mm; females, 5.9-7.9 mm; ovigerous females from 5.9 mm.

DESCRIPTION — Medium-sized species. *Carapace*: As long as broad; greatest breadth 1.3-1.5 × distance between anterolateral spines. Dorsal surface smooth and glabrous, slightly convex from anterior to posterior, with feeble depression between gastric and cardiac regions. Lateral margins medially convex (in small specimens) or convexly somewhat divergent posteriorly, bearing relatively small or moderate-sized anterolateral spine and another spine of same or slightly larger size at anterior end of branchial margin, followed by 5-7 tubercles or denticle-like small spines (often obsolescent). Rostrum broad triangular (distally narrower in small specimens), with interior angle of 25°-35°; dorsal surface concave; length 0.4-0.6 × that of carapace (longer in small specimens), breadth half carapace breadth at posterior carapace margin. Lateral orbital spine distinctly smaller than and situated anterior to level of anterolateral spine. Pterygostomian flap anteriorly roundish with very small spine on anterior margin, smooth on surface.

Sternum: Excavated sternum anteriorly ending in sharp spine, surface with small spine or low process in center. Sternal plastron very slightly shorter than broad, lateral extremities gently divergent posteriorly. Sternite 3 well depressed, anterior margin narrowly and deeply excavated, with 2 well-developed submedian spines nearly contiguous or separated by narrow notch, anterolaterally rounded. Sternite 4 having anterolateral margin relatively long, about 3 × as long as posterolateral margin, moderately convex, anteriorly ending in anteriorly or anterolaterally directed spine occasionally followed by a few posteriorly diminishing tubercles. Anterolateral margin of sternite 5 anteriorly convex and denticulate, 1.5 × longer than posterolateral margin of sternite 4.

Abdomen: Somites smooth and glabrous. Somite 1 slightly convex from anterior to posterior. Somite 2 tergite 2.4-2.6 × broader than long; pleuron anterolaterally blunt angular, posterolaterally angular, lateral margin concavely strongly divergent posteriorly. Pleuron of somite 3 with angular lateral terminus. Telson 0.60-0.65 × as long as broad; posterior plate 1.5-1.8 × longer than anterior plate, posterior margin emarginate.

Eye: 1.5-1.8 × longer than broad, overreaching midlength of, but not reaching apex of rostrum; mesial margins concave (strongly so in small specimens). Cornea dilated (more distinctly in small specimens), about half as long as remaining eyestalk.

Antennule and antenna: Distal 2 articles of antennular peduncle relatively bulky, ultimate article 1.7-2.1 × longer than high. Antennal peduncle terminating in corneal margin. Article 2 with small lateral spine. Antennal scale barely reaching distal end of article 5, breadth 1.7-1.8 × that of article 5. Article 5 with small distomesial spine, length 2.4-3.0 × that of article 4, breadth 0.4 × height of ultimate antennular article. Flagellum of 21-24 segments reaching or slightly falling short of distal end of P1 merus.

Mxp: Mxp1 with bases nearly contiguous. Mxp3 basis with 3 denticles on mesial ridge, proximal 2 often obsolete. Ischium with 18-25 denticles on crista dentata, flexor margin distally not rounded. Merus twice as long as ischium, unarmed or with vestigial distolateral spine; flexor margin well ridged, fringed with long setae; mesial face flattish. Carpus with very small or obsolescent distolateral spine.

P1: Massive in large males, nearly glabrous except for distal articles, length 4.5-4.8 × that of carapace. Ischium dorsally with depressed triangular spine, ventrally unarmed. Merus 1.1-1.2 × longer than carapace, granulose on ventral surface. Carpus granulose on ventral surface, 1.1-1.2 × longer than merus. Palm 1.9-3.0 × (males), 2.3-2.4 × (females) longer than broad, slightly shorter than or subequal to carpus; breadth subequal to distance between lateral orbital spines in females and small males, occasionally more than that in large males; ventral surface smooth. Fingers slightly or moderately gaping (opposable margins fitting to each other in distal third, with proximal process of movable finger bidentate, located proximal to level of opposite low process on fixed finger) or strongly gaping (opposable margins fitting to each other in distal fifth of length, movable finger with relatively small blunt process in gaping margin, no process on opposable margin of fixed finger), crossing distally; movable finger 0.5-0.6 × length of palm (relatively long in small specimens, irrespective of sex).

P2-4: Relatively slender, compressed mesio-laterally, with sparse long setae. Meri subequal on P2 and P3 or slightly longer (in large specimens) or slightly shorter (in small specimens) on P3 than on P2; length-breadth ratio, 4.9-6.0 on P2, 4.7-6.0 on P3, 4.2-5.5 on P4 (broader in large specimens); breadths subequal on P2 and P3 or slightly larger on P3 than on P2; dorsal margin unarmed, ventrolateral margin with small distal spine; P2 merus subequal to length of carapace, 1.4 × longer than P2 propodus; P3 merus 1.3 × length of P3 propodus; P4 merus 0.7-0.8 × length and 0.8 × breadth of P3 merus, subequal to length of P4 propodus. P2 carpus subequal to P3 carpus, P4 carpus 0.8-0.9 × length of P3 carpus; carpus-propodus length ratio, 0.6 on P2, 0.5 on P3 and P4, distinctly longer than dactyli (carpus-dactylus length ratio, 1.6-1.7 on P2, 1.5-1.7 on P3, 1.2-1.3 on P4); extensor margin with small distal spine at least on P2 (obsolescent in small specimens). Propodi shortest on P2, longest on P3; flexor margin nearly straight, ending in pair of spines (very close to juncture with dactylus) preceded by row of 9-10 relatively long movable spines on P2 and P3, 7-8 spines on P4 (located on distal three-quarters on P2, slightly more distal on P3 and P4). Dactyli 0.3-0.4 × as long as propodi on P2-4, shorter than carpi (dactylus-carpus length ratio, 0.6 on P2, 0.6-0.7 on P3, 0.8 on P4); flexor margin curving at proximal third, ending in strong spine preceded by 8-9 similar spines diminishing toward base of article, ultimate spine strongest, penultimate slightly closer to ultimate than to antepenultimate on P2, equidistant between on P3 and P4.

Eggs. Eggs carried 30 in number; size, 1.11 mm × 1.16 mm - 1.37 mm × 1.53 mm.

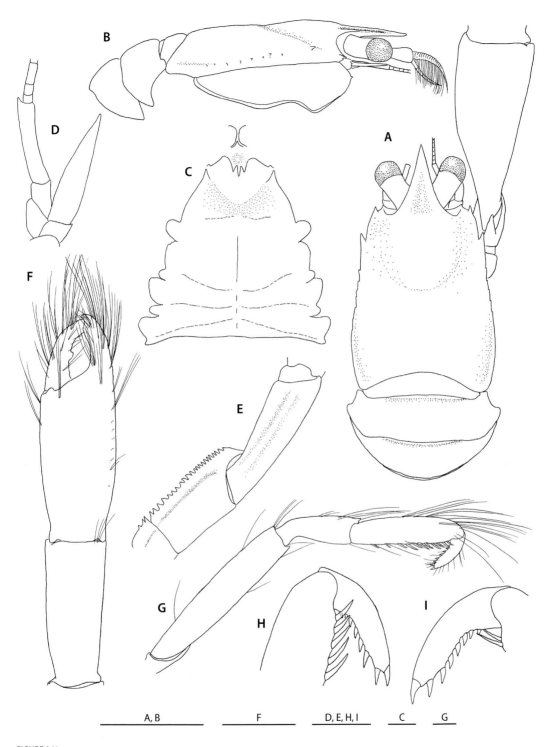

FIGURE 241

Uroptychus sibogae Van Dam, 1933, male 7.2 mm (MNHN-IU-2014-16950). **A**, carapace and anterior part of abdomen, proximal part of right P1 included, dorsal. **B**, same, lateral. **C**, sternal plastron, with excavated sternum and basal parts of Mxp1. **D**, left antenna, ventral. **E**, left Mxp3, setae omitted, ventral. **F**, right P1, dorsal. **G**, right P2, lateral. **H**, same, distal part, setae omitted, lateral. **I**, left P3, distal part, setae omitted, lateral. Scale bars: A, B, F, 5 mm; C, D, E, G, H, I, 1 mm.

REMARKS — The species strongly resembles *U. longicarpus* n. sp. and *U. nebulosus* n. sp. Characters distinguishing these species are outlined under the remarks of *U. nebulosus* (see above).

Uroptychus smib n. sp.

Figures 242, 243

TYPE MATERIAL — Holotype: **New Caledonia**, Norfolk Ridge. **SMIB** 5 Stn DW103, 23°16'S, 168°04'E, 300-315 m, 14.IX.1989, ov. ♀ 2.9 mm (MNHN-IU-2013-8520).

ETYMOLOGY — Named for the SMIB 5 cruise by which the new species was collected; used as a noun in apposition.

DISTRIBUTION — Norfolk Ridge; 315 m.

DESCRIPTION — Small species. *Carapace*: 1.1 × broader than long (0.9 × as long as broad); greatest breadth 2.0 × distance between anterolateral spines. Dorsal surface moderately convex from anterior to posterior, with weak depression between gastric and cardiac regions; very small tubercle-like spines or denticles scattered on anterior gastric and hepatic regions. Lateral margins convex and moderately divergent posteriorly; anterolateral spine well developed, directly lateral to ill-defined lateral limit of orbit; 6 small spines on branchial margin, last one situated at midlength of posterior branchial margin, followed by ridge leading to posterior end. Rostrum relatively broad triangular, with interior angle of 40°, length one-third that of remaining carapace, breadth slightly less than half carapace breadth at posterior carapace margin; dorsal surface nearly straight, horizontal, and moderately concave; lateral margin with small subapical spine on left side, absent on right side. Pterygostomian flap anteriorly somewhat roundish, with very small spine at anterior end and a few small spines on anterior surface.

Sternum: Excavated sternum anteriorly narrow triangular, surface cristate on anterior half in midline. Sternal plastron 1.4 × broader than long, lateral extremities straight divergent posteriorly. Sternite 3 shallowly depressed, relatively short; anterior margin moderately concave, with 2 small submedian spines separated by shallow notch. Anterolateral margin of sternite 4 somewhat convex, anteriorly produced to anterolaterally directed short spine, 1.4 × longer than posterolateral margin. Anterolateral margin of sternite 5 slightly convexly divergent posteriorly, slightly longer than posterolateral margin of sternite 4.

Abdomen: Glabrous and polished. Somite 1 without transverse ridge. Somite 2 tergite 1.9 × broader than long; pleuron blunt on anterolateral and posterolateral ends, lateral margin feebly concave and strongly divergent posteriorly. Pleuron of somite 3 with blunt lateral terminus. Telson about half as long as broad; posterior plate 1.8 × longer than anterior plate, posterior margin feebly concave.

Eye: Elongate, 2 × longer than broad at base, distally narrowed, reaching apex of rostrum. Cornea much less than half as long as remaining eyestalk.

Antennule and antenna: Ultimate article of antennule 2.7 × longer than high. Antennal peduncle overreaching rostrum, relatively slender. Article 2 with small distolateral spine. Antennal scale 1.4 × broader than article 5, ending in midlength of that article. Distal 2 articles each with distomesial spine. Article 5 2.7 × longer than article 4, breadth about half height of antennular ultimate article. Flagellum consisting of 12 segments.

Mxp: Mxp1 with bases nearly contiguous. Mxp3 barely setose on lateral surface. Basis with 3 denticles on mesial ridge. Ischium with 16 (left) or 18 (right) denticles on crista dentata, flexor margin distally not rounded. Merus 2.6 × longer than ischium, flexor margin not well ridged, lateral surface with 1 spine close to distal end of flexor margin and another spine at proximal third (between flexor and extensor margins); distolateral spine distinct. Carpus with distolateral spine and another spine on proximal part of extensor margin.

P1: Missing.

FIGURE 242

Uroptychus smib n. sp., holotype, ovigerous female 2.9 mm (MNHN-IU-2013-8520). **A**, carapace and anterior part of abdomen, dorsal. **B**, same, lateral. **C**, sternal plastron, with excavated sternum and basal parts of Mxp1. **D**, telson. **E**, right antenna, ventral. **F**, right Mxp3, setae omitted, ventral. **G**, same, setae omitted, lateral. Scale bars: 1 mm.

P2-4: Slender, compressed mesio-laterally, sparsely setose on meri and carpi, more setose on distal articles. Meri successively shorter posteriorly (P3 merus 0.9 × length of P2 merus, P4 merus 0.8 × length of P3 merus); P2-3 meri equally broad; P4 merus slightly narrower than P3 merus; length-breadth ratio, 5.2 on P2, 4.7 on P3, 4.2 on P4; P2 merus about as long as carapace, 1.1 × longer than P2 propodus; P3 merus about as long as P3 propodus; P4 merus 0.9 × length of P4 propodus; dorsal margins with a few very small spines on proximal portion on P2 and P3, smooth on P4. Carpi unarmed, successively shorter posteriorly; carpus-propodus length ratio, 0.6 on P2-4, much longer than dactyli (carpus-dactylus length ratio, 1.9 on P2, 1.7 on P3, 1.6 on P4). Propodi subequal in length on P2 and P3, shorter on P4; flexor margin somewhat convex distally in lateral view, with pair of terminal spines preceded by row of spines closely arranged distally, loosely arranged proximally, proximalmost situated at point proximal third and remotely equidistant between proximal second and proximal end of article. Dactyli subequal, about one-third as long as propodi, much shorter than carpi (dactylus-carpus length ratio, 0.5 on P2, 0.6 on P3 and P4), moderately curving; extensor margin with row of plumose setae; flexor margin ending in slender spine preceded by 8-9 proximally diminishing, somewhat inclined sharp spines, penultimate and antepenultimate subequal and distinctly larger than ultimate.

Eggs. Number of eggs carried, 5; size, 0.9-1.0 mm × 1.0-1.1 mm.

REMARKS — The spination of both the carapace and the P2-4 dactyli displayed by the new species is similar to that of *U. lanatus* n. sp. (see above). *Uroptychus smib* is differentiated from *U. lanatus* by the ill-defined lateral orbital angle, the elongate eyes, the short antennal scale terminating in midlength of the ultimate article, sternite 3 having a pair of median spines on the anterior margin, and P2-4 being more slender and propodi much longer relative to dactyli.

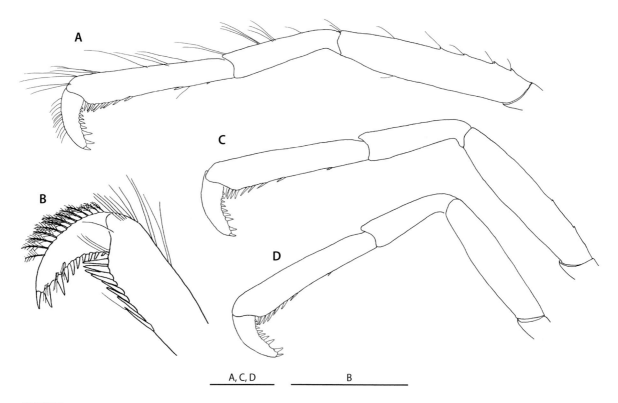

FIGURE 243

Uroptychus smib n. sp., holotype, ovigerous female 2.9 mm (MNHN-IU-2013-8520). **A**, left P2, lateral. **B**, same, distal part, lateral. **C**, left P3, setae omitted, lateral. **D**, left P4, setae omitted, lateral. Scale bars: 1 mm.

Uroptychus soyomaruae Baba, 1981

Figures 244A, B, 306D

Uroptychus soyomaruae Baba, 1981: 129, figs 12, 13. — Baba 1990: 948.
Not *Uroptychus soyomaruae* – Zarenkov & Khodkina 1983: 20, fig. 5 (= *U. sternospinosus* Tirmizi, 1964).

TYPE MATERIAL — Holotype: **Japan**, SE of Miyake-jima, Izu Islands, 33°55.1′N, 140°00.5′E, 860-870 m, ov. female (NSMT-Cr. 6178). [not examined].

MATERIAL EXAMINED — **New Caledonia**, Norfolk Ridge. HALIPRO 2 Stn BT 66, 24°45′S, 168°29′E, 885-1450 m, 19.XI.1996, 1 ov. ♀ 14.0 mm (MNHN-IU-2014-16954). **French Polynesia**, Marquesas Islands. MUSORSTOM 9 Stn CP1303, 8°50′S, 140°19′W, 705-794 m, with *Thouarella* sp. (Calcaxonia, Primnoidae), 9.IX.1997, 1 ♂ 13.0 mm, 1 ov. ♀ 12.5 mm, 2 ♀ 9.4, 14.0 mm (MNHN-IU-2014-16955). **French Polynesia**, Austral Islands. BENTHAUS Stn DW2020, 22°37′S, 152°49.1′W, 920-930 m, 25.XI.2002, 1 ♀ 9.8 mm (MNHN-IU-2014-16956).

DISTRIBUTION — Off Izu Islands (Japan), Madagascar, and now the Norfolk Ridge and French Polynesia; 705-1450 m.

SIZE — Male, 13.0 mm; females, 9.4-14.0 mm; ovigerous females from 12.5 mm.

DIAGNOSIS — Large species. Carapace 1.2 × longer than broad; greatest breadth 1.7 × distance between anterolateral spines. Dorsal surface with granules well developed on large specimens, less so on small specimens; gastric and cardiac regions well inflated; pair of strong gastric spines; lateral margins slightly convexly divergent posteriorly; anterolateral spine overreaching small lateral orbital spine. Rostrum narrow triangular, with interior angle of 20°; dorsal surface flattish or feebly convex from side to side and slightly upcurved. Pterygostomian flap granulose on surface, anteriorly somewhat angular, produced to distinct spine. Excavated sternum with broadly subtriangular anterior margin ending in blunt, occasionally obsolescent spine, surface with low central process; sternal plastron slightly shorter than broad, lateral extremities nearly straight divergent posteriorly. Sternite 3 having anterior margin broadly and deeply excavated, with small submedian spines; sternite 4 relatively short; anterolateral margin as long as posterolateral margin, with strong median spine directed anteroventrally, transverse ridge elevated and denticulate. Anterolateral margin of sternite 5 convexly divergent posteriorly, about half length of posterolateral margin of sternite 4. Abdomen smooth and glabrous; somite 1 tergite with antero-posteriorly convex transverse ridge; Somite 2 tergite 2.0-2.6 × broader than long; pleuron anterolaterally rounded, lateral margin somewhat convex, strongly divergent posteriorly ending in blunt tip. Pleuron of somite 3 bluntly angular on lateral end. Telson one-third to three-quarters as long as broad; posterior plate subsemicircular, length 1.6-1.9 × that of anterior plate, posterior margin convex or slightly emarginate. Eyes 1.5 × longer than broad, distally somewhat dilated, mesial margin proximal to cornea concave, lateral margin convex; cornea much more than half length of remaining eyestalk. Ultimate article of antennular peduncle 2.3-2.5 × longer than high. Antennal peduncle slender, overreaching cornea, unarmed on distal 2 articles; article 2 with tiny or obsolescent distolateral spine; antennal scale barely reaching midlength of article 5; article 5 2.0-2.3 × longer than article 4, breadth 0.3-0.4 × height of ultimate antennular article; flagellum of more than 17 segments far falling short of distal end of P1 merus. Mxp1 with bases nearly contiguous. Mxp3 basis with a few denticles, distal one distinct; ischium with flexor margin distally not rounded, crista dentata with obsolescent denticles; merus 2.6-2.8 × longer than ischium, rather thick mesio-laterally, flexor margin ridged but not well cristate; carpus unarmed. P1 slender, covered with granules; ischium with short, blunt distodorsal process, ventromesially without subterminal spine; merus 1.4-1.5 × longer than carapace, with well-developed distodorsal spine; carpus 1.3-1.4 × length of merus; palm 0.8 × length of carpus, twice length of movable finger; fingers spooned distally, not crossing. P2-4 also slender, meri and carpi subcylindrical; meri equally broad on P2-4; P2 merus slightly longer than or subequal to P3 merus, slightly shorter than carapace, 1.4-1.6 × longer than P2 propodus; P3 merus 1.2-1.3 × length of P3 propodus; P4 merus 0.9 × length of P3 merus, 1.1-1.2 × length of P4 propodus; carpi subequal, 0.7-0.8 × length of propodi on P2-4, much longer than dactyli (carpus-dactylus length ratio,

2.0-2.4 on P2, 1.8-2.8 on P3, 1.8-2.7 on P4); propodi subequal in length; flexor margin somewhat convex on distal half, with slender movable spines, terminal single, slightly distant from distal end; dactyli less than half length of propodi on P2-4, about half as long as carpi (dactylus-carpus length ratio, 0.4-0.5 on P2, 0.4-0.6 on P3 and P4); well curved, terminating in stout spine preceded by 7-12 moderately inclined spines diminishing toward base of article, often obscured by setae.

Eggs. More than 80 eggs carried; size, 1.4 mm × 1.5 mm - 1.6 mm × 1.7 mm.

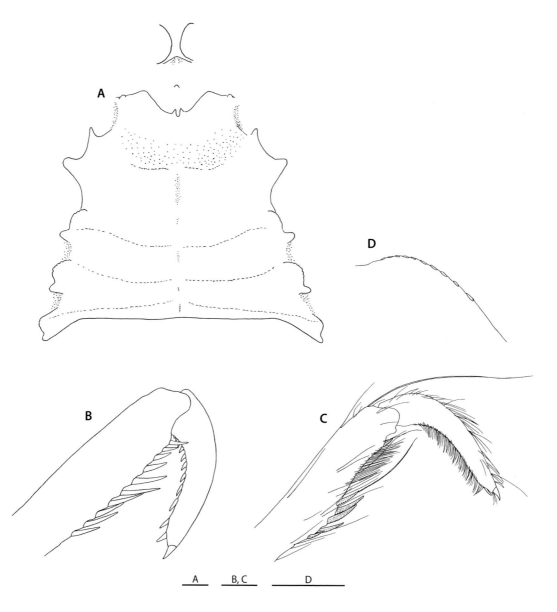

FIGURE 244

A, **B**, *Uroptychus soyomaruae* Baba, 1981, male 13.0 mm (MNHN-IU-2014-16955). **C**, **D**, *Uroptychus sternospinosus* Tirmizi, 1964, male syntype (BMNH 1966.2.3.21-22). **A**, sternal plastron, with excavated sternum and basal parts of Mxp1. **B**, distal part of right P2, setae omitted, lateral. **C**, same. **D**, dactylus, flexor margin, setae omitted, lateral. Scale bars: 1 mm.

Color. Ovigerous female (MNHN-IU-2014-16954) from HALIPRO 2 Stn BT 66, Norfolk Ridge: Orange red all over, including appendages.

PARASITES — The largest female from the Marquesas Islands (MNHN-IU-2014-16955) bears a bopyrid isopod on the left branchial region.

REMARKS — The granulose carapace, the slender pereopods, the strong lateral spine on the thoracic sternite 4, the elongate, unarmed Mxp3 merus, the short antennal scale, and the long carpi of P2-4, characteristic of the species, are also possessed by *U. sternospinosus* Tirmizi, 1964 and *U. thermalis* Baba & de Saint Laurent, 1992. However, the latter two species are distinctive in the spination of P2-4: the dactylus has the flexor margin almost smooth (though obliquely directed small spines are present on the proximal third but hardly visible under normal magnification and obscured by thick setae) except for the terminal and subterminal spines, and the propodus has the distal two flexor marginal spines remotely separated from each other (see Figures 244C, D, 263D-G). These characters of *U. sternospinosus* have been confirmed by examination of the male and female syntypes (BMNH 1966.2.3.21-22) now in the collection of the Natural History Museum, London. The material reported under *U. soyomaruae* by Zarenkov & Khodkina (1983) from the Marcus-Necker Rise is in all probability referable to *U. sternospinosus*.

Uroptychus spinirostris (Ahyong & Poore, 2004)

Figures 245, 306E

Gastroptychus spinirostris Ahyong & Poore, 2004: 9, fig. 1.
Uroptychus spinirostris – Baba 2005: 231 (synonymy and key). — Baba *et al.* 2008: 43, fig. 1H. — Poore *et al.* 2011: 330, pl. 8, fig. C. — McCallum & Poore 2013: 165, figs 8, 12A.

TYPE MATERIAL — Holotype: **Australia**, NE of Tweed Heads, Queensland, 28°02-05'S, 153°57'E, 364 m, male (AM P31418). [not examined].

MATERIAL EXAMINED — **Indonesia**, Kai Islands. KARUBAR Stn CP16, 05°17'S, 132°50'E, 315-349 m, 24.X.1991, 1 ♂ 7.4 mm, 1 ♀ 7.2 mm (MNHN-IU-2014-16957). **Vanuatu.** BOA 0 Stn CP2330, 15°44.4'S, 167°01.83'E, 295-890 m, 18.XI.2004, 1 ov. ♀ 7.5 mm (MNHN-IU-2014-16958). MUSORSTOM 8 Stn CP1083, 15°51.91'S, 167°19.42'E, 397-439 m, 5.X.1994, 1 ♂ 3.6 mm (MNHN-IU-2014-16959). **Vanuatu**, Monts Gemini. GEMINI Stn DW51, 20°58'S, 170°04'E, 450-360 m, 04.VII.1989, 1 ♂ 7.0 mm (MNHN-IU-2014-16960). **New Caledonia.** BATHUS 4 Stn DW903, 18°59.93'S, 163°13.55'E, 386-400 m, 4.VIII.1994, 1 ♂ 13.2 mm (MNHN-IU-2014-16961). **New Caledonia**, Norfolk Ridge. BATHUS 3 Stn CP812, 23°43'S, 168°16'E, 391-440 m, 28.XI.1993, 3 ov. ♀ 10.3 -11.7 mm, 1 ♀ 11.3 mm (MNHN-IU-2014-16962). NORFOLK 2 Stn CP2111, 23°48.56'S, 168°16.78'E, 500-1074 m, 31.X.2003, 1 ♀ 6.0 mm (MNHN-IU-2014-16963).

DISTRIBUTION — Queensland and western Australia, and now New Caledonia, the Norfolk Ridge, Monts Gemini, Vanuatu and the Kai Islands (Indonesia); 315-1074 m.

SIZE — Males, 3.6-13.2 mm; females, 6.0-11.7 mm; ovigerous females from 7.5 mm.

DIAGNOSIS — Large species. Body and appendages covered with sharp spines. Carapace as long as broad, greatest breadth 1.6-1.7 × distance between anterolateral spines; gastric and cardiac regions distinctly bordered by deep groove. Rostrum narrow and elongate, 2.1-2.3 × longer than broad, laterally with 2 pairs of dorso-anteriorly directed strong spines, breadth less than half carapace breadth at posterior carapace margin. Excavated sternum anteriorly produced, strongly ridged or cristate in midline on surface. Sternal plastron slightly broader than long; sternite 3 having anterior margin deeply excavated, with 2 submedian spines separated by deep notch; sternite 4 with 2 pairs of ventrally directed spines (anterior one smaller) on anterolateral margin and pair of submedian spines on posterior surface; posterolateral margin as long as anterolateral margin; anterolateral margin of sternite 5 with strong anterior spine, shorter (0.7) than posterolateral margin of sternite 4. Abdominal somite 1 with transverse row of 4 spines; somite 2 tergite 2.7-2.8 broader than long; pleuron anterolaterally blunt, posterolaterally strongly produced and tapering to sharp point; pleuron of somite 3 also tapering; somites 2-4 each with anterior row of 4-6 spines and posterior row of 6-9 spines; somite 5 with anterior row of 4-7 spines, posterior row of 4-9 spines, somite 6 with anterior row of 4-7 spines and posterior row of 4 spines and posterior marginal denticles; protopod of uropod with obsolescent protuberance on mesial margin; endopod 1.2-1.5 × longer than broad; telson half as long as broad, posterior plate 1.3-2.0 × longer than anterior plate, emarginate on posterior margin. Eyes proximally broadened in large specimens. Antennal peduncle extending far beyond cornea; article 2 with strong distolateral spine; antennal scale overreaching article 4, at most terminating in midlength of article 5; article 3 with distinct distomesial spine and small spine in ventral midline near juncture with article 4; articles 4 and 5 each with strong ventral distomesial spine; article 4 with additional spine at midlength of ventral surface in large specimens; article 5 as long as or slightly shorter than article 4, with small ventral spine about at midlength; flagellum consisting of 6-16 segments, falling short of distal end of P1 merus (in small specimens, slightly overreaching rostral tip). Mxp1 with bases close to each other. Mxp3 coxa with strong, ventrolaterally directed spine; basis unarmed or with obsolescent denticles on mesial ridge; ischium with strong distolateral spine (occasionally with small accompanying spine directly mesial to it), crista dentata with 25-30 denticles; merus twice (1.9-2.0 x) longer than ischium, with strong distolateral spine, flexor margin with laterally directed median spine, and 1-3 additional spines proximal to it, distolateral spine occasionally absent; carpus with strong distolateral spine and a few (usually 2) spines of small to good size on extensor surface. Pereopods spinose; P1 with spines roughly in 8 rows on merus, carpus and propodus; ischium with strong subterminal spine on ventromesial margin; merus slightly longer than carapace. P2-4 with dorsal (extensor), lateral and ventral (flexor) spines; meri successively shorter posteriorly (P3 merus 0.9 × length of P2 merus, P4 merus 0.8-0.9 × length of P3 merus), equally broad on P2-4; P2 merus as long as or slightly longer than carapace,

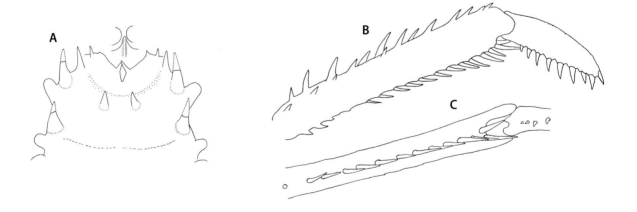

FIGURE 245

Uroptychus spinirostris (Ahyong & Poore, 2004), female 7.2 mm (MNHN-IU-2014-16957). **A**, anterior part of sternal plastron, with excavated sternum and basal parts of Mxp1. **B**, right P2, distal part, lateral. **C**, same, propodus, proximal part omitted, flexor. Scale bar: 1 mm.

0.8-1.1 × longer than P2 propodus; P3 merus as long as P3 propodus, P4 merus 0.9-1.1 × length of P4 propodus; carpi subequal or successively slightly shorter posteriorly, slightly less than half as long as propodi on P2-4, distinctly longer than dactyli; propodi successively longer posteriorly, flexor margin straight, with pair of terminal spines preceded by row of spines placed in straight arrangement; dactyli slightly less than half length of propodi on P2-4, shorter than carpi in large specimens, subequal in small specimens (dactylus-carpus length ratio, 0.6-1.0 on P2 and P3, 0.7-1.1 on P4); flexor margin with row of 9-13 proximally diminishing slender spines subperpendicular to margin, ultimate more slender than penultimate.

Eggs. Number of eggs carried, more than 100; size, 1.02 mm × 1.13 mm - 1.57 mm × 1.58 mm.

Color. Illustrated by McCallum & Poore (2013) for a specimen from Western Australia. Male from Vanuatu (MNHN-IU-2014-16959): uniformly translucent pale orange; spines on pereopods proximally reddish, distally pale or whitish.

REMARKS — The female (6.0 mm) from NORFOLK 2 Stn CP2111 (MNHN-IU-2014-16963) and another female (ovigerous, 7.5 mm) from BOA0 Stn CP2330 (MNHN-IU-2014-16958) bear an extra spine lateral to the submedian pair on the posterior surface of sternite 4, the feature as displayed by the larger of two western Australian specimens reported by McCallum & Poore (2013). Mentioning that this larger western Australian specimen differs from the holotype in having a more spinose carapace, with three instead of two strong rostral spines but the other smaller specimen agrees well with the holotype, McCallum & Poore believed these differences as allometric. All of our specimens, ranging from 3.6 to 13.2 mm, agree quite well with the holotype. However, the P2-4 dactyli in the specimens examined are shorter than the carpi in large specimens, subequal (or slightly longer on P4) in small specimens, with the dactylus-carpus length ratio, 0.6-1.0 on P2 and P3, 0.7-1.1 on P4, the feature apparent in the western Australian specimen of large size (McCallum & Poore 2013: fig. 8A).

Three closely related species, *U. numerosus* n. sp., *U. quartanus* n. sp. and *U. senarius* n. sp., are described in this paper. Their relationships are discussed under *U. senarius* (see below).

Uroptychus spinosior n. sp.

Figures 246, 247

TYPE MATERIAL — Holotype: **Wallis and Futuna Islands**. MUSORSTOM 7 Stn DW516, 14°13'S, 178°12'W, 441-550 m, 12.V.1992, ov. ♀ 3.5 mm (MNHN-IU-2011-5941). Paratypes: Wallis and Futuna Islands. MUSORSTOM 7 Stn DW514, 14°13'S, 178°11'W, 349-355 m, 12.V.1992, 1 ♂ 2.9 mm, 1 ov. ♀ 3.3 mm (MNHN-IU-2011-5943, MNHN-IU-2011-5944). – Stn CP517, 14°13'S, 178°10'W, 223-235 m, 12.V.1992, 1 ♂ 3.3 mm (MNHN-IU-2011-5945), 2 ov. ♀ 3.3, 4.2 mm (MNHN-IU-2011-5946, MNHN-IU-2011-5947). **Fiji Islands**. MUSORSTOM 10 Stn CP1389, 18°19'S, 178°05'E, 241-417 m, 19.VIII.1998, 1 ♂ 2.6 mm (MNHN-IU-2011-5939). BORDAU 1 Stn DW1454, 16°46'S, 179°59'E, 300-370 m, 4.III.1999, 1 ♂ 4.0 mm (MNHN-IU-2011-5940). **Tonga**. BORDAU 2 Stn CP1626, 23°20'S, 176°16'W, 220-249 m, 19.VI.2000, 1 ♂ 3.1 mm (MNHN-IU-2011-5942).

ETYMOLOGY — From the Latin *spinosior* (more thorny), alluding to more spinose lateral margins of the carapace, a character to separate the species from the close relatives *U. tridentatus* (Henderson, 1885) and *U. oxymerus* Ahyong & Baba, 2004.

DISTRIBUTION — Wallis and Futuna Islands (SW Pacific), Fiji Islands and Tonga; 220-550 m.

SIZE — Males, 2.6-4.0 mm; females, 3.3-4.2 mm; ovigerous females from 3.3 mm.

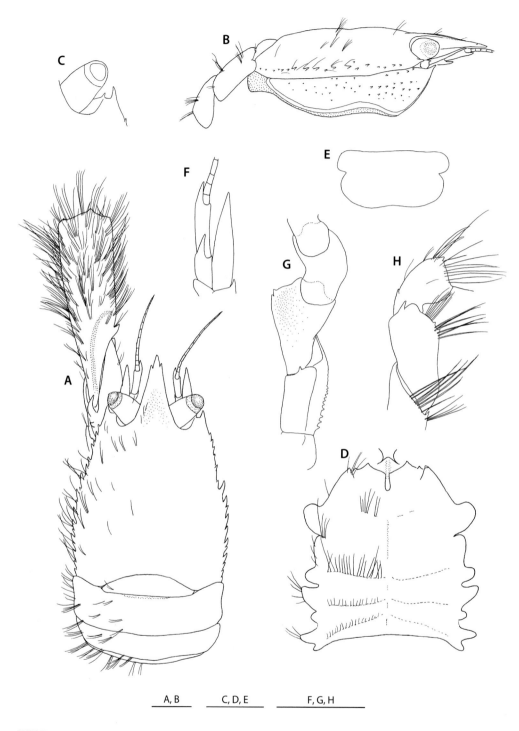

A, B   C, D, E   F, G, H

FIGURE 246

Uroptychus spinosior n. sp., holotype, ovigerous female 3.5 mm (MNHN-IU-2011-5941). **A**, carapace and anterior part of abdomen, proximal part of left P1 included, dorsal. **B**, same, lateral. **C**, anterolateral part of carapace showing eye, lateral orbital spine and anterolateral spine, dorsal. **D**, sternal plastron, with excavated sternum and basal parts of Mxp1. **E**, telson. **F**, left antenna, ventral. **G**, right Mxp3, setae omitted, ventral. **H**, same, lateral. Scale bars: 1 mm.

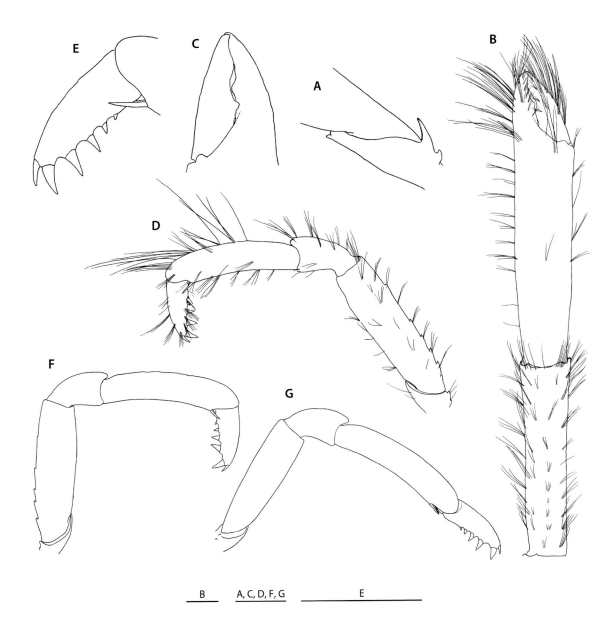

FIGURE 247

Uroptychus spinosior n. sp., holotype, ovigerous female 3.5 mm (MNHN-IU-2011-5941). **A**, left P1, proximal part, lateral. **B**, same, dorsal. **C**, same, distal part, setae omitted, ventral. **D**, left P2, lateral. **E**, same, distal part, setae omitted, lateral. **F**, right P3, setae omitted, lateral. **G**, right P4, setae omitted, lateral. Scale bars: 1 mm.

DESCRIPTION — Small species. *Carapace*: Slightly broader than long (0.9 × as long as broad); greatest breadth 1.5-1.6 × distance between anterolateral spines. Dorsal surface feebly convex or nearly horizontal from anterior to posterior, smoothly sloping down to rostrum without any border, with or without weak depression between gastric and cardiac regions; with sparse or moderately dense, relatively long setae. Lateral margins convexly divergent posteriorly, with row of 12-18 spines: those on hepatic and anterior branchial margins small except for anterolateral spine, those on posterior branchial margin successively smaller posteriorly, all arranged in same level in profile; first (anterolateral) spine situated directly lateral to tiny lateral orbital spine, distinctly overreaching it; second much more distant from first than from third, last situated near posterior end. Rostrum straight and horizontal, narrow triangular, with interior angle 25-30°; dorsal surface concave; lateral margin with small subapical spine; length varying from more than half to less than half that of remaining carapace, breadth half carapace breadth at posterior carapace margin. Pterygostomian flap anteriorly angular, produced to sharp spine; surface with numerous small spines.

Sternum: Excavated sternum with anterior margin convex or broad subtriangular between bases of Mxp1, surface with ridge in midline. Sternal plastron slightly longer than broad, lateral extremities subparallel between sternites 4 and 7. Sternite 3 shallowly or moderately depressed; anterior margin shallowly excavated, with narrow (rarely broad), deep median notch without flanking spines. Sternite 4 having anterolateral margin feebly convex or nearly straight, less than twice as long as posterolateral margin, anteriorly angular. Anterolateral margins of sternite 5 subparallel, anteriorly rounded, about as long as posterolateral margin of sternite 4.

Abdomen: Smooth and sparsely setose. Somite 1 without transverse ridge. Somite 2 tergite 2.2-2.3 × broader than long; pleuron with lateral margin slightly concave, slightly or barely divergent posteriorly, anterior and posterior lateral termini rounded. Pleuron of somite 3 laterally blunt. Telson half as long as broad; posterior plate slightly or somewhat concave on posterior margin, length 1.1-1.5 × that of anterior plate.

Eye: Relatively broad (1.5 × longer than broad), slightly narrowed distally, somewhat overreaching midlength of rostrum. Cornea much more than half length of remaining eyestalk.

Antennule and antenna: Ultimate article of antennular peduncle 2.5-3.3 × longer than high. Antennal peduncle reaching subapical spine of rostrum. Article 2 with distinct lateral spine. Antennal scale distally sharp, reaching or somewhat overreaching distal end of article 5, breadth 1.2-1.3 × that of article 5. Distal 2 articles each ventrally armed with strong distomesial spine; article 5 1.5-1.9 × longer than article 4, breadth 0.6-0.7 × height of ultimate antennular article. Flagellum consisting of 9-11 segments, not reaching distal end of P1 merus.

Mxp: Mxp1 with bases distinctly separated. Mxp3 basis lacking denticles on mesial ridge. Ischium with small distal spine and tuft of long setae lateral to rounded distal end of flexor margin; crista dentata with 6-8 distally diminishing denticles, proximal denticles loosely arranged. Merus 1.7 × longer than ischium, broad relative to length, flattish on mesial face, with distolateral spine and 1 or 2 small spines distal to point distal third of cristate flexor margin. Carpus with 1 distolateral and 1 proximal extensor marginal spine.

P1: 4.3-5.4 × longer than carapace, subcylindrical (somewhat massive on palm in large males), covered with soft fine setae. Ischium dorsally with strong spine (rarely 2, distal larger), ventromesially with short subterminal spine and a few proximal tubercles. Blunt distomesial and distolateral spines on ventral surface of merus and carpus. Merus with 2-4 additional spines on proximal part of ventromesial surface, length 1.2-1.4 × that of carapace. Carpus 1.2-1.3 × longer than merus. Palm 3.0-3.7 × (males), 3.5-4.4 × (females) longer than broad, 0.95-1.1 × length of carpus. Fingers gently incurved distally and slightly crossing when closed, slightly gaping in both sexes; movable finger 0.4-0.5 × length of palm, opposable margin with obtuse subtriangular proximal process (very low in small specimens) fitting to longitudinal groove proximal to median process or between 2 eminences (in largest male) on opposable margin of fixed finger when closed.

P2-4: Meri moderately compressed mesio-laterally, successively shorter posteriorly (P3 merus 0.8-0.9 × length of P2 merus, P4 merus 0.9 × length of P3 merus), subequally broad on P2-4; dorsal margin with a few eminences or small spines on proximal half or several spines on entire length on P2 and often on P3, absent on P4; length-breadth ratio, 3.7-4.6 on P2, 3.2-3.9 on P3, 3.3-3.8 on P4; P2 merus 0.9 × length of carapace, 1.2 × length of P2 propodus; P3 merus 0.9 × length of P3 propodus; P4 merus 0.8-0.9 × length of P4 propodus. Carpi subequal on P3 and P4, slightly longer on P2 or subequal on P2-4; carpus-propodus length ratio, 0.3-0.4 on P2, 0.3 on P3 and P4. Propodi successively slightly longer posteriorly; flexor margin nearly straight or slightly concavely curving in lateral view, with pair of distal spines (preceded by 1 spine on P2 only in some paratypes). Dactyli 1.1-1.2 × longer than carpi and 0.4 × as long as propodi on P2-4; flexor margin nearly straight, with 6-7 spines loosely arranged and perpendicular to margin (excepting 1 or 2 proximal spines moderately inclined), ultimate spine slender, penultimate slightly larger than antepenultimate, both stronger than remainder.

Eggs. Eggs carried, 6-23; size, 0.75 mm × 0.80 mm.

REMARKS — The combination of the following characters links the species to *U. annae* n. sp., *U. oxymerus* Ahyong & Baba, 2004 and *U. tridentatus* (Henderson, 1885): the rostrum bearing a subapical spine on each side, the sternal plastron with subparallel lateral extremities, the antenna bearing a strong spine on each of articles 4 and 5, the antennal scale overreaching article 5, Mxp3 with a small spine lateral to the rounded distal end of flexor margin, and the P2-4 dactyli bearing triangular spines perpendicular to the flexor margin. *Uroptychus spinosior* is readily distinguished from these congeners by having more numerous (12-18) spines along the entire lateral margin of the carapace, instead of at most 7 spines along the anterior three-quarters of the margin.

Uroptychus spinulus n. sp.
Figures 248, 249

TYPE MATERIAL — Holotype: **Vanuatu**. MUSORSTOM 8 Stn DW1030, 17°51.80'S, 168°30.44'E, 180-190 m, 29.IX.1994, ♂ 3.5 mm (MNHN-IU-2014-16964). Paratypes: **Vanuatu**. MUSORSTOM 8 Stn DW1030, collected with holotype, 1 ♀ 3.8 mm (MNHN-IU-2014-16965). – Stn CP1084, 15°50.29'S, 167°17.48'E, 207-280 m, 5.X.1994: 1 ♂ 3.8 mm, 2 ov. ♀ 3.3, 3.5 mm (MNHN-IU-2014-16966).

ETYMOLOGY — From the Latin *spinulus* (dim. of spina, small spine), referring to small spines (two in number) on the dorsal proximal margin of P2 merus, a character to help separate the species from related species.

DISTRIBUTION — Vanuatu; 180-280 m.

SIZE — Males, 3.5-3.8 mm; females, 3.3-3.8 mm; ovigerous females from 3.3 mm.

DESCRIPTION — Small species. *Carapace*: Slightly broader than long (0.9 × as long as broad); greatest breadth 1.6 × distance between anterolateral spines. Dorsal surface smooth and nearly glabrous; cardiac region very weakly convex; gastric region almost horizontal, preceded by depressed rostrum. Lateral margins slightly convexly divergent posteriorly, bearing 6 spines; first anterolateral spine, overreaching tiny lateral orbital spine and situated slightly posterior to level of and separated from that spine at most by its basal breadth; second remote from first, equidistant between first and third; fifth located at midlength of posterior branchial margin, sixth very small or absent, occasionally followed by a few tubercle-like processes. Rostrum sharp triangular, with interior angle of about 30°; dorsal surface straight horizontal and concave, ventral surface directed dorsally; length about half that of carapace, breadth half carapace breadth at posterior carapace margin. Pterygostomian flap with roundish anterior margin bearing small spine; surface smooth, bearing spine or tubercle on anterior portion.

Sternum: Excavated sternum with sharp ridge in midline, anterior margin triangular. Sternal plastron slightly broader than long, lateral extremities divergent posteriorly on sternites 4-7. Sternite 3 moderately depressed, anterior margin gently concave with U-shaped median notch without flanking spine, anterolateral corner angular. Sternite 4 having anterolateral margin entire and slightly concave, anterior end angular, posterolateral margin slightly shorter than anterolateral margin.

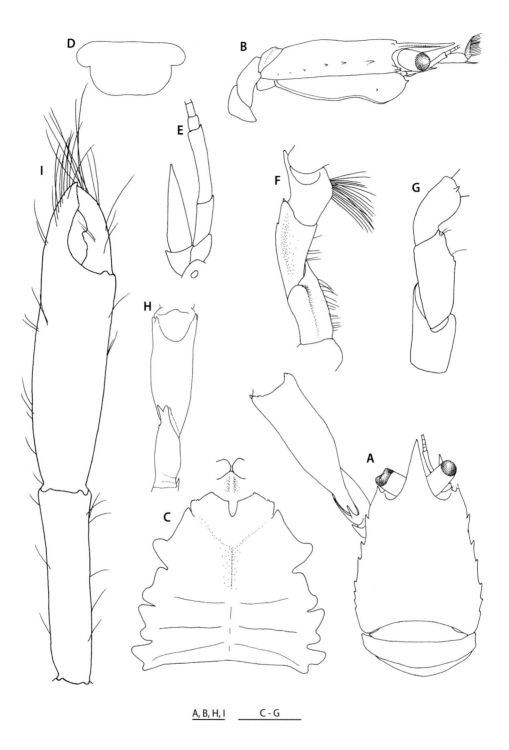

A, B, H, I   C - G

FIGURE 248

Uroptychus spinulus n. sp., holotype, male 3.5 mm (MNHN-IU-2014-16964). **A**, carapace and anterior part of abdomen, proximal part of left P1included, dorsal. **B**, same, lateral. **C**, sternal plastron, with excavated sternum and basal parts of Mxp1. **D**, telson. **E**, right antenna, ventral. **F**, right Mxp3, ventral. **G**, same, lateral, **H**, left P1, proximal part, setae omitted, ventral. **I**, same, dorsal. Scale bars: 1 mm.

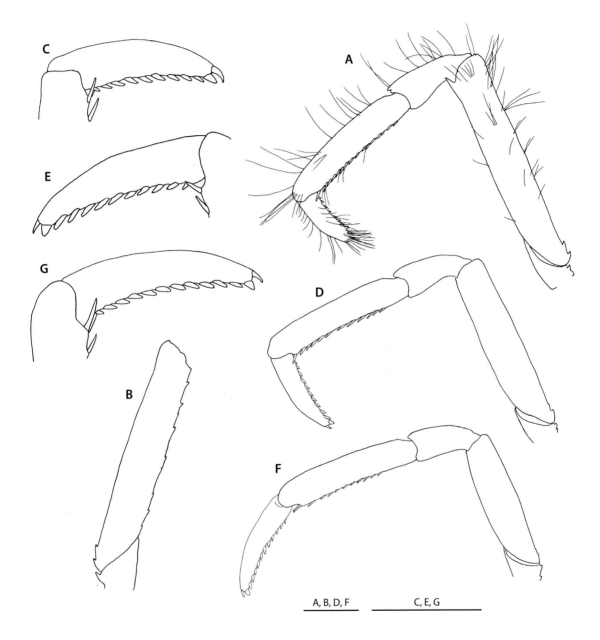

FIGURE 249

Uroptychus spinulus n. sp., holotype, male 3.5 mm (MNHN-IU-2014-16964). **A**, left P2, lateral. **B**, same, proximal part (ischium and merus), setae omitted, mesial. **C**, same, distal part, setae omitted, mesial. **D**, left P3, setae omitted, lateral. **E**, same, distal part, setae omitted, lateral. **F**, left P4, setae omitted, lateral. **G**, same distal part, setae omitted, mesial. Scale bars: 1 mm.

Anterolateral margin of sternite 5 anteriorly rounded, divergent posteriorly, length 0.7 × that of posterolateral margin of sternite 4.

Abdomen: Dorsal surface smooth with very sparse short setae. Somite 1 with somewhat elevated transverse ridge. Somite 2 tergite 2.6-2.8 × broader than long; pleuron posterolaterally blunt angular, lateral margin moderately concave and divergent posteriorly. Pleuron of somite 3 laterally blunt. Telson slightly less than half as long as broad; posterior plate slightly longer than anterior plate, feebly concave on posterior margin.

Eye: Elongate, 2 × as long as broad, overreaching midlength of rostrum, somewhat narrowed distally. Cornea about half as long as remaining eyestalk.

Antennule and antenna: Antennular ultimate article about 3 × longer than high. Antennal peduncle overreaching cornea, slightly falling short of rostral tip. Article 2 with small or obsolescent distolateral spine. Antennal scale 2 × as broad as article 5, sharply tapering, slightly overreaching midlength of article 5. Article 4 with distomesial spine. Article 5 with very small or obsolescent distomesial spine, length 1.5-1.7 × that of article 4, breadth less than half height of ultimate article of antennular peduncle. Flagellum of 11-14 segments not reaching distal end of P1 merus.

Mxp: Mxp1 with bases close to each other or nearly contiguous. Mxp3 scarcely setose on lateral surface. Basis without denticles on mesial ridge. Ischium with flexor margin distally rounded; crista dentata almost entire, without denticles. Merus twice as long as ischium, flexor margin moderately ridged, bearing 2 very small spines distal to point one-third from distal end; distolateral spine distinct. Carpus also with distolateral spine.

P1: More slender in females than in males, 5.3-5.8 × longer than carapace, smooth, bearing long setae on fingers, sparsely setose elsewhere. Ischium dorsally with long spine, ventromesially with well-developed subterminal spine. Merus 1.1-1.2 × longer than carapace, dorsally with very small distomesial spine, ventrally with distomesial and distolateral spines. Carpus with mesial and lateral distoventral spines and 2 tubercle-like processes distal to dorsal juncture with merus, length 1.3 × that of merus. Palm massive in males, 3.0-3.1 × (males), 4.6-4.9 × (females) longer than broad, 2.2-2.4 × (males), 2.8-2.9 × (females) longer than movable finger, 1.1-1.2 × longer than carpus. Fingers distally ending in incurved small spine, not spooned; in females, opposable margin of movable finger with strong median process distal to proximal process of fixed finger; in males, fingers gaping, opposable margins fitting to each other (or slightly crossing) in distal third of length when closed, fixed finger lacking proximal process, movable finger with prominent process on midlength of gaping margin; movable finger 0.3-0.4 × length of palm.

P2-4: With fine plumose setae, denser on dactyli. Ischium with 2 small distal spines on extensor margin. Meri successively diminishing posteriorly (P3 merus 0.7-0.8 × length of P2 merus, P4 merus 0.7-0.8 × length of P3 merus); breadths subequal on P2-4; length-breadth ratio, 6.1-7.3 on P2, 4.5-5.0 on P3, 3.7-3.8 on P4; dorsal margin with 1 or 2 (usually 2) spines distinct near proximal end on P2, occasionally obsolescent on P3, vestigial or absent on P4; row of tubercle-like small spines along entire ventromesial margin on P2 only, and distinct spine at distal end of ventrolateral margin (obsolescent on P3 and P4); P2 merus 1.2 × longer than carapace, 1.5-1.7 × longer than P2 propodus; P3 merus 0.9 × length of P3 propodus; P4 merus 0.9 × length of P4 propodus. Carpi successively slightly shorter posteriorly or subequal on P3-4; carpus-propodus length ratio, 0.4-0.5 on P2, 0.4 on P3, 0.3-0.4 on P4; extensor margin with small distal spine distinct or obsolescent on P2, absent on P3 and P4. Propodi subequal on P2 and P4, longest on P3, or subequal on P2-4; flexor margin straight, with pair of terminal spines preceded by 14-17 slender movable spines in zigzag arrangement along entire length. Dactyli as long as carpi on P2, 1.3 × longer on P3, 1.5 × longer on P4; dactylus-propodus length ratio, 0.4 on P2, 0.5 on P3, 0.5-0.6 on P4; flexor margin slightly curving, with 12-14 obliquely directed spines closely arranged and obscured by dense setae, ultimate spine slender, penultimate 2 × broader than antepenultimate, remaining spines slender, close to one another but not contiguous.

Eggs. Number of eggs carried, 6; size, 0.93 mm × 0.98 mm - 0.93 mm × 1.19 mm.

REMARKS — The arrangements of spines on the carapace and P2-4 dactyli are very much like those of *U. dissitus* n. sp. (see above). *Uroptychus spinulus* is distinguished from *U. dissitus* by the following: the pterygostomian flap is anteriorly roundish instead of angular, although ending in a small spine in both species; the antennal scale slightly overreaches the midlength of article 5, not reaching the end of that article, instead of slightly overreaching the distal end of article 5; the P1 merus is mesially smooth and unarmed instead of bearing strong spines; the P2 merus bears 1 or 2 (usually 2) dorsal spines on the proximal portion instead of being unarmed; and the P2-4 propodi bear a row of 14-17 flexor spines, other than the terminal pair, along the entire length, instead of at most 5 spines on the distal half.

Uroptychus squamifer n. sp.
Figures 250, 251

TYPE MATERIAL — Holotype: **New Caledonia**, South of Loyalty Ridge. BATHUS 3 Stn DW778, 24°43'S, 170°07'E, 750-760 m, 24.XI.1993, ov. ♀ 9.4 mm (MNHN-IU-2014-16967).

ETYMOLOGY — From the Latin *squamus* (scale) and *fer* (the suffix meaning bear), referring to scale-like ridges on the carapace dorsal surface.

DISTRIBUTION — South of Loyalty Ridge; 750-760 m.

DESCRIPTION — *Carapace*: About as long as broad; greatest breadth 1.7 × distance between anterolateral spines. Dorsal surface slightly convex from anterior to posterior, with weak depression between anterior and posterior branchial regions, and between anterior and posterior branchial regions, covered with setiferous, scale-like, granulate ridges; 2 epigastric spines preceded by depressed rostrum, flanking larger spine placed slightly posteriorly. Lateral margins convexly divergent posteriorly, ridged along posterior quarter, with 10 spines discernible in dorsal view; first 3 spines ventral to level of remaining spines; first anterolateral, well developed, directed forward, overreaching much smaller lateral orbital spine; second and third small, placed on hepatic margin; fourth to sixth on anterior branchial margin and seventh to tenth on posterior branchial margin; fourth slightly smaller than first, fifth and sixth small, seventh to ninth well developed, tenth last much smaller; extra small spine between ninth and tenth on left side, and another small spine between eighth and ninth on right side, last spine followed by ridge leading to posterior end. Rostrum somewhat deflected ventrally, narrow triangular, with interior angle of 20°; length half as long as remaining carapace, breadth much less than half carapace breadth at posterior carapace margin; lateral margin with 4 spinules; dorsal surface flattish, distinctly lowered from level of epigastric region. Lateral orbital spine slightly anterior to level of anterolateral spine, and separated by basal breadth of anterolateral spine. Pterygostomian flap feebly granulose on surface, anterior margin somewhat angular, produced to strong spine.

Sternum: Excavated sternum convexly produced anteriorly, surface with rounded ridge in midline. Sternal plastron 0.9 × as long as broad, lateral extremities moderately divergent posteriorly. Sternite 3 moderately depressed, anterior margin shallowly excavated, with 2 small submedian spines (with extra small spine lateral to left spine) separated by narrow notch. Sternite 4 with scattered denticulate ridges on surface; anterolateral margin slightly longer than posterolateral margin, denticulate, anteriorly with 2 processes (distally broken) placed side by side, mesial one smaller. Anterolateral margin of sternite 5 denticulate and convexly divergent posteriorly, slightly shorter than posterolateral margin of sternite 4.

Abdomen: Smooth. Somite 1 with well-developed setiferous transverse ridge. Somite 2 tergite 2.6 × broader than long; tergite with setiferous anterior transverse ridge, pleuron concavely divergent posteriorly, anterolateral and posterolateral termini angular. Pleuron of somite 3 tapering to sharp tip. Telson half as long as broad; posterior plate 1.5 × longer than anterior plate, distinctly emarginate on posterior margin.

Eye: Short relative to length, 1.5 × longer than broad, feebly narrowed proximally, barely reaching midlength of rostrum. Cornea slightly dilated, about as long as remaining eyestalk.

Antennule and antenna: Ultimate article of antennule 3.7 × as long as high. Antennal peduncle overreaching eye by full length of article 5. Article 2 with well-developed lateral spine. Antennal scale slightly falling short of distal end of

FIGURE 250

Uroptychus squamifer n. sp., holotype, male 9.4 mm (MNHN-IU-2014-16967). **A**, carapace and anterior part of abdomen, dorsal. **B**, same, lateral. **C**, sternal plastron, with excavated sternum, basal parts of Mxp1 included. **D**, telson. **E**, right antenna, ventral. **F**, left Mxp3, ventral. **G**, same, setae omitted, lateral. Scale bars: A, B, 5 mm; C-G, 1 mm.

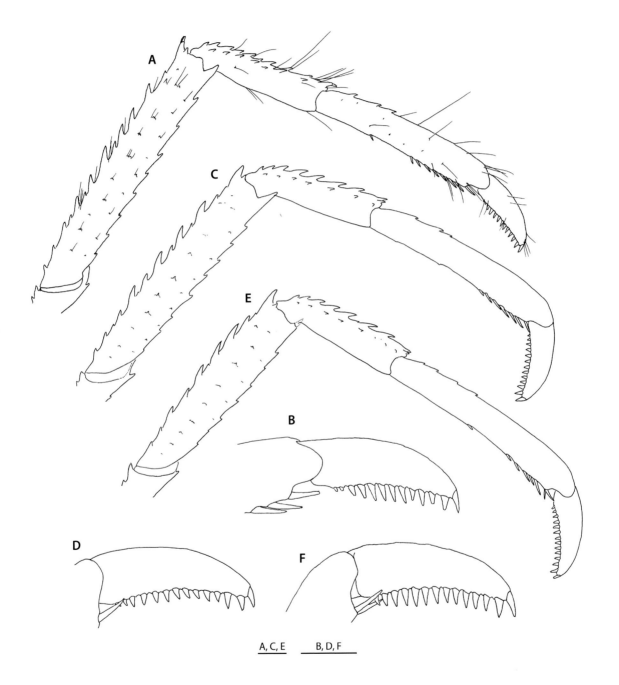

FIGURE 251

Uroptychus squamifer n. sp., holotype, male 9.4 mm (MNHN-IU-2014-16967). **A**, right P2, lateral. **B**, same, distal part, setae omitted, lateral. **C**, right P3, setae omitted, lateral. **D**, same, distal part, lateral. **E**, right P4, lateral. **F**, same, distal part, lateral. Scale bars: 1 mm.

peduncle, 2 × broader than article 5. Articles 4 and 5 each with distomesial spine; article 5 1.8 × longer than article 4, breadth 0.5 × height of antennular ultimate article. Flagella broken.

Mxp: Mxp1 with bases close to each other. Mxp3 basis with 4 denticles on mesial ridge, distal one distinct, others obsolescent. Ischium with very small denticles (about 35) on crista dentata, flexor margin not rounded distally. Merus twice as long as ischium, ridged along flexor margin with a few obsolescent denticles distal to midlength, distolateral spine well developed. Carpus with 1 distolateral and 1 proximo-lateral spine, and a few denticles between.

P1: Missing.

P2-4: Relatively slender, with very sparse, relatively short setae. Meri moderately compressed mesio-laterally, successively shorter posteriorly (P3 merus 0.9 × length of P2 merus, P4 merus 0.9 × length of P4 merus), equally broad on P2-4; length-breadth ratio, 7.0 on P2, 5.9 on P3, 4.7 on P4; dorsal margin with row of spines of irregular sizes (13-14 spines on P2 and P3, 9 spines on P4); ventrolateral margin with 9 spines on P2 and P3, 7 spines on P4; ventromesial margin with 6 or 7 small spines on P2 only; lateral surface sparsely denticulate. P2 merus as long as carapace, 1.3 × longer than P2 propodus; P3 merus 1.1 × longer than P3 propodus; P4 merus 0.9 × length of P4 propodus. Carpi subequal in length on P2-4; much longer than dactyli (carpus-dactylus length ratio, 1.9 on P2, 1.8 on P3, 1.5 on P4); extensor margin with row of spines (9 on P2, 8 on P3 and P4), with subparalleling row of 6-7 smaller spines laterally; carpus-propodus length ratio, 0.6 on P2 and P3, 0.5 on P4. Propodi successively longer posteriorly; extensor margin with a few small proximal spines (3 on P2, 2 on P3, 4 on P4); flexor margin nearly straight, with pair of terminal spines preceded by 7 slender movable spines on P2 and P3, 6 spines on P4, proximalmost spine remote from proximal second, situated at proximal third on P2 and P3, at midlength on P4. Dactyli successively longer posteriorly, length 0.5 × that of carpi on P2, 0.6 × on P3, 0.7 × on P4; one-third length of propodi on P2-4; flexor margin slightly curving, with 13 or 14 sharp spines perpendicular to margin and proximally diminishing, ultimate spine slightly longer than penultimate.

Eggs. Number of eggs carried, 8 (normal number probably more); size, 1.23 mm × 1.23 mm-1.40 mm × 1.19 mm.

REMARKS — *Uroptychus squamifer* resembles *U. strigosus* n. sp. in the carapace that bears scale-like ridges on the dorsal surface and a row of strong spines on the lateral margin, and in the abdominal somite 1 that bears a sharp transverse ridge. However, they differ from each other in many aspects as discussed under the remarks of *U. strigosus* (see below).

Uroptychus stenorhynchus n. sp.

Figures 252, 253, 306F

Uroptychus sp. — Poore *et al.* 2011: 330, pl. 8, fig. H.

TYPE MATERIAL — Holotype: **Vanuatu**. MUSORSTOM 8 Stn CP1051, 16°36.63'S, 167°59.90'E, 558-555 m, 1.X.1994, ♂ 7.2 mm (MNHN-IU-2014-16968). Paratypes: **Vanuatu**. MUSORSTOM 8 Stn CP974, 19°21.51'S, 169°28.26'E, 492-520 m, 22.IX.1994, 1 ♂ 5.8 mm (MNHN-IU-2014-16969). – Stn CP1051, station data as for the holotype, 1 ♂ 5.1, 2 ov. ♀ 5.7, 5.8 mm, 1 ♀ 5.0 mm (MNHN-IU-2014-16970). SANTO Stn AT73, 15°40.8'S, 167°00.5'E, 514-636 m, 07.X.2006, 1 ov. ♀ 6.0 mm (MNHN-IU-2014-16971). – Stn AT10, 15°41.1'S, 167°00.5'E, 509-659 m, 17.IX.2006, 1 ov. ♀ 7.1 mm (MNHN-IU-2014-16972). – Stn AT19, 15°40.8'S, 167°00.5'E, 503-600 m, 21.IX.2006, 1 ♂ 6.0 mm (MNHN-IU-2014-16973).

ETYMOLOGY — From the Greek *stenos* (narrow) and *rhynchus* (rostrum) referring to the narrow rostrum of the new species.

DISTRIBUTION — Vanuatu; 492-659 m.

SIZE — Males, 5.1-7.2 mm; females, 5.0-7.1 mm; ovigerous females from 5.7 mm.

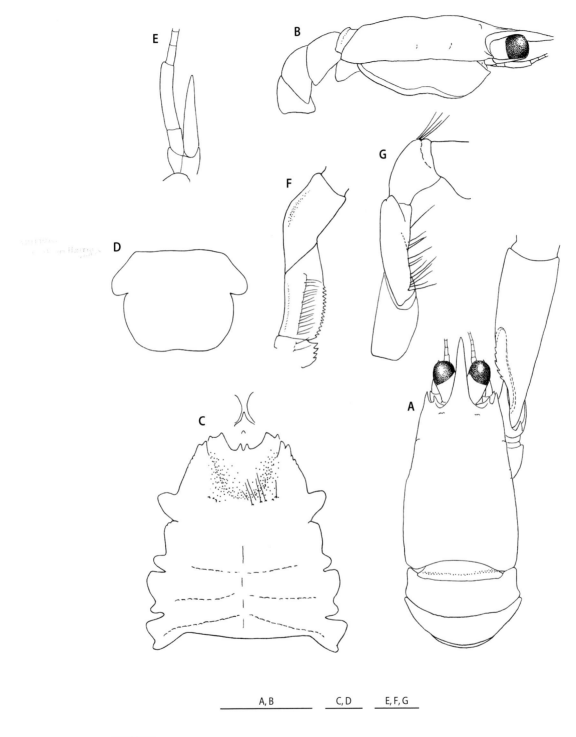

FIGURE 252

Uroptychus stenorhynchus n. sp., holotype, male 7.2 mm (MNHN-IU-2014-16968). **A**, carapace and anterior part of abdomen, proximal part of right P1 included, dorsal. **B**, same, lateral. **C**, sternal plastron, with excavated sternum, basal parts of Mxp1 included. **D**, telson. **E**, left antenna, ventral. **F**, right Mxp3, ventral. **G**, same, lateral. Scale bars: A, B, 5 mm; C-G, 1 mm.

DESCRIPTION — Medium-sized species. *Carapace*: 1.1-1.2 × longer than broad; greatest breadth 1.7 × distance between anterolateral spines. Dorsal surface smooth and glabrous, moderately convex from anterior to posterior, smoothly sloping down to rostrum, bearing weak depression between gastric and cardiac regions; epigastric region with pair of denticulate ridges (obsolescent in small paratypes) behind eyes. Lateral margins slightly convexly divergent posteriorly, occasionally with slight constriction about at midlength; anterolateral spine stout, directed straight forward, overreaching small lateral orbital spine; with 2 eminences or small spines, one on anterior end of anterior branchial margin and another on anterior end of posterior branchial margin. Rostrum horizontal, narrow triangular, with interior angle of 20°; length slightly less than half that of remaining carapace, breadth less than half carapace breadth at posterior carapace margin; dorsal surface flattened. Lateral orbital spine slightly anterior to level of, and separated from anterolateral spine by basal breadth of that spine. Pterygostomian flap smooth on surface, anteriorly angular and pointed.

Sternum: Excavated sternum produced to spine between bases of Mxp1, surface with low process or spine in center. Sternal plastron about as long as broad, lateral extremities gently divergent posteriorly. Sternite 3 strongly depressed, anterior margin broadly and deeply excavated, with pair of submedian spines without distinct notch between. Sternite 4 also depressed medially; anterolateral margin slightly convex, anteriorly ending in short spine followed by small or denticle-like, posteriorly diminishing spines, length about twice that of posterolateral margin. Anterolateral margin of sternite 5 nearly subparallel, anteriorly rounded, slightly longer than posterolateral margin of sternite 4.

Abdomen: Smooth and glabrous. Somite 1 with antero-posteriorly convex, well-elevated transverse ridge. Somite 2 tergite 2.3-2.5 × broader than long; pleuron posterolaterally blunt, lateral margin concavely moderately divergent posteriorly. Pleuron of somite 3 with bluntly angular lateral terminus. Telson 0.6-0.7 × as long as broad; posterior plate 1.2-1.4 × longer than anterior plate, posterior margin slightly emarginate.

Eye: Broad relative to length, 1.7 × longer than broad, reaching point distal third of rostrum; mesial and lateral margins subparallel. Cornea not inflated, length slightly less than half that of eyestalk.

Antennule and antenna: Ultimate article of antennule 2.0-2.5 × longer than high. Antennal peduncle terminating in or slightly overreaching corneal margin. Article 2 with well-developed distolateral spine. Antennal scale slightly falling short of distal end of peduncle, 1.2 × broader than article 5. Distal 2 articles unarmed; article 5 2.2-2.7 × longer than article 4; breadth at most half height of ultimate article of antennule. Flagellum of 11-15 segments slightly falling short of distal end of P1 merus.

Mxp: Mxp1 with bases close to each other. Mxp3 glabrous on lateral surface. Basis with 3-4 denticles on mesial ridge. Ischium having flexor margin not rounded distally, crista dentata with about 15 denticles. Merus twice as long as ischium, flattish on mesial face, flexor margin moderately ridged. No spine on merus and carpus.

P1: Massive in largest male (holotype), not so in all paratypes, 4.3-4.7 × longer than carapace in both sexes, bearing setae only on fingers. Ischium with short procurved dorsal spine; ventromesial margin with a few denticle-like proximal processes, subterminal spine vestigial (holotype) or absent (paratypes). Merus with row of 4-5 small spines on proximal mesial margin and small tubercle-like spines on ventral surface; length 1.0-1.1 × that of carapace. Carpus unarmed, 1.2-1.4 × longer than merus. Palm also spineless, relatively high dorsoventrally, 1.7-2.8 × (males), 2.5-3.1 × (females) longer than broad, as long as or slightly shorter than carpus. Fingers distally crossing, largely gaping in largest male (holotype), slightly so in other male paratypes and large females, not gaping in females; movable finger half as long as palm, opposable margin with obtuse proximal process (situated at midlength of gaping portion in largest male holotype).

P2-4: Moderately compressed mesio-laterally, setose on distal 2 articles, sparsely so on carpi, barely so on meri, setae on carpi and propodi long. Meri subequal in length and breadth on P2 and P3 (rarely slightly longer on P3), much shorter and narrower on P4 (P4 merus 0.6-0.7 × length of, 0.7 × breadth of P3 merus); length-breadth ratio, 4.2-4.7 on P2, 4.1-4.6 on P3, 3.7-3.9 on P4; dorsal margin roundly ridged and unarmed. P2 merus 0.9 × length of carapace, 1.0-1.2 × length of P2 propodus; P3 merus 1.1 × length of P3 propodus; P4 merus 0.9-1.0 × length of P4 propodus. Carpi subequal on P2-3, shorter on P4 (P4 carpus 0.7-0.8 × length of P3 carpus), distinctly longer than dactyli (carpus-dactylus length ratio, 1.6-1.7 on P2, 1.4-1.5 on P3, 1.1-1.2 on P4); carpus-propodus length ratio, 0.5-0.6 on P2, 0.5 on P3 and P4. Propodi longest on P3, shortest on P4 or subequal on P2 and P4; flexor margin nearly straight, with single terminal spine slightly distant

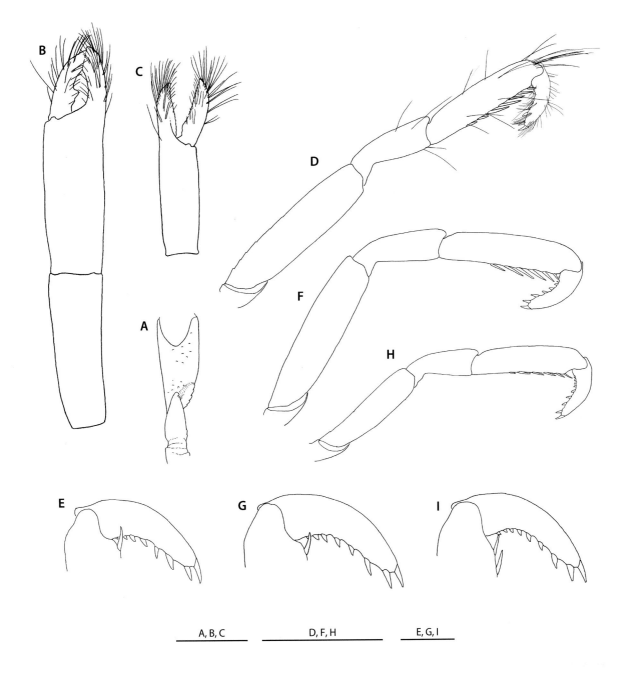

FIGURE 253

Uroptychus stenorhynchus n. sp., holotype, male 7.2 mm (MNHN-IU-2014-16968). **A**, right P1, proximal part, ventral. **B**, same, dorsal. **C**, left P1, distal part, dorsal. **D**, right P2, lateral. **E**, same, distal part, setae omitted, lateral. **F**, right P3, setae omitted, lateral. **G**, same, distal part, setae omitted, lateral. **H**, right P4, setae omitted, lateral. **I**, same, distal part, setae omitted, lateral. Scale bars: A, B, C, D, F, H, 5 mm; E, G, I, 1 mm.

from juncture with dactylus, preceded by 6-7 (P2 and P3) or 5 (P4) movable spines. Dactyli proportionately broad, strongly curving at proximal third; shorter than carpi, dactylus-propodus length ratio, 0.4 on P2 and P3, 0.4-0.5 on P4; flexor margin with 8-9 proximally diminishing, somewhat inclined sharp spines, ultimate largest or subequal to penultimate.

Eggs. Eggs carried, 16-28 in number; size, 1.05 mm × 0.98 mm - 1.16 mm × 1.18 mm.

Color. Male from SANTO AT19 (MNHN-IU-2014-16973), Vanuatu: Carapace and appendages pale pink; distal articles of P2-4 more or less orangish; abdomen translucent white. This specimen was shown in Poore *et al.* (2011).

REMARKS — The elongate carapace, shapes of the pterygostomian flap and sternum, the length ratio of P2-3 carpi to dactyli, and the spination of the P2-4 dactyli displayed by *U. stenorhynchus* resemble those of *U. lacunatus* n. sp. (see above). Their differences are very small but *U. stenorhynchus* is distinguished from that species by the more narrowly elongate rostrum, the antennal article 2 bearing a strong instead of small distolateral spine, and the P2-4 propodi bearing a single instead of paired terminal spines on the flexor margin.

Uroptychus stenorhynchus also resembles *U. dejouanneti* n. sp. (see above) in having the P2-4 propodi with the distalmost of the flexor terminal marginal spines single and not paired, and the dactyli with sharply triangular, obliquely directed flexor marginal spines, the distal two of which are subequal. *Uroptychus stenorhynchus* is more largely different from *U. dejouanneti* than from *U. lacunatus*: the rostrum is narrower, distinctly overreaching the eyes; the anterolateral spine of the carapace is well developed instead of small, distinctly overreaching instead of terminating at most in the tip of the lateral orbital spine; the antennal article 2 bears a well-developed instead of very tiny distolateral spine; and the P4 merus is 0.6-0.7 instead of 0.9 times as long as the P3 merus. In addition, P1-4 are broader relative to length.

Uroptychus strigosus n. sp.
Figures 254, 255

TYPE MATERIAL — Holotype: **Solomon Islands**. SALOMON 1 Stn DW1827, 9°59.1'S, 161°05.8'E, 804-936 m, 4.X.2001, ♂ 6.1 mm (MNHN-IU-2014-16974).

ETYMOLOGY — From the Latin *strigosus* (= full of striae), alluding to the transverse ridges on the carapace, which are interrupted and mostly scale-like, a rare character among *Uroptychus* species.

DISTRIBUTION — Solomon Islands; 804-936 m.

DESCRIPTION — *Carapace*: 1.1 × broader than long (0.9 × as long as broad); greatest breadth 1.7 × distance between anterolateral spines. Dorsal surface with elevated, scale-like setiferous ridges; distinct groove between gastric and cardiac regions; gastric region somewhat convex. Lateral margins divergent to point two-thirds from anterior end, then convergent, bearing 7 spines: first anterolateral, stout, somewhat posterior to level of small lateral orbital spine, and distinctly overreaching that spine; second and third small, placed on hepatic region and ventral to level of remainder; fourth to seventh on branchial margin, fourth to sixth well developed, seventh small, situated at midlength of posterior branchial margin, followed by ridge leading to posterior end. Rostrum narrow triangular, with interior angle of 20°; dorsal surface flattish, with small scale-like ridges in 2 longitudinal rows; length 0.6 × that of remaining carapace, breadth much less than half carapace breadth at posterior carapace margin. Pterygostomian flap anteriorly angular, produced to small spine; surface with scattered short setiferous ridges.

Sternum: Excavated sternum with anterior margin subtriangular between bases of Mxp1; surface with rounded ridge in midline. Sternal plastron 1.2 × broader than long, posteriorly broadened. Sternite 3 strongly depressed, anterior margin in broad V-shape lacking median notch and submedian spines. Sternite 4 having anterolateral margin 1.4 × longer than

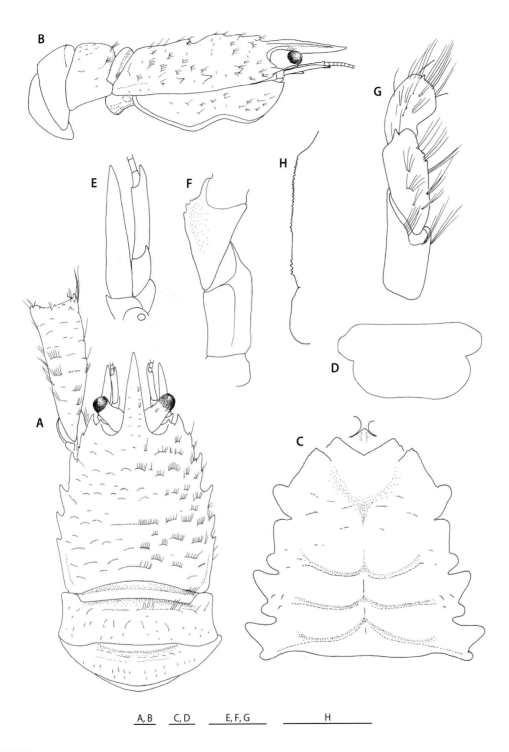

FIGURE 254

Uroptychus strigosus n. sp., holotype, male 6.1 mm (MNHN-IU-2014-16974). **A**, carapace and anterior part of abdomen, proximal part of left P1 included, dorsal. **B**, same, lateral. **C**, sternal plastron, with excavated sternum, basal parts of Mxp1 included. **D**, telson. **E**, right antenna, ventral. **F**, right Mxp3, setae omitted, ventral. **G**, same, lateral. **H**, left Mxp3, crista dentata, ventral. Scale bars: 1 mm.

posterolateral margin, anteriorly denticulate, with angular terminus. Sternite 5 with anterolateral margins subparallel, anteriorly rounded, about as long as posterolateral margin of sternite 4.

Abdomen: Somite 1 with well-elevated, setiferous transverse ridge. Somite 2 tergite 2.5 × broader than long; pleuron posterolaterally blunt, lateral margin slightly concave and slightly divergent posteriorly. Tergites of somites 2 and 3 covered with small scale-like striae, bearing interrupted, setiferous anterior transverse ridge. Pleuron of somite 3 laterally blunt angular. Telson half as long as broad; posterior plate slightly concave on posterior margin, length 1.2 × that of anterior plate.

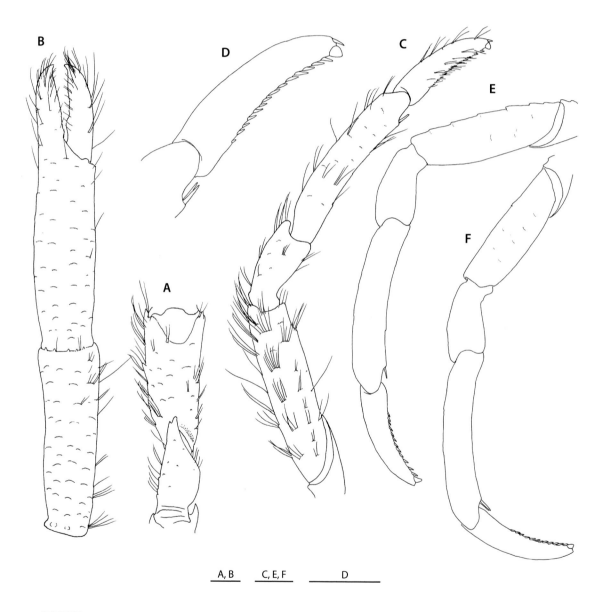

FIGURE 255

Uroptychus strigosus n. sp., holotype, male 6.1 mm (MNHN-IU-2014-16974). **A**, Left P1, proximal part, ventral. **B**, same, dorsal. **C**, right P2, lateral. **D**, same, distal part, setae omitted, lateral. **E**, left P3, lateral. **F**, left P4, lateral. Scale bars: 1 mm.

Eye: Elongate, 2 × as long as broad, distally narrowed. Cornea about half as long as remaining eyestalk.

Antennule and antenna: Antennular ultimate article 3.2 × longer than high, reaching midlength of rostrum. Antennal peduncle extending far beyond cornea. Article 2 with small distolateral spine. Antennal scale overreaching proximal segment of flagellum, breadth 1.8 × that of article 5. Articles 4 and 5 each with distinct distomesial spine; article 5 2.0 × longer than article 4, breadth 0.7 × height of antennular ultimate article. Flagellum consisting of 17-18 segments, overreaching distal end of P1 merus.

Mxp: Mxp1 with bases moderately separated. Mxp3 with long setae on lateral surface. Obsolescent denticles on mesial ridge of basis. Ischium with numerous small denticles on crista dentata, proximal group of 4 denticles remotely separated from distal group; flexor margin distally rounded. Merus twice as long as ischium, with short distolateral spine; flexor margin sharply ridged, bearing a few papilla-like spines on distal third of length. Carpus with small distolateral spine.

P1: Right P1 missing. Left P1 subcylindrical, palm somewhat depressed; covered with short, scale-like setiferous ridges. Ischium dorsally with lobe-like process, ventromesially with very small subterminal spine and obsolescent proximal tubercles. Merus slightly shorter than carapace, dorsally with very small terminal spines near juncture with carpus, mesially with 1 distinct spine at midlength, ventrolaterally with small distal spine. Carpus with 2 tubercle-like processes directly distal to juncture with merus, and a number of spinules on dorsal terminal margin, length subequal to that of merus. Palm as long as carpus, 2.9 × longer than broad, unarmed. Fingers distally ending in short incurved spine; movable finger half as long as palm, opposable margin with very low proximal process, that of fixed finger nearly straight.

P2-4: Sparingly setose. Meri moderately compressed mesio-laterally, successively shorter posteriorly (P3 merus 0.9 × length of P2 merus, P4 merus 0.9 × length of P3 merus), equally broad on P2 and P3, slightly narrower on P4 (P4 merus 0.9 × breadth of P3 merus); length-breadth ratio, 4.0 on P2, 3.4 on P3, 3.2 on P4; dorsal margin irregular, with low eminences on P2 and P3, nearly smooth on P4; P2 merus 0.8 × length of carapace, as long as P2 propodus; P3 merus 0.9 × length of P3 propodus; P4 merus 0.7 × length of P4 propodus. Carpi successively slightly shorter posteriorly (P3 carpus 0.95 × length of P2 carpus, P4 carpus 0.92 × length of P3 carpus); shorter than dactyli (carpus-dactylus length ratio, 0.8 on P2, 0.7 on P7, 0.6 on P4), about 0.4 × as long as propodi; extensor margin with low eminence on proximal portion on P2, smooth on P3 and P4. Propodi subequal in length; flexor margin slightly concavely curving in lateral view, with pair of terminal spines only. Dactyli successively slightly longer posteriorly, proportionately broad; dactylus-carpus length ratio, 1.2 on P2, 1.5 on P3, 1.6 on P4; dactylus-propodus length ratio, 0.5 on P2, 0.6 on P3 and P4; flexor margin somewhat curving, sharply crested, with slender, obliquely directed, closely arranged spines, but penultimate spine pronouncedly broad, more than 3 × as broad as ultimate and antepenultimate.

REMARKS — At first glance the species looks close to *U. squamifer* n. sp. in having the carapace with strong lateral spinens and covered with scale-like setiferous ridges and in having the abdominal somite 1 with a sharp transverse ridge, but it is closer to *U. occultispinatus* Baba, 1988, sharing the P2-4 dactyli that bear a prominent penultimate spine proximally preceded by a row of closely arranged, obliquely directed, slender spines on the flexor margin. *Uroptychus strigosus* differs from *U. occultispinatus* in having the carapace dorsal surface with more distinct scale-like ridges and in having sternite 3 with the anterior margin more deeply emarginate. In addition, the rostrum is narrow triangular with the interior angle of 20° instead of more than 30° as interpreted from the figures in the previous species accounts of *U. occultispinatus* (Balss 1913b: fig. 18; Baba 1988: fig. 14).

Uroptychus strigosus differs from *U. squamifer* in many aspects: the carapace dorsal surface is unarmed instead of bearing small spines on the anterior portion; the pterygostomian flap anteriorly ends in a small instead of strongly produced spine; sternite 3 represents a broad V-shaped anterior margin without submedian spines, instead of a shallowly concave anterior margin with two small submedian spines separated by a notch; the antennal scale overreaches the antennal peduncle instead of falling short of the distal end of the antennal article 5; the P2-4 meri, carpi and propodi are unarmed instead of bearing a row of spines on the dorsal or extensor margin; the P2-4 carpi are distinctly shorter instead of longer than the dactyli; and the flexor margin of the P2-4 dactyli bears a slender ultimate and a broad penultimate spine preceded by obliquely directed slender spines, instead of bearing all similar, sharp, slender spines subperpendicular to the margin.

Uroptychus tafeanus n. sp.

Figures 256, 257

TYPE MATERIAL — Holotype: **Vanuatu**. MUSORSTOM 8 Stn CP974, 19°22'S, 169°28'E, 492-520 m, 22.IX.1994, ov. ♀ 4.3 mm (MNHN-IU-2012-680). Paratype: Station data as for the holotype, 1 ♂ 4.7 mm (MNHN-IU-2013-8511).

ETYMOLOGY — From Tafea, one of the six provinces of Vanuatu from which the species was found, and the Latin suffix *anus* (belonging to).

DISTRIBUTION — Vanuatu, 492-520 m.

DESCRIPTION — Small species. *Carapace:* Broader than long (0.8 × as long as broad), broadest on posterior third; greatest breadth 2.0-2.1 × distance between anterolateral spines. Dorsal surface strongly convex from anterior to posterior, smoothly sloping down to rostrum, sparsely with short fine setae. Lateral margins convexly divergent posteriorly (divergent posteriorly to point one-third from posterior end, and then convergent), granulate, weakly ridged along posterior sixth of length; anterolateral spine strong, directly slightly mesially, overreaching lateral orbital spine. Rostrum somewhat deflected ventrally, relatively narrow triangular, with interior angle 20-22°; dorsal surface feebly concave, length slightly more than half (0.54-0.58) that of remaining carapace, breadth about one-third carapace breadth at posterior carapace margin. Lateral orbital spine small, situated at level of anterolateral spine and separated from that spine by its basal breadth. Pterygostomian flap smooth on surface, anterior margin angular, produced to distinct spine.

Sternum: Excavated sternum with broadly convex anterior margin; surface smooth, with or without weak low ridge in midline. Sternal plastron 0.7-0.8 × as long as broad, lateral extremities convexly divergent posteriorly; broadest on sternite 7. Sternite 3 shallowly depressed; anterior margin in broad V-shape without submedian spines, medially roundly excavated. Sternite 4 with anterolateral margin more than 2 × longer than posterolateral margin, anteriorly convex with obsolescent denticles. Anterolateral margin of sternite 5 convexly divergent posteriorly, 1.6-1.7 × longer than posterolateral margin of sternite 4.

Abdomen: Somite 1 weakly convex from anterior to posterior, without transverse ridge. Somite 2 tergite 3.2 × broader than long; pleural lateral margin concavely divergent posteriorly, posterolateral end blunt. Pleuron of somite 3 with blunt posterolateral terminus. Telson 0.4 × as long as broad; posterior plate about as long as anterior plate, distinctly emarginate on posterior margin.

Eye: Short, 1.5 × longer than broad, not reaching midlength of rostrum, distally narrowed, lateral and mesial margins convex. Cornea more than half as long as remaining eyestalk.

Antennule and antenna: Ultimate article of antennule 3.0-3.3 × longer than high. Antennal peduncle extending far beyond cornea, reaching midlength of rostrum. Article 2 with small lateral spine. Antennal scale tapering, 2.4 × broader than article 5, reaching distal end of proximal third segment of flagellum, extending forward far beyond eye. Articles 4 and 5 each with distinct distomesial spine, article 5 1.3-1.5 × longer than article 5, breadth about half height of antennular ultimate article. Flagellum of 13 segments overreaching apex of rostrum, not reaching distal margin of P1 merus.

Mxp: Mxp1 with bases broadly separated. Mxp3 with relatively long fine setae other than brushes on distal articles. Basis having mesial ridge lobe-like, smooth or with obsolescent denticle. Ischium with flexor margin not rounded distally; crista dentata with numerous small denticles. Merus twice as long as ischium, with small distolateral spine; flexor margin ridged, with 2 or 3 small spines on distal third. Carpus also with small distolateral spine.

P1: 4.7-4.8 × longer than carapace, covered with scattered short ridges supporting short fine soft setae. Ischium with short blunt dorsal spine, ventromesial margin denticulate, without subterminal spine. Merus with 2 small spines on proximal part of mesial margin, length subequal to or slightly more than that of carapace. Carpus 1.3 × longer than merus, dorsal surface with 4 small protuberances transversely arranged near juncture with merus. Palm 2.9-3.0 × longer than broad, 1.3 × longer than carpus. Fingers moderately depressed, distally incurved, gaping in distal half in male paratype, not gaping

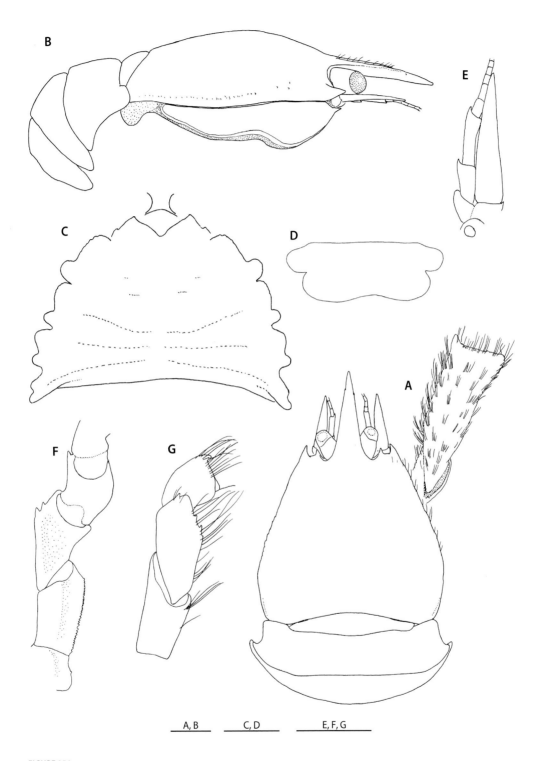

FIGURE 256

Uroptychus tafeanus n. sp., holotype, ovigerous female 4.3 mm (MNHN-IU-2012-680). **A**, carapace and anterior part of abdomen, proximal part of right P1 included, dorsal. **B**, same, lateral. **C**, sternal plastron, with excavated sternum and basal parts of Mxp1. **D**, telson. **E**, left antenna, ventral. **F**, right Mxp3, ventral. **G**, same, setae omitted, lateral. Scale bars: 1 mm.

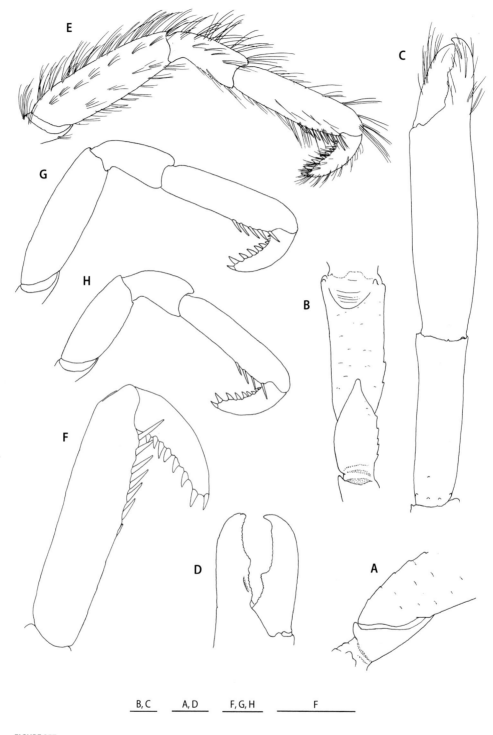

B, C   A, D   F, G, H   F

FIGURE 257

Uroptychus tafeanus n. sp., holotype, ovigerous female 4.3 mm (MNHN-IU-2012-680). **A**, right P1, proximal part, setae omitted, lateral. **B**, same, setae omitted, ventral. **C**, same, distal three articles, setiferous ridges on palm and carpus omitted, dorsal. **D**, same, fingers, setae omitted, ventral. **E**, right P2, lateral. **F**, same, distal articles, setae omitted. **G**, right P3, setae omitted, lateral. **H**, right P4, setae omitted, lateral. Scales bars: 1 mm.

in female holotype; fixed finger with blunt triangular process at midlength of opposable margin; movable finger 0.4 × as long as palm, opposable margin with median process proximal to position of opposite process of fixed finger; ventral face of fixed finger with longitudinal groove accommodating opposite process of movable finger when closed.

P2-4: Relatively broad and somewhat compressed mesio-laterally, with fine soft setae. Meri successively shorter posteriorly (P3 merus 0.8-0.9 × length of P2 merus, P4 merus 0.7 × length of P3 merus); breadths subequal on P2 and P3 or slightly smaller on P3 than on P2, smallest on P4 (P4 merus 0.8-0.9 × breadth of P3 merus); length-breadth ratio, 3.1-3.4 on P2, 2.9 on P3, 2.3-2.5 on P4; dorsal margin rounded, not crested and unarmed; P2 merus 0.7-0.8 × as long as carapace, subequal to length of P2 propodus; P3 merus 0.9-1.0 × length of P3 propodus; P4 merus 0.7 × length of P4 propodus. Carpi successively shorter posteriorly or subequal on P2 and P3, much shorter on P4; carpus-propodus length ratio, 0.5 on P2 and P3, 0.4 on P4; carpus-dactylus length ratio, 1.2 on P2, 1.1 on P3, 0.9 on P4. Propodi successively shorter posteriorly or subequal on P2 and P3 and shorter on P4; flexor margins slightly convex on distal portion, ending in pair of terminal spines preceded by 5 movable spines on P2, 4 or 5 on P3, 3 on P4 (restricted to distal third of length). Dactyli 0.4 × as long as propodi on P2 and P3, 0.5 × on P4; dactylus-carpus length ratio, 0.8 on P2, 0.9 on P3, 1.1 on P4; flexor margins slightly curving, with 9, 9 or 10, 9-11 somewhat obliquely directed, proximally diminishing, triangular spines on P2, P3, P4 respectively, ultimate slightly narrower than penultimate.

Eggs. Ten eggs carried; size, 1.00 × 1.22 mm - 1.00 × 1.27 mm.

Color (in preservative). Reddish stripe in midline of carapace as broad as rostrum, extending on to somite 6 of abdomen; similar stripe on each side of carapace but not clear on abdomen; in female, median stripe broadened on gastric region where connecting to side stripes.

REMARKS — The species resembles *U. plumella* n. sp., *U. senticarpus* n. sp., and *U. shanei* n. sp. (see above under the accounts of these species for their similarities). From all of these congeners *U. tafeanus* is readily distinguished by the ultimate flexor marginal spine of the P2-4 dactyli that is somewhat smaller instead of distinctly larger than the penultimate, hence these two groups are remote from each other in the key to species (see above). *Uroptychus tafeanus* is closer to *U. senticarpus*, from which it is distinguished by the anterolateral spine of the carapace directed anteromesially instead of anterolaterally; the distodorsal margin of the P1 carpus is unarmed instead of bearing a distinct spine at the midpoint of the distodorsal margin; and the P2-4 carpi are unarmed instead of bearing a small proximal spine on the extensor margin.

Uroptychus terminalis n. sp.

Figures 258-260

TYPE MATERIAL — Holotype: **New Caledonia**, Norfolk Ridge. BIOCAL Stn CP30, 23°09'S, 166°41'E, 1140 m, 29.VIII.1985, ♂ 8.4 mm (MNHN-IU-2014-16975). Paratypes: **New Caledonia**, Norfolk Ridge. Collected with holotype, 1 ♂ 7.5 mm, 2 ov. ♀ 7.0, 7.2 mm (MNHN-IU-2014-16976). BATHUS 3 Stn CP823, 23°23'S, 167°52'E, 980-1000 m, 29.XI.1993, 10 ♂ 2.9-8.6 mm, 9 ov. ♀ 5.8-6.8 mm, 3 ♀ 6.0-6.8 mm (MNHN-IU-2014-16977). – Stn CP831, 23°04'S, 166°56'E, 650-658 m, 30.XI.1993, 1 ♀ 6.7 mm (MNHN-IU-2014-16978). NORFOLK 2 Stn CP2131, 23°13.19'S, 168°11.21'E, 1114-1190 m, 2.XI.2003, 4 ♂ 5.4-7.9 mm, 3 ov. ♀ 5.8-6.0 mm, 1 ♀ 6.9 mm (MNHN-IU-2014-16979). – Stn CP2138, 23°00.56'S, 168°22.80'E, 396-405 m, 3.XI.2003, 2 ♀ 6.4, 6.6 mm (MNHN-IU-2014-16980). – Stn CP2139, 23°01'S, 168°23'E, 372-393 m, 03.XI.2003, 1 ♀ 6.5 mm (MNHN-IU-2014-16981). **New Caledonia**, Loyalty Basin. BIOGEOCAL Stn CP297, 20°37'S, 167°11'E, 1230-1240 m, 28.IV.1987, 1 ov. ♀ 5.7 mm, 3 ♀ 4.7-5.8 mm (MNHN-IU-2014-16982). **Vanuatu**. MUSORSTOM 8 Stn CP1111, 14°51.09'S, 167°14.00'E, 1210-1250 m, 8.X.1994, 1 ♂ 6.2 mm, 3 ov. ♀ 5.6-6.4 mm (MNHN-IU-2014-16983). – Stn CP1125, 15°57.63'S, 166°38.43'E, 1160-1220 m, 10.X.1994, 3 ♂ 3.6-8.0 mm, 3 ov. ♀ 4.1-6.7 mm, 2 ♀ 5.7, 6.2 mm (MNHN-IU-2014-16984). – Stn CP1126, 15°58.35'S, 166°39.98'E, 1210-1260 m, 10.X.1994, 1 ♂ 6.5 mm, 1 ov. ♀ 7.1 mm (MNHN-IU-2014-16985). – Stn CP1129, 16°00.73'S, 166°39.94'E, 1014-1050 m, 10.X.1994, 1 ♂ 6.0 mm, 1 ov. ♀ 6.4 mm (MNHN-IU-2014-16986). **Solomon Islands**. SALOMON 2 Stn CP2189, 8°19.6'S, 160°01.9'E, 660-854 m, 23.X.2004, 1 ov. ♀ 5.1 mm (MNHN-IU-2014-16987).

FIGURE 258

Uroptychus terminalis n. sp., holotype, male 8.4 mm (MNHN-IU-2014-16975). **A**, carapace and anterior part of abdomen, proximal part of left P1 included, dorsal. **B**, same, lateral. **C**, anterior part of sternal plastron, with excavated sternum and basal parts of Mxp1. **D**, telson. **E**, right antennule and antenna, ventral. **F**, right Mxp3, setae omitted, ventral. **G**, same, lateral. **H**, left P1, proximal part, ventral. **I**, same, proximal part omitted, dorsal. Scale bars: A, B, H, I, 5 mm; C-G, 1 mm.

OTHER MATERIAL EXAMINED — **Kermadec Islands**. CHALLENGER Stn 170, 29°55′S, 178°14′W, 520 fms (946 m), 1 ov. ♀ 5.9 mm (BMNH 1888:33) [syntype of *U. australis* (Henderson, 1885)]. – Stn 171, 28°33′S, 177°50′W, 600 fms (1098 m), 1 ov. ♀ 7.4 mm (BMNH 1888:33) [syntype of *U. australis* (Henderson, 1885)].

ETYMOLOGY — From the Latin *terminalis* (terminal), alluding to the terminal flexor marginal spine of the P2 propodus, which is very close to the distal end of the article, one of characters to separate the new species from *U. nigricapillis* Alcock, 1901.

DISTRIBUTION — Solomon Islands, Vanuatu, Loyalty Basin, Norfolk Ridge, and Kermadec Islands; 372-1260 m.

SIZE — Males, 2.9-8.6 mm; females, 4.1-7.4 mm; ovigerous females from 4.1 mm.

DESCRIPTION — Medium-sized species. *Carapace*: 1.1 × longer than broad; greatest breadth 1.7 × distance between anterolateral spines. Dorsal surface convex from anterior to posterior, with weak or feeble depression between gastric and cardiac regions; pair of epigastric spines distinct, often small. Lateral margins convexly divergent or with constriction about at midlength (between anterior and posterior branchial margin), ridged along posterior half of branchial margin; anterolateral spines small, not overreaching lateral orbital spine; tiny spine usually present at anterior end of branchial region, followed by at most 5 small, often obsolescent tubercle-like small spines on posterior branchial margin. Rostrum narrow triangular, with interior angle of 15-25°, straight horizontal, slightly upcurved distally or slightly deflected ventrally; dorsal surface flattish; length 0.4-0.5 × that of remaining carapace, breadth half or slightly less than half carapace breadth at posterior carapace margin. Pterygostomian flap anteriorly angular, produced to acute spine, smooth on surface.

Sternum: Excavated sternum anteriorly ending in sharp spine between bases of Mxp1, surface weakly ridged in midline, bearing small spine in center. Sternal plastron slightly broader than long; sternites successively broader posteriorly. Sternite 3 depressed well, anterior margin deeply excavated, with 2 submedian spines contiguous to each other or separated by narrow notch. Sternite 4 with transverse row of tubercles preceded by strong depression; anterolateral margin slightly convex, bearing posteriorly diminishing denticles, anteriorly produced to small spine; posterolateral margin more than half length of anterolateral margin. Anterolateral margin of sternite 5 strongly convex and denticulate, about as long as posterolateral margin of sternite 4.

Abdomen: Smooth and glabrous. Somite 1 with antero-posteriorly convex transverse ridge. Somite 2 tergite 2.3-2.5 × broader than long; pleuron posterolaterally blunt, lateral margin concavely divergent posteriorly. Pleuron of somite 3 with blunt lateral end. Telson two-thirds as long as broad; posterior plate 1.7-1.8 × longer than anterior plate, posterior margin distinctly emarginate.

Eye: Broad relative to length (1.5-1.6 × longer than broad), broadened distally, overreaching midlength of rostrum. Cornea slightly dilated, more than half length of remaining eyestalk.

Antennule and antenna: Ultimate article of antennular peduncle about twice as long as high. Antennal peduncle reaching distal margin of cornea. Article 2 with distinct distolateral spine. Antennal scale 1.3-1.7 × broader than article 5, reaching or overreaching midlength of, not reaching distal end of article 5. Distal 2 articles unarmed; article 5 2.1-2.4 × longer than article 4, breadth half height of ultimate article of antennule. Flagellum consisting of 14-16 segments, not reaching distal end of P1 merus.

Mxp: Mxp1 with bases close but not contiguous to each other. Mxp3 barely setose on lateral surface. Basis with 5-6 denticles often obsolescent, distal one distinct. Ischium with 10-14 denticles on crista dentata; flexor margin distally not rounded. Merus and carpus unarmed; merus 2.5 × longer than ischium, not flattened, moderately thick, rounded along flexor margin.

P1: Somewhat massive in males, length 4.3-4.6 × that of carapace; smooth and barely setose except for fingers. Ischium with short, depressed dorsal spine, ventrally with a few denticle-like spinules, ventromesially with vestigial subterminal spine. Merus with a few longitudinally arranged ventromesial spines and row of small or obsolescent spines in ventral midline (both occasionally obsolescent in females); length 1.1-1.3 × that of carapace. Carpus 1.2-1.3 × length of merus. Palm 2.8-3.3 × (males), 3.4-4.4 × (females) longer than broad, 0.8 × (rarely 0.9 x) length of carpus. Fingers proportionately broad, somewhat curving ventrally, distally crossing in males, barely crossing in females; gaping or not gaping in males, not gaping in females; opposable margins fitting to each other in distal half in non-gaping fingers, fitting in distal third in gaping fingers, fitting along entire length in females; movable finger half as long as palm, opposable margin in females with proximo-distally broad process proximal to position of median low eminence on fixed finger.

P2-4: Relatively broad in lateral view, well compressed mesio-laterally, bearing long setae numerous on distal articles. P2 merus slightly shorter than or as long as, and as broad as P3 merus, P4 merus shortest and much narrower than P2-3 meri, 0.6-0.7 × length of P3 merus; length-breadth ratio, 3.7-4.8 on P2 and P3, 3.5-4.0 on P4; P2 merus 0.8-0.9 × length of carapace, 1.1-1.3 × length of P2 propodus; P3 merus subequal to length of P3 propodus; P4 merus 0.9 × length of P4 propodus; dorsal margin with or without a few very small spines or eminences proximally on P2; ventrolateral margin with very small terminal spine distinct on P2 and P3, often obsolete on P4. Carpi slightly longer on P3 than on P2 or subequal on P2-3, shortest on P4; carpus-propodus length ratio, 0.6 on P2, 0.5 on P3, 0.4-0.5 on P4. Propodi longest on P3, shortest on P4; broad relative to length, flexor margin nearly straight, with 7-11 spines on P2, 6-8 on P3, 5-7 on P4, terminal spine single, very close to juncture with dactylus on P2, somewhat remote on P3 and P4. Dactyli curving at proximal third, shorter than carpi (dactylus-carpus length ratio, 0.7 on P2, 0.7-0.8 and P3, 0.9-1.0 on P4); dactylus-propodus length ratio, 0.4 on P2 and P3, 0.5 on P4; flexor margin strongly curving, with 8 or 9 (mostly 9) subtriangular spines obscured by setae, inclined but not oriented parallel to flexor margin, ultimate and penultimate close to each other, antepenultimate remotely equidistant between penultimate and distal quarter; ultimate distinctly larger than penultimate, remaining spines successively diminishing toward base of article.

Eggs. Eggs carried 6-14 in number (normal number probably more); size, 1.20 mm × 1.30 mm - 1.93 mm × 1.97 mm.

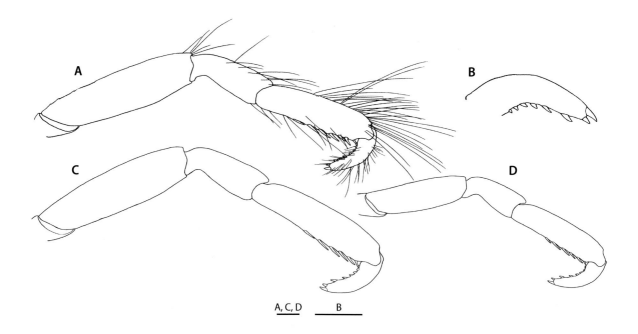

FIGURE 259

Uroptychus terminalis n. sp., holotype, male 8.4 mm (MNHN-IU-2014-16975). **A**, right P2, lateral. **B**, same, distal part, setae omitted, lateral. **C**, right P3, setae omitted, lateral. **D**, right P4, lateral. Scale bars: 1 mm.

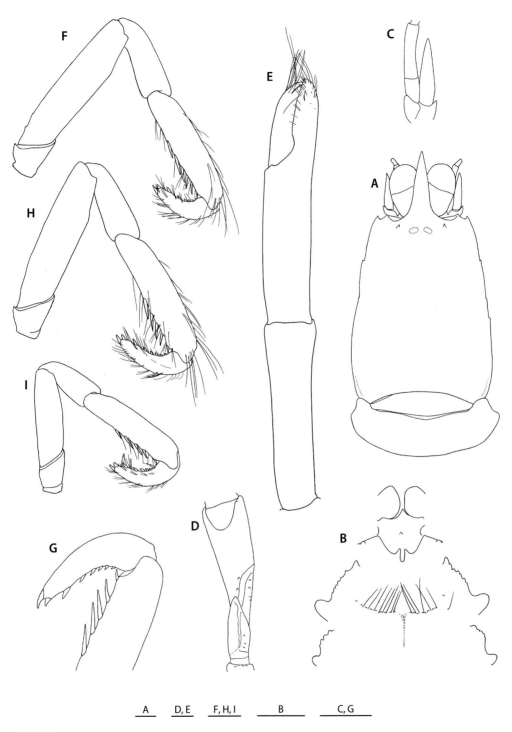

FIGURE 260

Uroptychus terminalis n. sp., ovigerous female 5.9 mm from *Challenger* Stn 170. **A**, carapace and anterior part of abdomen, dorsal.
B, anterior part of sternal plastron, with excavated sternum and basal parts of Mxp1. **C**, left antenna, ventral. **D**, right P1, proximal
part, ventral. **E**, same, distal articles, dorsal. **F**, right P2, lateral. **G**, same, distal part, setae omitted, mesial. **H**, right P3, lateral. **I**, right
P4, lateral. Scale bars: 1 mm.

PARASITES — Male from MUSORSTOM 8 Stn CP1129 (MNHN-IU-2014-16986) with bopyrid isopod on left branchial cavity.

REMARKS — As mentioned under *Uroptychus australis* (see above), the specimen (ovigerous female) from Challenger Stn 170 (see Figure 260) and another specimen (ovigerous female) from Challenger Stn 171, both syntypes of *Diptychus australis* Henderson, 1885 (currently *U. australis* (Henderson, 1885)), are referred to this new species (see above under *U. australis*). Also one of the specimens, ovigerous female, from Challenger Stn 194 proved to be identical with *U. empheres* Ahyong & Poore, 2004 (see above under *U. australis*). *Uroptychus terminalis* is distinguished from *U. australis* by having the P2-4 propodi with the terminal of the flexor marginal spines single, not paired, and the flexor marginal spines of the dactyli are obliquely directed, not oriented parallel to the margin. *Uroptychus empheres* has the P2-4 propodal terminal spine paired, not single as in this new species.

The general shape of the carapace and the flexor margin of the P2-4 propodi distally ending in a single spine link the species to *U. dejouanneti* n. sp., *U. gracilimanus* (Henderson, 1885), *U. brevisquamatus* Baba, 1988, and *U. nigricapillis* Alcock, 1901. *Uroptychus terminalis* is closer to *U. nigricapillis* than to the other species in having a pair of epigastric spines. However, *U. nigricapillis* (see above) has the terminal of the flexor marginal spines of P2-4 propodi rather remote from the juncture with the dactylus. This new species is distinguished from *U. dejouanneti* and *U. gracilimanus* by having the ultimate spine of the P2-4 dactyli distinctly larger than instead of smaller than (*U. gracilimanus*) or subequal to (*U. dejouanneti*) the penultimate; and the antennal article 2 bears a distinct instead of very small lateral spine. *Uroptychus terminalis* is distinguished from *U. brevisquamatus* by the anterolateral spine of the carapace that is small instead of stout, at most terminating in rather than distinctly overreaching the tip of the lateral orbital spine; the P4 merus is much shorter and narrower than the P3 merus; and the pterygostomian flap is anteriorly angular and produced to a sharp spine, instead of being roundish bearing a very small spine.

Uroptychus thermalis Baba & de Saint Laurent, 1992

Figures 261-263

Uroptychus thermalis Baba & de Saint Laurent, 1992: 324, fig. 2. — Chevaldonné & Olu 1996: 293. — Ahyong & Poore 2004a: 77, fig. 24. — Baba 2005: 231.

TYPE MATERIAL — Holotype: **North Fiji Basin**, 16°59.50'S, 173 °55.47'W, hydrothermal vent, 2000 m, male (MNHN-Ga 2351). [examined].

MATERIAL EXAMINED — **New Caledonia**. BIOCAL Stn CP72, 22°10'S, 167°33'E, 2100-2110 m, 4.IX.1985, 1 ♀ 8.1 mm (MNHN-IU-2014-16988).

DISTRIBUTION — North Fiji Basin (active thermal vent site), Queensland, and now New Caledonia; 1497-2110 m.

DIAGNOSIS — Medium-sized species. Carapace slightly longer than broad; greatest breadth 1.6 × distance between anterolateral spines. Dorsal surface with elevated, granulate, short ridges; gastric and cardiac regions distinctly bordered by deep groove, and both well inflated; anterolateral spine directed anterolaterally, overreaching angular lateral limit of orbit, followed by no spine but a row of short granulate ridges along branchial margin. Rostrum narrow triangular, with interior angle of 22-26°; breadth half length of posterior carapace margin; dorsal surface smooth, slightly convex from side to side. Pterygostomian flap with roundish anterior margin bearing small spine. Excavated sternum anteriorly convex with small spine, surface with small central spine; sternal plastron 1.2 × broader than long; lateral extremities divergent posteriorly; sternite 3 depressed, anterior margin moderately concave, with pair of small submedian spines; sternite 4 with strong spine about at midlength of anterolateral margin, posterolateral margin as long as anterolateral

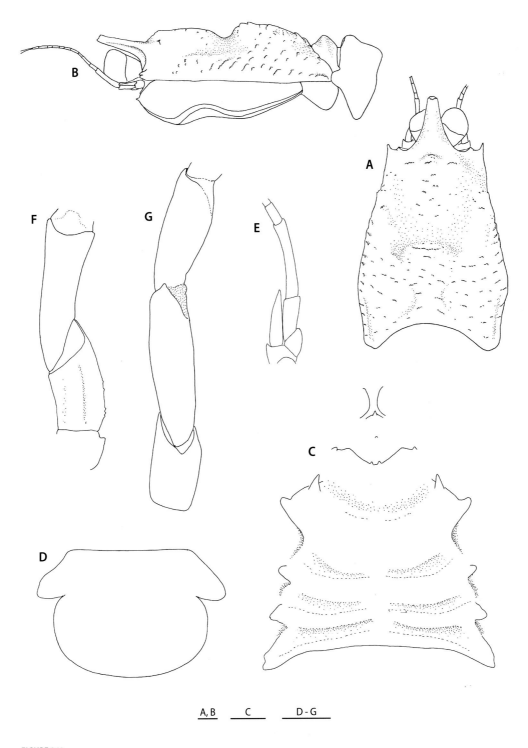

A, B   C   D - G

FIGURE 261

Uroptychus thermalis Baba & de Saint Laurent, 1992, female 8.1 mm (MNHN-IU-2014-16988). **A**, carapace, dorsal. **B**, same, anterior part of abdomen included, lateral. **C**, sternal plastron, with excavated sternum and basal parts of Mxp1. **D**, telson. **E**, right antenna, ventral. **F**, right Mxp3, ventral. **G**, same, lateral. Scale bars: 1 mm.

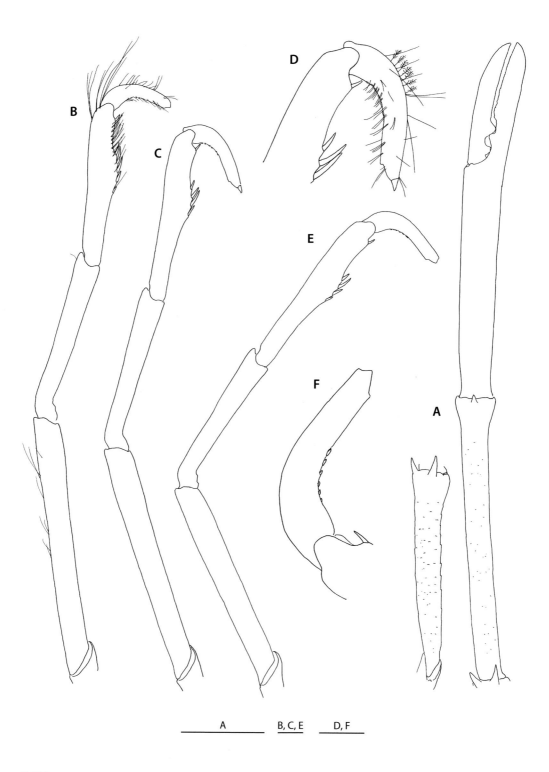

B

C

D

E

F

A

A   B, C, E   D, F

FIGURE 262

Uroptychus thermalis Baba & de Saint Laurent, 1992, female 8.1 mm (MNHN-IU-2014-16988). **A**, right P1, dorsal. **B**, right P2, lateral. **C**, right P3, setae omitted, lateral. **D**, same, distal part, lateral. **E**, right P4, lateral. **F**, same, distal part, lateral. Scale bars: A, 5 mm; B-F, 1 mm.

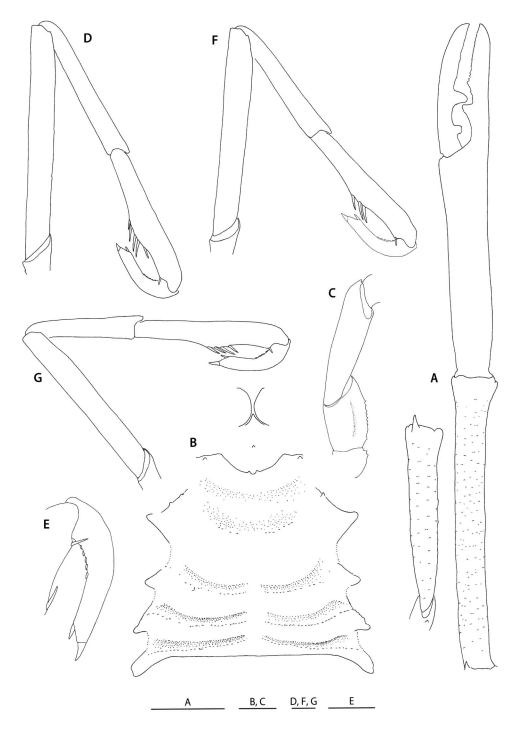

FIGURE 263

Uroptychus thermalis Baba & de Saint Laurent, 1992, male holotype 8.4 mm (MNHN-IU-2014-12825 [=MNHN-Ga 2351]). **A**, right P1, setae omitted, dorsal. **B**, sternal plastron, with excavated sternum and basal parts of Mxp1. **C**, right Mxp3, setae omitted, ventral. **D**, right P2, setae omitted, lateral. **E**, same, distal part, lateral. **F**, right P3, lateral. **G**, right P4, lateral. Scale bars: 1 mm.

margin; sternite 5 with anterolateral margin straightly divergent posteriorly, slightly more than half as long as posterolateral margin of sternite 4. Abdomen smooth and glabrous, somite 1 strongly convex from anterior to posterior; somite 2 tergite 2.0 × broader than long; pleuron having lateral margin concavely strongly divergent posteriorly, ending in blunt tip; pleuron of somite 3 blunt on lateral terminus. Telson three-fifths as long as broad; posterior plate subsemicircular, posteriorly rounded, length 1.7 × that of anterior plate. Eyes short relative to breadth; cornea inflated, as long as remaining eyestalk. Ultimate article of antennular peduncle 2.5 × longer than high. Antennal peduncle slender, overreaching cornea; article 2 with small lateral spine; antennal scale slightly overreaching article 4; distal 2 articles unarmed, article 5 twice as long as article 4, breadth less than half height of ultimate article of antennule; flagellum of 10 or 11 segments far falling short of distal end of P1 merus. Mxp1 with bases close to each other. Mxp3 slender and spineless; coxa with strong ventral spine; basis with 1 denticle on distal part of mesial ridge; ischium having flexor margin not rounded distally, crista dentata with obsolescent denticles; merus 2.8 × longer than ischium, relatively thick mesio-laterally; carpus longer than ischium. P1 slender; ischium with small dorsal process, ventrally unarmed; merus with 2 prominent distodorsal spines, length 1.4 × that of carapace; carpus 1.5 × longer than merus, with 1 distodorsal and 1 lateral distoventral spine; palm 0.8 × length of carpus; fingers spooned on prehensile face, not crossing distally when closed; movable finger 0.6 × as long as palm; fixed finger with median process opposite to between 2 proximal processes of movable finger. P2-4 slender, subcylindrical on meri and carpi, sparsely setose on distal articles; meri subequally broad on P2-4, unarmed; P2 merus slightly longer than carapace, 1.7 × longer than P2 propodus; P3 merus 0.9 × length of P2 merus; P4 merus 0.9 × length of P3 merus, slightly longer than P4 propodus; carpi long, subequal in length on P2 and P3, slightly shorter on P4 than on P2 and P3, carpus-propodus length ratio, 1.0 on P2, 0.8 on P3 and P4; propodi broadened medially; flexor margin with slender movable spines distal to midlength, distalmost single, slightly proximal to juncture with dactylus, distantly separated from proximal group of 4-5 spines; dactyli strongly curving, 0.4 × as long as propodi on P2-4, and 0.4 × as long as carpi on P2 and P3, 0.5 × on P5, flexor margin with terminal and subterminal spines of good size (subterminal spine absent in this specimen), both remotely separated from proximal group of 6 very small obliquely directed spines.

REMARKS — No clear differences were found in the relative lengths of articles of P2-4 between the type and the present material. Apparently this female specimen is identical with the male reported by Ahyong & Poore (2004) from a non-vent site in Queensland, sharing the presence of a strong lateral spine on sternite 4, two distodorsal spines on the P1 merus and one distodorsal spine on the P1 carpus. In the type material, the lateral spine of sternite 4 is very small, the P1 merus bears one instead of two spines, and the P1 carpus has no distodorsal spine. Ahyong & Poore (2004) considered these differences as age-related variations, but the present specimen is about as large as in the type material. The crista dentata of the Mxp3 ischium in this specimen bears denticles more obsolescent than in the type. Given that the type material was taken from the active thermal vent-site, the material from Queensland and New Caledonia may be referable to a different species. However, this is provisionally placed in *U. thermalis*, pending a discovery of additional material from the type locality and molecular analyses.

The spination of the P2 dactylus is illustrated here for the BIOCAL specimen; the small spines on the proximal portion are also discernible under high magnification in the type (Figure 263E). Dactylar and propodal spination of the P2-4 are similar between *U. thermalis* and *U. sternospinosus* Tirmizi, 1964, as also are the slender pereopods, short antennal scales, and the shape of the sternal plastron. In *U. sternospinosus*, however, the P2-4 carpi are consistently longer than the propodi, and the carapace bears a distinct longitudinal carina in midline and a pair of strong gastric spines, the characters confirmed by examination of the type material (BMNH 1966.2.3.21-22).

Uroptychus toka Schnabel, 2009

Figures 264, 265

Uroptychus toka Schnabel, 2009: 568, fig. 14.

TYPE MATERIAL — Holotype: **New Zealand**, L'Esperance Rock, Kermadec Ridge, 33°02.59'S, 179°34.60'W, 350-490 m, ov. female (NMNZ Cr. 012090). [not examined].

MATERIAL EXAMINED — **New Caledonia**, Loyalty Ridge. MUSORSTOM 6 Stn DW478, 21°08.96'S, 167°54.28'E, 400 m, 22.II.1989, 1 ♀ 3.7 mm (MNHN-IU-2014-16989). **New Caledonia**, Norfolk Ridge. CHALCAL 2 Stn DW82, 23°13.68'S, 168°04.27'E, 304 m, 31.X.1986, 1 ♂ 2.6 mm (MNHN-IU-2014-16991). NORFOLK 1 Stn DW1657, 23°28'S, 167°52'E, 305-332 m, 19.VI.2001, 1 ♀ 2.9 mm (MNHN-IU-2014-16992). NORFOLK 2 Stn CP2048, 23°43.82'S, 168°16.24'E, 380-389 m, unidentified host, 24.X.2003, 2 ov. ♀ 2.7, 2.8 mm (MNHN-IU-2014-16993). **Vanuatu**, Monts Gemini. GEMINI Stn DW51, 20°58'S, 170°04'E, 450-360 m, 04.VII.1989, 1 ♀ 2.9 mm (MNHN-IU-2014-16990).

DISTRIBUTION — Kermadec Ridge in 350-490 m, and now south of Vanuatu, Loyalty Ridge and Norfolk Ridge, in 304-450 m.

SIZE — Male, 2.6 mm; females, 2.8-3.7 mm; ovigerous females from 2.7 mm.

DESCRIPTION — Small species. Body and appendages very setose. *Carapace*: 0.85-0.95 × as long as broad; greatest breadth 1.7 × distance between anterolateral spines. Dorsal surface with long soft setae particularly numerous on branchial region, denticle-like small spines on hepatic and epigastric regions, and distinct depression bordering gastric and cardiac regions; gastric region slightly inflated, anteriorly elevated from level of rostrum. Lateral margins convex medially or on posterior third, with row of short, oblique, elevated setiferous ridges usually weak, often distinct; ridged along posterior third; anterolateral spine small, subequal to and very close to lateral orbital spine (in dorsal view, nearly contiguous at base). Rostrum nearly horizontal, triangular with interior angle of 35°, anteriorly ending in blunt tip; length 0.4 × that of remaining carapace, breadth half carapace breadth at posterior carapace margin; dorsal surface concave. Lateral orbital spine larger than anterolateral spine. Pterygostomian flap equally high on anterior and posterior halves, surface with scattered denticles and long setae, and row of tubercle-like small spines below linea anomurica; anterior margin angular, produced to small spine.

Sternum: Excavated sternum with somewhat rounded or bluntly triangular anterior margin, surface ridged in midline. Sternal plastron slightly longer than broad; lateral extremities subparallel between sternites 4 and 7. Sternite 3 weakly depressed; anterior margin shallowly emarginate, representing broad V-shape, with narrow U-shaped median sinus flanked by obsolescent spine, lateral end blunt. Sternite 4 having anterolateral margin nearly straight or slightly convex, anterior end blunt angular or rounded; posterolateral margin as long as anterolateral margin. Anterolateral margins of sternite 5 somewhat convex anteriorly, subparallel, slightly more than half as long as posterolateral margin of sternite 4.

Abdomen: Sparsely or thickly setose. Somite 2 tergite 2.2 × broader than long; pleuron posterolaterally rounded, lateral margin nearly straight or feebly concave. Pleuron of somite 3 blunt on lateral terminus. Telson about half as long as broad; posterior plate distinctly emarginate on posterior margin, slightly (1.1 x) longer than anterior plate.

Eye: Slightly falling short of apex of rostrum, 1.7 × longer than broad; mesial and lateral margins slightly convex or subparallel. Cornea not dilated, length half that of remaining eyestalk.

Antennule and antenna: Ultimate article of antennular peduncle 2.6-2.7 × longer than high. Antennal peduncle reaching apex of rostrum. Article 2 with strong distolateral spine. Antennal scale 1.5 × broader than article 5, slightly or moderately overreaching distal end of article 4, not reaching midlength of article 5. Article 4 distomesially produced to short spine. Article 5 unarmed, length 1.4 × that of article 4, breadth half height of antennular ultimate article. Flagellum of 10-11 segments not reaching distal end of P1 merus.

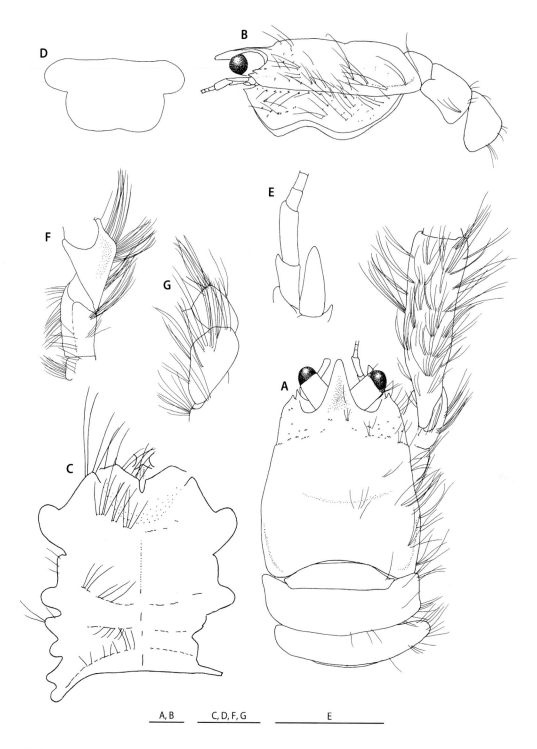

FIGURE 264

Uroptychus toka Schnabel, 2009, female 3.7 mm (MNHN-IU-2014-16989). **A**, carapace and anterior part of abdomen, proximal part of right P1 included, dorsal. **B**, same, lateral. **C**, sternal plastron, with excavated sternum and basal parts of Mxp1. **D**, telson. **E**, left antenna, ventral. **F**, left Mxp3, ventral. **G**, same, merus and carpus, lateral. Scale bars: 1 mm.

Mxp: Mxp1 with bases remotely separated. Mxp3 with long setae on lateral surface. Basis without denticles on mesial ridge. Ischium with obsolescent denticles on crista dentata, flexor margin rounded at distal end. Merus unarmed, twice as long as ischium, broad relative to length, well compressed and crested along flexor margin. Carpus unarmed.

P1: 5.4-5.8 × longer than carapace in females (missing in male); relatively slender, smooth with long setae. Ischium with short dorsal spine, ventromesially unarmed. Merus 1.3-1.4 × longer than carapace. Carpus 1.2-1.3 × longer than merus, nearly as long as or slightly longer than palm. Palm 3.9-4.8 × longer than broad, slightly broader than carpus. Fingers ventrally curved, ending in slightly incurved spine, slightly crossing when closed; movable finger 0.4-0.5 × length of palm, opposable margin with 2 low prominences in proximal half, that of fixed finger gently concave.

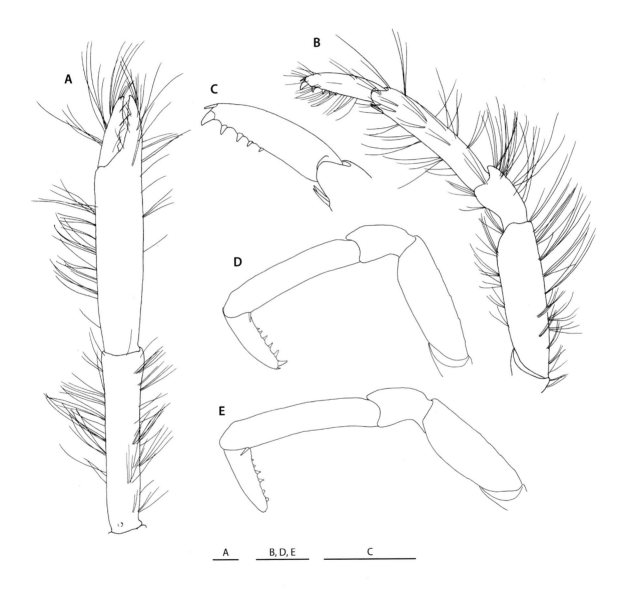

FIGURE 265

Uroptychus toka Schnabel, 2009, female 3.7 mm (MNHN-IU-2014-16989). **A**, right P1, dorsal. **B**, left P2, lateral. **C**, same, distal part, setae omitted, lateral. **D**, left P3, setae omitted, lateral. **E**, left P4, lateral. Scale bars: 1 mm.

P2-4: With long soft setae. Meri successively shorter posteriorly (P3 merus 0.8-0.9 × length of P2 merus, P4 merus 0.9 × length of P3 merus), equally broad on P2-4; rather thick mesio-laterally and unarmed; length-breadth ratio, 3.2-4.1 on P2, 3.1-3.7 on P3, 2.6-3.3 on P4; P2 merus 0.8-0.9 × length of carapace, about as long as or slightly longer than P2 propodus; P3 merus 0.8 × length of P3 propodus; P4 merus 0.7-0.8 × length of P4 propodus. Carpi subequal, relatively short, 0.3-0.4 × length of propodus on P2 and P3, 0.3 × on P4. Propodi successively longer posteriorly; flexor margin slightly curving, with pair of terminal spines only. Dactyli subequal, 1.3 (P2 and P3)-1.4 × (P4) longer than carpi, slightly less than half (0.44-0.49 on P2, 0.45-0.49 on P3, 0.41-0.47 on P4) length of propodi; flexor margin straight, with 6 or 7 loosely arranged spines nearly perpendicular to margin except for ultimate spine; ultimate very small and slender, close to penultimate; penultimate prominent; antepenultimate half as broad as penultimate, subequal to fourth; remainder smaller, diminishing toward base of article.

Eggs. Number of eggs carried, 4-5; size, 0.97 mm × 1.08 mm - 1.06 mm × 1.18 mm.

REMARKS — The specimen from NORFOLK 1 Stn DW1657 (MNHN-IU-2014-16992) is much more densely setose on the body and appendages than the other specimens. *Uroptychus toka* resembles *U. volsmar* n. sp. more closely than *U. bertrandi* n. sp., *U. sarahae*, *U. rutua*, in having the anterolateral spine of the carapace contiguous at base with the lateral orbital spine when viewed from dorsal side. The relationships with *U. volsmar* are discussed under that species (see below). For the relationships between the other congeners, see under the remarks of *U. bertrandi* and *U. sarahae* (see above).

Uroptychus triangularis Miyake & Baba, 1967

Uroptychus triangularis Miyake & Baba, 1967: 203, fig. 1. — Baba *et al.* 2009: 64, figs 54, 55. — Poore *et al.* 2011: 331, pl. 8, fig. D.

TYPE MATERIAL — Holotype: **Japan**, near Muko-jima, Ogasawara Island, ov. female (ZLKU 4883). [not examined].

MATERIAL EXAMINED — **Philippines**. MUSORSTOM 1 Stn CP03, 14°01'N, 120°15'E, 183-185 m, 19.III.1976, 1 ov. ♀ 3.0 mm (MNHN-IU-2014-16994).

DISTRIBUTION — Ogasawara Islands (Japan), Taiwan and now the Philippines; in 183-236 m.

DIAGNOSIS — Small species. Carapace subtriangular in dorsal view, 1.5 × broader than long; greatest breadth 2.2 × distance between anterolateral spines. Dorsal surface gently convex from anterior to posterior, nearly glabrous, with very small, scattered spines on anterior half; lateral margins strongly divergent to point one-third from posterior end, then convergent, bearing row of spines: first anterolateral, directed slightly anterolaterally, distinctly overreaching lateral orbital spine; second as long as first, located at anterior end of anterior branchial region; third strongest, basally broad and subtriangular, located at anterior end of posterior branchial region, followed by 3 or 4 posteriorly diminishing, short subtriangular spines. Rostrum narrow triangular, horizontal, with interior angle of 22°, length more than half that of remaining carapace, breadth slightly more than one-third carapace breadth at posterior carapace margin; dorsal surface concave. Lateral orbital spine small, directly mesial to anterolateral spine. Pterygostomian flap anteriorly angular, produced to sharp spine, surface with small spines. Excavated sternum sharply triangular on anterior margin, surface with rounded ridge in midline. Sternal plastron 1.7 × broader than long, lateral extremities convexly divergent posteriorly; sternite 3 deeply excavated on anterior margin, with pair of small submedian spines; sternite 4 having anterolateral margin twice as long as posterolateral margin, anteriorly ending in small spine; anterolateral margin of sternite 5 2 × longer than posterolateral margin of sternite 4. Somite 1 of abdomen without transverse ridge; somite 2 tergite 3.1 × broader than long; pleural lateral margin strongly concave and strongly divergent posteriorly, posterolateral corner angular; pleuron of somite 3 with blunt terminus. Telson 0.4 × as long as broad, posterior plate slightly longer than anterior plate, somewhat emarginate on posterior margin. Eyes slightly broadened proximally, elongate, slightly more than twice as long as broad, overreaching midlength of rostrum; cornea less than half length of remaining eyestalk. Ultimate article of antennular peduncle 2.4 × longer than high. Antennal peduncle overreaching cornea; article 2 with small distolateral spine; antennal scale overreaching proximal first segment of flagellum; distal 2 articles each with well-developed distomesial spine; article 5 slightly longer than article 4, breadth slightly more than half height of antennular ultimate article; flagellum consisting of 10 segments, not reaching distal end of P1 merus. Mxp1 with bases close to each other. Mxp3 basis with 3-4 denticles on mesial ridge; ischium with about 20 denticles on crista dentata, flexor margin not rounded distally; merus 1.9 × longer than ischium, distolateral spine well developed, flexor margin with 2 or 3 small spines distal to midlength, extensor margin with very small distal spine; carpus with distolateral spine and 2 extensor marginal spines. P1 massive; ischium with strong distodorsal spine, ventromesial margin with row of tubercle-like spines, without subterminal spine; merus as long as carapace; short stout spines on merus and carpus; carpus slightly longer than merus; palm 2.2 × longer than broad, 1.5 × longer than carpus, with a few spines along proximal part of mesial margin; fingers strongly incurved distally; opposable margin of fixed finger sinuous, that of movable finger with obtuse proximal process. P2 and P4 meri equally broad (P3 missing); P2 merus 3.0 × longer than broad, shorter than carapace, subequal to or slightly shorter than P2 propodus; P4 merus 2.4 × longer than broad, 0.8 × length of P2 merus, 0.8 × length of P4 propodus; dorsal margin with 4 spines on P2, 2 spines on P4, other than very small terminal spine; carpi 1.1 × as long as dactyli on P2, 0.8 × on P4, about half as long as propodi on P2, 0.4 × as long on P4; propodi shorter on P4 than on P2, extensor margin with plumose setae on distal portion, flexor margin slightly convex, with pair of terminal spines preceded by 5 spines on P2, 4 spines on P4; dactyli 0.9 × length of carpi on P2, 1.2 × on P4, slightly less than half as long as propodi on P2 and P4, flexor margin somewhat curving, bearing 12 sharp triangular spines, ultimate spine shorter and narrower than penultimate, subequally broad as antepenultimate as well as remaining spines.

Eggs. Number of eggs carried, 9; size, 0.90 mm × 1.1 mm - 1.05 mm × 1.16 mm.

Color. A specimen from Taiwan was illustrated by Baba *et al.* (2009) and Poore *et al.* (2011).

REMARKS — The second specimen of the species was recorded from Taiwan (Baba *et al.* 2009). This is the third specimen. No additional characters of significance are noted.

Uroptychus tridentatus (Henderson, 1885)

Figures 266, 267

Diptychus tridentatus Henderson, 1885: 421.
Uroptychus tridentatus — Henderson 1888: 181, pl. 6, figs 1, 1a. — Van Dam 1933: 30, figs 45, 46; 1937: 99. — Baba 2005: 61 (part, type material only; other specimens = *U. annae* n. sp.), figs 20.
?*Uroptychus tridentatus* — Baba 1990: 948.
Not *Uroptychus tridentatus* — Baba 1973: 117 (different species, see below).

TYPE MATERIAL — Holotype: **Indonesia**, East Indian Archipelago [Ambon], 15 fms (27 m), ov. female (BMNH 1888:33). [not examined].

MATERIAL EXAMINED — **Philippines**. MUSORSTOM 3 Stn CP134, 12°01'N, 121°57'E, 92-95 m, 5.VI.1985, 1 ov. ♀ 3.4 mm (MNHN-IU-2014-16995).

DISTRIBUTION — Banda Sea (Ambon and Kai Islands), N of the Sulu Islands, Solor Strait [between Solor and Alor Archipelagos, in 27-304 m; now Philippines (off NW of Panay), 92-95 m.

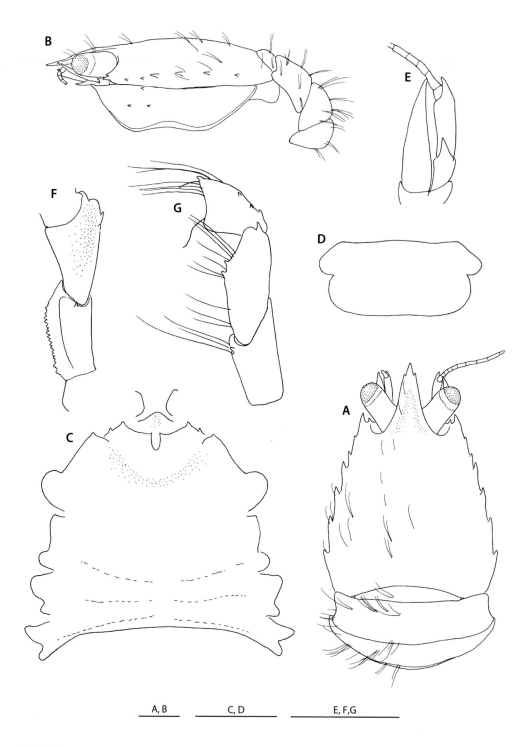

FIGURE 266

Uroptychus tridentatus (Henderson, 1885), ovigerous female 3.4 mm (MNHN-IU-2014-16995). **A**, carapace and anterior part of abdomen, dorsal. **B**, same, lateral. **C**, sternal plastron, with excavated sternum and basal parts of Mxp1. **D**, telson. **E**, right antenna, ventral. **F**, left Mxp3, ventral. **G**, same, lateral. Scale bars: 1 mm.

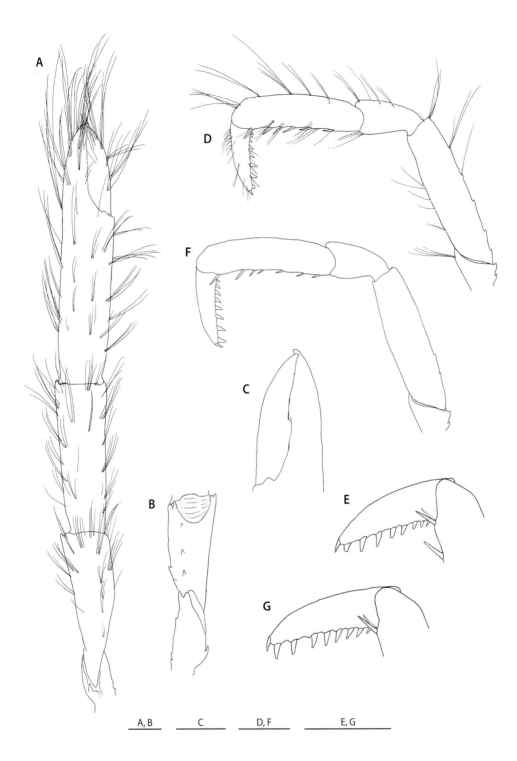

A, B   C   D, F   E, G

FIGURE 267

Uroptychus tridentatus (Henderson, 1885), ovigerous female 3.4 mm (MNHN-IU-2014-16995). **A**, left P1, dorsal. **B**, same, proximal part, setae omitted, ventral. **C**, same, distal part, ventral. **D**, left P2, lateral. **E**, same, distal part, setae omitted, lateral. **F**, left P3, lateral. **G**, same, distal part, lateral. Scale bars: 1 mm.

DESCRIPTION — Small species. *Carapace*: Slightly broader than long; greatest breadth 2.2 × distance between anterolateral spines. Dorsal surface with sparse, relatively long setae, unarmed, slightly convex from anterior to posterior. Lateral margins posteriorly divergent to point one-quarter from posterior end, then convergent, with 7 spines: first anterolateral, overreaching small lateral orbital spine; second and third much smaller, situated at same level as first spine in lateral view and ventral to level of posterior spines; fourth to seventh larger, placed on branchial region, seventh situated at point one-quarter from posterior end. Rostrum triangular, with interior angle of about 30°, bearing subterminal small spine on each side; dorsal surface concave; length slightly less than half that of remaining carapace, breadth about half carapace breadth at posterior carapace margin. Lateral orbital spine small, located directly mesial to anterolateral spine. Pterygostomian flap anteriorly sharp angular, produced to sharp spine, surface with small spines on anterior half, anterior half about as high as posterior half.

Sternum: Excavated sternum with convex anterior margin, ventral surface roundly ridged in midline. Sternal plastron shorter than broad (length 0.8 × breadth), lateral extremities slightly divergent posteriorly. Sternite 3 depressed well, anterior margin shallowly concave, with narrowly U-shaped median notch with small incurved flanking spine; posterolateral margin much more than half (0.8 x) length of anterolateral margin. Anterolateral margins of sternite 5 subparallel, about as long as posterolateral margin of sternite 4.

Abdomen: Somite 1 tergite well convex from anterior to posterior. Somite 2 tergite 2.3 × broader than long; pleuron with lateral margin moderately concave and moderately divergent posteriorly leading to blunt terminus. Pleuron of somite 3 posterolaterally bunt. Telson 0.4 × as long as broad; posterior plate 1.2 × longer than anterior plate, posterior margin feebly concave.

Eyes: 1.7 × longer than broad, distally slightly narrowed, overreaching distal third of rostrum; cornea more than half length of remaining eyestalk.

Antennule and antenna: Ultimate article of antennular peduncle 3.6 × longer than high. Antennal peduncle reaching apex of cornea. Article 2 with acute lateral spine. Antennal scale reaching second segment of flagellum. Distal 2 articles each with strong distomesial spine; article 5 1.7 × longer than article 4, breadth 0.8 × height of ultimate article of antennule. Flagellum of 10 or 11 segments not reaching distal end of P1 merus.

Mxps: Mxp1 with bases broadly separated. Mxp3 basis without denticles on mesial ridge. Ischium with small spine lateral to rounded distal end of flexor margin, crista dentata with about 17 denticles. Merus 2.0 × longer than ischium, flattish, with distolateral spine and 2 small spines around distal third of flexor margin. Carpus with distolateral spine and 3 small spines long extensor margin.

P1: Relatively massive, 4.5 × longer than carapace, with relatively long fine setae. Ischium with strong dorsal spine and well-developed subterminal spine on ventromesial margin. Merus with a few small ventral spines proximally, short ventromesial and ventrolateral spines distally, length 1.2 × that of carapace. Carpus 0.9 × length of merus. Palm 3.2 × longer than broad, length 1.2 × longer than carpus. Fingers distally slightly incurved, crossing when closed, not gaping and straightly fitting along opposable margins; movable finger 0.5 × length of palm, opposable margin with low blunt proximal process fitting to longitudinal groove on opposite face of fixed finger when closed.

P2-4: P4 missing. Meri compressed mesio-laterally and relatively broad, mesial face flattish, P2 merus subequally long and broad as P3 merus, with length-breadth ratio 3.4; dorsal margin with 4 small spines on P2, 2 proximal spines on P3, ventrolateral margin with small distal spine; P2 merus 0.8 × length of carapace, subequal to P2 propodus; P3 merus 0.9 × as long as P3 propodus. Carpi slightly shorter on P3 than on P2, unarmed, length 0.4 × that of propodus. Propodi slightly longer on P3 than on P2; flexor margin slightly concave, with pair of terminal spines preceded by 5 spines on P2, 6 spines on P3. Dactyli slightly longer on P3 than on P2, dactylus-carpus length ratio, 1.0 on P2, 1.3 on P3; flexor margin nearly straight, bearing slender terminal (ultimate) spine preceded by 8 triangular, somewhat obliquely directed, loosely arranged, proximally diminishing spines, penultimate subequal to antepenultimate.

Eggs. Number of eggs carried, 6; size, 0.75 × 0.90 mm - 0.76 × 0.93 mm.

REMARKS — The type material is now in poor condition, with all the pereopods missing. In my earlier paper (Baba, 2005), the specimens from New Caledonia (MNHN-IU-2014-17292, MNHN-IU-2014-17298) were thought to be identical with *U. tridentatus*. However, the present sole specimen has the carapace much more like that of the type, especially with the last lateral spine being located at a point in the posterior quarter rather than posterior third. Its occurrence is near to the type locality of *U. tridentatus* rather than being far south in New Caledonia and vicinity. In addition, their bathymetric range is shallow, unlike that of the specimens of Baba (2005), which were from below 290 m. The specimens of Baba (2005) are now described as *U. annae* n. sp. (see above). The present specimen has more numerous spines on the P2-4 propodi and dactyli (see above under *U. annae*).

The identification of the SIBOGA material from the Sulu Sea and Kai Islands (Van Dam, 1933) appear to be correct. The material reported by Van Dam (1937) from Solor Strait [between Solor and Alor Archipelago] may also be referable to this species, judging from her note that the specimen agrees well with her description of SIBOGA material.

The material reported from Japan (Baba, 1973) is also different from the present material in having only 3 spines on the branchial lateral margin with the last one located slightly posterior to the midlength of the carapace lateral margin, in having the P2-4 propodi with a pair of terminal spines preceded by 2, 0, 0 spine on P2, P3, P4 respectively, the features suggesting that it is close to *U. zezuensis*. However, *U. zezuensis* has the last branchial marginal spine distinctly more anterior in position. The Japanese material is now removed from the synonymy, requiring further study. The specimens from Madagascar (Baba, 1990) are very similar to the present material but differ in having more slender P2-4, especially the dactyli, and in having a few additional spines proximal to the distal spine on the meral ventrolateral margin. This is removed from the synonymy, pending extensive studies.

Uroptychus inclinis Baba, 2005 from the Kai Islands is now synonymized with *U. tridentatus*. The lateral carapace margins of *U. inclinis* seem to be more convex than those of both the holotype and the present material of *U. tridentatus*, but no additional difference worthy of note is found.

Uroptychus trispinatus n. sp.

Figures 268, 269, 306G

TYPE MATERIAL — Holotype: **New Caledonia**. BIOCAL Stn DW37, 23°00'S, 167°16'E, 350 m, 30.VIII.1985, ov. ♀ 4.0 mm (MNHN-IU-2014-16996). Paratypes: **New Caledonia**, Chesterfield Islands. MUSORSTOM 5 Stn DW337, 19°53.80'S, 158°38.00'E, 412-430 m, 15.X.1986, 1 ♂ 5.2 mm, 1 ov. ♀ 4.8 mm (MNHN-IU-2014-16997). New Caledonia, Norfolk Ridge. CHALCAL 2 Stn NORFOLK 1 Stn DW1654, 23°26'S, 167°51'E, 366-560 m, 19.VI.2001, 1 ov. ♀ 3.8 mm (MNHN-IU-2014-17002). NORFOLK 2 Stn CP2130, 23°15.90'S, 168°13.54'E, 375-427 m, 2.XI.2003, 1 ov. ♀ 3.5 mm (MNHN-IU-2014-17003). BATHUS 3 Stn DW830, 23°20'S, 168°01'E, 361-365 m, 29.XI.1993, 1 ov. ♀ 3.7 mm (MNHN-IU-2014-17004). New Caledonia. MUSORSTOM 4 Stn CP213, 22°51.3'S, 167°12.0'E, 405-430 m, 28.IX.1985, 1 ♀ 3.0 mm (MNHN-IU-2014-16998). SMIB 2 Stn DW05, 22°56'S, 167°15'E, 398-410 m, 17.IX.1986, 1 ♂ 3.0 mm (MNHN-IU-2014-16999). BIOCAL Stn DW37, collected with holotype, 2 ♂ 2.0, 3.0 mm (MNHN-IU-2014-17000). Vanuatu. MUSORSTOM 8 Stn CP1026, 17°50.35'S, 168°39.33'E, 437-504 m, 28.IX.1994, 1 ♂ 4.5 mm (MNHN-IU-2014-17001).

ETYMOLOGY — From the Latin *tri* (three) and *spinatus* (spined), in reference to three strong spines on the anterior part of the carapace lateral margin.

DISTRIBUTION — Chesterfield Islands, New Caledonia, Vanuatu, and Norfolk Ridge, 350-560 m.

SIZE — Males, 2.0-5.2 mm; females, 3.0-4.8 mm; ovigerous females from 3.5 mm.

DESCRIPTION — Small species. *Carapace*: Broader than long (0.7-0.8 × as long as broad); greatest breadth 1.6-1.8 × distance between anterolateral spines. Dorsal surface smooth, moderately convex from anterior to posterior, with shallow depression between gastric and cardiac regions; epigastric region with pair of well-developed spines. Lateral margins divided by constriction anterior to midlength (between anterior and posterior branchial regions) into anterior and posterior parts; anterior part divergent posteriorly, with 3 prominent spines, all directed anterolaterally, first anterolateral, overreaching base of antennal scale, second slightly smaller than first, third strongest, dorsal to level of preceding spines; posterior part strongly convex, with 4 or 5 (usually 5), rarely 3 sharp spines diminishing posteriorly, all directed anterolaterally. Rostrum narrow triangular, with interior angle of 25-28°, slightly upcurved; dorsal surface concave; length 0.3-0.4 × that of remaining carapace, breadth much less than half carapace breadth at posterior carapace margin. Lateral orbital spine small but distinct, slightly anterior to level of anterolateral spine. Pterygostomian flap anteriorly angular, produced to strong spine, followed by a few spines often small, occasionally pronounced, along lower margin and/or in midline of anterior part; height of posterior portion 0.4 × that of anterior portion.

Sternum: Excavated sternum with sharp ridge in midline leading to anterior margin produced triangularly between bases of Mxp1. Sternal plastron 1.5 × broader than long, lateral extremities convexly divergent behind sternite 4, sternite 7 narrower than sternite 6. Sternite 3 shallowly depressed; anterior margin shallowly concave, with deep narrow or V-shaped median notch without flanking spine. Sternite 4 short relative to breadth; anterolateral margin convex, anteriorly rounded or feebly angular, length twice that of posterolateral margin. Sternite 5 with medially convex, posteriorly divergent anterolateral margin distinctly longer than posterolateral margin of sternite 4.

Abdomen: Somites short relative to breadth. Somite 1 convex from anterior to posterior. Somite 2 tergite 3.2 × broader than long, pleuron concave on lateral margin, posterolaterally strongly produced. Pleura of somites 3-4 laterally tapering more strongly in females than in males. Telson 0.4 × as long as broad; posterior plate 1.5-2.0 × longer than anterior plate, posterior margin feebly or moderately concave, or distinctly emarginate.

Eye: Elongate, 2.0-2.1 × longer than broad, slightly falling short of apex of rostrum, somewhat narrowed distally. Cornea not dilated, less than half as long as remaining eyestalk.

Antennule and antenna: Ultimate article of antennular peduncle 1.8-2.1 × longer than high. Antennal peduncle reaching or slightly overreaching apex of rostrum. Article 2 with strong lateral spine. Antennal scale 1.3-1.6 × broader than article 5, usually slightly falling short of, rarely reaching distal end of article 5. Article 4 distomesially broadened and produced to short spine. Article 5 unarmed, length 1.4-1.5 × that of article 4, breadth 0.5-0.6 × height of ultimate article of antennule. Flagellum of 11-13 (rarely 9) segments not reaching distal end of P1 merus.

Mxp: Mxp1 with bases broadly separated. Mxp3 basis with 1 distal denticle on mesial ridge. Ischium with flexor margin distally rounded, crista dentata with about 20 distally diminishing denticles. Merus twice as long as ischium, mesial face flattish; flexor margin roundly ridged on proximal half, sharply ridged on distal half, bearing 1 or 2 blunt spines on distal third of length; lateral face with strong, blunt distal spine. Carpus unarmed.

P1: 6.3-6.8 × longer than carapace, smooth and barely setose except for fingers. Ischium with strong dorsal spine; ventromesial subterminal spine distally blunt and well developed, overreaching distal end of ischium, followed distantly by several protuberances on proximal portion. Merus mesially with blunt spines roughly in 3 rows (mesial, ventromesial, dorsomesial); length 1.3 × (1.6 × in largest male) that of carapace. Carpus 1.4-1.5 × length of merus, with several small or obsolescent spines in 2 rows along mesial margin, one of these somewhat dorsal. Palm unarmed, depressed, with lateral and mesial margins subparallel or somewhat convex; length 3.3-3.8 × (males), 3.6-4.1 × (females) breadth, subequal to or at most 1.1 × that of carpus. Fingers relatively slender distally, sparingly setose, moderately gaping in distal half, not spooned along opposable margins, barely incurved distally and slightly crossing when closed; fixed finger inclined somewhat laterally; movable finger 0.4-0.5 × as long as palm, opposable margin with prominent blunt process (distal margin of process perpendicular to opposable margin) fitting into narrow longitudinal groove on opposite ventromesial face of fixed finger when closed.

P2-4: Relatively slender, well compressed mesio-laterally, sparsely with simple setae, more setose on dactyli. Ischia with 2 short dorsal spines. Meri subequal in length on P2 and P3 or very slightly shorter on P3 than on P2, shortest on

A, B, D, E _____ C _____ F - I _____

FIGURE 268

Uroptychus trispinatus n. sp., **A-D**, **F-I**, holotype, ovigerous female 4.0 mm (MNHN-IU-2014-16996); **E**, paratype, male 5.2 mm (MNHN-IU-2014-16997). **A**, carapace and anterior part of abdomen, proximal part of right P1 included, dorsal. **B**, same, lateral. **C**, sternal plastron, with excavated sternum and basal parts of Mxp1. **D**, left pleura of abdominal somites 2-5, dorsolateral. **E**, same. **F**, telson. **G**, left antenna, ventral. **H**, right Mxp3, ventral. **I**, same, lateral. Scale bars: 1 mm.

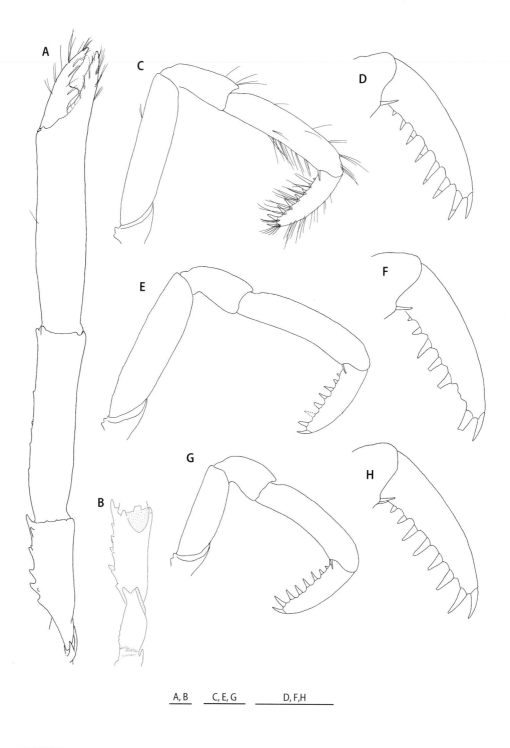

FIGURE 269

Uroptychus trispinatus n. sp., holotype, ovigerous female 4.0 mm (MNHN-IU-2014-16996). **A**, right P1, dorsal. **B**, same, proximal part, ventral. **C**, right P2, lateral. **D**, same, distal part, setae omitted, lateral. **E**, right P3, setae omitted, lateral. **F**, same, distal part, setae omitted, lateral. **G**, right P4, setae omitted, lateral. **H**, same, distal part, setae omitted, lateral. Scale bars: 1 mm.

P4 (P4 merus 0.6 × length of P3 merus); subequally broad on P2 and P3, narrower on P4 (P4 merus 0.8 × breadth of P3 merus); dorsal margin unarmed; length-breadth ratio, 3.6-4.3 on P2, 3.5-4.2 on P3, 2.5-2.8 on P4; P2 merus 0.8-0.9 × length of carapace, 1.1-1.2 × length of P2 propodus; P3 merus 0.9-1.1 × length of P3 propodus. P4 merus 0.7 × length of P4 propodus. Carpi subequal in length on P2 and P3 or slightly longer on P2 than on P3, shortest on P4 (P4 carpus 0.7-0.8 × length of P3 carpus), 0.4-0.5 × length of propodus on P2, 0.4 × on P3, slightly less than 0.4 × on P4. Propodi longest on P3, shortest on P4; flexor margin slightly concave in lateral view, with pair of slender movable distal spines only. Dactyli subequal in length on P2-4; dactylus-carpus length ratio, 1.1-1.3 on P2, 1.2-1.3 on P3, 1.6-1.8 on P4; dactylus-propodus length ratio, 0.5 on P2 and P3, 0.6-0.7 on P4; flexor margin slightly curving (slightly concave in lateral view), with 8-9 sharp spines loosely arranged, proximally diminishing and nearly perpendicular to margin, ultimate spine slightly more slender than penultimate.

Eggs. Number of eggs carried, 5-10; size, 0.97 × 1.06 mm - 1.30 × 1.00 mm.

Color. Based on male and ovigerous female from Musorstom 5 Stn DW337, Chesterfield Islands (MNHN-IU-2014-16997): ground color translucent seashell pink. Visceral parts under carapace pinkish. P1 with pinkish bands on distal portion of merus, on both distal and median portions of carpus and palm and on fingers; P2-4 totally translucent pinkish, faded proximally on meri and carpi, proximally and distally on propodi, and distally on dactyli.

REMARKS — The carapace shape including the presence of a pair of epigastric spines, the pterygostomian flap with spines on the anterior surface and the spination of P2-4, especially the arrangement of spines on dactyli link the species to *U. corbariae* n. sp., *U. paraplesius* n. sp., *U. mesodme* n. sp., *U. clarki* n. sp. and *U. defayeae* n. sp. Three of these, *i.e. Uroptychus trispinatus*, *U. paraplesius* and *U. mesodme*, share the carapace lateral margin bearing 3 anterior spines remotely separated from another group of 3-5 posterior spines, the antennal scale overreaching the midlength of article 5, and article 4 bearing a distomesial spine. The other three species, *U. corbariae*, *U. clarki* and *U. defayeae*, differ from the above three in having irregular spination on the anterior third of the carapace lateral margin, in having the antennal scale barely reaching the midlength of article 5, and in having the articles 4 and 5 unarmed. *Uroptychus trispinatus* is differentiated from *U. paraplesius* by the carapace lateral spines that are directed more laterally instead of more anteriorly; the antennal article 5 is unarmed instead of bearing a distinct distomesial spine; the epigastric spines are stronger; and the pleura of the abdominal somites 2-4 in females are more sharply tapering. *Uroptychus mesodme* can be distinguished from *U. trispinatus* by the anterolateral spine directed forward, the branchial marginal spines much smaller than the anterior three spines, the epigastric region bearing an additional pair of small spines between the usual epigastric spines, and the smaller subterminal spine of the P1 ischium.

This species is somewhat similar to *U. macrolepis* n. sp. (see above) in the spination of the carapace lateral margin and in having elongate eyes. *Uroptychus trispinatus*, however, is readily distinguished from that species by the P2-4 propodi that bears the flexor margins nearly straight instead of convex in lateral view, distally ending in a pair of terminal spines only instead of bearing a row of 6 or 7 single spines proximal to the terminal pair; sternite 3 bears a distinct instead of obsolescent median notch on the anterior margin; the antennal scale usually slightly falls short of (at most slightly overreaching) instead of extending far beyond the antennal article 5; the P2-4 ischia each bear 2 short spines instead of being unarmed on the dorsal margin; and the flexor marginal spines of P2-4 dactyli are elongate and perpendicularly directed, instead of being short and obliquely directed.

Uroptychus tuberculatus n. sp.

Figures 270, 271

TYPE MATERIAL — Holotype: **New Caledonia**, Chesterfield Islands. MUSORSTOM 5 Stn DW355, 19°36'S, 158°43'E, 580 m, 18.X.1986, ♀ 3.5 mm (MNHN-IU-2014-17005). Paratypes: Collected with holotype, 2 ♀ 3.6 mm (carapace in 1 ♀ broken) (MNHN-IU-2014-17006).

ETYMOLOGY — From the Latin *tuberculatus* (tuberculate), referring to a tuberculate dorsal surface of the carapace.

DISTRIBUTION — Chesterfield Islands, in 580 m.

DESCRIPTION — Small species. *Carapace*: Broader than long (0.8 × as long as broad); greatest breadth 1.6 × distance between anterolateral spines. Dorsal surface slightly convex from anterior to posterior, with feeble depression between gastric and cardiac regions, and between anterior and posterior branchial regions, sparingly tuberculose; epigastric region with tubercles roughly arranged in transverse row. Lateral margins subparallel; anterolateral spine well-developed, overreaching much smaller lateral orbital spine, followed by row of tubercle-like spines subparalleling another row of small spines above linea anomurica; ridged along posterior third. Rostrum narrow triangular, with interior angle of 22°; length slightly less than half that of remaining carapace, breadth much less than half carapace breadth at posterior carapace margin; dorsal surface deflected ventrally and concave, devoid of tubercles; ventral surface horizontal. Lateral orbital spine slightly anterior to level of anterolateral spine and separated from that spine by its basal breadth. Pterygostomian flap anteriorly angular, produced to well-developed spine; surface with row of spines directly below linea anomurica (spines on anterior half larger than lateral marginal spines of carapace, those in posterior half obsolescent), and scattered obsolescent spines further ventral to this row.

Sternum: Excavated sternum obtusely produced on anterior margin, surface with sharp, elevated ridge in midline. Sternal plastron slightly broader than long, lateral extremities gently divergent posteriorly on sternites 4-7. Sternite 3 moderately depressed; anterior margin strongly concave, bearing deep, broad median notch without flanking spine; anterolaterally sharp angular. Sternite 4 having anterolateral margin relatively short and slightly concave, anteriorly bearing 2 small processes placed side by side; posterolateral margin longer than anterolateral margin. Anterolateral margin of sternite 5 straight divergent posteriorly, slightly rounded on anterior end, about half as long as posterolateral margin of sternite 4.

Abdomen: Smooth and glabrous. Somite 1 dorsally convex from anterior to posterior, without transverse ridge. Somite 2 tergite 2.1-2.3 × broader than long; pleuron posterolaterally blunt, lateral margin feebly concave and weakly divergent posteriorly. Pleuron of somite 3 with blunt lateral end. Telson 0.4 × as long as broad; posterior margin not emarginate, nearly transverse, length 1.3 × that of anterior plate.

Eye: Relatively short, 1.4-1.5 × longer than broad, overreaching midlength of rostrum; mesial and lateral margins convex. Cornea distinctly (in holotype) or somewhat (paratypes) narrower than greatest breadth of remaining eyestalk, length subequal to or slightly less than that of remaining eyestalk.

Antennule and antenna: Ultimate article of antennule 3.4 × longer than high. Antennal peduncle overreaching cornea by full length of article 5, reaching apex of rostrum. Article 2 with small distolateral spine. Antennal scale 2 × broader than article 5, proportionately broad, overreaching midlength of article 5, barely reaching its distal end, with or without small lateral spine. Article 4 distomesially produced to short spine. Article 5 unarmed, 1.5 × longer than article 4; breadth 0.5 × height of antennular ultimate article. Flagellum of 6-7 segments falling short of distal end of P1 merus.

Mxp: Mxp1 with bases close to each other, slightly separated. Mxp3 barely setose on lateral surface. Coxa with blunt ventrolateral process. Basis with 3 or 4 obsolescent denticles on mesial ridge. Ischium having flexor margin not rounded distally, crista dentata with no distinct denticles. Merus twice as long as ischium, with blunt distolateral spine and 2 similar spines on distal third of flexor margin. Carpus with blunt distolateral spine.

P1: 6.2 × longer than carapace, relatively slender. Ischium with short blunt dorsal spine, ventromesially with several proximally diminishing spines (distalmost subterminal and well developed) subparalleling another row of 4 or 5 small mesial spines. Merus 1.3-1.4 × longer than carapace, with small mesial and ventral spines, in addition to short, blunt ventral distolateral and distomesial spines; dorsal small spines present or obsolescent. Carpus 1.5 × longer than merus, slightly shorter than palm, mesial margin with obsolescent tubercles or with accompanying row of small spines; dorsal surface sparsely or obsolescently tuberculose. Palm 6 × longer than broad, barely setose; dorsal surface with tubercles near juncture with carpus. Fingers gently incurved distally, crossing when closed, inclined slightly laterally, slightly gaping or

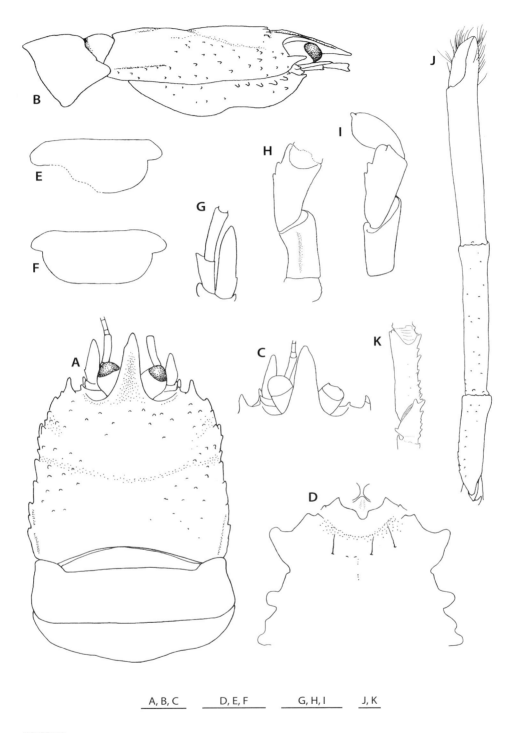

FIGURE 270

Uroptychus tuberculatus n. sp., **A**, **B**, **D**, **E**, **G-K**, holotype, female 3.5 mm (MNHN-IU-2014-17005); **C**, **F**, paratype, female 3. 6 mm (MNHN-IU-2014-17006). **A**, carapace and abdomen, dorsal. **B**, same, lateral. **C**, anterior part of carapace, including rostrum, eyes, antennal peduncle, dorsal. **D**, anterior part of sternal plastron, with excavated sternum and basal parts of Mxp1. **E**, telson. **F**, same. **G**, left antenna, ventral. **H**, right Mxp3, setae omitted, ventral. **I**, left Mxp3, setae omitted, lateral. **J**, right P1, dorsal. **K**, same, proximal part, ventral. Scale bars: 1 mm.

not gaping, sparingly with short setae; movable finger one-third length of palm, opposable margin with proximal process fitting to longitudinal groove on opposite ventromesial face of fixed finger when closed.

P2-4: Relatively broad, compressed mesio-laterally; sparsely setose on meri, carpi and propodi, moderately so on dactyli. Meri successively shorter posteriorly (P3 merus 0.9 × length of P2 merus, P4 merus 0.8-0.9 × length of P3 merus), successively slightly broader posteriorly; length-breadth ratio, 3.5-3.6 on P2, 3.0-3.2 on P3, 2.3-2.6 on P4; dorsal margin with a few denticle-like small spines on proximal third; ventromesial margin with 1 or 2 small spines on P2, ventrolateral margin with small distal spine; P2 merus 0.8-0.9 × length of carapace, 1.0-1.1 × length of P2 propodus; P3 merus 0.8-0.9 × length of P3 propodus; P4 merus 0.7-0.8 × length of P4 propodus. Carpi subequal, distinctly shorter than dactyli (carpus-dactylus length ratio, 0.7 on P2, 0.6 on P3 and P4); carpus-propodus length ratio, 0.4 on P2, 0.3 on P3 and P4.

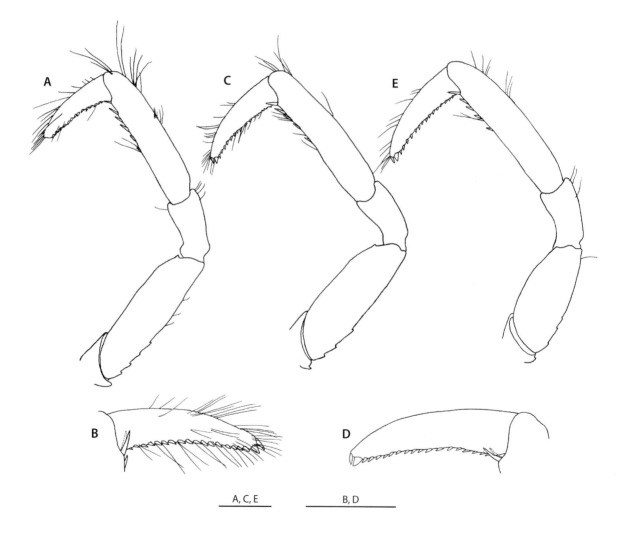

A, C, E   B, D

FIGURE 271

Uroptychus tuberculatus n. sp., holotype, female 3.5 mm (MNHN-IU-2014-17005). **A**, left P2, lateral. **B**, same, distal part, lateral. **C**, left P3, lateral. **D**, same, distal part, setae omitted, lateral. **E**, left P4, lateral. Scale bars: 1 mm.

Propodi subequal on P3 and P4, shorter on P2; flexor margin nearly straight, with pair of terminal spines preceded by 5 (on P2) or 4 (on P3 and P4) spines at most on distal half. Dactyli subequal in length on P3 and P4, shorter on P2, slightly more than half length of propodi (dactylus-propodus length ratio, 0.5-0.6 on P2, 0.6 on P3 and P4), longer than carpi (dactylus-carpus length ratio, 1.4 on P2, 1.6-1.7 on P3, 1.7-1.8 on P4); flexor margin slightly curving, with about 20 spines obscured by setae, ultimate spine slender, penultimate more than 2 × broader than antepenultimate, remaining spines small, shorter than and as broad as ultimate, obliquely directed and very close to one another.

REMARKS — The spination of the P2-4 and the shape of sternite 4, especially the posterolateral margin longer than the anterolateral margin, link the species to *U. floccus* n. sp., *U. dualis* n. sp. and *U. levicrustus* Baba, 1988. However, *U. tuberculatus* is readily distinguished from *U. floccus* and *U. dualis* by the carapace dorsally sparingly tuberculose instead of bearing a row of small epigastric spines, laterally with small instead of strong spines, and by the P1 merus bearing spines, without a field of three obliquely arranged spines on the proximal mesial surface. *Uroptychus levicrustus* is different from *U. tuberculatus* in having the carapace dorsum and the pterygostomian flap both smooth, with at least a well-developed spine on the lateral branchial margin.

Uroptychus turgidus n. sp.

Figures 272-274

TYPE MATERIAL — Holotype: **New Caledonia**, Chesterfield Islands. MUSORSTOM 5 Stn CP278, 24°10.80'S, 159°38.10'E, 265 m, coral unidentified, 10.X.1986, ♂ 2.0 mm (MNHN-IU-2014-17007). Paratype: Collected with holotype, 1 ov. ♀ 1.9 mm (MNHN-IU-2014-17008).

ETYMOLOGY — From the Latin *turgidus* (swollen), alluding to the dorsally swollen branchial region, characteristic of the species.

DISTRIBUTION — Chesterfield Islands; 265 m.

DESCRIPTION — Small species. *Carapace*: Broader than long (0.8 × as long as broad); greatest breadth 1.6 × distance between anterolateral spines. Dorsal surface with sparse tufts of long soft setae on posterior half and relatively small spines of irregular sizes along lateral margin; posterior branchial regions markedly inflated, other portions nearly horizontal in profile. Lateral margins convex on branchial region, with spines of irregular sizes; anterolateral spine small, distinctly posterior to level of, but not reaching apex of lateral orbital spine; 2 somewhat larger spines on posterior branchial region, anterior one dorsal in position. Rostrum broad triangular, with interior angle of 45°, distally blunt; dorsal surface concave; length less than half that of carapace, breadth much more than half carapace breadth at posterior carapace margin. Lateral orbital spine well developed, much larger than anterolateral spine. Pterygostomian flap covered with spinules, anterior margin somewhat angular, produced to distinct spine.

Sternum: Excavated sternum anteriorly convex, bearing ridge in midline. Sternal plastron as long as broad; lateral extremities subparallel between sternites 5-7. Sternite 3 shallowly depressed, anterior margin representing broad V-shape, with narrow deep median notch, lacking submedian spines. Sternite 4 broadest among sternites; anterolateral margin smooth, with blunt anterior end, length subequal to that of posterolateral margin. Anterolateral margins of sternite 5 subparallel, anteriorly rounded, length 0.7 × that of posterolateral margin of sternite 4.

Abdomen: Sparsely setose. Somite 1 without transverse ridge. Somite 2 tergite 2.5 × broader than long; pleuron posterolaterally blunt, lateral margins feebly concave and subparallel. Pleuron of somite 3 with rounded lateral margin. Telson half as long as broad; posterior plate medially not expanded laterally, length 1.3 × that of anterior plate, posterior margin feebly concave.

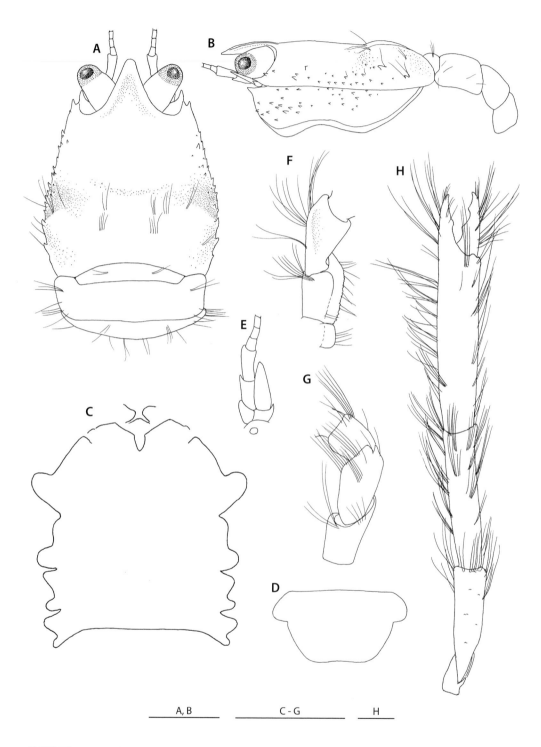

FIGURE 272

Uroptychus turgidus n. sp., holotype, male 2.0 mm (MNHN-IU-2014-17007). **A**, carapace and anterior part of abdomen, dorsal. **B**, same, lateral. **C**, sternal plastron, with excavated sternum and basal parts of Mxp1. **D**, telson. **E**, left antenna, ventral. **F**, right Mxp3, ventral. **G**, left Mxp3, lateral. **H**, left P1, dorsal. Scale bars: 1 mm.

Eye: 1.5 × longer than broad, distally narrowed, reaching apex of rostrum; lateral and mesial margins convex. Cornea about half as long as remaining eyestalk.

Antennule and antenna: Ultimate article of antennule 3.5-3.7 × longer than high. Antennal peduncle reaching apex of rostrum. Article 2 with strong distolateral spine. Antennal scale slightly overreaching midlength of article 5, 1.5 × broader than article 5. Article 4 with distomesial spine. Article 5 unarmed, slightly longer than article 4, breadth 0.6 × height of antennular ultimate article. Flagellum consisting of 13-14 segments, not reaching distal end of P1 merus.

Mxp: Mxp1 with bases moderately separated. Mxp3 with long soft setae on lateral surface. Basis without denticles on mesial ridge. Ischium with distally rounded flexor margin; crista dentata with obsolescent denticles. Merus 2.2 × longer than ischium, flattish, not strongly cristate but somewhat rounded on flexor margin, bearing well-developed distolateral spine. Carpus unarmed.

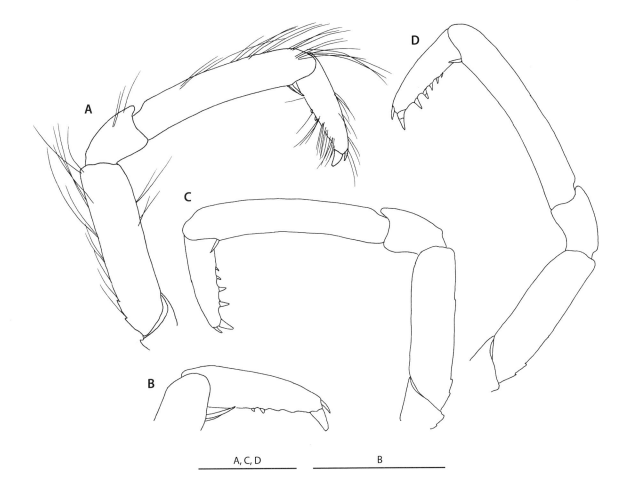

FIGURE 273

Uroptychus turgidus n. sp., holotype, male 2.0 mm (MNHN-IU-2014-17007). **A**, right P2, lateral. **B**, same, distal part, setae omitted, lateral. **C**, left P3, setae omitted, lateral. **D**, left P4, setae omitted, lateral. Scale bars: 1 mm.

P1: 7 × longer than carapace, relatively slender, sparingly with long soft setae. Ischium dorsally with sharp long spine, ventromesially unarmed. Merus, carpus and palm unarmed, Merus 1.5-1.6 × longer than carapace. Carpus 1.2-1.3 × longer than merus, 0.8 × (male) or 1.0 × (female) as long as palm. Palm 4.2 × (male) or 6.5 × (female) longer than broad, 1.1 × (female) or 1.3 × (male) longer than carpus. Fingers relatively narrowed distally, curving ventrally, somewhat incurved distally, crossing when closed; in female, opposable margins nearly straight, without processes in female; in male, opposable margin of movable finger with 2 low processes, distal one at distal third, proximal one at proximal third and slightly proximal to position of opposite low process on fixed finger; movable finger about one-third length of palm.

P2-4: Relatively thick mesio-laterally, bearing long soft setae. Meri successively slightly shorter posteriorly (P3 merus 0.96-0.97 × length of P2 merus, P4 merus 0.86-0.95 × length of P3 merus), breadths subequal on P2-4; length-breadth ratio, 3.7-3.8 on P2, 3.5-3.6 on P3, 3.1-3.6 on P4; dorsal margin with 2 small proximal spines distinct on P2-4 (male holotype) or obsolescent on P2-4 (ovigerous paratype); P2 merus 0.8 × length of carapace as well as P2 propodus; P3 merus 0.7 × length of P3 propodus; P4 merus 0.7 length of P4 propodus. Carpi subequal, length 0.2-0.3 × that of propodi on P2-4, 0.6-0.7 × that of dactyli on P2, 0.6 × on P3 and P4. Propodi shortest on P2, longer and subequal on P3 and P4; flexor margin slightly concave in lateral view, with pair of terminal spines only. Dactyli subequal in length on P2-4, slightly less than half length of propodi (dactylus-propodus length ratio, 0.40-0.45 on P2, 0.40-0.43 on P3, 0.42-0.43 on P4); 1.6-1.7 × longer than carpi on P2 and P3, 1.6 × longer on P4; flexor margin nearly straight, with 6 or 7 spines, ultimate spine very slender and close to penultimate spine; penultimate spine strong, much broader and longer than remainder, twice as broad as antepenultimate, remaining spines loosely arranged and proximally diminishing, antepenultimate and fourth spines perpendicular to margin, fifth and sixth (and seventh where present) somewhat inclined, antepenultimate more distant from penultimate than from fourth.

Eggs. Number of eggs carried, 3; size, 0.88 mm × 1.05 mm - 0.88 mm × 1.15 mm.

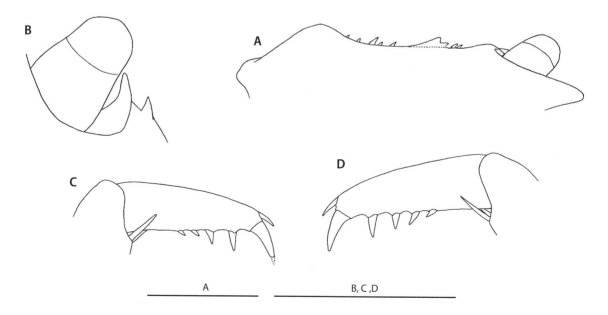

FIGURE 274

Uroptychus turgidus n. sp. **A**, **B**, holotype, male 2.0 mm (MNHN-IU-2014-17007). **C**, **D**, paratype, ovigerous female 1.9 mm (MNHN-IU-2014-17008). **A**, carapace, left part, dorsolateral. **B**, same, anterolateral part, right, showing eye and lateral orbital spine, dorsal. **C**, P2, distal part, setae omitted, lateral. **D**, P4, distal part, setae omitted, lateral. Scale bars: 1 mm.

REMARKS — The combination of the following characters links the species to *U. sarahae* n. sp., *U. toka* Schnabel, 2009 and *U. volsmar* n. sp.: the small anterolateral spine of the carapace subequal to or smaller than the lateral orbital spine, sternite 3 with a distinct median notch on the anterior margin, the short antennal scale ending at most in the midlength of article 5, and the P2-4 dactyli with perpendicularly directed spines proximal to the prominent penultimate spine on the flexor margin. *Uroptychus turgidus* is readily distinguished from these species by the presence of the pronounced inflation on the posterior branchial region and a few distinct spines along its lateral margin.

Uroptychus vandamae Baba, 1988

Figure 275

Uroptychus vandamae Baba, 1988: 49 (part), fig. 21. — McCallum & Poore 2013: 169, fig. 12H.
Uroptychus australis var. *indicus* — Van Dam 1933: 18 (part), figs 25, 27, 28.
Not *Uroptychus vandamae* — Baba 1990: 942, fig. 8c (= 2 new species, see below under the remarks).

TYPE MATERIAL — Holotype: **Indonesia**, Molucca Sea off west coast of Halmahera, 655 m, male (USNM 150316) (Not ov. female from Albatross Stn 5664 in Makassar Strait = *U. australis* (Henderson, 1885)). [not examined].

MATERIAL EXAMINED — **Indonesia**, Kai Islands. KARUBAR Stn CP19, 5°15'S, 133°01'E, 605-576 m, 25.X.1991, 1 ♂ 5.9 mm, 1 ov. ♀ 6.8 mm, 2 ♀ 4.5, 5.9 mm, 1 sp. (sex indet.) 3.1 mm (MNHN-IU-2014-17009). – Stn CP20, 5°15'S, 132°59'E, 769-809 m, with *Chrysogorgia* sp. (Calcaxonia, Chrysogorgiidae), 25.X.1991, 17 ♂ 6.0-7.8 mm, 14 ov. ♀ 5.8-7.1 mm, 10 ♀ 4.0-7.6 mm (MNHN-IU-2014-17010). – Stn CC21, 5°14'S, 133°00'E, 688-694 m, 25.X.1991, 1 ♂ 7.2 mm, 1 ♀ 6.4 mm (MNHN-IU-2014-17011); 4 ♂ 7.1-7.6 mm, 4 ov. ♀ 6.1-6.6 mm, 1 ♀ 6.4 mm (MNHN-IU-2014-17012). **Indonesia**, Makassar Strait. CORINDON Stn CH201, 01°11'S, 117°06'E, 21 m, 1 ♂ 6.3 mm (MNHN-IU-2014-17013). **Solomon Islands**. SALOMON 2 Stn CP2230, 6°27.8'S, 156°24.3'E, 837-945 m, 29.X.2004, 3 ♂ 4.5-6.7 mm, 3 ov. ♀ 5.3-6.4 mm, 1 ♀ 5.9 mm (MNHN-IU-2014-17014).

DISTRIBUTION — Molucca Sea off west coast of Halmahera, Timor Sea, off Roti, Makassar Strait and Western Australia, in 463-1289 m, and now Makassar Strait, Kai Islands and Solomon Islands, in 605-945 m. The male from the Makassar Strait (MNHN-IU-2014-17013) was taken in 21 m, but this depth record is questioned.

SIZE — Males, 4.5-7.8 mm; females, 4.0-7.6 mm; ovigerous females from 5.3 mm.

DIAGNOSIS — Medium-sized species. *Carapace*: 1.1-1.2 × longer than broad; greatest breadth 1.6 × distance between anterolateral spines. Dorsal surface smooth and glabrous, unarmed; lateral margins convexly moderately divergent posteriorly; anterolateral spine slightly overreaching lateral orbital spine. Rostrum narrow, elongate triangular, with interior angle of about 20°, dorsally somewhat convex from side to side, extending far beyond cornea, length more than half that of carapace, breadth half carapace breadth at posterior carapace margin. Lateral orbital spine small, often obsolete but acuminate, its base somewhat anterior to that of anterolateral spine. Pterygostomian flap with anterior margin somewhat roundish, bearing small spine. Excavated sternum with small sharp median process between bases of Mxp1, surface with small central spine. Sternal plastron slightly shorter than broad, lateral extremities straightly divergent posteriorly; sternite 3 well depressed, anterior margin deeply excavated, with 2 submedian spines separated by semicircular or narrow notch; sternite 4 having anterolateral margin anteriorly ending in well-developed process followed by posteriorly diminishing crenulations, posterolateral margin slightly more than half-length of anterolateral margin; anterolateral margin of sternite 5 convexly slightly divergent posteriorly, about as long as posterolateral margin of sternite 4. Abdominal somite 1 weakly ridged transversely; somite 2 tergite 2.4-2.6 × broader than long; pleuron posterolaterally blunt, lateral margin weakly concave and moderately divergent posteriorly; pleuron of somite 3 posterolaterally blunt. Telson half as long as broad or slightly more than so; posterior plate 1.7-

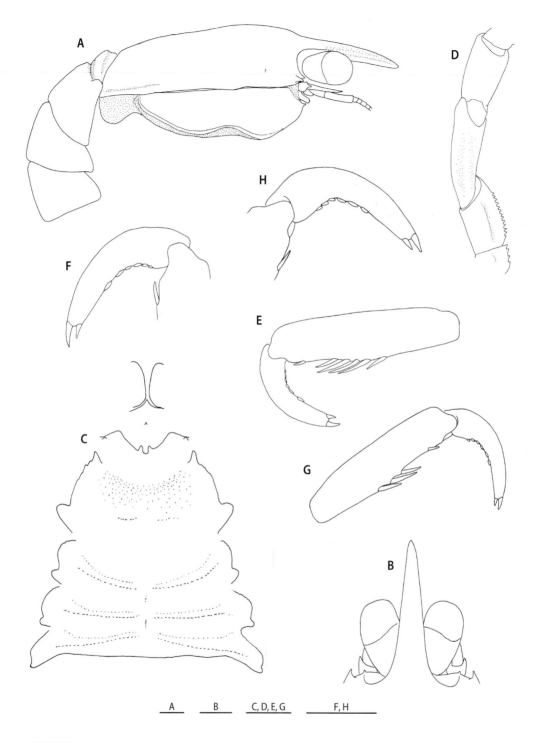

FIGURE 275

Uroptychus vandamae Baba, 1988, ovigerous female 6.9 mm (MNHN-IU-2014-17010). **A**, carapace and anterior part of abdomen, lateral. **B**, anterior part of carapace, dorsal. **C**, sternal plastron, with excavated sternum and basal parts of Mxp1. **D**, right Mxp3, setae omitted, ventral. **E**, left P2, distal articles, setae omitted, lateral. **F**, same. **G**, right P4, distal articles, setae omitted, lateral. **H**, same. Scale bars: 1 mm.

1.8 × longer than anterior plate, posterior margin emarginate. Eyes usually terminating in midlength of rostrum, relatively broad (1.6 × longer than broad), slightly narrowed proximally, lateral margins convex; cornea slightly longer than remaining eyestalk. Ultimate article of antennular peduncle twice as long as high. Antennal peduncle slightly overreaching cornea; article 2 with acute lateral spine; antennal scale slender, terminating in midlength of article 5; distal 2 articles unarmed; article 5 slightly more than twice as long as article 4, breadth less than half height of ultimate article of antennule; flagellum consisting of 13-15 segments, falling short of distal end of P1 merus. Mxp1 with bases close to each other. Mxp3 basis with 4 denticles on mesial ridge; ischium with flexor margin not rounded distally, crista dentata with 13-15 denticles; merus 2.2-2.5 × longer than ischium, distolateral spine obsolescent, flexor margin roundly ridged on proximal two-thirds. P1 slender, unarmed except for ischium, setose on fingers; ischium with small or obsolescent dorsal spine, ventromesially without subterminal spine; merus 1.1-1.2 × length of carapace; carpus 1.2 × length of merus; palm 0.8-0.9 × length of carpus; fingers proportionately broad, distally ending in tiny incurved spine, not distinctly crossing distally when closed, gaping in large males; movable finger with subtrapezoid proximal process (in dorsal view) in large males, low process in females and small males. P2-4 well compressed mesio-laterally, unarmed on meri and carpi, with long setae especially numerous on propodi; meri successively shorter posteriorly (P3 merus 0.9-1.0 × length of P2 merus, P4 merus 0.6-0.7 × length of P3 merus), subequally broad on P2 and P3, much narrower on P4 (P4 merus 0.8 × breadth of P3 merus); P2 merus 0.8-0.9 × length of carapace, 1.2 × length of P2 propodus, P4 merus 0.8-0.9 × length of P4 propodus; P2 carpus subequal to or slightly longer than P3 carpus, P4 carpus shortest; carpus-propodus length ratio, 0.6 on P2, 0.5-0.6 on P3 and P4; carpus-dactylus length ratio, 1.3-1.5 on P2, 1.2-1.5 on P3, 1.0 on P4; propodi slightly shorter on P2 than on P3 or subequal on P2 and P3, shortest on P4; flexor margin somewhat convex medially, with 5-6 spines on distal half, terminal (ultimate) spine single, somewhat distant from juncture with dactylus, penultimate spine remotely separated from ultimate; dactyli curving at proximal quarter, dactylus-carpus length ratio, 0.6-0.8 on P2, 0.7-0.9 on P3, 1.0 on P4; dactylus-propodus length ratio, 0.4 on P2 and P3, 0.4-0.5 on P4; flexor margin with 2 groups of spines remotely separated; distal group of 2 larger spines, terminal longer; proximal group of 6-8 smaller spines, all oriented parallel to flexor margin.

Eggs. Number of eggs carried, 6-25; size, 1.15 mm × 1.18 mm - 1.46 mm × 1.58 mm.

Color. Illustrated by McCallum & Poore (2013) for a western Australian specimen.

REMARKS — Reexamination of the type material of *U. vandamae* disclosed that the female paratype from Albatross Station 5664 is different from both the holotype and the male paratype from Albatross Station 5620 in having a small anterolateral spine situated distinctly posterior to the lateral orbital spine, in having the antennal scale reaching the distal end of antennal article 5, and in having the terminal of the flexor spines of P2-4 propodi paired, not single. This specimen is referred to *U. australis* (Henderson, 1885) (see above under *U. australis*).

The material reported under *U. vandamae* by Baba (1990) from Madagascar contains two species, both different from *U. vandamae*: one includes all of the specimens from Vauban Station 102 and the other includes the remaining material taken at 13 other stations. The former is characterized by having a pair of epigastric spines that are variable from distinct to obsolete, the antennal scale terminates in the midlength of the antennal article 5, the P2-4 propodi bear a pair of terminal spines preceded by nearly equidistantly arranged single spines on the flexor margin, and the P2-4 dactyli bear flexor spines arranged regularly, not separated into proximal and distal groups. The latter is much like *U. vandamae* but differs in having a pair of distinct epigastric spines, in having the antennal scale overreaching the antennal article 5, and in having the P2-4 dactyli bearing a distal group of three spines, not two spines as in *U. vandamae*. These two species are apparently new to science and will be described later elsewhere. The male from Valdivia Station 245 in Zanzibar Canal (SMF 4549) reported under *U. vandamae* by Baba (1990) is referable to the first of the above species.

As mentioned by Baba (1988), four lots of the SIBOGA material reported under *U. australis* var. *indicus* by Van Dam (1933) are referable to *U. vandamae*. The remaining two lots, one from SIBOGA Station 262 and the other from Station 266, both from the Kai Islands, are different from *U. vandamae* in having a longer antennal scale reaching the distal end of article 5, and in having flexor spines of the P2-4 dactyli arranged regularly, with no remote separation into proximal and distal groups. These materials have the ultimate of the flexor spines of P2-4 dactyli stronger than the penultimate, the feature not in agreement with that of *U. indicus* Alcock, 1901 (Figure 217). Reexamination of these specimens is recommended to establish their systematic status.

The combination of the following characters may characterize *U. vandamae*: the P2-4 dactyli bear distal and proximal groups of flexor marginal spines that are remotely separated from each other; and the ultimate of the flexor marginal spines of the P2-4 propodi is moderately remote from the juncture with dactyli and the penultimate spine is remarkably distant from the ultimate.

Uroptychus vegrandis n. sp.
Figures 276, 277

TYPE MATERIAL — Holotype: **Solomon Islands**. SALOMON 1 Stn DW1827, 9°59.1'S, 161°05.8'E, 804-936 m, 4.X.2001, ♂ 7.2 mm (MNHN-IU-2014-17015). Paratypes: Collected with holotype, 1 ♂ 6.0 mm (MNHN-IU-2014-17016). **Fiji Islands**. BORDAU 1 Stn DW1459, 17°18'S, 179°33'W, 820-863 m, 5.III.1999, 2 ♀ 4.1, 5.8 mm (MNHN-IU-2014-17017). MUSORSTOM 10 Stn CP1344, 16°45.3'S, 177°40.5'E, 588-610 m, 10.VIII.1998, 1 ov. ♀ 6.6 mm (MNHN-IU-2014-17018).

ETYMOLOGY — From the Latin *vegrandis* (not large), alluding to the anteriormost branchial marginal spine smaller than the anterolateral spine, a character to separate the species from the nearest congener *U. karubar* n. sp.

DISTRIBUTION — Solomon Islands and Fiji Islands; 588-936 m.

SIZE — Males, 6.0-7.2 mm; females, 4.1-6.6 mm, ovigerous female, 6.6 mm.

DESCRIPTION — Medium-sized species. *Carapace*: About as long as broad; greatest breadth 1.5 × distance between anterolateral spines. Dorsal surface moderately convex from anterior to posterior, with feeble, indistinct groove between cardiac and posterior branchial regions; granulose in large specimens, feebly or barely so in small specimens. Lateral margins slightly convexly divergent posteriorly, with 5 well-developed spines: first anterolateral, overreaching much smaller lateral orbital spine; second smaller than first, located at anterior end of anterior branchial region, preceded by 1 or 2 very small or obsolescent denticle-like spines; third to fifth on posterior branchial region, all acute, third larger than second, fifth situated at point one-fifth from posterior end, followed by ridge. Rostrum slightly deflected ventrally, narrow triangular, with interior angle of 24-25°, length 0.6 × that of remaining carapace, breadth slightly less than half carapace breadth at posterior carapace margin; dorsal surface feebly concave; lateral margin with 2-6 very small, denticle-like spines in distal two-thirds. Lateral orbital spine situated directly mesial to and distant somewhat from anterolateral spine. Pterygostomian flap granulose on surface, anteriorly angular, produced to distinct spine.

Sternum: Excavated sternum with convex anterior margin. Sternal plastron about as long as broad, lateral extremities gently divergent posteriorly. Sternite 3 well depressed, anterior margin moderately concave, with 2 small incurved submedian spines separated by semicircular or broad U-shaped sinus, anterolateral corner angular with tiny accompanying denticle on each side. Sternite 4 having anterolateral margins sharply produced anteriorly, weakly divergent posteriorly, with posteriorly diminishing spines; posterolateral margin slightly shorter than anterolateral margin. Anterolateral margins of sternite 5 subparallel, anteriorly convex or bluntly rectangular, about as long as posterolateral margin of sternite 4.

Abdomen: Smooth and glabrous. Somite 1 with distinct transverse ridge. Somite 2 tergite 2.6-2.8 × broader than long; pleuron angular on anterolateral and posterolateral termini, lateral margins well concave, feebly or somewhat divergent posteriorly. Pleuron of somite 3 with blunt (small specimens) or angular lateral tip. Telson half as long as broad, posterior plate 1.6 × longer than anterior plate, laterally somewhat convex, and posteriorly emarginate.

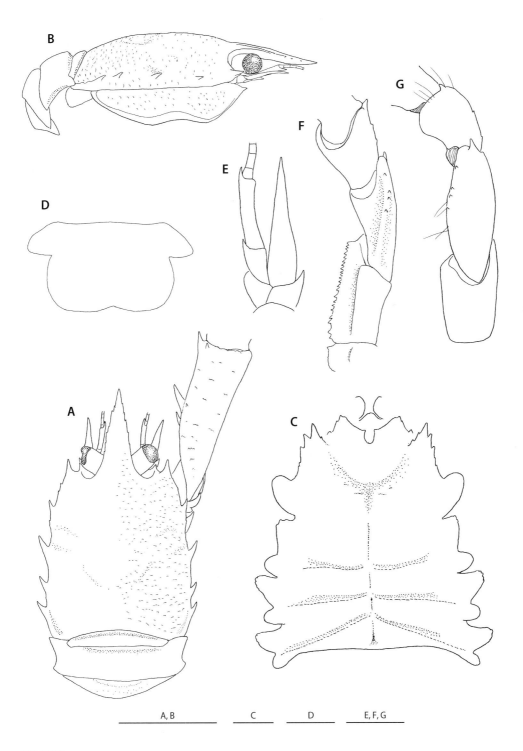

| A, B | C | D | E, F, G |

FIGURE 276

Uroptychus vegrandis n. sp., holotype, male 7.2 mm (MNHN-IU-2014-17015). **A**, carapace and anterior part of abdomen, proximal part of right P1 included, dorsal. **B**, same, lateral. **C**, sternal plastron, with excavated sternum and basal parts of Mxp1. **D**, telson. **E**, left antenna, ventral. **F**, left Mxp3, setae omitted, ventral. **G**, same, lateral. Scale bars: A, B, 5 mm; C-G, 1 mm.

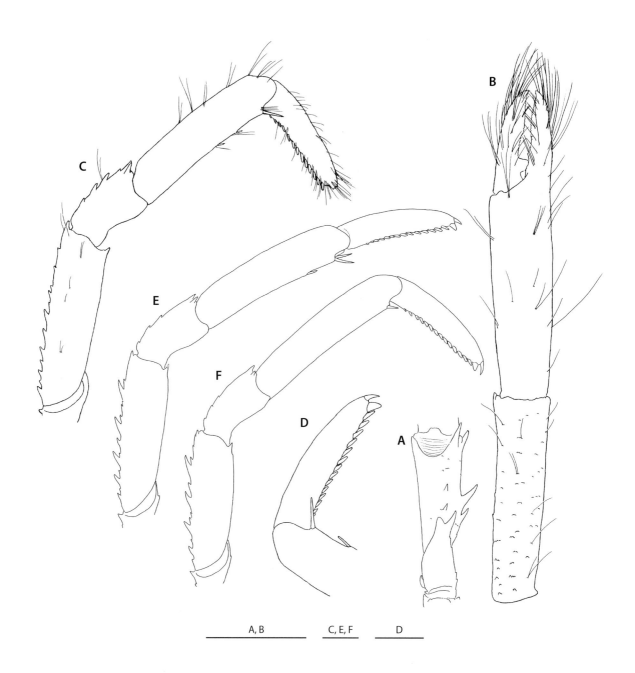

FIGURE 277

Uroptychus vegrandis n. sp., holotype, male 7.2 mm (MNHN-IU-2014-17015). **A**, right P1, proximal part, ventral. **B**, same, proximal part omitted, dorsal. **C**, right P2, lateral. **D**, same, distal part, setae omitted, lateral. **E**, right P3, setae omitted, lateral. **F**, right P4, setae omitted, lateral. Scale bars: A, B, 5 mm; C-F, 1 mm.

Eye: Relatively short (1.7 × longer than broad), ending at most in midlength of rostrum, lateral and mesial margins subparallel or feebly convex. Cornea not dilated, length much more than half that of remaining eyestalk.

Antennule and antenna: Ultimate article of antennule 2.6-2.9 × longer than high. Antennal peduncle extending far beyond cornea, far falling short of apex of rostrum. Article 2 with well-developed lateral spine. Antennal scale tapering, twice as broad as article 5, reaching proximal first or second segment of flagellum. Distal 2 articles each with distomesial spine. Article 5 2 × longer than article 4, breadth slightly more than half height of antennular ultimate article. Flagellum of 9-11 segments far falling short of distal end of P1 merus.

Mxp: Mxp1 with bases more or less close to each other. Mxp3 basis with 3-4 denticles, distalmost distinct, remainder obsolescent. Ischium with 20-23 denticles on crista dentata, flexor margin distally rounded. Merus 2.4 × longer than ischium, well ridged along flexor margin bearing 3 or 4 small spines distal to midlength, distolateral spine distinct. Carpus with small or moderate-sized distolateral spine and a few smaller spines along extensor margin.

P1: 3.9-4.5 × longer than carapace, sparsely setose but fingers more setose. Ischium with strong dorsal spine accompanying small spine proximally, ventromesial margin with well-developed subterminal spine. Merus as long as carapace, somewhat granulose; distomesial and distoventral (mesial) spines well developed; strong spine at midlength of mesial margin, with or without accompanying smaller spine proximal to it; ventral spines (1 or 2) small or obsolescent. Carpus 1.2-1.4 × longer than merus, granulose on surface. Palm slightly longer to slightly shorter than carpus, 3.1-3.6 × longer than broad. Fingers slightly or largely gaping in males, not gaping in females, distally ending small incurved spines, not distinctly crossing when closed (only distal spines crossing); movable finger 0.5-0.6 × length of palm, opposable margin with proximal process well developed in males, less so in females, fitting to longitudinal groove on opposite face of fixed finger when closed.

P2-4: Relatively broad, compressed mesio-laterally, sparsely setose. Meri successively shorter posteriorly (P3 merus 0.9 × length of P2 merus, P4 merus 0.9 × length of P3 merus), equally broad on P2-4; length-breadth ratio, 2.7-3.5 on P2, 2.3-3.3 on P3, 2.6-2.9 on P4; dorsal margin with 8-12 small spines on P2, 7-8 spines on P3, 4-8 spines on P4; P2 merus 0.6-0.7 × length of carapace, subequal to or slightly shorter than P2 propodus; P3 merus 0.8 × length of P3 propodus; P4 merus 0.7-0.8 × length of P4 propodus. Carpi subequal on P2-4, 0.3-0.4 × as long as propodi on P2-4, shorter than dactyli (carpus-dactylus length ratio, 0.6-0.7 on P2 and P3, 0.6 on P4); dorsal margin with 5-6 spines on P2, less numerous or obsolescent spines on P3 and P4. Propodi subequal in length on P3 and P4, shorter on P2, or subequal on P2 and P3, longer on P4; flexor margin nearly straight, ending in pair of spines preceded by 1 or 2 spines on P2, 1 spine on P3, 1 or 0 spine on P4. Dactyli subequal in length on P2-4, proportionately broad, longer than carpi (dactylus-carpus length ratio, 1.5-1.7 on P2, 1.5-1.8 on P3, 1.6-1.7 on P4), 0.6 × as long as propodi on P2-4; flexor margin straight, bearing row of spines, ultimate slender, as broad as antepenultimate, penultimate about 2 × broader than ultimate, remaining proximal spines slender, strongly inclined and nearly contiguous to one another, but not oriented parallel to flexor margin.

Eggs. Number of eggs carried, 25 (normal number probably more); size, 0.88 mm × 1.04 mm - 0.93 mm × 0.97 mm.

REMARKS — The ovigerous female from MUSORSTOM 10 Stn CP1344 is somewhat different from the others: the carapace bears a small spine dorsomesial to the second lateral spine; pereopods are more slender (length-breadth ratio, 5.0 on P1 palm, 4.1, 3.9, 3.4 on P2, P3, P4 merus respectively); the P1 merus is 1.2 times longer than the carapace and P1 carpus bears distinct spines along the mesial margin. Inasmuch as no other differences are found, this specimen is regarded as a variant of the species.

The carapace lateral marginal spination and the granulose dorsal surface in large specimens are very much like those of large specimens of *U. nanophyes* Alcock & McArdle, 1902. *Uroptychus vegrandis* differs from *U. nanophyes* in having four instead of five spines on the branchial margin, the anteriormost of which is smaller instead of larger than the anterolateral spine; the P2 merus has a terminal spine only instead of bearing a row of spines on each of the ventrolateral and ventromesial margins; and the P1 ischium bears on the ventromesial margin a subterminal spine only instead of a row of spines.

The species is closely related to *U. karubar* n. sp. (see above) in having the branchial lateral margin with four strong spines, the abdominal somite 1 well ridged transversely, and the antennal scale distinctly overreaching the antennal peduncle.

However, the first of the four branchial lateral spines in *U. karubar* is larger than the anterolateral spine followed by a distinct spine on the hepatic margin, whereas in *U. vegrandis* this spine is smaller than the anterolateral spine followed by one or two tiny denticle-like spines (usually absent); the epigastric region bears a row of transverse spines in *U. karubar*, unarmed in *U. vegrandis*; the Mxp3 ischium bears a small but distinct spine lateral to the distal end of flexor margin in *U. karubar*, which spine is absent in *U. vegrandis*; and the pterygostomian flap is granulose in *U. karubar*, whereas covered with small spines in *U. vegrandis*.

As mentioned under *U. quinarius* n. sp. (see above), that species also resembles this new species. *Uroptychus vegrandis* can be distinguished from *U. quinarius* by the granulose carapace surface and pterygostomian flap and by the lack of three closely arranged spines on the proximal mesial surface of P1.

Uroptychus vicinus n. sp.
Figures 278, 279

TYPE MATERIAL — Holotype: **New Caledonia**, Norfolk Ridge. BATHUS 3 Stn CP833, 23°03'S, 166°58'E, 441-444 m, 30.XI.1993, ♂ 3.3 mm (MNHN-IU-2014-17019). Paratypes: **Vanuatu**. MUSORSTOM 8 Stn CP1026, 17°50.35'S, 168°39.33'E, 437-504 m, 28.IX.1994, 1 ov. ♀ 3.3 mm (MNHN-IU-2014-17020). **New Caledonia**. MUSORSTOM 4 Stn CP180, 18°57'S, 163°18'E, 450 m, 18.IX.1985, 1 ♂ 3.1 mm, 1 ov. ♀ 2.9 mm (MNHN-IU-2014-17021). **New Caledonia**, Loyalty Ridge. MUSORSTOM 6 Stn DW428, 20°23.54'S, 166°12.57'E, 420 m, on *Thouarella* sp. (Alcyonacea: Primnoidae), 17.II.1989, 1 ov. ♀ 3.2 mm (MNHN-IU-2014-17022). **Philippines**. MUSORSTOM 3 Stn CP133, 11°58'N, 121°52'E, 334-390 m, 5.VI.1985, 4 ♂ 2.5-3.4 mm, 7 ov. ♀ 2.3-3.5 mm, 4 ♀ 2.1-2.7 mm (MNHN-IU-2014-17023).

ETYMOLOGY — From the Latin *vicinus* (near), alluding to a character of the close proximity of the the anterolateral and the orbital lateral spine of the carapace.

DISTRIBUTION — New Caledonia, Loyalty Islands, Norfolk Ridge, and Philippines; 334-504 m.

SIZE — Males, 2.5-3.4 mm; females, 2.1-3.5 mm; ovigerous females from 2.3 mm.

DESCRIPTION — Small species. *Carapace*: Broader than long (0.9 × as long as broad); greatest breadth 1.6 × distance between anterolateral spines. Dorsal surface smooth, with very sparse setae, slightly convex from anterior to posterior. Lateral margins straightly or slightly convexly divergent posteriorly, bearing 9 spines: first anterolateral, directly lateral to lateral orbital spine and reaching apex of that spine; second small, remote from first and close to third; third as large as first, situated at anterior end of anterior branchial region, dorsal to level of other spines, accompanying small spine dorsomesial to it; fourth small (obsolete on right side in holotype); fifth to ninth posteriorly diminishing, placed on posterior branchial region; fifth at anterior end of posterior branchial region, as large as third; ninth often obsolescent. Rostrum narrow triangular, with interior angle of 18-23°; ventral surface horizontal or slightly deflected ventrally; dorsal surface concave, lateral margin with subapical spine; length about half that of remaining carapace, breadth half carapace breadth at posterior carapace margin. Lateral orbital spine subequal to or very slightly smaller than closely subparalleling anterolateral spine. Pterygostomian flap with small spines arranged roughly in 2 rows, one in midline and another along anterior half of dorsal margin; anterior margin somewhat angular, produced to small spine.

Sternum: Excavated sternum with convex anterior margin, surface with obsolescent ridge in midline. Sternal plastron nearly as long as broad, lateral extremities slightly divergent posteriorly on sternites 4-5, subparallel on sternites 6-7. Sternite 3 shallowly depressed; anterior margin shallowly or moderately concave, medially with small V-shaped or deep narrow notch flanked by indistinct or very small spine discernible only under high magnification, anterolaterally sharp angular. Sternite 4 having anterolateral margin nearly straight or slightly convex, anteriorly ending in small spine; posterolateral

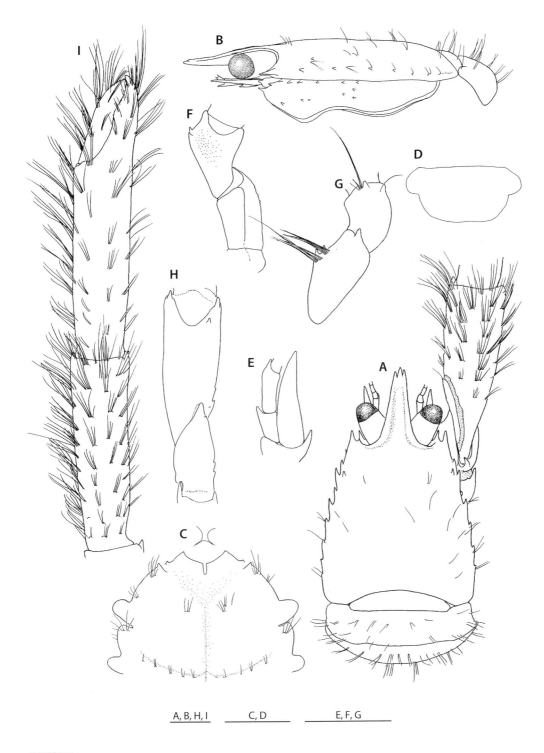

A, B, H, I   C, D   E, F, G

FIGURE 278

Uroptychus vicinus n. sp., holotype, male 3.3 mm (MNHN-IU-2014-17019). **A**, carapace and anterior part of abdomen, proximal part of right P1 included, dorsal. **B**, same, lateral. **C**, anterior part of sternal plastron, with excavated sternum and basal part of Mxp1. **D**, telson. **E**, left antenna, ventral. **F**, right Mxp3, setae omitted, ventral. **G**, left Mxp3, merus and carpus, lateral. **H**, right P1, proximal part, setae omitted, ventral. **I**, same, proximal part omitted, dorsal. Scale bars: 1 mm.

margin about two-thirds length of anterolateral margin. Anterolateral margin of sternite 5 anteriorly convex, about as long as posterolateral margin of sternite 4.

Abdomen: Polished, sparingly setose. Somite 1 without transverse ridge. Somite 2 tergite 2.5-2.8 × broader than long; pleural lateral margin feebly concave and slightly divergent posteriorly, posterolaterally bluntly angular. Pleuron of somite 3 with bluntly angular lateral terminus. Telson half as long as broad; posterior plate as long as anterior plate, feebly emarginate on posterior margin.

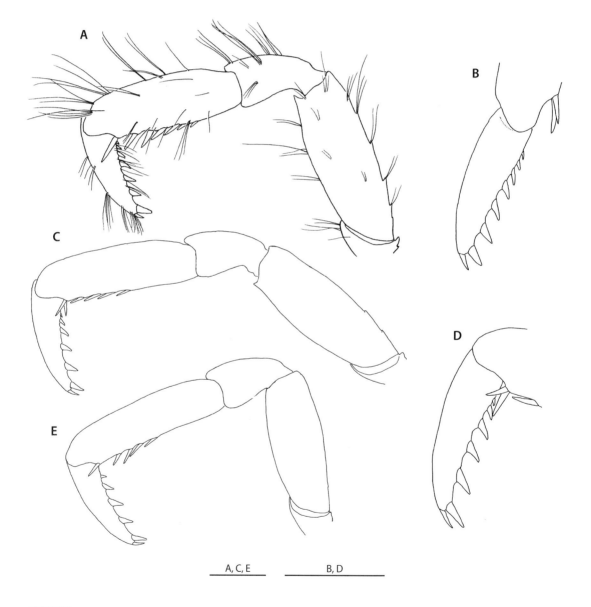

A, C, E   B, D

FIGURE 279

Uroptychus vicinus n. sp., holotype, male 3.3 mm (MNHN-IU-2014-17019). **A**, left P2, lateral. **B**, same, distal part, setae omitted, lateral. **C**, left P3, setae omitted, lateral. **D**, same, distal part, setae omitted, lateral. **E**, left P4, setae omitted, lateral. Scale bars: 1 mm.

Eye: Relatively short (1.6 × longer than broad), terminating in midlength of rostrum, distally somewhat narrowed. Cornea more than half length of remaining eyestalk.

Antennule and antenna: Ultimate article of antennular peduncle 2.5-3.2 × longer than high. Antennal peduncle overreaching cornea. Article 2 with strong distolateral spine. Antennal scale distally sharp, distinctly overreaching peduncle, occasionally with small spine on lateral proximal margin, breadth 1.8-2.0 × that of article 5. Distal 2 articles each with distomesial spine; article 5 1.5 × longer than article 4, breadth 0.5-0.6 × height of ultimate article of antennule. Flagellum consisting of 9-12 segments, not reaching distal end of P1 merus.

Mxp: Mxp1 with bases close to each other but not contiguous. Mxp3 basis without denticles on mesial ridge. Ischium with distally rounded flexor margin, crista dentata with very small or obsolescent denticles. Merus twice as long as ischium, well compressed mesio-laterally, somewhat concave on mesial face, bearing small or moderate-sized distolateral spine; flexor margin with 2 small, close spines distal to midlength, each accompanying long setae. Carpus with distolateral spine.

P1: 4.8-5.2 × longer than carapace, covered with fine setae. Ischium with moderate-sized dorsal spine, ventromesial margin with a few very small spines or tubercles, subterminal spine obsolescent. Merus 1.0-1.2 × longer than carapace, ventrally with 1 distomesial and 1 distolateral spine, and a few tubercles on proximal mesioventral surface. Carpus 1.0-1.2 × length of merus, with mesial and lateral distoventral spines. Palm unarmed, 3.0-3.4 × (males) or 2.9-5.9 × (females) longer than broad, as long as or slightly longer than carpus. Fingers inclined slightly laterally, distally incurved, crossing when closed; movable finger 0.4-0.5 × as long as palm, opposable margin with median process fitting to between opposite 2 low processes on fixed finger when closed.

P2-4: Moderately setose, broad and well compressed mesio-laterally. Meri successively slightly shorter posteriorly (P3 merus often subequal to P2 merus, P4 merus 0.9 × length of P3 merus), equally broad on P2-4; length-breadth ratio, 2.8-3.4 on P2, 2.2-3.1 on P3, 2.1-3.0 on P4; dorsal margin with several very small spines on P2, a few small spines or eminences proximally on P3, nearly smooth on P4; ventrolateral margin with distal spine distinct on P2 and P3, obsolete on P4. P2 merus 0.7-0.8 × length of carapace, 1.0-1.1 × length of P2 propodus; P3 merus subequal to length of P3 propodus; P4 merus 0.8-0.9 × length of P4 propodus. Carpi subequal; shorter than dactyli (carpus-dactylus length ratio, 0.9-0.95 on P2, 0.7-0.9 on P3 and P4); carpus-propodus length ratio, 0.4-0.5 on P2, 0.4 on P3 and P4. Propodi shorter on P2 than on P3 and P4, subequal on P3 and P4; flexor margin slightly convex or straight, ending in pair of spines preceded by 5-6 spines somewhat distant from one another at most along distal two-thirds on P2 and P3, 4 spines on P4. Dactyli longer than carpi on P2-4 (dactylus-carpus length ratio, 1.1 on P2, 1.2-1.4 on P3 and P4); dactylus-propodus length ratio, 0.5-0.6 on P2-4; flexor margin nearly straight on P2, slightly curved on P3, more curved on P4, with 7-8 (rarely 6) loosely arranged spines, ultimate spine slender, remaining spines sharp subtriangular, obliquely directed, loosely arranged and proximally diminishing, penultimate more than 2 × broader than ultimate, subequal to or slightly broader than antepenultimate, proximal 1 or 2 much smaller, ultimate very close to penultimate.

Eggs. Eggs carried 1-9 in number; size, 0.88 mm × 0.94 mm - 1.15 mm × 1.25 mm.

PARASITES — One of the males from MUSORSTOM 3 Stn CP133 (MNHN-IU-2014-17023) bears a rhizocephalan externa.

REMARKS — *Uroptychus vicinus* strongly resembles *U. beryx* n. sp. in the spination of the carapace, especially the lateral orbital spine being slightly larger than or subequal to the anterolateral spine and in having a median notch on the anterior margin of sternite 3. However, they can be separated by the following differences: the last lateral marginal spine of the carapace in *U. beryx* is closer to the posterior end, rather than to the preceding spine as in *U. vicinus*; the antennal scale is shorter in *U. beryx*, barely reaching instead of overreaching the distal end of article 5, and article 5 is unarmed instead of bearing a strong distomesial spine; and the P2-4 dactyli bear 9 or 10 closely arranged, perpendicularly directed spines proximal to the slender ultimate in *U. beryx*, 5-7 loosely arranged, obliquely directed spines in *U. vicinus*.

The characters listed below link the species to *U. annae* n. sp., *U. multispinosus* Ahyong & Poore, 2004, *U. perpendicularis* n. sp., *U. seductus* n. sp., *U. spinosior* n. sp., and *U. tridentatus* (Henderson, 1885): the carapace lateral margin with a row of spines, the rostrum with a subapical spine on each side, the pterygostomian flap with small spines on the surface, the antennal articles 4 and 5 each bearing a distomesial spine, the antennal scale overreaching article 5, the Mxp3 ischium rounded on the distal end of flexor margin, and the P2-4 dactyli with subtriangular flexor marginal spines.

Uroptychus annae, *U. perpendicularis*, *U. spinosior* and *U. tridentatus* are readily distinguished from the others by having the flexor marginal spines of P2-4 dactyli directed subperpendicularly instead of obliquely.

The lateral orbital spine is subequal to or slightly smaller than the anterolateral spine in *U. vicinus* and *U. multispinosus*, whereas in the other species it is much smaller. In *U. vicinus*, the anterolateral spine is situated directly lateral and very close to the lateral orbital spine, whereas it is distinctly posterior to and separated from that spine by its basal breadth in *U. multispinosus*. In addition, the lateral marginal spines of the carapace in *U. vicinus* are distinct rather than very small or obsolescent as in *U. multispinosus*, and an additional spine dorsomesial to the third spine is consistent in *U. vicinus*, absent in *U. multispinosus*. *Uroptychus vicinus* is distinguished from *U. seductus* by having a small dorsal spine mesial to the third carapace lateral spine, the P1 ischium bearing an obsolescent instead of strong subterminal spine on the ventromesial margin, the Mxp3 ischium lacking a spine near the rounded flexor distal margin. The last-mentioned spine is present in *U. seductus*, *U. spinosior* and *U. tridentatus*, absent in the other species. *Uroptychus spinosior* differs from *U. vicinus* in having more numerous spines on the carapace lateral margin and in having the anterolateral spine overreaching the much smaller lateral orbital spine.

Uroptychus perpendicularis has a small dorsal spine mesial to the third carapace lateral spine as in *U. vicinus*. It differs from *U. vicinus*, in addition to the arrangement of spines on the P2-4 dactyli (see above), in having the anterolateral spine of the carapace distinctly larger than instead of subequal to or slightly larger than the lateral orbital spine, and in having the antennal article 2 unarmed instead of bearing a strong distolateral spine.

Uroptychus volsmar n. sp.
Figures 280, 281

TYPE MATERIAL — Holotype: **New Caledonia**, Hunter and Matthew Islands. VOLSMAR Stn DW16, 22°25'S, 171°41'E, 420-500 m, 03.VI.1989, ♂ 3.0 mm (MNHN-IU-2011-5968). Paratypes: New Caledonia, Hunter and Matthew Islands. VOLSMAR, collected with holotype, 1 ♂ 2.7 mm (MNHN-IU-2011-5969). – Stn DW07, 22°26'S, 171°44'E, 325-400 m, 01.VI.1989, 1 ov. ♀ 2.5 mm (MNHN-IU-2011-5964). **New Caledonia**, Loyalty Ridge. MUSORSTOM 6 Stn DW406, 20°40.65'S, 167°06.80'E, 373 m, 15.II.1989, 1 ov. ♀ 2.5 mm (MNHN-IU-2014-17024).

ETYMOLOGY — Named for the cruise VOLSMAR, by which the holotype was collected; used as a noun in apposition.

DISTRIBUTION — Loyalty Islands and Hunter and Matthew Islands; 373-500 m.

SIZE — Males, 2.7, 3.0 mm; ovigerous females, 2.5 mm.

DESCRIPTION — Small species. *Carapace*: Slightly broader than long (0.9 × as long as broad); greatest breadth 1.5 × distance between anterolateral spines. Dorsal surface smooth, sparsely with soft setae, with distinct depression between gastric and cardiac regions; gastric region somewhat convex, cardiac region well convex, indistinctly separated from branchial region. Lateral margins slightly convexly divergent posteriorly; anterolateral spine short and basally stout, somewhat dorsal in position, very close to but distinctly posterior to level of lateral orbital spine; small spines along branchial margin. Rostrum broad triangular, with interior angle of 30-35°, nearly straight, directed slightly dorsally; dorsal surface concave; length slightly less than half that of remaining carapace, breadth half carapace breadth at posterior carapace margin. Lateral orbital spine relatively large, subequal to anterolateral spine. Pterygostomian flap with scattered denticles or small spines on surface, anterior margin somewhat angular, produced to small spine.

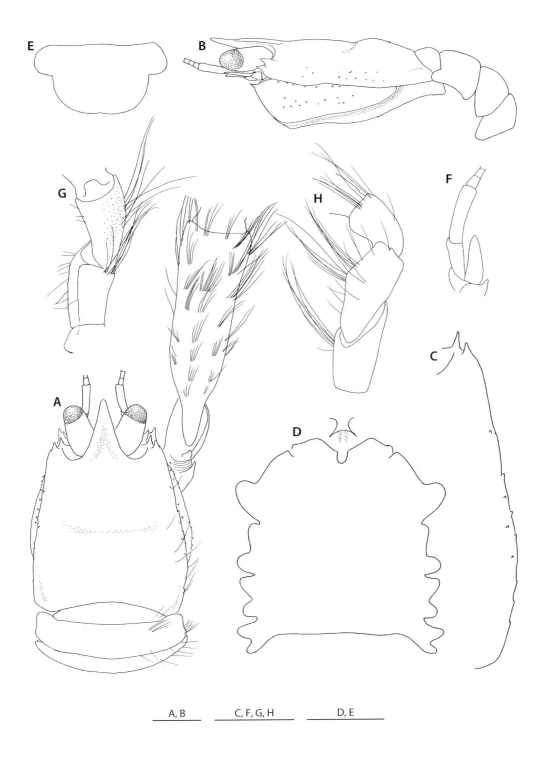

A, B   C, F, G, H   D, E

FIGURE 280

Uroptychus volsmar n. sp., holotype, male 3.0 mm (MNHN-IU-2011-5968). **A**, carapace and anterior part of abdomen, proximal articles of right P1 included, dorsal. **B**, same, lateral. **C**, lateral margin of carapace, right, setae omitted, dorsal. **D**, sternal plastron, with excavated sternum and basal parts of Mxp1. **E**, telson. **F**, left antenna, ventral. **G**, left Mxp3, ventral. **H**, same, lateral. Scale bars: 1 mm.

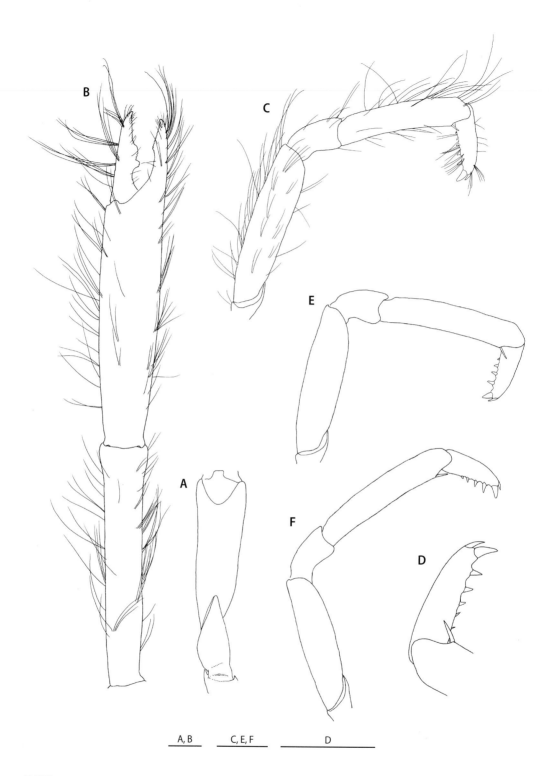

FIGURE 281

Uroptychus volsmar n. sp., holotype, male 3.0 mm (MNHN-IU-2011-5968). **A**, right P1, proximal part, setae omitted, ventral. **B**, same, dorsal. **C**, right P2, lateral. **D**, same, distal part, setae omitted, lateral. **E**, right P3, setae omitted, lateral. **F**, right P4, setae omitted, lateral. Scale bars: 1 mm.

Sternum: Excavated sternum with moderately or strongly convex anterior margin, bearing ridge in midline on surface. Sternal plastron about as long as broad, lateral extremities subparallel between sternites 4 and 7. Sternite 3 shallowly depressed; anterior margin gently concave with semioval median sinus flanked by very small or obsolescent spine. Sternite 4 having anterolateral margin as long as posterolateral margin, convex with bluntly angular anterior end. Anterolateral margins of sternite 5 subparallel but anteriorly convex, length 0.6-0.7 × that of posterolateral margin of sternite 4.

Abdomen: Smooth, with sparse long setae. Somite 2 tergite 2.8 × broader than long; pleuron posterolaterally blunt, lateral margin somewhat concave, moderately divergent posteriorly. Pleuron of somite 3 with blunt lateral terminus. Telson about half as long as broad; posterior plate semicircular and feebly emarginate on posterior margin, length 1.3 × that of anterior plate.

Eye: Elongate (2 × longer than broad), slightly falling short of rostral tip; distally narrowed, lateral margin strongly convex. Cornea not dilated, length about half or slightly less than half that of remaining eyestalk.

Antennule and antenna: Antennular ultimate article 3.2-3.4 × longer than high. Antennal peduncle overreaching rostrum. Article 2 with distolateral spine. Antennal scale 1.3 × broader than article 5, terminating in or slightly overreaching distal end of article 4. Article 4 with small distomesial spine. Article 5 unarmed, length 1.7-1.8 × that of article 4. Flagellum consisting of 12-14 segments, not reaching distal end of P1 merus.

Mxp: Mxp1 with bases distinctly separated from each other. Mxp3 basis lacking denticles on proximally lobe-like mesial ridge. Ischium with tuft of long setae lateral to rounded distal end of flexor margin; crista dentata with very small obsolescent denticles. Merus 2.3 × longer than ischium, distolateral spine short or obsolescent; flexor margin moderately ridged without spine. Carpus also spineless.

P1: 6.7-7.0 × longer than carapace, sparingly with long, soft, plumose setae. Ischium with short distodorsal spine, ventromesially unarmed. Merus unarmed, 1.5-1.7 × longer than carapace. Carpus 1.4 × longer than merus, unarmed. Palm 3.7-5.0 × longer than broad, about as long as palm. Fingers relatively narrow and slightly incurved distally, feebly crossing when closed, curving slightly ventrally, somewhat or greatly gaping; when strongly gaping, opposable margins touching each other at distal portion, crossing at tip when closed; movable finger 0.4 × length of palm, opposable margin with obtuse proximal process or with additional lower prominence distal to it, irrespective of sex; opposable margin of fixed finger with or without prominence distal to midlength.

P2-4: Sparingly with long, distally soft setae, moderately compressed mesio-laterally. Meri unarmed, successively shorter posteriorly (P3 merus 0.9 × length of P2, P4 merus 0.9 × length of P3 merus), equally broad on P2-4; length-breadth ratio, 4.0-4.4 on P2, 3.7-3.9 on P3, 3.5-3.8 on P4; P2 slightly shorter than carapace, slightly longer than P2 propodus; P3 merus 0.9 × length of P3 propodus; P4 merus 0.8-0.9 × length of P4 propodus. Carpi subequal; slightly less than one-third as long as propodus. Propodi subequal on P2-4, shorter on P2 than on P3 and P4 and subequal on P3 and P4 or longer on P3 than on P4; flexor margin nearly straight, with pair of terminal spines only. Dactyli subequal, 1.2 × longer than carpi and 0.3-0.4 × as long as propodi on P2-4; flexor margin straight, with 7 spines: ultimate slender; penultimate prominent, twice as broad as and very close to ultimate, remaining proximal spines loosely arranged, diminishing toward base of article, nearly perpendicular to margin (proximal 2 somewhat inclined), antepenultimate half as broad as penultimate, proximal-most very small and often obsolete.

Eggs. Number of eggs carried, 2-12; size, 0.88 mm × 0.92 mm - 1.10 mm × 1.20 mm.

REMARKS — The new species resembles *U. laurentae* n. sp., *U. paenultimus* Baba, 2005, and *U. toka* Schnabel, 2009 in having the anterolateral spine of the carapace close to the lateral orbital spine, the sternites 4-7 with subparallel lateral extremities, the antennal scale being short, ending at most in the midlength of antennal article 5, and the P2-4 dactyli bearing the penultimate flexor marginal spine fully twice as broad as the antepenultimate. *Uroptychus volsmar* is differentiated from *U. laurentae* and *U. paenultimus* by the P2-4 dactyli that bear flexor marginal spines directed perpendicularly instead of obliquely. In addition, the distal two of these spines are different in size: the penultimate spine is twice as broad as the ultimate spine in *U. volsmar*, 9 times broader in *U. laurentae*, 5 times broader in *U. pae-nultimus*. *Uroptychus volsmar* differs from *U. toka* in having the eyes distally narrowed rather than nearly uniformly broad from proximal to distal; the carapace dorsal surface is smooth instead of bearing denticle-like small spines on the gastric and epigastric regions; and the lateral orbital spine is subequal to instead of distinctly larger than the anterolateral spine of the carapace.

The shape of the sternal plastron, the spination of the P2-4 propodi and dactyli link this new species to *U. turgidus* n. sp. However, *U. volsmar* is readily distinguished from that species by the lateral orbital spine that is subequal to instead of much stronger than the anterolateral spine of the carapace, and by the P2-4 dactyli that are slightly (1.2 ×) instead of much (1.6 ×) longer than the carpi. In addition, the distinct lateral spines and pronounced inflation on the posterior branchial margin as displayed by *U. turgidus* are missing in *U. volsmar*.

Uroptychus vulcanus n. sp.
Figures 282, 283

TYPE MATERIAL — Holotype: **New Caledonia**, Loyalty Islands. BIOCAL Stn CP109, 22°11'S, 167°16'E, 495-515 m, with gorgonian corals of Primnoidae (Suborder Calcaxonia), 9.IX.1985, ov. ♀ 2.0 mm (MNHN-IU-2013-8534).

ETYMOLOGY — The specific name is derived from the Latin *Vulcanus* (the god of fire), in the Roman mythology.

DISTRIBUTION — Loyalty Islands; 495-515 m.

DESCRIPTION — Small species. *Carapace*: Much broader than long (0.6 × as long as broad); greatest breadth 1.7 × distance between anterolateral spines. Dorsal surface moderately convex from anterior to posterior, more so on posterior portion, with small spines scattered on hepatic and branchial regions. Lateral margins with very strong, anterolaterally produced spine at two-fifths from anterior end, bordering posteriorly convergent anterior part and subparallel posterior part; anterior part with well-developed anterolateral spine reaching tip of lateral orbital spine, followed by more than 10 small spines; posterior part well ridged, hanging over pterygostomian flap, bearing numerous very small spines. Rostrum directed slightly dorsally, relatively narrow, sharp triangular, with interior angle of 27°, ending in sharp tip; dorsal surface moderately concave; lateral margin with denticle-like small spines along entire length; length 0.85 × that of remaining carapace, breadth half carapace breadth at posterior carapace margin. Lateral orbital angle produced to well-developed spine situated slightly anterior to position of somewhat larger anterolateral spine. Pterygostomian flap anteriorly angular, produced to distinct spine, surface with spines in 2 rows on anterior portion, one directly below anterior part of linea anomurica, and another on lower part.

Sternum: Excavated sternum nearly transverse on anterior margin, surface smooth. Sternal plastron 1.4 × broader than long, lateral extremities divergent posteriorly. Sternite 3 slightly depressed from level of sternite 4; anterior margin moderately concave with obsolescent median notch, laterally angular with 2 small spines. Sternite 4 having anterolateral margin strongly convex anteriorly, weakly divergent posteriorly; posterolateral margin half as long as anterolateral margin. Anterolateral margin of sternite 5 anteriorly produced to rounded lobe, 1.3 × longer than posterolateral margin of sternite 4.

Abdomen: With sparse short setae. Somite 1 without transverse ridge. Somite 2 tergite 2.6 × broader than long; pleural lateral margins slightly concave, nearly subparallel in dorsal view, posterolaterally rounded. Pleuron of somite 3 with broadly rounded lateral margin. Telson slightly less than half as long as broad; posterior plate four-fifths length of anterior plate, moderately emarginate on posterior margin.

Eye: Relatively short and broad (1.3 × longer than broad), proximally slightly narrowed, slightly overreaching midlength of rostrum; lateral and mesial margins somewhat concave. Cornea slightly dilated, about as long as remaining eyestalk.

Antennule and antenna: Ultimate article of antennular peduncle 3.3 × longer than high. Antennal peduncle overreaching cornea. Article 2 laterally acuminate, without distinct spine. Antennal scale 1.8 × broader than article 5, terminating in midlength of article 5 (left one shorter, possibly regenerated). Distal 2 articles each with distomesial spine, that of article 5

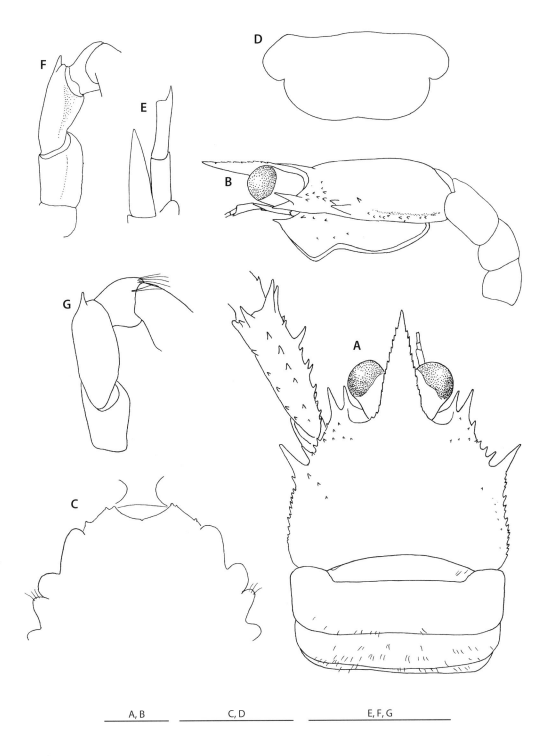

FIGURE 282

Uroptychus vulcanus n. sp., holotype, ovigerous female 2.0 mm (MNHN-IU-2013-8534). **A**, carapace and anterior part of abdomen, proximal part of left P1 included, dorsal. **B**, same, lateral. **C**, anterior part of sternal plastron, with excavated sternum and basal parts of Mxp1. **D**, telson. **E**, right antenna, ventral. **F**, right Mxp3, setae omitted, ventral. **G**, same, lateral. Scale bars: 1 mm.

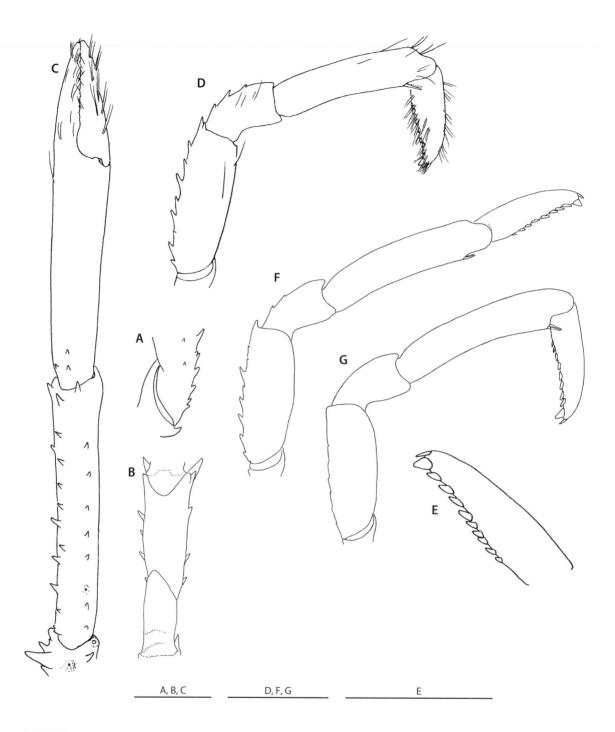

FIGURE 283

Uroptychus vulcanus n. sp., holotype, ovigerous female 2.0 mm (MNHN-IU-2013-8534). **A**, left P1, proximal part, lateral. **B**, same, ventral. **C**, same, proximal part omitted, dorsal. **D**, right P2, lateral. **E**, same, distal part, setae omitted, lateral. **F**, right P3, setae omitted, lateral. **G**, right P4, setae omitted, lateral. Scale bars: 1 mm.

stronger. Article 5 slightly longer than article 4, breadth 0.7 × height of ultimate article of antennule. Flagellum of 5 or 6 segments overreaching apex of rostrum by 2 or 3 segments, not reaching distal end of P1 merus.

Mxp: Mxp1 with bases broadly separated. Mxp3 basis without denticle on mesial ridge. Ischium having flexor margin not rounded distally, crista dentata with obsolescent denticles. Merus 2.5 × longer than ischium, with strong distolateral spine; flexor margin well ridged along entire length, mesial face concave. Carpus unarmed.

P1: About 5 × longer than carapace, slender, subcylindrical. Ischium with sharp dorsal spine, unarmed on ventral surface and ventromesial margin. Merus slightly longer than carapace, bearing lateral, dorsal and mesial rows of spines, distal spine of each row prominent; ventral surface unarmed. Carpus 1.6 × longer than merus, with dorsal spines arranged roughly in 2 rows and row of mesial spines. Palm 3.9 × longer than broad, 0.9 × length of carpus, slightly depressed, with a few small spines on proximal part of dorsal surface. Fingers distally slightly crossing when closed, not gaping, sparingly with relatively short setae; movable finger half as long as palm, opposable margin with low proximal process fitting to opposite longitudinal groove on ventral face of fixed finger when closed.

P2-4: Left P2 and P3 missing. P2-4 relatively short and sparsely setose, more setose on dactyli; P2 merus 0.8 × length of carapace, as long as P3 merus, 1.2 × longer than P2 propodus; P3 merus 0.7 × length of P3 propodus; P4 merus slightly shorter than P3 merus, 0.7 × length of P4 propodus; length-breadth ratio, 2.7 on P2, 2.6 on P3, 2.5 on P4; dorsal margin with row of 8-9 spines on P2 and P3, 5 obsolescent spines on P4. Carpi subequal, 0.4 × as long as propodi and 0.7 × as long as dactyli on P2-4; extensor margin with 2 or 3 spines small on P2, obsolescent on P3, absent on P4. Propodi shorter on P2 than on P3 and P4; flexor margin nearly straight, with pair of slender terminal spines only. Dactyli relatively stout, 1.5 × longer than carpi on P2-4, 0.50 × as long as propodi on P2 and P3, 0.55 × on P4; flexor margin nearly straight, with 10-11 spines obscured by setae, ultimate spine slender, narrower than antepenultimate, penultimate spine prominent but relatively short, 1.8-2.0 × broader than antepenultimate, remaining spines similar to antepenultimate, obliquely directed and close to one another, diminishing toward proximal end of article.

Eggs. Number of eggs carried, 7; size, 1.10 × 1.30 mm.

REMARKS — The new species somewhat resembles *U. obtusus* n. sp. (see above) in the spination of the P2-4, but they are largely different. In *U. vulcanus*, the carapace lateral margin bears a strong spine at the anterior two-fifths of length, which spine is absent in *U. obtusus*; the lateral orbital spine is well developed, paralleling the larger anterolateral spine, whereas this spine is very small in *U. obtusus*; the lateral margin of the rostrum bears denticle-like small spines along the entire length instead of along the distal third; the anterior margin of sternite 3 is deeply concave with a tiny median notch instead of being shallowly concave with a deep median sinus; the posterolateral margin of sternite 4 is distinctly shorter than instead of as long as the anterolateral margin; the anterior margin of the excavated sternum is nearly transverse rather than triangularly produced; and the Mxp3 merus is unarmed instead of bearing two distinct spines on the flexor margin. This species is readily distinguished from other known species of the genus by the combination of the following characters: the rostral lateral margin with numerous denticle-like small spines along the entire length; the carapace is much broader than long, with a prominent, anterolaterally directed spine at the anterior end of each of the subparallel branchial margins; and the lateral orbital spine is well developed but smaller than the anterolateral spine of the carapace.

Uroptychus yokoyai Ahyong & Poore, 2004

Figures 284, 306H

Uroptychus yokoyai Ahyong & Poore, 2004: 79, fig. 25. — Poore *et al*. 2011: 330, pl. 8, fig. E.

TYPE MATERIAL — Holotype: **Australia**, Gifford Guyot, E of Brisbane, 26°44.27'S, 159°28.93'E, 306 m, male (AM P65827). [not examined].

MATERIAL EXAMINED — **Philippines**. MUSORSTOM 1 Stn CP60, 14°05'N, 120°19'E, 129-124 m, 27.III.1976, 2 ♂ 2.6, 3.4 mm, 4 ov. ♀ 2.6-3.3 mm, 2 ♀ 2.5, 3.1 mm (MNHN-IU-2014-17025). MUSORSTOM 3 Stn CP131, 11°37'N, 121°43'E, 120-122 m, with Chrysogorgiidae gen. sp. (Calcaxonia), 5.VI.1985, 5 ♂ 2.5-3.7 mm, 4 ov. ♀ 3.3-3.9 mm, 2 ♀ 2.9, 3.3 mm (MNHN-IU-2014-17026). **Vanuatu**. MUSORSTOM 8 Stn CP1017, 17°52.80'S, 168°26.20'E, 294-295 m, 27.IX.1994, 1 ♂ 3.9 mm (MNHN-IU-2014-17027). – Stn DW964, 20°19.60'S, 169°49.00'E, 360-408 m, 21.IX.1994, 1 ♀ 3.0 mm (MNHN-IU-2014-17028). – Stn CP1133, 15°38.83'S, 167°03.06'E, 174-210 m, 11.X.1994, 2 ov. ♀ 2.9, 3.0 mm, 1 ♀ 2.3 mm (MNHN-IU-2014-17029). Santo Stn AT07, 15°38'S, 167°02'E, 180-223 m, 15.IX.2006, 1 ♀ 3.6 mm (MNHN-IU-2014-17030). – Stn AT17, 15°39.9'S, 167°02.0'E, 267-270 m, 21.IX.2006, 1 ♀ 3.1 mm (MNHN-IU-2014-17031). – Stn AT76, 15°38.7'S, 167°03.6'E, 105-135 m, 10.X.2006, 1 ♀ 3.2 mm (MNHN-IU-2014-17032). **New Caledonia**. LAGON Stn DW1151, 19°01'S, 163°27'E, 270-280 m, 28.X.1989, 1 ♂ 2.8 mm (MNHN-IU-2014-17034). – Stn DW389, 22°44'S, 167°04'E, 274-274 m, 22.I.1985, 1 ♀ 3.0 mm (MNHN-IU-2014-17035). – Stn DW396, 22°40'S, 167°09'E, 280-284 m, 23.I.1985, 1 ♀ 3.2 mm (MNHN-IU-2014-17036). **New Caledonia**, South of Chesterfield Islands. MUSORSTOM 5 Stn CP254, 25°10.29'S, 159°53.07'E, 280-290 m, with coral, ?Octocorallia, 7.X.1986, 1 ♂ 3.2 mm (MNHN-IU-2014-17037). – Stn DW273, 24°43.02'S, 159°43.26'E, 290 m, with Chrysogorgiidae gen. sp. (Calcaxonia), 9.X.1986, 1 ♂ 4.1 mm (MNHN-IU-2014-17038). – Stn CP289, 24°01.50'S, 159°38.40'E, 273 m, 10.X.1986, 1 ov. ♀ 3.4 mm (MNHN-IU-2014-17039). EBISCO, CP2494, 24°45'S, 159°42'E, 348-354 m, 06.X.2005, 1 ov. ♀ 3.6 mm (MNHN-IU-2014-17033). **New Caledonia**, Loyalty Ridge. MUSORSTOM 6 Stn DW399, 20°41.80'S, 167°00.20'E, 282 m, with Chrysogorgiidae gen. sp. (Calcaxonia), 14.II.1989, 1 ♂ 2.5 mm, 1 ov. ♀ 2.8 mm (MNHN-IU-2014-17040). – Stn CP400, 20°42.18'S, 167°00.40'E, 270 m, with Chrysogorgiidae gen. sp. (Calcaxonia), 14.II.1989, 2 ov. ♀ 2.9, 3.7 mm (MNHN-IU-2014-17041). – Stn DW417, 20°41.80'S, 167°03.65'E, 283 m, 16.II.1989, 1 spec. (sex indet.) 1.6 mm (MNHN-IU-2014-17042). **New Caledonia**, Norfolk Ridge. LITHIST Stn CP10, 24°48.4'S, 168°09.0' E, 245-261 m, 11.VIII.1999, 1 ♂ 2.6 mm, 4 ov. ♀ 2.3-3.2 mm, 1 ♀ 2.6 mm (MNHN-IU-2014-17043). – Stn DW11, 24°46.7'S, 168°08.3'E, 254-283 m, 11.VIII.1999, 1 ♂ 2.9 mm, 1 ov. ♀ 2.7 mm (MNHN-IU-2014-17044). – Stn DW13, 23°45.0'S, 168°16.7'E, 400 m, 12.VIII.1999, 1 ♂ 2.1 mm (MNHN-IU-2014-17045). BATHUS 3 Stn CP804, 23°41.40'S, 168°00.42'E, 244-278 m, 27.XI.1993, 1 ♀ 3.0 mm (MNHN-IU-2014-17046). SMIB 4 Stn DW41, 24°44'S, 168°09'E, 230-235 m, 08.III.1989, 1 ♀ 2.3 mm (MNHN-IU-2014-17047). – Stn DW45, 24°45'S, 168°08'E, 245-260 m, 08.III.1989, with Chrysogorgiidae gen. sp. (Calcaxonia), 3 ov. ♀ 2.2-2.6 mm (MNHN-IU-2014-17048). – Stn DW47, 24°46'S, 168°08'E, 250-280 m, 08.III.1989, 2 ov. ♀ 2.4, 2.5 mm (MNHN-IU-2014-17049). SMIB 5, Banc Alis, 250 m, rock, 1.IX.1985, 1 ♂ 3.3 mm, 1 ov. ♀ 3.2 mm (MNHN-IU-2014-17050). – Stn DW75, 23°41'S, 168°01'E, 250-270 m, 07.IX.1989, 1 ♀ 2.0 mm (MNHN-IU-2014-17051). – Stn DW95, 23°00'S, 168°19'E, 140-200 m, 14.IX.1989, 1 ♂ 2.3 mm, 2 ov. ♀ 2.4, 2.5 mm (MNHN-IU-2014-17052). – Stn DW97, 23°02'S, 168°18'E, 240-300 m, 14.IX.1989, 1 ♂ 3.1 mm, 1 ov. ♀ 3.3 mm (MNHN-IU-2014-17053). – Stn DW101, 23°21'S, 168°05'E, 285-270 m, 14.IX.1989, 1 ov. ♀ 2.3 mm, 1 ♀ 2.8 mm (MNHN-IU-2014-17054). SMIB 8 Stn DW155, 24°45.7'S, 168°08.2'E, 257-262 m, 28.I.1993, 2 ♂ 2.5, 2.6 mm, 1 ov. ♀ 2.8 mm, 2 ♀ 2.7, 2.8 mm (MNHN-IU-2014-17055); 2 ov. ♀ 2.5, 2.7 mm (MNHN-IU-2014-17056). – Stn DW157, 24°45.64'S, 168°08.23'E, 251-255 m, 28.I.1993, 1 ov. ♀ 2.5 mm (MNHN-IU-2014-17057). – Stn DW159, 24°45.8'S, 168°08.2'E, 241-245 m, 28.I.1993, 2 ♂ 2.5, 2.7 mm (MNHN-IU-2014-17058). NORFOLK 1 Stn DW1675, 24°45'S, 168°09'E, 231-233 m, 22.VI.2001, 3 ov. ♀ 2.7-3.1 mm, 2 ♀ 2.8, 2.8 mm (MNHN-IU-2014-17059). – Stn CP1676, 24°43'S, 168°09'E, 227-232 m, 22.VI.2001, 4 ♂ 2.5-3.3 mm, 4 ov. ♀ 2.3-2.8 mm, 1 ♀ 3.1 mm (MNHN-IU-2014-17060). – Stn CP1677, 24°43'S, 168°10'E, 233-259 m, 22.VI.2001, with coral, Hexacorallia?, 5 ♂ 2.3-2.9 mm, 10 ov. ♀ 2.1-3.0 mm, 1 ♀ 2.9 mm (MNHN-IU-2014-17061). – Stn CP1681, 24°44'S, 168°08'E, 240-228 m, 22.VI.2001, with Chrysogorgiidae gen. sp. (Calcaxonia), 1 ♂ 2.9 mm, 3 ov. ♀ 2.3-2.8 mm (MNHN-IU-2014-17062). – Stn CP1682, 24°42'S, 168°09'E, 331-379 m, 22.VI.2001, 1 ov. ♀ 2.8 mm (MNHN-IU-2014-17063). – Stn CP1683, 24°44'S, 168°07'E, 248-272 m, 22.VI.2001, 1 ♂ 3.1 mm, 3 ov. ♀ 2.7-2.8 mm (MNHN-IU-2014-17064). – Stn CP1714, 23°22'S, 168°03'E, 257-269 m, 26.VI.2001, 1 ♂ 3.2 mm (MNHN-IU-2014-17065). NORFOLK 2 Stn DW2093, 24°44'S, 168°08'E, 230 m, 29.X.2003, 1 ♂ 2.9 mm (MNHN-IU-2014-17066). – Stn CP2096, 24°43.91'S, 168°08.88'E, 230-240 m, 29.X.2003, 4 ♂ 2.6-2.9 mm, 4 ov. ♀ 2.3-2.8 mm, 2 ♀ 2.5, 2.6 mm (MNHN-IU-2014-17067). – Stn DW2158, 22°41.40'S, 167°14.01'E, 265-283 m, 6.XI.2003, 1 ♂ 2.4 mm (MNHN-IU-2014-17068). BERYX 11 Stn CP16, 24°47'S, 168°09'E, 240-250 m, 16.X.1992, 1 ov. ♀ 2.3 mm (MNHN-IU-2014-17069). – Stn CP17, 24°48'S, 168°09'E, 250-270 m, 16.X.1992, 1 ♂ 2.3 mm (MNHN-IU-2014-17070). – Stn DW18, 24°48'S, 168°09'E, 250-270 m, 16.X.1992, 1 ov. ♀ 2.6 mm, 1 ♀ 2.7 mm (MNHN-IU-2014-17071), 1 ♂ 2.6 mm, 1 ov. ♀ 2.2 mm (MNHN-IU-2014-17072). – Stn CP25, 24°43.52'S, 168°08.52'E, 230-235 m, 17.X.1992, 4 ov. ♀ 3.0-3.1 mm, 1 ♀ 2.7 mm (MNHN-IU-2014-17073). BIOCAL Stn DW65, 24°48'S, 168°09'E, 245-275 m, 3.IX.1985, 1 ov. ♀ 2.3 mm (MNHN-IU-2014-17074). CHALCAL 2 Stn CP19, 24°42.85'S, 168°09.73'E, with Chrysogorgiidae gen. sp. (Calcaxonia), 271 m, 27.X.1986, 1 ov. ♀ 2.6 mm (MNHN-IU-2014-17075), 2 ♂ 2.4, 2.6 mm, 11 ov. ♀ 2.1-2.6 mm (MNHN-IU-2014-17076). – Stn DW70, 24°46.0'S, 168°09.0'E, 232 m, with coral Chrysogorgiidae (Calcaxonia), 27.X.1986, 3 ♂ 2.3-2.9 mm, 4 ov. ♀ 2.4-2.7 mm, 1 ♀ 3.1 mm (MNHN-IU-2014-17077). – Stn DW71, 24°42.26'S, 168°09.52'E, 230 m, with Chrysogorgiidae gen. sp. (Calcaxonia), 27.X.1986, 1 ♂ 2.1 mm, 10 ov. ♀ 2.0-2.6 mm (MNHN-IU-2014-17078). MUSORSTOM 4 Stn CP152, 19°05'S, 163°23'E, 228 m, 14.IX.1985, 1 ♀ 3.0 mm (MNHN-IU-2014-17079). – Stn DW209, 22°41.8'S, 167°09.1'E, 310-315 m, 28.IX.1985, 1 ♀ 2.8 mm (MNHN-IU-2014-17080). **New Caledonia**, Hunter-Matthew Islands. VOLSMAR Stn DW39, 22°20'S, 168°44'E, 280-305 m, 08.VI.1989, 3 ♂ 2.5-2.9 mm, 6 ov. ♀ 2.7-3.3 mm (MNHN-IU-2014-17081). – Stn DW41, 22°19'S, 168°41'E, 195-250 m, 08.VI.1989, 1 ♂ 3.1 mm (MNHN-IU-2014-17082), 1 ♂ 3.1 mm, 1 ov. ♀ 3.0 mm (MNHN-IU-2014-17083), 1 ♂ 3.2 mm (MNHN-IU-2014-17084). – Stn DW42, 22°17'S, 168°42'E, 340-400 m, 08.VI.1989, 1 ♂ 3.0 mm, 1 ♀ 3.3 mm (MNHN-IU-2014-17085). **Tonga**. BORDAU 2 Stn CP1626, 23°20'S, 176°16'W, 220-249 m, 19.VI.2000, 1 ♀ 2.7 mm (MNHN-IU-2014-17086).

DISTRIBUTION — Previously known from Tasman Sea, in 295-306 m; now from the Philippines, Vanuatu, New Caledonia, south of the Chesterfield Islands, Norfolk Ridge, the Hunter-Matthew Islands, and Tonga, in 105-573 m.

SIZE — Males, 2.1-4.1 mm; females, 2.0-4.5 mm; ovigerous females from 2.0 mm; sex indet., 1.6 mm.

DIAGNOSIS — Small species. Carapace slightly broader than long; greatest breadth 1.3 × distance between anterolateral spines. Dorsal surface smooth and glabrous, slightly convex from anterior to posterior. Lateral margins usually strongly convex, particularly on posterior two-thirds; anterolateral spine small and subequal to lateral orbital spine; pronounced spine at anterior end of branchial margin followed by additional 0-3 (usually 2) small spines. Rostrum very broad, nearly equilateral triangular, with interior angle of 50-57°; breadth more than two-thirds carapace breadth at posterior carapace margin; dorsal surface concave. Lateral orbital spine very close to anterolateral spine. Pterygostomian flap smooth on surface, anterior margin roundish with small or obsolescent spine. Excavated sternum anteriorly produced to small

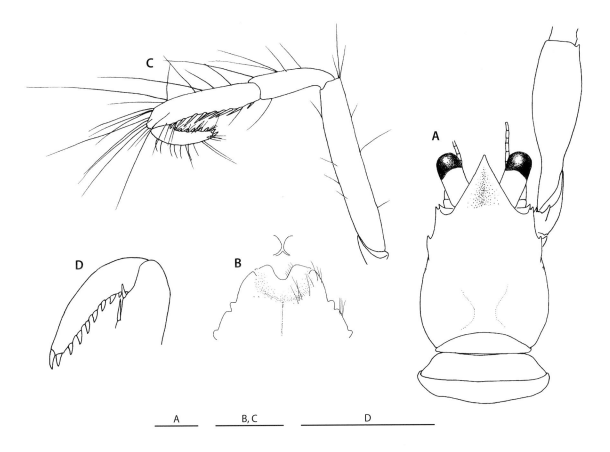

FIGURE 284

Uroptychus yokoyai Ahyong & Poore, ovigerous female 2.6 mm (MNHN-IU-2014-17075). **A**, carapace and anterior part of abdomen, proximal part of right P1 included, dorsal. **B**, anterior part of sternal plastron, with excavated sternum and basal parts of Mxp1. **C**, left P2, lateral. **D**, same, distal part, setae omitted, lateral. Scale bars: 1 mm.

sharp spine, smooth on surface. Sternal plastron 1.2-1.3 × broader than long, lateral extremities convexly divergent posteriorly, sternites 6 and 7 equally broad. Sternite 3 strongly depressed; anterior margin deeply or shallowly semicircular, occasionally medially deeply emarginate, with no submedian spines; sternite 4 with anterolateral margin relatively long, convex and anteriorly roundish; posterolateral margin one-quarter length of anterolateral margin; anterolateral margin of sternite 5 more than 3 × longer than posterolateral margin of sternite 4. Abdomen smooth and glabrous; somite 1 without transverse ridge; somite 2 tergite 3.1-3.4 × broader than long; pleuron posterolaterally blunt, lateral margin feebly concave, strongly divergent posteriorly; pleuron of somite 3 with blunt posterolateral terminus; telson 0.7-0.8 × as long as broad; posterior plate trianguloid, with rounded posterior margin, length subequal to greatest breadth, 2.3-3.0 × that of anterior plate. Eye reaching or slightly overreaching apex of rostrum, 2 × longer than broad, with subparallel mesial and lateral margins, cornea half as long as remaining eyestalk. Ultimate antennular article about 2 × as long as high. Antennal peduncle relatively slender, not reaching corneal margin; article 2 without distinct distolateral spine; antennal scale reaching or slightly overreaching midlength of article 5; distal 2 articles each with distomesial spine, that of article 4 often obsolescent; article 5 2.0-2.2 × longer than article 4, breadth 0.4 × height of ultimate antennular article; flagellum of about 20 segments overreaching distal end of P1 merus. Mxp1 with bases very close to each other. Mxp3 basis with obsolescent denticles on mesial ridge; ischium having flexor margin not rounded distally, crista dentata with about 20 denticles; merus 2.4-2.9 × longer than ischium, flexor margin not cristiform but rounded; distolateral spine distinct, very small or obsolete; carpus with small distolateral spine. P1 with long setae on fingers and distal portion of palm, glabrous elsewhere; ischium with strong, curved dorsal spine, subterminal spine on ventromesial margin obsolescent; merus 1.3-1.5 × longer than carapace, proximally and distally narrowed especially in small specimens (hence representing reverse shape of "bowling pin"); carpus proximally narrowed, 1.4-1.6 × (males), 1.6-1.7 × (females) longer than merus, finely granulose on mesial and lateral faces; palm 3.0-3.4 (males), 4.4-5.3 × (females) longer than broad, subequal to or slightly shorter than carpus; laterally and often dorsally with fine tubercles; fingers ending in incurved tip, crossing when closed; fixed finger with convex, finely dentate cutting edge; opposing margin of movable finger with disto-proximally broad proximal process low in females and small males, often sharply pointed (distal margin perpendicular to opposable margin) in males; movable finger 0.4 × length of palm in both sexes.P2-4 slender, bearing sparse long setae, especially on carpi and meri, unarmed on meri and carpi; meri successively shorter posteriorly (P3 merus 0.9 × length of P2 merus, P4 merus 0.8-0.9 × length of P3 merus), equally broad on P2-4; length-breadth ratio, 6.0 on P2, 5.0 on P3, 4.0-5.0 on P4; P2 merus subequal to or slightly longer than carapace, 1.5-1.6 × length of P2 propodus; P3 merus 1.3 × length of P3 propodus; P4 merus 1.1 × length of P4 propodus; carpi subequal or successively slightly shorter posteriorly, more than half as long as propodus on P2 and P3, nearly half as long on P4, usually with long setae distally; propodi subequal in length on P3 and P4, slightly shorter on P2, flexor margin concavely curving, ending in pair of spines preceded by 7 or 8 slender spines on P2 and P3, 6 spines on P4; dactyli slightly shorter than carpi (dactylus-carpus length ratio, 0.7-0.8 on P2, 0.8-0.9 on P3 and P4), about half or slightly less than half length of propodi, moderately curved; flexor margin with 9-10 (usually 10) subtriangular spines somewhat inclined and proximally diminishing.

Color. The female (3.6 mm) (MNHN-IU-2014-17030) from SANTO Stn AT07 was illustrated in Poore *et al.* (2011). Base color pale pink, cephalothorax and abdomen more pinkish along midline, abdominal pleura translucent.

Eggs. Number of eggs carried, 4 (in small specimens)-20 (in large specimen); size, 0.93 mm × 1.01 mm - 0.99 mm × 1.01 mm.

PARASITES — The male from MUSORSTOM 5 Stn CP254 (MNHN-IU-2014-17037) bears a rhizocephalan externa on the abdomen.

REMARKS — The species strongly resembles *Uroptychus latirostris* Yokoya, 1933 in the carapace ornamentation, in having a broad triangular rostrum, and in having no distinct submedian spines on the sternite 3 anterior margin, but it is distinctive in the following particulars: The P1 merus is proximally narrowed, representing a unique shape like a bowling pin, especially in small specimens; the anterolateral spine of the carapace is very close to instead of somewhat distant from the lateral orbital spine; and the posterior plate of the telson is elongate and narrowed posteriorly (hence, triangu-loid) instead of semicircular.

This is one of the most common species in the vicinity of New Caledonia, taken at 58 sampling stations.

Uroptychus zezuensis Kim, 1972

Uroptychus zezuensis Kim, 1972: 53, figs 1, 2. — Kim 1973: 171, fig. 17, pl. 64, figs 4a, 4b. — Baba 2005: 64, fig. 23. — Baba *et al.* 2009: 66, figs 56-58. — Poore *et al.* 2011: 330, pl. 8, figs F, G.

TYPE MATERIAL — Holotype: **Korea**, south of Seogwipo, Jeju Island, ov. female, (SNU). [not examined].

MATERIAL EXAMINED — **Philippines**. MUSORSTOM 1 Stn CP03, 14°01′N, 120°15′E, 183-185 m, 19.III.1976, 1 ♂ 4.2 mm, 1 ♀ 3.2 mm (MNHN-IU-2014-17087). – Stn CP27, 14°00′N, 120°16′E, 192-188 m, 22.III.1976, 1 ♂ 5.0 mm (MNHN-IU-2014-17088 [=MNHN-Ga 4614]) (incorporated in Baba 2005). – Stn CP63, 14°00′N, 120°16′E, 191-195 m, 27.III.1976, 1 ♂ 4.9 mm, 1 ov. ♀ 5.2 mm (MNHN-IU-2014-17089 [=MNHN-Ga 4615]) (incorporated in Baba 2005). MUSORSTOM 2 Stn CP01, 14°00′N, 120°18′E, 198-188 m, 20.XI.1980, 1 ov. ♀ 4.7 mm (MNHN-IU-2014-17090). – Stn CP19, 14°01′N, 120°18′E, 189-192 m, 22.XI.1980, with *Chironephthya* sp. (Alcyoniina: Nidaliidae), 1 ♀ 5.1 mm (MNHN-IU-2014-17091). MUSORSTOM 3 Stn CP108, 14°01′N, 120°18′E, 188-195 m, with *Chironephthya* sp. (Alcyoniina: Nidaliidae), 2.VI.1985, 1 ♂ 4.5 mm, 1 ♀ 2.4 mm (MNHN-IU-2014-17092).

DISTRIBUTION — Nagasaki (Japan), Jeju Island (Korea), Taiwan, and the Philippines; 60-311 m.

SIZE — Males, 4.2-5.0 mm; females, 2.4-5.2 mm; ovigerous females from 3.4 mm.

DIAGNOSIS — Small species. Carapace slightly broader than long, greatest breadth 1.5-1.6 × distance between ante-rolateral spines. Dorsal surface slightly convex from anterior to posterior; lateral margins convex, with 5 spines on anterior half: first anterolateral, moderate in size, overreaching smaller lateral orbital spine; second and third small; fourth and fifth acute, subequal, both much larger than first; fifth situated slightly anterior to midlength. Rostrum narrowly triangular, with interior angle 20-30°, slightly deflected ventrally, deeply concave dorsally, lateral margin with or without small subapical spine on each side, breadth slightly more than half carapace breadth at posterior carapace margin. Pterygostomian flap with spinules on surface, anteriorly angular, produced to sharp spine. Excavated sternum with distinct ridge in midline on surface, anterior margin broadly subtriangular between broadly separated bases of Mxp1. Sternite 3 shallowly depressed, anterior margin shallowly concave, with narrow or relatively broad U-shaped median notch; sternite 4 with anterolateral margin slightly convex and anteriorly angular, posterolateral margin 0.8 × as long as anterolateral margin; anterolateral margin of sternite 5 about as long as posterolateral margin of sternite 4. Abdominal somite 2 tergite 2.3 × broader than long; pleuron posterolaterally bluntly angular, lateral margins sub-parallel or weakly concave and weakly divergent posteriorly. Pleuron of somite 3 with bluntly angular posterolateral terminus. Telson with posterior margin feebly or more or less concave, not distinctly emarginated. Eyes 1.8 × longer than broad, slightly narrowed distally, reaching midlength of rostrum; cornea about half as long as remaining eyestalk. Antennal peduncle overreaching cornea; article 2 with acute lateral spine; antennal scale overreaching article 5 but not its distal spine; distal 2 articles each with very strong distomesial spine, article 5 about twice as long as article 4; flagellum not reaching distal end of P1 merus. Mxp3 having ischium with small spine directly lateral to rounded distal end of flexor margin, crista dentata with very small denticles; merus 1.9-2.0 × longer than ischium, with 1 distolateral spine and 1-3 small spines distal to midlength of flexor margin; carpus with 1 distolateral and 1 extensor marginal spine. Pereopods with long non-plumose setae. P1 ischium dorsally with strong spine often with accompanying smal-ler spine proximal to it, ventromesially with strong subterminal spine; merus and carpus usually with distomesial and distolateral spines ventrally, merus with row of 3 spines arranged obliquely on proximal part of ventromesial surface and row of 3 spines on ventral surface, length 1.2-1.3 × that of carapace; fingers more or less tapering, gently incurved distally, about half as long as palm. P2-4 moderately broad in lateral view, with long, distally softened setae; meri successively shorter posteriorly, equally broad on P2-4; dorsal margin smooth, without spines; P2 merus slightly shorter than carapace, 1.1 × length of P2 propodus; P3 merus 0.9-1.0 × length of P3 propodus; P4 merus 0.8 × length of P4 propodus; carpi subequal; length 0.4 × that of propodi on P2-4; propodus with straight flexor margin bearing pair of terminal spines preceded by 1 spine on P2, none on P3 and P4; dactyli slightly less than half as long as propodi, about as long as carpi (dactylus-carpus length ratio, 0.9-1.0 on P2, 1.0 on P3, 1.1 on P4); flexor margin with slender ultimate spine preceded by 5 or 6 spines on flexor margin, penultimate and antepenultimate strong, triangular, subequal and perpendicular to flexor margin, remaining proximal spines diminishing toward base of article, proximal-most spine inclined.

Eggs. Number of eggs carried up to 18; size, 0.83 mm × 0.85 mm - 0.95 mm × 1.05 mm.

Color. A specimen from Taiwan was illustrated by Baba *et al.* (2009) and Poore *et al.* (2011).

REMARKS — The part of the series from the Philippines (MNHN-Ga 4614, 4615) was incorporated in Baba (2005), in which the relationships between *U. zezuensis* and *U. tridentatus* (Henderson, 1885) were discussed.

This species is very similar to *U. joloensis* Van Dam, 1939 in morphology, coloration and host preference. Morphologically, *U. zezuensis* differs from *U. joloensis* in having the carapace lateral margin with five instead of three spines including the anterolateral spine (the last two being larger and sharper), in having the antennal scale distinctly overreaching instead of barely reaching the end of article 5, and in having the sternite 3 anterior margin more deeply excavated. The coloration was illustrated by Baba *et al.* (2009). It differs from that of *U. joloensis* (=*U. kudayagi*) in having a red spot around the hepatic region rather than around the rostrum.

The alcyonacean *Siphonogorgia* sp. (Nidaliidae) is known as a host associate (Baba *et al.* 2009). Two lots from the Philippines here examined (MNHN-IU-2014-17091; MNHN-IU-2014-17092) were taken with *Chironephthya* sp. (Nidaliidae). The red spots on both the carapace and P1 well harmonize with the color of the host's polyps, just as displayed by *U. joloensis* (see above).

Uroptychus zigzag n. sp.

Figures 285, 286

TYPE MATERIAL — Holotype: **Vanuatu**. MUSORSTOM 8 Stn CP983, 19°21.61'S, 169°27.76'E, 480-475 m, 23.IX.1994, ♂ 4.3 mm (MNHN-IU-2014-17093). Paratypes: **Vanuatu**. Collected with holotype, 1 ♂ 4.0 mm, 1 ov. ♀ 5.1 mm (MNHN-IU-2014-17094). **New Caledonia**, Norfolk Ridge. NORFOLK 2 Stn DW2063, 24°41'S, 168°40'E, 624-724 m, 25.X.2003, 1 ♂ 3.4 mm (MNHN-IU-2014-17095). **Indonesia**, Kai Islands. KARUBAR Stn CP16, 5°17'S, 132°50'E, 315-349 m, 24.X.1991, 1 ♀ 4.7 mm (MNHN-IU-2014-17096).

ETYMOLOGY — The specific name is a noun in apposition from the French *zigzag* (zigzag), referring to the zigzag arrangement of flexor marginal spines of P2-4 propodi displayed by the new species.

DISTRIBUTION — Vanuatu, Norfolk Ridge and Kai Islands (Indonesia); 315-724 m.

SIZE — Males, 3.4-4.3 mm; females, 4.7, 5.1 mm; ovigerous female, 5.1 mm.

DESCRIPTION — Medium-sized species. *Carapace*: Broader than long (0.7-0.8 × as long as broad); greatest breadth 1.8 × distance between anterolateral spines. Dorsal surface polished, moderately convex from side to side, more distinctly so from anterior to posterior, without depression; small spines mesial to fourth lateral marginal spine, another smaller

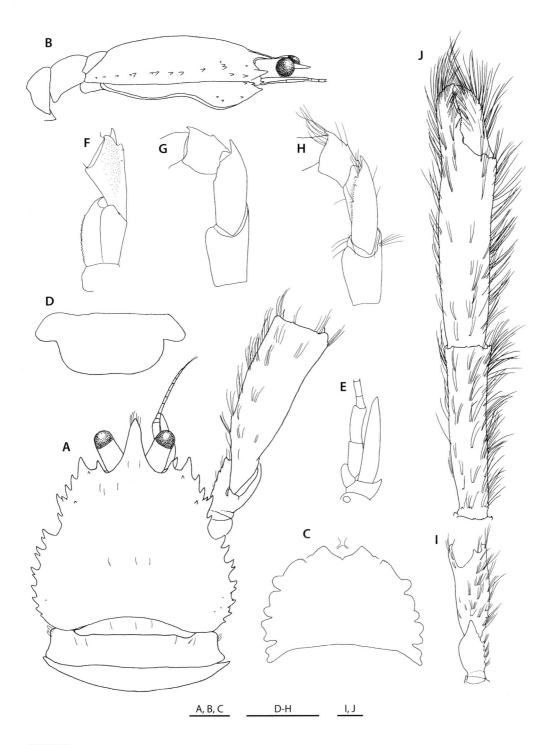

A, B, C   D-H   I, J

FIGURE 285

Uroptychus zigzag n. sp., holotype, male 4.3 mm (MNHN-IU-2014-17093). **A**, carapace and anterior part of abdomen, proximal part of right P1 included, dorsal. **B**, same, lateral. **C**, sternal plastron, with excavated sternum and basal parts of Mxp1. **D**, telson. **E**, left antenna, ventral. **F**, left Mxp3, ventral. **G**, same, lateral. **H**, same, ventromesial. **I**, left P1, proximal part, ventral. **J**, same, proximal articles omitted, dorsal. Scale bars: 1 mm.

spine directly behind each anterolateral spine and a few denticles on posterior lateral portion of branchial region; rarely several tiny spines in epigastric row (female, MNHN-IU-2014-17096). Lateral margins convexly divergent posteriorly, somewhat constricted at level of insertion of P1, bearing 12 spines (5 in front of and 7 behind constriction): first spine (anterolateral) strong, directed somewhat anterolaterally, overreaching smaller lateral orbital spine; second and third subequal and smaller than first, both situated ventral to level of other spines; fourth subequal in size to first, occasionally bifurcate dorsoventrally, closer to third than to fifth; remaining spines behind constriction or on posterior branchial margin nearly contiguous at base to one another, posteriorly diminishing, some of these often laciniate or bifurcate, last spine situated near posterior end of margin. Rostrum narrow triangular, with interior angle of 27-35°; length about half or less than half that of remaining carapace, breadth much less than half carapace breadth at posterior carapace margin; dorsal surface somewhat concave; ventral surface horizontal. Lateral orbital spine slightly anterior to level of anterolateral spine, and separated by basal breadth of that spine. Pterygostomian flap anteriorly sharp angular, produced to distinct spine, bearing 1-3 spines near anterior end of linea anomurica (absent in male paratype MNHN-IU-2014-17095) and additional few spines on anterior lower surface; height of posterior portion 0.5 × that of anterior portion.

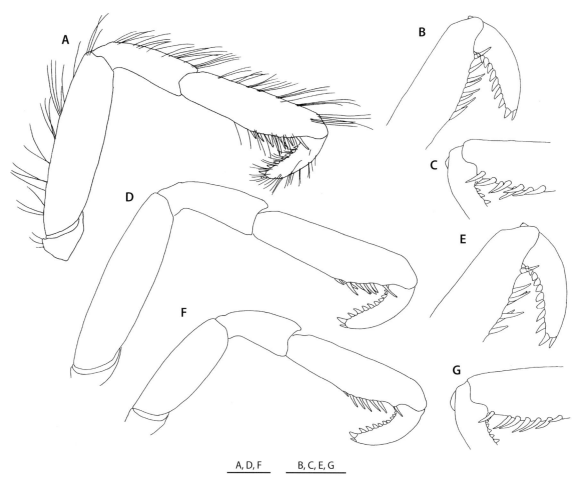

A, D, F   B, C, E, G

FIGURE 286

Uroptychus zigzag n. sp., holotype, male 4.3 mm (MNHN-IU-2014-17093). **A**, right P2, lateral. **B**, same, distal part, setae omitted, lateral. **C**, same, mesial. **D**, right P3, setae omitted, lateral. **E**, same, distal part, setae omitted, lateral. **F**, right P4, setae omitted, lateral. **G**, same, distal part, setae omitted, mesial. Scale bars: 1 mm.

Sternum: Excavated sternum with longitudinal ridge in midline on surface, anteriorly bluntly produced. Sternal plastron 1.4 × broader than long (three-quarters as long as broad), convexly divergent posteriorly. Sternite 3 weakly depressed, anterior margin gently concave, with small V-shaped median notch lacking flanking spines. Sternite 4 with convex anterolateral margin anteriorly ending in a few tiny spines; posterolateral margin about half length of anterolateral margin. Anterolateral margin of sternite 5 strongly convexly divergent posteriorly, 1.3 × longer than posterolateral margin of sternite 4.

Abdomen: Smooth and glabrous. Somite 1 without transverse ridge. Somite 2 tergite 3.2-3.5 × broader than long; pleuron posterolaterally bluntly angular, lateral margin moderately concavely divergent posteriorly. Pleuron of somite 3 posterolaterally angular. Telson about half as long as broad; posterior plate 1.2-1.4 × longer than anterior plate, posterior margin feebly concave.

Eye: 2.0-2.4 × longer than broad, slightly falling short of rostral tip, distally somewhat narrowed. Cornea about half as long as remaining eyestalk.

Antennule and antenna: Ultimate article of antennular peduncle 2.3-3.0 × longer than high. Antennal peduncle overreaching cornea, barely reaching rostral tip. Article 2 with short lateral spine. Antennal scale 1.4-1.5 × as broad as article 5, distinctly overreaching peduncle, reaching apex of rostrum. Distal 2 articles unarmed; article 5 1.0-1.3 × longer than article 4, breadth 0.5-0.7 × height of ultimate article of antennule. Flagellum consisting of 12-13 segments, far falling short of distal end of P1 merus.

Mxp: Mxp1 with bases very close to each other. Mxp3 with sparse soft setae on ischium, merus and carpus. Basis without denticles on mesial ridge. Ischium having flexor margin roundly ridged distally, crista dentata with small denticles. Merus 1.8 × longer than ischium, with well-developed distolateral spine and 2 small flexor marginal spines (obsolete in ovigerous female paratype) distal to midlength; mesial face somewhat concave; flexor margin sharply ridged. Carpus with distolateral spine and occasionally a few eminences on extensor surface.

P1: 5.7-6.1 × (males), 4.8 × (ovigerous female) longer than carapace; with soft setae, especially numerous along mesial margin. Ischium dorsally with short, basally broad, depressed spine, ventromesially with 5 or 6 short spines, without distinct subterminal spine. Merus 1.2-1.3 × (1.0 × in ovigerous female paratype) longer than carapace; a few to a number of denticle-like processes along proximal half of mesial margin and a few small spines on proximal ventral surface; dorsal terminal margin with 4 tubercle-like small spines, ventral terminal margin with obtuse mesial and lateral spines. Carpus 1.3-1.4 × longer than merus, dorsally with a few tubercles proximally and a few obsolescent terminal tubercles, ventrally with distomesial and distolateral processes, both blunt and short. Palm 3.3-3.8 × longer than broad in both sexes, 1.1-1.3 × longer than carpus; surface smooth. Fingers proportionately broad, not gaping, distally incurved, crossing when closed; opposable margins sinuous, without strong process; movable finger 0.4-0.5 × length of palm.

P2-4: Broad relative to length, moderately depressed mesio-laterally, with soft fine setae more numerous along dorsal and extensor margins. Meri unarmed; length-breadth ratio, 3.2-3.9 on P2, 3.1-3.9 on P3, 2.2-3.1 on P4; P2 merus slightly narrower than, and as long as or slightly shorter than P3 merus, 0.8-1.0 × length of carapace, 1.1-1.2 × longer than P2 propodus; P3 merus subequal to P3 propodus; P4 merus 0.6-0.7 × length of and 0.7-0.9 × breadth of P3 merus, 0.8 × length of P4 propodus. Carpi subequal on P2-3, P4 carpus 0.7-0.8 × length of P3 carpus, about half length of propodus on P2 and P3, slightly less than so on P4. Propodi shorter on P2 than on P3 and P4, P3 propodus usually longest; flexor margin medially convex, ending in pair of spines preceded by 6-7 slender spines on P2, 5 or 6 spines on P3, 4-6 spines on P4, those on P2 and P3 in zigzag arrangement, placed on distal half, ultimate of these spines more remote from distal pair than from penultimate. Dactyli slightly longer on P3 and P4 than on P2, slightly shorter than carpus (dactylus-carpus length ratio, 0.8 on P2 and P3, 0.9-1.0 on P4), half as long as propodus; flexor margin curving at proximal quarter, ending in slender spine preceded by 10-12 relatively broad, subtriangular, somewhat inclined spines diminishing toward base of article.

Eggs. Number of eggs carried, 17; size, 1.23 mm × 1.37 mm - 1.27 mm × 1.53 mm.

REMARKS — The new species resembles *U. duplex* n. sp. and *U. macrolepis* n. sp. (see above) in having a spinose carapace lateral margin, the antenna unarmed on articles 4 and 5, with the antennal scale reaching article 5, the Mxp3 ischium with rounded distal end of flexor margin, the P2-4 propodi distally broadened, and in the spination of the P2-4 dactyli. *Uroptychus zigzag* differs from *U. macrolepis* in having 12 instead of 8 or 9 spines on the carapace lateral margin; the spines on the posterior branchial margin are broad at base (distally laciniate or bifurcate) instead of much smaller (narrowed at base); the eyes are shorter relative to breadth (2.0-2.4 instead of 2.6-3.2 times longer than broad); and the anterolateral spine of the carapace is much larger instead of slightly larger than the lateral orbital spine. *Uroptychus duplex* is distinguished from *U. zigzag* by the carapace being much broader than long (breadth 1.7 versus 1.3 times length); the posteriormost of the lateral carapace spines is situated on the posterior quarter rather than near the posterior end; the sternal plastron has the lateral extremities strongly rather than gently divergent posteriorly; the anterior margin of sternite 3 is transverse on the median third instead of concave; and the lateral orbital spine is much smaller compared with that of *U. zigzag*.

Uroptychus zigzag also resembles *U. megistos* n. sp. (see above) in having the spinose carapace lateral margin and the inflated flexor margin of the P2-4 propodi, from which it can be readily distinguished by the following: the distal two articles of the antennal peduncle are unarmed instead of bearing a strong distomesial spine; the P2 merus is subequal to or slightly shorter instead of consistently longer than the P3 merus; and the ultimate of the flexor marginal spines of the P2-4 dactyli is much more slender instead of stronger than the penultimate.

HETEROPTYCHUS n. gen.

ETYMOLOGY — From the Greek *heteros* (different), plus the last syllables of *Uroptychus* (*ptychos* = plate). The gender is masculine.

Type species. *Uroptychus scambus* Benedict, 1902.

DIAGNOSIS — Carapace dorsal surface smooth and glabrous, lateral margin with anterolateral spine only, rarely with 1 or 2 processes along branchial region. Rostrum narrowly or broadly triangular. Pterygostomian flap very low on posterior half, height of posterior half 0.1-0.3 × that of anterior half. Sternal plastron different in sexes, posterior margin in females strongly excavated, with median parts of sternites 5-7 absorbed into sternite 4 (left and right parts of sternites 5-7 discontinuous, interrupted by loss of median parts). Antennal scale articulated or fused with article 2, not reaching distal end of article 4, articles 4 and 5 unarmed. P1 ischium with anterior dorsal process lobe-like or spiniform, posterior process usually lobe-like (rarely obsolescent). P4 very short, especially carpus 0.3-0.5 × length of P3 carpus. Distal two articles of P2-4 with long prehensile margins thickly fringed with setae, dactyli with slender spines perpendicular to flexor margin. G1 and G2 present.

REMARKS — The new genus is established on the basis of the following characters: the female sternal plastron is strongly excavated on the posterior margin, with sternites 5-7 medially discontinuous; the pterygostomian flap is very low in the posterior half (the posterior height at most 0.3 times the anterior height, but usually slightly lower in the majority of *Uroptychus* species); and the P4 carpus is very short, at most half as long as the P3 carpus, whereas it is subequal to, slightly shorter or rarely longer in *Uroptychus*. In addition, it is noteworthy that the antenna is inclined laterally or posteriorly in most of preserved specimens.

The present proposal of the new genus is supported by molecular data that showed that *Uroptychus* is not monophyletic (Roterman *et al.* 2013; Bracken-Grissom *et al.* 2013). Apparently *U. scambus* resides in a clade separate from other species of *Uroptychus* species, although the specific identity of *U. scambus* reported by these papers remains to be confirmed.

The unique shape of the sternal plastron, different in sexes in *Uroptychus scambus*, was first mentioned by Baba *et al.* (2009). This is also applicable to *U. brevis* Benedict, 1902 from the western Atlantic Ocean, the features confirmed by examination of the type material (holotype, ovigerous female, poc 5.2 mm, USNM 20566) of that species. *Heteroptychus scambus* is definitely a deep-sea species and has been believed to show a wide distribution from the western Indian Ocean to the western Pacific. However, molecular analyses (L. Corbari, personal comm.) suggest that there are some cryptic species, so that an attempt was made to detect morphological differences. Accordingly, slight differences that had been regarded as individual variations were validated as discriminating characters: the presence or absence of articulation between the antennal scale and article 2 and in the carapace shape, especially the rostrum and anterolateral spine. These characters delineate nine species including six new species, but no clear differences are found on other characters in all the material examined. However, there exist unstable characters in *H. claudeae* n. sp., *i.e.* the presence or absence of the lateral orbital spine, and the presence or absence of median processes of the anterior margin of sternite 3. In addition, color differences are found in morphologically identical specimens from Taiwan (Baba *et al.* 2009). It is not unlikely that this species may contain cryptic species, for which extensive studies are required.

Two classical names pose a problem for establishing this new genus because of the brief species descriptions: *U. scambus* Benedict, 1902 and *U. glyphodactylus* MacGilchrist, 1905. Examination of the holotype of *Uroptychus scambus* from Japan now in the collection of the Smithsonian Institution has elaborated on its systematic status. However, the type material of *U. glyphodactylus* from the Andaman Sea now housed in the collection of the Zoological Survey of India could not have been examined, despite repeated queries to the director of the institution. Some of the morphological features of this species that were not described in MacGilchrist (1905) are interpreted from the illustrations given by Alcock & MacGilchrist (1905: pl. 70, fig. 4, pl. 71, figs 1, 1a, 1b, 1c, 1d) albeit not in sufficient detail: the antennal scale in the figure appears to be articulated with article 2. As shown in the key to species provided below, these two species key out in the same couplet. Given that these species are known from disjunct localities, it is not unlikely that they are different. However, they are treated as synonymous for the time being. *Uroptychus edwardi* Kensley, 1981 from off the east coast of South Africa was synonymized with *U. scambus* (see Baba, 1988), but it is validated in this paper, on the basis of differences in the rostrum and anterolateral spine of the carapace (see below under the remarks of *H. colini* n. sp.).

The genus now contains nine species, including six new species: eight from the Indo-West Pacific and one from the western Atlantic. Females of three species have not been collected (*H. apophysis* n. sp., *H. colini* n. sp. and *H. lemaitrei* n. sp.).

KEY TO SPECIES

1. Antennal scale articulated with antennal article 2 .... **2**
– Antennal scale fused with antennal article 2 .... **7**

2. Branchial margin with 2 processes .... *H. apophysis* n. sp.
– Branchial margin smooth **3**

3. Anterolateral spine of carapace slightly overreaching apex of rostrum .... *H. anouchkae* n. sp.
– Anterolateral spine of carapace not overreaching apex of rostrum **4**

4. Rostrum not reaching apex of eye ....
.... *H. scambus* (Benedict, 1902) (= *H. glyphodactylus* (MacGilchrist, 1905))
– Rostrum overreaching eye **5**

5. Rostrum broad triangular, distally blunt .... *H. paulae* n. sp.
– Rostrum narrow triangular or distally spiciform **6**

6. Rostrum narrow triangular. Sternite 3 with pair of median processes .... *H. colini* n. sp.
– Rostrum spiciform. Sternite 3 with median notch, with no median processes .... *H. edwardi* (Kensley, 1981)

7. Rostrum and anterolateral spines distally blunt. P1 ischium with posterior dorsal lobe obsolescent ....
.... *H. brevis* (Benedict, 1902) [Western Atlantic; holotype examined]
– Rostrum and anterolateral spines sharply pointed. P1 ischium with posterior dorsal lobe overhanging basis **8**

8. Carapace lateral margin smooth. Rostrum narrow, more or less depressed distally .... *H. claudeae* n. sp.
– Carapace lateral margin with small protuberance at anterior end of branchial margin. Rostrum spiniform .
.... *H. lemaitrei* n. sp.

Heteroptychus anouchkae n. sp.
Figures 287, 288, 306I

TYPE MATERIAL — Holotype: **New Caledonia**, Norfolk Ridge. BATHUS 3 Stn CC841, 23°03'S, 166°53'E, 640-680 m, 30.XI.1993, ov. ♀ 4.7 mm (MNHN-IU-2014-17097). Paratypes: **New Caledonia**. HALIPRO 1 Stn CP854, 21°45.37'S, 166°38.34'E, 650-780 m, 19.III.1994, 1 ♂ 4.6 mm (MNHN-IU-2014-17098). – Stn CH872, 23°03'S, 166°53'E, 620-700 m, 30.III.1994, 1 ♂ 4.6 mm, 3 ov. ♀ 4.0-4.3 mm (MNHN-IU-2014-17099). – Stn CH873, 23°01'S, 166°53'E, 640-680 m, 30.III.1994, 1 ov. ♀ 4.5 mm (MNHN-IU-2014-17100). **New Caledonia**, Norfolk Ridge. BATHUS 3 Stn CP831, 23°04'S, 166°56'E, 650-658 m, 30.XI.1993, 1 ♂ 4.7 mm (MNHN-IU-2014-17101), 1 ov. ♀ 4.5 mm (MNHN-IU-2013-12292). – Stn CC841, collected with holotype, 2 ov. ♀ 4.3, 4.5 mm (MNHN-IU-2014-17102). – Stn CP846, 23°03'S, 166°58'E, 500-514 m, 1.XII.1993, 1 ov. ♀ 3.4 mm (MNHN-IU-2014-17103). – Stn CC848, 23°02'S, 166°53'E, 680-700 m, 1.XII.1993, 1 ♂ 4.1 mm (MNHN-IU-2013-12291). **New Caledonia**, Loyalty Ridge. MUSORSTOM 6 Stn DW483, 21°19.80'S, 167°47.80'E, 600 m, 23.II.1989, 1 ov. ♀ 4.0 mm (MNHN-IU-2013-12294).

ETYMOLOGY — Named for Anouchka Sato who helped me in various ways during my stays at the Paris Museum.

DISTRIBUTION — Loyalty Ridge, New Caledonia and Norfolk Ridge, in 500-780 m.

SIZES — Males, 4.1-4.7 mm; females, 3.4-4.7 mm, ovigerous females from 3.4 mm.

DESCRIPTION — Small species. *Carapace*: 0.6-0.7 × as long as broad, greatest breadth 2.1-2.4 × distance between anterolateral spines. Dorsal surface polished, moderately convex from anterior to posterior, ridged along posterior half of lateral margin; greatest breadth measured at posterior quarter. Lateral margins strongly convexly divergent posteriorly; anterolateral spine prominent, directed straight forward but somewhat dorsally, slightly overreaching rostral tip. Rostrum distally narrow triangular, with interior angle of 23-37°, distally sharply pointed, straight horizontal, reaching or slightly falling short of tip of eye; length 0.6-0.8 × breadth, 0.2-0.3 × length of remaining carapace; dorsal surface flattish but feebly concave basally. Lateral limit of orbit with small but distinct spine. Pterygostomian flap very low in posterior half (posterior height 0.1 × anterior height); anteriorly produced to sharp strong spine; anterior surface well inflated, with longitudinal ridge bearing 2 or 3 obsolescent (females and small males) or distinct, blunt (males) processes, occasionally smooth.

Sternum: Excavated sternum with subtriangular anterior margin, surface with sharp ridge in midline on anterior half. Sternal plastron in males 0.5-0.6 × as long as broad; in females, left and right parts of sternites 5-7 discontinuous, interrupted by loss of median parts. Sternite 3 having anterior margin broadly and deeply excavated, with pair of small obtuse median processes. Sternite 4 with rounded anterolateral margin.

Abdomen: Smooth and polished. Dorsal surface of somite 1 gently convex from anterior to posterior. Somite 2 tergite 2.3-2.6 × (females), 3.0-3.4 × (males) broader than long; pleural lateral margin moderately concave, strongly divergent posteriorly, ending in blunt tip. Pleura of somites 3-4 strongly tapering in females, less so in males, ending in rounded

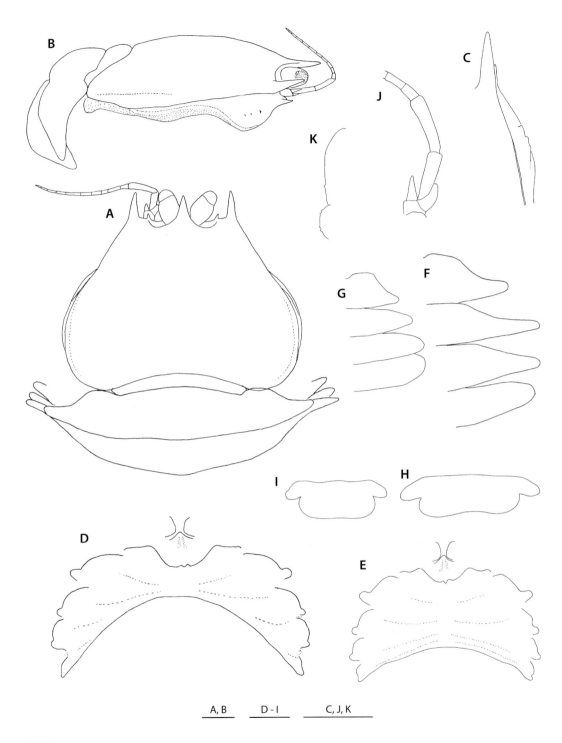

A, B   D - I   C, J, K

FIGURE 287

Heteroptychus anouchkae n. sp. **A-D**, **F**, **H**, **J**, **K**, holotype, ovigerous female 4.7 mm (MNHN-IU-2014-17097); **E**, **G**, **I**, male paratype 4.1 mm (MNHN-IU-2013-12291). **A**, carapace and anterior part of abdomen, dorsal. **B**, same, lateral. **C**, same, anterior part showing pterygostomian flap, dorsolateral. **D**, sternal plastron, with excavated sternum and basal parts of Mxp1. **E**, same. **F**, pleura of abdominal somite 2-5. **G**, same. **H**, telson. **I**, same. **J**, right antenna, ventral. **K**, left Mxp3, crista dentata, ventral. Scale bars: 1 mm.

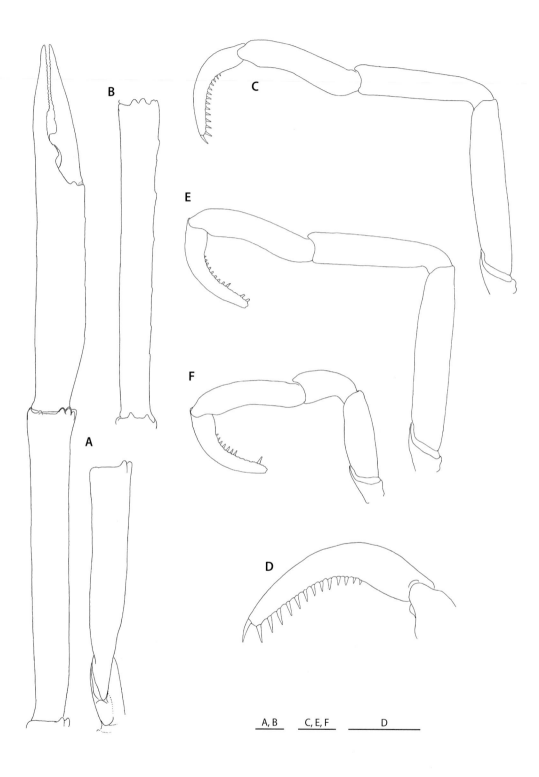

FIGURE 288

Heteroptychus anouchkae n. sp., holotype, ovigerous female 4.7 mm (MNHN-IU-2014-17097). **A**, left P1, dorsal. **B**, same, carpus, dorsomesial. **C**, left P2, setae omitted, lateral. **D**, same, distal part, lateral. **E**, left P3, lateral. **F**, left P4, lateral. Scale bars: 1 mm.

margin. Telson 0.3-0.4 × as long as broad; posterior plate moderately concave on posterior margin, length subequal to or slightly more than that of anterior plate.

Eyes: About 1.8 × as long as broad, subovate, medially inflated, distally and proximally narrowed; cornea 0.4 × length of remaining eyestalk.

Antennule and antenna: Ultimate article of antennular peduncle 3.5 × longer than high. Article 2 of antennal peduncle without lateral spine. Antennal scale narrow triangular, ending in or barely reaching midlength of article 4, distinctly articulated. Distal 2 articles unarmed; article 5 1.1-1.3 × longer than article 4, breadth 0.44 × height of ultimate antennular article. Flagellum of 9-12 segments directed laterally or posteriorly.

Mxp: Mxp1 with bases broadly separated. Mxp3 basis with 3-4 denticles on mesial ridge. Ischium relatively short, 0.4 × as long as merus, crista dentata with 1-3 obsolescent denticles on proximal third of length, flexor margin not rounded distally. Merus relatively thick mesio-laterally, with rounded ridge along flexor margin, unarmed. No spine on carpus.

P1: Smooth and nearly glabrous except for fingers, 7.5-8.7 × (males), 6.8-7.1 × (females) longer than carapace. Ischium with basally broad, distally blunt distodorsal process and lobe-like proximal process overhanging basis. Merus with 2 distodorsal spines mesially, length 1.8-1.9 × (males), 1.4-1.7 × (females) that of carapace. Carpus 1.5-1.6 × (males), 1.3-1.4 × (females) longer than merus, dorsally with 3 or 4 terminal spines, ventromesially with several obsolescent spines. Palm 3.5-3.7 × (males), 3.6-4.3 × (females) longer than broad, 0.6-0.7 × as broad as high, 0.7-0.8 × as long as carpus, subequally broad medially and distally or very slightly broader medially, proximally narrowed; mesial margin sharply ridged, and shallowly or obsolescently serrated in males, more obsolescently so in females. Fingers directed straight forward or somewhat inclined laterally, gaping in proximal half, straightly fitting to each other in distal half when closed, relatively slender, distally spooned; movable finger 0.5-0.6 × (usually 0.6 x) length of palm, opposable margin with prominent blunt (truncate) process at midlength of gaping portion, distal to position of opposite process on fixed finger.

P2-4: Setose along prehensile margins of propodi and dactyli. Meri subequal on P2 and P3, P4 merus 0.4-0.5 × length of P3 merus; length-breadth ratio, 4.5-4.8 on P2, 4.7-4.8 on P3, 2.6-2.9 on P4; P2 merus subequal to length of carapace, 1.3 × length of P2 propodus, P3 merus 1.3 × length of P3 propodus, P4 merus 0.7 × length of P4 propodus. Carpi subequal in length on P2 and P3, P4 carpus 0.4 × length of P3 carpus; carpus-dactylus length ratio, 1.4-1.5 on P2, 1.4 on P3, 0.6 on P4. Propodi subequal on P2 and P3, P4 propodus 0.9 × length of P3 propodus; propodus-dactylus length ratio, 1.5 on P2, 1.6 on P3, 1.3-1.4 on P4; flexor margin concavely curving in lateral view, unarmed. Dactyli subequal on P2-4, dactylus-carpus length ratio, 0.7 on P2 and P3, 2.0 on P4; flexor margin with row of 13 or 14 (rarely 15) sharp slender spines proximally diminishing and subperpendicular to margin, ultimate and penultimate spines subequal.

Eggs. Number of eggs carried, 30; size, 1.00 × 1.24 mm - 1.12 × 1.35 mm (MNHN-IU-2014-17097).

Color. Male 4.7 mm from Bathus 3 Stn CP831, Norfolk Ridge (MNHN-IU-2014-17101): Reddish all over surface, abdomen translucent, P2-4 dactyli pale. Ovigerous female 4.6 mm from HalipRo 1 Stn CP854 (MNHN-IU-2014-17098): Reddish; abdomen, P2-3 dactyli and entire P4 pale.

REMARKS — The species of *Heteroptychus* are classified into two groups by the presence or absence of articulation between the antennal scale and article 2: articulation present in *H. anouchkae* n. sp., *H. apophysis* n. sp., *H. colini* n. sp., *H. edwardi* (Kensley, 1981), *H. paulae* n. sp., and *H. scambus* (Benedict, 1902); articulation absent in *H. brevis* (Benedict, 1902), *H. claudeae* n. sp. and *H. lemaitrei* n. sp.

In the first group, *H. anouchkae*, *H. apophysis* and *H. paulae* share a long anterolateral spine of the carapace fully reaching or slightly falling short of the apex of the rostrum, whereas in the other species of the group, it is much shorter. *Heteroptychus anouchkae* differs from *H. apophysis* and *H. paulae* in having a much narrower, distally sharp pointed rostrum; the carapace lateral margin is more strongly convex around the posterior quarter rather than around posterior third; the anterolateral spine is directed somewhat dorsally instead of straight horizontal, the lateral orbital spine is consistently present; the anterior margin of sternite 3 bears a pair of submedian spines instead of a median notch.

Heteroptychus apophysis n. sp.

Figures 289-291

TYPE MATERIAL — Holotype: **Solomon Islands**. SALOMON 2 Stn CP2269, 7°45.1'S, 156°56.3'E, 768-890 m, 4.XI.2004, 1 ♂ 5.1 mm, (MNHN-IU-2014-17104). Paratype: **Solomon Islands**. SALOMON 2 Stn DW2238, 06°53'S, 156°21'E, 470-443 m, 30.X.2004, 1 ♂ 5.1 mm (MNHN-IU-2014-10139).

ETYMOLOGY — The specific name is a noun in apposition from the Greek *apophysis* (process, projection) for two projections on the branchial lateral margin, by which the species is distinguished from *U. paulae* n. sp. (see below).

DISTRIBUTION — Solomon Islands, 632-890 m.

DESCRIPTION — Medium-sized species. *Carapace*: 0.8 × as long as broad, greatest breadth 1.9 × distance between anterolateral spines. Dorsal surface polished, moderately convex from anterior to posterior, ridged along posterior two-fifths of lateral margin. Lateral margin strongly convex around posterior third, with rounded process on anterior end of branchial margin, followed by smaller one at anterior end of posterior branchial margin; anterolateral spine prominent, horizontally directed straight forward reaching apex of rostrum. Rostrum broad triangular, with interior angle of 36-40°, straight horizontal, reaching tip of eye; length 0.7 × breadth, 0.25-0.28 × that of remaining carapace; dorsal surface flattish. Lateral limit of orbit without spine. Pterygostomian flap anteriorly produced to sharp spine; anterior surface well inflated, convex from dorsal to ventral, with 3 or 4 blunt processes of good size; posterior height 0.2 × anterior height.

Sternum: Excavated sternum with bluntly subtriangular anterior margin, surface with sharp ridge in midline on anterior half. Sternal plastron 0.6 × as long as broad, lateral extremities convexly divergent posteriorly, sternite 6 broadest. Sternite 3 anterior margin excavated in broad V-shape with median notch lacking flanking spine. Sternite 4 with convex anterolateral margin.

Abdomen: Smooth and polished. Somite 1 in profile gently convex from anterior to posterior. Somite 2 2.8-3.0 × broader than long; pleural lateral margin slightly concave, strongly divergent posteriorly, ending in blunt tip. Pleura of somites 3-4 ending in rounded margin. Telson 0.4 × as long as broad; posterior plate feebly concave on posterior margin, length 1.3 × that of anterior plate.

Eyes: 1.6 × as long as broad, subovate. Cornea 0.6 × length of remaining eyestalk.

Antennule and antenna: Ultimate article of antennular peduncle 3.2-3.8 × longer than high. Antennal peduncle relatively short and slender. Article 2 without lateral spine. Antennal scale ending in proximal third or midlength of article 4, distinctly articulated. Distal 2 articles unarmed; article 5 1.2-1.4 × longer than article 4, breadth 0.4-0.5 × height of ultimate antennular article. Flagellum of 11 segments directed posterolaterally.

Mxp: Mxp1 with bases broadly separated. Mxp3 basis with 2 obsolescent denticles or unarmed on mesial ridge. Ischium relatively short, 0.4 × as long as merus, crista dentata with 2-4 denticles on proximal half, flexor margin not rounded distally. Merus relatively thick mesio-laterally, with rounded ridge along flexor margin, unarmed. No spine on carpus.

P1: Smooth and nearly glabrous except for fingers, 7.2-7.8 × longer than carapace. Ischium with basally broad, distally blunt distodorsal process and lobe-like proximal process overhanging basis. Merus with 3 spines on distal margin (1 small median, 2 large mesial) and 1 or 2 small mesial marginal protuberances proximal to distal end, length 1.7-1.9 × that of carapace. Carpus 1.5-1.7 × longer than merus, with 4 terminal spines and 2 rows of short blunt spines along mesial margin. Palm 4.0-4.8 × longer than broad, 0.5-0.6 × broader than high, 0.7-0.8 × as long as carpus, mesial and lateral margins subparallel, mesial margin sharply ridged, with 9 or 10 short spines. Fingers half as long as palm, gaping in proximal two-thirds, straightly fitting to each other in distal half when closed, relatively slender, distally spooned; opposable margin of movable finger with prominent truncate process at midlength of gaping portion or distal to position of somewhat shorter truncate process on fixed finger.

P2-4: Meri subequal on P2 and P3, P4 merus 0.5 × length of P3 merus; length-breadth ratio, 6.3-6.7 on P2, 6.2-6.5 on P3, 3.6-3.7 on P4; P2 merus 0.9-1.0 × length of carapace,1.4 × length of P2 propodus, P3 merus 1.3-1.4 × length of P3

A, B   C, E   D   F, G

FIGURE 289

Heteroptychus apophysis n. sp., holotype, male 5.1 mm (MNHN-IU-2014-17104). **A**, carapace and anterior part of abdomen, dorsal. **B**, same, lateral. **C**, sternal plastron, with excavated sternum and basal parts of Mxp1. **D**, right pleura of abdominal somites 2-5, dorsolateral. **E**, telson. **F**, left antenna, ventral. **G**, left Mxp3, setae omitted, ventral. Scale bars: 1 mm.

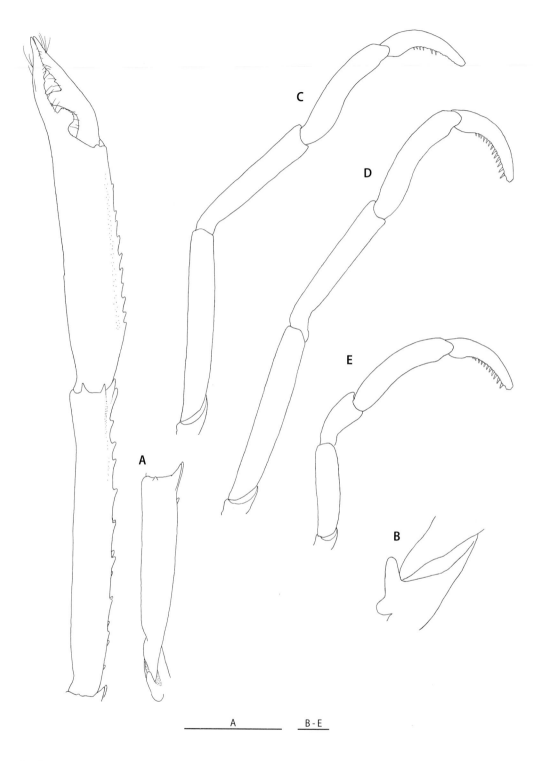

FIGURE 290

Heteroptychus apophysis n. sp., holotype, male 5.1 mm (MNHN-IU-2014-17104). **A**, left P1, dorsal. **B**, same, proximal part, lateral. **C**, right P2, setae omitted, lateral. **D**, right P3, lateral. **E**, right P4, lateral. Scale bars: A, 5 mm; B-E, 1 mm.

propodus, P4 merus 0.7-0.8 × length of P4 propodus. Carpi subequal in length on P2 and P3, P4 carpus one-third length of P3 carpus; slightly (1.1 x) longer than propodus on P2 and P3, much shorter (0.4 x) on P4; carpus-dactylus length ratio, 1.5-1.7 on P2 and P3, 0.6-0.7 on P4. Propodi subequal on P2 and P3, P4 propodus 0.9 × length of P3 propodus; propodus-dactylus length ratio, 1.6 on P2, 1.4-1.7 on P3 and P4; flexor margin concavely curving in lateral view, unarmed. Dactyli subequal on P2 and P3, slightly shorter on P4 than on P2 and P3; dactylus-carpus length ratio, 0.6 on P2 and P3, 1.5-1.7 on P4; flexor margin with row of 12-14 slender, proximally diminishing spines, ultimate and penultimate spines subequal.

REMARKS — This species strongly resembles *U. paulae* n. sp. in nearly all aspects. Their relationships are discussed under that species (see below).

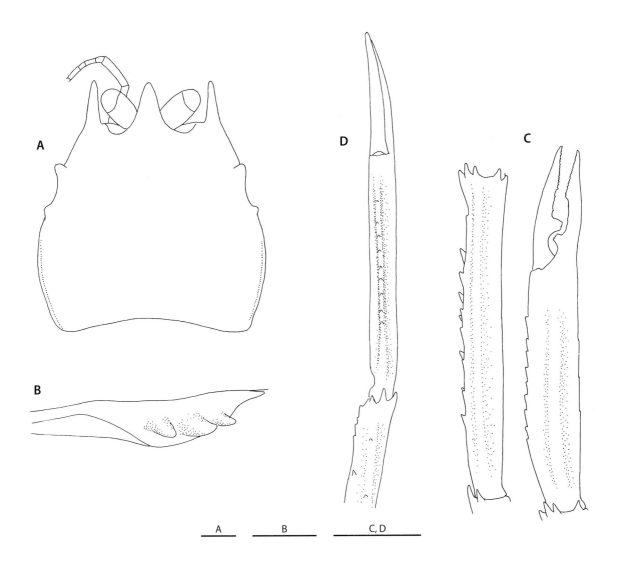

FIGURE 291

Heteroptychus apophysis n. sp., paratype, male 5.1 mm (MNHN-IU-2014-10139). **A**, carapace and anterior part of abdomen, dorsal. **B**, right pterygostomian flap, lateral. **C**, right P1, proximal articles omitted, dorsal. **D**, same, mesial. Scale bars: A, B, 1 mm; C, D, 5 mm.

Heteroptychus claudeae n. sp.

Figures 292-294

Uroptychus scambus — Baba 1981: 120; 1988: 43. — Baba *et al.* 2009: 59 (part), figs 49-51.

TYPE MATERIAL — Holotype: **Wallis and Futuna Islands**. MUSORSTOM 7 Stn CP564, 11°46'S, 178°27'W, 1015-1020 m, with Chrysogorgiidae gen. sp. (Calcaxonia), 20.V.1992, ♂ 5.0 mm (MNHN-IU-2013-8571). Paratypes: **New Caledonia**, Chesterfield Islands. MUSORSTOM 5 Stn CP387, 20°53.41'S, 160°52.14'E, 650-660 m, with Chrysogorgiidae gen. sp. (Calcaxonia), 22.X.1986, 1 ov. ♀ 4.5 mm (MNHN-IU-2013-8573). **Solomon Islands**. SALOMON 2 Stn CP2189, 8°19.6'S, 160°01.9'E, 660-854 m, 23.X.2004, 1 ov. ♀ 3.9 mm (MNHN-IU-2014-17105); 2 ♂ 3.8, 4.0 mm, 2 ov. ♀ 3.9, 4.1 mm, 1 ♀ 3.7 mm (MNHN-IU-2014-17106). – Stn DW2190, 08°24'S, 159°27'E, 140-263 m, 24.X.2004, 1 ♀ 4.0 mm (MNHN-IU-2014-17107). – Stn CP2250, 7°29.2'S, 156°16.7'E, 845-970 m, 2.XI.2004, 1 ov. ♀ 4.5 mm (MNHN-IU-2014-17108). – Stn CP2253, 7°26.5'S, 156°15.0'E, 1200-1218 m, 2.XI.2004, 1 ov. ♀ 4.9 mm (MNHN-IU-2014-17109). **Wallis and Futuna Islands**. MUSORSTOM 7 Stn CP564, collected with holotype, 7 ♂ 4.2-5.5 mm, 8 ov. ♀ 4.2-5.6 mm (MNHN-IU-2014-17110). – Stn CP567, 11°47'S, 178°27'W, 1010-1020 m, 20.V.1992, 4 ♂ 4.6-4.9 mm, 6 ov. ♀ 4.2-5.5 mm (MNHN-IU-2013-12286, MNHN-IU-2013-12284, MNHN-IU-2013-11272), 1 ♂ 5.8 mm, (MNHN-IU-2013-12285). – Stn DW589, 12°16'S, 174°41'W, 400 m, 23.V.1992, 1 ov. ♀ 4.1 mm (MNHN-IU-2014-17111). – Stn CP551, 12°15'S, 177°28'W, 791-795 m, 18.V.1992, 2 ♂ 4.2, 4.3 mm, 1 ov. ♀ 4.3 mm (MNHN-IU-2013-8572). – Stn CP550, 12°15'S, 177°28'W, 800-810 m, 18.V.1992, 1 ♂ 4.1 mm, 1 ov. ♀ 4.5 mm (MNHN-IU-2014-17112). **New Caledonia**, Loyalty Ridge. BIOGEOCAL Stn CP297, 20°37'S, 167°11'E, 1230-1240 m, 28.IV.1987, 1 ♂ 3.8 mm (MNHN-IU-2014-17113). MUSORSTOM 6 Stn CP438, 20°23'S, 166°20'E, 800 m, 18.II.1989, 1 ♂ 4.4 mm, 1 ov. ♀ 3.6 mm (MNHN-IU-2014-17114). **New Caledonia**. HALIPRO 2 Stn BT96, 23°59'S, 161°55'E, 1034-1056 m, 25.XI.1996, 1 ♂ 4.2 mm (MNHN-IU-2014-17115). – Stn BT101, 24°19'S, 161°43'E, 970-1063 m, 26.XI.1996, 1 ♂ 4.8 mm (MNHN-IU-2014-17116). **New Caledonia**, Norfolk Ridge. BIO-CAL Stn CP31, 23°08'S, 166°51'E, 850 m, 29.VIII.1985, 1 ♂ 3.5 mm (MNHN-IU-2014-17117). – Stn CP32, 23°07'S, 166°51'E, 825 m, 29.VIII.1985, 1 ov. ♀ 3.7 mm (MNHN-IU-2014-17118). – Stn CP54, 23°10'S, 167°43'E, 950-1000 m, 1.IX.1985, 1 ♂ 4.5 mm, 2 ov. ♀ 4.7, 4.8 mm, 1 ♀ 3.8 mm (MNHN-IU-2014-17119). – Stn CP61, 24°11'S, 167°32'E, 1070 m, 2.IX.1985, 1 ♂ 3.4 mm, 3 ov. ♀ 3.7-4.2 mm, 1 sp., sex indet., 2.4 mm (MNHN-IU-2013-12287, MNHN-IU-2013-12288, MNHN-IU-2013-11271). BATHUS 3 Stn CP823, 23°23'S, 167°52'E, 980-1000 m, 29.XI.1993, 1 ♂ 5.0 mm (MNHN-IU-2014-17120). **Vanuatu**. MUSORSTOM 8 Stn CP992, 18°52.34'S, 168°55.16'E, 775-748 m, 24.IX.1994, 1 ov. ♀ 4.7 mm (MNHN-IU-2014-17121). – Stn CP1125, 15°57.63'S, 166°38.43'E, 1160-1220 m, 10.X.1994, 1 ov. ♀ 4.4 mm (MNHN-IU-2014-17122). – Stn CP1036, 18°01.00'S, 168°48.20'E, 920-950 m, 29.IX.1994, 1 ♂ 4.6 mm (MNHN-IU-2014-17123). **Tonga**. BORDAU 2 Stn CP1613, 23°03'S, 175°47'W, 331-352 m, 17.VI.2000, 1 ♂ 4.1 mm (MNHN-IU-2014-17124). – Stn CP1625, 23°28'S, 176°22'W, 824 m, 19.VI.2000, 1 ♀ 4.1 mm (MNHN-IU-2014-17125).

ETYMOLOGY — Named for Claude Crosnier for her hospitality during my stays in Paris.

DISTRIBUTION — Japan, Taiwan, Indonesia, in 1175-1184 m, and now Wallis and Futuna Islands, Chesterfield Islands, Solomon Islands, New Caledonia, Vanuatu, Loyalty Ridge, Norfolk Ridge, and Tonga, in 331-1240 m.

SIZE — Males, 3.4-5.5 mm; females, 3.7-5.6 mm; ovigerous females from 3.7 mm.

DESCRIPTION — *Carapace*: Broader than long (0.7 × as long as broad), greatest breadth 2.1-2.3 × distance between anterolateral spines. Dorsal surface polished, moderately convex in profile on anterior half, feebly so on posterior half, ridged along posterior half of lateral margin. Lateral margins strongly convex around posterior third; anterolateral spine prominent, horizontally inclined anteromesially, occasionally directed straight forward, not reaching apex of rostrum. Rostrum distally narrow triangular, laterally concave, with interior angle of 20-30°, nearly horizontal or directed somewhat

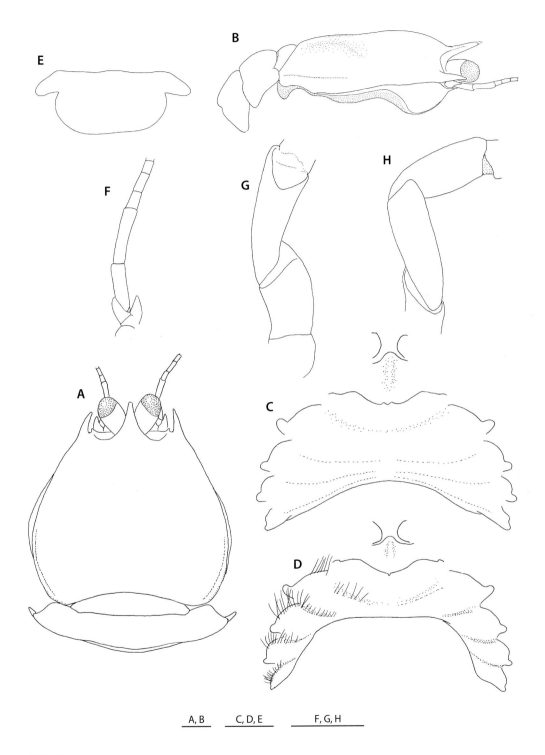

FIGURE 292

Heteroptychus claudeae n. sp., **A-C**, **E-H**, holotype, male 5.0 mm (MNHN-IU-2013-8571); **D**, paratype, ovigerous female 4.5 mm (MNHN-IU-2014-17110). **A**, carapace and anterior part of abdomen, dorsal. **B**, same, lateral. **C**, sternal plastron, with excavated sternum and basal parts of Mxp1. **D**, same. **E**, telson. **F**, left antenna, ventral. **G**, right Mxp3, ventral. **H**, same, lateral. Scale bars: 1 mm.

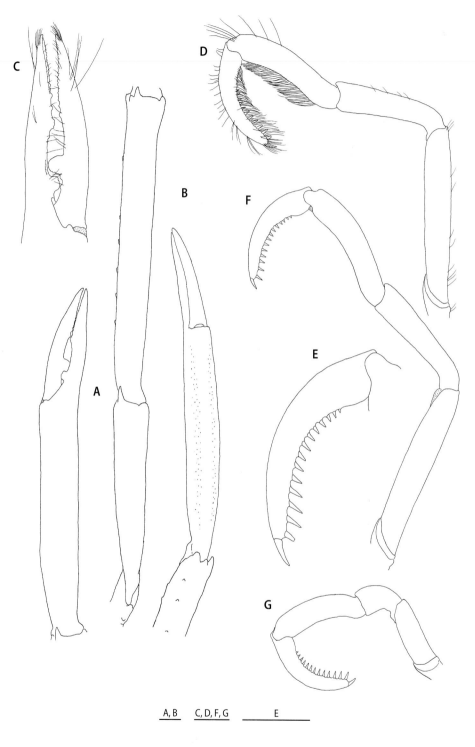

A, B C, D, F, G E

FIGURE 293

Heteroptychus claudeae n. sp., holotype, male 5.0 mm (MNHN-IU-2013-8571). **A**, right P1, dorsal. **B**, same, proximal part omitted, mesial. **C**, same, fingers, ventral. **D**, left P2, lateral. **E**, same, distal part, setae omitted, lateral. **F**, left P3, lateral. **G**, left P4, lateral. Scale bars: 1 mm.

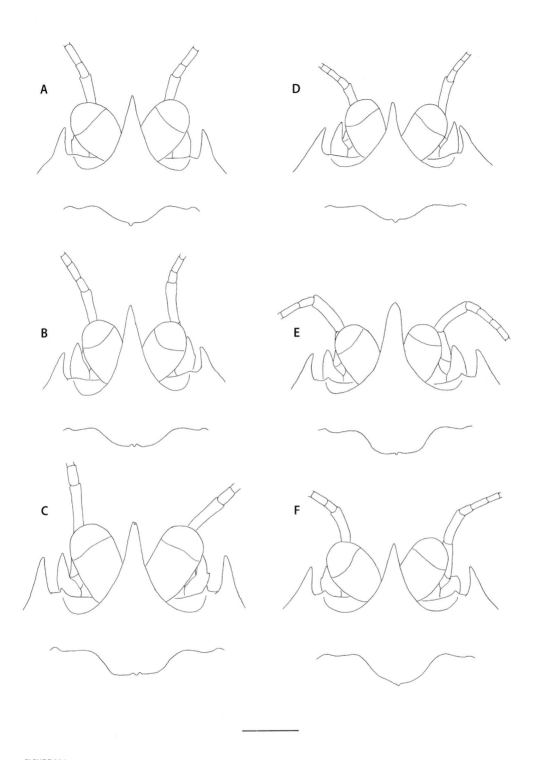

FIGURE 294

Heteroptychus claudeae n. sp., paratypes, showing anterior part of carapace and anterior part of sternal plastron (MNHN-IU-2014-17110). **A**, male 4.4 mm. **B**, male 4.6 mm. **C**, male 5.7 mm. **D**, ovigerous female 5.1 mm. **E**, ovigerous female 5.2 mm. **F**, ovigerous female 5.6 mm. Scale bar: 1 mm.

or moderately dorsally, reaching or slightly overreaching tip of eye; length 0.9 × breadth, 0.3 × length of remaining carapace; dorsal surface flattish but feebly concave basally. Lateral limit of orbit unarmed, acuminate or with small spine. Pterygostomian flap anteriorly produced to sharp strong spine; anterior surface well inflated, smooth or with feeble or distinct ridge in lateral midline on anterior portion; posterior height 0.1 × anterior height.

Sternum: Excavated sternum with roundly produced anterior margin, surface bearing sharp ridge in midline on anterior half. Sternal plastron in males 0.5-0.6 × as long as broad; in females, left and right parts of sternites 5-7 discontinuous, interrupted by loss of median parts. Sternite 3 having anterior margin broadly and deeply excavated, with small (rarely obsolescent) median notch, with or without small submedian processes. Sternite 4 with rounded anterolateral margin. Sternites 5 and 7 subequal in breadth, sternite 6 broadest.

Abdomen: Smooth and polished. Dorsal surface of somite 1 gently convex from anterior to posterior. Somite 2 tergite 2.4-2.8 × (males), 2.1-2.6 × (females) broader than long; pleural lateral margin slightly concave, strongly divergent posteriorly, ending in blunt tip. Pleura of somites 3-4 somewhat more narrowed in females than in males, ending in rounded margin. Telson 0.4 × as long as broad; posterior plate feebly concave on posterior margin, length 1.4-1.8 × that of anterior plate.

Eyes: 1.3-1.4 × as long as broad, subovate. Cornea 0.5-0.7 × length of remaining eyestalk.

Antennule and antenna: Ultimate article of antennular peduncle 2.7-2.8 × longer than high. Article 2 of antennal peduncle fused with antennal scale. Antennal scale narrow triangular, reaching or barely reaching midlength of article 4, laterally with or without small spine. Distal 2 articles unarmed; article 5 1.2-1.3 × as long as article 4, breadth less than half height of ultimate antennular article. Flagellum of 9-11 segments.

Mxp: Mxp1 with bases broadly separated. Mxp3 basis unarmed on mesial ridge. Ischium relatively short, 0.4 × as long as merus, crista dentata without distinct denticles, flexor margin not rounded distally. Merus relatively thick mesio-laterally, with rounded ridge along flexor margin, unarmed. No spine on carpus.

P1: Smooth and nearly glabrous except for fingers, length 6.3-8.7 × (males), 6.3-6.9 × (females) that of carapace. Ischium with basally broad, blunt distodorsal process (elongate in large specimens) and lobe-like proximal process overhanging basis, unarmed elsewhere. Merus mesially with 1 well-developed distodorsal and 1 small or obsolescent distoventral spine, length 1.6-1.9 × (males), 1.6-1.8 × (females) that of carapace. Carpus 1.4-1.5 × longer than merus, with 2 terminal spines dorsomesially and several obsolescent spines ventromesially. Palm 5.1-5.4 × (males), 3.8-5.9 × (females) longer than broad, 0.8-0.9 × length of carpus, subequally broad medially and distally, proximally narrowed; mesial margin variably ridged from roundly to sharply, more sharply in large specimens, usually with no spine or serration (with obsolescent blunt spines in large specimens). Fingers directed somewhat anterolaterally, gaping in proximal half, straightly fitting to each other in distal half when closed, relatively slender, distally spooned; movable finger 0.5-0.6 × length of palm, opposable margin with prominent blunt process at midlength of gaping portion, distal to position of opposite truncate process on fixed finger.

P2-4: Setose along prehensile margins of propodi and dactyli. Meri subequal on P2 and P3, P4 merus 0.3-0.4 × length of P3 merus; length-breadth ratio, 5.1-5.8 on P2, 5.1-5.5 on P3, 2.5-2.8 on P4; P2 merus 0.9 × (1.1 × in largest male MNHN-IU-2013-12285) length of carapace, 1.1-1.2 × length of P2 propodus, P3 merus 1.1-1.3 × length of P3 propodus, P4 merus 0.6-0.7 × length of P4 propodus. Carpi subequal in length on P2 and P3, P4 carpus 0.3 × length of P3 carpus; carpus-dactylus length ratio, 1.1-1.2 on P2 and P3, 0.4 on P4. Propodi subequal on P2 and P3, P4 propodus 0.8 × length of P3 propodus; propodus-dactylus length ratio, 1.4 on P2, 1.3-1.4 on P3, 1.2 on P4; flexor margin concavely curving in lateral view, unarmed. Dactyli subequal on P2 and P3, slightly shorter on P4; dactylus-carpus length ratio, 0.9 on P2 and P3, 2.2-2.6 on P4; flexor margin with row of 13-16 sharp slender spines proximally diminishing and perpendicular to margin, ultimate somewhat larger than penultimate.

Eggs. Number of eggs carried, 20; size, 1.33 × 1.04 mm - 1.33 × 1.08 mm (MNHN-IU-2013-8573); 25 eggs; size, 1.23 × 1.49 mm - 1.41 × 1.67 mm (MNHN-IU-2013-12285).

Color. Two specimens in different color of *Uroptychus scambus* from Taiwan (now referred to *H. claudeae*; see below) were illustrated by Baba *et al.* (2009: Figures 49, 50).

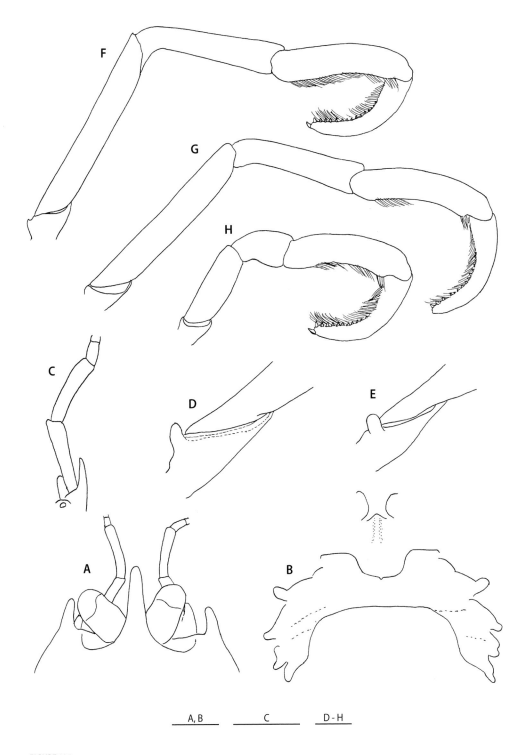

A, B   C   D - H

FIGURE 295

Heteroptychus brevis (Benedict, 1902), holotype of *Uroptychus brevis*, ovigerous female 5.2 mm, USNM 20566. **A**, anterior part of carapace, dorsal. **B**, sternal plastron, with excavated sternum and basal parts of Mxp1. **C**, left antenna, ventral. **D**, right P1, proximal part, lateral. **E**, same, dorsal. **F**, right P2, lateral. **G**, right P3, lateral. **H**, right P4, lateral. Scales: 1 mm.

REMARKS — The antennal scale is fused with article 2 in *H. claudeae*, *H. lemaitrei* n. sp. and the western Atlantic *H. brevis* (Benedict, 1902), whereas in all the other species it is articulated. *Heteroptychus brevis* has both the rostrum and the anterolateral spines distally blunt and the P1 ischium bearing an obsolescent posterior dorsal lobe (see Figure 295), the features differentiating it from *H. claudeae* and *H. lemaitrei*. *Heteroptychus claudeae* differs from *H. lemaitrei* in having the carapace lateral margin smooth instead of bearing a small but distinct protuberance at the anterior end of branchial margin (see below under the remarks of *H. lemaitrei*).

Unlike all the other species, the presence or absence of lateral orbital spine and the shape of the anterior margin of sternite 3 are unstable in *H. claudeae* (Figure 294).

The specimens reported under *U. scambus* in my earlier paper (Baba, 1981) from Japan are referable to *H. claudeae*, as also are the specimens from Taiwan (Baba *et al.* 2009) except for the male from TAIWAN Stn CD141 (= *H. scambus*; see below) and the specimens from Station PCP342 (not examined).

One of the specimens, the largest of the type series (1 ♂ 5.8 mm, MNHN-IU-2013-12285) collected together with the holotype from Wallis and Futuna, resembles *H. lemaitrei* in having a spiniform rostrum and in having the P1 ischium with a spiniform distodorsal spine. Unfortunately, the molecular analyses of this lot were unsuccessful (L. Corbari, personal comm.). Also, one of the Japanese specimens collected at Soyo-Maru Stn B4, 1700 m (Baba, 1981) has a spiniform rostrum but the distodorsal process of the P1 ischium is distally obtuse, not spiniform. These two specimens are provisionally placed in *U. claudeae*.

The variability noted above, as well as the color difference in Taiwan specimens (Baba *et al.* 2009: figs 49, 50) suggest that the species may likely contain cryptic species, for which further study is required.

Heteroptychus colini n. sp.
Figures 296, 297

TYPE MATERIAL — Holotype: **Wallis and Futuna Islands**. MUSORSTOM 7 Stn CP552, 12°16'S, 177°28'W, 786-800 m, 18.V.1992, ♂ 4.3 mm (MNHN-IU-2014-17126). Paratypes: **New Caledonia**, Norfolk Ridge. BIOCAL Stn CP75, 22°20'S, 167°23'E, 825-860 m, 04-05.IX.1985, 1 ♂ 3.6 mm (MNHN-IU-2014-17127). **Fiji Islands**. BORDAU 1 Stn DW1417, 16°27'S, 178°55'W, 353 m, 27.II.1999, 1 ♂ 4.5 mm (MNHN-IU-2014-17128).

ETYMOLOGY — Named for Colin L. McLay of the University of Canterbury for his friendship and help in reading many of my manuscripts.

DISTRIBUTION — Fiji Islands, Wallis and Futuna Islands and Norfolk Ridge; 353-860 m.

SIZE — Males, 3.6-4.5 mm.

DESCRIPTION — Small species. Carapace: 0.7 × as long as broad, greatest breadth measured at posterior third, 2.2 × distance between anterolateral spines. Dorsal surface polished, somewhat convex from anterior to posterior or slightly depressed between gastric and cardiac regions, ridged along posterior half of lateral margin. Lateral margins convexly divergent posteriorly; anterolateral spine prominent, horizontally directed straight forward or slightly inclined mesially, ending in distal quarter of rostral length. Rostrum distally narrow triangular, with interior angle of 27-29°, straight, directed somewhat dorsally, reaching tip of eye; length 0.7 × breadth, 0.3 × length of remaining carapace; dorsal surface flattish. Lateral limit of orbit angular and acuminate. Pterygostomian flap anteriorly produced to sharp strong spine; anterior surface smooth and well inflated; posterior height 0.1 × anterior height.

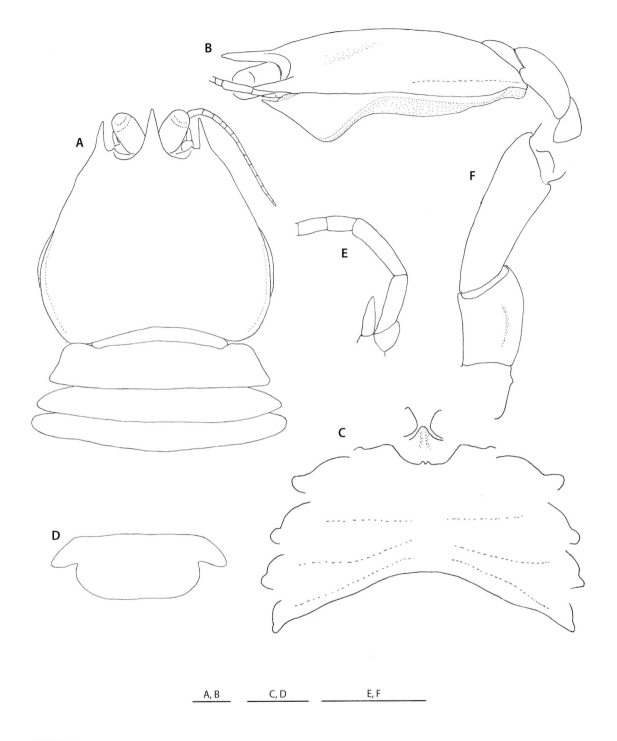

A, B   C, D   E, F

FIGURE 296

Heteroptychus colini n. sp., holotype, male 4.3 mm (MNHN-IU-2014-17126). **A**, carapace and anterior part of abdomen, dorsal. **B**, same, lateral. **C**, sternal plastron, with excavated sternum and basal parts of Mxp1. **D**, telson. **E**, right antenna, ventral. **F**, right Mxp3, ventral. Scale bars: 1 mm.

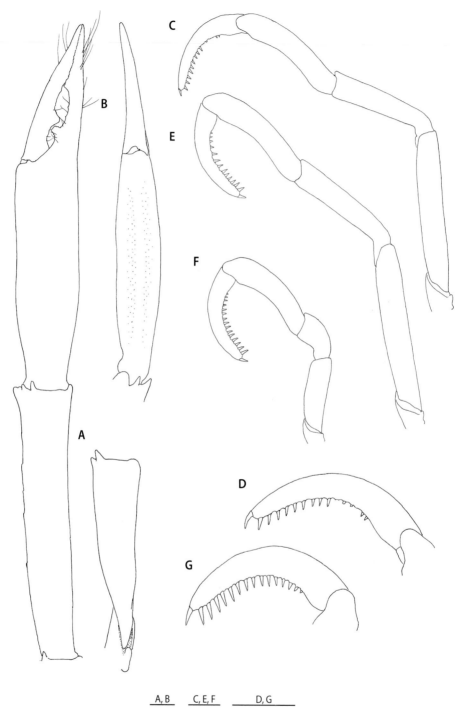

FIGURE 297

Heteroptychus colini n. sp., holotype, male 4.3 mm (MNHN-IU-2014-17126). **A**, right P1, dorsal. **B**, same, distal part, mesial. **C**, left P2, setae omitted, lateral. **D**, same, distal part, lateral. **E**, left P3, lateral. **F**, left P4, lateral. **G**, same, distal part, lateral. Scale bars: 1 mm.

Sternum: Excavated sternum with subtriangular anterior margin, surface with sharp ridge in midline on anterior half. Sternal plastron half as long as broad; lateral extremities of sternites 4-7 convexly divergent posteriorly; sternite 6 broadest. Sternite 3 having anterior margin broadly and deeply excavated, with pair of blunt submedian processes.

Abdomen: Smooth and polished. Somite 1 gently convex from anterior to posterior. Somite 2 tergite 2.7 × broader than long; pleural lateral margin slightly concave, strongly divergent posteriorly, ending in blunt tip. Pleura of somites 3-4 ending in rounded margin. Telson 0.4 × as long as broad; posterior plate feebly concave on posterior margin, length 1.2-1.3 × that of anterior plate.

Eyes: 1.8-1.9 × as long as broad, subovate. Cornea 0.5 × length of remaining eyestalk.

Antennule and antenna: Ultimate article of antennular peduncle 2.9-3.0 × longer than high. Antennal article 2 without lateral spine (with spine on left side in male MNHN-IU-2014-17128). Antennal scale articulated, relatively narrow, terminating in midlength of article 4. Distal 2 articles unarmed; article 5 1.2 × longer than article 4, breadth 0.4-0.5 × height of ultimate antennular article. Flagellum of 12 segments.

Mxp: Mxp1 with bases broadly separated. Mxp3 basis with 1 denticle near distal end of mesial ridge. Ischium 0.4 × as long as merus, crista dentata without denticles, flexor margin not rounded distally. Merus relatively thick mesio-laterally, with rounded ridge along flexor margin, unarmed. No spine on carpus.

P1: Smooth and nearly glabrous except for fingers, length 5.9-7.1 that of carapace. Ischium with lobe-like distodorsal process and short blunt proximal process overhanging basis. Merus with 2 distomesial spines (distodorsal strong, distoventral small), length 1.5-1.6 × that of carapace. Carpus 1.4 × longer than merus, dorsally with 3 or 4 terminal spines, mesially with row of 7-9 short blunt spines subparalleling ventromesial row of smaller spines, ventral distomesial and distolateral spines small. Palm 2.7-3.2 × longer than broad, 0.7 × as broad as high, 0.8-0.9 × length of carpus; mesial margin not cristiform but weakly ridged. Fingers inclined laterally, gaping in proximal two-thirds, straightly fitting to each other in distal third when closed; relatively slender, distally spooned; movable finger 0.6-0.7 × length of palm, opposable margin with prominent distally blunt process at midlength of gaping portion, distal to position of opposite truncate process on fixed finger.

P2-4: Sparsely setose on meri and carpi, thickly so along prehensile margins of propodi and dactyli. Meri slightly longer on P3 than on P2, P4 merus 0.4 × length of P3 merus; length-breadth ratio, 4.8-5.7 on P2, 4.7-5.7 on P3, 2.3-2.9 on P4; P2 merus 0.9-1.0 × length of carapace, 1.2 × length of P2 propodus, P3 merus 1.2 × length of P3 propodus, P4 merus 0.6-0.7 × length of P4 propodus. Carpi subequal in length on P2 and P3, P4 carpus 0.3-0.4 × length of P3 carpus; carpus-dactylus length ratio, 1.1-1.2 on P2, 1.1-1.3 on P3, 0.4-0.5 on P4. Propodi subequal on P2 and P3, P4 propodus 0.8 × length of P3 propodus; propodus-dactylus length ratio, 1.3-1.4 on P2 and P3, 1.2 on P4; flexor margin concavely curving in lateral view, unarmed. Dactyli subequal on P2-4 or slightly shorter on P4, dactylus-carpus length ratio, 0.8-0.9 on P2 and P3, 2.1-2.3 on P4; flexor margin with row of sharp spines (16 on P2, 14-18 on P3, 15-16 on P4) proximally diminishing and perpendicular to margin, ultimate spine subequal to or slightly longer than penultimate spine.

REMARKS — The male from the Fiji Islands (MNHN-IU-2014-17128) is different from the others in the shape of sternite 3: the excavated anterior margin is medially more produced anteriorly bearing pair of median spines.

Among the species possessing the articulated antennal scale, *H. colini* n. sp., *H. anouchkae* n. sp. and *H. edwardi* (Kensley, 1981) are grouped together by having a distally narrowed rostrum, whereas *H. apophysis* n. sp., *H. paulae* n. sp. and *H. scambus* (Benedict, 1902) have a broad rostrum. *Heteroptychus colini* differs from *H. anouchkae* in having the rostrum distinctly overreaching instead of falling short of the eye, in having the anterolateral spine of the carapace horizontal rather than dorsally deflected, barely reaching instead of overreaching the apex of the rostrum, and in having an angular or acuminate lateral orbital angle instead of a small but distinct spine. The original description of *H. edwardi* (Kensley, 1981) from the western Indian Ocean off South Africa is rather brief, but the following characters may discriminate it from *H. colini*. The rostrum is distally more slender (Kensley used "spiciform"); the anterior margin of sternite 3 bears a median notch, lacking a pair of median processes; the lateral limit of orbit is rounded (Kensley described "anterior margin sinuous between rostrum and anterolateral spine") instead of acuminate; and according to the description, the pterygostomian flap anteriorly ends in a short spine instead of being strongly produced.

Heteroptychus lemaitrei n. sp.

Figures 298, 299

Uroptychus scambus — Baba 2005: 58 (not *H. scambus* (Benedict, 1902)).

TYPE MATERIAL — Holotype: **New Caledonia**, Norfolk Ridge. BIOCAL Stn CP27, 23°07'S, 166°26'E, 1850-1900 m, 28-29.VIII.1985, ♂ 5.4 mm (MNHN-IU-2013-12289). Paratypes: **Wallis and Futuna Islands**. MUSORSTOM 7 Stn CP623, 12°34'S, 178°15'W, 1280-1300 m, 28.V.1992, 1 ♂ 6.2 mm (MNHN-IU-2014-17129). **Solomon Islands**. SALOMON 1 Stn CP1761, 8°46.5'S, 160°01.6'E, 191-290 m, 27.IX.2001, 1 ♂ 5.3 mm (MNHN-IU-2014-17130).

ETYMOLOGY — Named for Rafael Lemaitre of the Smithsonian Institution, for his help and friendship.

DISTRIBUTION — Solomon Islands, Wallis and Futuna Islands, Norfolk Ridge and Makassar Strait; 1280-2084 m.

DESCRIPTION — Medium-sized species. *Carapace*: 0.7-0.8 × as long as broad, greatest breadth 1.9-2.1 × distance between anterolateral spines. Dorsal surface polished, moderately convex from anterior to posterior, ridged along posterior half of lateral margin. Lateral margins convexly divergent posteriorly but feebly concave at anterior third, with small protuberance at anterior end of branchial margin; greatest breadth of carapace measured around posterior third of length; anterolateral spine prominent, horizontally directed more or less anteromesially, ending in or somewhat overreaching midlength of rostrum. Rostrum distally narrow triangular, with interior angle of 15-17°, straight, directed moderately dorsally, overreaching eyes by one-quarter length of rostrum; length subequal to breadth, 0.35 length of remaining carapace; dorsal surface flattish. Lateral limit of orbit acuminate or with small spine. Pterygostomian flap anteriorly produced to sharp strong spine; anterior surface well inflated, with no longitudinal ridge in midline; posterior height 0.3 × anterior height.

Sternum: Excavated sternum with triangular anterior margin, surface with sharp ridge in midline on anterior half. Sternal plastron 0.63-0.65 as long as broad; lateral extremities of sternites 4-7 slightly convex, sternite 6 broadest. Sternite 3 having anterior margin broadly and deeply excavated, medially slightly or somewhat produced with pair of short blunt processes.

Abdomen: Smooth and polished. Somite 1 gently convex from anterior to posterior. Somite 2 tergite 2.5-2.7 × broader than long; pleural lateral margin slightly concave, strongly divergent posteriorly, ending in blunt tip. Pleura of somites 3-4 ending in rounded margin. Telson 0.4 × as long as broad; posterior plate feebly concave on posterior margin, length 1.3-1.4 × that of anterior plate.

Eyes: About 1.6 × as long as broad, subovate. Cornea 0.6 × length of remaining eyestalk.

Antennule and antenna: Ultimate article of antennular peduncle 2.6-3.3 × longer than high. Antennal scale fused with article 2, as broad as or slightly broader than article 5, ending in or barely reaching midlength of article 4. Distal 2 articles unarmed; article 5 1.2-1.4 × as long as article 4, breadth 0.4 × height of ultimate antennular article. Flagellum of 11-13 segments.

Mxp: Mxp1 with bases broadly separated. Mxp3 basis with 1 denticle near distal end of mesial ridge. Ischium relatively short, 0.4 × as long as merus, crista dentata with 1 or 2 obsolescent denticle proximally, flexor margin not rounded distally. Merus relatively thick mesio-laterally, with rounded ridge along flexor margin, unarmed. No spine on carpus.

P1: Smooth and nearly glabrous except for fingers, 7.6-8.1 × longer than carapace. Ischium with well-developed distodorsal spine and short, distally blunt proximal process overhanging basis. Merus with 2 distal spines mesially (distodorsal strong, distoventral small), length 1.7-1.9 × that of carapace. Carpus 1.4-1.5 × longer than merus, dorsally with 3 or 4 terminal spines, ventromesially with 6-8 spines. Palm 3.4-3.6 × longer than broad, 0.5-0.6 × as broad as high, 0.8 × length of carpus; mesial margin sharply ridged and cristiform with obsolescent serration. Fingers directed anterolaterally, gaping in proximal half, straightly fitting to each other in distal half when closed, relatively slender, distally spooned; movable finger 0.5-0.6 × length of palm, opposable margin with prominent blunt process at midlength of gaping portion, distal to position of opposite truncate process on fixed finger.

FIGURE 298

Heteroptychus lemaitrei n. sp., holotype, male 5.4 mm (MNHN-IU-2013-12289). **A**, carapace and anterior part of abdomen, dorsal. **B**, same, lateral. **C**, pterygostomian flap, anterior part, dorsolateral. **D**, sternal plastron, with excavated sternum and basal parts of Mxp1. **E**, telson. **F**, right antennal peduncle, ventral. **G**, right Mxp3, ventral. Scale bars: 1 mm.

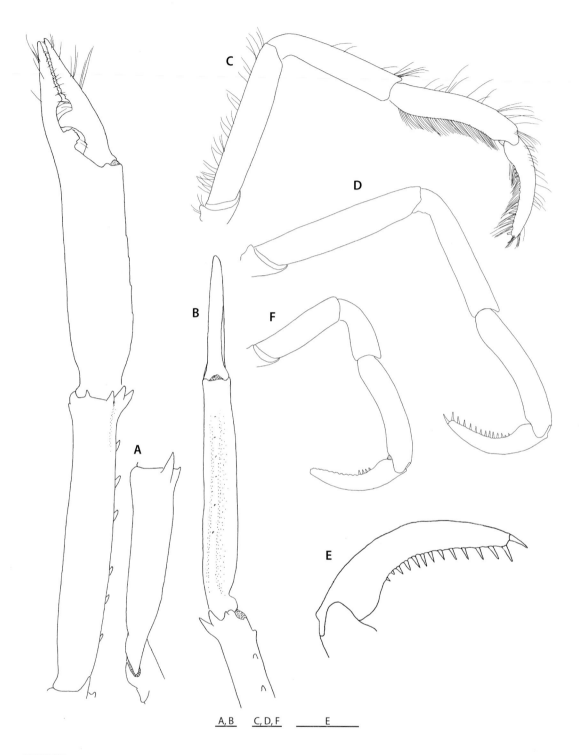

FIGURE 299

Heteroptychus lemaitrei n. sp., holotype, male 5.4 mm (MNHN-IU-2013-12289). **A**, left P1, dorsal. **B**, same, proximal part omitted, mesial. **C**, right P2, lateral. **D**, right P3, setae omitted, lateral. **E**, same, distal part, lateral. **F**, right P4, lateral. Scale bars: 1 mm.

P2-4: Sparsely setose on meri and carpi, thickly so along prehensile margins of propodi and dactyli. Meri subequal on P2 and P3, P4 merus 0.5-0.6 × length of P3 merus; length-breadth ratio, 5.1-5.6 on P2, 5.2-6.2 on P3, 3.5-3.9 on P4; P2 merus 0.9-1.1 × length of carapace, 1.2 × length of P2 propodus, P3 merus 1.2-1.3 × length of P3 propodus, P4 merus 0.8 × length of P4 propodus. Carpi subequal in length on P2 and P3, P4 carpus 0.4-0.5 × length of P3 carpus; carpus-dactylus length ratio, 1.4 on P2, 1.4-1.5 on P3, 0.6-0.7 on P4. Propodi subequal on P2 and P3, P4 propodus 0.9 × length of P3 propodus; propodus-dactylus length ratio, 1.5 on P2, 1.5-1.6 on P3, 1.4 on P4; flexor margins concavely curving in lateral view, unarmed. Dactyli subequal on P2-4, dactylus-carpus length ratio, 0.7 on P2 and P3, 1.5-1.6 on P4; flexor margins with row of 14 or 15 sharp spines proximally diminishing and slightly inclined or subperpendicular to margin, ultimate spine subequal to or somewhat longer than penultimate spine.

REMARKS — The fusion of the antennal scale with article 2 is shared by *H. claudeae* n. sp., *H. lemaitrei* n. sp. and the western Atlantic *H. brevis* (Benedict, 1902) (see above under *H. claudeae*). *Heteroptychus lemaitrei* differs from *H. claudeae* in having a small but distinct protuberance at the anterior end of the carapace branchial margin and in having the rostrum distally spiniform rather than narrow triangular.

The material reported under *Uroptychus scambus* from the Makassar Strait (Baba, 2005) was reexamined. The larger of the two male specimens can definitely be referred to *H. lemaitrei*, bearing a distinct protuberance, characteristic of this species, at the anterior branchal margin. In the smaller male, the protuberance is reduced to a very tiny denticle that is discernible only under high magnification; in addition, the P1 has less pronounced distal spines on the merus and carpus, with no distinct spines along the mesial margin of carpus. These differences appear to be size-related.

Heteroptychus paulae n. sp.
Figures 300, 301

TYPE MATERIAL — Holotype: **Solomon Islands**. SALOMON 2 Stn CP2269, 7°45.1'S, 156°56.3'E, 768-890 m, 4.XI.2004, ov. ♀ 4.7 mm (MNHN-IU-2014-17131). Paratypes: **Solomon Islands**. SALOMON 2 Stn CP2215, 7°44.3'S, 157°42.3'E, 718-880 m, 26.X.2004, 1 ov. ♀ 4.1 mm (MNHN-IU-2014-17132). – Stn CP2246, 7°42.6'S, 156°24.6'E, 664-682 m, 1.XI.2004, 1 ov. ♀ 4.8 mm (MNHN-IU-2013-12293). – Stn CP2247, 7°44.9'S, 156°24.7'E, 686-690 m, 1.XI.2004, 1 ♂ 4.2 mm (MNHN-IU-2014-17133). – Stn CP2248, 7°45.212'S, 56°25.606'E, 673-650 m, 1.XI.2004, 1 ov. ♀ 5.2 mm (MNHN-IU-2013-12290). – Stn CP2289, 08°36'S, 157°28'E, 627-623 m, 07.XI.2004, 1 ♂ 3.8 mm, 1 ov. ♀ 4.2 mm (MNHN-IU-2014-17134).

ETYMOLOGY — Name for Paula Martin-Lefevre of the Muséum national d'Histoire naturelle for her contribution to collection and data management on crustaceans.

DISTRIBUTION — Solomon Islands, in 660-890 m.

SIZE — Males, 3.8-5.1 mm; females, 4.2-5.2 mm, ovigerous females from 4.2 mm.

DESCRIPTION — Small to medium-sized species. *Carapace*: 0.7-0.8 × as long as broad, greatest breadth 1.9-2.2 × distance between anterolateral spines. Dorsal surface polished, moderately convex from anterior to posterior, ridged along posterior third of lateral margin. Lateral margin strongly convex around posterior third; anterolateral spine prominent, horizontally directed forward but slightly inclined mesially, reaching or slightly falling short of rostral tip. Rostrum relatively broad triangular, with interior angle of 36-48°, straight horizontal, fully reaching or slightly falling short of tip of eye; length 0.6-0.7 × breadth, 0.20-0.27 × length of remaining carapace; distally blunt; dorsal surface flattish but feebly concave basally. Lateral limit of orbit rounded, without spine. Pterygostomian flap very low on posterior half (posterior height 0.2 ×

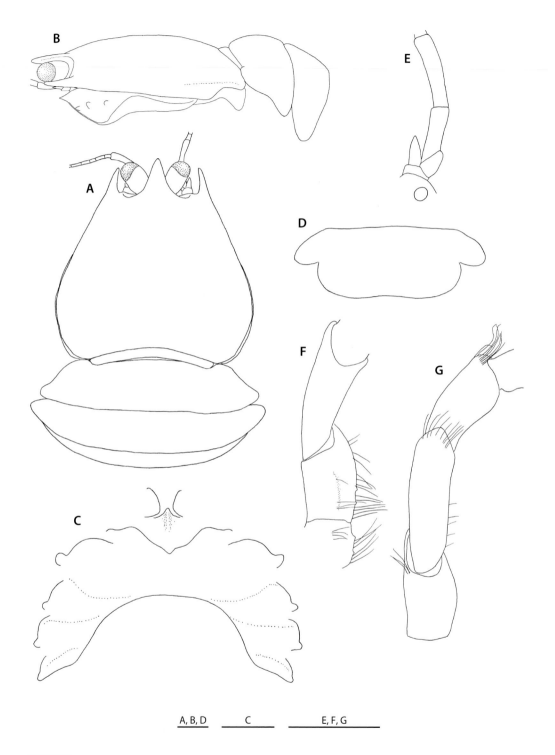

A, B, D   C   E, F, G

FIGURE 300

Heteroptychus paulae n. sp., holotype, ovigerous female 4.7 mm (MNHN-IU-2014-17131). **A**, carapace and anterior part of abdomen, dorsal. **B**, same, lateral. **C**, sternal plastron, with excavated sternum and basal parts of Mxp1. **D**, telson. **E**, right antenna, ventral. **F**, right Mxp3, setae on merus omitted, ventral. **G**, same, lateral. Scale bars: 1 mm.

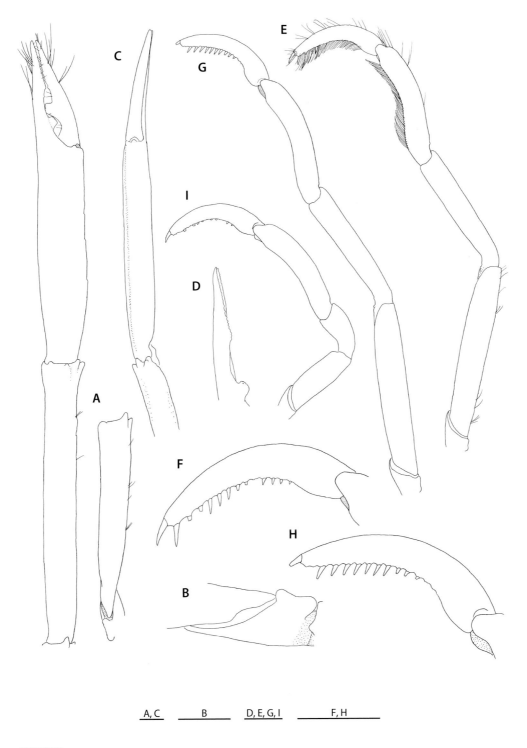

A, C   B   D, E, G, I   F, H

FIGURE 301

Heteroptychus paulae n. sp., holotype, ovigerous female 4.7 mm (MNHN-IU-2014-17131). **A**, left P1, dorsal. **B**, same, proximal part, lateral. **C**, same, distal part, setae omitted, mesial. **D**, same, fixed finger, dorsomesial. **E**, left P2, lateral. **F**, same, distal part, setae omitted, lateral. **G**, left P3, setae omitted, lateral. **H**, same, distal part, lateral. **I**, left P4, lateral. Scale bars: 1 mm.

anterior height), anteriorly produced to sharp spine; anterior surface well inflated (convex from dorsal to ventral) with 3 or 4 distally blunt coniform processes of small to good size or ridged longitudinally with 2 or 3 small or low processes.

Sternum: Excavated sternum with subtriangular anterior margin, surface with sharp ridge in midline on anterior half. Sternal plastron in males 0.6 × as long as broad; in females, left and right parts of sternites 5-7 discontinuous, interrupted by loss of median parts; lateral extremities of sternites 4-7 slightly convex and slightly divergent posteriorly; sternite 6 broadest. Sternite 3 anterior margin excavated in broad V-shape or semicircular shape, with small blunt (rarely obsolete) median notch lacking flanking spine or process. Sternite 4 with rounded anterolateral margin. Sternite 6 broadest.

Abdomen: Smooth and polished. Somite 1 in profile gently convex from anterior to posterior. Somite 2 tergite 2.2-2.8 × broader than long; pleural lateral margins slightly concave, strongly divergent posteriorly, ending in blunt tip. Pleura of somites 3-4 ending in rounded margin. Telson 0.4 × as long as broad; posterior plate slightly or very feebly concave on posterior margin, length 1.0-1.3 × that of anterior plate.

Eyes: 1.4-1.6 × as long as broad, subovate. Cornea 0.5-0.7 × length of remaining eyestalk.

Antennule and antenna: Ultimate article of antennular peduncle 3.2-3.8 × longer than high. Antennal peduncle relatively short and slender. Article 2 without lateral spine. Antennal scale distinctly articulated, ending in proximal third or midlength of article 4. Distal 2 articles unarmed; article 5 1.2-1.4 × longer than article 4, breadth 0.4-0.5 × height of ultimate antennular article. Flagellum of 11 or 12 segments directed posterolaterally.

Mxp: Mxp1 with bases broadly separated. Mxp3 basis with 2 obsolescent denticles or unarmed on mesial ridge. Ischium 0.4 × as long as merus, crista dentata with 2-4 denticles on proximal half, flexor margin not rounded distally. Merus relatively thick mesio-laterally, with rounded ridge along flexor margin, unarmed. No spine on carpus.

P1: Smooth and nearly glabrous except for fingers, 6.9-7.3 × (males), 5.8-7.4 × (females) longer than carapace. Ischium with basally broad, distally blunt distodorsal process and lobe-like proximal process overhanging basis. Merus with 2-4 blunt distodorsal spines (rarely obsolescent, leaving 1 short blunt spine), length 1.4-1.6 × (males), 1.8 × (females) that of carapace. Carpus 1.3-1.4 × longer than merus, unarmed or with several obsolescent blunt spines (in both sexes) along mesial margin. Palm 4.1-4.5 × (males), 4.3-4.7 × (females) longer than broad, 0.6 × (rarely 0.7 x) as broad as high, 0.8 × length of carpus, mesial and lateral margins subparallel, mesial margin roundly or moderately ridged, not cristiform, with or without obsolescent blunt spines. Fingers half as long as palm, gaping in proximal two-thirds, straightly fitting to each other when closed in distal half, relatively slender, distally spooned; opposable margin of movable finger with prominent blunt (truncate) process at midlength of gaping portion, distal to position of opposite process on fixed finger.

P2-4: Setose along prehensile margins of propodi and dactyli. Meri subequal on P2 and P3, P4 merus 0.4-0.5 × length of P3 merus; length-breadth ratio, 5.2-6.4 on P2, 5.3-6.4 on P3, 2.8-3.4 on P4; P2 merus 0.9-1.0 × length of carapace, 1.2-1.3 × length of P2 propodus, P3 merus 1.2-1.3 × length of P3 propodus, P4 merus 0.6-0.7 × length of P4 propodus. Carpi subequal in length on P2 and P3, P4 carpus one-third length of P3 carpus; subequal to or slightly shorter than propodus on P2 and P3, much shorter (0.31-0.36) on P4; carpus-dactylus length ratio, 1.3-1.5 on P2 and P3, 0.5 on P4. Propodi subequal on P2 and P3 or shorter on P3 than on P2, P4 propodus 0.8-0.9 × length of P3 propodus; propodus-dactylus length ratio, 1.5-1.6 on P2, 1.4-1.7 on P3, 1.3-1.4 on P4; flexor margin concavely curving in lateral view, unarmed. Dactyli subequal on P2-4, dactylus-carpus length ratio, 0.7-0.8 on P2, 0.6-0.8 on P3, 2.0-2.2 on P4; flexor margin with row of 12-14 sharp spines proximally diminishing and slightly inclined, ultimate and penultimate spines subequal or ultimate slightly larger.

Eggs. Number of eggs carried, 13-26; size, 1.00 × 1.08 mm - 1.10 × 1.12 mm (holotype); 1.25 × 1.21 mm - 1.46 × 1.29 mm (paratype, MNHN-IU-2014-17134).

REMARKS — The female holotype of *H. paulae* was collected together with the male holotype of *H. apophysis* n. sp. Initially, these two specimens were thought to be identical, because of the similarity in most characters including the relatively broad rostrum, the absence of lateral orbital spine and lack of paired median spines on the anterior margin of sternite 3. However, the following characters are not observed in all the paratypes including males of *H. paulae*: the carapace branchial margin with two rounded processes, the anterolateral spine stronger and directed straight forward, and the P1 carpus bearing two rows instead of one row of spines along the mesial margin and the palm mesially cristiform instead of roundly ridged. It is not unlikely that these are size-related, representing grown male characters, but the differences seem too large. Unfortunately, genetic analyses of this and related material failed (L. Corbari, pers. comm.), but here I propose to treat them as different species.

The antennal scale articulated with article 2 and the relatively long anterolateral spine of the carapace link *H. paulae* to *H. anouchkae* n. sp. However, *H. paulae* differs from *H. anouchkae* in having the rostrum that is distally blunt instead of sharply pointed and that is relatively broad and more trianguloid with the interior angle of at least 45°, instead of relatively narrow with the interior angle of less than 35°; the greatest breadth of the carapace is measured at the posterior quarter of the lateral margin rather than posterior third; the anterolateral spine of the carapace is horizontal instead of deflected dorsally; the lateral orbital angle is rounded instead of bearing small distinct spine; the anterior margin of sternite 3 is medially notched instead of bearing pair of small processes; and the P1 palm is mesially rounded or moderately ridged, not sharply ridged.

Heteroptychus scambus (Benedict, 1902)

Figures 302-304

Uroptychus scambus Benedict, 1902: 297, fig. 41. — Baba 2009: 59 (part; see below under Remarks).
Uroptychus glyphodactylus MacGilchrist, 1905: 249 — Alcock & MacGilchrist 1905: pl. 70, fig. 4; pl. 71, figs 1, 1a, 1b, 1c, 1d.
Not *Uroptychus scambus* — Baba 2005: 58 (= *H. lemaitrei* n. sp.). — Baba *et al.* 2009: 59, figs 49-50. (part = *H. claudeae* n. sp.).

Identification questioned:
Uroptychus scambus — Doflein & Balss 1913: 134. — Van Dam 1937: 100, fig. 1. — Schnabel 2009: 567.

TYPE MATERIAL — Holotype: **Japan**, off Honshu, 617 m [2 miles off Entr. Port Heda, N 86° E], ov. female (USNM 26165). [examined].

MATERIAL EXAMINED — Holotype: **Japan**, off Honshu. ALBATROSS Stn 3706, 617 m [2 miles off Entr. Port Heda, N 86° E], ov. ♀ 3.8 mm (USNM 26165). **Taiwan**. TAIWAN Stn CD141, 22°12.04′N, 119°59.96′E, 1110-985 m, 24.IX.2001, 1 ♂ 4.3 mm (NTOU).

DISTRIBUTION — Japan, 617 m, and Taiwan, 1110-985 m.

DESCRIPTION OF HOLOTYPE (female) — *Carapace*: 0.7 × as long as broad, greatest breadth mesured at posterior third, 2.5 × distance between anterolateral spines. Dorsal surface polished, moderately convex from anterior to posterior, ridged along posterior half of lateral margin. Lateral margin strongly convex posteriorly; anterolateral spine prominent, directed anteromesially, reaching distal quarter of rostrum. Rostrum broad triangular, with interior angle of 53°, half as long as broad, less than 0.2 × length of remaining carapace, straight horizontal, barely reaching tip of eye, dorsal surface flattish but feebly concave basally. Lateral limit of orbit with very small spine. Pterygostomian flap very low on posterior half (posterior height 0.3 × anterior height), anteriorly produced to sharp spine; anterior surface well inflated (convex from dorsal to ventral), with 3 small, low, blunt processes or protuberances.

Sternum: Excavated sternum bluntly produced anteriorly, surface with sharp ridge in midline on anterior half. Sternal plastron half as long as broad, lateral extremities convexly divergent posteriorly, sternite 6 broadest; left and right parts of sternites 5-7 discontinuous, interrupted by loss of median parts; anterior margin of sternite 3 deeply excavated in semicircular shape, with small median notch.

Abdomen: Smooth and polished. Somite 1 tergite gently convex dorsally from anterior to posterior. Somite 2 tergite 3 × broader than long; pleural lateral margin moderately concave, strongly divergent posteriorly, ending in blunt tip. Pleura of somites 3-4 ending in rounded margin. Telson 0.34 × as long as broad; posterior plate nearly transverse on posterior margin, length subequal to that of anterior plate.

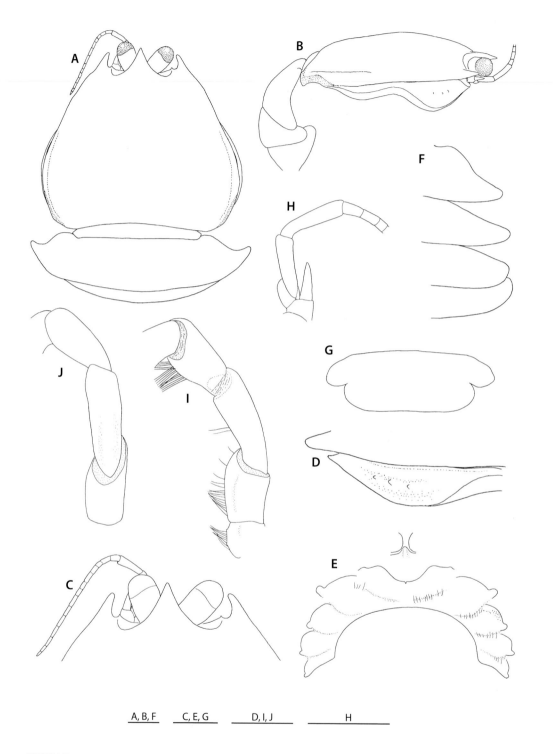

A, B, F   C, E, G   D, I, J   H

FIGURE 302

Heteroptychus scambus (Benedict, 1902), holotype of *Uroptychus scambus*, ovigerous female 5.7 mm (USNM 26165). **A**, carapace and anterior part of abdomen, dorsal. **B**, same, lateral. **C**, carapace, anterior part, dorsal. **D**, left pterygostomian flap, dorsolateral. **E**, sternal plastron, with excavated sternum and basal parts of Mxp1. **F**, right pleura of abdominal somites 2-5, dorsolateral. **G**, telson. **H**, left antenna, ventral. **I**, left Mxp3, ventral. **J**, same, setae omitted, lateral. Scale bars: 1 mm.

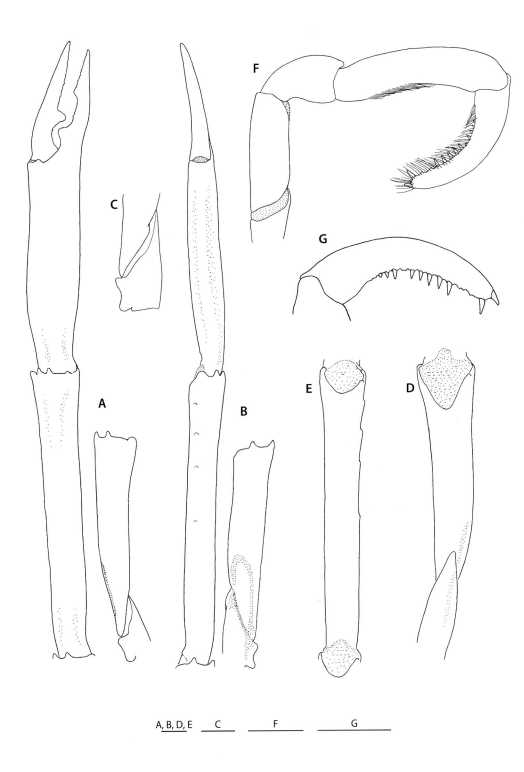

A, B, D, E ___ C ___ F ___ G ___

FIGURE 303

Heteroptychus scambus (Benedict, 1902), holotype of *Uroptychus scambus*, ovigerous female 5.7 mm (USNM 26165). **A**, right P1, dorsal. **B**, same, mesial. **C**, same, proximal part, lateral. **D**, same, ischium and merus, ventral. **E**, same, carpus, ventral. **F**, right P4, lateral. **G**, same, distal part, setae omitted. Scale bars: 1 mm.

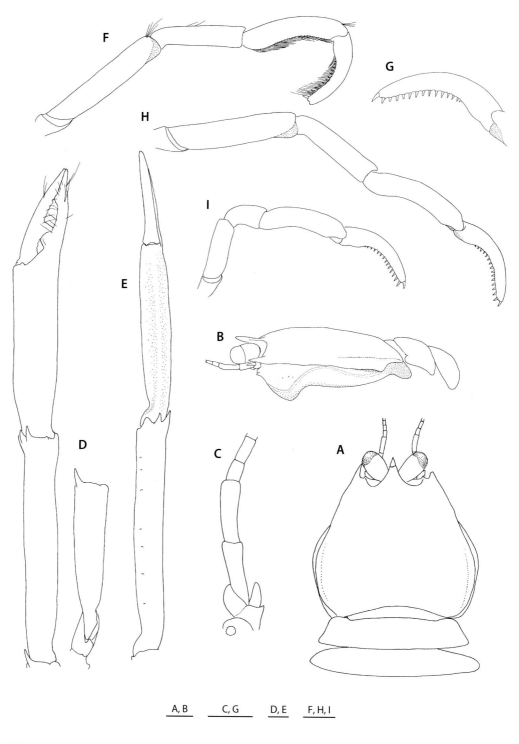

FIGURE 304

Heteroptychus scambus (Benedict, 1902), male 4.3 mm, from *Taiwan* Stn CD141. **A**, carapace and anterior part of abdomen, dorsal. **B**, same, lateral. **C**, left antenna, ventral. **D**, right P1, dorsal. **E**, same, mesial. **F**, right P2, lateral. **G**, left P2, distal part, setae omitted, lateral. **H**, right P3, setae omitted, lateral. **I**, right P4, lateral. Scale bars: 1 mm.

Eyes: About 2 × as long as broad, subovate. Cornea slightly shorter than (0.8 x) remaining eyestalk.

Antennule and antenna: Ultimate article of antennular peduncle 3.3 × longer than high. Antennal peduncle relatively short and slender. Article 2 without lateral spine. Antennal scale articulated, slightly narrower than article 5, slightly overreaching midlength of article 4. Distal 2 articles unarmed; article 5 as long as article 4, breadth 0.6 × height of ultimate antennular article. Flagellum of 14 segments directed posteriorly.

Mxp: Mxp1 with bases moderately separated. Mxp3 basis with 2 obsolescent denticles on mesial ridge. Ischium 0.4 × as long as merus, crista dentata unarmed, flexor margin not rounded distally. Merus relatively thick mesio-laterally, with rounded ridge along flexor margin, unarmed. No spine on carpus.

P1: Left P1 missing. Right P1 smooth and glabrous except for fingers, 5.9 × longer than carapace. Ischium with basally broad, distally blunt distodorsal process and lobe-like proximal process overhanging basis. Merus with 2 distodorsal spines, length 1.4 × that of carapace. Carpus 1.4 × longer than merus, with row of several obsolescent blunt spines along distal half of mesial margin. Palm 3.8 × longer than broad, 0.6 × as high as broad, 0.8 × length of carpus; mesial margin roundly ridged, not cristiform. Fingers gaping proximally, straightly fitting to each other in distal third when closed, relatively slender, distally spooned; opposable margin of movable finger with prominent, distally blunt subtriangular process at proximal third, distal to position of opposite process of on fixed finger.

P2-4: Only right P4 available, setose along prehensile margins of propodus and dactylus. Merus 2.6 × longer than broad, 0.6 × length of propodus. Carpus slightly less than 0.4 × length of propodus. Propodus 1.4 × longer than dactylus, flexor margin concavely curving in lateral view. Dactylus 2 × length of carpus, 0.7 × length of propodus, flexor margin with row of 14 sharp spines proximally diminishing and nearly perpendicular to margin, ultimate and penultimate spines subequal.

Eggs. Number of eggs carried, 8; size, 1.26 × 1.46 mm - 1.23 × 1.50 mm.

REMARKS — This species is characterized by the antennal scale articulated with the antennal article 2 and the broad triangular rostrum far falling short of the tip of the eye, both of which are possessed only by *H. glyphodactylus* (MacGilchrist, 1905) from east of Andaman Islands. Because of the brevity of the original description of *H. glyphodactylus* and because its type material is hardly accessible, further comparison between the two nominal species is impossible. Therefore these species are treated as identical for the time being, although it is highly probable that they are different because of their disjunct distribution.

Uroptychus edwardi Kensley, 1977 from the western Indian Ocean off South Africa (holotype, ovigerous female, SAF A16033) was synonymized with *U. scambus* in my earlier paper (Baba, 1988). However, the following characters are distinctive in *H. edwardi*: the spiciform rostrum overreaching the eyes, the anterolateral spine of the carapace directed straight forward, reaching the tip of the eye, and the frontal margin somewhat sinuous between the spiciform rostrum and the anterolateral spine [with no distinct lateral orbital spine]. Hence, the species is now resurrected.

The specimens reported under *Uroptychus scambus* from Japan (Baba 1981) are referable to *H. claudeae* n. sp., as also are the female from Albatross Station 5605 (Indonesia) and the two specimens from Station 5083 (Japan) (Baba 1988); the other specimen from the Albatross Station 4959 (Japan) requires reexamination. The two specimens taken at Galathea Station 453, Makassar Strait (Baba 2005) are referred to *H. lemaitrei* n. sp. (see above).

The NTOU material from Taiwan (Baba *et al.* 2009) was reexamined. The male collected at Taiwan Station CD141 (222°12.04'N, 119°59.96'E, 1110-985 m) may be referable to *H. scambus*, and the other specimens (except for the material from Station PCP342 that was not located) are identified with *H. claudeae* n. sp. (see above). This male (Figure 304) is not in complete agreement with the holotype in having the rostrum distally narrower and directed somewhat dorsally, and in having the antennal article 2 with a small distolateral spine and the scale barely reaching the midlength of article 4. These subtle differences are regarded as individual variations. The terminal spines of P1 merus and carpus are stronger, probably sex-related. P2-4 are as illustrated (P2 and P3 are missing in the holotype), with no clear difference from those of all the other species.

Records of *U. scambus* reported by Doflein & Balss (1913: 134), Van Dam (1937: 100, fig. 1), and Schnabel (2009: 567) require verification.

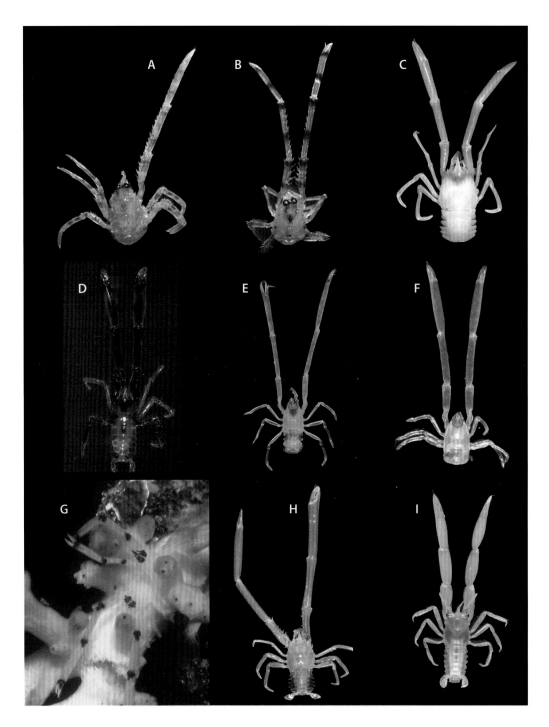

FIGURE 305

A, *Uroptychus abdominalis* n. sp., paratype, ovigerous female 3.0 mm, MNHN-IU-2014-17140, Norfolk Ridge. **B**, *U. anoploetron* n. sp., paratype, ovigerous female 2.7 mm, MNHN-IU-2014-17302, Norfolk Ridge. **C**, *U. babai* Ahyong & Poore, 2004, male 9.2 mm, MNHN-IU-2014-16300, Chesterfield Islands. **D**, *U. elongatus* n. sp., holotype, ovigerous female 4.3 mm, MNHN-IU-2014-16440, Vanuatu. **E**, *U. joloensis* Van Dam, 1933, ovigerous female, 3.0 mm, MNHN-IU-2014-16595, Vanuatu. **F**, *U. joloensis* Van Dam, 1933, ovigerous female, MNHN-IU-2014-16602, Wallis and Futuna Islands. **G**, *U. joloensis* Van Dam, 1933, male 2.3 mm and ovigerous female 2.1 mm, in situ habitat, associated with nidaliid corals, QM W25105, Papua New Guinea. **H**, *U. nanophyes* McArdle, 1901, male, 10.7 mm, MNHN-IU-2014-16757, Norfolk Ridge. **I**, *U. nebulosus* n. sp., paratype, male, 8.1 mm, MNHN-IU-2014-16812, Vanuatu.

FIGURE 306

A, *Uroptychus psilus* n. sp., paratype, male 9.1 mm, MNHN-IU-2014-16888, Norfolk Ridge. **B**, *U. senarius* n. sp., paratype, ovigerous female 4.9 mm, MNHN-IU-2014-16946, Vanuatu. **C**, *U. setifer* n. sp., holotype, ovigerous female 2.2 mm, MNHN-IU-2010-5436, Vanuatu. **D**, *U. soyomaruae* Baba, 1981, ovigerous female 14.0 mm, MNHN-IU-2014-16954, Norfolk Ridge. **E**, *U. spinirostris* (Ahyong & Poore, 2004), 3.6 mm, MNHN-IU-2014-16959, Vanuatu. **F**, *U. stenorhynchus* n. sp., paratype, male 6.0 mm, MNHN-IU-2014-16973, Vanuatu. **G**, *U. trispinatus* n. sp., paratypes, male 5.2 mm (left), ovigerous female 4.8 mm (right), MNHN-IU-2014-16997, Chesterfield Islands. **H**, *U. yokoyai* Ahyong & Poore, 2004, female 3.2 mm, MNHN-IU-2014-17032, Vanuatu. **I**, *Heteroptychus anouchkae* n. sp., paratype, male 4.7 mm, MNHN-IU-2014-17101, Norfolk Ridge.

ACKNOWLEDGMENTS

The main material in this paper originates from various deep sea cruises, conducted by MNHN and Institut de Recherche pour le Développement (IRD) as part of the *Tropical Deep-Sea Benthos* program. I thank Alain Crosnier of the Muséum national d'Histoire naturelle, Paris (MNHN), for his friendship, help and support. Thanks are also due to the following colleagues for loans of type and/or comparative materials under their care: Tin-Yam Chan of the National Taiwan Ocean University, Keelung; Paul F. Clark and Miranda Lowe of the Natural History Museum, London; Régis Cleva, Paula Martin-Lefevre and Laure Corbari (MNHN); Rafael Lemaitre and Karen Reed of the Smithsonian Institution, Washington, D.C.; Charles Oliver Coleman of the Museum für Naturkunde an der Humboldt-Universität zu Berlin, Berlin; Hironori Komatsu of the National Museum of Nature and Science, Tsukuba; Junji Okuno of the Coastal Branch of the Natural History Museum and Institute, Chiba; Jørgen Olesen of the Natural History Museum of Denmark, Copenhagen; Michitaka Shimomura of the Kitakyushu Museum of Natural History and Human History, Kitakyushu. My thanks also go to Laure Corbari, Julien Brisset and Danielle Defaye of MNHN, Rafael Lemaitre and Karen Reed, and Elisabeth Lang of the Musée Zoologique, Strasbourg for laboratory facilities during my visits to their institutions. Laure Corbari, Sarah Samadi and Marie-Catherine Boisselier of MNHN helped me with molecular analyses. Peter J.F. Davie of the Queensland Museum, Brisbane sent me specimens from Papua New Guinea along with a photograph, which have been incorporated in the present paper. At my request Anna McCallum of the Museum Victoria examined her material of *Uroptychus scandens* from Western Australia. The manuscript benefited from discussions with Enrique Macpherson of the Centro de Estudios AVanzados de Blanes, Girona, Shane T. Ahyong of the Australian Museum, Sydney, Kareen E. Schnabel of the National Institute of Water & Atmospheric Research, Wellington, and Anna McCallum of the Museum Victoria. My best thanks to Enrique Macpherson, Shane Ahyong and Anna McCallum for their patient time to read the lengthy manuscript, without their help this paper could not have been completed. The station data have been verified by Laure Corbari and Julien Brisset. Color photographs used in this paper are due to the talent of Tin-Yam Chan, Pierre Laboute, Jean-Louis Menou, Gérard Moutham and Neville Coleman. Some of the comparative material used in this paper were made available by Sandy Bruce. Commensal host corals were identified by Yukimitsu Imahara of the Biological Institute of Kuroshio, Wakayama.

This work is a part of the Tropical Deep Sea Benthos (ex MUSORSTOM) program carried out during my visits to the MNHN, supported by grants from the Institut de Recherche pour le Développement (IRD) in 1998 and 1999, and from the MNHN in 2000, 2001, 2003, 2004. My visit to the MNHN in 2005 was supported partly by the courtesy of Philippe Bouchet of the MNHN. Tim O'Hara of Museum Victoria, Melbourne is thanked for a financial support for my visit to the MNHN in 2013. One month stay in the Paris Museum in 2014 was funded by the SPN, Service du Patrimoine Naturel (for conservation of natural heritage).

ADDENDUM

Since this paper went to press, four new species of *Uroptychus*, three from the Indo-West Pacific (Ahyong & Baba 2017; Dong *et al.* 2017; Schnabel *et al.* 2017) and one from the eastern Atlantic (Baba & Macpherson 2017) have been described.

Of the three Indo-West Pacific species described in the aforementioned papers:

(1) *Uroptychus michaeli* Ahyong & Baba, 2017 was described as suggested under the Remarks of *U. nigricapillis* Alcock, 1901 in this paper.

(2) *Uroptychus inaequipes* Dong, Li, Lu & Wang, 2017 differs from *U. dejouanneti* n. sp. in having the P2-4 dactyli with the ultimate spine distinctly stronger than, instead of subequally as broad as, the penultimate, and the P2-4 carpi more than twice as long instead of subequally as long as the dactyli.

(3) *Uroptychus macquariae* Schnabel, Burghardt & Ahyong, 2017 is distinguished from *U. insignis* (Henderson, 1885) by the carapace and P1 being less setose, in having larger, more pronounced epigastric spines in large specimens, especially the largest epigastric spines being as large as instead of smaller than adjacent branchial marginal spines, eyes relatively long (1.8-2.0 × versus 1.4-1.5 × longer than broad), and the antennal scale never reaching, instead of reaching or overreaching the end of antennal article 5.

REFERENCES

AHYONG S. T. & BABA K. 2004 — Chirostylidae from north-western Australia (Crustacea: Decapoda: Anomura). *Memoirs of Museum Victoria* 61: 57-64.

AHYONG S. T. & BABA K. 2017 — *Uroptychus michaeli* (Decapoda, Chirostylidae), a new species of deep-water squat lobster from northwestern Australia and Taiwan. *Crustaceana* 90: 799-806.

AHYONG S. T. & POORE G. C. B. 2004 — The Chirostylidae of southern Australia (Crustacea: Decapoda: Anomura). *Zootaxa* 436: 1-88.

AHYONG S. T., SCHNABEL K. E. & BABA K. 2015— Southern high latitude squat lobsters: Galatheoidea and Chirostyloidea from Marcquarie Ridge with description of a new species of *Uroptychus*. *Records of the Australian Museum* 67: 109-128.

ALCOCK A. 1901 — A Descriptive Catalogue of the Indian Deep-Sea Crustacea Decapoda Macrura and Anomala in the Indian Museum. Being a Revised Account of the Deep-Sea Species collected by the Royal Indian Marine Survey Ship Investigator. *Trustees of the Indian Museum Calcutta*, iv + 286 pp., 3 pls.

ALCOCK A. & ANDERSON A. R. S. 1894 — Natural history notes from H.M. Royal Indian Marine Survey Steamer "Investigator" commander C.F. Oldham R.N. commanding. Series II No. 14. An account of a recent collection of deep sea Crustacea from the Bay of Bengal and Laccadive Sea. *Journal of the Asiatic Society of Bengal (2) (Natural History)* 63: 141-185, pl. 9.

ALCOCK A. & ANDERSON A. R. S. 1899 — Natural history notes from H. M. Royal Indian Marine Survey Ship «Investigator» commander T.H. Heming R.N. commanding. Series III No. 2. An account of the deep-sea Crustacea dredged during the surveying season of 1897-98. *Annals and Magazine of Natural History Series* 7(3): 1-27.

ALCOCK A. & MCARDLE A. F. 1902 — Crustacea Part X. Illustrations of the Zoology of the Royal Indian Marine Surveying Steamer Investigator. *Office of the Superintendent of Government Printing Calcutta*, pls 56-67.

ANDERSON A.R.S. 1896 — An account of the deep-sea Crustacea from the Bay of Bengal and Laccadive Sea. *Journal of the Asiatic Society of Bengal (2) (Natural History)* 65: 88-106.

ANONYMOUS 1914 — Biological collections of the R.I.M.S. "Investigator." List of stations. 1884-1913. *Trustees of the Indian Museum Calcutta*, 33 pp.

BABA K. 1973 — Remarkable species of the Chirostylidae (Crustacea Anomura) of Japanese waters. *Memoirs of the Faculty of Education Kumamoto University Section 1 (Natural Science)* 22: 117-124, pl. 4.

BABA K. 1974 — Four new species of Galatheidean Crustacea from New Zealand waters. *Journal of the Royal Society of New Zealand* 4: 381-393.

BABA K. 1977 — Five new species of chirostylid crustaceans (Decapoda Anomura) from off Midway Island. *Bulletin of the National Science Museum Tokyo* Series A (Zoology) 3: 141-156.

BABA K. 1979 — First records of chirostylid and Galatheid crustaceans (Decapoda Anomura) from New Caledonia. *Bulletin du Muséum national d'Histoire naturelle Série 4 Section A* 1: 521-529.

BABA K. 1981 — Deep-sea Galatheidean Crustacea (Decapoda Anomura) taken by the R/V Soyo-Maru in Japanese waters. I. Family Chirostylidae. *Bulletin of the National Science Museum Tokyo* Series A (Zoology) 7: 111-134.

BABA K. 1986a — Two new anomuran Crustacea (Decapoda: Anomura) from North-West Australia. *The Beagle* 3: 1-5.

BABA K. 1986b — Two new species of anomuran crustaceans (Decapoda: Chirostylidae and Galatheidae) from the Andaman Sea. *Journal of Crustacean Biology* 6: 625-632.

BABA K. 1988 — Chirostylid and Galatheid crustaceans (Decapoda: Anomura) of the "Albatross" Philippine Expedition 1907-1910. *Researches on Crustacea* Special Number 2: 203 pp.

BABA K. 1990 — Chirostylid and Galatheid crustaceans of Madagascar (Decapoda: Anomura). *Bulletin du Muséum national d'Histoire naturelle Série 4 Section A* (1989) 11: 921-975.

BABA K. 2000 — Two new species of chirostylids (Decapoda: Anomura: Chirostylidae) from Tasmania. *Journal of Crustacean Biology* 20 spec. no. 2: 246-252.

BABA K. 2004 — *Uroptychodes* new genus of Chirostylidae (Crustacea: Decapoda: Anomura) with description of three new species. *Scientia marina* 68: 97-116.

BABA K. 2005 — Deep-sea chirostylid and Galatheid crustaceans (Decapoda: Anomura) from the Indo-Pacific with a list of species. *Galathea Report* 20: 1-317.

BABA K. & LIN C.-W. 2008 — Five new species of chirostylid crustaceans (Crustacea: Decapoda: Anomura: Chirostylidae) from Taiwan. *Zootaxa* 1919: 1-24.

BABA K. & MACPHERSON E. 2012 — A new squat lobster (Crustacea: Decapoda: Anomura: Chirostylidae) from off NW Spain. *Zootaxa* 3224: 49-56.

BABA K. & MACPHERSON E. 2017 — *Uroptychus tuerkayi* sp. nov. (Anomura, Chirostylidae), a new squat lobster from the Atlantis-Great Meteor Seamount Chain in the eastern Atlantic. *Crustaceana* 90: 807-817.

BABA K. & DE SAINT LAURENT M. 1992 — Chirostylid and Galatheid crustaceans (Decapoda: Anomura) from active thermal vent areas in the southwest Pacific. *Scientia marina* 56: 321-332.

BABA K. & TIRMIZI N. M. 1979 — A new chirostylid (Crustacea Decapoda Anomura) from deeper parts of the Japanese waters and off the east coast of Africa. *Proceedings of the Japanese Society of Systematic Zoology* 17: 52-57.

BABA K. & WICKSTEN M. 2015 — *Uroptychus minutus* Benedict 1902 and a closely related new species (Crustacea: Anomura: Chirostylidae) from the western Atlantic Ocean. *Zootaxa* 3957: 215-225.

BABA K. & WICKSTEN M. 2017a — *Uroptychus nitidus* (A. Milne-Edwards 1880) and related species (Crustacea: Decapoda: Anomura: Chirostylidae) from the western Atlantic. *Zootaxa* 4221: 251-290.

BABA K. & WICKSTEN M. 2017b — *Uroptychus atlanticus* a new species of squat lobster (Crustacea: Decapoda: Anomura: Chirostylidae) from the western Atlantic Ocean. *Zootaxa* 4227: 295-300.

BABA K. & WILLIAMS A. B. 1998 — New Galatheoidea (Crustacea Decapoda Anomura) from hydrothermal systems in the West Pacific Ocean: Bismarck Archipelago and Okinawa Trough. *Zoosystema* 20: 143-156.

BABA K., MACPHERSON E., POORE G. C. B., AHYONG S. T., BERMUDEZ A., CABEZAS P., LIN C.-W., NIZINSKI M., RODRIGUES C. & SCHNABEL K. E. 2008 — Catalogue of squat lobsters of the world (Crustacea: Decapoda: Anomura families Chirostylidae Galatheidae and Kiwaidae). *Zootaxa* 1905: 1-220.

BABA K., MACPHERSON E., LIN C.-W. & CHAN T.-Y. 2009 — Crustacean Fauna of Taiwan: Squat Lobsters (Chirostylidae and Galatheidae). National Taiwan Ocean University Keelung, ix + 311 pp.

BABA K., AHYONG S. T. & MACPHERSON E. 2011 — Morphology of marine squat lobsters. Pp. 1-37, *in* POORE G. C. B, AHYONG S. T. & TAYLOR J. (eds), *The Biology of Squat Lobsters*. CSIRO Publishing Melbourne, xviii + 364 pp.

BAEZA A. J. 2011 — Squat lobsters as symbionts and in chemo-autotrophic environments. Pp. 249-270, *in* POORE G. C. P, AHYONG S. T. & TAYLOR J. (eds), *The Biology of Squat Lobsters*. CSIRO Publishing Melbourne, xviii + 364 pp.

BALSS H. 1913a — Neue Galatheiden aus der Ausbeute der deutschen Tiefsee-Expedition 'Valdivia.' *Zoologischer Anzeiger* 41: 221-226.

BALSS H. 1913b — Ostasiatische Decapoden I. Die Galatheiden und Paguriden. In: Doflein F. (ed.) Beiträge zur Naturgeschichte Ostasiens. *Abhandlungen der Mathematisch-Physikalischen Klasse der Königlich Bayerischen Akademie der Wissenschaften* 2: 1-85, pls 1, 2.

BALSS H. 1957 — Decapoda. *Bronn's Klassen und Ordnungen des Tierreichs* 5 (Abteilung 1 Buch 7 Lieferung 12), 1505-1672.

BARNARD K. H. 1950 — Descriptive catalogue of South African decapod Crustacea (crabs and shrimps). *Annals of the South African Museum* 38: 1-837.

BENEDICT J. E. 1902 — Description of a new genus and forty-six new species of crustaceans of the family Galatheidae with a list of the known marine species. *Proceedings of the Biological Society of Washington* 26: 243-334.

BORRADAILE L. A. 1916 — Crustacea. Part 1. - Decapoda. *British Antarctic ("Terra Nova") Expedition 1910. Natural History Report Zoology* 3: 75-110.

BRACKEN-GRISSOM H. D., CANNON M. E., CABEZAS P., FELDMANN R. M., SCHWEITZER C. E., AHYONG S. T., FELDER D. L., LEMAITRE R. & CRANDALL K. A. 2013 — A comprehensive and integrative reconstruction of evolutionary history for Anomura (Crustacea: Decapoda). *BMC Evolutionary Biology* 13: 128.

CABEZAS P., LIN C.-W. & CHAN T.-Y. 2011 — A new species of *Uroptychus* (Decapoda: Anomura: Chirostylidae) from Taiwan. *Bulletin of Marine Science* 88: 113-118.

CHEVALDONNÉ P. & OLU K. 1996 — Occurrence of anomuran crabs (Crustacea: Decapoda) in hydrothermal vent and cold-seep communities: a review. *Proceedings-Biological society of Washington* 109: 286-298.

DOFLEIN F. & BALSS H. 1913 — Die Galatheiden der Deutschen Tiefsee-Expedition. *Wissenschaftliche Ergebnisse der Deutschen Tiefsee-Expedition auf dem Dampfer "Valdivia" 1898-1899* 20: 125-184, pls 12-17.

DONG D. & LI X. 2015 — Galatheid and chirostylid crustaceans (Decapoda: Anomura) from a cold seep environment in the northeastern South China Sea. *Zootaxa* 4057: 91-105.

DONG D., LI X., LU B. & WANG C. 2017 — Three squat lobsters (Crustacea: Decapoda: Anomura) from tropical West Pacific seamounts, with description of a new species of *Uroptychus* Henderson, 1888. *Zootaxa* 4311: 389-398.

FROGLIA C. 1987 — Redescription of *Uroptychus ensirostris* Parisi 1917 (Decapoda Anomura Chirostylidae). *Crustaceana* 53: 148-151.

HAIG J. 1974 — The anomuran crabs of Western Australia: Their distribution in the Indian Ocean and adjacent seas. *Journal of the Marine Biological Association of India* 14: 443-451.

HENDERSON J. R. 1885 — Diagnoses of new species of Galatheidae collected during the "Challenger" expedition. *Annals and Magazine of Natural History Series 5* 16: 407-421.

HENDERSON J. R. 1888 — Report on the Anomura collected by H.M.S. *Challenger* during the years 1873-76. *Report on the Scientific Results of the Voyage of H.M.S. Challenger during the years 1873-76 Zoology* 27: 1-221, 21 pls.

KENSLEY B. 1977 — The South African Museum's *Neiring Naude* cruises. Part 2. Crustacea Decapoda Anomura and Brachyura. *Annals of the South African Museum* 72: 161-188.

KENSLEY B. 1981 — The South African Museum's *Meiring Naude* cruises. Part 12. Crustacea Decapoda of the 1977 1978 1979 cruises. *Annals of the South African Museum* 83: 49-78.

KIM H. S. 1972 — A new species of family Chirostylidae (Crustacea: Anomura) from Jeju Island Korea. *Korean Journal of Zoology* 15: 53-56.

KIM H. S. 1973 — Anomura Brachyura, *in Ilustrated Encyclopedia of Fauna and Flora of Korea. Seoul*, 694 pp.

KIM H. S. & CHOE B. L. 1976 — A report on four unrecorded anomuran species (Crustacea Decapoda) from Korea. *Korean Journal of Zoology* 19: 43-49.

LAURIE R. D. 1926 — Reports of the Percy Sladen Trust Expedition to the Indian Ocean in 1905, under the leadership of Mr J. Stanley Gardiner, M.A. Vol. 8, No. VI - Anomura collected by Mr J. Stanley Gardiner in the western Indian Ocean in H.M.S. Sealark. *Transactions of the Zoological Society of London (ser. 2, Zoology)* 19: 121-167, pls 8, 9.

MACPHERSON E. & BABA K. 2011 — Taxonomy of squat lobsters, *in* POORE G. C. B. AHYONG S. T. & TAYLOR J. (eds), *The Biology of Squat Lobsters*. CSIRO Publishing Melbourne and CRC Press Boca Raton, pp. 39-71.

MACPHERSON E. & ROBAINAS-BARCIA A. 2015 — Species of the genus *Galathea* Fabricius 1793 (Crustacea Decapoda Galatheidae) from the Indian and Pacific Oceans with descriptions of 92 new species. *Zootaxa* 3913: 1-335.

MARTIN J.W. & HANEY T.A. 2005 — Decapod crustaceans from hydrothermal vents and cold seeps: a review through 2005. *Zoological Journal of the Linnean Society* 145: 445-522.

MCARDLE A. F. 1901 — Natural history notes from the R.I.M.S. Ship 'Investigator.' — Series III No. 5. An account of the trawling operations during the surveying-season of 1900-1901. *Annals and Magazine of Natural History Series 7* 8: 517-526.

MCCALLUM A. W. & POORE G. C. B. 2013 — Chirostylidae of Australia's western continental margin (Crustacea: Decapoda: Anomura) with the description of five new species. *Zootaxa* 3664: 149-175.

MIYAKE S. & NAKAZAWA K. 1947 — Crustacea Anomura, *in* UCHIDA S. (ed.), *Illustrated Encyclopedia of the Fauna of Japan (exclusive of insects) Revised edition*. Hokuryukan Tokyo, pp. 731-750.

MIYAKE S. 1960 — Decapod Crustacea Anomura, *in* OKADA Y. K. & UCHIDA T. (eds), *Encyclopedia Zoologica Illustrated in Colours*. Hokuryukan Tokyo, pp. 89-97, pls 44-48.

MIYAKE S. 1965 — Crustacea Anomura, *in* OKADA Y. K. & UCHIDA T. (eds), *New illustrated Encyclopedia of the Fauna of Japan*. Hokuryukan Tokyo, pp. 630-652.

MIYAKE S. 1982 — *Japanese Crustacean Decapods and Stomatopods in Color. Vol. 1. Macrura Anomura and Stomatopoda*. Hoikusha Osaka 261 pp. (first edition; second printing in 1991 including some name changes of some species).

MIYAKE S. & BABA K. 1967 — Descriptions of new species of Galatheids from the Western Pacific. *Journal of the Faculty of Agriculture Kyushu University* 14: 203-212.

ORTMANN A. E. 1892 — Die Decapoden-Krebse des Strassburger Museums mit besonderer Berücksichtigung der von Herrn Dr. Döderlein bei Japan und bei den Liu-Kiu-Inseln gesammelten und zur Zeit im Strassburger Museum aufbewahrten Formen. IV. Die Abtheilungen Galatheidea und Paguridea. *Zoologische Jahrbücher Abtheilung für Systematik Geographie und Biologie der Thiere* 6: 241-326, pls 11-12.

PARISI B. 1917 — I Decapoda giapponesi des Museo di Milano. V. Galatheidea e Reptantia. *Atti della Societa' Italiana di Scienze Naturali e del Museo Civico di Storia Naturale Milano* 56: 1-24.

POORE G. C. B. & ANDREAKIS N. 2011 — Morphological molecular and biogeographic evidence support two new species in the *Uroptychus naso* complex (Crustacea: Decapoda: Chirostylidae). *Molecular Phylogenetics and Evolution* 60: 152-169.

POORE G. C. B., MCCALLUM A. W. & TAYLOR J. 2008 — Decapod Crustacea of the continental margin of southwestern and central Western Australia: preliminary identifications of 524 species from FRV *Southern Surveyor* voyage SS10-2005. Museum Victoria Science Report 11: 1-106.

POORE G. C. B., AHYONG S. T. & TAYLOR J. (eds) 2011 — *The Biology of Squat Lobsters*. CSIRO Publishing Melbourne, xviii: 364 pp.

ROTERMAN C. N., COPLEY J. T., LINSE K. T., TYLER P. A. & ROGERS A. D. 2013 — The biogeography of the yeti crabs (Kiwaidae) with notes on the phylogeny of the Chirostyloidea (Decapoda: Anomura). *Proceedings of the Royal Society B* 280 (1764): 20130718.

SCHNABEL K. E. 2009 — A review of the New Zealand Chirostylidae (Anomura: Galatheoidea) with description of six new species from the Kermadec Islands. *Zoological Journal of the Linnean Society* 155: 542-582.

SCHNABEL K. E. & AHYONG S. T. 2010 — A new classification of the Chirostyloidea (Crustacea: Decapoda: Anomura). *Zootaxa* 2687: 56-64.

SCHNABEL K. E., AHYONG S. T. & MAAS E. L. 2011 — Galatheoidea are not monophyletic - molecular and morphological phylogeny of the squat lobsters (Decapoda: Anomura) with recognition of a new superfamily. *Molecular Phylogenetics and Evolution* 58: 157-168.

SCHNABEL K. E., BURGHARDT I. & AHYONG S. T., 2017 — Southern high latitude squat lobsters II: description of *Uroptychus macquariae* sp. nov. from Macquarie Ridge. *Zootaxa* 4353: 327-338.

TAKEDA M. 1982 — *Keys to the Japanese and foreign crustaceans fully illustrated in colors*. Hokuryukan Tokyo. 284 pp

THOMSON G. M. 1899 — A revision of the Crustacea Anomura of New Zealand. *Transactions and Proceedings of the New Zealand Institute 1898* 31: 169-197, pls 20 21.

TIRMIZI N. M. 1964 — Crustacea: Chirostylidae (Galatheidea). *The John Murray Expedition 1933-34 Scientific Reports* 10: 385-415.

VAN DAM A. J. 1933 — Die Decapoden der SIBOGA-Expedition. VIII. Galatheidea: Chirostylidae. *SIBOGA-Expeditie Monographie* 39a (7): 1-46.

VAN DAM A. J. 1937 — Einige neue Fundorte von Chirostylidae. *Zoologischer Anzeiger* 120: 99-103.

VAN DAM A. J. 1939 — Über einige *Uroptychus*-Arten des Museums zu Kopenhagen. *Bijdrgen tot de Dierkunde Uitgegeven door het Koninklijk zoologisch genootschap Natur Artis Magistra de Amsterdam* 27: 392-407.

VAN DAM A. J. 1940 — Anomura gesammelt vom Dampfer "Gier" in der Java-See. *Zoologischer Anzeiger* 129: 95-104.

YOKOYA Y. 1933 — On the distribution of decapod crustaceans inhabiting the continental shelf around Japan chiefly based upon the materials collected by S.S. Soyo-Maru during the years 1923-30. *Journal of the College of Agriculture Tokyo Imperial University* 12: 1-226.

ZARENKOV N. A. & KHODKINA I. V. 1983 — [Decapod crustaceans] (in Russian), *in Benthos of the submarine mountains Marcus-Necker and adjacent Pacific regions*. Academija Nauk SSSR Moscow, pp. 83-93.

INDEX

Page numbers in *italics* refer to figures.

ACHEVÉ D'IMPRIMER
EN SEPTEMBRE 2018
SUR LES PRESSES
DE
L'IMPRIMERIE F. PAILLART
À ABBEVILLE

Date de distribution : 14 septembre 2018
Dépôt légal : septembre 2018
N° d'impression : 16012

RECENTLY PUBLISHED MEMOIRS

Tome 211 : Marie-Claude DURETTE-DESSET, María Celina DIGIANI, Mohamed KILANI & Didier GEFFARD-KURIYAMA 2017 — Critical revision of the Heligmonellidae (Nematoda: Trichostrongylina: Heligmosomoidea). 290 pp. (ISBN: 978-2-85653-801-2) 59 €.

Tome 210 : Sylvain CHARBONNIER, Denis AUDO, Alessandro GARASSINO & Matúš HYŽNÝ 2017 — Fossil Crustacea of Lebanon. 252 pp. (ISBN : 978-2-85653-785-5) 69 €.

Tome 209 : Tony ROBILLARD, Frédéric LEGENDRE, Claire VILLEMANT & Maurice LEPONCE (eds) 2016 — Insects of Mount Wilhelm. Papua New Guinea. 573 pp. (ISBN : 978-2-85653-784-8) 99 €.

Tome 208 : Virginie HÉROS, Ellen STRONG & Philippe BOUCHET (eds) 2016 — Tropical Deep-Sea Benthos, volume 29. 463 pp. (ISBN : 978-2-85653-774-9) 89 €.

Tome 207 : Stephen D. CAIRNS 2015 — Stylasteridae (Cnidaria: Hydrozoa: Anthoathecata) of the NewCaledonian Region, *in* Tropical Deep-Sea Benthos, volume 28. 362 pp. (ISBN : 978-2-85653-767-1) 89 €.

Tome 206 : Éric GUILBERT, Tony ROBILLARD, Hervé JOURDAN & Philippe GRANDCOLAS (eds) 2014 — Zoologia Neocaledonica 8. Biodiversity studies in New Caledonia. 315 pp. (ISBN : 978-2-85653-707-7) 89 €.

Tome 205 : Sylvain CHARBONNIER, Alessandro GARASSINO, Günter SCHWEIGERT & Martin SIMPSON (eds) 2013 — A worldwide review of fossil and extant glypheid and lotogastrid lobsters (Crustacea, Decapoda, Glypheoidea). 304 pp. (ISBN: 978-2-85653-706-0) 69 €.

Tome 204 : Shane T. AHYONG, Tin-Yam CHAN, Laure CORBARI & Peter K. L. NG (eds) 2013 — Tropical Deep-Sea Benthos, volume 27. 501 pp. (ISBN: 978-2-85653-692-6) 99 €.

Tome 203 : Stéphane PEIGNÉ & Sevket SEN (eds) 2012 — Mammifères de Sansan. 709 pp. (ISBN: 978-2-85653-681-0) 119 €.

Tome 202 : Tyson R. ROBERTS 2012 — Systematics, Biology, and Distribution of the Species of the Oceanic Oarfish Genus *Regalecus* (Teleostei, Lampridiformes, Regalecidae). 268 pp. (ISBN: 978-2-85653-677-3) 69 €.

Prix TTC, frais de port en sus.
Prices in €uros, postage not included.

http://sciencepress.mnhn.fr/fr